Calculus of Vector Functions

Third Edition

Calculus of Vector Functions

RICHARD E. WILLIAMSON

Department of Mathematics
Dartmouth College

RICHARD H. CROWELL

Department of Mathematics
Dartmouth College

HALE F. TROTTER

Department of Mathematics
Princeton University

Prentice-Hall, Inc., *Englewood Cliffs, New Jersey*

10 9 8 7

ISBN: 0-13-112367-X

Library of Congress Catalog Card Number 75-167788

Printed in the United States of America

PRENTICE-HALL INTERNATIONAL, INC., *London*
PRENTICE-HALL OF AUSTRALIA, PTY. LTD., *Sydney*
PRENTICE-HALL OF CANADA, LTD., *Toronto*
PRENTICE-HALL OF INDIA PRIVATE LIMITED, *New Delhi*
PRENTICE-HALL OF JAPAN, INC., *Tokyo*

Preface

This book is an introduction to the calculus of functions of several variables and vector calculus, following the unifying idea that calculus deals with linear approximations to functions. A working knowledge of one-variable calculus is the only prerequisite. The necessary linear algebra is presented in the first two chapters.

The emphasis in this third edition is on learning and understanding mathematical techniques, together with their applications to specific problems. The framework of linear algebra is used both to clarify concepts and to emphasize algebraic techniques as useful tools. We have given precise statements of all definitions and theorems, and have included complete proofs of practically everything in the book. The proofs are, however, meant to be studied only insofar as they help understanding. With careful attention to the examples, a person can go through the text intelligently without studying any proofs at all. The book is designed to be flexible, thereby allowing an instructor to select a level of theoretical emphasis appropriate to the interests and abilities of his class. It is unlikely that anyone would follow either extreme course: covering all proofs or including none.

In this edition, the material on linear algebra has been expanded and reorganized as two chapters. The first covers vectors and linear functions in $\mathcal{R}^n$ and contains nearly all the basic algebraic ideas used in the calculus. Chapter 2 includes a complete discussion of the solution of linear systems with a section on applications, presents the fundamental ideas of abstract vector spaces and dimension, and contains a brief but complete treatment of linear differential equations with constant coefficients; a section on complex vector spaces shows their usefulness in this connection. Other topics such as orthonormal bases and eigenvectors are also treated briefly.

Material involving numerical calculation has been added in several places. There are sections on Newton's method for functions of several variables, on numerical estimation of implicitly defined functions, and on numerical approximation of definite integrals. This material goes particularly well with the use of an automatic computer. We have included some exercises in these sections which cannot reasonably be done by hand, but which students with access to a computer can do by writing fairly simple programs. Of course we have also included enough numerical exercises that do not require a computer.

We want to thank the people at many universities who have made suggestions for improving the book. Their helpful interest has been most welcome. Thanks are also due to Mrs. Dorothy Krieger and to Mr. Arthur Wester of Prentice-Hall, whose efforts have helped to make the book more attractive and useful.

R. E. W.

R. H. C.

H. F. T.

Possible Courses of Study

We have tried to organize the text so that each section leads naturally to the next, but it is by no means necessary to include all sections or to take them up in strictly consecutive order. In particular, it is not necessary to complete the linear algebra before starting on the calculus. (Students who have already taken linear algebra can, of course, start with Chapter 3 and use the first two chapters for reference and review.) Everything essential for Chapters 3 through 7 is contained in the first seven sections of Chapter 1. (An exception is the use of facts about dimension in the last section of Chapter 4.) The study of determinants can be postponed until the section on change of variable in multiple integrals and the last three sections on vector analysis, where they are really needed. On the other hand, determinants can be used in practice for inverting small (2-by-2 or 3-by-3) matrices or solving small systems of linear equations, so that some may prefer to take them up earlier and postpone (or omit) the row-reduction method given in the first section of Chapter 2. At the cost of avoiding a few higher dimensional exercises later on, one may even restrict discussion of matrix inversion to the trivial 2-by-2 case.

Other changes of order, such as taking up multiple integration early in the course, should also cause no difficulty.

The sequences of section numbers listed below are minimal in the sense that they contain the prerequisite material for later entries, but do not contain everything that might be desirable in a typical course. Additional sections can be added to make up a full one-year course suitable for particular needs. Experience has shown that for ordinary class use two meetings should be used to cover an average section. An occasional short section merits only one meeting, and some longer ones, in order to be covered entirely, require three.

Ch. 1	*Ch. 2*	*Ch. 3*	*Ch. 4*	*Ch. 5*	*Ch. 6*	*Ch. 7*	*Comments*
1–5, 7		1–8	1–3, 5	1, 5	1–3	1, 3–5	General survey
1–5, 7, 8	1	1–9	1–3, 5	1	1–3, 6		Numerically oriented
1–5, 7	4, 5	1–8	1–3, 5, 7		1–4		Emphasizes dimension
1–5, 7	4, 5, 8, 9	1–8	1–3	5–7			No multiple integrals
1–5,7		1–8	1–3, 5	1	1–3		One-term course
1–5	1	1–8	1–3	1	1, 2, 7	1	No determinants

Contents

Calculus of Vector Functions

Introduction

A first course in calculus deals with real-valued functions of one variable, that is, functions defined on all or part of the real number line $\mathcal{R}$, and having values in $\mathcal{R}$. For example, a formula such as

$$y = x^2 + 3$$

yields a real number y for any real number x and so defines a function f from $\mathcal{R}$ to $\mathcal{R}$ with (for instance) $f(0) = 3$, $f(-2) = 7$, $f(\sqrt{3}) = 6$, etc. In this book we are concerned with functions of several variables whose values may be real numbers or, more generally, may be m-tuples of real numbers. For example, a pair of formulas

$$y_1 = \sqrt{x_1^2 + x_2^2 + x_3^2}$$
$$y_2 = x_1x_2 + 5x_3$$

yields a pair of numbers (y_1, y_2) for any triple of numbers (x_1, x_2, x_3) and so defines a function g from "3-dimensional space" to "2-dimensional space." In particular, the formulas above give

$$g(0, 0, 0) = (0, 0),$$
$$g(1, 2, 3) = (\sqrt{14}, 17),$$
$$g(3, 2, 1) = (\sqrt{14}, 11).$$

We shall use $\mathcal{R}^n$ to stand for the set of all n-tuples of real numbers. ($\mathcal{R}^1$ is thus the same as $\mathcal{R}$.) The **domain** of a function is the set on which it is defined, and the **range** or **image** is the set of values assumed by the function. We speak of functions "from $\mathcal{R}^n$ to $\mathcal{R}^m$" and write

$$\mathcal{R}^n \xrightarrow{f} \mathcal{R}^m$$

to indicate that f is a function whose domain is a subset of $\mathcal{R}^n$ and whose range is a subset of $\mathcal{R}^m$. $\mathcal{R}^n$ is then called the **domain space** of the function and $\mathcal{R}^m$ is called its **range space.** The terms "transformation" or "mapping" are sometimes used instead of "function."

While functions of one variable are basic and very useful, there are many situations whose mathematical formulation requires the more general functions we consider here. For example, just as certain curves in the plane can be represented as graphs of functions from $\mathcal{R}^1$ to $\mathcal{R}^1$, so certain surfaces in 3-space can be represented as graphs of functions from $\mathcal{R}^2$ to $\mathcal{R}^1$. The picture below illustrates the graph of a function $\mathcal{R}^2 \xrightarrow{f} \mathcal{R}^1$ defined by a formula

$$f(x, y) = z.$$

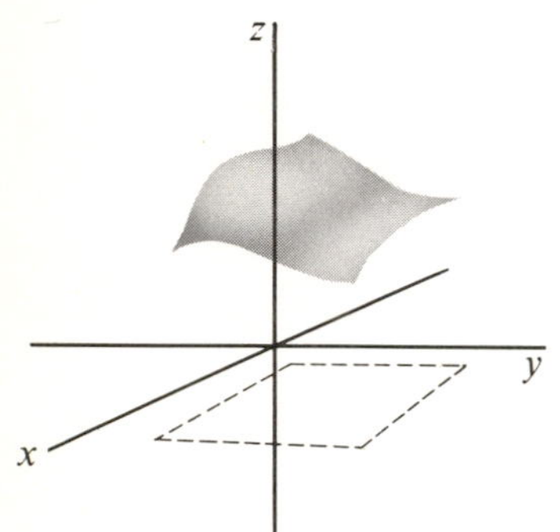

Other examples showing how curves and surfaces can be described by functions are given at the beginning of Chapter 3.

Most of the problems and examples in this book involve functions from $\mathcal{R}^n$ to $\mathcal{R}^m$ with values of m and n not more than 2 or 3, since higher-dimensional problems are difficult to visualize and often require inordinate amounts of computation. In the theoretical development we nevertheless provide formulations valid for arbitrary dimensions. This is not an empty generality. An economist may wish to consider a mathematical model in which the prices of a number of commodities are determined by a number of other variables such as costs of production and demands. A civil engineer may want to study the displacements produced in the many joints of a complex structure by various combinations of loads applied at many points. Now that automatic computers have made the arithmetic calculations feasible, problems involving dozens or even hundreds of variables are being attacked and solved. Thus it is important to have techniques and theorems that apply to functions from $\mathcal{R}^n$ to $\mathcal{R}^m$ for all values of m and n.

Most of the ideas studied in one-variable calculus reappear in a more general form in multivariable calculus. For example, the problem of finding a tangent line to the graph $y = f(x)$ of a function of one variable becomes the problem of finding a tangent plane to the graph $z = g(x, y)$ of a function of two variables. The tangent line has an equation of the form $y = mx + b$, and the coefficient m is found by computing the derivative of f. In the higher-dimensional case, the tangent plane has an equation of the form $z = px + qy + c$, and, as we shall see, the coefficients p and q are given by the higher-dimensional derivative of g.

Differential calculus can be described as the technique of studying general functions by approximating them with functions of the simple form exemplified by $y = mx + b$ or $z = px + qy + c$. In one dimension, these "linear" functions are very simple indeed, but in higher dimensions they are more complicated. The first two chapters of the book are concerned with linear functions and some closely related topics that have applications in many fields other than calculus.

The main purpose of the book, to which the later chapters are largely devoted, is to study generalizations of derivative and integral to higher dimensions. We shall be concerned with the relations between these ideas and with how they can be used to solve a variety of problems. The formal techniques of differentiation and integration from one-variable calculus continue to be directly applicable in multivariable calculus as methods of calculation. However the interpretations, and often the underlying ideas, may be quite different. To get geometric insight into these interpretations it is worth learning to visualize three-dimensional pictures. Two dimensional ones come fairly readily because they are easy to draw on paper. Learning to make perspective drawings is a big help in understanding three-dimensional problems and relating them to the physical world.

The boldface letters **x**, **a**, etc., that are used to distinguish vectors from numbers can be written longhand in several ways. Some possibilities are $\bar{\mathrm{x}}$, $\vec{\mathrm{x}}$, or $\underline{\mathrm{x}}$; capital letters or ordinary small letters can also be used in a context in which they will not be confused with the usual notations for matrices and numbers. The printing of a word in boldface type indicates that the word is being defined at that point in the text.

1

Vectors and Linearity

SECTION 1

VECTORS

We denote by $\mathcal{R}^n$ the set of all n-tuples $(x_1, x_2, \ldots, x_n)$ of real numbers. Boldface letters $\mathbf{x}$, $\mathbf{y}$, $\mathbf{z}$, etc., will stand for n-tuples, while ordinary lightface letters will stand for single real numbers. In particular we may write $\mathbf{x} = (x, y)$ or $\mathbf{x} = (x, y, z)$ for general pairs and triples in order to save writing subscripts.

For any two elements $\mathbf{x} = (x_1, x_2, \ldots, x_n)$ and $\mathbf{y} = (y_1, y_2, \ldots, y_n)$ in $\mathcal{R}^n$, we define the **sum** $\mathbf{x} + \mathbf{y}$ to be the n-tuple

$$(x_1 + y_1, x_2 + y_2, \ldots, x_n + y_n).$$

For any real number r and n-tuple $\mathbf{x} = (x_1, x_2, \ldots, x_n)$, we define the **numerical multiple** $r\mathbf{x}$ to be the n-tuple

$$(rx_1, rx_2, \ldots, rx_n).$$

For example, if $\mathbf{x} = (2, -1, 0, 3)$ and $\mathbf{y} = (0, 7, -2, 3)$, then

$$\mathbf{x} + \mathbf{y} = (2, 6, -2, 6)$$

and

$$3\mathbf{x} = (6, -3, 0, 9).$$

We write $-\mathbf{x}$ for the numerical multiple $(-1)\mathbf{x}$, and $\mathbf{x} - \mathbf{y}$ as an abbreviation for $\mathbf{x} + (-\mathbf{y})$. We use 0 to denote an n-tuple consisting entirely of zeros. The zero notation is ambiguous since, for example, 0 may stand for $(0, 0)$ in one formula and for $(0, 0, 0)$ in another. The ambiguity seldom causes any confusion since in most contexts only one interpretation makes sense. For instance, if $\mathbf{z} = (-2, 0, 3)$, then in the formula $\mathbf{z} + 0$, the 0 must stand for $(0, 0, 0)$ since addition is defined only between n-tuples with the same number of entries.

The following formulas hold for arbitrary $\mathbf{x}$, $\mathbf{y}$, and $\mathbf{z}$ in $\mathcal{R}^n$ and arbitrary numbers r, s. They express laws for our new operations of addition and numerical multiplication very closely analogous to the familiar distributive, commutative, and associative laws for ordinary addition and multiplication of numbers.

1. $r\mathbf{x} + s\mathbf{x} = (r + s)\mathbf{x}$.
2. $r\mathbf{x} + r\mathbf{y} = r(\mathbf{x} + \mathbf{y})$.
3. $r(s\mathbf{x}) = (rs)\mathbf{x}$.
4. $\mathbf{x} + \mathbf{y} = \mathbf{y} + \mathbf{x}$.
5. $(\mathbf{x} + \mathbf{y}) + \mathbf{z} = \mathbf{x} + (\mathbf{y} + \mathbf{z})$.
6. $\mathbf{x} + \mathbf{0} = \mathbf{x}$.
7. $\mathbf{x} + (-\mathbf{x}) = \mathbf{0}$.

These laws are all quite obvious consequences of the definitions of our new operations and the laws of arithmetic. For illustration, we give a formal proof of law 2.

Let $\mathbf{x} = (x_1, x_2, \ldots, x_n)$ and $\mathbf{y} = (y_1, y_2, \ldots, y_n)$, and let r be a real number. Then

$$r\mathbf{x} = (rx_1, rx_2, \ldots, rx_n) \qquad \text{[definition of numerical multiplication]}$$

$$r\mathbf{y} = (ry_1, ry_2, \ldots, ry_n) \qquad \text{[definition of numerical multiplication]}$$

and so

$$r\mathbf{x} + r\mathbf{y} = (rx_1 + ry_1, rx_2 + ry_2, \ldots, rx_n + ry_n) \qquad \text{[definition of addition].}$$

On the other hand,

$$\mathbf{x} + \mathbf{y} = (x_1 + y_1, x_2 + y_2, \ldots, x_n + y_n) \qquad \text{[definition of addition]}$$

and so

$$r(\mathbf{x} + \mathbf{y}) = \big(r(x_1 + y_1), r(x_2 + y_2), \ldots, r(x_n + y_n)\big) \qquad \text{[definition of numerical multiplication].}$$

By the distributive law of ordinary arithmetic, $r(x_1 + y_1) = rx_1 + ry_1$, $r(x_2 + y_2) = rx_2 + ry_2$, etc., and therefore the n-tuples $r\mathbf{x} + r\mathbf{y}$ and $r(\mathbf{x} + \mathbf{y})$ are the same, as was to be proved.

Any set with operations of addition and multiplication by real numbers defined in such a way that the laws 1–7 hold is called a **vector space,** and its elements are called **vectors.** We shall discuss some other vector spaces in Chapter 2, but for the present "vector" may be taken to mean "element of $\mathcal{R}^n$" for some n. Numbers are sometimes called **scalars** when emphasis on the distinction between numbers and vectors is wanted. In physics, for example, mass and energy may be referred to as scalar quantities in distinction to vector quantities such as velocity or momentum. The term scalar multiple is synonymous with what we have called numerical multiple.

The vectors

$$\mathbf{e}_1 = (1, 0, \dots, 0)$$
$$\mathbf{e}_2 = (0, 1, 0, \dots, 0)$$
$$\vdots$$
$$\mathbf{e}_n = (0, \dots, 0, 1)$$

have the property that, if $\mathbf{x} = (x_1, \dots, x_n)$, then

$$\mathbf{x} = x_1\mathbf{e}_1 + \dots + x_n\mathbf{e}_n.$$

Because every element of $\mathcal{R}^n$ can be so simply represented in this way, the set of vectors $\{\mathbf{e}_1, \dots, \mathbf{e}_n\}$ is called the **natural basis** for $\mathcal{R}^n$. For example, in $\mathcal{R}^3$ we have $(1, 2, -7) = \mathbf{e}_1 + 2\mathbf{e}_2 - 7\mathbf{e}_3$. The entries in

$$\mathbf{x} = (x_1, \dots, x_n)$$

are then called the **coordinates** of $\mathbf{x}$ relative to the natural basis.

A sum of numerical multiples $x_1\mathbf{e}_1 + \dots + x_n\mathbf{e}_n$ is called a linear combination of the vectors $\mathbf{e}_1, \dots, \mathbf{e}_n$. More generally, a sum of multiples $a_1\mathbf{x}_1 + \dots + a_n\mathbf{x}_n$ is called a **linear combination** of the vectors $\mathbf{x}_1, \dots, \mathbf{x}_n$. Thus, for example, the equation

$$(2, 3, 4) = 4(1, 1, 1) - 1(1, 1, 0) - 1(1, 0, 0)$$

shows the vector $(2, 3, 4)$ represented as a linear combination of the vectors $(1, 1, 1)$, $(1, 1, 0)$, and $(1, 0, 0)$.

EXERCISES

1. Given $\mathbf{x} = (3, -1, 0)$, $\mathbf{y} = (0, 1, 5)$, and $\mathbf{z} = (2, 5, -1)$ compute $3\mathbf{x}$, $\mathbf{y} + \mathbf{z}$, and $4\mathbf{x} - 2\mathbf{y} + 3\mathbf{z}$. [*Ans.* (18, 9, −13).]

2. Find numbers a and b such that $a\mathbf{x} + b\mathbf{y} = (9, -1, 10)$, where $\mathbf{x}$ and $\mathbf{y}$ are as in Problem 1. Is there more than one solution?

3. Show that no choice of numbers a and b can make $a\mathbf{x} + b\mathbf{y} = (3, 0, 0)$, where $\mathbf{x}$ and $\mathbf{y}$ are as in Problem 1. For what value(s) of c (if any) can the equation $a\mathbf{x} + b\mathbf{y} = (3, 0, c)$ be satisfied?

4. Write out proofs for (a) law 3 and (b) law 4 on page 5, giving precise justification for each step.

5. Verify that the set $C[a, b]$ of all continuous real-valued functions defined on the interval $a \leq x \leq b$ is a vector space, with addition and numerical multiplication defined by $(f + g)(x) = f(x) + g(x)$ and $(rf)(x) = rf(x)$.

6. Prove that the representation of a vector $\mathbf{x}$ in $\mathcal{R}^n$ in terms of the natural basis is unique. That is, show that if

$$x_1\mathbf{e}_1 + \ldots + x_n\mathbf{e}_n = y_1\mathbf{e}_1 + \ldots + y_n\mathbf{e}_n,$$

then $x_k = y_k$ for $k = 1, \ldots, n$.

7. Represent the first vector below as a linear combination of the remaining vectors, either by inspection or by solving an appropriate system of equations.

(a) $(2, 3, 4)$; $(1, 1, 1)$, $(1, 2, 1)$, $(-1, 1, 2)$.

(b) $(2, -7)$; $(1, 1)$, $(1, -1)$.

(c) $(-2, 3)$; $\mathbf{e}_1$, $\mathbf{e}_2$.

SECTION 2

GEOMETRIC INTERPRETATIONS

Geometric representations of $\mathcal{R}^1$ as a line, of $\mathcal{R}^2$ as a plane, and of $\mathcal{R}^3$ as 3-dimensional space may be obtained by using coordinates. To represent $\mathcal{R}^1$ as a line, one must first specify a point on the line to be called the **origin,** a unit of distance, and a direction on the line to be called positive. (The opposite direction is then called negative.) Then a positive number x corresponds to the point which is a distance x in the positive direction from the origin. A negative number x corresponds to the point which is a distance $|x|$ from the origin in the negative direction. The number zero of course corresponds to the origin. The number line is most often thought of as horizontal with the positive direction to the right. With this standard convention, we obtain the familiar Fig. 1, in which the arrow indicates the positive direction.

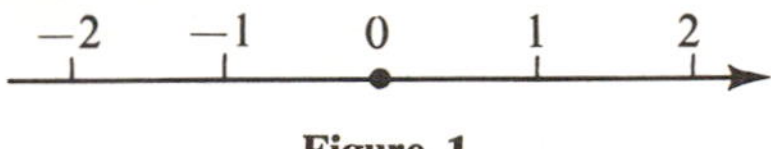

Figure 1

In the plane, one takes an origin, a unit of distance, a pair of perpendicular lines (called the **axes**) through the origin, and a positive direction on each axis. Given a vector in $\mathcal{R}^2$, that is, a pair of numbers (x_1, x_2), the procedure described in the preceding paragraph determines a point p_1 on the first axis corresponding to the number x_1 and a point p_2 on the second axis corresponding to the number x_2. Then the vector (x_1, x_2) corresponds to the point p in the plane whose projection on the first axis is p_1 and whose projection on the second axis is p_2. The (perpendicular) **projection** of a point p on a line L is defined as the foot of the perpendicular from p to L if p is not on L. If p is on L, then the projection of p on L is p itself.

The conventional choice is to take the first axis horizontal with the positive direction to the right, and the second axis vertical with the positive direction upwards. This leads to the usual picture shown in Fig. 2.

Representing a vector by an arrow from the origin to the corresponding point, as we have done in Fig. 2, often makes a better picture than simply marking the point.

An obvious extension of the procedure works in three dimensions. One takes an origin, three perpendicular axes through it, and a positive direction on each. A vector in $\mathcal{R}^3$ is a triple of numbers (x_1, x_2, x_3) and gives points p_1, p_2, and p_3 on the three axes. Then the point p, corresponding to (x_1, x_2, x_3), is the one whose projections on the three axes are p_1, p_2, and p_3.

There is no universally accepted convention for labeling the axes in 3-dimensional figures. Figure 3 illustrates the convention followed in this book, but several other schemes are also in common use.

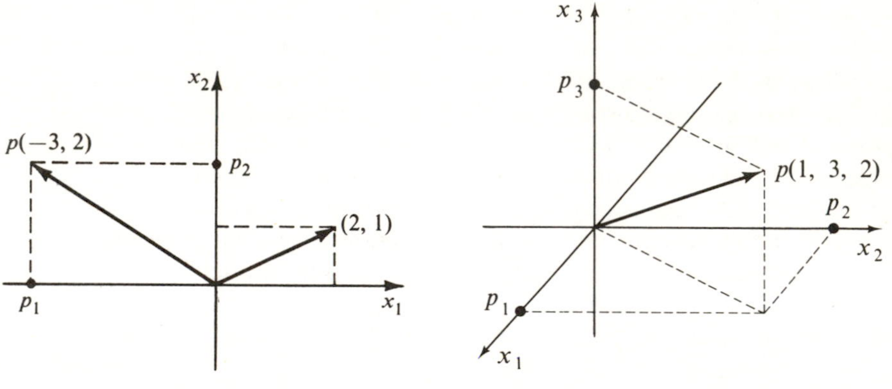

Figure 2

Figure 3

We have described how to set up a correspondence between vectors (n-tuples of numbers) and points. We now consider the geometrical interpretation of the vector operations of addition and numerical multiplication. This time we represent a vector as an arrow from the origin to the point with given coordinates. Figure 4 shows two vectors $\mathbf{u} = (u_1, u_2)$ and $\mathbf{v} = (v_1, v_2)$. The sum in $\mathcal{R}^2$ is $\mathbf{u} + \mathbf{v} = (u_1 + v_1, u_2 + v_2)$, so the arrow representing $\mathbf{u} + \mathbf{v}$ must be drawn as shown.

As Fig. 4 suggests, $\mathbf{u} + \mathbf{v}$ is represented by an arrow from the origin to the opposite corner of the parallelogram whose sides are the arrows representing $\mathbf{u}$ and $\mathbf{v}$. This geometric rule for adding vectors is often referred to as the **parallelogram law of addition.** (To prove that this

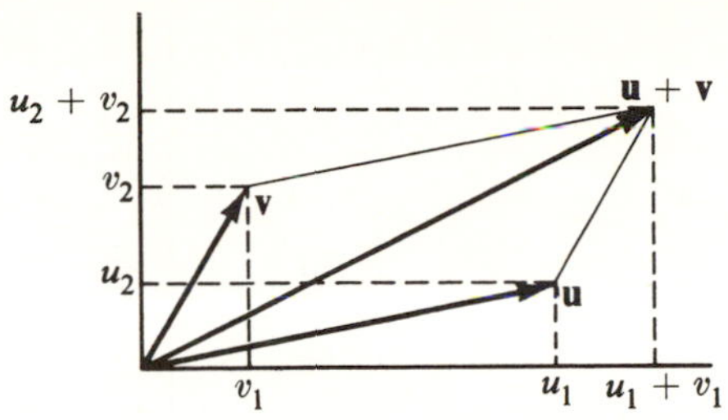

Figure 4

law is a consequence of our definitions, one of course has to use some theorems of geometry. For an outline of a proof, see Exercise 6(a).)

Another rule for adding vectors geometrically is illustrated in Fig. 5. Starting at the end of the arrow representing **u**, draw an arrow equal in length and parallel to the arrow representing **v**. (In other words, translate the arrow representing **v** from the origin to the end of **u**.) Then **u** + **v** is represented by an arrow from the origin to the end of the translation of **v**. This rule can be applied to any pair of vectors, whereas the parallelogram law does not (strictly speaking) apply if the arrows representing **u** and **v** lie in the same straight line.

If we write **w** for **u** + **v**, then **v** = **w** − **u**. Figure 6 (which is simply Fig. 5 relabeled) illustrates the useful fact that the difference of two vectors **w** and **u** is represented by the arrow from the end of **u** to the end of **w**, appropriately translated.

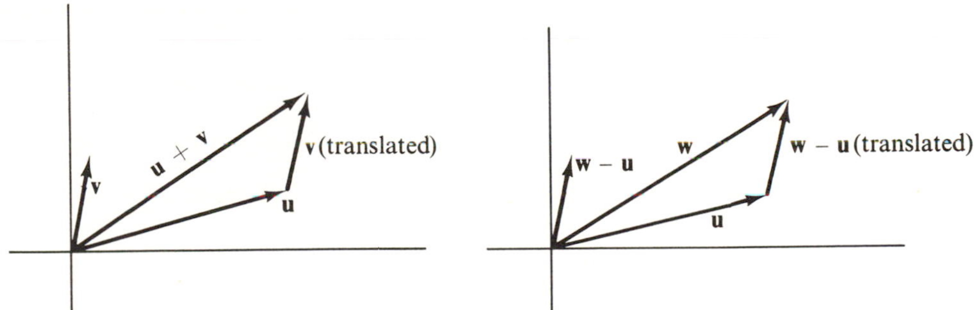

Figure 5 **Figure 6**

Figure 7 illustrates numerical multiplication, both by a positive number a and a negative number b. For a positive number a, the arrow representing a**u** has the same direction as the arrow representing **u** and is a times as long. For a negative number b, b**u** points in exactly the opposite direction to **u** and is $|b|$ times as long. (See Exercise 6(b).)

The question of whether to represent a vector geometrically as a

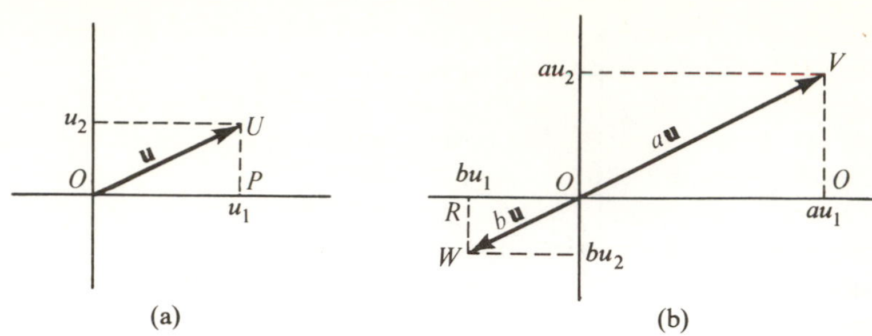

Figure 7

point, or as an arrow, or perhaps as an arrow translated, has different answers depending on the geometric situation in which the vector occurs. Logically we could restrict ourselves to using only points, or only arrows, and thus avoid the necessity of making a decision. However, there is a considerable gain in geometric insight from being able to use both interpretations, and the examples will show which is more useful in a particular kind of problem.

So far we have used lines in $\mathcal{R}^2$ and $\mathcal{R}^3$ informally to get pictures of vector addition and numerical multiplication. Having done this, however, we can formally define what we shall mean by a line in $\mathcal{R}^n$. To begin, we fix a *nonzero* vector $\mathbf{x}_1$ and call the set of all numerical multiples $t\mathbf{x}_1$ a line L_0 through the origin. Then we define a **line** L to be the set of all points representable in the form $t\mathbf{x}_1 + \mathbf{x}_0$, where $\mathbf{x}_0$ is some fixed vector. The relationship between L_0 and L is shown in Fig. 8 (for vectors $\mathbf{x}_1$ and $\mathbf{x}_0$ in $\mathcal{R}^3$).

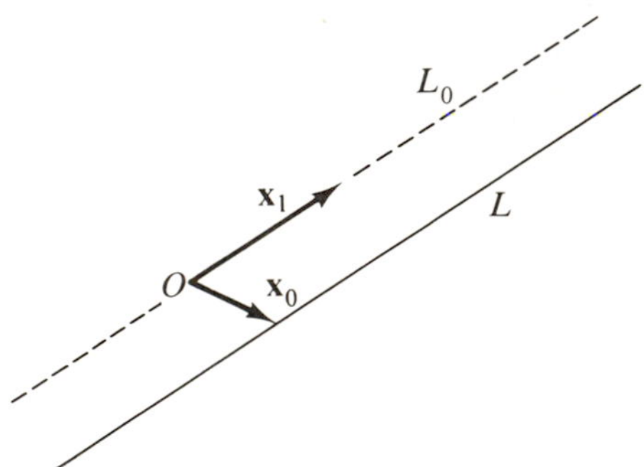

Figure 8

We define two lines, one consisting of points of the form $s\mathbf{u}_1 + \mathbf{u}_0$ and the other consisting of points of the form $t\mathbf{v}_1 + \mathbf{v}_0$ to be **parallel** if $\mathbf{v}_1$ is a numerical multiple of $\mathbf{u}_1$. Any line $s\mathbf{u}_1 + \mathbf{u}_0$ is parallel to a line through the origin, namely, $s\mathbf{u}_1$.

Example 1. To find a representation for the line in $\mathcal{R}^3$ parallel to the vector $(1, 1, 1)$ and passing through the point $(-1, 3, 6)$ we form all multiples $t(1, 1, 1)$ to get a line through the origin. Then the set of all points $t(1, 1, 1) + (-1, 3, 6)$ is a line containing the point $(-1, 3, 6)$, as we see by setting $t = 0$.

To determine a plane in $\mathcal{R}^3$ it is natural to start with two *noncollinear* vectors $\mathbf{x}_1$ and $\mathbf{x}_2$ (that is, such that neither is a multiple of the other) and consider all points $u\mathbf{x}_1 + v\mathbf{x}_2$, where u and v are numbers. The geometric interpretation of numerical multiplication and addition of vectors shows that the points $u\mathbf{x}_1 + v\mathbf{x}_2$ constitute what we would like to call a plane P_0 through the origin. We then define a **plane** P parallel to P_0 to be the set of all points $u\mathbf{x}_1 + v\mathbf{x}_2 + \mathbf{x}_0$, where $\mathbf{x}_0$ is some fixed vector. P_0 and P are related as shown in Fig. 9.

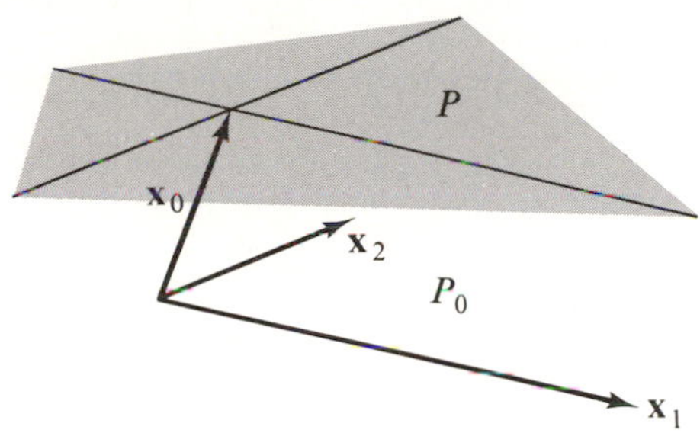

Figure 9

Example 2. We represent a plane in $\mathcal{R}^3$ parallel to the two vectors $(1, 1, 1)$ and $(2, -1, 3)$, and passing through the point $(0, 1, -1)$. The set of all linear combinations $u(1, 1, 1) + v(2, -1, 3)$ is a plane P_0 through the origin, and the set of all points $u(1, 1, 1) + v(2, -1, 3) + (0, 1, -1)$ is a parallel plane P. That P contains the point $(0, 1, -1)$ becomes evident on setting $u = v = 0$.

The set of all linear combinations of a set S of vectors is called the **span** of S, and we speak of the set of linear combinations as being spanned by S. For example, if S consists of one vector $\mathbf{x}$, then the line $\mathcal{L}$ consisting of all numerical multiples $t\mathbf{x}$ is the span of S. If $S = \{\mathbf{x}_1, \mathbf{x}_2\}$, the set spanned by S is the set P of all linear combinations $u\mathbf{x}_1 + v\mathbf{x}_2$. For P to be a plane, we require that $\mathbf{x}_1$ and $\mathbf{x}_2$ not lie on the same line. Another way to state this condition is that no multiple of $\mathbf{x}_1$ should equal a multiple of $\mathbf{x}_2$ except for the zero multiples. More generally, we say that a set of vectors

$$\mathbf{x}_1, \mathbf{x}_2, \ldots, \mathbf{x}_n$$

is **linearly independent** if, whenever

$$c_1\mathbf{x}_1 + \ldots + c_n\mathbf{x}_n = 0$$

for some numbers $c_1, \ldots, c_n$, then all the c's must be zero. If on the other hand it is possible to have a linear combination of vectors equal to the zero vector, but with not all the coefficients equal to zero, then those vectors are said to be **linearly dependent.**

Example 3. The set of three vectors

$$(1, 2), (-1, 1), (1, -3)$$

in $\mathcal{R}^2$ is linearly dependent. For, the equation

$$c_1(1, 2) + c_2(-1, 1) + c_3(1, -3) = 0$$

is equivalent to the two equations

$$\begin{aligned} c_1 - c_2 + c_3 &= 0 \\ 2c_1 + c_2 - 3c_3 &= 0. \end{aligned}$$

If we now let c_3 have any nonzero value, we can then solve for c_1 and c_2. For example, if we set $c_3 = 1$, then

$$\begin{aligned} c_1 - c_2 &= -1 \\ 2c_1 + c_2 &= 3, \end{aligned}$$

and we solve the two equations. Adding them gives $3c_1 = 2$, or $c_1 = \frac{2}{3}$, while subtracting the second from two times the first gives $-3c_2 = -5$, or $c_2 = \frac{5}{3}$. Thus

$$(\tfrac{2}{3})(1, 2) + (\tfrac{5}{3})(-1, 1) + (1)(2, -3) = 0,$$

so the vectors are linearly dependent. Linear independence can be checked similarly by solving a vector equation and showing that it has only zero solutions.

A linearly independent set of vectors that spans a subset $\mathcal{S}$ of a vector space $\mathcal{V}$ is called a **basis** for $\mathcal{S}$. A basis for $\mathcal{S}$ is useful because, if $\mathbf{x}$ is any vector in $\mathcal{S}$, then $\mathbf{x}$ can be represented as a linear combination of basis elements with uniquely determined coefficients. Thus if $\mathbf{x}_1, \ldots, \mathbf{x}_n$ is a basis for $\mathcal{S}$ and $\mathbf{x}$ is in $\mathcal{S}$, we have, by the spanning property,

$$\mathbf{x} = c_1\mathbf{x}_1 + \ldots + c_n\mathbf{x}_n,$$

for some constants $c_1, \ldots, c_n$. But the c's are completely determined. For if we had also

$$\mathbf{x} = d_1\mathbf{x}_1 + \ldots + d_n\mathbf{x}_n,$$

then subtraction of one equation from the other gives

$$0 = (c_1 - d_1)\mathbf{x}_1 + \ldots + (c_n - d_n)\mathbf{x}_n.$$

It follows from the linear independence of the $\mathbf{x}$'s that $c_k - d_k = 0$ for each k, so that $c_k = d_k$.

Example 4. The natural basis for $\mathcal{R}^3$ is the set of vectors

$$\mathbf{e}_1 = (1, 0, 0), \quad \mathbf{e}_2 = (0, 1, 0), \quad \mathbf{e}_3 = (0, 0, 1).$$

It is easy to check that these vectors are linearly independent and that they span $\mathcal{R}^3$. For if

$$c_1\mathbf{e}_1 + c_2\mathbf{e}_2 + c_3\mathbf{e}_3 = 0,$$

then $(c_1, c_2, c_3) = (0, 0, 0)$. On the other hand, any vector (x, y, z) in $\mathcal{R}^3$ can be written

$$(x, y, z) = x\mathbf{e}_1 + y\mathbf{e}_2 + z\mathbf{e}_3.$$

EXERCISES

1. For each pair $\mathbf{u}$, $\mathbf{v}$ of vectors in $\mathcal{R}^2$ given below, draw the arrows representing $\mathbf{u}$, $\mathbf{v}$, $\mathbf{u} + \mathbf{v}$, $\mathbf{u} - \mathbf{v}$, and $\mathbf{u} + 2\mathbf{v}$.

(a) $\mathbf{u} = (1, 0)$, $\mathbf{v} = (0, 1)$.
(b) $\mathbf{u} = (-2, 1)$, $\mathbf{v} = (1, 2)$.
(c) $\mathbf{u} = (-1, 1)$, $\mathbf{v} = (\frac{1}{2}, \frac{1}{2})$.

2. Let $\mathbf{x} = (1, 1)$ and $\mathbf{y} = (0, 1)$. Draw the arrows representing $t\mathbf{x} + \mathbf{y}$ for the following values of t: $-1, \frac{1}{2}, 1, 2$.

3. Let $\mathbf{u} = (2, 1)$ and $\mathbf{v} = (-1, 2)$. Draw the arrows representing $t\mathbf{u} + (1 - t)\mathbf{v}$ for the following values of t: $-1, 0, \frac{1}{2}, 1, 2$.

4. (a) Let $\mathbf{u}_1 = (0, 1)$, $\mathbf{v}_1 = (-1, 1)$, $\mathbf{u}_2 = (-3, 2)$, and $\mathbf{v}_2 = (2, 1)$. Sketch the lines $\mathbf{u}_1 + t\mathbf{v}_1$ and $\mathbf{u}_2 + s\mathbf{v}_2$. Find the vector $\mathbf{w}$ at the point of intersection of the lines by finding values of t and s for which $\mathbf{w} = \mathbf{u}_1 + t\mathbf{v}_1 = \mathbf{u}_2 + s\mathbf{v}_2$. [*Ans.* $\mathbf{w} = (-\frac{5}{3}, \frac{8}{3})$.]

(b) Let $\mathbf{u}_1$, $\mathbf{v}_1$, and $\mathbf{u}_2$ be as in part (a), but take $\mathbf{v}_2 = (2, -2)$. (Note that $\mathbf{v}_2$ is then a numerical multiple of $\mathbf{v}_1$.) Sketch the lines $\mathbf{u}_1 + t\mathbf{v}_1$ and $\mathbf{u}_2 + s\mathbf{v}_2$. Show algebraically that the lines do not intersect.

5. Show that in $\mathcal{R}^n$, the lines represented by $s\mathbf{u}_1 + \mathbf{u}_0$ and $t\mathbf{v}_1 + \mathbf{v}_0$ are the same if and only if both $\mathbf{v}_1$ and $\mathbf{v}_0 - \mathbf{u}_0$ are numerical multiples of $\mathbf{u}_1$.

6.

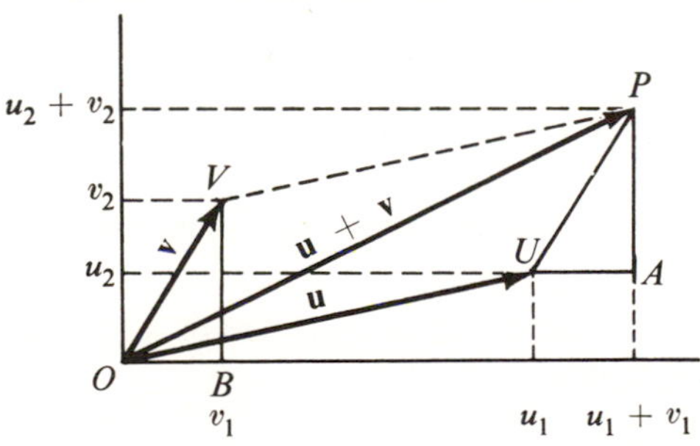

(a) In the figure above, the points U, V, and P are constructed to have coordinates as shown. By using plane geometry, show that the triangles

OVB and *UPA* are congruent. From this deduce that *OV* and *UP* are parallel and of equal length. Then *OVPU* is a parallelogram, and the parallelogram law follows.

(b) In Figs. 7(a) and 7(b) the points *U*, *V*, and *W* are constructed to have coordinates as shown. Prove that the triangles *OUP*, *OVQ*, and *OWR* are similar. Show that the angles *UOP*, *VOQ*, and *WOR* are therefore equal and that the lengths *OV* and *OW* are proportional to the length of *OU* as stated in the text.

7. It is not essential to use perpendicular axes in setting up coordinates in the plane. The same procedure can be used if the projection of a point on an axis is defined as the point of intersection of that axis and the line through the given point and parallel to the other axis. (If the axes are perpendicular, this is equivalent to our previous definition. Why?) It is also possible to choose different units of distance along the two axes. Show that the geometric interpretation of the vector operations in two dimensions remains the same in this more general setting. How would you extend the definition of projection to make the same generalization in three dimensions?

8. (a) Show that if $\mathbf{u}$ and $\mathbf{v}$ are two distinct vectors, then the vectors $t\mathbf{u} + (1 - t)\mathbf{v}$ form a line through the points corresponding to $\mathbf{u}$ and $\mathbf{v}$.

(b) If $\mathbf{x}_1$ and $\mathbf{x}_2$ are two vectors in $\mathcal{R}^n$, then the set of all vectors $t\mathbf{x}_1 + (1 - t)\mathbf{x}_2$, where $0 \leq t \leq 1$, is the **line segment** joining $\mathbf{x}_1$ and $\mathbf{x}_2$. A set S in $\mathcal{R}^n$ is **convex** if, whenever S contains two points, it also contains the line segment joining them. Prove that the intersection of any collection of convex sets is convex.

9. Represent the following lines in the form $t\mathbf{x}_1 + \mathbf{x}_0$, where t runs over all real numbers. Sketch each line.

(a) The line in $\mathcal{R}^3$ parallel to (1, 2, 0) and passing through the point (1, 1, 1).
(b) The line in $\mathcal{R}^2$ joining the points (1, 0) and (0, 1).
(c) The line in $\mathcal{R}^3$ joining the points (1, 0, 0) and (0, 0, 1).

10. Represent the following planes in $\mathcal{R}^3$ in the form $u\mathbf{x}_1 + v\mathbf{x}_2 + \mathbf{x}_0$, where u and v run over all real numbers. Sketch each plane.

(a) The plane parallel to the vectors (1, 1, 0) and (0, 1, 1) and passing through the origin.
(b) The plane parallel to the vectors $\mathbf{e}_1$ and $\mathbf{e}_2$ and passing through the point (0, 0, 1).
(c) The plane passing through the three points (1, 0, 0), (0, 1, 0), and (0, 0, 1).

11. Determine whether each of the following sets of vectors is linearly dependent or linearly independent.

(a) (1, 2), (2, 1).
(b) (1, 2), (−3, −6).
(c) (1, −1), (0, 1), (2, 5).
(d) (1, 2, 0), (−1, 1, 1), (−1, 0, 0).

12. Determine whether the first vector below is in the set spanned by the set S. In each case give a geometric interpretation: the first point does (or does not) lie on the plane (or line) spanned by S.

(a) $(1, 1)$; $S = \{(-2, -2)\}$.
(b) $(-1, 1, 3)$; $S = \{(1, 1, 1), (-1, 0, 1)\}$.
(c) $(-1, -2, 1)$; $S = \{(3, 6, -3), (1, 2, -1)\}$.

13. Which of the following sets form bases for $\mathcal{R}^2$ or $\mathcal{R}^3$?

(a) $(1, 1), (-1, 1)$.
(b) $(1, 2, 1), (-1, 2, 1), (0, 4, 2)$.
(c) $(1, 1, 1), (1, 1, 0), (1, 0, 0)$.
(d) $(1, 0, 0), (0, 2, 0), (0, 0, 3)$.

14. (a) Prove that two nonzero vectors $\mathbf{x}$ and $\mathbf{y}$ are linearly dependent if and only if they lie on the same line through the origin.
(b) Prove that three nonzero vectors $\mathbf{x}$, $\mathbf{y}$, and $\mathbf{z}$ are linearly dependent if and only if they all lie on the same plane through the origin.

SECTION 3

MATRICES

A set of equations such as

$$\begin{aligned} y_1 &= 2x_1 + 3x_2 - 4x_3 \\ y_2 &= x_1 - x_2 + 2x_3, \end{aligned}$$

in which each y_i is given as a sum of constant multiples of the x_j, defines a function (in this example from $\mathcal{R}^3$ to $\mathcal{R}^2$) of a particularly simple kind. It is an example of what we shall call a *linear* function. (The precise definition of "linear" appears in the next section.) An understanding of these functions is basic to the study of more general functions. Linear functions from $\mathcal{R}^1$ to $\mathcal{R}^1$ are so simple that they can be taken for granted in studying the calculus for functions of one variable. In higher dimensions, however, linear functions can be more complicated, and the main business of this first chapter is to develop their basic properties and to present notation and methods of calculation for dealing with them.

In the foregoing example, the x's and y's can be thought of as place-holders. The function is described completely by the array of coefficients

$$\begin{pmatrix} 2 & 3 & -4 \\ 1 & -1 & 2 \end{pmatrix}.$$

Any such rectangular array of numbers is called a **matrix.** Thus

$$\begin{pmatrix} 0 & 5 \\ -1 & \frac{1}{2} \\ 0 & 4 \end{pmatrix}, \quad \begin{pmatrix} 1 & 0.7 & 3 \\ 0.9 & 0 & 2.8 \end{pmatrix}, \quad \begin{pmatrix} 1 & 0 \\ 0 & -1 \end{pmatrix}, \quad (\tfrac{1}{2}, \tfrac{1}{3}, 0), \quad \begin{pmatrix} 0.325 \\ 0.007 \\ 0.579 \\ 3.142 \end{pmatrix}$$

are all examples of matrices. The horizontal lines of numbers in a matrix are called its **rows** and the vertical lines are called its **columns.** The number of rows and number of columns of a matrix determine its **shape.** Thus the five examples above have shapes 3-by-2, 2-by-3, 2-by-2, 1-by-3, and 4-by-1. Note that the number of rows always comes *before* the number of columns. The 1-by-n matrices are called **n-dimensional row vectors,** and n-by-1 matrices are called **n-dimensional column vectors.** A matrix is **square** if it has the same number of rows as it has columns.

The number in the ith row and jth column of a matrix is called the ijth **entry** of that matrix. Once more the row index is always put before the column index. Two matrices are equal if and only if they have the same shape and the entries in corresponding positions in the two matrices are equal.

We use capital letters to denote matrices, and often use the corresponding small letters with appropriate subscript to denote their entries. Thus we write

$$A = \begin{pmatrix} a_{11} & a_{12} & a_{13} & a_{14} \\ a_{21} & a_{22} & a_{23} & a_{24} \\ a_{31} & a_{32} & a_{33} & a_{34} \end{pmatrix} \quad \text{or} \quad A = (a_{ij}),$$

where $i = 1, 2, 3$, $j = 1, 2, 3, 4$. We usually write simply $x_1, x_2, \ldots, x_n$ for the entries of an n-dimensional column vector $\mathbf{x}$ rather than $x_{11}, x_{21}, \ldots, x_{n1}$.

The operations of addition and numerical multiplication which were defined in the last section for vectors in $\mathcal{R}^n$ can be extended to matrices. If A and B have the same shape, then their **sum** $A + B$ is defined as the matrix with the same shape and ijth entry equal to $a_{ij} + b_{ij}$. For example

$$\begin{pmatrix} 2 & 3 \\ -2 & \frac{1}{2} \\ 0 & 4 \end{pmatrix} + \begin{pmatrix} 1 & -1 \\ 2 & \frac{2}{3} \\ 0 & 0 \end{pmatrix} = \begin{pmatrix} 3 & 2 \\ 0 & \frac{7}{6} \\ 0 & 4 \end{pmatrix}.$$

Addition is not defined between matrices of different shapes. Recall that we did not define addition between elements of $\mathcal{R}^n$ and $\mathcal{R}^m$ with $m \neq n$.

For any matrix A and number r, the **numerical multiple** rA is defined as the matrix with the same shape as A and ijth entry equal to ra_{ij}. For example

$$2\begin{pmatrix} 3 & 4 \\ 5 & 6 \end{pmatrix} = \begin{pmatrix} 6 & 8 \\ 10 & 12 \end{pmatrix}$$

and

$$5\begin{pmatrix} 1 & 0 \\ 3 & 2 \\ -1 & 0 \end{pmatrix} + 3\begin{pmatrix} -2 & 2 \\ -4 & 0 \\ 0 & 1 \end{pmatrix}$$

$$= \begin{pmatrix} 5 & 0 \\ 15 & 10 \\ -5 & 0 \end{pmatrix} + \begin{pmatrix} -6 & 6 \\ -12 & 0 \\ 0 & 3 \end{pmatrix} = \begin{pmatrix} -1 & 6 \\ 3 & 10 \\ -5 & 3 \end{pmatrix}.$$

As with vectors in $\mathcal{R}^n$, we write $-A$ for $(-1)A$ and $A - B$ for $A + (-1)B$. For every shape there is a **zero matrix** which has all its entries equal to zero. We use 0 to denote any zero matrix; the shape intended will always be clear from the context.

It is easy to see that if the matrices X, Y, and Z all have the same shape, and r and s are any numbers, then the formulas 1–7 in Section 1 all hold. (The proofs are just the same as when X, Y, and Z are all in $\mathcal{R}^n$.) In other words, according to the definition in Section 2, for any fixed m and n, the set of m-by-n matrices forms a vector space with the operations of addition and numerical multiplication that we have just defined.

Another operation between matrices is suggested by the way they may be used to describe functions from $\mathcal{R}^m$ to $\mathcal{R}^n$. For example, suppose we have formulas

$$\begin{aligned} z_1 &= 3y_1 - y_2 \\ z_2 &= 5y_1 + 2y_2, \end{aligned}$$

and

$$\begin{aligned} y_1 &= 2x_1 + 3x_2 + 4x_3 \\ y_2 &= x_1 - x_2 + 2x_3, \end{aligned}$$

defining functions from $\mathcal{R}^2$ to $\mathcal{R}^2$ and from $\mathcal{R}^3$ to $\mathcal{R}^2$, respectively. These functions are described by the matrices A and B where

$$A = \begin{pmatrix} 3 & -1 \\ 5 & 2 \end{pmatrix} \quad \text{and} \quad B = \begin{pmatrix} 2 & 3 & 4 \\ 1 & -1 & 2 \end{pmatrix}.$$

If we express the z's directly in terms of the x's we obtain

$$\begin{aligned} z_1 &= 3(2x_1 + 3x_2 + 4x_3) - (x_1 - x_2 + 2x_3) \\ z_2 &= 5(2x_1 + 3x_2 + 4x_3) + 2(x_1 - x_2 + 2x_3). \end{aligned}$$

Rearranging terms gives

$$\begin{aligned} z_1 &= (3 \cdot 2 - 1 \cdot 1)x_1 + (3 \cdot 3 - 1 \cdot (-1))x_2 + (3 \cdot 4 - 1 \cdot 2)x_3 \\ z_2 &= (5 \cdot 2 + 2 \cdot 1)x_1 + (5 \cdot 3 + 2 \cdot (-1))x_2 + (5 \cdot 4 + 2 \cdot 2)x_3 \end{aligned}$$

and we see that the resulting function is described by a matrix C with

$$C = \begin{pmatrix} 3 \cdot 2 - 1 \cdot 1 & 3 \cdot 3 - 1 \cdot (-1) & 3 \cdot 4 - 1 \cdot 2 \\ 5 \cdot 2 + 2 \cdot 1 & 5 \cdot 3 + 2 \cdot (-1) & 5 \cdot 4 + 2 \cdot 2 \end{pmatrix} = \begin{pmatrix} 5 & 10 & 10 \\ 12 & 13 & 24 \end{pmatrix}.$$

We say that C is obtained from A and B by matrix multiplication and write $C = AB$.

To see how the general definition should be made, note that the ijth entry of C is the sum of products of entries in the ith row of A and the jth column of B. Thus $c_{21} = a_{21}b_{11} + a_{22}b_{21} = 5 \cdot 2 + 2 \cdot 1 = 12$.

We give the definition of matrix multiplication in two stages. First suppose

$$A = (a_1, a_2, \ldots, a_k) \quad \text{and} \quad B = \begin{pmatrix} b_1 \\ b_2 \\ \cdot \\ \cdot \\ \cdot \\ b_k \end{pmatrix}$$

are row and column vectors of the same dimension. Then the product AB is defined to be the number $a_1b_1 + a_2b_2 + \ldots + a_kb_k$. Now let A be an m-by-k matrix and B be an k-by-n matrix. (It is important that the number of columns of A be equal to the number of rows of B.) Then each row of A is a k-dimensional row vector and each column of B is a k-dimensional column vector. We define the **matrix product** AB as the m-by-n matrix whose ijth entry is the product (in the sense just defined) of the ith row of A and the jth column of B. The product AB always has the same number of rows as A and the same number of columns as B. For instance, in our example

$$\begin{pmatrix} 3 & -1 \\ 5 & 2 \end{pmatrix}\begin{pmatrix} 2 & 3 & 4 \\ 1 & -1 & 2 \end{pmatrix} = \begin{pmatrix} 5 & 10 & 10 \\ 12 & 13 & 24 \end{pmatrix},$$

the entry in the second row and third column of the result is obtained by the calculation

$$(5 \quad 2)\begin{pmatrix} 4 \\ 2 \end{pmatrix} = 5 \cdot 4 + 2 \cdot 2 = 24.$$

You should check that the other entries in the product can be obtained by the rule stated above. Schematically the entries in a matrix product are found by the mechanism illustrated below. The process is sometimes called row-by-column multiplication of matrices.

$$\begin{pmatrix} * & * & * & * \\ \boxed{* \;\; * \;\; * \;\; *} \\ * & * & * & * \end{pmatrix}\begin{pmatrix} * & * & * & \boxed{*} & * \\ * & * & * & \boxed{*} & * \\ * & * & * & \boxed{*} & * \\ * & * & * & \boxed{*} & * \end{pmatrix} = \begin{pmatrix} * & * & * & * & * \\ * & * & * & \boxed{*} & * \\ * & * & * & * & * \end{pmatrix}$$

The following remark is an obvious consequence of the way matrix multiplication is defined. We state it formally for emphasis and because we shall refer to it later.

3.1 Theorem

The ith row of a matrix product AB is equal to the ith row of A times B. The jth column of AB is equal to A times the jth column of B.

There are several important laws relating matrix multiplication and the operations of matrix addition and numerical multiplication. They hold for any number t and matrices A, B, C for which the indicated operations are defined. (Addition is defined only between matrices of the same shape. Multiplication is defined only if the left factor has exactly as many columns as the right factor has rows.)

1. $(A + B)C = AC + BC$.
2. $A(B + C) = AB + AC$.
3. $(tA)B = t(AB) = A(tB)$.
4. $A(BC) = (AB)C$.

According to the last law, it makes sense to talk of *the* product of three matrices and simply write ABC, since the result is independent of how the factors are grouped. In fact this 3-term associative law implies that the result of multiplying together any finite sequence of matrices is independent of how they are grouped. Not all the laws that hold for multiplication of numbers hold for multiplication of matrices. In particular, the value of a matrix product depends on the order of the factors, and AB is usually different from BA. It is also possible for the product of two matrices to be a zero matrix, without either of the factors being zero. Exercise 5 at the end of this section illustrates these points.

The laws stated above are easily proved by writing out what they mean, using the definitions of the operations, and then applying the associative, distributive, and commutative laws of arithmetic. Number 4 is the most complicated to prove, and we give its proof in full below. The other proofs are left as exercises.

To prove that $A(BC) = (AB)C$, let A, B, and C have respective shapes p-by-q, q-by-r, and r-by-s. Let $U = BC$ and $V = AB$. (Then U has shape q-by-s, and V has shape p-by-r.) We have to show that $AU = VC$. The ijth element of AU is (by definition of matrix multiplication) equal to $\sum_{k=1}^{q} a_{ik}u_{kj}$. The kjth element of U is $\sum_{l=1}^{r} b_{kl}c_{lj}$. Thus the ijth element of AU equals $\sum_{k=1}^{q} a_{ik}\left(\sum_{l=1}^{r} b_{kl}c_{lj}\right)$. Similarly, the ijth element of VC is equal to $\sum_{l=1}^{r} v_{il}c_{lj} = \sum_{l=1}^{r} \left(\sum_{k=1}^{q} a_{ik}b_{kl}\right) c_{lj}$. Both these expressions are equal to the sum

$$\sum_{\substack{1 \le l \le r \\ 1 \le k \le q}} a_{ik}b_{kl}c_{lj}$$

and hence are equal to each other. Thus corresponding entries of AU and VC are equal and the matrices are the same. This completes the proof.

A square matrix of the form

$$I = \begin{pmatrix} 1 & 0 \\ 0 & 1 \end{pmatrix} \quad \text{or} \quad I = \begin{pmatrix} 1 & 0 & 0 \\ 0 & 1 & 0 \\ 0 & 0 & 1 \end{pmatrix} \quad \text{or} \quad I = \begin{pmatrix} 1 & 0 & \dots & 0 & 0 \\ 0 & 1 & \dots & 0 & 0 \\ \cdot & \cdot & & \cdot & \cdot \\ \cdot & \cdot & & \cdot & \cdot \\ \cdot & \cdot & & \cdot & \cdot \\ 0 & 0 & \dots & 0 & 1 \end{pmatrix}$$

that has 1's on its main diagonal and zeros elsewhere is called an **identity matrix.** It has the property that

$$IA = A, \quad BI = B$$

for any matrices A, B such that the products are defined. Thus it is an identity element for matrix multiplication just as the number 1 is an identity for multiplication of numbers. There is an $n \times n$ identity matrix for every value of n, but, as with the zero matrices, it is almost always clear from the context what the dimension of an identity matrix must be.

If A is a square matrix and there is a matrix B (of the same size) such that

$$AB = BA = I,$$

then we say that A is **invertible** and that B is an **inverse** of A. As we show below in Theorem 3.3, there is at most one matrix B satisfying these conditions, so we are justified in speaking of *the* inverse of a matrix. If A is an invertible matrix, we write A^{-1} for its inverse. For example, it is easy to check that

$$\begin{pmatrix} 1 & 2 \\ 3 & 7 \end{pmatrix}\begin{pmatrix} 7 & -2 \\ -3 & 1 \end{pmatrix} = \begin{pmatrix} 7 & -2 \\ -3 & 1 \end{pmatrix}\begin{pmatrix} 1 & 2 \\ 3 & 7 \end{pmatrix} = \begin{pmatrix} 1 & 0 \\ 0 & 1 \end{pmatrix},$$

and according to the definition this shows that

$$\begin{pmatrix} 1 & 2 \\ 3 & 7 \end{pmatrix} \quad \text{is invertible and that} \quad \begin{pmatrix} 1 & 2 \\ 3 & 7 \end{pmatrix}^{-1} = \begin{pmatrix} 7 & -2 \\ -3 & 1 \end{pmatrix}.$$

Many matrices, on the other hand, are not invertible. No zero matrix can have an inverse, and several less obvious examples are given in the exercises.

It is usually not easy to tell whether a large matrix is invertible, and it can take a lot of work to compute its inverse if it has one. Determinants can be used to give a formula for the inverse of a matrix (Theorem 8.3), and a more effective way to compute inverses is given in Section 1 of Chapter 2. Two-by-two matrices and a few other easy cases are discussed in Exercises 9 to 13 of this section. The rule for finding the inverse of a 2-by-2 matrix is as follows.

3.2
$$\begin{pmatrix} a & b \\ c & d \end{pmatrix}^{-1} = \frac{1}{ad - bc}\begin{pmatrix} d & -b \\ -c & a \end{pmatrix}, \quad \text{if} \quad ad - bc \neq 0.$$

Thus, for example

$$\begin{pmatrix} 1 & 3 \\ -1 & 2 \end{pmatrix}^{-1} = \tfrac{1}{5}\begin{pmatrix} 2 & -3 \\ 1 & 1 \end{pmatrix} = \begin{pmatrix} \frac{2}{5} & -\frac{3}{5} \\ \frac{1}{5} & \frac{1}{5} \end{pmatrix}.$$

It is easy to verify that

$$\begin{pmatrix} 1 & 3 \\ -1 & 2 \end{pmatrix}\begin{pmatrix} \frac{2}{5} & -\frac{3}{5} \\ \frac{1}{5} & \frac{1}{5} \end{pmatrix} = \begin{pmatrix} \frac{2}{5} & -\frac{3}{5} \\ \frac{1}{5} & \frac{1}{5} \end{pmatrix}\begin{pmatrix} 1 & 3 \\ -1 & 2 \end{pmatrix} = \begin{pmatrix} 1 & 0 \\ 0 & 1 \end{pmatrix},$$

so that we have indeed found the inverse.

Two important properties of invertible matrices are easily proved directly from the definition. The first one ensures that, no matter how an inverse to a matrix A is computed, the resulting matrix A^{-1} is always the same.

3.3 Theorem

A matrix A has at most one inverse.

Proof. Suppose there are two matrices, B and C, such that both

$$AB = BA = I \quad \text{and} \quad AC = CA = I.$$

Then $AB = AC$, because both products equal I. Multiplying by B on the left gives

$$BAB = BAC.$$

But since $BA = I$, substitution gives $IB = IC$, from which it follows that $B = C$, as was to be proved.

The next theorem shows how to compute the inverse of a product of matrices each of which is invertible.

3.4 Theorem

If $A = A_1A_2 \ldots A_n$, and all of $A_1, \ldots, A_n$ are invertible, then A is invertible and $A^{-1} = A_n^{-1}A_{n-1}^{-1} \ldots A_2^{-1}A_1^{-1}$.

Proof. In the product $(A_n^{-1} \ldots A_2^{-1}A_1^{-1})(A_1A_2 \ldots A_n)$, the terms $A_1^{-1}A_1$ combine to give I, which may then be dropped. Then A_2^{-1} and A_2 cancel, and so on, until I is obtained as the final result. Similarly $(A_1 \ldots A_n)(A_n^{-1} \ldots A_1^{-1})$ reduces to I, and this shows that the two products are inverses of each other.

An identity matrix is a special case of what is called a diagonal matrix. A square matrix A is **diagonal** if its entries off the "main diagonal" are all zero, that is, if $a_{ij} = 0$ whenever $i \neq j$. The notation **diag** $(t_1, t_2, \ldots, t_n)$ is convenient for the $n \times n$ diagonal matrix which has entries $t_1, t_2, \ldots, t_n$ on the diagonal. For example, **diag** $(2, 0, -1, 3)$ is a notation for

$$\begin{pmatrix} 2 & 0 & 0 & 0 \\ 0 & 0 & 0 & 0 \\ 0 & 0 & -1 & 0 \\ 0 & 0 & 0 & 3 \end{pmatrix}.$$

Problems 7 and 8 show that matrix operations with diagonal matrices are particularly simple.

EXERCISES

1. Given the matrices

$$A = \begin{pmatrix} 1 & 3 \\ -4 & 2 \end{pmatrix}, \quad B = \begin{pmatrix} 0 & -2 & 1 \\ -1 & 3 & 0 \end{pmatrix}, \quad C = \begin{pmatrix} -2 & 0 & 1 \\ 0 & 3 & 0 \\ 2 & 3 & -1 \end{pmatrix},$$

$$D = \begin{pmatrix} 2 & -4 \\ 0 & 0 \\ 3 & 3 \end{pmatrix}, \quad G = \begin{pmatrix} 1 & -1 & 2 \\ 1 & 0 & 3 \end{pmatrix},$$

determine which of the following expressions are defined, and compute those that are.

(a) $2B - 3G$. (f) $CD + 3DB$.
(b) AB. (g) $2AB - 5G$.
(c) BA. (h) $2GC - 4AB$.
(d) BD. (i) CDC.
(e) DB. (j) DCD.

$$\left[Ans.\text{ (b) } \begin{pmatrix} -3 & 7 & 1 \\ -2 & 14 & -4 \end{pmatrix}.\right]$$

2. Show that for any matrix A and zero matrices of appropriate shapes,

$$A0 = 0 \quad \text{and} \quad 0A = 0.$$

If A is m-by-n, for what possible shapes of zero matrices is $A0$ defined? For what shapes is $0A$ defined? What are the shapes of the products?

3. With A, B, C, D as in Problem 1, determine what shapes X and Y would have to have for each of the following equations to be possible. (In some cases there may be no possible shape; in some cases there may be more than one.)

(a) $AX = B + Y$. (d) $CX + DY = 0$.
(b) $(D + 2X)YC = 0$. (e) $AX = YC$.
(c) $AX = YD$. (f) $AX = CY$.

[*Ans.* (d) X is 3-by-n, Y is 2-by-n.]

4. Prove the distributive law

$$A(B + C) = AB + AC$$

for matrix multiplication.

5. Let $U = \begin{pmatrix} -1 & 2 \\ 2 & -4 \end{pmatrix}$, $V = \begin{pmatrix} 2 & 6 \\ 1 & 3 \end{pmatrix}$. Compute UV and VU. Are they the same? It is possible for the product of two matrices to be zero without either factor being zero?

6. Let $X = \begin{pmatrix} 1 \\ 1 \\ 1 \end{pmatrix}$, $P = \begin{pmatrix} 0 & -1 \\ 4 & 3 \\ 2 & 0 \end{pmatrix}$, $Q = \begin{pmatrix} 1 & 2 & 0 \\ 3 & -4 & -1 \\ -1 & 2 & 0 \end{pmatrix}$.

Let $D = \mathbf{diag}\,(1, 2, 3) = \begin{pmatrix} 1 & 0 & 0 \\ 0 & 2 & 0 \\ 0 & 0 & 3 \end{pmatrix}$.

Compute DX, DP, and DQ.

7. Show that, as is illustrated by Exercise 6, the product DR, where D is a diagonal matrix $\mathbf{diag}\,(d_1, \ldots, d_n)$ and R is any n-rowed matrix, is obtained by multiplying the ith row of R by d_i, for all i. Suppose S has n columns.

How may the product SD be described? (Computing the product QD using the matrices of Exercise 6 should suggest the general rule.)

8. Using the matrices B, C, and G of Exercise 1, compute $B\mathbf{e}_1$, $C\mathbf{e}_1$, $G\mathbf{e}_1$, $B\mathbf{e}_3$, $C\mathbf{e}_3$, $G\mathbf{e}_3$. Prove the general rule that for any matrix M and column vector $\mathbf{e}_j$ with appropriate dimension, the product $M\mathbf{e}_j$ is the jth column of M.

9. What is the product $\mathbf{diag}\,(a_1, \ldots, a_n)\ \mathbf{diag}\,(b_1, \ldots, b_n)$? When is the result the identity matrix? Show that $\mathbf{diag}\,(a_1, \ldots, a_n)$ has an inverse provided none of the numbers a_i is zero.

10. (a) Let $A = \begin{pmatrix} a & b \\ c & d \end{pmatrix}$ be a 2-by-2 matrix with $ad \neq bc$, and let A^{-1} be given by Formula 3.2. Show that $AA^{-1} = A^{-1}A = I$. (This proves that A is invertible if $ad - bc$, called the *determinant* of A, is not zero.)

(b) Try to find the inverses of the following 2-by-2 matrices using Formula 3.2.

$$\begin{pmatrix} -1 & 1 \\ 2 & 1 \end{pmatrix}, \quad \begin{pmatrix} 0 & 1 \\ 1 & 0 \end{pmatrix}, \quad \begin{pmatrix} 2 & 6 \\ 1 & 3 \end{pmatrix}.$$

What is wrong with the last one?

11. Show that if there is any matrix $X \neq 0$ such that $AX = 0$, then A cannot be invertible. [*Hint.* Suppose $B = A^{-1}$, and consider $(BA)X = B(AX)$.] Use this result to show that $\mathbf{diag}\,(a_1, \ldots, a_n)$ is not invertible if any of the a_i is zero.

12. Show that if $ad = bc$, then $\begin{pmatrix} a & b \\ c & d \end{pmatrix}$ is not invertible.

13. If A is a square matrix, it can be multiplied by itself, and we can define $A^2 = AA$, $A^3 = AAA = A^2A$, $A^n = AA \ldots A$ (n factors). These powers of A all have the same shape as A. Find A^2 and A^3 if

$$\text{(a) } A = \begin{pmatrix} 2 & 1 \\ 0 & 1 \end{pmatrix} \qquad \text{(b) } A = \begin{pmatrix} 1 & 0 & -1 \\ -1 & 0 & 1 \\ 2 & 1 & -1 \end{pmatrix}.$$

$$\left[\textit{Ans.}\ \text{(a)} \quad A^2 = \begin{pmatrix} 4 & 3 \\ 0 & 1 \end{pmatrix}. \right]$$

(Note that 0 is the only number whose cube is 0. Part (b) of this problem thus illustrates another difference between the arithmetic of numbers and of matrices.)

14. The numerical equation $a^2 = 1$ has $a = 1$ and $a = -1$ as its only solutions.

(a) Show that if $A = I$ or $-I$, then $A^2 = I$, where I is an identity matrix of any dimension.

(b) Show that $\begin{pmatrix} a & b \\ c & -a \end{pmatrix}^2 = \begin{pmatrix} 1 & 0 \\ 0 & 1 \end{pmatrix}$ if $a^2 + bc = 1$; so the equation

$A^2 = I$ has infinitely many different solutions in the set of 2-by-2 matrices.

(c) Show that every 2-by-2 matrix A for which $A^2 = I$ is either I, $-I$, or one of the matrices described in (b).

SECTION 4

LINEAR FUNCTIONS

The product of an m-by-n matrix and n-dimensional column vector (n-by-1 matrix) is an m-dimensional column vector. An n-tuple in $\mathcal{R}^n$ obviously corresponds to a unique n-dimensional column vector, and vice versa. From here on we shall often simply consider elements of $\mathcal{R}^n$ to be column vectors. With this convention, multiplication by any given m-by-n matrix defines a function from $\mathcal{R}^n$ to $\mathcal{R}^m$. Indeed a matrix equation such as

$$\begin{pmatrix} y_1 \\ y_2 \end{pmatrix} = \begin{pmatrix} 2 & 3 & -4 \\ 1 & -1 & 2 \end{pmatrix} \begin{pmatrix} x_1 \\ x_2 \\ x_3 \end{pmatrix}$$

is equivalent to a set of numerical equations

$$\begin{aligned} y_1 &= 2x_1 + 3x_2 - 4x_3 \\ y_2 &= x_1 - x_2 + 2x_3, \end{aligned}$$

as may be seen by simply writing out the result of the matrix multiplication. Thus the vector function described by a matrix amounts to multiplication of a domain vector by the matrix.

Functions given by matrix multiplication can be characterized by some very simple properties. Note that the definitions given below apply to functions between any two vector spaces, although we are at present concerned only with the spaces $\mathcal{R}^n$.

A function f with domain a vector space $\mathcal{V}$ and range a subset of a vector space $\mathcal{W}$ is a **linear** function if the equations

$$f(\mathbf{x} + \mathbf{y}) = f(\mathbf{x}) + f(\mathbf{y})$$
$$f(r\mathbf{x}) = rf(\mathbf{x})$$

hold for all vectors $\mathbf{x}$, $\mathbf{y}$ in $\mathcal{V}$, and all numbers r.

4.1 Theorem

A function from $\mathcal{R}^n$ to $\mathcal{R}^m$ is linear if and only if it coincides with multiplication by some m-by-n matrix.

Proof. We show first that if f is defined by $f(\mathbf{x}) = A\mathbf{x}$ for a fixed matrix A, then f is linear. We must show that $f(\mathbf{x} + \mathbf{y}) = f(\mathbf{x}) + f(\mathbf{y})$ and $f(r\mathbf{x}) = rf(\mathbf{x})$ for any $\mathbf{x}$, $\mathbf{y}$, and r. But by the definition of f, these equations amount to $A(\mathbf{x}+\mathbf{y}) = A\mathbf{x} + A\mathbf{y}$ and $A(r\mathbf{x}) = r(A\mathbf{x})$, which hold by properties 2 and 3 for matrix multiplication.

The proof of the converse is a little more complicated. Suppose we are given a linear function g from $\mathcal{R}^n$ to $\mathcal{R}^m$. We have to find an m-by-n matrix A such that $g(\mathbf{x}) = A\mathbf{x}$ for every column vector $\mathbf{x}$ in $\mathcal{R}^n$. Let $\mathbf{e}_j$ be the column vector in $\mathcal{R}^n$ that has 1 for its jth entry and 0 for all other entries. Now let A be the matrix whose jth column is the m-dimensional vector $g(\mathbf{e}_j)$ for $j = 1, 2, \ldots, n$. (Thus A has m rows and n columns, as required.) Any column vector $\mathbf{x}$ with entries $x_1, x_2, \ldots, x_n$ can be written as $x_1\mathbf{e}_1 + \ldots + x_n\mathbf{e}_n$. Since g is linear,

$$
\begin{aligned}
g(\mathbf{x}) &= g(x_1\mathbf{e}_1 + \ldots + x_n\mathbf{e}_n) \\
&= x_1 g(\mathbf{e}_1) + \ldots + x_n g(\mathbf{e}_n).
\end{aligned}
$$

By the definition of A we have

$$
g(\mathbf{x}) = x_1 \begin{pmatrix} a_{11} \\ a_{21} \\ \cdot \\ \cdot \\ \cdot \\ a_{m1} \end{pmatrix} + \ldots + x_n \begin{pmatrix} a_{1n} \\ a_{2n} \\ \cdot \\ \cdot \\ \cdot \\ a_{mn} \end{pmatrix}
$$

$$
= \begin{pmatrix} a_{11}x_1 + \ldots + a_{1n}x_n \\ a_{21}x_1 + \ldots + a_{2n}x_n \\ \cdot \\ \cdot \\ \cdot \\ a_{m1}x_1 + \ldots + a_{mn}x_n \end{pmatrix}
$$

Finally, by the definition of matrix multiplication,

$$
g(\mathbf{x}) = A\mathbf{x}.
$$

The proof given above actually shows how to construct the matrix corresponding to a linear function. This construction is important and we summarize it as a theorem for emphasis.

4.2 Theorem

Let $\mathcal{R}^n \xrightarrow{f} \mathcal{R}^m$ be a linear function, and let A be the matrix whose jth column is $f(\mathbf{e}_j)$. Then $f(\mathbf{x}) = A\mathbf{x}$ for every $\mathbf{x}$ in $\mathcal{R}^n$.

The result is easy to remember if we consider, for example, the case of a function f from $\mathcal{R}^2$ to $\mathcal{R}^2$. The matrix A, for which $f(\mathbf{x}) = A\mathbf{x}$, is then 2-by-2, and its columns are the result of applying A successively to $\mathbf{e}_1 = \begin{pmatrix} 1 \\ 0 \end{pmatrix}$ and $\mathbf{e}_2 = \begin{pmatrix} 0 \\ 1 \end{pmatrix}$:

$$\begin{pmatrix} a & c \\ b & d \end{pmatrix}\begin{pmatrix} 1 \\ 0 \end{pmatrix} = \begin{pmatrix} a \\ b \end{pmatrix}, \qquad \begin{pmatrix} a & c \\ b & d \end{pmatrix}\begin{pmatrix} 0 \\ 1 \end{pmatrix} = \begin{pmatrix} c \\ d \end{pmatrix}.$$

Example 1. Consider the function $\mathcal{R}^2 \xrightarrow{f} \mathcal{R}^2$ described in terms of the geometric representation of $\mathcal{R}^2$ as a counterclockwise rotation of 30°. Any rotation leaves the origin fixed, preserves distances, and carries figures such as parallelograms onto congruent figures. From these properties, and the way in which addition and numerical multiplication of vectors can be done by geometrical constructions (Section 2), it can be shown that a rotation is a linear function. Figure 10(a) shows the vectors $\mathbf{e}_1$, $\mathbf{e}_2$, $f(\mathbf{e}_1)$, and $f(\mathbf{e}_2)$. By trigonometry we see that

$$f(\mathbf{e}_1) = \begin{pmatrix} \dfrac{\sqrt{3}}{2} \\ \dfrac{1}{2} \end{pmatrix} \quad \text{and} \quad f(\mathbf{e}_2) = \begin{pmatrix} -\dfrac{1}{2} \\ \dfrac{\sqrt{3}}{2} \end{pmatrix}.$$

By Theorem 4.2,

$$f\begin{pmatrix} x_1 \\ x_2 \end{pmatrix} = \begin{pmatrix} \dfrac{\sqrt{3}}{2} & -\dfrac{1}{2} \\ \dfrac{1}{2} & \dfrac{\sqrt{3}}{2} \end{pmatrix}\begin{pmatrix} x_1 \\ x_2 \end{pmatrix} = \begin{pmatrix} \dfrac{\sqrt{3}}{2}x_1 - \dfrac{1}{2}x_2 \\ \dfrac{1}{2}x_1 + \dfrac{\sqrt{3}}{2}x_2 \end{pmatrix}$$

for any vector $\begin{pmatrix} x_1 \\ x_2 \end{pmatrix}$ in $\mathcal{R}^2$.

Example 2. Define $\mathcal{R}^2 \xrightarrow{f} \mathcal{R}^2$ to be the linear function which multiplies horizontal distances by 3 and vertical distances by 2. Then $f(\mathbf{e}_1) = 3\mathbf{e}_1$ and $f(\mathbf{e}_2) = 2\mathbf{e}_2$, and f has the diagonal matrix

$$\begin{pmatrix} 3 & 0 \\ 0 & 2 \end{pmatrix}.$$

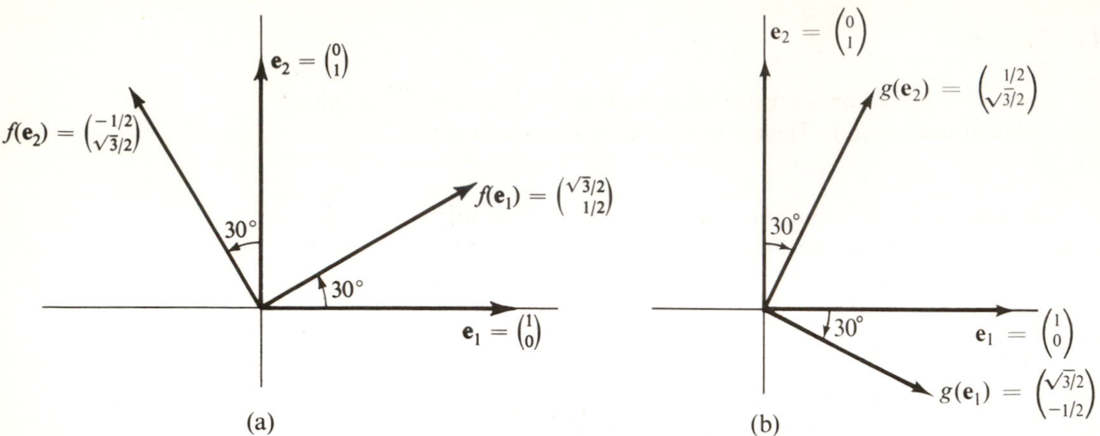

Figure 10

The geometrical effect of f is illustrated in Fig. 11, in which C is the unit circle $x_1^2 + x_2^2 = 1$, and $f(C)$ is the image of C under f. If $\begin{pmatrix} u_1 \\ u_2 \end{pmatrix}$ is the image of $\begin{pmatrix} x_1 \\ x_2 \end{pmatrix}$ under f, then $x_1 = \frac{1}{3}u_1$ and $x_2 = \frac{1}{2}u_2$. Hence, if $\begin{pmatrix} u_1 \\ u_2 \end{pmatrix}$ is in $f(C)$, then $\frac{u_1^2}{9} + \frac{u_2^2}{4} = 1$; this can be recognized as the equation of an ellipse with semimajor axis 3 and semiminor axis 2.

Theorem 4.2 says that, if f is a linear function from $\mathcal{R}^n$ to $\mathcal{R}^m$, then the jth column of the matrix of f is $f(\mathbf{e}_j)$. Because we can write any vector $\mathbf{x}$ in $\mathcal{R}^n$ as a linear combination

$$\mathbf{x} = c_1\mathbf{e}_1 + \ldots + c_n\mathbf{e}_n,$$

$f(e_2) = 2e_2$
e_2
e_1
$f(e_1) = 3e_1$
C
$f(C)$

Figure 11

we can use the linearity of f to write

$$f(\mathbf{x}) = c_1 f(\mathbf{e}_1) + \ldots + c_n f(\mathbf{e}_n).$$

Since $\mathbf{x}$ is an arbitrary vector in the domain of f, the last equation expresses an arbitrary vector in the range of f as a linear combination of the vectors $f(\mathbf{e}_1), \ldots, f(\mathbf{e}_n)$. But these vectors are just the columns of the matrix of f so we have the following.

4.3 Theorem

Let $\mathcal{R}^n \xrightarrow{f} \mathcal{R}^m$ be a linear function with matrix A. Then the columns of A, looked at as vectors of $\mathcal{R}^m$, span the range of f.

Example 3. If P is a plane through the origin in $\mathcal{R}^3$, then P consists of all linear combinations of two vectors $\mathbf{y}_1$ and $\mathbf{y}_2$ in $\mathcal{R}^3$. The function $\mathcal{R}^2 \xrightarrow{f} \mathcal{R}^3$ defined by

$$f\begin{pmatrix} u \\ v \end{pmatrix} = u\mathbf{y}_1 + v\mathbf{y}_2$$

is linear (why?), and Theorem 4.2 says that the matrix of f has as columns the two vectors

$$f\begin{pmatrix} 1 \\ 0 \end{pmatrix} = \mathbf{y}_1 \quad \text{and} \quad f\begin{pmatrix} 0 \\ 1 \end{pmatrix} = \mathbf{y}_2.$$

For example, if

$$\mathbf{y}_1 = \begin{pmatrix} 1 \\ 2 \\ 1 \end{pmatrix} \quad \text{and} \quad \mathbf{y}_2 = \begin{pmatrix} -1 \\ 0 \\ 1 \end{pmatrix},$$

then

$$f\begin{pmatrix} u \\ v \end{pmatrix} = \begin{pmatrix} 1 & -1 \\ 2 & 0 \\ 1 & 1 \end{pmatrix}\begin{pmatrix} u \\ v \end{pmatrix},$$

and indeed the range of f is spanned by the columns of the 3-by-2 matrix.

Example 4. If L is a line consisting of all points of the form $t\mathbf{u}_1 + \mathbf{u}_0$, with $\mathbf{u}_1 \neq 0$, and f is a linear function, then

$$\begin{aligned} f(t\mathbf{u}_1 + \mathbf{u}_0) &= tf(\mathbf{u}_1) + f(\mathbf{u}_0) \\ &= t\mathbf{v}_1 + \mathbf{v}_0. \end{aligned}$$

Thus, unless $f(\mathbf{u}_1) = 0$, the image $f(L)$ is also a line. If $f(\mathbf{u}_1) = 0$, then of course $f(L)$ is the single point $f(\mathbf{u}_0)$. The equation $f(\mathbf{u}_1) = 0$ cannot hold for a nonzero vector $\mathbf{u}_1$ if the linear function f is one-to-one, as defined below.

A function f is said to be **one-to-one** if each point in the range of f corresponds to exactly one point in the domain of f. In other words, f is one-to-one if the equation $f(\mathbf{x}_1) = f(\mathbf{x}_2)$ always implies that $\mathbf{x}_1 = \mathbf{x}_2$. For example, of the two real-valued functions $f(x) = x$ and $g(x) = x^2$, it is obvious that f is one-to-one. But g is not one-to-one because $g(x) = g(-x)$ for every real number x. For *linear* functions we have the following criterion.

4.4 Theorem

A linear function f is one-to-one if and only if $f(\mathbf{x}) = 0$ implies $\mathbf{x} = 0$.

Proof. For suppose that $f(\mathbf{x}) = 0$ always implies $\mathbf{x} = 0$. If $f(\mathbf{x}_1) = f(\mathbf{x}_2)$, the linearity of f shows that $f(\mathbf{x}_1 - \mathbf{x}_2) = 0$. But then by assumption $\mathbf{x}_1 - \mathbf{x}_2 = 0$; so $\mathbf{x}_1 = \mathbf{x}_2$, and therefore f must be one-to-one. Conversely, suppose that $f(\mathbf{x}_1) = f(\mathbf{x}_2)$ always implies $\mathbf{x}_1 = \mathbf{x}_2$. Then, because $f(0) = 0$ for a linear function, the equation $f(\mathbf{x}) = 0$ can be written $f(\mathbf{x}) = f(0)$. But then our assumption implies $\mathbf{x} = 0$.

Example 5. The counterclockwise rotation through 30° in Example 1 is certainly one-to-one. For only one vector is carried into any other vector by the rotation. Alternatively, the zero vector is the only vector carried into the zero vector. In algebraic terms, the rotation is proved in Example 1 to be representable by

$$f\begin{pmatrix} x \\ y \end{pmatrix} = \begin{pmatrix} \dfrac{\sqrt{3}}{2}x - \dfrac{1}{2}y \\ \dfrac{1}{2}x + \dfrac{\sqrt{3}}{2}y \end{pmatrix}$$

Then Theorem 4.4 shows that the one-to-one property of f is equivalent to the equations

$$\frac{\sqrt{3}}{2}x - \frac{1}{2}y = 0$$

$$\frac{1}{2}x + \frac{\sqrt{3}}{2}y = 0$$

having only the solution $(x, y) = (0, 0)$. This of course can be verified directly by solving the equations.

If f and g are any two functions (not necessarily linear) such that the range space of f is the same as the domain space of g we define the **composition** $g \circ f$ to be the function obtained by applying first f and then g. More explicitly, if $\mathbf{x}$ is in the domain of f and $f(\mathbf{x})$ is in the domain of g, then $g \circ f(\mathbf{x})$ is defined as $g(f(\mathbf{x}))$. If $f(\mathbf{x})$ or $g(f(\mathbf{x}))$ is not defined, then $g \circ f(\mathbf{x})$ is not defined.

Composition of linear functions lies behind matrix multiplication. In introducing the concept of matrix multiplication, we considered a function from $\mathcal{R}^2$ to $\mathcal{R}^2$ given by

$$\begin{aligned} z_1 &= 3y_1 - y_2 \\ z_2 &= 5y_1 + 2y_2 \end{aligned}$$

and another from $\mathcal{R}^3$ to $\mathcal{R}^2$ given by

$$\begin{aligned} y_1 &= 2x_1 + 3x_2 + 4x_3 \\ y_2 &= x_1 - x_2 + 2x_3. \end{aligned}$$

We then computed that the composition of the two functions was given by the formulas

$$\begin{aligned} z_1 &= 5x_1 + 10x_2 + 10x_3 \\ z_2 &= 12x_1 + 13x_2 + 24x_3. \end{aligned}$$

The definition of matrix multiplication was set up to give the matrix $\begin{pmatrix} 5 & 10 & 10 \\ 12 & 13 & 24 \end{pmatrix}$ of the composite function as the product of the matrices

$$\begin{pmatrix} 3 & -1 \\ 5 & 2 \end{pmatrix}\begin{pmatrix} 2 & 3 & 4 \\ 1 & -1 & 2 \end{pmatrix}$$

for the original functions. The following theorem states the important fact that the composition of linear functions is always given by matrix multiplication.

4.5 Theorem

Let f and g be linear functions with $\mathcal{R}^n \xrightarrow{f} \mathcal{R}^m$, $\mathcal{R}^m \xrightarrow{g} \mathcal{R}^p$ given by matrices A, B. Thus $f(\mathbf{x}) = A\mathbf{x}$ and $g(\mathbf{y}) = B\mathbf{y}$ for all $\mathbf{x}$ in $\mathcal{R}^n$ and $\mathbf{y}$ in $\mathcal{R}^m$. Then the composition $g \circ f$ is given by the matrix BA, so that $g \circ f(\mathbf{x}) = BA\mathbf{x}$ for all $\mathbf{x}$ in $\mathcal{R}^n$.

Proof. Note that A is an m-by-n matrix and B a p-by-m matrix, so the matrix product BA is defined and has the appropriate shape, namely, p-by-n. By the definition of $g \circ f$, $g \circ f(\mathbf{x}) = g(f(\mathbf{x})) = B(A\mathbf{x})$ for all $\mathbf{x}$ in $\mathcal{R}^n$. By the associative law of matrix multiplication this is equal to $(BA)\mathbf{x}$, as was to be proved.

Example 6. Consider the function $\mathcal{R}^2 \xrightarrow{g} \mathcal{R}^2$ defined as $f \circ f \circ f$, where f is the function of Example 1. By Theorem 4.5, $g(\mathbf{x}) = B\mathbf{x}$, where

$$B = \begin{pmatrix} \frac{\sqrt{3}}{2} & -\frac{1}{2} \\ \frac{1}{2} & \frac{\sqrt{3}}{2} \end{pmatrix} \begin{pmatrix} \frac{\sqrt{3}}{2} & -\frac{1}{2} \\ \frac{1}{2} & \frac{\sqrt{3}}{2} \end{pmatrix} \begin{pmatrix} \frac{\sqrt{3}}{2} & -\frac{1}{2} \\ \frac{1}{2} & \frac{\sqrt{3}}{2} \end{pmatrix}$$

$$= \begin{pmatrix} \frac{1}{2} & -\frac{\sqrt{3}}{2} \\ \frac{\sqrt{3}}{2} & \frac{1}{2} \end{pmatrix} \begin{pmatrix} \frac{\sqrt{3}}{2} & -\frac{1}{2} \\ \frac{1}{2} & \frac{\sqrt{3}}{2} \end{pmatrix} = \begin{pmatrix} 0 & -1 \\ 1 & 0 \end{pmatrix}.$$

Thus $g(\mathbf{e}_1) = B\mathbf{e}_1 = \begin{pmatrix} 0 \\ 1 \end{pmatrix} = \mathbf{e}_2$ and $g(\mathbf{e}_2) = B\mathbf{e}_2 = \begin{pmatrix} -1 \\ 0 \end{pmatrix} = -\mathbf{e}_1$. As Fig. 12 shows, g amounts geometrically to a counterclockwise rotation of 90°, which of course is what the result of three successive 30° rotations ought to be.

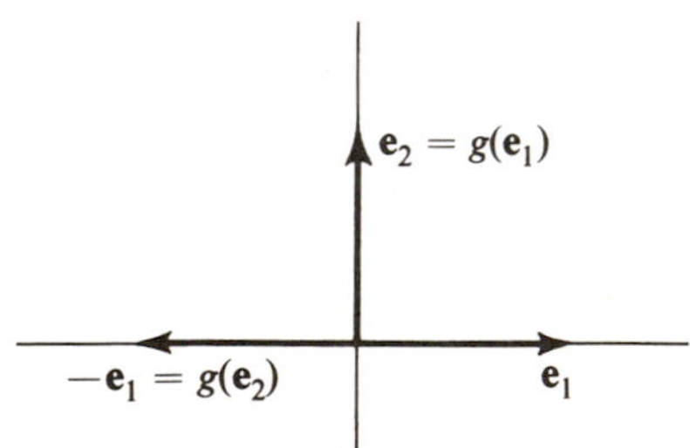

Figure 12

Example 7. The identity function $\mathcal{R}^n \xrightarrow{h} \mathcal{R}^n$, such that $h(\mathbf{x}) = \mathbf{x}$, is a linear function. Its matrix is the n-by-n identity matrix I. Two functions $\mathcal{R}^n \xrightarrow{f} \mathcal{R}^n$ and $\mathcal{R}^n \xrightarrow{g} \mathcal{R}^n$ are said to be **inverses** of each other if $f \circ g$ and $g \circ f$ are both equal to the identity function. For example, if $n = 2$, suppose that f is the counterclockwise rotation through 30° discussed in Example 1, and suppose that g is a *clockwise* rotation through 30°. The functions f and g are obviously inverses of each other. The matrices of

f and g are found from Fig. 12 to be

$$\begin{pmatrix} \frac{\sqrt{3}}{2} & -\frac{1}{2} \\ \frac{1}{2} & \frac{\sqrt{3}}{2} \end{pmatrix} \quad \text{and} \quad \begin{pmatrix} \frac{\sqrt{3}}{2} & \frac{1}{2} \\ -\frac{1}{2} & \frac{\sqrt{3}}{2} \end{pmatrix},$$

respectively. It is easy to verify that multiplying these matrices in either order gives the 2-by-2 identity matrix. Thus the matrices of f and g are inverses of each other. More generally, linear functions from $\mathcal{R}^n$ to $\mathcal{R}^n$ are inverses of each other if and only if their matrices are inverses of each other.

In dealing with functions of one variable, it is customary to call any function of the form $f(x) = ax + b$ a linear function. This terminology is not consistent with our definition of linearity. For example, if $f(x) = 2x + 1$, then $f(x + y) = 2(x + y) + 1$, while $f(x) + f(y) = 2x + 2y + 2$; so this function fails to satisfy the condition $f(x + y) = f(x) + f(y)$ that is required for linearity. Since our definition of linear functions between vector spaces is the standard one, we shall not consider functions such as $2x + 1$ to be linear. Functions of this more general type are important, however, and there is a standard term for them. They are called *affine* functions, and are characterized as having the form of a linear function (such as $2x$) plus a constant (such as 1). The formal definitions follow.

If $\mathbf{b}$ is an element of a vector space $\mathcal{V}$, then the function $t_{\mathbf{b}}$ from $\mathcal{V}$ to itself defined by $t_{\mathbf{b}}(\mathbf{x}) = \mathbf{x} + \mathbf{b}$ for all $\mathbf{x}$ in $\mathcal{V}$ is called the **translation** of $\mathcal{V}$ induced by $\mathbf{b}$.

Example 8. Let $\mathbf{b}$ be the vector $(1, 0)$ in $\mathcal{R}^2$. Then t_b sends any vector (x, y) into the vector $(x + 1, y)$. In geometrical language, every point of the plane is moved one unit to the right.

A function from a vector space $\mathcal{V}$ to a vector space $\mathcal{W}$ is an **affine** function if it is the composition of a linear function from $\mathcal{V}$ to $\mathcal{W}$ with a translation of $\mathcal{W}$.

Example 9. Let f from $\mathcal{R}^2$ to $\mathcal{R}^2$ have the matrix $\begin{pmatrix} 1 & 0 \\ 0 & -1 \end{pmatrix}$ and let $\mathbf{b}$ be the vector $(0, 2)$. The affine function $t_{\mathbf{b}} \circ f$ then sends any vector (x, y) into $(x, -y + 2)$. Geometrically, this amounts to a reflection in the x-axis followed by a motion of 2 units straight up. It is easy to see that all the points of the line $y = 1$ remain fixed under this function, and that it may be described as reflection in the line $y = 1$.

The following theorem is an immediate consequence of the definitions and of Theorem 4.1.

4.6 Theorem

> A function f from $\mathcal{R}^n$ to $\mathcal{R}^m$ is affine if and only if $f(\mathbf{x}) = A\mathbf{x} + \mathbf{b}$ for some fixed m-by-n matrix A and m-dimensional column vector $\mathbf{b}$.

Finding the matrix A and vector $\mathbf{b}$ needed to describe an affine transformation is easy. If $f(\mathbf{x}) = A\mathbf{x} + \mathbf{b}$, then $\mathbf{b} = f(0)$. Then the function $g(\mathbf{x}) = f(\mathbf{x}) - f(0)$ is linear and the matrix A is found by using Theorem 4.2.

Example 10. Let $\mathcal{R}^2 \xrightarrow{f} \mathcal{R}^2$ be defined in terms of the standard geometric representation as a counterclockwise rotation of 90° about the center $\begin{pmatrix} 1 \\ 0 \end{pmatrix} = \mathbf{e}_1$. We see that $f(0) = \begin{pmatrix} 1 \\ -1 \end{pmatrix}$, $f(\mathbf{e}_1) = \begin{pmatrix} 1 \\ 0 \end{pmatrix}$ (since the center of a rotation stays fixed), and $f\begin{pmatrix} 0 \\ 1 \end{pmatrix} = \begin{pmatrix} 0 \\ -1 \end{pmatrix}$. Introducing the linear function $g(\mathbf{x}) = f(\mathbf{x}) - f(0)$, we have $g(\mathbf{e}_1) = \begin{pmatrix} 1 \\ 0 \end{pmatrix} - \begin{pmatrix} 1 \\ -1 \end{pmatrix} = \begin{pmatrix} 0 \\ 1 \end{pmatrix}$, and

$$g(\mathbf{e}_2) = \begin{pmatrix} 0 \\ -1 \end{pmatrix} - \begin{pmatrix} 1 \\ -1 \end{pmatrix} = \begin{pmatrix} -1 \\ 0 \end{pmatrix}.$$

Therefore $g(\mathbf{x}) = \begin{pmatrix} 0 & -1 \\ 1 & 0 \end{pmatrix} \mathbf{x}$ for all $\mathbf{x}$ and

$$f(\mathbf{x}) = \begin{pmatrix} 0 & -1 \\ 1 & 0 \end{pmatrix}\mathbf{x} + \begin{pmatrix} 1 \\ -1 \end{pmatrix}.$$

As a check, let us compute

$$f\begin{pmatrix} 2 \\ 0 \end{pmatrix} = \begin{pmatrix} 0 & -1 \\ 1 & 0 \end{pmatrix}\begin{pmatrix} 2 \\ 0 \end{pmatrix} + \begin{pmatrix} 1 \\ -1 \end{pmatrix} = \begin{pmatrix} 0 \\ 2 \end{pmatrix} + \begin{pmatrix} 1 \\ -1 \end{pmatrix} = \begin{pmatrix} 1 \\ 1 \end{pmatrix},$$

which agrees with the geometric description of f. See Fig. 13.

Linear and affine functions are often called linear and affine **transformations,** though we more often use the term function in this book.

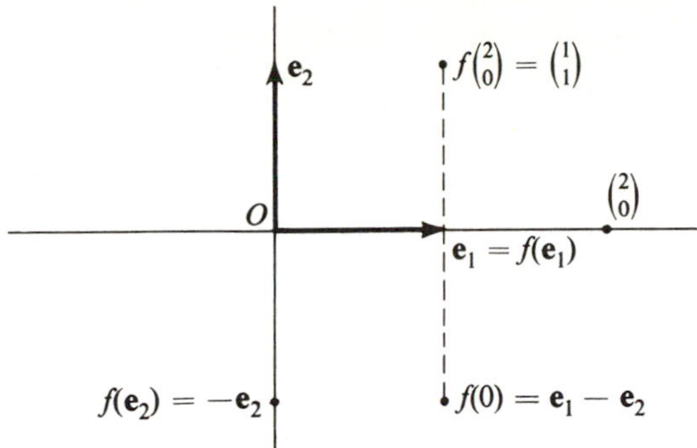

Figure 13

EXERCISES

1. (a) Show that the two conditions on a linear function given in the text are equivalent to the single condition that

$$f(r\mathbf{x} + s\mathbf{y}) = rf(\mathbf{x}) + sf(\mathbf{y})$$

hold for all $\mathbf{x}$, $\mathbf{y}$ in the domain of f and all numbers r, s.

(b) Give a proof by induction on n that if f is linear, then

$$f(r_1\mathbf{x}_1 + \ldots + r_{n-1}\mathbf{x}_{n-1} + r_n\mathbf{x}_n) = r_1 f(\mathbf{x}_1) + \ldots + r_{n-1}f(\mathbf{x}_{n-1}) + r_n f(\mathbf{x}_n)$$

for any numbers $r_1, \ldots, r_n$ and vectors $\mathbf{x}_1, \ldots, \mathbf{x}_n$.

2. Show that $f(x) = ax + b$ is a linear function in the terminology of this book if and only if $b = 0$.

3. Let f and g be linear functions from $\mathcal{R}^2$ to $\mathcal{R}^2$ such that

(a) $f\begin{pmatrix}1\\0\end{pmatrix} = \begin{pmatrix}2\\1\end{pmatrix}$, $f\begin{pmatrix}0\\1\end{pmatrix} = \begin{pmatrix}1\\-1\end{pmatrix}$.

(b) $g\begin{pmatrix}1\\1\end{pmatrix} = \begin{pmatrix}2\\1\end{pmatrix}$, $g\begin{pmatrix}-1\\1\end{pmatrix} = \begin{pmatrix}1\\-1\end{pmatrix}$.

In each case, find the matrix that represents the function.

4. Show that a counterclockwise rotation in $\mathcal{R}^2$ through an angle α is described by the matrix $R_\alpha = \begin{pmatrix}\cos\alpha & -\sin\alpha\\ \sin\alpha & \cos\alpha\end{pmatrix}$. Let β be another angle, and compute the product $R_\alpha R_\beta$. The composition of a rotation through angle α with one through angle β is a rotation through the angle $\alpha + \beta$. What is the relation between $R_\alpha R_\beta$ and $R_{\alpha+\beta}$? What is the inverse of R_α?

5. (a) Show that the matrix $\begin{pmatrix}0 & 1\\ 1 & 0\end{pmatrix}$ gives a linear function from $\mathcal{R}^2$ to $\mathcal{R}^2$ which corresponds geometrically to reflection in the line through the origin 45° counterclockwise from the horizontal.

(b) What matrix corresponds to reflection in the line through the origin 135° counterclockwise from the horizontal?

(c) Compute the product of the matrices in (a) and (b) and interpret the result geometrically.

$$\left[\textit{Ans.}\ \text{(b)}\ \begin{pmatrix} 0 & -1 \\ -1 & 0 \end{pmatrix}.\right]$$

6. (a) Find the 2-by-2 matrix M_α corresponding to reflection in the line through the origin at an angle α from the horizontal. Check your result against Exercise 5 for $\alpha = 45°$, $\alpha = 135°$. What is M_α^2?

(b) Let β be another angle and compute the product $M_\alpha M_\beta$. Show that this represents a rotation, and identify the angle of rotation. When does $M_\alpha M_\beta = M_\beta M_\alpha$?

7. (a) Show that

$$U = \begin{pmatrix} 1 & 0 & 0 \\ 0 & 0 & -1 \\ 0 & 1 & 0 \end{pmatrix} \quad \text{and} \quad V = \begin{pmatrix} 0 & 0 & 1 \\ 0 & 1 & 0 \\ -1 & 0 & 0 \end{pmatrix}$$

represent 90° rotations of $\mathcal{R}^3$ about the x_1-axis and x_2-axis, respectively. Find the matrix W which represents a 90° rotation about the x_3-axis. Also find U^{-1} and V^{-1} (which represent rotations in the opposite direction).

(b) Compute UVU^{-1} and VUV^{-1} and interpret the results geometrically. (You may find it helpful to manipulate an actual 3-dimensional model.)

8. Show that a function f from one vector space to another is affine if and only if

$$f(r\mathbf{x} + (1 - r)\mathbf{y}) = rf(\mathbf{x}) + (1 - r)f(\mathbf{y})$$

for all numbers r and all $\mathbf{x}$, $\mathbf{y}$ in the domain of f. [*Hint.* Consider the function $g(\mathbf{x}) = f(\mathbf{x}) - f(0)$.]

9. Let f be a linear function from a vector space $\mathcal{V}$ to a vector space $\mathcal{W}$. Show that if $\mathbf{x}_1, \ldots, \mathbf{x}_n$ are linearly independent vectors in $\mathcal{V}$ and f is one-to-one, then the vectors $f(\mathbf{x}_1), \ldots, f(\mathbf{x}_n)$ are linearly independent in $\mathcal{W}$. [*Hint.* If f is one-to-one, and $f(\mathbf{x}) = 0$, then $\mathbf{x} = 0$.]

10. (a) Prove that if f is a linear function from $\mathcal{R}^3$ to $\mathcal{R}^3$ and f is one-to-one, then the image $f(L)$ of a line L by f is also a line.

(b) Show by example that, if the linear function of part (a) fails to be one-to-one, then the image of a line by f may reduce to a point.

(c) Show that if L_1 and L_2 are parallel lines and f is a linear function, then $f(L_1)$ and $f(L_2)$ are parallel lines, provided that f is one-to-one.

11. Show that the composition of two affine functions from a space into itself is affine. Suppose $f(\mathbf{x}) = A\mathbf{x} + \mathbf{b}$, $g(\mathbf{x}) = C\mathbf{x} + \mathbf{d}$. Suppose $(f \circ g)(\mathbf{x}) = P\mathbf{x} + \mathbf{q}$. Express P and $\mathbf{q}$ in terms of A, $\mathbf{b}$, C, and $\mathbf{d}$. When is $f \circ g$ the same function as $g \circ f$?

12. (a) Let $\mathbf{a} = \begin{pmatrix} 1 \\ 0 \end{pmatrix}$, $\mathbf{b} = -\mathbf{a} = \begin{pmatrix} -1 \\ 0 \end{pmatrix}$, and let $\mathcal{R}^2 \xrightarrow{g} \mathcal{R}^2$ be a counterclockwise rotation of 90° with center at the origin. Find the matrix corresponding to the linear function g and compute the affine function $f = t_{\mathbf{a}} \circ g \circ t_{\mathbf{b}}$, where $t_{\mathbf{a}}$ and $t_{\mathbf{b}}$ are the translations induced by **a** and **b**.
(b) Give a geometric interpretation of the composition $f = t_{\mathbf{a}} \circ g \circ t_{\mathbf{b}}$, where **a** is any vector in $\mathcal{R}^2$, $\mathbf{b} = -\mathbf{a}$, and g is any rotation about the origin. [*Hint.* What happens to the point corresponding to **a** under the function f?]

13. Show that an affine function A is one-to-one if and only if $A(\mathbf{x}) = A(0)$ always implies $\mathbf{x} = 0$.

14. Which of the following linear functions are one-to-one?

(a) $f(x_1, x_2) = (x_1 + 2x_2, 2x_1 + x_2)$.
(b) $g(x_1, x_2, x_3) = (x_2 - x_3, x_3 - x_1, x_1 - x_2)$.

SECTION 5 DOT PRODUCTS

To allow the full application of vector ideas to Euclidean geometry, we must have a means of introducing concepts such as length, angle, and perpendicularity. We will show in this section that all these concepts can be defined if we introduce a new operation on vectors. If $\mathbf{x} = (x_1, \ldots, x_n)$ and $\mathbf{y} = (y_1, \ldots, y_n)$ are vectors in $\mathcal{R}^n$, we define the **dot product** or **inner product** of **x** and **y** to be the number

$$\mathbf{x} \cdot \mathbf{y} = x_1y_1 + \ldots + x_ny_n.$$

It is easy to verify (see Exercise 4) that the dot product of vectors in $\mathcal{R}^n$ has the following properties.

5.1 Positivity: $\mathbf{x} \cdot \mathbf{x} > 0$ except that $0 \cdot 0 = 0$.

Symmetry: $\mathbf{x} \cdot \mathbf{y} = \mathbf{y} \cdot \mathbf{x}$.

Additivity: $(\mathbf{x} + \mathbf{y}) \cdot \mathbf{z} = \mathbf{x} \cdot \mathbf{z} + \mathbf{y} \cdot \mathbf{z}$.

Homogeneity: $(r\mathbf{x}) \cdot \mathbf{y} = r(\mathbf{x} \cdot \mathbf{y})$.

Because of the symmetry of the dot product, it follows immediately that additivity and homogeneity hold for the second vector also, that is,

$$\mathbf{x} \cdot (\mathbf{y} + \mathbf{z}) = \mathbf{x} \cdot \mathbf{y} + \mathbf{x} \cdot \mathbf{z}$$

and

$$\mathbf{x} \cdot (r\mathbf{y}) = r(\mathbf{x} \cdot \mathbf{y}).$$

Let us first of all consider the length of a given vector **x** in $\mathcal{R}^3$. If $\mathbf{x} = (x_1, x_2, x_3)$, we think of the length of the vector as the distance from the origin to the point with coordinates (x_1, x_2, x_3). In $\mathcal{R}^3$ the Pythagorean theorem gives us a simple formula for the distance (see Fig. 14). Letting $|\mathbf{x}|$ stand for the length of the vector $\mathbf{x}$, we have

$$|\mathbf{x}|^2 = x_1^2 + x_2^2 + x_3^2. \tag{1}$$

Thus we see that the **length** of the vector $\mathbf{x}$ can be expressed in terms of the dot product as $\sqrt{\mathbf{x} \cdot \mathbf{x}} = |\mathbf{x}|$. Note that we use the same symbol for the length of a vector as for the absolute value of a number. Indeed, if we think of a number as a one-coordinate vector, its length is its absolute value. In $\mathcal{R}^n$ we define the **length** of a vector by the same formula that works in $\mathcal{R}^3$: $|\mathbf{x}| = \sqrt{\mathbf{x} \cdot \mathbf{x}}$.

Next we would like to express the angle between two nonzero vectors in $\mathcal{R}^n$. The usual convention is to take this angle θ to be in the interval $0 \leq \theta \leq \pi$ (see Exercise 1). The solution to the problem is provided by the following theorem.

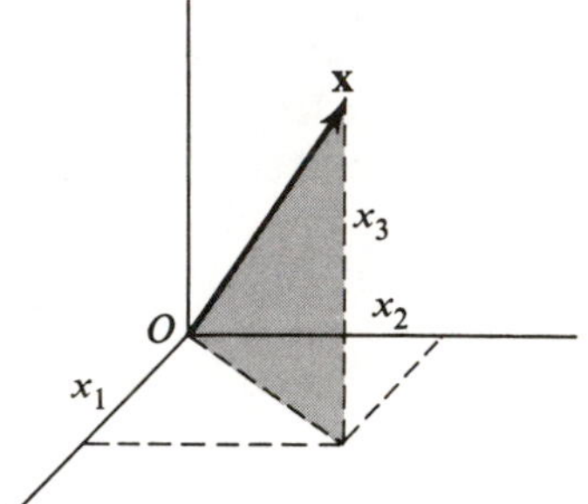

Figure 14

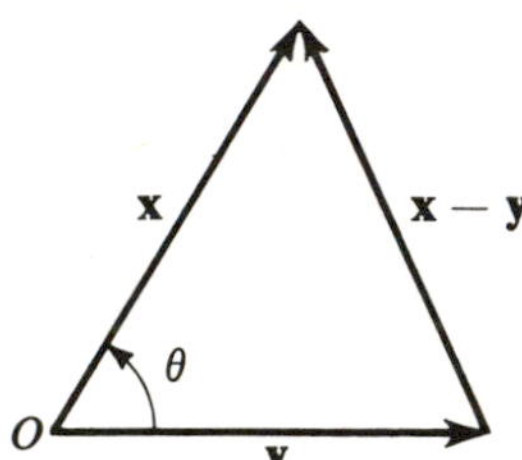

Figure 15

5.2 Theorem

If θ is the angle between $\mathbf{x}$ and $\mathbf{y}$, then

$$\cos \theta = \frac{\mathbf{x} \cdot \mathbf{y}}{|\mathbf{x}|\,|\mathbf{y}|}.$$

Proof. Let us apply the law of cosines to the triangle shown in Fig. 15. It states that

$$|\mathbf{x} - \mathbf{y}|^2 = |\mathbf{x}|^2 + |\mathbf{y}|^2 - 2\,|\mathbf{x}|\,|\mathbf{y}| \cos \theta,$$

which we can rewrite, using $|\mathbf{x}|^2 = \mathbf{x} \cdot \mathbf{x}$, as

$$(\mathbf{x} - \mathbf{y}) \cdot (\mathbf{x} - \mathbf{y}) = \mathbf{x} \cdot \mathbf{x} + \mathbf{y} \cdot \mathbf{y} - 2\,|\mathbf{x}|\,|\mathbf{y}| \cos \theta.$$

Expanding the left-hand member, we obtain

$$\mathbf{x} \cdot \mathbf{x} - \mathbf{x} \cdot \mathbf{y} - \mathbf{y} \cdot \mathbf{x} + \mathbf{y} \cdot \mathbf{y} = \mathbf{x} \cdot \mathbf{x} + \mathbf{y} \cdot \mathbf{y} - 2\,|\mathbf{x}|\,|\mathbf{y}| \cos \theta.$$

Hence,

$$2\mathbf{x} \cdot \mathbf{y} = 2\,|\mathbf{x}|\,|\mathbf{y}| \cos \theta,$$

and the theorem follows by dividing by $2\,|\mathbf{x}|\,|\mathbf{y}|$.

We see from this theorem that $\mathbf{x} \cdot \mathbf{y}$ in absolute value is at most $|\mathbf{x}|\,|\mathbf{y}|$, i.e., $|\mathbf{x} \cdot \mathbf{y}| \leq |\mathbf{x}|\,|\mathbf{y}|$. This is known as the Cauchy-Schwarz inequality, proved more generally in Theorem 5.4.

Example 1. What is the angle θ between $\mathbf{x} = (1, 3)$ and $\mathbf{y} = (-1, 1)$? We easily compute that

$$\mathbf{x} \cdot \mathbf{x} = 1^2 + 3^2 = 10,$$

$$\mathbf{y} \cdot \mathbf{y} = (-1)^2 + 1^2 = 2,$$

and

$$\mathbf{x} \cdot \mathbf{y} = -1 + 3 = 2.$$

Hence

$$|\mathbf{x}| = \sqrt{10}, \quad |\mathbf{y}| = \sqrt{2}, \quad \text{and} \quad \cos \theta = \frac{1}{\sqrt{5}}.$$

By consulting a trigonometric table we find that $\theta = 1.1$ radians approximately (or about 63°).

The theorem also provides a simple test for perpendicularity of two vectors. They are perpendicular if and only if $\theta = \pi/2$, and hence $\cos \theta = 0$. Thus the condition for perpendicularity is simply

$$\mathbf{x} \cdot \mathbf{y} = 0. \tag{2}$$

Example 2. Let us find a vector $\mathbf{x}$ of length 2 perpendicular to $(1, 2, 3)$ and to $(1, 0, -1)$. From geometric considerations we see that there will be two solutions since, if $\mathbf{x}$ is a solution, so is $-\mathbf{x}$. (See Fig. 16.) We

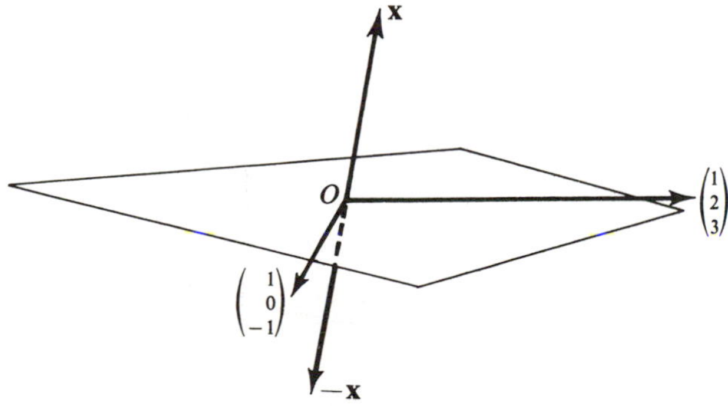

Figure 16

have to write down three conditions, two for the perpendicularity requirements, using (2), and a third condition to assure length 2:

$$\begin{aligned} x_1 + 2x_2 + 3x_3 &= 0, \\ x_1 \qquad\quad - x_3 &= 0, \\ x_1^2 + x_2^2 + x_3^2 &= 4. \end{aligned}$$

These equations have the pair of solutions $\mathbf{x} = \pm(\sqrt{\frac{2}{3}}, -\sqrt{\frac{8}{3}}, \sqrt{\frac{2}{3}})$.

If $\mathbf{n}$ is a unit vector, that is, a vector of length 1, then the dot product $\mathbf{n} \cdot \mathbf{x}$ is called the **coordinate** of $\mathbf{x}$ in the direction of $\mathbf{n}$. The geometric interpretation of $\mathbf{n} \cdot \mathbf{x}$ is shown in Fig. 17. For since $\cos\theta = (\mathbf{n} \cdot \mathbf{x})/|\mathbf{x}|$, it follows that $\mathbf{n} \cdot \mathbf{x}$ is either the length of the perpendicular projection on the line containing $\mathbf{n}$, or else its negative. The vector $(\mathbf{n} \cdot \mathbf{x})\mathbf{n}$ is called the **component** of $\mathbf{x}$ in the direction of $\mathbf{n}$ and is sometimes denoted $\mathbf{x}_\mathbf{n}$.

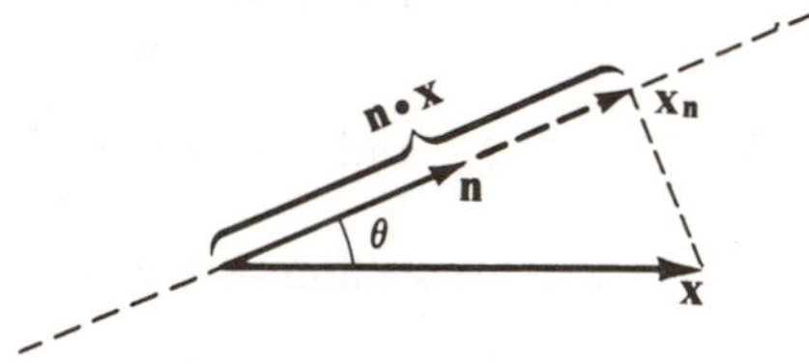

Figure 17

Example 3. Suppose that the vector $(1, -1, 2)$ represents a force $\mathbf{F}$ acting at a point in the sense that the direction of the force is in the direction of the vector and the magnitude of the force is equal to the length of the vector, namely, $\sqrt{6}$. In such an interpretation it is customary to picture the vector not as an arrow from the origin to the point with coordinates $(1, -1, 2)$, but as that arrow translated parallel to itself and with the tail moved to the point of application of the force. Figure 18

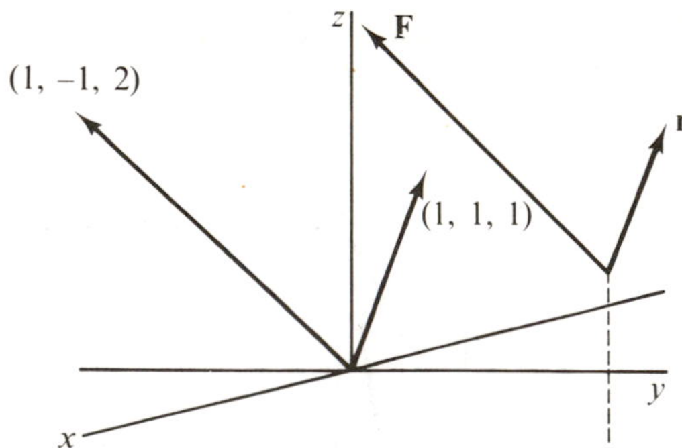

Figure 18

illustrates one possibility. The component of the force $\mathbf{F}$ in the direction of a unit vector $\mathbf{n}$ is then interpreted as the force resulting from the application of F at a point that can move only along a line parallel to $\mathbf{n}$. Thus to find the component of $\mathbf{F} = (1, -1, 2)$ in the direction of $(1, 1, 1)$ we first find the unit vector in that direction. Since $|(1, 1, 1)| = \sqrt{3}$, the desired unit vector is $\mathbf{n} = (1/\sqrt{3}, 1/\sqrt{3}, 1/\sqrt{3})$. The component of $\mathbf{F}$ in the direction of $\mathbf{n}$ is then

$$(\mathbf{F} \cdot \mathbf{n})\mathbf{n} = (1, -1, 2) \cdot \left(\frac{1}{\sqrt{3}}, \frac{1}{\sqrt{3}}, \frac{1}{\sqrt{3}}\right)\mathbf{n}$$

$$= \frac{2}{\sqrt{3}}\mathbf{n} = \left(\frac{2}{3}, \frac{2}{3}, \frac{2}{3}\right).$$

Any vector space on which there is defined a product with the properties 5.1 is called an **inner product space.** Thus $\mathcal{R}^n$ is an inner product space, and some other examples are given in Problems 8 and 9. Inner products in spaces other than $\mathcal{R}^n$ are used in this book only in Chapter 2 Section 4 and Chapter 5 Section 5.

In terms of the inner product we can always define the **length** or **norm** of a vector by

$$|\mathbf{x}| = \sqrt{\mathbf{x} \cdot \mathbf{x}}.$$

Then length has the following properties.

5.3 Positivity: $|\mathbf{x}| > 0$ except that $|0| = 0$.

Homogeneity: $|r\mathbf{x}| = |r|\,|\mathbf{x}|$.

Triangle inequality: $|\mathbf{x} + \mathbf{y}| \leq |\mathbf{x}| + |\mathbf{y}|$.

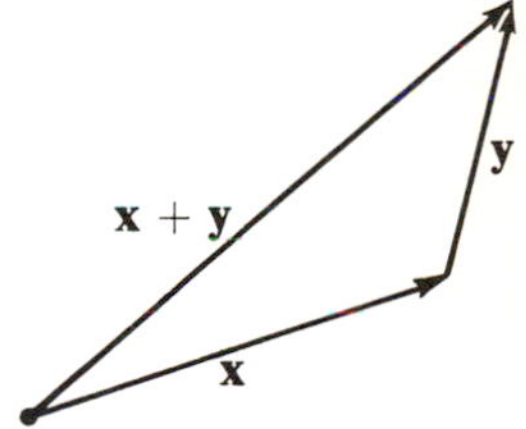

Figure 19

The proofs of the first two are easy and are left for the reader to check. The proof of the third is harder and will be taken up later, though we remark here on its geometric significance, illustrated by Fig. 19.

The fact that length has been defined in terms of an inner product leads to some properties of length that are not derivable from those already listed in 5.3. First we prove the following.

5.4 Cauchy-Schwarz inequality

$$|\mathbf{x} \cdot \mathbf{y}| \leq |\mathbf{x}|\,|\mathbf{y}|.$$

Proof. We assume first that $\mathbf{x}$ and $\mathbf{y}$ are unit vectors, that is, that $|\mathbf{x}| = |\mathbf{y}| = 1$. Then,

$$0 \leq |\mathbf{x} - \mathbf{y}|^2 = (\mathbf{x} - \mathbf{y}) \cdot (\mathbf{x} - \mathbf{y})$$

$$= |\mathbf{x}|^2 - 2\mathbf{x} \cdot \mathbf{y} + |\mathbf{y}|^2 = 2 - 2\mathbf{x} \cdot \mathbf{y},$$

or

$$\mathbf{x} \cdot \mathbf{y} \leq 1.$$

Assuming that neither **x** nor **y** is zero (for the inequality obviously holds if one of them *is* zero), we can replace **x** and **y** by the unit vectors $\mathbf{x}/|\mathbf{x}|$ and $\mathbf{y}/|\mathbf{y}|$, getting

$$\mathbf{x} \cdot \mathbf{y} \leq |\mathbf{x}|\,|\mathbf{y}|.$$

Now replace **x** by $-\mathbf{x}$ to get

$$-\mathbf{x} \cdot \mathbf{y} \leq |\mathbf{x}|\,|\mathbf{y}|.$$

The last two inequalities imply the Cauchy-Schwarz inequality.

Notice that the Cauchy-Schwarz inequality may be written as

$$\frac{|\mathbf{x} \cdot \mathbf{y}|}{|\mathbf{x}|\,|\mathbf{y}|} \leq 1,$$

so there will always be an angle θ such that

$$\cos \theta = \frac{\mathbf{x} \cdot \mathbf{y}}{|\mathbf{x}|\,|\mathbf{y}|}.$$

Defining the cosine of the angle θ between **x** and **y** is sometimes more satisfactory than defining θ itself, though it doesn't show us how to tell whether the angle is

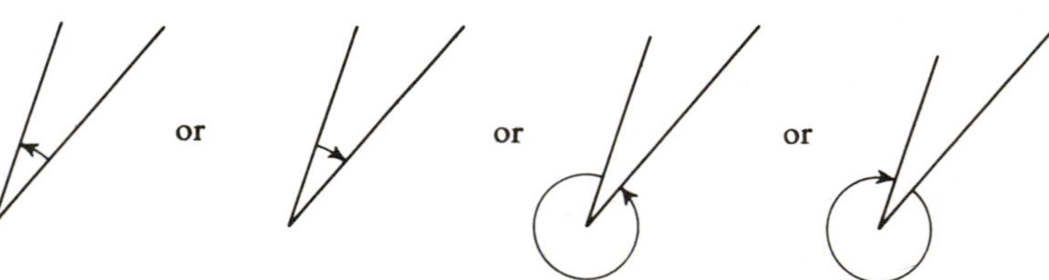

Using the Cauchy-Schwarz inequality, it is easy to give the deferred proof of the triangle inequality, for from

$$|\mathbf{x} \cdot \mathbf{y}| \leq |\mathbf{x}|\,|\mathbf{y}|,$$

we get

$$\begin{aligned} |\mathbf{x} + \mathbf{y}|^2 &= (\mathbf{x} + \mathbf{y}) \cdot (\mathbf{x} + \mathbf{y}) = |\mathbf{x}|^2 + 2\mathbf{x} \cdot \mathbf{y} + |\mathbf{y}|^2 \\ &\leq |\mathbf{x}|^2 + 2\,|\mathbf{x}|\,|\mathbf{y}| + |\mathbf{y}|^2 = (|\mathbf{x}| + |\mathbf{y}|)^2, \end{aligned}$$

from which follows

$$|\mathbf{x} + \mathbf{y}| \leq |\mathbf{x}| + |\mathbf{y}|.$$

Two vectors **x** and **y** that are perpendicular with respect to an inner product are sometimes called **orthogonal.** Furthermore, a set S of vectors that are mutually orthogonal and have length 1 is called an **orthonormal set.** The idea has a useful application to finding the inverses of certain matrices. Any square matrix whose rows, or whose columns, looked at as vectors in $\mathcal{R}^n$, form an orthonormal set with respect to the Euclidean dot product is called an **orthogonal matrix.** It is very easy to find the inverse of such a matrix. Suppose the columns of

$$A = \begin{pmatrix} a_{11} & a_{12} & \cdot & \cdot & \cdot \\ a_{21} & a_{22} & \cdot & \cdot & \cdot \\ \cdot & \cdot & & & \\ \cdot & \cdot & & & \\ \cdot & \cdot & & & \end{pmatrix}$$

form an orthonormal set. Thus, for example $a_{11}^2 + a_{21}^2 + \ldots = 1$, $a_{12}^2 + a_{22}^2 + \ldots = 1$ and $a_{11}a_{12} + a_{21}a_{22} + \ldots = 0$. We form the matrix A^t, called the **transpose** of A, by reflecting A across its main diagonal. Thus

$$A^t = \begin{pmatrix} a_{11} & a_{21} & \cdot & \cdot & \cdot \\ a_{12} & a_{22} & \cdot & \cdot & \cdot \\ \cdot & \cdot & & & \\ \cdot & \cdot & & & \\ \cdot & \cdot & & & \end{pmatrix}.$$

Then the fact that the columns of A are orthonormal shows that $A^tA = I$. We shall see later (Theorem 8.3) that this implies also that $AA^t = I$. A similar argument replacing orthonormal columns by orthonormal rows would show that $AA^t = I$ and then (by Theorem 8.3) that $A^tA = I$. Thus we have proved the following

5.5 Theorem

Every orthogonal matrix A is invertible and $A^{-1} = A^t$.

Example 4. It is easy to verify that the matrix

$$A = \begin{pmatrix} \frac{\sqrt{3}}{2} & -\frac{1}{2} \\ \frac{1}{2} & \frac{\sqrt{3}}{2} \end{pmatrix}$$

has columns (and rows) orthonormal. The transposed matrix is

$$A^t = \begin{pmatrix} \frac{\sqrt{3}}{2} & \frac{1}{2} \\ -\frac{1}{2} & \frac{\sqrt{3}}{2} \end{pmatrix},$$

and so $A^{-1} = A^t$.

The second of the above two square matrices is obtained from the first by interchanging rows and columns. When two matrices are so related we still say that one is the **transpose** of the other, even if they are not square. Thus if

$$A = \begin{pmatrix} 1 & 2 & 3 \\ 4 & 5 & 6 \end{pmatrix},$$

then

$$A^t = \begin{pmatrix} 1 & 4 \\ 2 & 5 \\ 3 & 6 \end{pmatrix}.$$

Obviously,

$$(A^t)^t = A \qquad \text{for any matrix } A.$$

EXERCISES

1. Show that the natural basis vectors satisfy $\mathbf{e}_i \cdot \mathbf{e}_i = 1$, and $\mathbf{e}_i \cdot \mathbf{e}_j = 0$ if $i \neq j$.

2. (a) Find the angle between the vectors (1, 1, 1) and (1, 0, 1).
(b) Find a vector of length 1 perpendicular to both vectors in part (a).

3. Find the distance between (1, 2) and (0, 5).

4. Prove that the dot product has the properties listed in 5.1.

5. Prove the positivity and homogeneity properties of length listed in 5.3.

6. Find the *coordinate* of the vector $\mathbf{x} = (1, -1, 2)$: (a) in the direction of $\mathbf{n} = (1/\sqrt{3}, 1/\sqrt{3}, 1/\sqrt{3})$ and (b) in the direction of the nonunit vector (1, 1, 3). (c) What is the *component* of $\mathbf{x}$ in the direction of $\mathbf{n}$?

7. (a) Prove that if $\mathbf{x}$ is any vector in $\mathcal{R}^n$ and $\mathbf{n}$ is a unit vector, then $\mathbf{x}$ can be written as $\mathbf{x} = \mathbf{y} + \mathbf{z}$, where $\mathbf{y}$ is a multiple of $\mathbf{n}$ and $\mathbf{z}$ is perpendicular to $\mathbf{n}$. [*Hint.* Take $\mathbf{y}$ to be the component of $\mathbf{x}$ in the direction of $\mathbf{n}$.]
(b) Show that the vectors $\mathbf{y}$ and $\mathbf{z}$ of part (a) are uniquely determined. The vector $\mathbf{z}$ so determined is called the component of $\mathbf{x}$ perpendicular to $\mathbf{n}$

8. (a) Consider the vector space $\mathcal{C}[0, 1]$ consisting of all continuous real-valued functions defined on the interval $0 \leq x \leq 1$. [The sum of f and g is defined by $(f + g)(x) = f(x) + g(x)$, and the numerical multiple rf by $(rf)(x) = rf(x)$.] Show that the product $\langle f, g \rangle$ defined by

$$\langle f, g \rangle = \int_0^1 f(x)g(x)\, dx$$

is an inner product on $\mathcal{C}[0, 1]$.

(b) Define the norm of f by $|f| = \langle f, f \rangle^{1/2}$. What is the norm of $f(x) = x$?

9. (a) Show that the product of $\mathbf{x} = (x_1, x_2)$ and $\mathbf{y} = (y_1, y_2)$ defined by

$$\mathbf{x} * \mathbf{y} = x_1 y_1 + 2x_2 y_2$$

is an inner product.

(b) With length defined by $|\mathbf{x}|_* = (\mathbf{x} * \mathbf{x})^{1/2}$, sketch the set of points satisfying $|\mathbf{x}|_* = 1$.

10. Show that if $|\mathbf{x}| = |\mathbf{y}|$ in an inner product space, then $\mathbf{x} + \mathbf{y}$ is perpendicular to $\mathbf{x} - \mathbf{y}$.

11. Show that the matrix

$$\begin{pmatrix} \cos\theta & -\sin\theta \\ \sin\theta & \cos\theta \end{pmatrix}$$

is orthogonal, and find its inverse.

12. Derive the inequality

$$|\mathbf{x}| - |\mathbf{y}| \leq |\mathbf{x} - \mathbf{y}|$$

from the triangle inequality, and then show that

$$\big||\mathbf{x}| - |\mathbf{y}|\big| \leq |\mathbf{x} - \mathbf{y}|.$$

13. (a) Show that if A is a 2-by-2 orthogonal matrix, then $|A\mathbf{x}| = |\mathbf{x}|$ for all vectors $\mathbf{x}$ in $\mathcal{R}^2$.

(b) Prove the result of part (a) for n-by-n matrices.

(c) Prove that, if A is n-by-n and $|A\mathbf{x}| = |\mathbf{x}|$ for all $\mathbf{x}$, then $(A\mathbf{x} \cdot A\mathbf{y}) = \mathbf{x} \cdot \mathbf{y}$ for all $\mathbf{x}$ and $\mathbf{y}$ in $\mathcal{R}^n$.

14. A real-valued linear function $\mathcal{R}^n \xrightarrow{f} \mathcal{R}$ is sometimes called a **linear functional**. Show that, if f is a linear functional defined on $\mathcal{R}^n$, then there is a fixed vector $\mathbf{y}$ in $\mathcal{R}^n$ such that $f(\mathbf{x}) = \mathbf{y} \cdot \mathbf{x}$ for all $\mathbf{x}$ in $\mathcal{R}^n$. [*Hint.* What is the matrix of f?]

15. Let

$$A = \begin{pmatrix} 0 & 2 & -1 \\ 3 & -5 & 2 \end{pmatrix} \quad \text{and} \quad B = \begin{pmatrix} -3 & -1 \\ 1 & 0 \\ 2 & -2 \end{pmatrix}.$$

Compute AB, BA, A^tB^t, and B^tA^t.

16. (a) For any two matrices A and B, show that if AB is defined, so is B^tA^t, and that $B^tA^t = (AB)^t$.
(b) Show that if A is invertible, then so is A^t, and that $(A^t)^{-1} = (A^{-1})^t$.

SECTION 6 EUCLIDEAN GEOMETRY

In this section we develop some facts and formulas about lines and planes in $\mathcal{R}^2$ and $\mathcal{R}^3$. Some of the ideas will reappear in Section 1 of Chapter 3.

Recall that if $\mathbf{x}_1$ is a nonzero vector in $\mathcal{R}^2$ or $\mathcal{R}^3$, then the set of all numerical multiples $t\mathbf{x}_1$ is a line L_0 passing through the origin, and any line parallel to L_0 consists of the set of all points $t\mathbf{x}_1 + \mathbf{x}_0$, where $\mathbf{x}_0$ is some fixed vector. An alternative way to say the same thing is: a line is the range of a function of the form $f(t) = t\mathbf{x}_1 + \mathbf{x}_0$, where $\mathbf{x}_1 \neq 0$.

Example 1. To describe a line L passing through two distinct points $\mathbf{a}_1$ and $\mathbf{a}_2$, take $\mathbf{x}_1 = \mathbf{a}_2 - \mathbf{a}_1$. Then all multiples $t(\mathbf{a}_2 - \mathbf{a}_1)$ make up a line parallel to L, and the points $t(\mathbf{a}_2 - \mathbf{a}_1) + \mathbf{a}_1$ make up L itself. For example, when $t = 0$ we get $\mathbf{a}_1$, and when $t = 1$ we get $\mathbf{a}_2$. See Fig. 20. Alternatively, we can write the points on L in the form $t\mathbf{a}_2 + (1 - t)\mathbf{a}_1$.

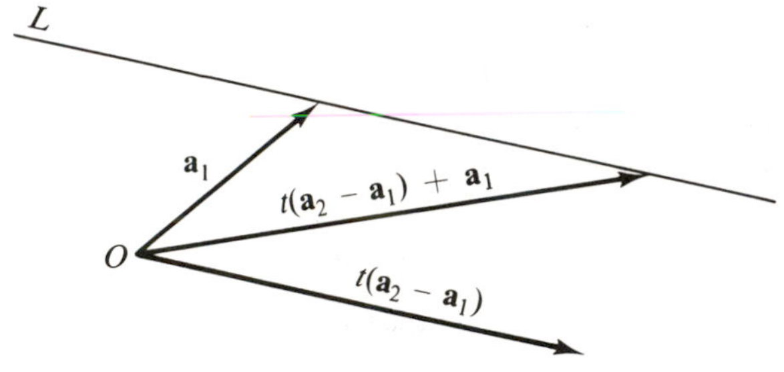

Figure 20

To determine a plane in $\mathcal{R}^3$, we take two noncollinear vectors $\mathbf{x}_1$ and $\mathbf{x}_2$ (that is, so that neither is a multiple of the other) and consider all points $u\mathbf{x}_1 + v\mathbf{x}_2$ where u and v are numbers. These points form a plane P_0 through the origin. A parallel plane will then consist of all points $u\mathbf{x}_1 + v\mathbf{x}_2 + \mathbf{x}_0$, where $\mathbf{x}_0$ is some fixed vector. We can restate the definition by saying that a plane is the range of a function $\mathcal{R}^2 \xrightarrow{g} \mathcal{R}^3$, where $g(u, v) = u\mathbf{x}_1 + v\mathbf{x}_2 + \mathbf{x}_0$, and the vectors $\mathbf{x}_1$, $\mathbf{x}_2$ do not lie on a line.

Example 2. For three points $\mathbf{a}_1$, $\mathbf{a}_2$, and $\mathbf{a}_3$ to determine a unique plane passing through them, they must not lie on a line. (Then the vectors $\mathbf{a}_3 - \mathbf{a}_1$ and $\mathbf{a}_2 - \mathbf{a}_1$ are not collinear. For otherwise $\mathbf{a}_3 - \mathbf{a}_1 = t(\mathbf{a}_2 - \mathbf{a}_1)$, so $\mathbf{a}_3 = t\mathbf{a}_2 + (1 - t)\mathbf{a}_1$ and $\mathbf{a}_3$ would lie on the line determined by $\mathbf{a}_2$ and $\mathbf{a}_1$.) Now take $\mathbf{x}_1 = \mathbf{a}_3 - \mathbf{a}_1$ and $\mathbf{x}_2 = \mathbf{a}_2 - \mathbf{a}_1$. The desired plane consists of all points

$$\mathbf{w} = u(\mathbf{a}_3 - \mathbf{a}_1) + v(\mathbf{a}_2 - \mathbf{a}_1) + \mathbf{a}_1.$$

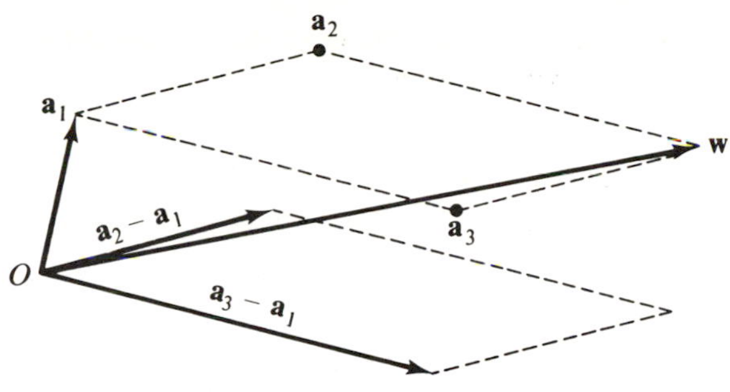

Figure 21

To get $\mathbf{a}_1$, take $u = v = 0$. To get $\mathbf{a}_2$, take $u = 0$, $v = 1$. To get $\mathbf{a}_3$, take $u = 1$, $v = 0$. See Fig. 21.

Using the dot product gives an alternative way to describe a plane P. Suppose $\mathbf{x}_0$ is an arbitrary point on P and that $\mathbf{p}$ is a nonzero vector perpendicular to both $\mathbf{x}_1$ and $\mathbf{x}_2$ (so $\mathbf{p} \cdot \mathbf{x}_1 = \mathbf{p} \cdot \mathbf{x}_2 = 0$) where $\mathbf{x}_1$ and $\mathbf{x}_2$ are vectors parallel to P. Then the equation

$$\mathbf{p} \cdot (\mathbf{x} - \mathbf{x}_0) = 0 \tag{1}$$

is satisfied by all points $\mathbf{x} = u\mathbf{x}_1 + v\mathbf{x}_2 + \mathbf{x}_0$ on P because $\mathbf{p} \cdot (u\mathbf{x}_1 + v\mathbf{x}_2) = u\mathbf{p} \cdot \mathbf{x}_1 + v\mathbf{p} \cdot \mathbf{x}_2 = 0$. It is also easy to check that Equation (1) can be solved for $\mathbf{x}$ to get $\mathbf{x} = u\mathbf{y}_1 + v\mathbf{y}_2 + \mathbf{x}_0$, and we leave the solution as an exercise. Figure 22 shows the relationship between the vectors. To find a vector $\mathbf{p}$ perpendicular to two vectors $\mathbf{x}_1$ and $\mathbf{x}_2$, it is convenient to use the **cross-product** of $\mathbf{x}_1$ and $\mathbf{x}_2$, defined for $\mathbf{x}_1 = (u_1, u_2, u_3)$ and $\mathbf{x}_2 = (v_1, v_2, v_3)$ by

$$\mathbf{p} = (u_2v_3 - u_3v_2, u_3v_1 - u_1v_3, u_1v_2 - u_2v_1). \tag{2}$$

It is routine to check that $\mathbf{p} \cdot \mathbf{x}_1 = 0$ and $\mathbf{p} \cdot \mathbf{x}_2 = 0$. For example,

$$\mathbf{p} \cdot \mathbf{x}_1 = u_1u_2v_3 - u_1u_3v_2 + u_2u_3v_1 - u_2u_1v_3 + u_3u_1v_2 - u_3u_2v_1 = 0.$$

The cross-product is taken up in more detail in the next section. In the meantime we will simply use it to find a vector perpendicular to two given vectors. See Problem 13.

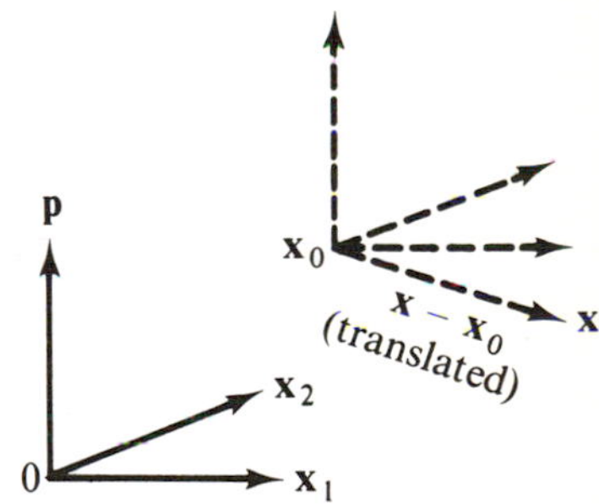

Figure 22

Example 3. Suppose we are given a plane parallel to the two vectors $\mathbf{x}_1 = (1, 2, -3)$ and $\mathbf{x}_2 = (2, 0, 1)$, and containing the point $\mathbf{x}_0 = (1, 1, 1)$. A vector $\mathbf{p}$ perpendicular to $\mathbf{x}_1$ and $\mathbf{x}_2$ is given by their cross-product:

$$\begin{aligned}\mathbf{p} &= \big((2)(1) - (0)(-3), (2)(-3) - (1)(1), (1)(0) - (2)(2)\big) \\ &= (2, -7, -4).\end{aligned}$$

Now writing $\mathbf{x} = (x, y, z)$, we require that $\mathbf{p} \cdot (\mathbf{x} - \mathbf{x}_0) = 0$, that is,

$(2, -7, -4) \cdot (x - 1, y - 1, z - 1) = 0$. According to the definition of the dot product, this last equation is

$$2(x - 1) - 7(y - 1) - 4(z - 1) = 0$$

or

$$2x - 7y - 4z + 9 = 0. \tag{3}$$

In other words, the given plane consists of all points (x, y, z) with coordinates satisfying that equation.

The simplest way to sketch the plane satisfying an equation like (3) is to pick three points with simple coordinates that satisfy it, say $(0, 0, \frac{9}{4})$, $(0, \frac{9}{7}, 0)$, and $(-\frac{9}{2}, 0, 0)$. Then locate the points relative to coordinate axes and sketch the plane using the three points as references. We have purposely chosen in our example points that lie on the coordinate axes. See Fig. 23.

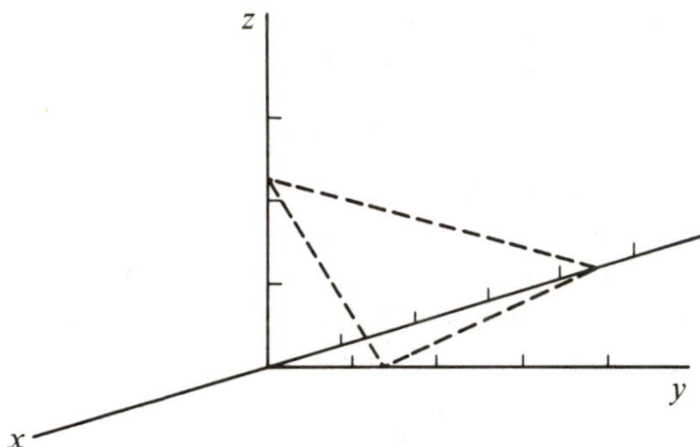

Figure 23

If $\mathbf{p}$ is a nonzero vector in $\mathcal{R}^2$ and $\mathbf{x}_0$ is a point in $\mathcal{R}^2$, we can determine the line perpendicular to $\mathbf{p}$ and containing $\mathbf{x}_0$ by the equation $\mathbf{p} \cdot (\mathbf{x} - \mathbf{x}_0) = 0$. If $\mathbf{p} = (p_1, p_2)$ and $\mathbf{x}_0 = (x_0, y_0)$, the equation becomes

$$(p_1, p_2) \cdot (x - x_0, y - y_0) = 0$$

or

$$p_1(x - x_0) + p_2(y - y_0) = 0.$$

This is one of the several forms for the equation of a line in the xy-plane. The slope of the line is evidently $-p_1/p_2$.

The representation of a plane by an equation $\mathbf{p} \cdot (\mathbf{x} - \mathbf{x}_0) = 0$ is not unique because any nonzero multiple of $\mathbf{p}$ can replace it, leaving the set of vectors $\mathbf{x}$ satisfying the equation unchanged. However, it is sometimes useful to normalize the equation of a plane by requiring that $\mathbf{p}$ be a unit vector. The normalized equation then becomes $\mathbf{n} \cdot (\mathbf{x} - \mathbf{x}_0) = 0$, where $\mathbf{n} = \mathbf{p}/|\mathbf{p}|$. Alternatively we can write $\mathbf{n} \cdot \mathbf{x} = c$, where $c = \mathbf{n} \cdot \mathbf{x}_0$.

6.1 Theorem

If $\mathbf{n} \cdot \mathbf{x} = c$ is the normalized equation of a plane P and $\mathbf{y}$ is any point, then the distance from $\mathbf{y}$ to P is the absolute value of $c - \mathbf{n} \cdot \mathbf{y}$.

Proof. By definition the distance is to be measured along a line from $\mathbf{y}$ perpendicular to P. This line can be represented by $t\mathbf{n} + \mathbf{y}$, and the intersection with P will occur for some $t = t_0$. The desired distance is then $|(t_0\mathbf{n} + \mathbf{y}) - \mathbf{y}| = |t_0\mathbf{n}|$, which is simply the absolute value of t_0, since $\mathbf{n}$ has length 1. Since $t_0\mathbf{n} + \mathbf{y}$ lies in P, then $\mathbf{n} \cdot (t_0\mathbf{n} + \mathbf{y}) = c$. But $\mathbf{n} \cdot \mathbf{n} = 1$, so we obtain $t_0 + \mathbf{n} \cdot \mathbf{y} = c$, from which the theorem follows.

Example 4. We shall find the distance from $(1, 1, 1)$ to the plane $(3, 0, -4) \cdot \mathbf{x} = -3$. The normalized equation of the plane is given by $(\frac{3}{5}, 0, -\frac{4}{5}) \cdot \mathbf{x} = -\frac{3}{5}$. Then

$$\begin{aligned} c - \mathbf{n} \cdot \mathbf{y} &= -\tfrac{3}{5} - (\tfrac{3}{5}, 0, -\tfrac{4}{5}) \cdot (1, 1, 1) \\ &= -\tfrac{2}{5}. \end{aligned}$$

Hence the distance is $\frac{2}{5}$. Notice that the equation of the plane could also be written $3x - 4z = -3$ and, in normalized form, $(\frac{3}{5})x - (\frac{4}{5})z = -\frac{3}{5}$.

EXERCISES

1. Write a formula in the form $f(t) = t\mathbf{x}_1 + \mathbf{x}_0$ for a line containing $\mathbf{x}_0$ and parallel to $\mathbf{x}_1$. In each case determine whether the point $\mathbf{a}$ lies on the line, that is, whether $\mathbf{a}$ is in the range of f.

   (a) $\mathbf{x}_0 = (1, 1)$, $\mathbf{x}_1 = (-2, 3)$; $\mathbf{a} = (-3, 7)$.
   (b) $\mathbf{x}_0 = (0, 0, 1)$, $\mathbf{x}_1 = (1, 1, 1)$; $\mathbf{a} = (5, 5, 4)$.
   (c) $\mathbf{x}_0 = (1, 0, 2)$, $\mathbf{x}_1 = (2, 0, 3)$; $\mathbf{a} = (3, 0, 5)$.

2. Write a formula in the form $g(u, v) = u\mathbf{x}_1 + v\mathbf{x}_2 + \mathbf{x}_0$ for a plane containing $\mathbf{x}_0$ and parallel to $\mathbf{x}_1$ and $\mathbf{x}_2$. In each case determine whether the point $\mathbf{a}$ lies on the plane, that is, whether $\mathbf{a}$ is in the range of g.

   (a) $\mathbf{x}_0 = (0, 0, 0)$, $\mathbf{x}_1 = (1, 0, 1)$, $\mathbf{x}_2 = (2, 3, 0)$; $\mathbf{a} = (4, 3, 1)$.
   (b) $\mathbf{x}_0 = (1, -1, 1)$, $\mathbf{x}_1 = (1, 1, 0)$, $\mathbf{x}_2 = (-1, 2, 0)$; $\mathbf{a} = (1, 1, 2)$.
   (c) $\mathbf{x}_0 = (0, 0, 1)$, $\mathbf{x}_1 = (1, 2, 1)$, $\mathbf{x}_2 = (2, 1, 2)$; $\mathbf{a} = (3, 0, 4)$.

3. Sketch the lines determined in $\mathcal{R}^2$ or $\mathcal{R}^3$ by

   (a) $t(1, 2) + (-1, -1)$.
   (b) $t(1, 0, 1) + (1, 1, 0)$.
   (c) $(1, 2) \cdot \mathbf{x} = 2$.

4. Sketch the plane determined in $\mathcal{R}^3$ by

(a) $u(1, 2, 0) + v(2, 0, 1) + (1, 0, 0)$.
(b) $(-1, -1, 1) \cdot \mathbf{x} = 1$.
(c) $2x + y + z = 0$.

5. Find an equation for the line in $\mathcal{R}^2$ that is perpendicular to the line $t\mathbf{x}_1 + \mathbf{x}_0$ and passes through the point $\mathbf{y}_2$.

6. Find the cosine of the angle between the planes $\mathbf{x}_1 \cdot \mathbf{x} = c_1$ and $\mathbf{y}_1 \cdot \mathbf{y} = c_2$.

7. Prove that if $\mathbf{x}_1 \cdot \mathbf{x} = c_1$ and $\mathbf{y}_1 \cdot \mathbf{y} = c_2$ are normalized equations of two planes, then the cosine of the angle between them is $\mathbf{x}_1 \cdot \mathbf{y}_1$.

8. For each of the points and planes or lines listed below, find the distance from the point to the plane or line.

(a) $(1, 0, -1)$; $(1, 1, 1) \cdot \mathbf{x} = 1$.
(b) $(1, 0, -1)$; $x + 2y + 3z = 1$.
(c) $(1, 2)$; $(3, 4) \cdot \mathbf{x} = 0$.

9. (a) Verify that the cross-product of $\mathbf{x}_1$ and $\mathbf{x}_2$ [formula (2) in the text] is perpendicular to $\mathbf{x}_2$.

(b) Find a representation for the line perpendicular to the plane consisting of all points $u(1, 2, 1) + v(-1, 0, 1) + (1, 1, 1)$ and passing through the origin.

10. (a) Find a representation for $\mathbf{x} = (x, y, z)$ satisfying

$$2x + 3y + z = 1$$

by finding vectors $\mathbf{x}_1$, $\mathbf{x}_2$, and $\mathbf{x}_0$ such that $\mathbf{x} = u\mathbf{x}_1 + v\mathbf{x}_2 + \mathbf{x}_0$. [*Hint.* Let $u = y$, $v = z$.]

(b) Do the same as in part (a) for the general equation $\mathbf{p} \cdot (\mathbf{x} - \mathbf{x}_0) = 0$, when $\mathbf{p} \neq 0$. [*Hint.* If $\mathbf{p} \neq 0$, then $\mathbf{p}$ has a nonzero coordinate.]

11. (a) If $\mathbf{x}$ is any vector in $\mathcal{R}^3$, show that

$$\mathbf{x} = (\mathbf{x} \cdot \mathbf{e}_1)\mathbf{e}_1 + (\mathbf{x} \cdot \mathbf{e}_2)\mathbf{e}_2 + (\mathbf{x} \cdot \mathbf{e}_3)\mathbf{e}_3.$$

(b) If the vector $\mathbf{x}$ in part (a) is a unit vector, that is, a vector $\mathbf{u}$ of length 1, show that $\mathbf{u} \cdot \mathbf{e}_i = \cos \alpha_i$, where α_i is the angle between $\mathbf{u}$ and $\mathbf{e}_i$. The coordinates $\cos \alpha_i$ are called the **direction cosines** of $\mathbf{u}$ relative to the natural basis vectors $\mathbf{e}_i$. If $\mathbf{x}$ is any nonzero vector, the direction cosines of $\mathbf{x}$ are defined to be the direction cosines of the unit vector $\mathbf{x}/|\mathbf{x}|$.
(c) Find the direction cosines of $(1, 2, 1)$.
(d) Show that parts (a) and (b) generalize to $\mathcal{R}^n$.

12. Let $\mathbf{u}$ and $\mathbf{v}$ be points in $\mathcal{R}^n$. Show that the point $\frac{1}{2}\mathbf{u} + \frac{1}{2}\mathbf{v}$ is the midpoint of the line segment joining $\mathbf{u}$ and $\mathbf{v}$.

13. Let $\mathbf{u}$ and $\mathbf{v}$ be noncollinear vectors in $\mathcal{R}^3$. To find a vector $\mathbf{x}$ perpendicular to both $\mathbf{u}$ and $\mathbf{v}$, we solve the equations $\mathbf{u} \cdot \mathbf{x} = 0$ and $\mathbf{v} \cdot \mathbf{x} = 0$, that is, solve

$$u_1x + u_2y + u_3z = 0$$
$$v_1x + v_2y + v_3z = 0, \qquad (*)$$

where $\mathbf{u} = (u_1, u_2, u_3)$, $\mathbf{v} = (v_1, v_2, v_3)$, and $\mathbf{x} = (x, y, z)$.

(a) Show that

$$(u_2v_1 - u_1v_2)y = (u_1v_3 - u_3v_1)z$$
$$(u_1v_2 - u_2v_1)x = (u_2v_3 - u_3v_2)z.$$

(b) Show that if $u_1v_2 - u_2v_1 \neq 0$, then the equations $(*)$ have a solution

$$(x, y, z) = (u_2v_3 - u_3v_2, u_3v_1 - u_1v_3, u_1v_2 - u_2v_1).$$

(c) Show how to solve $(*)$ if $u_1v_2 - u_2v_1 = 0$.

SECTION 7

DETERMINANTS

In this section we define and study a certain numerical-valued function defined on the set of all square matrices. The value of this function for a square matrix M is called the **determinant** of M and is written det M. Another common way to denote the determinant of a matrix in displayed form is to replace the parentheses enclosing the array of entries by vertical bars. Thus the notations

$$\begin{vmatrix} 1 & 4 & 5 \\ 6 & 7 & -3 \\ -2 & 1 & 0 \end{vmatrix}, \qquad \begin{vmatrix} a & b \\ c & d \end{vmatrix}$$

mean the same as

$$\det \begin{pmatrix} 1 & 4 & 5 \\ 6 & 7 & -3 \\ -2 & 1 & 0 \end{pmatrix}, \qquad \det \begin{pmatrix} a & b \\ c & d \end{pmatrix}.$$

Our definition of determinant will be inductive, that is, we shall define det M first for 1-by-1 matrices, and then for each n define the determinant of an n-by-n matrix in terms of determinants of certain $(n - 1)$-by-$(n - 1)$ matrices. We first need a notation for certain submatrices of a given matrix.

For any matrix A, the matrix obtained by deleting the ith row and jth column of A is called the ijth **minor** of A and is denoted by A_{ij}. (Recall that we use the small letter a_{ij} to denote the ijth entry of a matrix A.)

Example 1. Let

$$A = \begin{pmatrix} -5 & -6 & 7 \\ 8 & -9 & 0 \\ -3 & 4 & 2 \end{pmatrix}, \qquad B = \begin{pmatrix} 1 & 2 \\ 3 & 4 \end{pmatrix}.$$

Then

$$a_{11} = -5, \qquad A_{11} = \begin{pmatrix} -9 & 0 \\ 4 & 2 \end{pmatrix}$$

$$a_{23} = 0, \qquad A_{23} = \begin{pmatrix} -5 & -6 \\ -3 & 4 \end{pmatrix}$$

$$b_{11} = 1, \qquad B_{11} = (4), \quad b_{12} = 2, \quad B_{12} = (3).$$

We can now make the definition of **determinant:**

For a 1-by-1 matrix $A = (a)$, we define

$$\det A = a.$$

For an n-by-n matrix $A = (a_{ij})$, $i, j = 1, \ldots, n$, we define

7.1
$$\det A = \sum_{j=1}^{n} (-1)^{j+1} a_{1j} \det A_{1j}$$
$$= a_{11} \det A_{11} - a_{12} \det A_{12} + \ldots - (-1)^n a_{1n} \det A_{1n}.$$

In words, the formula says that det A is the sum with alternating signs, of the elements of the first row of A, each multiplied by the determinant of its corresponding minor. For this reason the numbers

$$\det A_{11}, -\det A_{12}, \ldots, (-1)^{n+1} \det A_{1n}$$

are called the *cofactors* of the corresponding elements of the first row of A. In general, the **cofactor** of the entry a_{ij} in A is defined to be $(-1)^{i+j} \det A_{ij}$. Thus in Example 1 the entry $a_{21} = 8$ in the matrix A has cofactor

$$(-1)^{2+1} \det \begin{pmatrix} -6 & 7 \\ 4 & 2 \end{pmatrix} = 40.$$

The factor $(-1)^{i+j}$ associates plus and minus signs with det A_{ij} according to the pattern

$$\begin{pmatrix} + & - & + & - & \cdots \\ - & + & - & + & \cdots \\ + & - & + & - & \cdots \\ - & + & - & + & \cdots \\ \cdot & \cdot & \cdot & \cdot & \\ \cdot & \cdot & \cdot & \cdot & \\ \cdot & \cdot & \cdot & \cdot & \end{pmatrix}.$$

Example 2.

(a) $\det\begin{pmatrix}1 & 2\\ 3 & 4\end{pmatrix} = 1(4) - 2(3) = 4 - 6 = -2.$

(b)
$$\det\begin{pmatrix}-5 & -6 & 7\\ 8 & -9 & 0\\ -3 & 4 & 2\end{pmatrix} = -5\det\begin{pmatrix}-9 & 0\\ 4 & 2\end{pmatrix} - (-6)\det\begin{pmatrix}8 & 0\\ -3 & 2\end{pmatrix}$$
$$+ 7\det\begin{pmatrix}8 & -9\\ -3 & 4\end{pmatrix}$$
$$= -5(-18 - 0) + 6(16 - 0)$$
$$+ 7(32 - 27)$$
$$= 90 + 96 + 35 = 221.$$

(c) $\det\begin{pmatrix}a & b\\ c & d\end{pmatrix} = ad - bc.$

The result of the last example is worth remembering as a rule of calculation. *The determinant of a 2-by-2 matrix is the product of the entries on the main diagonal minus the product of the other two entries.* Thus 2-by-2 determinants can usually be computed mentally, and 3-by-3 determinants in one or two lines. In principle, any determinant can be calculated from the definition, but this involves formidable amounts of arithmetic if the dimension is at all large. Some of the theorems we prove will justify other methods of calculation, which involve less arithmetic than that required in working directly from the definition for $n > 3$.

Determinants were originally invented (in the middle of the eighteenth century) as a means of expressing the solutions of systems of linear equations. To see how this works for two equations in two unknowns, consider the general system

$$a_{11}x_1 + a_{12}x_2 = r_1$$
$$a_{21}x_1 + a_{22}x_2 = r_2.$$

The variable x_2 can be eliminated by multiplying the first equation by a_{22} and the second by a_{12}, and then taking the difference. The result is the equation

$$(a_{22}a_{11} - a_{12}a_{21})x_1 = (a_{22}r_1 - a_{12}r_2).$$

This equation may be written as

$$x_1 \det A = \det B^{(1)}$$

where $A = \begin{pmatrix} a_{11} & a_{12} \\ a_{21} & a_{22} \end{pmatrix}$ is the matrix of coefficients and $B^{(1)} = \begin{pmatrix} r_1 & a_{12} \\ r_2 & a_{22} \end{pmatrix}$ is the result of replacing the first column of A by $\begin{pmatrix} r_1 \\ r_2 \end{pmatrix}$. The reader can easily derive the equation $x_2 \det A = \det B^{(2)}$, where $B^{(2)} = \begin{pmatrix} a_{11} & r_1 \\ a_{21} & r_2 \end{pmatrix}$ is the result of substituting $\begin{pmatrix} r_1 \\ r_2 \end{pmatrix}$ for the second column of A. As we shall see in Theorem 8.4, a similar result holds for systems of n linear equations in n unknowns, for all values of n.

Since our definition of determinants by 7.1 is inductive, most of the proofs have the same character. That is, to prove a theorem about determinants of all square matrices, we verify it for 1-by-1 (or in some cases, 2-by-2) matrices, and also show that if it is true for $(n-1)$-by-$(n-1)$ matrices, then it holds for n-by-n matrices. In the proofs, we give only the argument for going from step $n-1$ to step n. The reader should verify the propositions directly for 1-by-1 and 2-by-2 matrices; the verification is in all cases quite trivial. A, B, and C will always denote n-by-n matrices. We write $\mathbf{a}_j$ for the jth column of a matrix A. If A has n rows, then $\mathbf{a}_j$ is a vector in $\mathcal{R}^n$.

7.2 Theorem

If B is obtained from A by multiplying some column by a number r, then $\det B = r \det A$.

Proof. Suppose $\mathbf{b}_j = r\mathbf{a}_j$, while $\mathbf{b}_k = \mathbf{a}_k$ for $k \neq j$. Then in particular $b_{1j} = ra_{1j}$. For $k \neq j$, B_{1k} is obtained from A_{1k} by multiplying a column by r; since B_{1k} and A_{1k} are $(n-1)$-by-$(n-1)$, we have $\det B_{1k} = r \det A_{1k}$ by the inductive assumption. On the other hand, $B_{1j} = A_{1j}$, and $b_{1k} = a_{1k}$ for $k \neq j$. Thus, whether $k = j$ or not, $b_{1k} \det B_{1k} = ra_{1k} \det A_{1k}$. Therefore

$$\det B = \sum_{k=1}^{n} (-1)^{k+1} b_{1k} \det B_{1k}$$

$$= \sum_{k=1}^{n} (-1)^{k+1} r a_{1k} \det A_{1k} = r \det A.$$

7.3 Corollary

If a matrix has a zero column, then its determinant is zero.

Proof. If $\mathbf{a}_j = 0$, then $\mathbf{a}_j = 0\mathbf{a}_j$. Then by Theorem 7.2, $\det A = 0 \cdot \det A = 0$.

Example 3. Let

$$A = \begin{pmatrix} 1 & 2 & 3 \\ -1 & 2 & 4 \\ 0 & 1 & 2 \end{pmatrix}, \qquad B = \begin{pmatrix} 1 & 6 & 3 \\ -1 & 6 & 4 \\ 0 & 3 & 2 \end{pmatrix}.$$

B is obtained from A by multiplying the second column by 3.

$$\begin{aligned} \det A &= (1)(4-4) - 2(-2-0) + 3(-1+0) \\ &= 0 + 4 - 3 = 1 \\ \det B &= (1)(12-12) - 6(-2-0) + 3(-3+0) \\ &= 0 + 12 - 9 = 3 = 3 \det A. \end{aligned}$$

7.4 Theorem

Let A, B, and C be identical except in the jth column, and suppose that the jth column of C is the sum of the jth columns of A and B. Then $\det C = \det A + \det B$.

Proof. We have $c_{1j} = a_{1j} + b_{1j}$, and also $C_{1j} = A_{1j} = B_{1j}$. For $k \neq j$, $c_{1k} = a_{1k} = b_{1k}$, and C_{1k} is identical with A_{1k} and B_{1k} except for one column which is the sum of the corresponding columns of A_{1k} and B_{1k}. Thus for $k \neq j$, $\det C_{1k} = \det A_{1k} + \det B_{1k}$, by the inductive assumption. For $k \neq j$ we have

$$\begin{aligned} c_{1k} \det C_{1k} &= c_{1k} \det A_{1k} + c_{1k} \det B_{1k} \\ &= a_{1k} \det A_{1k} + b_{1k} \det B_{1k}, \end{aligned}$$

while

$$\begin{aligned} c_{1j} \det C_{1j} &= a_{1j} \det C_{1j} + b_{1j} \det C_{1j} \\ &= a_{1j} \det A_{1j} + b_{1j} \det B_{1j}. \end{aligned}$$

Hence

$$\begin{aligned} \det C &= \sum_{k=1}^{n} (-1)^{k+1} c_{1k} \det C_{1k} \\ &= \sum_{k=1}^{n} (-1)^{k+1} a_{1k} \det A_{1k} + \sum_{k=1}^{n} (-1)^{k-1} b_{1k} \det B_{1k} \\ &= \det A + \det B. \end{aligned}$$

Example 4. Let

$$A = \begin{pmatrix} 1 & 2 & 3 \\ -1 & 2 & 4 \\ 0 & 1 & 2 \end{pmatrix}, \qquad B = \begin{pmatrix} 1 & 3 & 3 \\ -1 & 1 & 4 \\ 0 & -2 & 2 \end{pmatrix},$$

$$C = \begin{pmatrix} 1 & 5 & 3 \\ -1 & 3 & 4 \\ 0 & -1 & 2 \end{pmatrix}.$$

A, B, and C are identical except in the second column, and the second column of C is the sum of the second columns of A and B.

$$\begin{aligned} \det A &= (1)(4 - 4) - 2(-2 - 0) + 3(-1 - 0) \\ &= 0 + 4 - 3 = 1 \\ \det B &= (1)(2 + 8) - 3(-2 - 0) + 3(2 - 0) \\ &= 10 + 6 + 6 = 22 \\ \det C &= (1)(6 + 4) - 5(-2 - 0) + 3(1 - 0) \\ &= 10 + 10 + 3 = 23 = \det A + \det B. \end{aligned}$$

Theorems 7.2 and 7.4 have an important interpretation if the determinant of a matrix is viewed as a function of the columns of that matrix. Thus if $\mathbf{a}_1, \ldots, \mathbf{a}_n$ are the columns of an n-by-n matrix, we may write $\det A = \det(\mathbf{a}_1, \ldots, \mathbf{a}_n)$. Now suppose the $n - 1$ vectors $\mathbf{a}_2, \ldots, \mathbf{a}_n$ are held fixed and $f(\mathbf{x}) = \det(\mathbf{x}, \mathbf{a}_2, \ldots, \mathbf{a}_n)$ is considered as a function of $\mathbf{x}$. Then Theorems 7.2 and 7.4 assert that $f(r\mathbf{x}) = rf(\mathbf{x})$ and $f(\mathbf{x}_1 + \mathbf{x}_2) = f(\mathbf{x}_1) + f(\mathbf{x}_2)$. The same would hold if the "variable" vector were in the jth place instead of the first. In other words, we have

7.5 Theorem

If all but one of $\mathbf{a}_1, \ldots, \mathbf{a}_n$ are held fixed, then $\det(\mathbf{a}_1, \ldots, \mathbf{a}_n)$ is a linear function of the remaining vector.

To see how this works out for the first column of a 3-by-3 matrix, let

$$\mathbf{x} = \begin{pmatrix} x_1 \\ x_2 \\ x_3 \end{pmatrix}, \qquad \mathbf{b} = \begin{pmatrix} b_1 \\ b_2 \\ b_3 \end{pmatrix}, \qquad \mathbf{c} = \begin{pmatrix} c_1 \\ c_2 \\ c_3 \end{pmatrix}.$$

To exhibit det $(\mathbf{x}, \mathbf{b}, \mathbf{c})$ as a linear function of $\mathbf{x}$, we need merely calculate

$$\begin{aligned}
\det(\mathbf{x}, \mathbf{b}, \mathbf{c}) &= \det\begin{pmatrix} x_1 & b_1 & c_1 \\ x_2 & b_2 & c_2 \\ x_3 & b_3 & c_3 \end{pmatrix} \\
&= x_1 \det\begin{pmatrix} b_2 & c_2 \\ b_3 & c_3 \end{pmatrix} - b_1 \det\begin{pmatrix} x_2 & c_2 \\ x_3 & c_3 \end{pmatrix} + c_1 \det\begin{pmatrix} x_2 & b_2 \\ x_3 & b_3 \end{pmatrix} \\
&= x_1(b_2c_3 - b_3c_2) - b_1(x_2c_3 - x_3c_2) + c_1(x_2b_3 - x_3b_2) \\
&= x_1(b_2c_3 - b_3c_2) - x_2(b_1c_3 - b_3c_1) + x_3(b_1c_2 - c_1b_2),
\end{aligned}$$

which is a linear function of $\mathbf{x}$.

Another important property of determinants is that if any two columns (or rows) of a matrix are interchanged, then its determinant changes sign. We first prove the result for *adjacent* columns.

7.6 Lemma

If B is obtained from A by exchanging two adjacent columns, then $\det B = -\det A$.

Proof. Suppose A and B are the same, except that $\mathbf{a}_j = \mathbf{b}_{j+1}$ and $\mathbf{a}_{j+1} = \mathbf{b}_j$. For $k \neq j$ or $j + 1$, we have $b_{1k} = a_{1k}$ and $\det B_{1k} = -\det A_{1k}$ by the inductive hypothesis, so $(-1)^{k+1}b_{1k} \det B_{1k} = -(-1)^{k+1}a_{1k} \det A_{1k}$. On the other hand $b_{1j} = a_{1,j+1}$ and $B_{1j} = A_{1,j+1}$ so $(-1)^{j+1}b_{1j} \det B_{1j} = (-1)^{j+1}a_{1,j+1} \det A_{1,j+1} = -(-1)^{j+2} a_{1,j+1} \det A_{1,j+1}$. Similarly $(-1)^{j+2}b_{1,j+1} \det B_{1,j+1} = (-1)^{j+1}a_{1j} \det A_{1j}$. Thus each term in the expansion of $\det B$ by 7.1 is matched by a term equal to its negative in the expansion of A, and it follows that $\det B = -\det A$.

Example 5. Let

$$A = \begin{pmatrix} 1 & 3 & -2 \\ 2 & -4 & 1 \\ 3 & 5 & -2 \end{pmatrix}, \qquad B = \begin{pmatrix} 1 & -2 & 3 \\ 2 & 1 & -4 \\ 3 & -2 & 5 \end{pmatrix}.$$

Then

$$\begin{aligned}
\det A &= (1)(8 - 5) - (3)(-4 - 3) + (-2)(10 - (-12)) \\
&= 3 + 21 - 44 = -20 \\
\det B &= (1)(5 - 8) - (-2)(10 - (-12)) + (3)(-4 - 3) \\
&= -3 + 44 - 21 = 20 \\
&= -\det A
\end{aligned}$$

7.7 Theorem

If B is obtained from A by exchanging any two columns, then $\det B = -\det A$.

Proof. Suppose there are k columns between the two columns in question (so $k = 0$ if they are adjacent). The first column can be brought next to the second by k exchanges of adjacent columns. Then the two columns can be exchanged, and with another k exchanges of adjacent columns the second column can be put back in the original place of the first. There are $2k + 1$ steps in all, and by Lemma 7.6 each step changes the sign of the determinant. Since $2k + 1$ is an odd number, $\det B = -\det A$.

7.8 Theorem

If any two columns of A are identical, then $\det A = 0$.

Proof. Exchanging the two columns gives A again. Therefore $\det A = -\det A$, and so $\det A = 0$.

Multiplication by an n-by-n matrix gives a linear function from $\mathcal{R}^n$ to $\mathcal{R}^n$, and it is natural to ask whether the determinant of the matrix is related to some geometric property of the corresponding linear function. It turns out that the determinant describes how the function affects volumes in $\mathcal{R}^n$. In $\mathcal{R}^2$, of course, "volume" is area.

Example 6. Multiplication by $\begin{pmatrix} 3 & 0 \\ 0 & 2 \end{pmatrix}$ gives a function $\mathcal{R}^2 \xrightarrow{f} \mathcal{R}^2$ which multiplies lengths in the x-direction by 3 and in the y-direction by 2. Areas are magnified by a factor of 6, as illustrated in Fig. 24(a), which shows a unit square S and its image $f(S)$. Note that $\det \begin{pmatrix} 3 & 0 \\ 0 & 2 \end{pmatrix} = 6$. For another example, consider the function g given by the matrix $\begin{pmatrix} 1 & 2 \\ 0 & 1 \end{pmatrix}$, which has determinant 1. Its effect is illustrated in Fig. 24(b). The unit square is mapped into a parallelogram with the same base and altitude, so the area remains unchanged. The composition $g \circ f$ multiplies areas by 6 [since $f(S)$ has 6 times the area of S and $g(f(S))$ has the same area as $f(S)$]. The matrix of $g \circ f$ is given by the matrix product

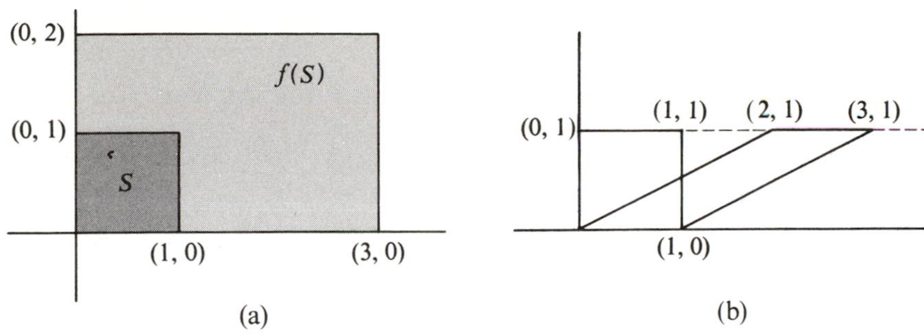

Figure 24

$$\begin{pmatrix} 1 & 2 \\ 0 & 1 \end{pmatrix}\begin{pmatrix} 3 & 0 \\ 0 & 2 \end{pmatrix} = \begin{pmatrix} 3 & 4 \\ 0 & 2 \end{pmatrix},$$

and we see that its determinant is 6.

These examples suggest that a linear function of $\mathcal{R}^2$ into itself multiplies areas by a factor equal to the determinant of the associated matrix. This is not quite right. For example, reflection in the y-axis (sending the point (x, y) into $(-x, y)$) obviously leaves areas unchanged, while its matrix $\begin{pmatrix} -1 & 0 \\ 0 & 1 \end{pmatrix}$ has determinant -1. To get a correct statement we must use the absolute value of the determinant. In fact, the following general statement is true.

7.9 Theorem

A linear function from $\mathcal{R}^n$ to $\mathcal{R}^n$ with matrix A multiplies volumes by the factor $|\det A|$.

The theorem cannot be proved in this general form without a rigorous definition of volume. The proof is therefore deferred to Chapter 4, Theorem 3.2, where it is obtained as a special case of the theorem proved in Section 9 of the Appendix.

If one function affects volumes by a factor u, and another affects them by a factor v, then the composition of the functions obviously affects volumes by the factor uv. Hence Theorem 7.9 implies that for n-by-n matrices A, B the relation $|\det (AB)| = |\det A| \cdot |\det B|$ holds. In fact the following theorem is true.

7.10 Product Rule

If A and B are any two square matrices of the same size, then $\det(AB) = (\det A)(\det B)$.

Proof. Let

$$L(\mathbf{x}_1, \ldots, \mathbf{x}_n) = \det A \det(\mathbf{x}_1, \ldots, \mathbf{x}_n) - \det(A\mathbf{x}_1, \ldots, A\mathbf{x}_n),$$

where $\mathbf{x}_1, \ldots, \mathbf{x}_n$ are vectors in $\mathcal{R}^n$. Clearly L is linear as a function of each vector $\mathbf{x}_j$. Furthermore, $L(\mathbf{e}_{i_1}, \ldots, \mathbf{e}_{i_n}) = 0$ for any set $\{\mathbf{e}_{i_1}, \ldots, \mathbf{e}_{i_n}\}$ of natural basis vectors. The reason is that if any of $\mathbf{e}_{i_1}, \ldots, \mathbf{e}_{i_n}$ are the same then both $\det(\mathbf{e}_{i_1}, \ldots, \mathbf{e}_{i_n})$ and $\det(A\mathbf{e}_{i_1}, \ldots, A\mathbf{e}_{i_n})$ are zero by Theorem 7.8. Otherwise $\mathbf{e}_{i_1}, \ldots, \mathbf{e}_{i_n}$ are just $\mathbf{e}_1, \ldots, \mathbf{e}_n$ in some order, and by Theorem 7.7

$$\begin{aligned}\det A \det(\mathbf{e}_{i_1}, \ldots, \mathbf{e}_{i_n}) &= \pm\det A \det(\mathbf{e}_1, \ldots, \mathbf{e}_n) = \pm\det A \det I \\ &= \pm\det A = \pm\det(A\mathbf{e}_1, \ldots, A\mathbf{e}_n) \\ &= \det(A\mathbf{e}_{i_1}, \ldots, A\mathbf{e}_{i_n}).\end{aligned}$$

But then $L(\mathbf{b}_i, \ldots, \mathbf{b}_n) = 0$, where

$$\mathbf{b}_j = \begin{pmatrix} b_{1j} \\ \cdot \\ \cdot \\ \cdot \\ b_{nj} \end{pmatrix}$$

is the jth column of B. For, using the linearity of L,

$$\begin{aligned}L(\mathbf{b}_1, \ldots, \mathbf{b}_n) &= L\left(\sum_{i=1}^{n} b_{i1}\mathbf{e}_i, \ldots, \sum_{i=1}^{n} b_{in}\mathbf{e}_i\right) \\ &= \sum_{i_1=1}^{n} \cdots \sum_{i_n=1}^{n} b_{i_1 1} \ldots b_{i_n n} L(\mathbf{e}_{i_1}, \ldots, \mathbf{e}_{i_n}) = 0.\end{aligned}$$

Hence,

$$\begin{aligned}\det A \det B - \det AB &= \det A \det(\mathbf{b}_1, \ldots, \mathbf{b}_n) \\ &\quad - \det(A\mathbf{b}_1, \ldots, A\mathbf{b}_n) \\ &= L(\mathbf{b}_1, \ldots, \mathbf{b}_n) = 0.\end{aligned}$$

Note that the proof just given uses only Theorems 7.5, 7.7, and 7.8 and does *not* use Theorem 7.9. (The point is important because the product rule is used in the proof in Section 7 of the Appendix, on which the proof of Theorem 7.9 depends.)

The natural unit of area in $\mathcal{R}^2$ is given by the unit square with edges (1, 0) and (0, 1), and the natural unit of volume in $\mathcal{R}^3$ is given by the unit cube with edges (1, 0, 0), (0, 1, 0), (0, 0, 1). In general we take the unit of volume in $\mathcal{R}^n$ to be that of the cube whose edges are the natural basis vectors $\mathbf{e}_1, \ldots, \mathbf{e}_n$ that form the columns of the n-by-n identity matrix. Moreover, we can take any n vectors $\mathbf{x}_1, \ldots, \mathbf{x}_n$ in $\mathcal{R}^n$ and form all linear combinations $t_1\mathbf{x}_1 + \ldots + t_n\mathbf{x}_n$, where each of the real numbers $t_1, \ldots, t_n$ satisfies the condition $0 \leq t_i \leq 1$. The resulting set of points is called the **parallelepiped** determined by its **edges** $\mathbf{x}_1, \ldots, \mathbf{x}_n$. If we choose only two vectors $\mathbf{x}_1$, $\mathbf{x}_2$, then we speak of the *parallelogram* determined by $\mathbf{x}_1$ and $\mathbf{x}_2$. Figure 25 shows a parallelepiped.

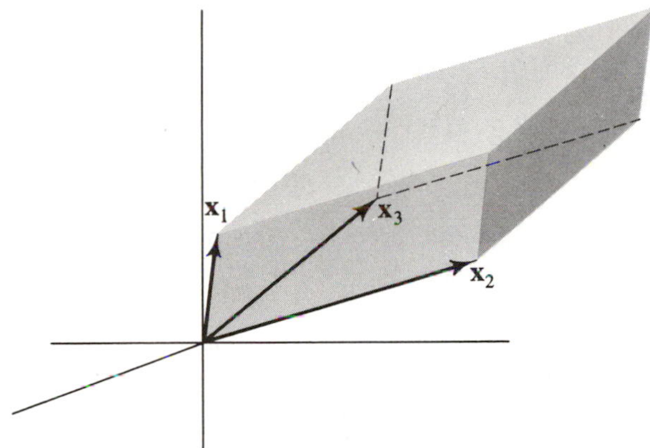

Figure 25

7.11 Theorem

Let $\mathbf{a}_1, \mathbf{a}_2, \ldots, \mathbf{a}_n$ be n vectors in $\mathcal{R}^n$. Then the volume of the parallelepiped with edges $\mathbf{a}_1, \ldots, \mathbf{a}_n$ is $|\det(\mathbf{a}_1, \ldots, \mathbf{a}_n)|$.

Proof. The linear function f whose matrix has columns $\mathbf{a}_1, \ldots, \mathbf{a}_n$ carries $\mathbf{e}_j$ into $\mathbf{a}_j$, by Theorem 4.2. Hence, f transforms the unit cube into the parallelepiped with edges a_j. Since it multiplies volumes by the factor $|\det(\mathbf{a}_1, \ldots, \mathbf{a}_n)|$, and the cube has unit volume, the volume of the parallelepiped is $|\det(\mathbf{a}_1, \ldots, \mathbf{a}_n)|$.

Example 7. Let $\mathbf{a}_1 = \begin{pmatrix} r_1 \cos\theta_1 \\ r_1 \sin\theta_1 \end{pmatrix}$, $\mathbf{a}_2 = \begin{pmatrix} r_2 \cos\theta_2 \\ r_2 \sin\theta_2 \end{pmatrix}$. The vectors have lengths r_1 and r_2, and make angles θ_1 and θ_2 with the x-axis as shown in Fig. 26. Then $\det(\mathbf{a}_1, \mathbf{a}_2)$ is

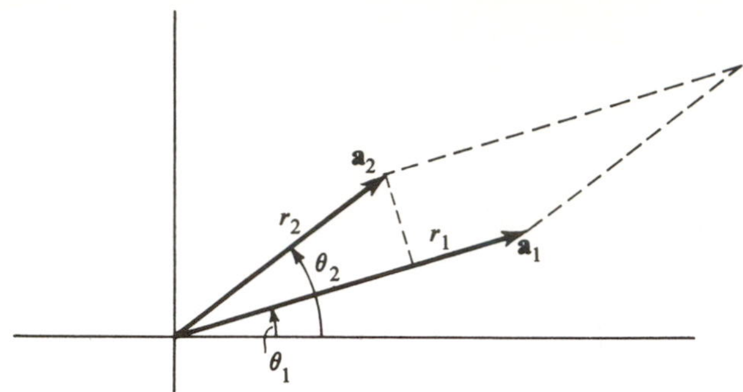

Figure 26

$$\det \begin{pmatrix} r_1 \cos \theta_1 & r_2 \cos \theta_2 \\ r_1 \sin \theta_1 & r_2 \sin \theta_2 \end{pmatrix} = r_1 r_2(\cos \theta_1 \sin \theta_2 - \sin \theta_1 \cos \theta_2)$$

$$= r_1 r_2 \sin (\theta_2 - \theta_1)$$

This number may be interpreted as the product of the base r_1 by the perpendicular height $r_2 \sin (\theta_2 - \theta_1)$.

We have seen that the absolute value of $\det (\mathbf{a}_1, \ldots, \mathbf{a}_n)$ can be interpreted as a volume. The sign of this determinant also has a geometric interpretation. We say that an ordered set of vectors $(\mathbf{a}_1, \ldots, \mathbf{a}_n)$ in $\mathcal{R}^n$ has **positive orientation** (or is **positively oriented**) if $\det (\mathbf{a}_1, \ldots, \mathbf{a}_n) > 0$, and has **negative orientation** if $\det (\mathbf{a}_1, \ldots, \mathbf{a}_n) < 0$. If the determinant is equal to zero, the orientation is not defined.

Example 7 shows that in $\mathcal{R}^2$, the sign of $\det (\mathbf{a}_1, \mathbf{a}_2)$ is the same as the sign of $\sin \theta$, where $\theta = \theta_2 - \theta_1$ is the angle from $\mathbf{a}_1$ to $\mathbf{a}_2$. Thus the orientation of $(\mathbf{a}_1, \mathbf{a}_2)$ is positive if some counterclockwise rotation of less than 180° will turn $\mathbf{a}_1$ to the direction of $\mathbf{a}_2$. The orientation is negative if a clockwise rotation is required; it is not defined if $\mathbf{a}_1$ and $\mathbf{a}_2$ lie in the same line. Thus in $\mathcal{R}^2$, orientation corresponds to a direction of rotation. (Note that the orientation of $(\mathbf{a}_1, \mathbf{a}_2)$ is opposite to that of $(\mathbf{a}_2, \mathbf{a}_1)$.) Property 7.7 of determinants, of course, implies that the orientation of a set of vectors is always reversed if two vectors of the set are exchanged.

The interpretation of orientation in $\mathcal{R}^3$ is less obvious. The sets of vectors $(\mathbf{x}, \mathbf{y}, \mathbf{z})$ and $(-\mathbf{x}, \mathbf{y}, \mathbf{z})$ shown in Fig. 27(a) and 27(b) have opposite orientations since, by Theorem 7.2, $\det (-\mathbf{x}, \mathbf{y}, \mathbf{z}) = -\det (\mathbf{x}, \mathbf{y}, \mathbf{z})$. The ordered set of vectors $(\mathbf{x}, \mathbf{y}, \mathbf{z})$ is said to form a right-handed system because, when the thumb and index finger of the right hand are made to point in the $\mathbf{x}$- and $\mathbf{y}$-directions, the middle finger will point in the $\mathbf{z}$-direction. Similarly, $(-\mathbf{x}, \mathbf{y}, \mathbf{z})$ form a left-handed system. In this book we have chosen to draw pictures in 3-space so that the vectors

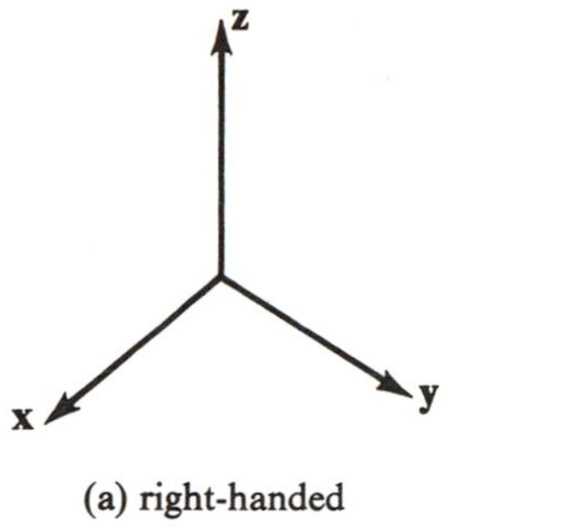

(a) right-handed

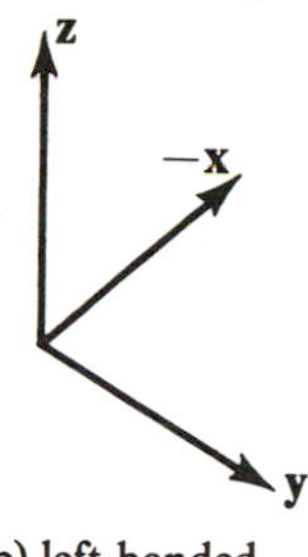

(b) left-handed

Figure 27

$\mathbf{e}_1, \mathbf{e}_2, \mathbf{e}_3$ form a right-handed system. Since $\det(\mathbf{e}_1, \mathbf{e}_2, \mathbf{e}_3) = \det I = 1$, this implies that our right-handed system has positive orientation, and a left-handed system would have negative orientation.

Let $\mathbf{u} = (u_1, u_2, u_3)$ and $\mathbf{v} = (v_1, v_2, v_3)$ be vectors in $\mathcal{R}^3$. The vector with coordinates

$$\begin{vmatrix} u_2 & u_3 \\ v_2 & v_3 \end{vmatrix}, \quad \begin{vmatrix} u_3 & u_1 \\ v_3 & v_1 \end{vmatrix}, \quad \begin{vmatrix} u_1 & u_2 \\ v_1 & v_2 \end{vmatrix}$$

is called the **cross-product** of $\mathbf{u}$ and $\mathbf{v}$, and is written $\mathbf{u} \times \mathbf{v}$. Note that interchanging $\mathbf{u}$ and $\mathbf{v}$ changes the sign of each component, so $\mathbf{u} \times \mathbf{v} = -(\mathbf{v} \times \mathbf{u})$. The significance of the cross product lies in the equation

7.12 $$\mathbf{x} \cdot (\mathbf{u} \times \mathbf{v}) = \det(\mathbf{x}, \mathbf{u}, \mathbf{v}),$$

which holds for any $\mathbf{x}$ in $\mathcal{R}^3$. To prove it we let $\mathbf{x} = (x_1, x_2, x_3)$. Then

$$\begin{aligned}
\det(\mathbf{x}, \mathbf{u}, \mathbf{v}) &= \det \begin{pmatrix} x_1 & u_1 & v_1 \\ x_2 & u_2 & v_2 \\ x_3 & u_3 & v_3 \end{pmatrix} \\
&= x_1 \begin{vmatrix} u_2 & v_2 \\ u_3 & v_3 \end{vmatrix} - u_1 \begin{vmatrix} x_2 & v_2 \\ x_3 & v_3 \end{vmatrix} + v_1 \begin{vmatrix} x_2 & u_2 \\ x_3 & u_3 \end{vmatrix} \\
&= x_1(u_2v_3 - u_3v_2) - u_1(x_2v_3 - x_3v_2) + v_1(x_2u_3 - x_3u_2) \\
&= x_1(u_2v_3 - u_3v_2) + x_2(u_3v_1 - u_1v_3) + x_3(u_1v_2 - u_2v_1) \\
&= x_1 \begin{vmatrix} u_2 & u_3 \\ v_2 & v_3 \end{vmatrix} + x_2 \begin{vmatrix} u_3 & u_1 \\ v_3 & v_1 \end{vmatrix} + x_3 \begin{vmatrix} u_1 & u_2 \\ v_1 & v_2 \end{vmatrix} \\
&= \mathbf{x} \cdot (\mathbf{u} \times \mathbf{v}).
\end{aligned}$$

A convenient way to remember the formula for the cross-product is to think of the formal "determinant"

$$\begin{vmatrix} \mathbf{e}_1 & \mathbf{e}_2 & \mathbf{e}_3 \\ u_1 & u_2 & u_3 \\ v_1 & v_2 & v_3 \end{vmatrix} = \begin{vmatrix} u_2 & u_3 \\ v_2 & v_3 \end{vmatrix}\mathbf{e}_1 - \begin{vmatrix} u_1 & u_3 \\ v_1 & v_3 \end{vmatrix}\mathbf{e}_2 + \begin{vmatrix} u_1 & u_2 \\ v_1 & v_2 \end{vmatrix}\mathbf{e}_3,$$

which expresses $\mathbf{u} \times \mathbf{v}$ in terms of the standard basis $\mathbf{e}_1, \mathbf{e}_2, \mathbf{e}_3$.

If either $\mathbf{u}$ or $\mathbf{v}$ is zero, or if one is a numerical multiple of the other, $\mathbf{u} \times \mathbf{v}$ is obviously zero. Otherwise $\mathbf{u}$ and $\mathbf{v}$ determine a plane. Substituting $\mathbf{u}$ and $\mathbf{v}$ for $\mathbf{x}$ in 7.12 gives

$$\mathbf{u} \cdot (\mathbf{u} \times \mathbf{v}) = \det(\mathbf{u}, \mathbf{u}, \mathbf{v}) = 0, \qquad \mathbf{v} \cdot (\mathbf{u} \times \mathbf{v}) = \det(\mathbf{v}, \mathbf{u}, \mathbf{v}) = 0$$

by Theorem 7.8. Thus $\mathbf{u} \times \mathbf{v}$ is perpendicular to the plane of $\mathbf{u}$ and $\mathbf{v}$. Let $\mathbf{x}$ be a vector of unit length in the direction of $\mathbf{u} \times \mathbf{v}$. Then $\mathbf{x} \cdot (\mathbf{u} \times \mathbf{v}) = |\mathbf{u} \times \mathbf{v}|$. On the other hand $\mathbf{x} \cdot (\mathbf{u} \times \mathbf{v}) = \det(\mathbf{x}, \mathbf{u}, \mathbf{v})$, and has absolute value equal to the volume of the parallelepiped with edges $\mathbf{x}, \mathbf{u}, \mathbf{v}$. Since $\mathbf{x}$ has unit length and is perpendicular to the plane of $\mathbf{u}$ and $\mathbf{v}$, this volume is equal to the area of the parallelogram with edges $\mathbf{u}, \mathbf{v}$. Thus the length of $\mathbf{u} \times \mathbf{v}$ is equal to the area of the parallelogram with edges $\mathbf{u}, \mathbf{v}$. Finally,

$$\det(\mathbf{u} \times \mathbf{v}, \mathbf{u}, \mathbf{v}) = (\mathbf{u} \times \mathbf{v}) \cdot (\mathbf{u} \times \mathbf{v}) = |\mathbf{u} \times \mathbf{v}|^2 \geq 0.$$

Then, unless $\mathbf{u} \times \mathbf{v} = 0$, the triple $(\mathbf{u} \times \mathbf{v}, \mathbf{u}, \mathbf{v})$ has positive orientation and forms a right-handed system.

We can summarize what we have just proved as:

7.13 Theorem

If $\mathbf{u}$ and $\mathbf{v}$ are noncollinear vectors in $\mathcal{R}^3$, then the cross-product $\mathbf{u} \times \mathbf{v}$ is a vector perpendicular to both $\mathbf{u}$ and $\mathbf{v}$, with length equal to the area of the parallelogram with edges $\mathbf{u}$ and $\mathbf{v}$. The ordered triple $(\mathbf{u} \times \mathbf{v}, \mathbf{u}, \mathbf{v})$ is a right-handed system.

Figure 28 shows the relation between $\mathbf{u}$, $\mathbf{v}$, and $\mathbf{u} \times \mathbf{v}$.

Example 8. Find the area of the parallelogram with edges $\mathbf{u} = (1, 2, 3)$, $\mathbf{v} = (3, 2, 1)$. We have

$$\begin{aligned} \mathbf{u} \times \mathbf{v} &= \big((2)(1) - (3)(2), -(1)(1) + (3)(3), (1)(2) - (2)(3)\big) \\ &= (-4, 8, -4). \end{aligned}$$

$$\begin{aligned} \text{Area} &= |\mathbf{u} \times \mathbf{v}| = (16 + 64 + 16)^{1/2} \\ &= \sqrt{96} = 4\sqrt{6}. \end{aligned}$$

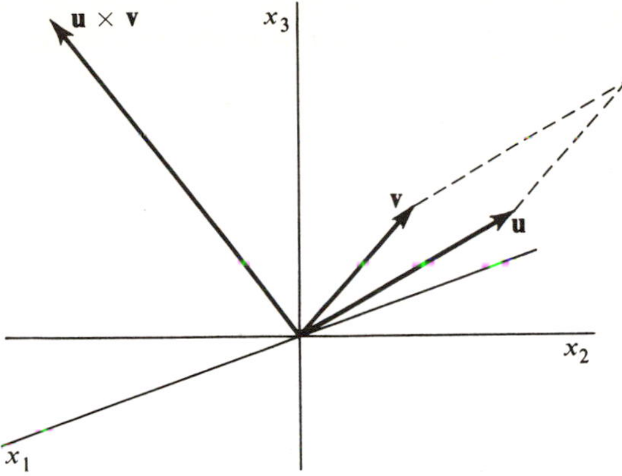

Figure 28

It is sometimes appropriate to combine the ideas of volume and orientation in a single concept. We define the **oriented volume** determined by an n-tuple of vectors to be the ordinary volume if the orientation of the n-tuple is positive and to be the negative of the volume if the orientation is negative. Then the oriented volume of the ordered set $(\mathbf{a}_1, \ldots, \mathbf{a}_n)$ is equal to $\det(\mathbf{a}_1, \ldots, \mathbf{a}_n)$. The relation between oriented volume and ordinary volume is very much like the relation between directed distance on a line and ordinary distance. Indeed, oriented volume may be considered a generalization of directed distance, and we use the idea in Chapter 7, Section 7.

EXERCISES

1. Find AB, BA, and the determinants of A, B, AB, and BA when

(a) $A = \begin{pmatrix} 1 & -2 \\ 3 & 1 \end{pmatrix}, \quad B = \begin{pmatrix} 0 & 1 \\ 2 & -3 \end{pmatrix}.$ [*Ans.* $\det AB = -14$.]

(b) $A = \begin{pmatrix} 2 & 0 & 0 \\ 0 & 3 & 0 \\ 0 & 0 & 4 \end{pmatrix}, \quad B = \begin{pmatrix} -1 & 0 & 1 \\ 2 & -1 & -3 \\ 0 & 3 & 5 \end{pmatrix}.$

2. Find the coefficients needed to express each of the following as a linear function of the x's.

(a) $\begin{vmatrix} x_1 & -1 & 6 \\ x_2 & 4 & -3 \\ x_3 & 2 & 5 \end{vmatrix}.$ (b) $\begin{vmatrix} 1 & x_1 & -3 & 2 \\ 0 & x_2 & 0 & 0 \\ 3 & x_3 & -1 & 5 \\ -2 & x_4 & 0 & 1 \end{vmatrix}.$

3. Show that if D is the diagonal matrix $\mathbf{diag}\,(r_1, \ldots, r_n)$, then $\det D = r_1 r_2 \ldots r_n$.

4. What is the relation between

(a) $\det A$ and $\det(-A)$?

(b) $\det(\mathbf{a}_1, \mathbf{a}_2, \ldots, \mathbf{a}_{n-1}, \mathbf{a}_n)$ and $\det(\mathbf{a}_n, \mathbf{a}_{n-1}, \ldots, \mathbf{a}_2, \mathbf{a}_1)$?

5. Verify the product rule for the pairs of matrices (a) and (b) in Problem 1.

6. Apply the product rule to show that, if A is invertible, then $\det A \neq 0$ and $(\det A^{-1}) = (\det A)^{-1}$.

7. Let A be an m-by-m matrix and B an n-by-n matrix. Consider the $(m + n)$-by-$(m + n)$ matrix $\begin{pmatrix} A & 0 \\ 0 & B \end{pmatrix}$ which has A in the upper left corner, B in the lower right corner, and zeros elsewhere. Show that its determinant is equal to $(\det A)(\det B)$. [*Suggestion.* First consider the case where one of A or B is an identity matrix, and derive the general result from the product rule.]

8. It is geometrically clear that a rotation in $\mathcal{R}^2$ preserves orientations, that a reflection reverses them, and that both leave areas unchanged. Verify this by finding the determinants of the associated matrices. (See Problems 4 and 6 in Section 4.)

9. By interpreting the determinant as a volume, show that $|\det(\mathbf{x}_1, \mathbf{x}_2, \mathbf{x}_3)| \leq |\mathbf{x}_1| \cdot |\mathbf{x}_2| \cdot |\mathbf{x}_3|$ for any three vectors in $\mathcal{R}^3$, and that equality holds if and only if the vectors are mutually orthogonal.

10. Find the volume, and the area of each side, of the parallelepiped with edges $(1, 1, 0)$, $(0, 1, 2)$, $(-3, 5, -1)$.

[*Ans.* volume $= 17$, areas $= \sqrt{166}$, $\sqrt{66}$, 3.]

11. If $\mathbf{u} = (2, 1, 3)$, $\mathbf{v} = (0, 2, -1)$, and $\mathbf{w} = (1, 1, 1)$, compute $\mathbf{u} \times \mathbf{v}$, $\det(\mathbf{u}, \mathbf{v}, \mathbf{w})$, $(\mathbf{u} \times \mathbf{v}) \cdot \mathbf{w}$, $(\mathbf{u} \times \mathbf{v}) \times \mathbf{w}$, and $\mathbf{u} \times (\mathbf{v} \times \mathbf{w})$.

12. Prove that $\mathbf{u} \times \mathbf{v} = -\mathbf{v} \times \mathbf{u}$, that $\mathbf{u} \times (\mathbf{v} + \mathbf{w}) = (\mathbf{u} \times \mathbf{v}) + (\mathbf{u} \times \mathbf{w})$, and that $(a\mathbf{u}) \times \mathbf{v} = \mathbf{u} \times (a\mathbf{v}) = a(\mathbf{u} \times \mathbf{v})$, for a real.

13. Find a representation for a line perpendicular to $(2, 1, 3)$ and $(0, 2, -1)$, and passing through $(1, 1, 1)$.

14. Let P be a parallelogram determined by two vectors in $\mathcal{R}^3$. Let P_x, P_y, and P_z be the projections of P on the yz-plane, the zx-plane, and the xy-plane, respectively. If $A(P)$ is the area of P, show that $A^2(P) = A^2(P_x) + A^2(P_y) + A^2(P_z)$.

15. (a) Verify by direct coordinate computation that $|\mathbf{u} \times \mathbf{v}|^2 = |\mathbf{u}|^2\,|\mathbf{v}|^2 - (\mathbf{u} \cdot \mathbf{v})^2$.

(b) Use the result of part (a) to show that $|\mathbf{u} \times \mathbf{v}| = |\mathbf{u}|\,|\mathbf{v}| \sin\theta$, where θ is the angle between $\mathbf{u}$ and $\mathbf{v}$ such that $0 \leq \theta \leq \pi$.

(c) Show that $|\mathbf{u}|\,|\mathbf{v}| \sin\theta$ is the area of the parallelogram with edges $\mathbf{u}$ and $\mathbf{v}$.

16. The complex numbers can be extended to the quaternion algebra $\mathcal{H}$, which is a four-dimensional vector space with natural basis $\{1, i, j, k\}$. Thus a typical quaternion is written $q = a_1 + a_2i + a_3j + a_4k$, where the a's are real numbers. A product is defined in $\mathcal{H}$ by requiring $i^2 = j^2 = k^2 = -1$ and $ij = -ji = k$, $jk = -kj = i$, $ki = -ik = j$. The product of two quaternions is got by multiplying out and using the above rules for products of basis vectors. $\mathcal{R}^3$ can be looked at as the vector subspace $\mathcal{G}$ of $\mathcal{H}$ consisting of quaternions with "real part" equal to zero and thus with natural basis $\{i, j, k\}$.

(a) Show that the quaternion product of two elements of $\mathcal{G}$ is not necessarily in $\mathcal{G}$.

(b) Define a product on $\mathcal{G}$ by first forming the quaternion product and then replacing its real part by zero. Show that the resulting product is the same as the cross-product in $\mathcal{R}^3$.

17. Prove the identity $\mathbf{a} \times (\mathbf{b} \times \mathbf{c}) = (\mathbf{a} \cdot \mathbf{c})\mathbf{b} - (\mathbf{a} \cdot \mathbf{b})\mathbf{c}$ for vectors $\mathbf{a}$, $\mathbf{b}$, and $\mathbf{c}$ in $\mathcal{R}^3$. [*Hint.* Choose an orthonormal set of vectors $(\mathbf{u}_1, \mathbf{u}_2, \mathbf{u}_3)$ for $\mathcal{R}^3$ so that $(\mathbf{u}_1, \mathbf{u}_2, \mathbf{u}_3)$ is positively oriented and

$$\mathbf{a} = a_1\mathbf{u}_1 + a_2\mathbf{u}_2 + a_3\mathbf{u}_3$$

$$\mathbf{b} = b_1\mathbf{u}_1 + b_2\mathbf{u}_2$$

$$\mathbf{c} = c_1\mathbf{u}_1.]$$

SECTION 8

DETERMINANT EXPANSIONS

In the previous section we defined the determinant of an n-by-n matrix $A = (a_{ij})$ by

$$\det A = \sum_{j=1}^{n} (-1)^{j+1} a_{1j} \det A_{1j},$$

where, in general, A_{ij} denotes the $(n-1)$-by-$(n-1)$ minor corresponding to a_{ij}. In the present section we prove more general formulas of the same kind. These formulas, which apply to any n-by-n matrix A, are

8.1 $$\det A = \sum_{i=1}^{n} (-1)^{i+j} a_{ij} \det A_{ij}$$

8.1R $$\det A = \sum_{j=1}^{n} (-1)^{i+j} a_{ij} \det A_{ij}$$

Formula 8.1 holds for each integer j between 1 and n, while Formula 8.1R holds for each integer i between 1 and n. (For $i = 1$, Formula 8.1R coincides with Formula 7.1 used earlier in defining determinants.) For a given j, the matrix elements a_{ij} on the right side of 8.1 are in the jth column of A, so 8.1 is called the expansion of det A by the jth column. Similarly, the right side of 8.1R is called the expansion of det A by the ith row. We postpone the proof of the formulas to the end of the section, first showing some of their consequences.

Example 1. Let

$$A = \begin{pmatrix} 2 & 3 & 4 \\ 5 & 6 & 7 \\ 8 & 9 & 0 \end{pmatrix}.$$

The expansion of det A by the second row is

$$\begin{aligned} &-5 \det \begin{pmatrix} 3 & 4 \\ 9 & 0 \end{pmatrix} + 6 \det \begin{pmatrix} 2 & 4 \\ 8 & 0 \end{pmatrix} - 7 \det \begin{pmatrix} 2 & 3 \\ 8 & 9 \end{pmatrix} \\ &= (-5)(-36) + (6)(-32) - (7)(-6) \\ &= 180 - 192 + 42 = 30. \end{aligned}$$

The expansion by the third column is

$$\begin{aligned} &4 \det \begin{pmatrix} 5 & 6 \\ 8 & 9 \end{pmatrix} - 7 \det \begin{pmatrix} 2 & 3 \\ 8 & 9 \end{pmatrix} + 0 \det \begin{pmatrix} 2 & 3 \\ 5 & 6 \end{pmatrix} \\ &= (4)(-3) - (7)(-6) \\ &= -12 + 42 = 30. \end{aligned}$$

Formulas 8.1 and 8.1R can be useful in evaluating a determinant, but some of their theoretical consequences are more important. Let the matrix A be given and consider the expression

$$\sum_{i=1}^{n} (-1)^{i+j} x_i \det A_{ij},$$

where $x_1, \ldots, x_n$ may be any set of n numbers. From 8.1 we see that this is equal to a certain determinant; in fact it is the expansion by the jth column of the matrix obtained from A by replacing the jth column with $(x_1, \ldots, x_n)$. Now consider what happens if we take $x_1, \ldots, x_n$ equal to the elements $a_{1k}, a_{2k}, \ldots, a_{nk}$ of the kth column of A. If $k = j$, of course we simply have the expansion of det A by the jth column. If $k \neq j$, we have the determinant of a matrix with two columns (the jth and kth) identical, and by Theorem 7.8 the result is 0. We have proved:

8.2 Theorem

For any n-by-n matrix A,

$$\sum_{i=1}^{n} (-1)^{i+j} a_{ik} \det A_{ij} = \begin{cases} \det A & \text{if } k = j \\ 0 & \text{if } k \neq j \end{cases}$$

An exactly similar argument, using 8.1R instead of 8.1, gives the "row" form:

8.2R For any n-by-n matrix A,

$$\sum_{j=1}^{n}(-1)^{i+j}a_{kj}\det A_{ij} = \begin{cases}\det A & \text{if } k = i\\ 0 & \text{if } k \neq i\end{cases}$$

The number $(-1)^{i+j}\det A_{ij}$ is called the ijth **cofactor** of the matrix A. We shall abbreviate it as $\tilde{a}_{ij}$ and write $\tilde{A}$ for the matrix with entries $\tilde{a}_{ij}$. Theorem 8.2 can be formulated as a statement about the matrix product $\tilde{A}^tA$. The jkth entry in the product is the product of the jth row of $\tilde{A}^t$ (i.e., the jth column of $\tilde{A}$) and the kth column of A. That is, it is the sum $\sum_{i=1}^{n}\tilde{a}_{ij}a_{ik}$. By Theorem 8.2 this is $\det A$ if $j = k$, and 0 otherwise. Hence $\tilde{A}^tA$ is equal to $(\det A)I$, a numerical multiple of the identity matrix. A similar calculation using 8.2 shows that $A\tilde{A}^t$ is also equal to $(\det A)I$. If $\det A \neq 0$, we may divide $\tilde{A}^t$ by it and obtain a matrix B such that $AB = BA = I$. Thus we have proved:

8.3 Theorem

If $\det A \neq 0$, then A is invertible, and $A^{-1} = (\det A)^{-1}\tilde{A}^t$, where $\tilde{A}$ is the matrix of cofactors of A.

It is an important consequence of Theorem 8.3 that, if a matrix A is known only to have a "right inverse" ($BA = I$) or a "left inverse" ($AB = I$), then A is invertible and $A^{-1} = B$. This is true because the product rule for determinants applied to $AB = I$, or to $BA = I$, gives

$$(\det A)(\det B) = \det I = 1.$$

But then $\det A \neq 0$, so A is invertible. That $A^{-1} = B$ follows immediately. (Why?)

Example 2. We shall compute the inverse of the matrix

$$A = \begin{pmatrix}2 & 3 & 4\\ 5 & 6 & 7\\ 8 & 9 & 0\end{pmatrix}$$

used in Example 1. Write b_{ij} as an abbreviation for det A_{ij}; the matrix B is then easily calculated to be

$$\begin{pmatrix} -63 & -56 & -3 \\ -36 & -32 & -6 \\ -3 & -6 & -3 \end{pmatrix}.$$

To obtain the matrix of cofactors, insert the factors $(-1)^{i+j}$, changing the sign of every second entry and giving

$$\begin{pmatrix} -63 & 56 & -3 \\ 36 & -32 & 6 \\ -3 & 6 & -3 \end{pmatrix}.$$

Finally, transpose and divide by det A, which was found to equal 30 in Example 1. The result is

$$A^{-1} = \tfrac{1}{30}\begin{pmatrix} -63 & 36 & -3 \\ 56 & -32 & 6 \\ -3 & 6 & -3 \end{pmatrix},$$

as can be verified by computing AA^{-1} and $A^{-1}A$.

A system of n linear equations in n unknowns

$$\begin{aligned} a_{11}x_1 + a_{12}x_2 + \ldots + a_{1n}x_n &= b_1 \\ a_{21}x_1 + a_{22}x_2 + \ldots + a_{2n}x_n &= b_2 \\ \cdot \qquad\qquad\qquad \cdot \quad &\quad \cdot \\ \cdot \qquad\qquad\qquad \cdot \quad &\quad \cdot \\ \cdot \qquad\qquad\qquad \cdot \quad &\quad \cdot \\ a_{n1}x_1 + a_{n2}x_2 + \ldots + a_{nn}x_n &= b_n \end{aligned} \tag{1}$$

may be written in matrix form as $A\mathbf{x} = \mathbf{b}$, where A is the n-by-n coefficient matrix with entries a_{ij}, and $\mathbf{x}$ and $\mathbf{b}$ are vectors in $\mathcal{R}^n$. Doing the matrix multiplication in the equation

$$\begin{pmatrix} a_{11} & \ldots & a_{1n} \\ \cdot & & \cdot \\ \cdot & & \cdot \\ \cdot & & \cdot \\ a_{n1} & \ldots & a_{nn} \end{pmatrix} \begin{pmatrix} x_1 \\ \cdot \\ \cdot \\ \cdot \\ x_n \end{pmatrix} = \begin{pmatrix} b_1 \\ \cdot \\ \cdot \\ \cdot \\ b_n \end{pmatrix}$$

shows at once that it is equivalent to the system (1).

If A is invertible, then $A\mathbf{x} = \mathbf{b}$ implies $A^{-1}A\mathbf{x} = A^{-1}\mathbf{b}$ or $\mathbf{x} = A^{-1}\mathbf{b}$; on the other hand, $A(A^{-1}\mathbf{b}) = (AA^{-1})\mathbf{b} = \mathbf{b}$. In other words, the equations have a unique solution, and it is $A^{-1}\mathbf{b}$. The jth entry in the column vector $A^{-1}\mathbf{b}$ is the matrix product of the jth row of A^{-1} and the vector $\mathbf{b}$. If $\det A \neq 0$, we may express the elements of A^{-1} in terms of cofactors of A and obtain

$$(\det A)^{-1} \sum_{i=1}^{n} (-1)^{i+j} (\det A_{ij}) b_i$$

for this product. From 8.1 this may be recognized as $(\det A)^{-1} \det B^{(j)}$ where $B^{(j)}$ is the result of replacing the jth column of A by $\mathbf{b}$. We have proved:

8.4 Cramer's Rule

If the determinant of the matrix of coefficients of a system of n linear equations in n unknowns $x_1, \ldots, x_n$ is different from zero, then there is a unique solution and it is given by

$$x_j = \frac{\det B^{(j)}}{\det A},$$

where A is the matrix of coefficients and $B^{(j)}$ is the result of replacing the jth column of A by the column of numbers that make up the right side of the equations.

Example 3. Solve the system

$$\begin{aligned} x_1 \quad -2x_2 \quad +4x_3 &= 1 \\ -x_1 \quad +x_2 \quad -x_3 &= 2 \\ 2x_1 \quad +3x_2 \quad -x_3 &= 3. \end{aligned}$$

We have

$$A = \begin{pmatrix} 1 & -2 & 4 \\ -1 & 1 & -1 \\ 2 & 3 & -1 \end{pmatrix}, \qquad B^{(1)} = \begin{pmatrix} 1 & -2 & 4 \\ 2 & 1 & -1 \\ 3 & 3 & -1 \end{pmatrix},$$

$$B^{(2)} = \begin{pmatrix} 1 & 1 & 4 \\ -1 & 2 & -1 \\ 2 & 3 & -1 \end{pmatrix}, \qquad B^{(3)} = \begin{pmatrix} 1 & -2 & 1 \\ -1 & 1 & 2 \\ 2 & 3 & 3 \end{pmatrix}.$$

Expanding the determinants by their first rows gives

$$\det A = (1)(2) - (-2)(3) + (4)(-5) = 2 + 6 - 20 = -12$$

$$\det B^{(1)} = (1)(2) - (-2)(1) + (4)(3) = 2 + 2 + 12 = 16$$

$$\det B^{(2)} = (1)(1) - (1)(3) + (4)(-7) = 1 - 3 - 28 = -30$$

$$\det B^{(3)} = (1)(-3) - (-2)(-7) + (1)(-5) = -3 - 14 - 5 = -22.$$

Then $x_1 = -\frac{16}{12} = -\frac{4}{3}$, $x_2 = \frac{30}{12} = \frac{5}{2}$, $x_3 = \frac{22}{12} = \frac{11}{6}$.

We have not made any assertions in this section about what happens if the determinant of the coefficient matrix of a system of equations is zero. It is an easy consequence of the product rule that a matrix with zero determinant cannot be invertible (see Problem 1 in Section 7), and it will be shown in Chapter 2, where we discuss the solution of linear systems in detail, that in this case the system of equations has either no solution or infinitely many. While Cramer's rule and the formula for A^{-1} in terms of cofactors are important as theoretical results and are quite useful for solving systems of two or three linear equations, they are less efficient for larger systems than the methods of Section 1, Chapter 2.

Since 8.1 with $j = 1$ is exactly like 7.1 except that it refers to the first column instead of the first row of a matrix, it is clear that transposing a matrix (which just exchanges the roles of rows and columns) should not affect the value of the determinant. The formal statement and proof follow.

8.5 Theorem

For any square matrix A, $\det A^t = \det A$.

Proof. Let $B = A^t$. By definition of the transpose, $b_{ij} = a_{ji}$, and it is easy to see that $B_{ij} = A^t_{ji}$. Thus by definition

$$\begin{aligned}\det B &= b_{11} \det B_{11} - b_{12} \det B_{12} + \ldots + (-1)^{n+1} b_{1n} \det B_{1n} \\ &= a_{11} \det A^t_{11} - a_{21} \det A^t_{21} + \ldots + (-1)^{n+1} a_{n1} \det A^t_{n1}.\end{aligned}$$

The A_{ij} are $(n - 1)$-by-$(n - 1)$ matrices, and by the inductive hypothesis we may replace $\det A^t_{i1}$ by $\det A_{i1}$ in the formula above. The result is equal to $\det A$ by 8.1 (with $j = 1$), and we have proved $\det A^t = \det B = \det A$.

We may now take any theorem about determinants that involves columns of matrices and immediately derive a corresponding theorem involving rows instead, by applying the given theorem to the transposes of the matrices. We shall not always bother to write out these corresponding theorems, but shall refer to the "row" version of a numbered statement by using the same number with an R after it. (We have already numbered 8.1R and 8.2R to conform to this convention.) For example, Theorem 7.2R would read: If B is obtained from A by multiplying some row by the number r, then $\det B = r \det A$.

Theorem 8.5 implies that, from any theorem about determinants that involves columns of matrices, we can derive a corresponding theorem involving rows instead, by applying the given theorem to the transposes of the matrices. In particular, 8.1R *is a consequence of* 8.1 *and* 8.5. We shall not bother to write out the row versions of the other statements but may refer to them by the original statement number with an R after it. The particularly important fact that a determinant is a linear function of each of its columns was proved in the previous section. It now follows that a determinant is also a linear function of each of its rows.

The following theorem (and the corresponding theorem for rows) leads to an efficient procedure for computing the determinants of large matrices.

8.6 Theorem

If C and A are n-by-n matrices and C is obtained from A by adding a numerical multiple of one column to another, then $\det C = \det A$.

Proof. Suppose C is the same as A except that $\mathbf{c}_j = \mathbf{a}_j + r\mathbf{a}_i$. Let B be the result of replacing $\mathbf{a}_j$ in A with $r\mathbf{a}_i$. By Theorem 7.4, $\det C = \det A + \det B$. By Theorem 7.2, $\det B$ is r times the determinant of a matrix with two identical columns. Therefore $\det B = 0$ and $\det C = \det A$.

Example 4. (a) Let

$$A = \begin{pmatrix} 1 & 3 & -2 \\ 2 & -4 & 1 \\ 3 & 5 & -2 \end{pmatrix}, \qquad C = \begin{pmatrix} 1 & 3 & 0 \\ 2 & -4 & 5 \\ 3 & 5 & 4 \end{pmatrix}.$$

The third column of C is equal to the third column of A plus 2 times the first column. As in Example 5, Section 7, $\det A = -20$. Then $\det C = -20$. As a check,

$$\begin{aligned}\det C &= (1)(-16 - 25) - (3)(8 - 15) + (0)(10 - (-12)) \\ &= -41 + 21 + 0 = -20.\end{aligned}$$

(b) Let

$$A = \begin{pmatrix} 2 & 4 & -1 & 0 \\ 3 & 0 & 2 & 3 \\ -1 & 2 & 3 & 1 \\ 0 & 1 & -2 & -1 \end{pmatrix}.$$

By adding 2 times column 3 to column 1, and 4 times column 3 to column 2, we obtain

$$B = \begin{pmatrix} 0 & 0 & -1 & 0 \\ 7 & 8 & 2 & 3 \\ 5 & 14 & 3 & 1 \\ -4 & -7 & -2 & -1 \end{pmatrix},$$

and by Theorem 8.6, det A = det B. The expansion of B has only one nonzero term and we get

$$\det B = (-1) \det \begin{pmatrix} 7 & 8 & 3 \\ 5 & 14 & 1 \\ -4 & -7 & -1 \end{pmatrix}$$

$$= -\det \begin{pmatrix} 7 & 1 & 3 \\ 5 & 9 & 1 \\ -4 & -3 & -1 \end{pmatrix} \qquad \text{[subtract column 1 from column 2]}$$

$$= -\det \begin{pmatrix} 0 & 1 & 0 \\ -58 & 9 & -26 \\ 17 & -3 & 8 \end{pmatrix} \qquad \text{[subtract 7 times column 2 from column 1 and 3 times column 2 from column 3].}$$

Then det $B = -(-1)\big((-58)(8) - (17)(-26)\big) = -22$.

This last example illustrates a way of computing determinants which is usually more efficient than direct calculation by Formula 8.1 or 8.1R for matrices of size larger than 3-by-3.

The following lemma is a preliminary step in the postponed proof of 8.1. Recall that $\mathbf{e}_i$ denotes the ith natural basis vector in $\mathcal{R}^n$ and is 0 except for a 1 in the ith entry.

8.7 Lemma

If the first column of the matrix A is $\mathbf{e}_i$, then

$$\det A = (-1)^{i+1} \det A_{i1}.$$

Proof. By Definition 7.1,

$$\det A = \sum_{j=1}^{n} (-1)^{j+1} a_{1j} \det A_{1j}.$$

If the first column of A is $\mathbf{e}_1$, then $a_{11} = 1$, while for $j > 1$ the first column of A_{1j} is 0, and so $\det A_{1j} = 0$ by Theorem 7.3. Thus the expression for $\det A$ reduces to one term, $\det A_{11}$, and Theorem 8.7 holds for the case $i = 1$. For $i > 1$ we need to use the inductive hypothesis that Lemma 8.7 holds for $(n-1)$-by-$(n-1)$ matrices. In this case $a_{11} = 0$ and

$$\det A = \sum_{j=2}^{n} (-1)^{j+1} a_{1j} \det A_{1j}.$$

Each minor A_{1j} has $\mathbf{e}_{i-1}$ for its first column. (Removing the top entry from the vector $\mathbf{e}_i$ in $\mathcal{R}^n$ gives $\mathbf{e}_{i-1}$ in $\mathcal{R}^{n-1}$.) By the inductive hypothesis,

$$\det A_{1j} = (-1)^i \det A_{1j,i1},$$

where we write $A_{1j,i1}$ for the matrix obtained from A by deleting the first row, the jth column, the ith row, and the first column. Let $B = A_{i1}$. Then (since the first column of B is formed from the second column of A, etc.) $A_{1j,i1} = B_{1,j-1}$ and $a_{1j} = b_{1,j-1}$. Combining the equations we have derived so far gives

$$\det A = \sum_{j=2}^{n} (-1)^{j+1} b_{1,j-1} (-1)^i \det B_{1,j-1}.$$

By Formula 7.1 the right side of this equation is $(-1)^{i+1} \det B$, which is $(-1)^{i+1} \det A_{i1}$, as was to be proved.

Proof of 8.1. We first prove 8.1 for the special case $j = 1$. The first column of A can be written as $\mathbf{a}_1 = a_{11}\mathbf{e}_1 + a_{21}\mathbf{e}_2 + \ldots + a_{n1}\mathbf{e}_n$; so by linearity of det and Lemma 8.7,

$$\begin{aligned}\det A &= \det(\mathbf{a}_1, \mathbf{a}_2, \ldots, \mathbf{a}_n)\\ &= a_{11} \det(\mathbf{e}_1, \mathbf{a}_2, \ldots, \mathbf{a}_n) + a_{21} \det(\mathbf{e}_2, \mathbf{a}_2, \ldots, \mathbf{a}_n)\\ &\quad + a_{n1} \det(\mathbf{e}_n, \mathbf{a}_2, \ldots, \mathbf{a}_n)\\ &= \sum_{i=1}^{n} (-1)^{i+1} a_{i1} \det A_{i1}.\end{aligned}$$

To prove the general case, consider the matrix B obtained from A by moving the jth column into the first position. This move requires a series of $j - 1$ exchanges of adjacent columns; so by Theorem 7.6, $\det A = (-1)^{j-1} \det B$. For all i, we have $b_{i1} = a_{ij}$ and $B_{i1} = A_{ij}$. Thus we obtain

$$\begin{aligned}\det A &= (-1)^{j-1} \det B \\ &= (-1)^{j-1} \sum_{j=1}^{n} (-1)^{i+1} b_{i1} \det B_{i1} \\ &= \sum_{i=1}^{n} (-1)^{i+j} a_{ij} \det A_{ij},\end{aligned}$$

as was to be proved.

EXERCISES

1. Using appropriate cases of 8.1 or 8.1R, express each of the following as a linear function of the x's.

(a) $\begin{pmatrix} 0 & 3 & 6 \\ -2 & -1 & 6 \\ x_1 & x_2 & x_3 \end{pmatrix}.$

(b) $\begin{pmatrix} -2 & 3 & x_1 & 1 \\ 0 & -1 & x_2 & 0 \\ -1 & 2 & x_3 & 1 \\ 1 & 0 & x_4 & 0 \end{pmatrix}.$

2. Prove that a product of square matrices $P_1P_2 \ldots P_k$ is invertible if and only if each of the P_i is invertible. Hence show that, if A and B are square matrices and $AB = I$, then A and B are invertible and are inverses of each other.

3. Find the inverses of

(a) $\begin{pmatrix} 1 & 2 \\ 3 & 4 \end{pmatrix}$, (b) $\begin{pmatrix} 0 & -1 & 3 \\ 2 & 5 & -4 \\ -3 & 7 & 1 \end{pmatrix}$,

(c) $\begin{pmatrix} 1 & 0 & 1 & 0 \\ 0 & 2 & 3 & 0 \\ -1 & 0 & 0 & 2 \\ -3 & 0 & -2 & 0 \end{pmatrix}.$ $\left[\textit{Ans.}\ \text{(b)}\ \frac{1}{77}\begin{pmatrix} 33 & 22 & -11 \\ 10 & 9 & 6 \\ 29 & 3 & 2 \end{pmatrix}. \right]$

Check your answers by multiplication.

4. Solve the systems

(a) $7x + 6y = 5$
$6x + 5y = -3.$

(b) $2x + y = 0$
$3y + z = 1$
$4z + x = 2.$ [*Ans.* $x = -\frac{2}{25}, y = \frac{4}{25}, z = \frac{13}{25}$.]

(c)
$$\begin{aligned} x_1 + x_2 + x_3 + x_4 &= -1 \\ x_1 - x_2 \qquad + 2x_4 &= 0 \\ 3x_2 - x_3 \qquad &= 3 \\ x_2 \qquad - x_4 &= 0. \end{aligned}$$

5. Use the method of Example 4(b) of the text to evaluate

(a) $\det \begin{pmatrix} -1 & 0 & 1 & 2 \\ 0 & 1 & 2 & -1 \\ 1 & 2 & -1 & 0 \\ 2 & -1 & 0 & 1 \end{pmatrix}.$ [*Ans.* (a) 32.]

(b) $\det \begin{pmatrix} 1 & 1 & 1 & 1 \\ 1 & 2 & 4 & 8 \\ 1 & 3 & 9 & 27 \\ 1 & 4 & 16 & 64 \end{pmatrix}.$

6. (a) Compute

$$\det \begin{pmatrix} 1 & 2 & 3 & 4 \\ 0 & -1 & 5 & 6 \\ 0 & 0 & 3 & -1 \\ 0 & 0 & 0 & 4 \end{pmatrix}.$$

(b) A matrix A, like the one in part (a) in which every element below the diagonal is 0, is said to be triangular. Show that if A is any triangular matrix, then det A is equal to the product of the diagonal elements.

7. (a) A matrix of the form

$$A = \begin{pmatrix} a_1 & b_1 & 0 & 0 & 0 \\ c_1 & a_2 & b_2 & 0 & 0 \\ 0 & c_2 & a_3 & b_3 & 0 \\ 0 & 0 & c_3 & a_4 & b_4 \\ 0 & 0 & 0 & c_4 & a_5 \end{pmatrix}$$

which is zero except for the entries on or adjacent to the diagonal, is called a tridiagonal matrix. Let d_k be the determinant of the k-by-k minor formed from the first k rows and columns of A, so, for example, $d_1 = a_1$ and $d_2 = a_1a_2 - b_1c_1$. Show that, for $k \geq 3$, $d_k = a_kd_{k-1} - b_{k-1}c_{k-1}d_{k-2}$.

(b) Consider tridiagonal matrices in which the entries on the diagonal all have the value 2 and the entries next to the diagonal all have the value 1. Let d_n be the determinant of an n-by-n matrix of this type. Find a formula for d_n. [*Suggestion.* Start out by seeing what happens for $n = 2, 3, 4$.]

2

Linear Algebra

SECTION 1

LINEAR EQUATIONS, INVERSE MATRICES

A **system of linear equations** is a finite set of equations

1.1
$$\begin{array}{c} a_{11}x_1 + \dots + a_{1n}x_n = b_1 \\ \vdots \qquad\qquad \vdots \quad\ \ \vdots \\ a_{m1}x_1 + \dots + a_{mn}x_n = b_m \end{array}$$

where the a's and b's are given and the x's are to be determined. The whole system can be written in matrix form as $A\mathbf{x} = \mathbf{b}$, where A is the m-by-n **coefficient matrix** with entries a_{ij}, $\mathbf{b}$ is a column vector in $\mathcal{R}^m$, and $\mathbf{x}$ is a column vector in $\mathcal{R}^n$. Doing the matrix multiplication in the equation

$$\begin{pmatrix} a_{11} & \dots & a_{1n} \\ \cdot & & \cdot \\ \cdot & & \cdot \\ \cdot & & \cdot \\ a_{m1} & \dots & a_{mn} \end{pmatrix} \begin{pmatrix} x_1 \\ \cdot \\ \cdot \\ \cdot \\ x_n \end{pmatrix} = \begin{pmatrix} b_1 \\ \cdot \\ \cdot \\ \cdot \\ b_m \end{pmatrix}$$

shows at once that it is equivalent to the system 1.1.

It frequently happens in applications that there are as many equations as there are unknowns, so that $m = n$. Then, if the determinant of the coefficient matrix is not zero, the solution can be obtained by Cramer's rule of Section 8, Chapter 1. The methods of the present section do not use determinants, and we do not assume $m = n$; even for systems that can be solved by Cramer's rule, these methods are more efficient when n is greater than 3.

Any vector $\mathbf{c}$ in $\mathcal{R}^n$ such that $A\mathbf{c} = \mathbf{b}$ is a solution of the system. As we shall show, some systems have no solution, some have exactly one, and some have infinitely many solutions.

We say that two systems are **equivalent** if they have exactly the same set of solutions. Our procedure will be to take a given system and alter it in a sequence of steps to obtain an equivalent system for which the solutions are obvious. We illustrate the process with an example before giving a general description.

Example 1.

$$\begin{aligned} 3x + 12y + 9z &= 3 \\ 2x + 5y + 4z &= 4 \\ -x + 3y + 2z &= -5. \end{aligned}$$

Multiply the first equation by $\frac{1}{3}$, which makes the coefficient of x equal to 1 and gives

$$\begin{aligned} x + 4y + 3z &= 1 \\ 2x + 5y + 4z &= 4 \\ -x + 3y + 2z &= -5. \end{aligned}$$

Add (-2) times the first equation to the second, and replace the second equation by the result. This makes the coefficient of x in the second equation equal to 0 and gives

$$\begin{aligned} x + 4y + 3z &= 1 \\ -3y - 2z &= 2 \\ -x + 3y + 2z &= -5. \end{aligned}$$

Add the first equation to the third, and replace the third equation by the result, to get

$$\begin{aligned} x + 4y + 3z &= 1 \\ -3y - 2z &= 2 \\ 7y + 5z &= -4. \end{aligned}$$

Multiply the second equation by $-\frac{1}{3}$, to get

$$\begin{aligned} x + 4y + 3z &= 1 \\ y + \tfrac{2}{3}z &= -\tfrac{2}{3} \\ 7y + 5z &= -4. \end{aligned}$$

Add (-4) times the second equation to the first, and (-7) times the second equation to the third, to get

$$\begin{aligned} x + \tfrac{1}{3}z &= \tfrac{11}{3} \\ y + \tfrac{2}{3}z &= -\tfrac{2}{3} \\ \tfrac{1}{3}z &= \tfrac{2}{3}. \end{aligned}$$

Multiply the third equation by 3 to get

$$\begin{aligned} x + \tfrac{1}{3}z &= \tfrac{11}{3} \\ y + \tfrac{2}{3}z &= -\tfrac{2}{3} \\ z &= 2. \end{aligned}$$

Add $(-\frac{1}{3})$ times the third equation to the first and $(-\frac{2}{3})$ times the third equation to the second to get

$$\begin{aligned} x &= 3 \\ y &= -2 \\ z &= 2. \end{aligned}$$

Clearly, this sytem has just one solution, namely, the column vector

$$\begin{pmatrix} 3 \\ -2 \\ 2 \end{pmatrix}.$$

It is easy to verify by substitution in a system of equations that we have found a solution for them. This verification of course does not rule out the theoretical possibility that the original equations might have other solutions as well. In fact the final system is equivalent to the original system and has the same set of solutions. The same is true for any pair of systems where one is obtained from the other by steps such as were used in this example. Before we can prove this, we must first state exactly what operations are allowed and then investigate their properties.

The operations used were "multiplying an equation by a number," and "adding a multiple of one equation to another." We prefer to give the formal definitions in terms of matrices and, accordingly, consider the general matrix equation $A\mathbf{x} = \mathbf{b}$.

We define three types of **elementary operations** which can be applied to any matrix M:

An **elementary multiplication** replaces a row of M by a numerical multiple of the row, where the multiplier is different from 0.

An **elementary modification** replaces a row of M by the sum of that row and a numerical multiple of some other row.

An **elementary transposition** interchanges two rows of M. We did not use any transpositions in Example 1, but they are sometimes useful.

It is important to understand that, if an elementary operation is applied to both sides of an equation $A\mathbf{x} = \mathbf{b}$, the result is a new matrix equation that is satisfied by every vector $\mathbf{x}$ that satisfied the original equation. Equally important is the fact that each elementary operation has an **inverse elementary operation** by which the original operation can be undone or reversed. For example, multiplication of a row by a number $r \neq 0$ is reversed by multiplying the row by $1/r$. Similarly, an elementary modification is reversed by subtracting the same numerical multiple of the same row instead of adding it. Finally, a transposition is reversed by interchanging the two rows again.

Written as a matrix equation, the original system of equations in Example 1 becomes

$$\begin{pmatrix} 3 & 12 & 9 \\ 2 & 5 & 4 \\ -1 & 3 & 2 \end{pmatrix} \mathbf{x} = \begin{pmatrix} 3 \\ 4 \\ -5 \end{pmatrix}.$$

If we apply the elementary operation of multiplying the first row by $\frac{1}{3}$ to the matrices on both sides, we obtain

$$\begin{pmatrix} 1 & 4 & 3 \\ 2 & 5 & 4 \\ -1 & 3 & 2 \end{pmatrix} \mathbf{x} = \begin{pmatrix} 1 \\ 4 \\ 5 \end{pmatrix},$$

which is the matrix form of the system of equations at the second step in Example 1. In the same way, each operation on the system of equations amounts to applying an elementary operation to the matrices on both sides of the equivalent matrix equation. At the final stage, the matrix equation becomes

$$\begin{pmatrix} 1 & 0 & 0 \\ 0 & 1 & 0 \\ 0 & 0 & 1 \end{pmatrix} \mathbf{x} = \begin{pmatrix} 3 \\ -2 \\ 2 \end{pmatrix}$$

which, since the matrix on the left is the identity matrix, simply amounts to saying that $\mathbf{x}$ is equal to the vector on the right.

The theorem that justifies our method of solving linear equations is as follows.

1.2 Theorem

If the system $A_1\mathbf{x} = \mathbf{b}_1$ is converted to a system $A_2\mathbf{x} = b_2$ by applying a sequence of elementary operations to A_1 to get A_2 and the *same* sequence of elementary operations to $\mathbf{b}_1$ to get $\mathbf{b}_2$, then the two systems are equivalent in the sense that they have the same solutions.

Proof. If $\mathbf{x}$ is a solution of $A_1\mathbf{x} = \mathbf{b}_1$, then applying an elementary operation to both sides produces a new matrix equation that is still satisfied by $\mathbf{x}$. So $\mathbf{x}$ also satisfies the equation $A_2\mathbf{x} = \mathbf{b}_2$ that results from a finite sequence of such operations. Conversely, having transformed $A_1\mathbf{x} = \mathbf{b}_1$ into $A_2\mathbf{x} = \mathbf{b}_2$ by a sequence of elementary operations, we can find a sequence of inverse elementary operations which transforms the new equation back to the original one. But then the same argument that applied in the first part of the proof allows us to conclude that every solution of $A_2\mathbf{x} = \mathbf{b}_2$ is also a solution of $A_1\mathbf{x} = \mathbf{b}_1$.

Example 2. We now exhibit a system of equations with infinitely many solutions. Consider the matrix equation

$$\begin{pmatrix} 1 & -2 & -3 \\ \frac{1}{2} & -2 & -\frac{13}{2} \\ -3 & 5 & 4 \end{pmatrix} \mathbf{x} = \begin{pmatrix} 2 \\ 7 \\ 0 \end{pmatrix}.$$

Add $(-\frac{1}{2})$ times the first row to the second row, and then add 3 times the first row to the third row to produce zeros in the second and third entries of the first column, and obtain

$$\begin{pmatrix} 1 & -2 & -3 \\ 0 & -1 & -5 \\ 0 & -1 & -5 \end{pmatrix} \mathbf{x} = \begin{pmatrix} 2 \\ 6 \\ 6 \end{pmatrix}.$$

Multiply the second row by (-1) to obtain

$$\begin{pmatrix} 1 & -2 & -3 \\ 0 & 1 & 5 \\ 0 & -1 & -5 \end{pmatrix} \mathbf{x} = \begin{pmatrix} 2 \\ -6 \\ 6 \end{pmatrix}.$$

Add 2 times the second row to the first and then add 1 times the second row to the third to obtain

$$\begin{pmatrix} 1 & 0 & 7 \\ 0 & 1 & 5 \\ 0 & 0 & 0 \end{pmatrix} \mathbf{x} = \begin{pmatrix} -10 \\ -6 \\ 0 \end{pmatrix}.$$

At the corresponding stage in Example 1, we performed an elementary multiplication to make the third entry in the third row equal to 1. We were then able to obtain the identity matrix by further elementary operations. Obviously the row of zeros prevents us from following this procedure. Let us put $\mathbf{x} = \begin{pmatrix} x \\ y \\ z \end{pmatrix}$, and translate the matrix equation back into a system of linear equations. The result is

$$\begin{aligned} x \qquad + 7z &= -10 \\ y + 5z &= -6 \\ 0x + 0y + 0z &= \quad 0. \end{aligned}$$

The third equation is satisfied for any values of x, y, z. The first two equations may be rewritten as $x = -10 - 7z$ and $y = -6 - 5z$. Thus, for any value of z,

$$\begin{pmatrix} -7z - 10 \\ -5z - 6 \\ z \end{pmatrix} = z \begin{pmatrix} -7 \\ -5 \\ 1 \end{pmatrix} + \begin{pmatrix} -10 \\ -6 \\ 0 \end{pmatrix}$$

is a solution, and every solution has this form for some value of z. We have now described the set of solutions of

$$\begin{pmatrix} 1 & 0 & 7 \\ 0 & 1 & 5 \\ 0 & 0 & 0 \end{pmatrix} \mathbf{x} = \begin{pmatrix} -10 \\ -6 \\ 0 \end{pmatrix},$$

and by Theorem 1.2 we know that this is the same as the set of solutions of the matrix equation we started with.

Example 3. Consider the matrix equation

$$\begin{pmatrix} 1 & -2 & -3 \\ \frac{1}{2} & -2 & -\frac{13}{2} \\ -3 & 5 & 4 \end{pmatrix} \mathbf{x} = \begin{pmatrix} 2 \\ 7 \\ 2 \end{pmatrix}.$$

The matrix on the left is the same as the one in Example 2. Carrying out the same sequence of elementary operations yields

$$\begin{pmatrix}1 & 0 & 7\\ 0 & 1 & 5\\ 0 & 0 & 0\end{pmatrix}\mathbf{x} = \begin{pmatrix}-10\\ -6\\ 2\end{pmatrix}.$$

Whatever $\mathbf{x}$ is, the third row in the product $\begin{pmatrix}1 & 0 & 7\\ 0 & 1 & 5\\ 0 & 0 & 0\end{pmatrix}\mathbf{x}$ will be zero because the third row of the left factor is zero. Thus no value of $\mathbf{x}$ can give a column vector with 2 in the third row, and we conclude that the equation has *no* solution.

(If we put $\mathbf{x} = \begin{pmatrix}x\\ y\\ z\end{pmatrix}$ and write the matrix equation out as a system of equations, we obtain

$$\begin{aligned} x \qquad + 7z &= -10 \\ y + 5z &= -6 \\ 0x + 0y + 0z &= 2. \end{aligned}$$

The last equation obviously cannot be satisfied for any values of x, y, z.)

In these examples we used elementary operations to transform the original systems of equations into equivalent systems for which the solutions were easy to find. The property of the final set of equations that made the solutions obvious was that each equation involved a variable that did not appear in any of the other equations. Thus in Example 2 we found $x = -10 - 7z$, $y = -6 - 5z$, and because the first equation involved x but not y, and the second involved y but not x, we could find the values of (x, y, z) satisfying both equations by considering the equations separately. In practice, there is no difficulty (beyond the labor of doing the arithmetic) in reducing any system of equations to a system that has this "noninterference" property. For such a reduced system, it is either obvious that no solution exists, as in Example 3, or else the solution or set of solutions can be described explicitly as in Example 1 or Example 2. We leave for Section 3 the proof that this procedure can always be carried out.

Suppose we have a square matrix A and are able to convert it to the identity matrix I by some sequence of elementary operations, as in Example 1. Theorem 1.2 then asserts that, if we carry out the same sequence of operations on a vector $\mathbf{b}$ and get $\mathbf{c}$ as a result, then $A\mathbf{x} = \mathbf{b}$

if and only if $I\mathbf{x} = \mathbf{c}$. In other words $\mathbf{c}$ is the (unique) solution of $A\mathbf{x} = \mathbf{b}$. Carrying out these same operations on the identity matrix gives a matrix C, whose jth column $\mathbf{c}_j$ is the solution of $A\mathbf{c}_j = \mathbf{e}_j$, where $\mathbf{e}_j$ is the jth column of I. Hence $AC = I$. This suggests that $C = A^{-1}$ and that we have arrived at a method for computing matrix inverses. This is so, and we state the result formally.

1.3 Theorem

If an n-by-n matrix A can be converted to the n-by-n identity matrix I by a sequence of elementary operations, then A is invertible and A^{-1} is equal to the result of applying the same sequence of operations to I.

Proof. The preceding discussion showed that if C is the result of applying to I some elementary operations that convert A to I, then $AC = I$. To show that A is invertible and $C = A^{-1}$, we must show that $CA = I$. The argument depends on the fact that elementary operations are reversible. Now since C was constructed by applying elementary operations to I, it follows that C can be converted back to I by elementary operations. Applying to C the argument that we previously applied to A shows that there is some matrix D such that $CD = I$. We now have

$$A = AI = A(CD) = (AC)D = ID = D;$$

so $D = A$. Thus $CA = I$, as was to be proved.

Example 4. We shall find the inverse of

$$A = \begin{pmatrix} 2 & 4 & 8 \\ 1 & 0 & 0 \\ 1 & -3 & -7 \end{pmatrix}.$$

We start with

$$\begin{pmatrix} 2 & 4 & 8 \\ 1 & 0 & 0 \\ 1 & -3 & -7 \end{pmatrix}, \quad \begin{pmatrix} 1 & 0 & 0 \\ 0 & 1 & 0 \\ 0 & 0 & 1 \end{pmatrix}.$$

Add -2 times the second row to the first and -1 times the second row to the third to get

$$\begin{pmatrix} 0 & 4 & 8 \\ 1 & 0 & 0 \\ 0 & -3 & -7 \end{pmatrix}, \quad \begin{pmatrix} 1 & -2 & 0 \\ 0 & 1 & 0 \\ 0 & -1 & 1 \end{pmatrix}.$$

Multiply the first row by $\frac{1}{4}$ and then add 3 times the first row to the third to get

$$\begin{pmatrix} 0 & 1 & 2 \\ 1 & 0 & 0 \\ 0 & 0 & -1 \end{pmatrix}, \quad \begin{pmatrix} \frac{1}{4} & -\frac{1}{2} & 0 \\ 0 & 1 & 0 \\ \frac{3}{4} & -\frac{5}{2} & 1 \end{pmatrix}.$$

Multiply the third row by -1 and then add -2 times the third row to the first to get

$$\begin{pmatrix} 0 & 1 & 0 \\ 1 & 0 & 0 \\ 0 & 0 & 1 \end{pmatrix}, \quad \begin{pmatrix} \frac{7}{4} & -\frac{11}{2} & 2 \\ 0 & 1 & 0 \\ -\frac{3}{4} & \frac{5}{2} & -1 \end{pmatrix}.$$

Transpose the first and second rows to get

$$\begin{pmatrix} 1 & 0 & 0 \\ 0 & 1 & 0 \\ 0 & 0 & 1 \end{pmatrix}, \quad \begin{pmatrix} 0 & 1 & 0 \\ \frac{7}{4} & -\frac{11}{2} & 2 \\ -\frac{3}{4} & \frac{5}{2} & -1 \end{pmatrix}.$$

The last matrix on the right is A^{-1}, as may be verified by multiplying by A.

EXERCISES

1. Solve the following systems of equations:

(a) $$\begin{pmatrix} 1 & 1 & 1 \\ -1 & 2 & -4 \\ 1 & 3 & 9 \end{pmatrix} \mathbf{x} = \begin{pmatrix} 2 \\ 2 \\ 0 \end{pmatrix}.$$

(b) $$\begin{aligned} x_1 + x_2 &= 1 \\ x_2 + x_3 &= 2 \\ x_3 + x_4 &= 3 \\ x_4 + x_1 &= 4. \end{aligned}$$

(c) $$\begin{pmatrix} 1 & 2 & 3 \\ 4 & 5 & 6 \\ 7 & 8 & 9 \end{pmatrix} \mathbf{x} = \begin{pmatrix} 10 \\ 11 \\ 12 \end{pmatrix}.$$

2. Show that Theorem 1.2 implies the following more general statement: If an elementary operation converts a matrix A_1 to A_2 and the same operation converts B_1 to B_2, then the set of matrices X such that $A_1X = B_1$ is the same as the set of X such that $A_2X = B_2$.

3. Solve the matrix equations:

(a) $\begin{pmatrix} 1 & 2 \\ 5 & 6 \end{pmatrix} X = \begin{pmatrix} 0 & -3 & 4 \\ 1 & 2 & 0 \end{pmatrix}.$

(b) $\begin{pmatrix} 1 & 2 & 3 \\ 1 & 0 & 0 \\ -4 & 1 & 2 \end{pmatrix} X = \begin{pmatrix} 2 & 1 \\ -5 & 4 \\ 0 & -3 \end{pmatrix}.$

4. For each of the following matrices A, find A^{-1} if A is invertible, and find a nonzero solution of $A\mathbf{x} = 0$ if A is not invertible.

(a) $\begin{pmatrix} 1 & 0 & 0 \\ 3 & 1 & 5 \\ -2 & 0 & 1 \end{pmatrix}$ $\left[\textit{Ans.} \begin{pmatrix} 1 & 0 & 0 \\ -13 & 1 & -5 \\ 2 & 0 & 1 \end{pmatrix}. \right]$

(b) $\begin{pmatrix} 1 & 2 & 3 \\ -1 & 1 & 0 \\ 0 & 3 & 3 \end{pmatrix}$

(c) $\begin{pmatrix} 4 & -1 & 0 & 0 \\ 0 & 5 & -2 & 0 \\ 0 & 0 & 6 & -3 \\ 0 & 0 & 0 & 7 \end{pmatrix}$

(d) $\begin{pmatrix} 1 & 1 & 2 & 3 \\ 0 & 5 & 4 & 2 \\ -1 & -3 & 1 & 0 \\ 0 & 3 & 7 & 5 \end{pmatrix}$

SECTION 2

SOME APPLICATIONS

In this section we shall look at some applications of vectors and linear equations. The selection of examples is made so as to avoid technical complications from the fields of application. Several of our examples involve the notion of a **network**, which we define to be a finite collection of points, or nodes, some of which may be joined to some others by line segments. It is theoretically unimportant whether a network is visualized as lying in 2-dimensional space or 3-dimensional space; we choose whichever is pictorially more convenient. Some networks are illustrated in Fig. 1.

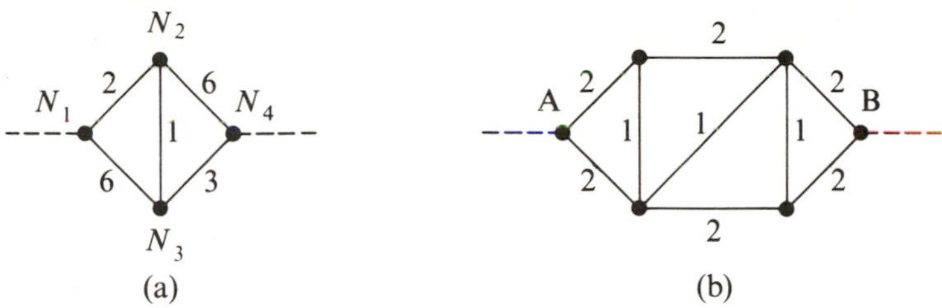

Figure 1

Example 1. Some electrical circuits can be considered as networks in which each line represents a connecting wire with a given electrical resistance. When such a circuit is connected to a battery or similar power source, currents flow in the connections, and a value of electrical potential is established at each node. For many types of connecting wires, the current flow in a wire is proportional to the difference in the potential at its two ends. In quantitative terms,

$$c_{ij} = \frac{(v_i - v_j)}{r_{ij}} \tag{1}$$

where c_{ij} is the current flowing from node i to node j, r_{ij} is the resistance of the connection between nodes i and j, and v_i and v_j are the values of the electrical potential at nodes i and j. (The appropriate units of measurement are amperes for current, ohms for resistance, and volts for potential.) A negative value for the current from i to j is to be interpreted as a current from j to i.

Figure 1(a) shows a circuit with four nodes and five segments, with the resistance of each segment indicated beside it. Suppose an external power source is connected at nodes 1 to 4 to maintain values $v_1 = 12$ and $v_4 = 0$. Since node 2 has no external connection, the current flowing in must balance the current flowing out, so that if signs are taken into account, the sum of the currents out of node 2 must be zero. Using (1), we get the equation

$$\tfrac{1}{2}(v_2 - v_1) + (v_2 - v_3) + \tfrac{1}{6}(v_2 - v_4) = 0.$$

When rewritten in the form

$$(\tfrac{1}{2} + 1 + \tfrac{1}{6})v_2 = \tfrac{1}{2}v_1 + v_3 + \tfrac{1}{6}v_4,$$

we see that v_2 is a weighted average of v_1, v_3, and v_4, with coefficients which are the reciprocals of the resistances in the lines joining node 2 to the others. A similar equation will hold at any node that does not have an external connection. Thus at node 3 we get

$$(\tfrac{1}{6} + 1 + \tfrac{1}{3})v_3 = \tfrac{1}{6}v_1 + v_2 + \tfrac{1}{3}v_4.$$

If we put in the assumed values $v_1 = 12$ and $v_4 = 0$, and rearrange terms, we get

$$\tfrac{5}{3}v_2 - v_3 = 6$$
$$-v_2 + \tfrac{3}{2}v_3 = 2.$$

Solving this system gives $v_2 = \frac{22}{3} \approx 7.33$ and $v_3 = \frac{56}{9} \approx 6.22$. Once the potentials are known, other quantities can be calculated directly. For example, the current from node 1 to node 2 is calculated as $(v_1 - v_2)/r_{12} = \frac{1}{2}(12 - 7.33) = 2.33$. Similarly, the current from node 1 to node 3 is $\frac{1}{6}(12 - 6.22) = 0.96$. The total current flowing from node 1 into the rest of the network is then $2.33 + 0.96 = 3.29$, which must of course be equal to the current flowing into node 1 from the external source.

Example 2. We can consider vectors in $\mathcal{R}^2$ or $\mathcal{R}^3$ as representing forces acting at some point which, for convenience, we take to be the origin. The direction of the arrow is the direction in which the force acts, and the length of the arrow is the magnitude of the force. Our fundamental physical assumption here is that, if more than one force acts at a point, then the **resultant force** acting at the point is represented by the *sum* of the separate force vectors acting there. In Fig. 2 we have shown two different pictures. The resultant arrow **r** is shown only in Fig. 2(a). For example, suppose that the force vectors in Fig. 2(a) lie in a plane, which we take to be $\mathcal{R}^2$ with the origin at the point of action. If we have

$$\mathbf{f}_1 = (-1, 3), \qquad \mathbf{f}_2 = (4, 3), \qquad \mathbf{f}_3 = (-2, -4), \tag{2}$$

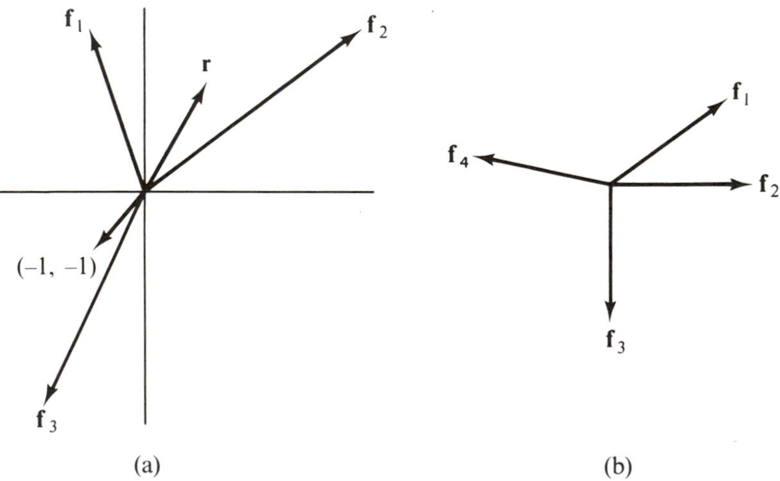

Figure 2

then by definition

$$\mathbf{r} = \mathbf{f}_1 + \mathbf{f}_2 + \mathbf{f}_3 = (1, 2).$$

But suppose we are given the directions of the three force vectors and are asked to find constants of proportionality that will produce a given resultant, say, $\mathbf{r} = (-1, -1)$. In other words, suppose we want to find nonnegative numbers c_1, c_2, c_3 such that

$$c_1\mathbf{f}_1 + c_2\mathbf{f}_2 + c_3\mathbf{f}_3 = (-1, -1).$$

(A negative value for some one of the c's would reverse the direction of the corresponding force.) This vector equation is equivalent to the system of equations we get by substituting the given vectors (1) that determine the force directions. We find

$$c_1\begin{pmatrix}-1\\3\end{pmatrix} + c_2\begin{pmatrix}4\\3\end{pmatrix} + c_3\begin{pmatrix}-2\\-4\end{pmatrix} = \begin{pmatrix}-1\\-1\end{pmatrix} \tag{3}$$

or

$$\begin{aligned}-c_1 + 4c_2 - 2c_3 &= -1,\\ 3c_1 + 3c_2 - 4c_3 &= -1.\end{aligned}$$

Since we have two equations and three unknowns, we would expect, in general, to be able to specify one of the c's and then solve for the others. However, recall that the c's are to be nonnegative. (In particular, a glance at Fig. 2(a) shows that we could not get a resultant equal to $(-1, -1)$ unless c_3 is actually positive.) Hence, we try $c_3 = 1$. This choice leads to the pair of equations

$$\begin{aligned}-c_1 + 4c_2 &= 1\\ c_1 + c_2 &= 1,\end{aligned}$$

which has the solution $c_1 = \frac{3}{5}$, $c_2 = \frac{2}{5}$. Then the triple $(c_1, c_2, c_3) = (\frac{3}{5}, \frac{2}{5}, 1)$ is one possible solution, and the three force vectors are

$$c_1\mathbf{f}_1 = (-\tfrac{3}{5}, \tfrac{9}{5}), \qquad c_2\mathbf{f}_2 = (\tfrac{8}{5}, \tfrac{6}{5}), \qquad c_3\mathbf{f}_3 = (-2, -4),$$

with magnitudes

$$|c_1\mathbf{f}_1| = \tfrac{3}{5}\sqrt{10}, \qquad |c_2\mathbf{f}_2| = 2, \qquad |c_3\mathbf{f}_3| = 2\sqrt{5}.$$

We could equally well have asked for an assignment of force magnitudes that would put the system in **equilibrium**, that is, so that the resultant is the zero vector. We would then have replaced the vector $(-1, -1)$ on the right side of Equation (2) by $(0, 0)$, and solved the new system in a similar way.

Example 3. In this example we consider the idea of a **random walk** in which an object always moves among finitely many positions along specified paths, each path having a certain probability or likelihood of being used. To be specific, we assume that the probability of leaving a certain position along some particular path is the same for all paths leading away from that position. Some sample configurations are shown in Fig. 3, and it is clear that they form networks.

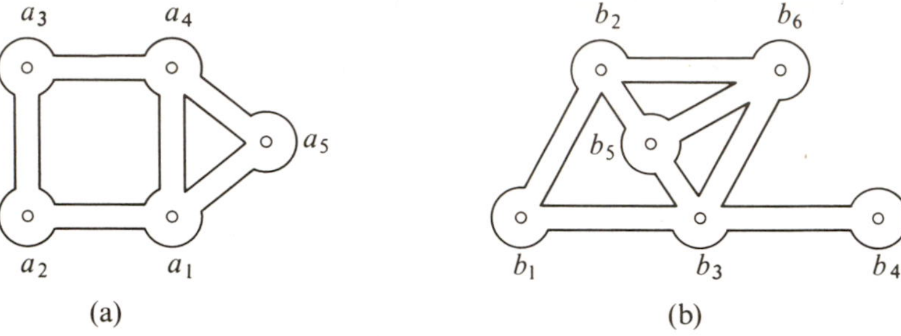

Figure 3

It is understood that a probability is always a number between 0 and 1 inclusive, and that the probability of a particular event is equal to the sum of the probabilities of the various distinct ways in which that event can occur. Thus in Fig. 3(a), since we assume that all paths away from a_5 are equally likely, it will happen that the probability of leaving a_5 along each of the two possible paths is $\frac{1}{2}$. Similarly, each of the three paths from a_4 has probability $\frac{1}{3}$. We further assume that transition from one position to another is such that the probability of two successive events is equal to the product of their respective probabilities. Thus going from a_2 to a_1 to a_4 has the probability $(\frac{1}{2})(\frac{1}{3}) = \frac{1}{6}$.

We can now ask a question such as the following: What is the probability p_k of starting at a_k and arriving at the specified position a_5, *without going to* a_4? We see that, starting at a_1, we can go to a_5 directly with probability $\frac{1}{3}$, or we can go to a_2 with probability $\frac{1}{3}$ and *then* go to a_5 with probability p_2. Thus

$$p_1 = \tfrac{1}{3} + (\tfrac{1}{3})p_2$$

Similarly, because going to a_4 does not occur in the events we are watching,

$$p_2 = (\tfrac{1}{2})p_1 + (\tfrac{1}{2})p_3$$

$$p_3 = (\tfrac{1}{2})p_2.$$

We can rewrite these equations as

$$\begin{aligned} p_1 - (\tfrac{1}{3})p_2 &= \tfrac{1}{3} \\ -(\tfrac{1}{2})p_1 + p_2 - (\tfrac{1}{2})p_3 &= 0 \\ -(\tfrac{1}{2})p_2 + p_3 &= 0 \end{aligned}$$

and solve them by routine methods. We get

$$p_1 = \tfrac{3}{7}, \qquad p_2 = \tfrac{2}{7}, \qquad p_3 = \tfrac{1}{7}.$$

It appears that, the nearer we start to a_5, the more likely we are to get to a_5 without going to a_4. However, it is important to understand that, in general, the values of the probabilities depend on the entire configuration.

An analysis like the one just given would be useful in distinguishing completely random behavior of a creature in a maze from purposeful or conditioned behavior.

Example 4. A system of interconnected water pipes can be thought of as a network in which the nodes are joints. It is usual in such a network to assign each pipe a natural direction of flow, indicated by arrows in Fig. 4.

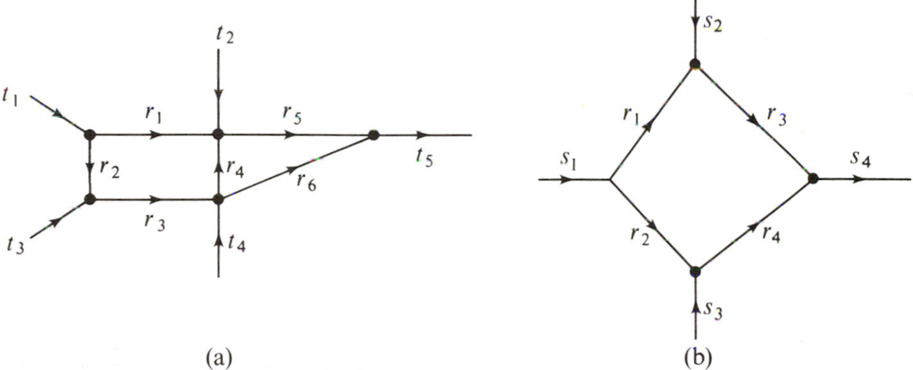

Figure 4

Then a positive number r will represent a rate of flow in the assigned direction, while a negative number $-r$ will represent a flow of equal rate in the opposite direction. We shall separate the flow rates into internal rates r_k and external rates denoted by t_k. (Specifying an external rate 0 at any joint has the effect of closing off the external pipe at that joint.) We assume that the inflow at any joint must equal the outflow. Thus at the upper left corner in Fig. 4(a) we find $t_1 = r_1 + r_2$, while at the lower left we find $t_3 + r_2 = r_3$. For the network of Fig. 4(a), the complete set of equations relating the r's and the t's can be put in the form

$$\begin{aligned}
r_1 + r_2 \qquad\qquad\qquad\qquad &= t_1 \\
-r_1 \qquad -r_4 + r_5 \qquad\qquad &= t_2 \\
-r_2 + r_3 \qquad\qquad\qquad &= t_3 \\
-r_3 + r_4 \qquad + r_6 &= t_4 \\
r_5 + r_6 &= t_5
\end{aligned} \tag{4}$$

In vector form we can write $A\mathbf{r} = \mathbf{t}$, where A is the 5-by-6 matrix

$$\begin{pmatrix} 1 & 1 & 0 & 0 & 0 & 0 \\ -1 & 0 & 0 & -1 & 1 & 0 \\ 0 & -1 & 1 & 0 & 0 & 0 \\ 0 & 0 & -1 & 1 & 0 & 1 \\ 0 & 0 & 0 & 0 & 1 & 1 \end{pmatrix}.$$

The vector equation shows that there is a linear relation between $\mathbf{t}$ and $\mathbf{r}$, namely, that $\mathbf{t}$ is a linear function of $\mathbf{r}$. This implies that, if the flows r_k are all multiplied by a number c, then so are the flows t_k. Similarly, if two sets of flows r_k and r_k' are superimposed, then the resulting external flows are given by $\mathbf{t} = A(\mathbf{r} + \mathbf{r}')$. This phenomenon is called the **superposition principle** for linear systems. Of course, specifying the r's will completely determine what the t's must be.

Turning the problem around, we can ask to what extent specifying the external flows at the joints will specify the flows r_k in the pipes. In particular, we can try specifying that the exterior flow t_k at each joint should be zero. This leads to the system of five equations in six unknowns

$$\begin{aligned} r_1 + r_2 \qquad\qquad\qquad\qquad &= 0 \\ -r_1 \qquad\qquad -r_4 + r_5 \qquad &= 0 \\ -r_2 + r_3 \qquad\qquad\qquad &= 0 \\ -r_3 + r_4 \qquad r_6 &= 0 \\ r_5 + r_6 &= 0. \end{aligned}$$

We can let $r_6 = a$ be any number; then we get $r_5 = -a$ from the last equation. Now let $r_4 = b$ be any number. The remaining four equations are easily solved in terms of a and b, and we find the solution

$$\mathbf{r}_0 = (-a - b, a + b, a + b, b, -a, a).$$

A simple check shows that these assignments for r_1 through r_6 will satisfy the above system. It follows that there are infinitely many different pipe flows that will produce zero external flow at each joint. Similarly, if we have a solution $\mathbf{r}$ of the original system (4) for some given vector $\mathbf{t}$ on the right side, then by the superposition principle any of the solutions $\mathbf{r}_0$ can be added to $\mathbf{r}$ to give a new solution $\mathbf{r}_0 + \mathbf{r}$; we have

$$\begin{aligned} A(\mathbf{r}_0 + \mathbf{r}) &= A\mathbf{r}_0 + A\mathbf{r} \\ &= 0 + \mathbf{t} = \mathbf{t}. \end{aligned}$$

Example 5. The derivation of Simpson's rule for approximate integration is based on the requirement that it should give exact results when applied to quadratic polynomials. The rule gives an approximation to the integral of a function over an interval $(a - h, a + h)$ in terms of the value of the function at the end-points and mid-point of the interval, in the form

$$\int_{a-h}^{a+h} f(x)\,dx \approx uf(a-h) + vf(a) + wf(a+h)$$

where u, v, and w are constants. If the formula is to be correct for all polynomials of degree less than or equal to 2, it must in particular be correct for the polynomials $f_0(x) = 1$, $f_1(x) = x$, and $f_2(x) = x^2$. Each of these gives an equation for u, v, and w. For instance, for $f_0(x) = 1$ we have

$$\int_{a-h}^{a+h} f_0(x)\,dx = 2h \qquad \text{and} \qquad f_0(a-h) = f_0(a) = f_0(a+h) = 1$$

so $u + v + w = 2h$. Using $f_1(x) = x$ and $f_2(x) = x^2$ similarly gives the equations

$$(a - h)u + av + (a + h)w = 2ah$$

and

$$(a^2 + h^2 - 2ah)u + a^2v + (a^2 + h^2 + 2ah)w = 2a^2h + \tfrac{2}{3}h^3.$$

These equations are easily solved (see Exercise 13) and give the result $u = w = \frac{1}{3}h$, $v = \frac{4}{3}h$.

We have obtained a rule which is correct for the particular polynomials f_0, f_1, and f_2. Its correctness for any quadratic polynomial follows readily. Let us write $E(f)$ for the error committed when the rule is applied to a general function f, so

$$E(f) = \int_{a-h}^{a+h} f(x)\,dx - \tfrac{1}{3}hf(a-h) - \tfrac{4}{3}f(a) - \tfrac{1}{3}f(a+h).$$

It is easy to see that E is a linear function from the vector space of polynomials to $\mathcal{R}^1$, and of course $E(f_0) = E(f_1) = E(f_2) = 0$ by construction. If f is any quadratic polynomial, so $f(x) = px^2 + qx + r$, then $f = pf_2 + qf_1 + f_0$ and by linearity

$$E(f) = pE(f_2) + qE(f_1) + rE(f_0) = 0.$$

The same method can be used to derive a wide variety of formulas useful in numerical analysis. (See Exercise 14.)

EXERCISES

1. Figure 1(b) shows an electrical network with the resistance (in ohms) of each edge marked on it. Suppose an external power supply maintains node A at a potential of 10 volts and node B at a potential of 4 volts. Following the procedure of Example 1 in the text, set up equations for the potential at the other nodes and solve them. From the results, calculate the current flowing into the network at A.

2. The edges and vertices of a 3-dimensional cube form a network with 8 nodes and 12 edges. Suppose each edge is a wire of resistance 1 ohm, and that two of the vertices have external connections that maintain one of them at a potential of 1 and the other at a potential of 0. Find the value of the potential at the other vertices, and the current flowing in the external connections if
(a) the two vertices with external connections are at opposite corners of the cube
(b) they are at the two ends of an edge
(c) they are at opposite corners of a face of the cube.

3. (a) Suppose that three forces acting at the origin in $\mathcal{R}^3$ have the same directions as (1, 0, 0), (1, 1, 0), and (1, 1, 1). Find magnitudes for the forces acting in these directions so that the resultant force vector will be (−1, 2, 4).
(b) Can *any* force vector be the resultant of forces acting in directions specified in part (a)?

4. Show that, if three linearly independent vectors in $\mathcal{R}^3$ are used to specify the directions of three forces, then magnitudes can be assigned to these forces to make the resultant force equal to any given vector.

5. If forces act in $\mathcal{R}^2$ in the directions of (2, 1), (2, 2), and (−3, −1), show that magnitudes can be assigned so that the system is in equilibrium.

6. Suppose that a random walk traverses the paths shown in Fig. 3(a). What is the probability p_k, $k = 1, 2, 3$, that a walk starting at a_k goes to a_4 without passing through a_5?

7. (a) Suppose that a particle traces a random walk on the paths shown in Fig. 3(b). Letting p_k be the probability of going from b_k to b_6 without going through b_5, find p_k for $k = 1, 2, 3, 4$.
(b) How is the result of part (a) modified if b_4 and the path leading to it are eliminated altogether?
(c) How is the result of part (a) modified if a new path is introduced between b_4 and b_6?

8. Suppose that various mixtures can be made of substances S_1, S_2, S_3, S_4 having densities 2, 17, 3, and 1 respectively, measured in grams per cubic centimeter. Suppose also that the price of each substance in cents per gram is 4, 51, 3 and 1 respectively. Is it possible to make a mixture weighing 10 grams, with a volume of 20 cubic centimeters and costing 1 dollar?

9. Suppose that a linear function f from $\mathcal{R}^3$ to $\mathcal{R}^3$ is such that $f(\mathbf{e}_1) = (1, 2, -1)$, $f(\mathbf{e}_2) = (0, 1, 1)$, and $f(\mathbf{e}_3) = (1, 2, 1)$. Find a vector $\mathbf{x}_1$, $\mathbf{x}_2$, $\mathbf{x}_3$ such that $f(\mathbf{x}_k) = \mathbf{e}_k$, for $k = 1, 2, 3$.

10. Let f be the linear function from $\mathcal{R}^3$ to $\mathcal{R}^3$ with matrix

$$\begin{pmatrix} 1 & 2 & -1 \\ 0 & 1 & 3 \end{pmatrix}.$$

Show that the points $\mathbf{x}$ in $\mathcal{R}^3$ such that $f(\mathbf{x}) = 0$ all lie on a line.

11. (a) Suppose the vector $\mathbf{t} = (t_1, t_2, t_3, t_4, t_5)$ in Fig. 4(a) is specified to be $\mathbf{t} = (-1, 0, 1, 2, 1)$. Find a vector $\mathbf{r}$ that determines consistent internal flow rates.
    (b) Verify that the vector $\mathbf{r}_0 = (-a - b, a + b, a + b, b, -a, 0)$, for any a and b, is consistent with external flow $\mathbf{t} = 0$ in Fig. 4(a).

12. (a) Let the external flow vector in Fig. 4(b) be given by $\mathbf{t} = (1, 1, 2, 4)$. Show that there is more than one consistent internal flow vector $\mathbf{r}$, and find two of them.
    (b) Let the external flow vector in Fig. 4(b) be given by $\mathbf{t} = (1, 0, 1, 1)$. Show that there is no consistent internal flow vector.

13. Carry out the solution of the equations for u, v, w given in Example 5 of the text. [Suggestion: begin by subtracting a times the first equation from the second and a^2 times the first equation from the third.]

14. By using the method of Example 5, find constants t, u, v, w such that the formula

$$\int_a^{a+3h} f(x)\,dx = tf(a) + uf(a + h) + vf(a + 2h) + wf(a + 3h)$$

is exact whenever f is a polynomial of degree less than or equal to 3.

SECTION 3

THEORY OF LINEAR EQUATIONS

In this section we are concerned with formalizing the method of solving linear equations that has been illustrated in the two preceding sections, and in proving that it always works. We begin by giving a precise definition for the "noninterference" property discussed informally after Example 3 of Section 1. At that point we expressed the idea by saying that each equation of a system contained a variable that did not occur in any other equation. The following definitions express this "noninterference" property in terms of the coefficient matrix.

An entry in a matrix is called a **leading entry** if it is the first nonzero entry in its row. Thus in

$$\begin{pmatrix} 3 & 0 & 2 & 4 \\ 0 & 0 & 4 & -10 \\ 0 & 0 & 0 & 0 \\ 2 & -2 & 3 & 1 \end{pmatrix},$$

the leading entries are the first entry in the first row, the third entry in the second row, and the first entry in the fourth row. There is no leading entry in the third row since it consists entirely of zeros. We say that a matrix is **reduced** if:

3.1 (a) Every column containing a leading entry is zero except for the leading entry.
(b) Every leading entry is 1.

In discussing reduced matrices it is frequently necessary to distinguish the columns that contain leading entries from those that do not. (Of course a row contains a leading entry if and only if it is nonzero.) We shall say that the columns that do contain leading entries are **pivotal.** In a reduced matrix, each leading entry belongs to one nonzero row and one pivotal column; we shall say that the row and column are **associated.** This gives a one-to-one correspondence between nonzero rows and pivotal columns, and it establishes the following fact which will be referred to later.

3.2 In a reduced matrix, the number of pivotal columns equals the number of nonzero rows.

Example 1. Consider the matrices

$$A = \begin{pmatrix} 1 & 2 & 0 & 5 \\ 0 & 0 & 1 & -3 \\ 0 & 0 & 0 & 0 \end{pmatrix}, \qquad B = \begin{pmatrix} 0 & 0 & 1 & 0 \\ 1 & 2 & 0 & 0 \\ 0 & 0 & 0 & 1 \end{pmatrix},$$

$$C = \begin{pmatrix} 2 & 4 & 0 & 0 \\ 0 & 0 & 1 & 0 \\ 0 & 0 & 0 & 1 \end{pmatrix}, \qquad D = \begin{pmatrix} 1 & 0 & 3 & 5 \\ 0 & 1 & 0 & -2 \\ 0 & 0 & 0 & 1 \end{pmatrix}.$$

A is reduced; its pivotal columns are the first (associated with the first row) and the third (associated with the second row). B is also reduced; its pivotal columns are the first, third, and fourth, associated with the second, first, and third rows, respectively. The matrix C fails to satisfy condition (b), but multiplying the first row by $\frac{1}{2}$ gives a reduced matrix. D does not satisfy condition (a), but subtracting 5 times the third row from the first and adding 2 times the third row to the second gives a reduced matrix.

We now present some theorems on the solutions of systems of equations for which the coefficient matrix is reduced. The first theorem shows how to tell whether an equation has any solutions at all.

3.3 Theorem

Suppose A is a reduced matrix. If A has a row of zeros for which the corresponding entry in $\mathbf{b}$ is not zero, then the equation $A\mathbf{x} = \mathbf{b}$ has no solution. Otherwise it has at least one solution.

Proof. If any row of A is zero, then the corresponding entry in $A\mathbf{x}$ will be zero, whatever $\mathbf{x}$ may be. Thus if the corresponding entry in $\mathbf{b}$ is not zero, there can be no solution.

To prove the other half of the theorem, we assume that every zero row of A (if there is any) is matched by a zero entry in $\mathbf{b}$, and we show how to write down a vector $\mathbf{x}$ that is a solution. The vector $\mathbf{x}$ must of course have entries $x_1, \ldots, x_n$ corresponding to the n columns of A, while $\mathbf{b}$ has entries $b_1, \ldots, b_m$ corresponding to the m rows of A. If the jth column of A is nonpivotal set $x_j = 0$. If the jth column of A is pivotal, let i be the number of the associated row, and set $x_j = b_i$. We claim that the vector $\mathbf{x}$ constructed in this way satisfies $A\mathbf{x} = \mathbf{b}$. The ith entry in the product is $\sum_{k=1}^{n} a_{ik}x_k$. We must show that it equals b_i. It is zero if the ith row of A is zero, and by assumption b_i is then also zero. If the ith row is not zero, then $a_{ij} = 1$, where j is the associated pivotal column. By construction, $x_j = b_i$, so $a_{ij}x_j = b_i$. The terms $a_{ik}x_k$ with $k \neq j$ are all zero because $a_{ik} = 0$ if the kth column is pivotal (because A is reduced) and $x_k = 0$ if the kth column is not pivotal (by construction).

Example 2. Consider

$$A = \begin{pmatrix} 0 & 0 & 1 & 5 & 0 \\ 1 & 2 & 0 & 3 & 0 \\ 0 & 0 & 0 & 0 & 0 \\ 0 & 0 & 0 & 0 & 1 \end{pmatrix}, \qquad \mathbf{b}_1 = \begin{pmatrix} 1 \\ 4 \\ 2 \\ 3 \end{pmatrix}, \qquad \mathbf{b}_2 = \begin{pmatrix} 1 \\ 4 \\ 0 \\ 3 \end{pmatrix}.$$

The third row of A is zero and the third entry of $\mathbf{b}_1$ is not; so according to Theorem 3.3 the equation $A\mathbf{x} = \mathbf{b}_1$ has no solution. On the other hand, the third entry of $\mathbf{b}_2$ is zero. The pivotal columns of A are the first, third, and fifth, with associated rows the second, first, and fourth. The proof of the theorem shows that, if we construct

$$\mathbf{x} = \begin{pmatrix} 4 \\ 0 \\ 1 \\ 0 \\ 3 \end{pmatrix}$$

by making the first, third, and fifth entries equal to the second, first, and fourth entries of $\mathbf{b}_2$, and making the other rows zero, then $A\mathbf{x}$ will be equal to $\mathbf{b}_2$. This is easily verified by doing the matrix multiplication.

The next problem is how to tell whether an equation that has a solution has more than one. Theorem 3.4 and its corollary show that the question can be reduced to a special case, and theorem 3.5 deals with this special case.

3.4 Theorem

Suppose $\mathbf{x}_0$ is a solution of $A\mathbf{x} = \mathbf{b}$. Then $\mathbf{x}_1$ is also a solution if and only if $\mathbf{x}_1 - \mathbf{x}_0$ is a solution of $A\mathbf{x} = 0$.

Proof. We have $A(\mathbf{x}_1 - \mathbf{x}_0) = A\mathbf{x}_1 - A\mathbf{x}_0 = \mathbf{b} - \mathbf{b} = 0$. Conversely, if $A(\mathbf{x}_1 - \mathbf{x}_0) = 0$, then $A\mathbf{x}_1 = A\mathbf{x}_0 = \mathbf{b}$.

The equation $A\mathbf{x} = 0$ is often called the **homogeneous** equation associated with $A\mathbf{x} = \mathbf{b}$. Observe that the homogeneous equation always has at least one solution, namely, $\mathbf{x} = 0$. From 3.4 we immediately obtain the following.

Corollary

Suppose $A\mathbf{x} = \mathbf{b}$ has *at least* one solution. Then it has *exactly* one solution if and only if $\mathbf{x} = 0$ is the only solution of the associated homogeneous equation.

3.5 Theorem

Suppose A is reduced. If every column of A is pivotal, then $\mathbf{x} = 0$ is the only solution of the homogeneous equation $A\mathbf{x} = 0$. Otherwise the equation has solutions with $\mathbf{x} \neq 0$.

Proof. Suppose every column of A is pivotal. Let i be the number of the row associated with column j. Then $a_{ij} = 1$ and $a_{ik} = 0$ for $k \neq j$ (because A is reduced and all columns are pivotal). Thus the ith entry in $A\mathbf{x}$ is $\sum_{k=1}^{n} a_{ik}x_k = x_j$; so if $A\mathbf{x} = 0$, $x_j = 0$ for all j.

Conversely, if r is the number of a nonpivotal column, we can construct a nonzero solution of $A\mathbf{x} = 0$ as follows. Take $x_r = 1$

(which guarantees $\mathbf{x} \neq 0$) and take $x_j = 0$ if j is the number of any other nonpivotal column of A. If the jth column is pivotal, take $x_j = -a_{ir}$ where i is the number of the row associated with column j. As in the proof of Theorem 3.4, we look at the product $A\mathbf{x}$ a row at a time. Zero rows of A, of course, give zero entries in the product. If the ith row of A is nonzero, we get the sum $\sum_{k=1}^{n} a_{ik}x_k$. If k is the number of any pivotal column except the jth (where column j is associated with row i), we have $a_{ik} = 0$. If it is the number of any nonpivotal column except the rth, we have $x_k = 0$. Thus the sum reduces to $a_{ij}x_j + x_{ir}x_r = 1 \cdot (-a_{ir}) + a_{ir} \cdot 1 = 0$, as required.

Example 3. Consider the homogeneous equation $A\mathbf{x} = 0$, where A is the matrix of Example 2. The second and fourth columns of A are nonpivotal. The construction given in the proof of Theorem 3.5 can thus be applied to give a solution with $x_2 = 1$, $x_4 = 0$, and one with $x_4 = 1$, $x_2 = 0$. The vectors obtained are

$$\mathbf{y} = \begin{pmatrix} -2 \\ 1 \\ 0 \\ 0 \\ 0 \end{pmatrix} \quad \text{and} \quad \mathbf{z} = \begin{pmatrix} -3 \\ 0 \\ -5 \\ 1 \\ 0 \end{pmatrix}.$$

Any combination $r\mathbf{y} + s\mathbf{z}$ is also a solution. (Why?) Using Theorem 3.4 to combine this information with the result of Example 2, we see that all vectors of the form

$$\begin{pmatrix} 4 \\ 0 \\ 1 \\ 0 \\ 3 \end{pmatrix} + r\begin{pmatrix} -2 \\ 1 \\ 0 \\ 0 \\ 0 \end{pmatrix} + s\begin{pmatrix} -3 \\ 0 \\ -5 \\ 1 \\ 0 \end{pmatrix}$$

are solutions of

$$\begin{pmatrix} 0 & 0 & 1 & 5 & 0 \\ 1 & 2 & 0 & 3 & 0 \\ 0 & 0 & 0 & 0 & 0 \\ 0 & 0 & 0 & 0 & 1 \end{pmatrix}\mathbf{x} = \begin{pmatrix} 1 \\ 4 \\ 0 \\ 3 \end{pmatrix}$$

We have answered most of the questions about solutions of linear systems that have a reduced coefficient matrix. If we can convert any given system into a reduced one by elementary operations, then we have a general method of finding solutions of linear systems. The examples of Section 1 illustrate a reduction process that can in fact be applied successfully to any matrix.

Suppose a matrix is not reduced. Then there must be some column containing a leading entry such that either 3.1(a) or 3.1(b) (or both) is violated. If the column contains the leading entry r for the ith row, multiplying the ith row by r^{-1} will make the leading entry 1. (Since r was a leading entry, it could not be zero. Of course it might be 1 to begin with, and the multiplication would be unnecessary.) If any other entries in the column are nonzero, they can be made zero by adding suitable multiples of the ith row to the rows they are in. Any column that was "correct" before these operations must have a zero for its ith entry, and therefore is unaltered by them. We have just described a process that can be applied to any unreduced matrix and that increases the number of columns that satisfy the Conditions 3.1. If the resulting matrix is not reduced, the process can be repeated. A reduced matrix will be obtained within at most n steps, where n is the number of columns in the matrix. We have proved

3.6 Theorem

Given any matrix A, a sequence of elementary operations can be found which converts A to a reduced matrix.

Example 4. We shall apply the method given in the proof of Theorem 3.6 to reduce the matrix

$$A = \begin{pmatrix} 1 & 3 & -2 & 0 \\ 2 & 6 & 0 & -1 \\ -1 & -3 & 4 & 1 \end{pmatrix}.$$

Column 1 does not satisfy 3.1(a), but has a leading entry of 1 in the first row; so no elementary multiplication is necessary. Subtracting 2 times row 1 from row 2, and adding row 1 to row 3, clears the other two entries to zero and gives

$$A_2 = \begin{pmatrix} 1 & 3 & -2 & 0 \\ 0 & 0 & 4 & -1 \\ 0 & 0 & 2 & 1 \end{pmatrix}.$$

Column 2 does not contain a leading entry. The 4 in column 3 can be converted to a 1 by multiplying the second row by $\frac{1}{4}$, which gives

$$A_3 = \begin{pmatrix} 1 & 3 & -2 & 0 \\ 0 & 0 & 1 & -\frac{1}{4} \\ 0 & 0 & 2 & 1 \end{pmatrix}.$$

Adding 2 times row 2 to row 1 and (—2) times row 2 to row 3 clears out the other entries in column 3 to give

$$A_4 = \begin{pmatrix} 1 & 3 & 0 & -\frac{1}{2} \\ 0 & 0 & 1 & -\frac{1}{4} \\ 0 & 0 & 0 & \frac{3}{2} \end{pmatrix}.$$

Multiplying row 3 by $\frac{2}{3}$ and next adding $\frac{1}{2}$ times row 3 to row 1 and $\frac{1}{4}$ times row 3 to row 2 gives the reduced matrix

$$\begin{pmatrix} 1 & 3 & 0 & 0 \\ 0 & 0 & 1 & 0 \\ 0 & 0 & 0 & 1 \end{pmatrix}$$

as the final result.

In working this example we used the standard procedure given in the proof of 3.6. This of course is not the only sequence of steps that will give a reduced matrix, and a different sequence may require less arithmetic. For instance, consider the matrix A_2. Adding row 3 to row 1 and subtracting 2 times row 3 from row 2 gives

$$\begin{pmatrix} 1 & 3 & 0 & 1 \\ 0 & 0 & 0 & -3 \\ 0 & 0 & 2 & 1 \end{pmatrix}.$$

Then adding $(\frac{1}{3})$ times row 2 to row 1 and $(\frac{1}{3})$ times row 2 to row 3 gives

$$\begin{pmatrix} 1 & 3 & 0 & 0 \\ 0 & 0 & 0 & -3 \\ 0 & 0 & 2 & 0 \end{pmatrix}.$$

Finally, multiplying row 2 by $(-\frac{1}{3})$ and row 3 by $(\frac{1}{2})$ gives the reduced matrix

$$\begin{pmatrix} 1 & 3 & 0 & 0 \\ 0 & 0 & 0 & 1 \\ 0 & 0 & 1 & 0 \end{pmatrix}.$$

We now turn to the special case of systems of n equations in n unknowns, that is, systems with square coefficient matrices. The following theorem is the one most important in applications.

3.7 Theorem

If A is a square matrix and the only solution of the homogeneous equation $A\mathbf{x} = 0$ is $\mathbf{x} = 0$, then the equation $A\mathbf{x} = \mathbf{b}$ has exactly one solution for every vector $\mathbf{b}$.

Proof. By Theorem 3.6 we can convert A to a reduced matrix C and, using the same elementary operations, convert $\mathbf{b}$ to a vector $\mathbf{d}$ so that $C\mathbf{x} = 0$ has the same solution set as $A\mathbf{x} = 0$ and $C\mathbf{x} = \mathbf{d}$ has the same solution set as $A\mathbf{x} = \mathbf{b}$. The hypothesis therefore implies that $C\mathbf{x} = 0$ has no solution except $\mathbf{x} = 0$. Since C is reduced, it follows from Theorem 3.5 that every column of C is pivotal. The number of nonzero rows of C is equal to the number of pivotal columns. Since C is square, there can be no zero rows, therefore, by Theorem 3.3 $C\mathbf{x} = \mathbf{d}$ (and hence $A\mathbf{x} = \mathbf{b}$) has at least one solution. That there is exactly one solution follows from the corollary of Theorem 3.4.

We can now also prove the converse of Theorem 1.3, namely:

3.8 Theorem

If A is invertible, then it can be converted to I by elementary operations.

Proof. If there is an $\mathbf{x} \neq 0$ with $A\mathbf{x} = 0$, A cannot have an inverse, since then $A\mathbf{x} = 0$ would imply $A^{-1}A\mathbf{x} = A^{-1}0$, i.e., $\mathbf{x} = 0$, which would be a contradiction. Otherwise A can be converted to a reduced matrix C, all of whose columns are pivotal. Every column of C then contains one 1 and has all other entries 0, and the 1's in different columns belong to different rows. Appropriately changing the order of the rows (which can be done by elementary transpositions) will give the identity matrix. Thus A can be converted to I by elementary operations.

We thus see that it is sensible to apply the method used for computing inverses given by Theorem 1.3 to any square matrix A. If the method succeeds, it shows that the matrix is invertible and computes A^{-1}. If the method fails, then one obtains a reduced matrix with at least one non-pivotal column and, hence, by Theorem 3.5, a nonzero $\mathbf{x}$ such that $A\mathbf{x} = 0$, which demonstrates that A is not invertible.

We conclude by observing that every elementary operation, when applied to a column vector in $\mathcal{R}^n$, is a linear function, called an **elementary transformation.** Indeed, it is obvious that (a) multiplication of a coordinate by $r \neq 0$, (b) addition of a multiple of one coordinate to another, and (c) transposition of two coordinates all have the properties of linearity. It follows that each of these operations can be performed by multiplying on the left by some n-by-n **elementary matrix.** (The precise forms of such matrices are described in Exercise 7 at the end of this section.) This fact enables us to prove the next theorem.

3.9 Theorem

An invertible matrix A can be written as a product of elementary matrices. It follows that the linear function defined by

$$f(\mathbf{x}) = A\mathbf{x}$$

can be expressed as a composition of elementary transformations.

Proof. By Theorem 3.8, the matrix A can be reduced to I by applying a product

$$Q = A_1 \ldots A_k$$

of elementary matrices to A. Thus $QA = I$. But since each elementary operation is reversible, each matrix A_i has an inverse A_i^{-1}. Then Q has an inverse

$$Q^{-1} = A_k^{-1} \ldots A_1^{-1},$$

and so $A = Q^{-1}I = A_k^{-1} \ldots A_1^{-1}$. This expresses A as the desired product.

EXERCISES

1. Determine which of the following matrices are reduced. For those that are not, state exactly how they violate the conditions. For those that are reduced, list the pivotal columns and their associated rows.

$$A = \begin{pmatrix} 1 & -1 & 0 & 1 \\ 0 & 0 & 1 & 2 \\ 0 & 0 & 0 & 0 \end{pmatrix}, \qquad B = \begin{pmatrix} 1 & 0 & 1 \\ 0 & 1 & 0 \\ 1 & 0 & 0 \end{pmatrix},$$

$$C = \begin{pmatrix} 1 & 3 & 0 & 5 \\ 0 & 0 & 2 & 4 \\ 0 & 0 & 0 & 0 \end{pmatrix}, \qquad D = \begin{pmatrix} 1 & 2 & 0 \\ 0 & 0 & 0 \\ 0 & 0 & 1 \end{pmatrix}.$$

2. For each matrix of Exercise 1 which is not already reduced, find an elementary operation which changes it to a reduced matrix.

3. Let

$$\mathbf{r} = \begin{pmatrix} 0 \\ 1 \\ 2 \end{pmatrix}, \quad \mathbf{s} = \begin{pmatrix} 2 \\ 0 \\ 1 \end{pmatrix}, \quad \mathbf{t} = \begin{pmatrix} 1 \\ 2 \\ 0 \end{pmatrix}.$$

For each of the equations $A\mathbf{x} = \mathbf{r}$, $A\mathbf{x} = \mathbf{s}$, $A\mathbf{x} = \mathbf{t}$, where A is the matrix of Problem 1, determine whether the equation has no solutions, exactly one solution, or more than one solution. If there is one solution, give it; if there are more than one, give two different solutions.

4. Repeat Problem 3 using the matrices B, C, and D instead of A from Problem 1. (Remember to get the coefficient matrix in reduced form.)

5. Show that if a square n-by-n matrix is reduced and has no all-zero row, then every row and column contains $n - 1$ zeros and one 1. Hence show that it can be converted to an identity matrix by elementary transpositions.

6. Determine the solution sets of the following systems of equations.

(a) $$\begin{pmatrix} 1 & 2 & 3 \\ 4 & 5 & 6 \\ 7 & 8 & 9 \end{pmatrix} \mathbf{x} = \begin{pmatrix} 10 \\ 11 \\ 12 \end{pmatrix}.$$

[*Ans.* $r(1, -2, 1) + (0, -9, \frac{28}{3})$.]

(b)
$$\begin{aligned} x_1 + x_2 &= 1 \\ x_2 + x_3 &= 2 \\ x_3 + x_4 &= 3 \\ x_4 + x_1 &= 4. \end{aligned}$$

(c) $$\begin{pmatrix} 1 & 1 & 1 \\ -1 & 2 & -4 \\ 1 & 3 & 9 \end{pmatrix} \mathbf{x} = \begin{pmatrix} 0 & 2 \\ 6 & 2 \\ 0 & 0 \end{pmatrix}.$$

(d)
$$\begin{aligned} x_1 + 2x_2 + x_3 - x_4 &= 1 \\ x_2 - x_3 + x_4 &= 0 \\ x_1 + 3x_3 - 3x_4 &= 3. \end{aligned}$$

(e)
$$\begin{aligned} x_1 + 2x_2 + x_3 - x_4 &= 1 \\ x_2 - x_3 + x_4 &= 0 \\ x_1 + 3x_3 - 3x_4 &= 1. \end{aligned}$$

7. (a) Denote by $D_i(r)$ a matrix which is the same as the identity matrix except for having r in place of 1 in the ith diagonal entry. Show that the matrix product

$$D_i(r)M$$

is the elementary modification of M that results from multiplying the ith row of M by r.

(b) Denote a matrix with 1 for its ijth entry and 0's elsewhere by E_{ij}. For example, for 3-by-3 matrices,

$$E_{13} = \begin{pmatrix} 0 & 0 & 1 \\ 0 & 0 & 0 \\ 0 & 0 & 0 \end{pmatrix} \quad \text{and} \quad E_{21} = \begin{pmatrix} 0 & 0 & 0 \\ 1 & 0 & 0 \\ 0 & 0 & 0 \end{pmatrix}.$$

Show that, if M is an m-by-n matrix and E_{ij} and I are both m-by-m, then

$$(I + rE_{ij})M$$

is the elementary modification of M which results from adding r times the jth row to the ith row.

(c) Denote by T_{ij} the matrix obtained from I by exchanging the ith and jth rows. Show that exchanging the ith and jth rows of M gives

$$T_{ij}M.$$

(d) Any matrix of the form $(I + rE_{ij})$, $D_i(r)$, or T_{ij} is called an **elementary matrix.** Show that each of the three types of elementary matrices is invertible by verifying that, for $i \neq j$,

$$(I + rE_{ij})^{-1} = (I - rE_{ij}), \qquad D_i(r)^{-1} = D_i\left(\frac{1}{r}\right), \quad \text{and} \quad T_{ij}^{-1} = T_{ij}.$$

8. Prove that if there are more unknowns than there are equations in a linear system, then the system has either no solutions or infinitely many.

9. Suppose A is a square matrix. Prove that A is invertible if and only if every matrix equation $A\mathbf{x} = \mathbf{b}$ has *at least* one solution.

10. Let $p(x) = a_0 + a_1x + \ldots + a_nx^n$ be a polynomial of degree $\leq n$. It is a well-known theorem of algebra that if there are more than n values of x which make $p(x) = 0$, then all of its coefficients are zero. Use this theorem to show that if $x_0, \ldots, x_n$ are any $n + 1$ different numbers, and $b_0, \ldots, b_n$ are any $n + 1$ numbers, then there is exactly one polynomial of degree $\leq n$ such that $p(x_0) = b_0, \ldots, p(x_n) = b_n$. [*Hint.* Show that the problem leads to a system of linear equations with $a_0, \ldots, a_n$ as unknowns.]

11. A reduced matrix in which the first pivotal column (starting from the left) is associated with the first row, the second pivotal column is associated with the second row, etc., is said to be in **echelon form.** Show that:

(a) Any reduced matrix can be put in echelon form by a sequence of elementary transpositions.

(b) If a matrix is in echelon form, then the zero rows (if there are any) come last.

(c) A square matrix in echelon form is either an identity matrix or has at least one zero row.

SECTION 4

VECTOR SPACES, SUBSPACES, DIMENSION

In the earlier parts of the book we have restricted ourselves to vectors in $\mathcal{R}^n$. In this section we consider more general vector spaces, though some of the ideas have already been introduced with $\mathcal{R}^2$ and $\mathcal{R}^3$ as the main examples.

Recall that a vector $\mathbf{x}$ is a **linear combination** of the vectors $\mathbf{x}_1, \ldots, \mathbf{x}_n$ if there are numbers $r_1, \ldots, r_n$ such that

$$\mathbf{x} = r_1\mathbf{x}_1 + \ldots + r_n\mathbf{x}_n.$$

Example 1. Let

$$\mathbf{x}_1 = (1, 0, 0), \qquad \mathbf{x}_2 = (0, 1, 0), \qquad \mathbf{x}_3 = (0, 0, 1), \qquad \mathbf{x}_4 = (1, 1, 1).$$

Then $\mathbf{y} = (2, 2, 0)$ is a linear combination of $\mathbf{x}_1$ and $\mathbf{x}_2$ because it is equal to $2\mathbf{x}_1 + 2\mathbf{x}_2$; it is a linear combination of $\mathbf{x}_3$ and $\mathbf{x}_4$ because it is equal to $2\mathbf{x}_4 - 2\mathbf{x}_3$. On the other hand, it is not a linear combination of $\mathbf{x}_2$ and $\mathbf{x}_3$ because $r\mathbf{x}_2 + s\mathbf{x}_3$ has a first entry of 0 whatever the values of r and s; so $\mathbf{y} = r\mathbf{x}_2 + s\mathbf{x}_3$ is impossible.

Since $0 = 0\mathbf{x}_1 + \ldots + 0\mathbf{x}_n$, the zero vector is a linear combination of any set of vectors. The linear combinations of a single vector $\mathbf{x}_1$ are just the numerical multiples $r\mathbf{x}_1$.

If a set of vectors lies in a plane through the origin in 3-space, then every linear combination of them lies in the same plane—recall that if $\mathbf{x}$ and $\mathbf{y}$ are in a plane through the origin, the parallelogram rule makes $\mathbf{x} + \mathbf{y}$ a vector in the same plane. Any numerical multiple of $\mathbf{x}$ lies in the same plane because it lies in the line containing $\mathbf{x}$. Any linear combination of $\mathbf{x}_1, \ldots, \mathbf{x}_n$ is built up by multiplications and additions, and if the vectors $\mathbf{x}_1, \ldots, \mathbf{x}_n$ lie in a plane, so do all linear combinations of them.

Similarly, if $\mathbf{x}_1, \ldots, \mathbf{x}_n$ all lie in one line through the origin (so they are all multiples of some one vector), any linear combination of them lies in the same line.

These remarks suggest the following generalization which includes lines and planes as special cases: A subset $\mathcal{S}$ of a vector space $\mathcal{V}$ is called a **linear subspace** (or, frequently, simply a **subspace**) if every linear combination of elements of $\mathcal{S}$ is also in $\mathcal{S}$. We assume $\mathcal{S}$ is non-empty.

Example 2. We list some examples of subspaces.

(a) The set of all vectors in $\mathcal{R}^n$ with first entry equal to 0 is a subspace, since any linear combination of such vectors will also have a 0 for its first entry.

(b) For any vector space $\mathcal{V}$, $\mathcal{V}$ itself is a subspace. The term *proper* subspace is often used to refer to those subspaces that are not the whole space. In any vector space the zero vector forms a subspace all by itself. This subspace is called the *trivial* subspace (because it is).

(c) For any linear function $\mathcal{V} \xrightarrow{f} \mathcal{W}$, the set $\mathcal{N}$ of vectors $\mathbf{x}$ in $\mathcal{V}$ with $f(\mathbf{x}) = 0$ is a subspace of $\mathcal{V}$ called the **null space** of f. If $\mathbf{x}_1, \ldots, \mathbf{x}_k$ are in $\mathcal{N}$ and $\mathbf{x} = \sum_{i=1}^{k} r_i\mathbf{x}_i$, then $f(\mathbf{x}) = \sum_{i=1}^{k} r_i f(\mathbf{x}_i) = 0$, because f is linear and all the $f(\mathbf{x}_i)$ are 0. Hence $\mathbf{x}$ is also in $\mathcal{N}$, and so $\mathcal{N}$ is a subspace of $\mathcal{V}$. In particular, the set of vectors (x, y, z) in $\mathcal{R}^3$ such that

$$x + 2y + 3z = 0,$$

or equivalently,

$$\begin{pmatrix} 1 & 2 & 3 \end{pmatrix}\begin{pmatrix} x \\ y \\ z \end{pmatrix} = 0,$$

is a subspace because it is the null space of the linear function from $\mathcal{R}^3$ to $\mathcal{R}$ defined by the preceding 1-by-3 matrix.

(d) The range of a linear function f defined on a vector space is a subspace of the range space of f. The reason is that for $\mathbf{y}_1, \ldots, \mathbf{y}_k$ to be in the range of f means that there are vectors $\mathbf{x}_1, \ldots, \mathbf{x}_k$ in the domain of f such that $f(\mathbf{x}_i) = \mathbf{y}_i$ for $i = 1, \ldots, k$. But then, by the linearity of f, an arbitrary linear combination of the $\mathbf{y}_i$ has the form

$$\begin{aligned} r_1\mathbf{y}_1 + \ldots + r_k\mathbf{y}_k &= r_1 f(\mathbf{x}_1) + \ldots + r_k f(\mathbf{x}_k) \\ &= f(r_1\mathbf{x}_1 + \ldots + r_k\mathbf{x}_k). \end{aligned}$$

Because the domain of f is a vector space, $r_1\mathbf{x}_1 + \ldots + r_k\mathbf{x}_k$ is in it, and so $r_1\mathbf{y}_1 + \ldots + r_k\mathbf{y}_k$ is in the range of f. In particular, the set of all vectors in $\mathcal{R}^3$ of the form

$$\begin{pmatrix} 1 & 4 \\ 2 & 5 \\ 3 & 6 \end{pmatrix}\begin{pmatrix} x \\ y \end{pmatrix} = x\begin{pmatrix} 1 \\ 2 \\ 3 \end{pmatrix} + y\begin{pmatrix} 4 \\ 5 \\ 6 \end{pmatrix}$$

is a subspace because it is the range of the linear function just defined by the above 3-by-2 matrix.

We define the **span** of a set S of vectors to be the set consisting of all linear combinations of elements of S. It is easy to show that the span of any set is a subspace. Suppose $\mathbf{x} = \sum_{i=1}^{k} r_i\mathbf{x}_i$ is a linear combination of vectors $\mathbf{x}_1, \ldots, \mathbf{x}_k$, which are in the span of S, that is, $\mathbf{x}_i = \sum_{j=1}^{n} s_{ij}\mathbf{u}_j$ for some vectors $\mathbf{u}_j$ in S. Then $\mathbf{x} = \sum_{i=1}^{k} r_i\left(\sum_{j=1}^{n} s_{ij}\mathbf{u}_j\right) = \sum_{j=1}^{n} t_j\mathbf{u}_j$, where $t_j = \sum_{i=1}^{k} r_i s_{ij}$. This shows that $\mathbf{x}$ is itself in the span of S.

We recall from Section 1 of Chapter 1 the general definition of a vector space as a set $\mathcal{V}$ of elements with operations of addition and numerical multiplication such that for any elements $\mathbf{x}$, $\mathbf{y}$, $\mathbf{z}$ of $\mathcal{V}$ and real numbers r, s:

1. $r\mathbf{x} + s\mathbf{x} = (r + s)\mathbf{x}$.
2. $r\mathbf{x} + r\mathbf{y} = r(\mathbf{x} + \mathbf{y})$.
3. $r(s\mathbf{x}) = (rs)\mathbf{x}$.
4. $\mathbf{x} + \mathbf{y} = \mathbf{y} + \mathbf{x}$.
5. $(\mathbf{x} + \mathbf{y}) + \mathbf{z} = \mathbf{x} + (\mathbf{y} + \mathbf{z})$.
6. There exists an element 0 in $\mathcal{V}$ such that $\mathbf{x} + 0 = \mathbf{x}$.
7. For any $\mathbf{x}$ in $\mathcal{V}$, $\mathbf{x} + (-1)\mathbf{x} = 0$.

Now if $\mathcal{S}$ is any subspace of a vector space $\mathcal{V}$, then the operations of addition and numerical multiplication, as given for $\mathcal{V}$, always yield a result in $\mathcal{S}$ when they are applied to elements of $\mathcal{S}$, because $\mathbf{x} + \mathbf{y} = 1\mathbf{x} + 1\mathbf{y}$ and $r\mathbf{x} = r\mathbf{x} + 0\mathbf{y}$ are linear combinations of $\mathbf{x}$ and $\mathbf{y}$. The laws 1 through 5 certainly hold for $\mathbf{x}$, $\mathbf{y}$, $\mathbf{z}$ in $\mathcal{S}$, since they hold for all elements of $\mathcal{V}$. The zero vector belongs to $\mathcal{S}$, since $0 = 0\mathbf{x}$ for any $\mathbf{x}$ in $\mathcal{S}$; also, if $\mathbf{x}$ is in $\mathcal{S}$, then so is $(-1)\mathbf{x} = -\mathbf{x}$. Thus laws 6 and 7 hold as well. In other words, we have proved the next theorem.

4.1 Theorem

Any subspace of a vector space is a vector space, with the operations inherited from the original space.

While subspaces of $\mathcal{R}^n$ provide important examples of vector spaces, there are many others. We list some below.

Example 3. (a) Let $\mathcal{V}$ consist of all continuous real-valued functions of a real variable. Define $f + g$ and rf in the obvious way as the functions whose values for any number x are $f(x) + g(x)$ and $rf(x)$, respectively. (Of course, we are using the theorems that $f + g$ and rf are continuous if

f and g are.) It is easy to verify that the laws for a vector space are satisfied.

(b) Let P be the subspace of $\mathcal{V}$ consisting of all polynomials, i.e., all functions f that can be expressed by a formula $f(x) = a_0 + a_1x + \ldots + a_kx^k$ for some constants $a_0, \ldots, a_k$. (What needs to be checked to verify that this is a subspace?)

(c) Let P_n be the subspace of polynomials of degree less than or equal to n, i.e., those that require no power of x higher than the nth for their expression. For $k \leq n$, P_k is a subspace of P_n, and all P_n are subspaces of P.

(d) Let $C^{(k)}$ be the vector space of real-valued functions $f(x)$ whose first k derivatives are continuous for all real x. Then $C^{(k+1)}$ is a subspace of $C^{(k)}$. If we denote by $C^{(\infty)}$, the vector space of functions having derivatives of *all* orders, then $C^{(\infty)}$ is a subspace of $C^{(k)}$ for every k.

The description of lines, planes, and ordinary space as 1-, 2-, and 3-dimensional is familiar. It is possible to define the dimension of any vector space. The examples of lines and planes suggest that the span of k vectors should have dimension k. This is not quite right, since, for example, the span of two vectors that happen to lie in the same line will be only a line instead of a plane. To handle the question properly requires the concept of linear independence introduced in Chapter 1. We recall the following definition.

A set of vectors $\{\mathbf{x}_1, \ldots, \mathbf{x}_k\}$ is **(linearly) independent** if the only set of numbers $r_1, \ldots, r_k$ such that $r_1\mathbf{x}_1 + \ldots + r_k\mathbf{x}_k = 0$ is the set $r_1 = r_2 = \ldots = r_k = 0$. A set of vectors is **(linearly) dependent** if it is not independent. For $\{\mathbf{x}_1, \ldots, \mathbf{x}_k\}$ to be dependent therefore means that there are numbers $r_1, \ldots, r_k$ *not* all zero, such that $r_1\mathbf{x}_1 + \ldots + r_k\mathbf{x}_k = 0$.

Example 4. (a) The four vectors $\mathbf{x}_1 = (2, 0, 0)$, $\mathbf{x}_2 = (0, -2, 0)$, $\mathbf{x}_3 = (0, 0, 3)$, $\mathbf{x}_4 = (2, -2, 3)$ are linearly dependent since $\mathbf{x}_1 + \mathbf{x}_2 + \mathbf{x}_3 - \mathbf{x}_4 = 0$. The set of three vectors $\mathbf{x}_1$, $\mathbf{x}_2$, $\mathbf{x}_3$ is independent since $r\mathbf{x}_1 + s\mathbf{x}_2 + t\mathbf{x}_3 = 0$ only if $r = s = t = 0$.

(b) A set of two vectors $\mathbf{x}$, $\mathbf{y}$ is independent only if neither is a numerical multiple of the other. For example, if $\mathbf{y} = 3\mathbf{x}$, then $3\mathbf{x} - \mathbf{y} = 0$, and the vectors are dependent.

A set of vectors which is linearly independent and spans a space $\mathcal{V}$ is called a **basis** for $\mathcal{V}$.

Example 5. (a) The natural basis vectors $\mathbf{e}_1, \ldots, \mathbf{e}_n$, where $\mathbf{e}_i$ has 1 for its ith entry and 0 for all other entries, form a basis for $\mathcal{R}^n$. Verification that the $\mathbf{e}_i$ are in fact linearly independent and span $\mathcal{R}^n$ is left to the reader.

(b) In the space P_n of Example 3(c), the polynomials $\mathbf{x}_0 = 1$, $\mathbf{x}_1 = x, \ldots, \mathbf{x}_n = x^n$ form a basis. Obviously, if $f(x) = a_0 + a_1x + \ldots + a_nx^n$ is a polynomial of degree less than or equal to n, it is the linear combination $a_0\mathbf{x}_0 + \ldots + a_n\mathbf{x}_n$ of the $\mathbf{x}$'s. If $a_0\mathbf{x}_0 + \ldots + a_n\mathbf{x}_n$ is the zero function, then (since a polynomial of degree less than or equal to n cannot have more than n roots unless its coefficients are all zero) $a_0 = a_1 = \ldots = a_n = 0$.

While it is true that every vector space has a basis, we shall usually consider only those which have a basis with a finite number of elements. The next theorem implies that if a space is spanned by a finite set (which is perhaps not linearly independent), then it has a finite basis.

4.2 Theorem

Let $\mathcal{V}$ be the span of the vectors $\mathbf{x}_1, \ldots, \mathbf{x}_n$. Either $\mathcal{V}$ consists of the zero vector alone, or some subset of the $\mathbf{x}$'s is a basis for $\mathcal{V}$.

Proof. If the set $\mathbf{x}_1, \ldots, \mathbf{x}_n$ is independent, then it is itself a basis for $\mathcal{V}$. Otherwise some relation $r_1\mathbf{x}_1 + \ldots + r_n\mathbf{x}_n = 0$ holds, with at least one r, say r_k, different from 0. Then we can divide by r_k and obtain $\mathbf{x}_k = -(r_1/r_k)\mathbf{x}_1 - \ldots - (r_n/r_k)\mathbf{x}_n$, where $\mathbf{x}_k$ does not appear on the right side. By substituting the right side for $\mathbf{x}_k$ in any linear combination of all the $\mathbf{x}$'s, a linear combination is obtained that does not involve $\mathbf{x}_k$. In other words, $\mathbf{x}_k$ can be dropped and the span of the remaining vectors will still be all of $\mathcal{V}$. If the resulting subset is not independent, the process can be repeated. It must end in a finite number of steps either because a basis has been obtained or because all the vectors have been discarded. The latter is possible only if the space contains only the zero vector.

The following theorem is the most important step in developing the theory of dimension.

4.3 Theorem

If a vector space is spanned by n vectors $\mathbf{x}_1, \ldots, \mathbf{x}_n$, and $\mathbf{y}_1, \ldots, \mathbf{y}_k$ is a linearly independent set of k vectors in the space, then $k \leq n$.

Proof. The statement is easy to prove if $n = 1$. In this case, the space spanned by $\mathbf{x}_1$ consists of all numerical multiples of $\mathbf{x}_1$. Then for any two vectors $\mathbf{y}_1$ and $\mathbf{y}_2$ in the space, one must be a multiple of the other, and so $\mathbf{y}_1$ and $\mathbf{y}_2$ cannot be independent. We proceed by induction and assume that, in any space spanned by $n - 1$ vectors, no independent subset can have more than $n - 1$ vectors in it. Suppose we now are given a space spanned by n vectors $\mathbf{x}_1, \ldots, \mathbf{x}_n$ and containing k vectors $\mathbf{y}_1, \ldots, \mathbf{y}_k$ with $k > n$. If we can show that the $\mathbf{y}$'s are dependent, then we will have shown that the statement of the theorem holds for a spanning set of n elements, and the inductive proof will be complete. Each of the vectors $\mathbf{y}_1, \ldots, \mathbf{y}_{n+1}$ can be written as a linear combination of $\mathbf{x}_1, \ldots, \mathbf{x}_n$, which means that there are numbers a_{ij} such that $\mathbf{y}_i = \sum_{j=1}^{n} a_{ij}\mathbf{x}_j$, for $i = 1, \ldots, n + 1$. If the $n + 1$ numbers a_{i1} are all zero, then the $\mathbf{y}$'s all lie in the space spanned by the $n - 1$ vectors $\mathbf{x}_2, \ldots, \mathbf{x}_n$; thus by the inductive assumption the $\mathbf{y}$'s would then be dependent, as we want to show. Otherwise (by renumbering, if necessary) we may suppose that a_{11} is not zero. Then define n vectors $\mathbf{z}_2, \ldots, \mathbf{z}_{n+1}$ by setting $\mathbf{z}_i = \mathbf{y}_i - a_{11}^{-1}a_{i1}\mathbf{y}_1$. By using the equations giving $\mathbf{y}_i$ in terms of the $\mathbf{x}$'s, it is easy to see that

$$\mathbf{z}_i = \sum_{j=2}^{n}(a_{ij} - a_{11}^{-1}a_{i1}a_{1j})\mathbf{x}_j,$$

so that the $\mathbf{z}$'s are linear combinations of the $n - 1$ vectors $\mathbf{x}_2, \ldots, \mathbf{x}_n$. By the inductive assumption, there are numbers $r_2, \ldots, r_{n+1}$, not all zero, such that $r_2\mathbf{z}_2 + \ldots + r_{n+1}\mathbf{z}_{n+1} = 0$. Using the definition of the $\mathbf{z}$'s in terms of the $\mathbf{y}$'s, this last relation becomes

$$-a_{11}^{-1}(r_2a_{21} + \ldots + r_{n+1}a_{n+11})\mathbf{y}_1 + r_2\mathbf{y}_2 + \ldots + r_{n+1}\mathbf{y}_{n+1} = 0.$$

But since not all the r's are zero, this implies that $\mathbf{y}_1, \ldots, \mathbf{y}_{n+1}$ are dependent, as we wanted to show.

4.4 Theorem

Let $\mathcal{V}$ be a vector space with a basis of n elements. Then every basis for $\mathcal{V}$ has n elements.

Proof. Let $\{\mathbf{x}_1, \ldots, \mathbf{x}_n\}$ and $\{\mathbf{y}_1, \ldots, \mathbf{y}_k\}$ be two bases for $\mathcal{V}$. Since both sets are independent, and both are spanning sets, Theorem 4.3 implies that $k \leq n$ and $n \leq k$.

The **dimension** of a vector space that has a finite spanning set is the number of elements in any basis for the space. (The dimension of the space consisting of the zero vector alone is defined to be 0.) We write dim ($\mathcal{V}$) for the dimension of the vector space $\mathcal{V}$. Note that Theorem 4.2

guarantees the existence of a basis, and that Theorem 4.4 guarantees that the dimension does not depend on which basis is taken.

Example 6. (a) By Example 5(a), dim $(\mathcal{R}^n) = n$.

(b) By Example 5(b), dim $(P_n) = n + 1$.

(c) The space P of all polynomials (Example 4(b)) does not have any finite spanning set. If it did have one with k elements, then the fact that $1, x, x^2, \ldots, x^k$ are $k + 1$ linearly independent elements of P would contradict Theorem 4.3.

A vector space with a finite basis is said to be **finite-dimensional**. As we have just seen in Example 6(c), there are spaces which are not finite-dimensional.

Theorem 4.2 asserts that we can get a basis from a finite spanning set by deleting some of its members. The next theorem shows that, in a finite-dimensional space, we can get a basis from a linearly independent set by adding vectors to it.

4.5 Theorem

Let $S = \{\mathbf{x}_1, \ldots, \mathbf{x}_k\}$ be a linearly independent set in a vector space $\mathcal{V}$. If S is not already a basis, either it can be extended to a finite basis for $\mathcal{V}$, or else it can be extended to an infinite sequence of independent vectors, and $\mathcal{V}$ is not finite-dimensional.

Proof. Suppose $\mathbf{x}_1, \ldots, \mathbf{x}_k$ are linearly independent but do not span all of $\mathcal{V}$. Then there is some vector $\mathbf{y}$ that is not a linear combination of $\mathbf{x}_1, \ldots, \mathbf{x}_k$. Take $\mathbf{x}_{k+1} = \mathbf{y}$. We claim that the set $\mathbf{x}_1, \ldots, \mathbf{x}_k, \mathbf{x}_{k+1}$ is linearly independent. Suppose $r_1\mathbf{x}_1 + \ldots + r_{k+1}\mathbf{x}_{k+1} = 0$. We must show that all the r's are 0. If r_{k+1} were not 0, we could write $\mathbf{x}_{k+1} = -(r_1/r_{k+1})\mathbf{x}_1 - \ldots - (r_k/r_{k+1})\mathbf{x}_k$, which is impossible because $\mathbf{x}_{k+1}$ is not a linear combination of the other $\mathbf{x}$'s. Therefore we have $r_{k+1} = 0$ and $r_1\mathbf{x}_1 + \ldots + r_k\mathbf{x}_k = 0$. Since $\mathbf{x}_1, \ldots, \mathbf{x}_k$ are independent, the last equation implies $r_1 = \ldots = r_k = 0$. In other words, if a linearly independent set does not span a space, then a vector can be added to it so that the resulting set is also independent.

This process of adding vectors can be repeated unless a spanning set is reached. If a spanning set is reached, then it is a basis and $\mathcal{V}$ is finite-dimensional. Otherwise an arbitrarily large independent set can be found, and $\mathcal{V}$ cannot be finite-dimensional. This completes the proof of the theorem.

4.6 Theorem

If $\mathcal{S}$ is a subspace of a finite-dimensional space $\mathcal{V}$ with dim $(\mathcal{V}) = n$, then $\mathcal{S}$ is finite-dimensional and dim $(\mathcal{S}) \leq n$. Any basis for $\mathcal{S}$ can be extended to a basis for $\mathcal{V}$, and if dim $(\mathcal{S}) = n$, then $\mathcal{S} = \mathcal{V}$.

Proof. If $\mathcal{S}$ consists of 0 alone it has dimension 0. Otherwise start with a nonzero vector $\mathbf{x}_1$ in $\mathcal{S}$ and apply Theorem 4.5. Since no independent subset of $\mathcal{V}$ can contain more than n elements, the construction must end with a finite basis for $\mathcal{S}$. This basis for $\mathcal{S}$ is a linearly independent subset of $\mathcal{V}$ and can if necessary be extended (Theorem 4.5 again) to a basis for $\mathcal{V}$. Thus we have a basis for $\mathcal{S}$ that is a subset of a basis for $\mathcal{V}$. If dim $(\mathcal{S}) = n$, the subset must be the whole set; so the same set is a basis for both $\mathcal{S}$ and $\mathcal{V}$, and $\mathcal{S} = \mathcal{V}$.

The following theorem states an important property of linear functions.

4.7 Theorem

Let f be a linear function defined on a finite-dimensional vector space. Then

$$\text{dim (null space of } f) + \text{dim (range of } f) = \text{dim (domain of } f).$$

Proof. As with any theorem about dimension, the trick to proving this is to find suitable bases for the spaces involved. Let us write $\mathcal{V}$ for the domain of f, $\mathcal{N}$ for its null space, and $\mathcal{W}$ for its range. Let $\mathbf{v}_1, \ldots, \mathbf{v}_k$ be a basis for $\mathcal{N}$, and extend it to a basis $\mathbf{v}_1, \ldots, \mathbf{v}_k$, $\mathbf{u}_1, \ldots, \mathbf{u}_r$ for $\mathcal{V}$. (Theorem 4.6 guarantees the possibility of this construction.) Then dim $(\mathcal{N}) = k$ and dim $(\mathcal{V}) = k + r$. Let $\mathbf{w}_1 = f(\mathbf{u}_1), \ldots, \mathbf{w}_r = f(\mathbf{u}_r)$. We claim that $\mathbf{w}_1, \ldots, \mathbf{w}_r$ is a basis for $\mathcal{W}$, which implies that dim $(\mathcal{W}) = r$ and proves the theorem. It is obvious that the vectors $\mathbf{w}_1, \ldots, \mathbf{w}_r$ span $\mathcal{W}$, for if $\mathbf{y} = f(\mathbf{x})$ is any vector in the range of f, we may write

$$\mathbf{x} = \sum_{i=1}^{k} a_i \mathbf{v}_i + \sum_{i=1}^{r} b_i \mathbf{u}_i,$$

and then

$$\mathbf{y} = \sum_{i=1}^{k} a_i f(\mathbf{v}_i) + \sum_{i=1}^{r} b_i f(\mathbf{u}_i) = 0 + \sum_{i=1}^{r} b_i \mathbf{w}_i,$$

which shows that $\mathbf{y}$ is a linear combination of the $\mathbf{w}$'s. It remains to be shown that $\mathbf{w}_1, \ldots, \mathbf{w}_r$ are linearly independent. Suppose

$\sum_{i=1}^{r} b_i\mathbf{w}_i = 0$. This means that the vector $\sum_{i=1}^{r} b_i\mathbf{u}_i$ is in the null space of f and is therefore equal to some linear combination $\sum_{i=1}^{r} a_i\mathbf{v}_i$, so $a_1\mathbf{v}_1 + \ldots + a_k\mathbf{v}_k - b_1\mathbf{u}_1 - \ldots - b_r\mathbf{u}_r = 0$. Since $\mathbf{v}_1, \ldots \mathbf{v}_k, \mathbf{u}_1, \ldots, \mathbf{u}_r$ are linearly independent, this is possible only if all the b's (and all the a's) are zero, which shows that the $\mathbf{w}$'s are linearly independent.

Example 7. Suppose $m \leq n$, and define f from $\mathcal{R}^n$ to $\mathcal{R}^m$ by letting $f(\mathbf{x})$ be the m-dimensional column vector consisting of the first m entries of $\mathbf{x}$. The range of f is all of $\mathcal{R}^m$, and the null space, of dimension $n - m$, consists of the vectors in $\mathcal{R}^n$ whose first m components are all 0.

The definitions of line and plane can be unified in terms of subspaces of $\mathcal{R}^n$. Because a subspace always contains the zero vector, we define a **line through the origin** to be a 1-dimensional subspace and a **plane through the origin** to be a 2-dimensional subspace. Examples in $\mathcal{R}^3$ are shown in Fig. 5(a). A line or a plane that does not pass through the origin can be obtained from one that does by a translation, that is, by adding a fixed vector to every vector in the subspace. We say that a subset $\mathcal{S}$ of a vector space is an **affine subspace** if it can be expressed in the form

$$\mathcal{V} + \mathbf{b},$$

where $\mathcal{V}$ is a linear subspace and $\mathbf{b}$ is a single vector. The dimension of an affine subspace is of course defined to be the dimension of $\mathcal{V}$, and a 1-dimensional affine subspace is usually called a **line,** while a 2-dimensional affine subspace is called a **plane.** Examples in $\mathcal{R}^3$ are shown in Fig. 5(b). Two affine subspaces are **parallel** if they are translates of one another.

Two of the most important ways of describing subspaces, namely, as range or null space of a linear function, provide standard ways of describing lines and planes. The parametric representation of planes discussed in

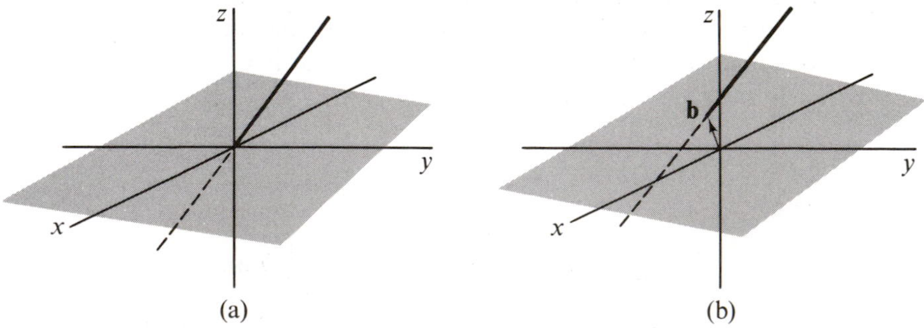

Figure 5

Section 2 of Chapter 1 is obtained by starting with a linear function $\mathcal{R}^2 \xrightarrow{L} \mathcal{R}^3$ having a 2-dimensional range with basis $\mathbf{x}_1, \mathbf{x}_2$. We write

$$L(u, v) = u\mathbf{x}_1 + v\mathbf{x}_2$$

and then form an affine function

$$A(u, v) = u\mathbf{x}_1 + v\mathbf{x}_2 + \mathbf{x}_0$$

by adding a fixed vector $\mathbf{x}_0$. The representation of a line takes a similar form,

$$A(t) = t\mathbf{x} + \mathbf{x}_0,$$

in which the vectors $t\mathbf{x}_1$ form a 1-dimensional subspace as the range of a linear function $\mathcal{R} \xrightarrow{L} \mathcal{R}^3$.

To represent a line or plane through the origin as the null space of a linear function is to consider all solution vectors $\mathbf{x}$ of a *homogeneous* equation

$$L(\mathbf{x}) = 0,$$

where L is linear. An affine subspace is then obtained by adding a fixed vector $\mathbf{x}_0$ to the result or, alternatively, by solving instead

$$L(\mathbf{x} - \mathbf{x}_0) = 0.$$

Because L is linear, this last equation can be written as a *nonhomogeneous* equation

$$L(\mathbf{x}) = L(\mathbf{x}_0). \tag{1}$$

Methods for solving such equations are discussed in the earlier sections of this chapter. If $\mathbf{x} = (x, y)$ is 2-dimensional then equation (1) would take the form

$$ax + by = c,$$

which is a familiar way to represent a line. If $\mathbf{x} = (x, y, z)$ is 3-dimensional, a single equation

$$a_1x + b_1y + c_1z = d_1$$

in general has as solution a plane in $\mathcal{R}^3$, while a system of two equations

$$\begin{aligned} a_1x + b_1y + c_1z &= d_1 \\ a_2x + b_2y + c_2z &= d_2 \end{aligned}$$

has as solution the intersection of two planes, which is either a line or a plane.

Example 8. If we try to solve the equations

$$\begin{aligned} x - y + 2z &= 1 \\ x + y + 3z &= 0 \end{aligned}$$

by row operations, we add the second equation to the first to get

$$\begin{aligned} 2x \qquad + 5z &= 1 \\ x + y + 3z &= 0. \end{aligned}$$

Multiplying the first equation by $\frac{1}{2}$ and subtracting from the second gives

$$\begin{aligned} x \qquad + \tfrac{5}{2}z &= \tfrac{1}{2} \\ y + \tfrac{1}{2}z &= -\tfrac{1}{2}. \end{aligned}$$

We can represent the set of all solutions of this pair of equations parametrically as an affine subspace by setting $z = t$. We find

$$\begin{aligned} x &= -\tfrac{5}{2}t + \tfrac{1}{2} \\ y &= -\tfrac{1}{2}t - \tfrac{1}{2} \\ z &= t \end{aligned}$$

or

$$\begin{pmatrix} x \\ y \\ z \end{pmatrix} = t\begin{pmatrix} -\frac{5}{2} \\ -\frac{1}{2} \\ 1 \end{pmatrix} + \begin{pmatrix} \frac{1}{2} \\ -\frac{1}{2} \\ 0 \end{pmatrix}.$$

Thus we have found a line of solutions; it can be described by starting with a line through the origin having the direction of $\mathbf{x}_1 = (-\frac{5}{2}, -\frac{1}{2}, 1)$, and then translating by adding the vector $\mathbf{x}_0 = (\frac{1}{2}, -\frac{1}{2}, 0)$. The result is shown in Fig. 6.

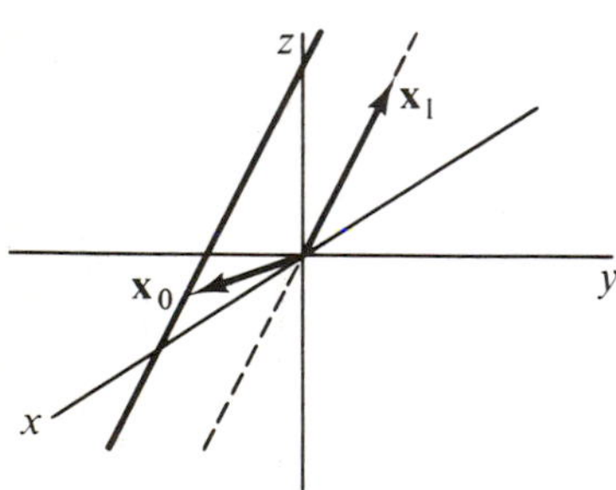

Figure 6

If we have a system of three equations, its solution set is the intersection of the solution sets of the three individual equations. The usual situation is that the three planes intersect in one point. Changing the right side of the equations shifts the planes parallel to themselves, and in this case they will always intersect in just one point. Another possibility is for all three planes to be parallel. The associated homogeneous system will have a two-dimensional solution set; the inhomogeneous system will also have a 2-dimensional solution set if the three planes happen to coincide. Otherwise there will be no solution. Another possibility (which has no analogue in two dimensions) is that no two of the planes are parallel, but that the planes representing the solutions of the *homogeneous* equations all pass through one line. Then (unless they all pass through one line) each pair of the planes representing the solution of the *in*homogeneous equations will intersect in a line parallel to the third plane, and there will be no common solution for all three equations.

EXERCISES

1. Which of the following subsets of $\mathcal{R}^3$ are subspaces? In each case either show that the subset is a subspace or find some linear combination of elements of the subset that is not in the subset.

(a) All vectors (x_1, x_2, x_3) with $x_1 + x_2 = 0$.
(b) All vectors with $x_3 = 0$.
(c) All vectors satisfying (a) *and* (b).
(d) All vectors satisfying (a) *or* (b).
(e) All vectors with $x_1 = (x_2)^3$.
(f) All vectors satisfying (a) *and* (e).

2. Let $\mathbf{x}_1 = (1, 2, 3)$, $\mathbf{x}_2 = (-1, 2, 1)$, $\mathbf{x}_3 = (1, 1, 1)$, and $\mathbf{x}_4 = (1, 1, 0)$.

(a) Show that $\mathbf{x}_1, \mathbf{x}_2, \mathbf{x}_3, \mathbf{x}_4$ is a linearly dependent set by solving an appropriate system of equations.
(b) Express $\mathbf{x}_1$ as a linear combination of $\mathbf{x}_2, \mathbf{x}_3, \mathbf{x}_4$ by a method similar to that used in part (a). [*Ans.* $\mathbf{x}_1 = \frac{1}{3}\mathbf{x}_2 + \frac{8}{3}\mathbf{x}_3 - \frac{4}{3}\mathbf{x}_4$.]

3. Let $C[a, b]$ be the vector space of continuous real-valued functions defined on the interval $[a, b]$. Let $C_0[a, b]$ be the set of functions f in $C[a, b]$ such that $f(a) = f(b) = 0$.

(a) Show that $C_0[a, b]$ is a proper subspace of $C[a, b]$.
(b) What if the condition $f(a) = f(b) = 0$ is replaced by $f(a) = 1, f(b) = 0$?

4. Show that the intersection of two subspaces is always a subspace.

5. Part (d) of Exercise 1 shows that the union of two subspaces is not always a subspace. Show that the union of two subspaces is a subspace if and only if one of them is contained in the other.

6. Show that the range of a linear function may be a proper subspace of its range space.

7. Let $\mathcal{R}^n \xrightarrow{f} \mathcal{R}^m$ be the linear function defined by multiplication by the m-by-n matrix A. Show that its range is the span of the columns of A (considered as elements of $\mathcal{R}^m$).

8. For any two subsets $\mathcal{A}$ and $\mathcal{B}$ of a vector space, let $\mathcal{A} + \mathcal{B}$ be the set of all vectors that can be expressed as a sum $\mathbf{a} + \mathbf{b}$ with $\mathbf{a}$ in $\mathcal{A}$ and $\mathbf{b}$ in $\mathcal{B}$. Show that if $\mathcal{A}$ and $\mathcal{B}$ are subspaces then so is $\mathcal{A} + \mathcal{B}$.

9. Show that if 0 is the only element in the intersection of two subspaces $\mathcal{S}$, $\mathcal{T}$, then

$$\dim(\mathcal{S} + \mathcal{T}) = \dim(\mathcal{S}) + \dim(\mathcal{T})$$

[*Hint.* Show that a basis for $\mathcal{S}$ together with a basis for $\mathcal{T}$ gives a basis for $\mathcal{S} + \mathcal{T}$.]

10. Show that for any two subspaces $\mathcal{S}$, $\mathcal{T}$,

$$\dim(\mathcal{S} + \mathcal{T}) = \dim(\mathcal{S}) + \dim(\mathcal{T}) - \dim(\mathcal{S} \cap \mathcal{T}).$$

[*Hint.* Start with a basis for $\mathcal{S} \cap \mathcal{T}$ and extend it (Theorem 4.6) to a basis for $\mathcal{S}$ and a basis for $\mathcal{T}$.]

11. Let $\mathcal{R}^3 \xrightarrow{f} \mathcal{R}^2$ be defined by the matrix

$$\begin{pmatrix} 1 & -3 & 2 \\ -2 & 6 & -4 \end{pmatrix}.$$

Find a basis for the null space of f, and one for the range of f. Verify that Theorem 4.7 holds.

12. Describe the solution set of each of the equations or systems of equations below as an affine subspace, that is, as a translate by a specified vector of a linear subspace with a specified basis.

(a) $x + y = 1.$

(b) $\begin{aligned} 2x + y \quad &= 1 \\ 2x - 3y + z &= 2. \end{aligned}$

(c) $\begin{aligned} x + 2y + 3z &= 10 \\ 4x + 5y + 6z &= 11 \\ 7x + 8y + 9z &= 12. \end{aligned}$ [*Ans.* $t(1, -2, 1) + (0, -9, \frac{28}{3})$.]

(d) $x - y + z = 2.$

13. For each of the following sets of three equations in three unknowns, determine which of the geometric possibilities discussed at the end of this section hold.

(a) $\begin{aligned} x + 2y - 3z &= 2 \\ 3x + 6y - 9z &= 0 \\ -2x - 4y + 6z &= 3. \end{aligned}$

(b) $\begin{aligned} x + y - z &= 1 \\ -x + y + z &= 3 \\ x - y + z &= 5. \end{aligned}$

(c) $\begin{aligned} x - y + z &= 0 \\ -x + 2y - 3z &= 2 \\ 2x - y \quad &= 6. \end{aligned}$

SECTION 5

LINEAR FUNCTIONS

In Chapter 1 we saw that matrices obey some of the same rules for addition and multiplication that ordinary numbers do. Furthermore, we have seen that given a linear function f from $\mathcal{R}^n$ to $\mathcal{R}^m$, there is an m-by-n matrix A such that $f(x) = A\mathbf{x}$ for all $\mathbf{x}$ in $\mathcal{R}^n$. Using these facts we could prove that linear functions from $\mathcal{R}^n$ to $\mathcal{R}^m$ obey algebraic rules just as their matrices do. However, it turns out that these same rules apply to a wider class of linear functions and not just those representable by matrices. For this reason we shall prove the rules in general form and then apply them in Section 6 to a systematic analysis of some linear differential equations.

We begin by describing the operations of addition, scalar multiplication, and composition of functions. Let f and g be functions with the same domain and having the same vector space as range space. Then the function $f + g$ is the **sum** of f and g defined by

$$(f + g)(\mathbf{x}) = f(\mathbf{x}) + g(\mathbf{x})$$

for all $\mathbf{x}$ in the domain of both f and g. Similarly, if r is a number, then rf is the numerical multiple of f by r and is defined by

$$rf(\mathbf{x}) = r(f(\mathbf{x})).$$

We have already defined the composition of functions in Chapter 1, but we repeat the definition here. We require now that the range of f be contained in the domain of g. Then $g \circ f$ is the **composition** of f and g and is defined by

$$g \circ f(\mathbf{x}) = g(f(\mathbf{x})).$$

It is an easy exercise to prove that sums, numerical multiples, and compositions of linear functions are linear.

Example 1. Suppose $\mathcal{R}^2 \xrightarrow{f} \mathcal{R}^2$ and $\mathcal{R}^2 \xrightarrow{g} \mathcal{R}^2$ are given by

$$f\begin{pmatrix} x \\ y \end{pmatrix} = \begin{pmatrix} x + y \\ x - y \end{pmatrix} = \begin{pmatrix} 1 & 1 \\ 1 & -1 \end{pmatrix}\begin{pmatrix} x \\ y \end{pmatrix}$$

and

$$g\begin{pmatrix} x \\ y \end{pmatrix} = \begin{pmatrix} 2x + y \\ x + 3y \end{pmatrix} = \begin{pmatrix} 2 & 1 \\ 1 & 3 \end{pmatrix}\begin{pmatrix} x \\ y \end{pmatrix}.$$

Then

$$\begin{aligned} (f + g)\begin{pmatrix} x \\ y \end{pmatrix} &= f\begin{pmatrix} x \\ y \end{pmatrix} + g\begin{pmatrix} x \\ y \end{pmatrix} \\ &= \begin{pmatrix} 3x + 2y \\ 2x + 2y \end{pmatrix} = \begin{pmatrix} 3 & 2 \\ 2 & 2 \end{pmatrix}\begin{pmatrix} x \\ y \end{pmatrix}. \end{aligned}$$

Also, $3f$ is given by

$$\begin{aligned} 3f\begin{pmatrix} x \\ y \end{pmatrix} &= 3\left(f\begin{pmatrix} x \\ y \end{pmatrix}\right) \\ &= \begin{pmatrix} 3x + 3y \\ 3x - 3y \end{pmatrix} = \begin{pmatrix} 3 & 3 \\ 3 & -3 \end{pmatrix}\begin{pmatrix} x \\ y \end{pmatrix}. \end{aligned}$$

Finally, $g \circ f$ is given, according to Theorem 4.4 of Chapter 1, by matrix multiplication as

$$\begin{aligned} g \circ f\begin{pmatrix} x \\ y \end{pmatrix} &= g\left(f\begin{pmatrix} x \\ y \end{pmatrix}\right) \\ &= \begin{pmatrix} 2 & 1 \\ 1 & 3 \end{pmatrix}\begin{pmatrix} 1 & 1 \\ 1 & -1 \end{pmatrix}\begin{pmatrix} x \\ y \end{pmatrix} \\ &= \begin{pmatrix} 3 & 1 \\ 4 & -2 \end{pmatrix}\begin{pmatrix} x \\ y \end{pmatrix}. \end{aligned}$$

Notice that in the above example the matrix of $f + g$ is equal to the matrix of f plus the matrix of g, and that the matrix of $3f$ is 3 times the matrix of f. Of course we have already proved in Chapter 1, Theorem 4.5, that the matrix of the composition $g \circ f$ is the product of the matrices of g and f in the same order. We can state the general result as follows.

5.1 Theorem

Let f and g be linear fu nctions from $\mathcal{R}^n$ to $\mathcal{R}^m$ with matrices A and B respectively. Then:

1. The matrix of $f + g$ is $A + B$.
2. For any number r, the matrix of rf is rA.
3. If $g \circ f$ is defined, its matrix is BA.

Proof. To prove statement 1 we write

$$\begin{aligned}(f + g)(\mathbf{x}) &= f(\mathbf{x}) + g(\mathbf{x}) \\ &= A\mathbf{x} + B\mathbf{x} = (A + B)\mathbf{x}.\end{aligned}$$

The first equality holds by the definition of $f + g$, the second by the relation of matrix to function, and the last by the general rule $AC + BC = (A + B)C$ for matrix multiplication and addition.

To prove statement 2 we write, similarly,

$$rf(\mathbf{x}) = r(A\mathbf{x}) = (rA)\mathbf{x},$$

where the last equality follows from the rule $B(AC) = (BA)C$ for matrix multiplication.

Statement 3 is just the result of Theorem 4.5, Chapter 1.

Example 2. Suppose f and g are linear functions such that

$$f\begin{pmatrix}1\\0\end{pmatrix} = \begin{pmatrix}2\\3\end{pmatrix}, \quad f\begin{pmatrix}0\\1\end{pmatrix} = \begin{pmatrix}-1\\0\end{pmatrix}$$

$$g\begin{pmatrix}1\\0\end{pmatrix} = \begin{pmatrix}1\\0\end{pmatrix}, \quad g\begin{pmatrix}0\\1\end{pmatrix} = \begin{pmatrix}0\\-1\end{pmatrix}.$$

Then the matrices of f and g are

$$\begin{pmatrix}2 & -1\\3 & 0\end{pmatrix} \quad \text{and} \quad \begin{pmatrix}1 & 0\\0 & -1\end{pmatrix}$$

respectively, by Theorem 4.2 of Chapter 1. By Theorem 5.1, the matrix of $f + g$ is the sum of the matrices of f and g:

$$\begin{pmatrix} 2 & -1 \\ 3 & 0 \end{pmatrix} + \begin{pmatrix} 1 & 0 \\ 0 & -1 \end{pmatrix} = \begin{pmatrix} 3 & -1 \\ 3 & -1 \end{pmatrix}$$

Thus $f + g = h$ is a function such that $h\begin{pmatrix} 1 \\ 0 \end{pmatrix} = \begin{pmatrix} 3 \\ 3 \end{pmatrix}$ and $h\begin{pmatrix} 0 \\ 1 \end{pmatrix} = \begin{pmatrix} -1 \\ -1 \end{pmatrix}$.

Similarly, the matrix of $f - 2g$ is

$$\begin{pmatrix} 2 & -1 \\ 3 & 0 \end{pmatrix} - 2\begin{pmatrix} 1 & 0 \\ 0 & -1 \end{pmatrix} = \begin{pmatrix} 0 & -1 \\ 3 & 2 \end{pmatrix}.$$

Finally, the matrix of $g \circ f$ is the product

$$\begin{pmatrix} 2 & -1 \\ 3 & 0 \end{pmatrix}\begin{pmatrix} 1 & 0 \\ 0 & -1 \end{pmatrix} = \begin{pmatrix} 2 & 1 \\ 3 & 0 \end{pmatrix}.$$

The close connection between combinations of matrices and combinations of functions from $\mathcal{R}^n$ to $\mathcal{R}^m$ suggests that the rules for matrix algebra should hold for operations with functions also. In fact, these rules hold for any linear functions for which the operations are defined.

5.2 Theorem

Let r be a number and let f, g, and h be linear functions for which both sides of the following equations are defined. Then

(1) $(f + g) \circ h = f \circ h + g \circ h$
(2) $f \circ (g + h) = f \circ g + f \circ h$
(3) $(rf) \circ g = r(f \circ g) = f \circ (rg)$
(4) $h \circ (g \circ f) = (h \circ g) \circ f.$

Proof. These formulas are all proved by applying the function on either side of the equality to an arbitrary domain vector $\mathbf{x}$, and then writing out the meaning of each side. We shall prove only Equation (4), leaving the others as exercises. To prove Equation (4) we observe that, by definition,

$$\begin{aligned} h \circ (g \circ f)(\mathbf{x}) &= h(g \circ f(\mathbf{x})) \\ &= h(g(f(\mathbf{x}))). \end{aligned}$$

Similarly,

$$\begin{aligned} (h \circ g) \circ f(\mathbf{x}) &= h \circ g(f(\mathbf{x})) \\ &= h(g(f(\mathbf{x}))). \end{aligned}$$

The results of the two computations are the same, so the formulas we started with must be the same. Since $\mathbf{x}$ is arbitrary, $h \circ (g \circ f) = (h \circ g) \circ f$.

Example 3. Let $C^{(\infty)}(\mathcal{R})$ be the vector space of infinitely often differentiable real-valued functions y of the real variable x. If we let D stand for differentiation with respect to x, then

$$Dy = y',$$

and D is a function with domain $C^{(\infty)}(\mathcal{R})$. The familiar rules about differentiation stating that $(f + g)' = f' + g'$ and $(cf)' = cf'$ imply that D is linear. Because the meaning will always be clear, we can omit the little circle in writing compositions, so that DD will stand for $D \circ D$.

The compositions $D^2 = DD$, $D^3 = DDD$, etc., are all meaningful, and we have for example

$$D^2y = y'', \qquad D^3y = y'''.$$

We can even define

$$D^0y = y$$

for occasional convenience. Combining powers of D by addition and numerical multiplication leads to formulas like

$$D^2 - D - 2, \qquad (D + 1)(D - 2),$$

and applying the rules (1) and (2) of Theorem 5.2 we get

$$(D + 1)(D - 2) = D^2 - D - 2.$$

If we apply either side of this last equation to a function y, we of course get the same thing:

$$y'' - y' - 2y.$$

Differential operators such as these will be studied in more detail in the following section.

We have seen that addition and multiplication of matrices are related to operations on linear functions. Similarly related to the idea of an inverse matrix is that of an inverse function. A function f has an **inverse function,** denoted f^{-1}, if

$$f^{-1} \circ f(\mathbf{x}) = \mathbf{x}$$

for every $\mathbf{x}$ in the domain of f. Applying f to both sides of this equation gives $f \circ f^{-1}(f(\mathbf{x})) = f(\mathbf{x})$; therefore, we also have

$$f \circ f^{-1}(\mathbf{y}) = \mathbf{y},$$

for every $\mathbf{y} = f(\mathbf{x})$ in the image of f. We leave as an exercise the proof that:

5.3 If f is linear, then so is f^{-1}.

Because we have derived the techniques of computing matrix inverses in several stages, we summarize here what we have proved so far. At the end of Chapter 1, we showed that

5.4 If A is an orthogonal matrix, then $A^{-1} = A^t$, where A^t is the transpose of A across its main diagonal.

In Theorem 8.3 of Chapter 1, it is proved that:

5.5 If $\det A \neq 0$, then A is invertible, and then $A^{-1} = (1/\det A)\,\tilde{A}^t$, where $\tilde{A}$ is the matrix of cofactors of elements of A.

Finally, the row reduction method of the first section of this chapter is an efficient way to find the inverse of an invertible matrix, that is, to find A^{-1} such that

$$A^{-1}A = AA^{-1} = I,$$

where I is the identity matrix.

The relationship between inverse linear function and inverse matrix is not quite so straightforward as in the case of the other operations on functions and matrices. The following example shows why.

Example 4. Let A be an n-by-n invertible matrix. Then the linear function f from $\mathcal{R}^n$ to $\mathcal{R}^n$, defined by

$$f(\mathbf{x}) = A\mathbf{x},$$

always has an inverse function f^{-1} given by

$$f^{-1}(\mathbf{x}) = A^{-1}\mathbf{x}.$$

Indeed,

$$f^{-1} \circ f(\mathbf{x}) = A^{-1}A\mathbf{x} = \mathbf{x},$$

for all $\mathbf{x}$ in $\mathcal{R}^n$. However, consider the linear function g from $\mathcal{R}$ to $\mathcal{R}^2$ defined by

$$g(x) = \begin{pmatrix} x \\ x \end{pmatrix}.$$

It is easy to check that g is linear and that g has an inverse function g^{-1} defined on the subspace $\mathfrak{L}$ of $\mathcal{R}^2$ consisting of vectors having both coordinates equal. We define

$$g^{-1}\begin{pmatrix} x \\ x \end{pmatrix} = x, \qquad \text{for} \qquad \begin{pmatrix} x \\ x \end{pmatrix} \quad \text{in } \mathfrak{L}.$$

But the matrix of g is

$$\begin{pmatrix} 1 \\ 1 \end{pmatrix},$$

which is not invertible because it isn't even a square matrix. The difficulty is that g^{-1} is not defined on all of $\mathcal{R}^2$, but only on the subspace $\mathfrak{L}$.

In the above example the crucial characteristic of the function g was the dimension of its range or image. The idea is important enough to have a name: the **rank** of a linear function f is the dimension of the range of f. Thus the linear function

$$f\begin{pmatrix} x \\ y \end{pmatrix} = \begin{pmatrix} 2 & 0 \\ 0 & 3 \end{pmatrix}\begin{pmatrix} x \\ y \end{pmatrix} = \begin{pmatrix} 2x \\ 3y \end{pmatrix}$$

has its rank equal to 2 because its range is all of $\mathcal{R}^2$. The function

$$g\begin{pmatrix} x \\ y \end{pmatrix} = \begin{pmatrix} 1 & 0 \\ 0 & 0 \end{pmatrix}\begin{pmatrix} x \\ y \end{pmatrix} = \begin{pmatrix} x \\ 0 \end{pmatrix}$$

has its rank equal to 1 because its range is just the x-axis in $\mathcal{R}^2$. For a linear function f from $\mathcal{R}^n$ to $\mathcal{R}^m$ with matrix A, we have the following useful criterion.

5.6 Theorem

Let f be a linear function with matrix A. Then the rank of f is equal to the number of linearly independent columns in A.

Proof. The columns of A span the range of f because

$$f(\mathbf{x}) = x_1 f(\mathbf{e}_1) + \ldots + x_n f(\mathbf{e}_n),$$

where $\mathbf{x} = (x_1, \ldots, x_n)$ and the column vectors $f(\mathbf{e}_1), \ldots, f(\mathbf{e}_n)$ are just the columns of A; hence, these columns span the range of f. The number of independent columns is then the dimension of the range of f.

Because of Theorem 5.6, we can speak of the **rank** of a matrix A and define it to be either the number of linearly independent columns or else the rank of the function $f(\mathbf{x}) = A\mathbf{x}$.

Example 5. To find the rank of the matrix

$$\begin{pmatrix} 1 & 0 & 7 \\ 0 & 1 & 5 \\ 2 & -1 & 9 \end{pmatrix},$$

we apply elementary row operations to reduce the matrix to echelon form. At each stage the new matrix is the matrix of a different linear function, but because the row operations are all invertible, the dimension of the range remains the same. We find that adding the second row to the third row gives

$$\begin{pmatrix} 1 & 0 & 7 \\ 0 & 1 & 5 \\ 2 & 0 & 14 \end{pmatrix}.$$

Then subtracting 2 times the first row from the third row gives

$$\begin{pmatrix} 1 & 0 & 7 \\ 0 & 1 & 5 \\ 0 & 0 & 0 \end{pmatrix}.$$

No further reduction is possible; so clearly the matrix has just two linearly independent columns. Thus the given matrix has rank 2, and the associated linear function from $\mathcal{R}^3$ to $\mathcal{R}^3$ has a 2-dimensional range.

Finally, we complete the discussion of the relationship between inverse function and inverse matrix in $\mathcal{R}^n$. If f is a function, then f is said to be **one-to-one** if each element of the range of f corresponds to exactly one element of the domain of f. Clearly, f is one-to-one if and only if f^{-1} exists. For *linear* functions we have the following simple fact.

5.7 Theorem

A linear function f is one-to-one if and only if $\mathbf{x} = 0$ whenever $f(\mathbf{x}) = 0$.

Proof. First assume that $f(\mathbf{x}) = 0$ implies $\mathbf{x} = 0$. If $f(\mathbf{x}_1) = f(\mathbf{x}_2)$, then, by the linearity of f, $f(\mathbf{x}_1 - \mathbf{x}_2) = 0$. But then by assumption $\mathbf{x}_1 - \mathbf{x}_2 = 0$, so $\mathbf{x}_1 = \mathbf{x}_2$. Thus f must be one-to-one. Conversely, suppose $f(\mathbf{x}_1) = f(\mathbf{x}_2)$ always implies $\mathbf{x}_1 = \mathbf{x}_2$. Then, because $f(0) = 0$ for a linear function, the equation $f(\mathbf{x}) = 0$ can be written $f(\mathbf{x}) = f(0)$. But then $\mathbf{x} = 0$, by assumption.

It follows from Theorem 5.7 that, if f is a one-to-one linear function from $\mathcal{R}^n$ to $\mathcal{R}^m$, then the rank of f equals n. The reason is that the null space of f has dimension 0 (that is, the dimension of the origin) and so, by Theorem 4.7, rank $(f) + 0 = n$. This idea is used in the proof of the following theorem. For linear functions from $\mathcal{R}^n$ to $\mathcal{R}^n$, that is, for functions representable by square matrices, we have:

5.8 Theorem

Let $\mathcal{R}^n \xrightarrow{f} \mathcal{R}^n$ be linear, with square matrix A. Then f has an inverse if and only if A is invertible.

Proof. We have seen at the beginning of Example 4 that if A^{-1} exists, then so does f^{-1}. Conversely, if f has an inverse f^{-1}, then f is one-to-one. By Theorem 5.7 the null space of f has dimension 0; so the range of f is all of $\mathcal{R}^n$. But this means that the domain of f^{-1} is all of $\mathcal{R}^n$. It follows that f^{-1} has a square matrix B, such that $f^{-1}(\mathbf{x}) = B\mathbf{x}$ for all $\mathbf{x}$ in $\mathcal{R}^n$. Then the equations

$$BA\mathbf{x} = f^{-1} \circ f(\mathbf{x}) = \mathbf{x}$$

$$AB\mathbf{x} = f \circ f^{-1}(\mathbf{x}) = \mathbf{x}$$

both hold. It follows that $AB = BA$; therefore A is invertible and $A^{-1} = B$.

EXERCISES

1. Suppose that f and g are linear functions such that

$$f\begin{pmatrix}1\\0\end{pmatrix} = \begin{pmatrix}1\\1\end{pmatrix}, \quad f\begin{pmatrix}0\\1\end{pmatrix} = \begin{pmatrix}1\\2\end{pmatrix},$$

$$g\begin{pmatrix}1\\0\end{pmatrix} = \begin{pmatrix}-1\\0\end{pmatrix}, \quad g\begin{pmatrix}0\\1\end{pmatrix} = \begin{pmatrix}-2\\3\end{pmatrix}.$$

Find the matrices of the following functions:

(a) f.
(b) g.
(c) $f + g$.
(d) $2f - g$.
(e) $g \circ f$.
(f) $f \circ (f + g)$.

2. If

$$A = \begin{pmatrix} 1 & 1 \\ 0 & 2 \\ 1 & 3 \end{pmatrix}, \quad B = \begin{pmatrix} -1 & 0 \\ 2 & 1 \\ 3 & 1 \end{pmatrix}, \quad C = \begin{pmatrix} 2 & 1 \\ 1 & 3 \end{pmatrix},$$

compute $(A + B)C$ and verify that it equals $AC + BC$.

3. (a) Prove rules (1), (2), and (3) of Theorem 5.2.
(b) Prove that linear combinations and compositions of linear functions are linear.

4. Find the inverse of each of the following matrices that has an inverse.

(a) $\begin{pmatrix} 1 & 3 \\ -1 & 2 \end{pmatrix}$.

(b) $\begin{pmatrix} \dfrac{1}{\sqrt{2}} & \dfrac{1}{\sqrt{2}} \\ \dfrac{-1}{\sqrt{2}} & \dfrac{1}{\sqrt{2}} \end{pmatrix}$.

(c) $\begin{pmatrix} 1 & 0 & 2 \\ 0 & 1 & 2 \\ -3 & 3 & 0 \end{pmatrix}$.

(d) $\begin{pmatrix} 2 & -1 & 3 \\ 1 & 2 & 1 \\ 2 & -2 & 1 \end{pmatrix}$.

5. Show that, if f is a linear function and has an inverse f^{-1}, then f^{-1} is also linear.

6. Let f be a linear function from $\mathcal{R}^n$ to $\mathcal{R}^m$.

(a) Show that if $n > m$, then the dimension of the null space of f is positive. Then show that f cannot have an inverse function.
(b) Suppose that $n = m$ and that f has an inverse function f^{-1}. Show that, if A is the matrix of f, then A is invertible and A^{-1} is the matrix of f^{-1}.
(c) Show that, if $n < m$, then even if f^{-1} exists, the matrix of f is not invertible.

7. (a) Show that $(D + a)(D + b) = (D + b)(D + a)$, where a and b are constants.

(b) Define $D + f(x)$ by

$$(D + f(x))y = y' + f(x)y.$$

Show that

$$(D + 1)(D + x) \neq (D + x)(D + 1)$$

by applying both sides to a twice-differentiable function $y(x)$.

8. Find the rank of f, where f has matrix

(a) $\begin{pmatrix} 0 & 1 \\ 1 & 0 \end{pmatrix}$ (b) $\begin{pmatrix} 1 & 2 \\ 2 & 4 \end{pmatrix}$

(c) $\begin{pmatrix} 0 & 0 & 1 \\ 0 & 1 & 0 \\ 1 & 0 & 0 \end{pmatrix}$ (d) $\begin{pmatrix} 2 & 1 & 3 \\ 2 & -2 & 2 \\ 4 & -1 & 5 \end{pmatrix}$.

9. Let $\mathcal{V}$ and $\mathcal{W}$ be finite dimensional vector spaces and let f be a linear function with domain $\mathcal{V}$ and with range equal to $\mathcal{W}$. (Thus $\dim(\mathcal{W}) = \text{rank}(f)$.)

(a) Show that $\mathcal{V}$ contains a largest subspace $\mathcal{S}$ such that f, when restricted to $\mathcal{S}$, becomes one-to-one from $\mathcal{S}$ to $\mathcal{W}$. [*Hint*. Let $\mathcal{N}$ be the null space of f with basis $\mathbf{n}_1, \ldots, \mathbf{n}_k$. Extend this basis to a basis for $\mathcal{V}$ by adding vectors $\mathbf{s}_1, \ldots, \mathbf{s}_l$.]

(b) Show that $\dim(\mathcal{S}) = \text{rank}(f)$, where $\mathcal{S}$ is the subspace of part (a).

10. If f is a function and $\mathbf{y}$ is an element of the image of f, then the subset S of the domain of f consisting of those $\mathbf{x}$ for which $f(\mathbf{x}) = \mathbf{y}$ is called the **inverse image** or **pre-image** of $\mathbf{y}$. Show that, if f is linear, then the inverse image of a fixed vector is either a subspace of the domain of f or else is a subspace translated by a fixed vector, that is, an affine subspace.

11. What is the simplest way to find the rank of a diagonal matrix $\text{diag}(a_1, \ldots, a_n)$?

12. Show that if A is an m-by-n matrix, then the rank of A is equal to $n - k$, where k is the dimension of the solution set of the equation $A\mathbf{x} = 0$.

SECTION 6

DIFFERENTIAL OPERATORS

In this section we look at some vector space ideas that arise in studying differential equations. The equations we shall treat will be like the following:

$$y' - 2y = 0 \tag{1}$$

$$y' - 3y = e^x \tag{2}$$

$$y'' + 3y' + y = \sin x, \tag{3}$$

where the primes denote differentiation with respect to x. Equations of this kind are important because they express a relation between the value of y, its first derivative (velocity), and its second derivative (acceleration). The problem posed by each equation is to find all functions $y(x)$ such that replacing y by $y(x)$ satisfies the equation identically. Before beginning a systematic treatment of the problem, we shall consider some examples whose results will be useful later.

Example 1. The differential equation $y' - ry = 0$, where r is a constant, can be written

$$y' = ry \tag{4}$$

and so specifies that the rate of change of y is proportional to the value of y for every value of the variable x. The growth of a population is sometimes assumed to obey such a law. To find solutions we use the fact that if $y = e^{rx}$, then $y' = re^{rx}$. It follows that Equation (4) is satisfied if we take $y = e^{rx}$. More generally, if c is an arbitrary constant, then Equation (4) is satisfied if we take

$$y = ce^{rx}, \tag{5}$$

because the c will cancel on both sides. In fact, Equation (5) gives the most general solution to (4), for observe that we can write (5) in the form

$$e^{-rx}y = c.$$

Differentiating with respect to x gives

$$(e^{-rx}y)' = 0$$

or, using the product rule for derivatives,

$$e^{-rx}y' - re^{-rx}y = e^{-rx}(y' - ry) = 0.$$

Dividing by e^{-rx} leaves $y' - ry = 0$, which is the given Equation (4) rewritten. But now we can reverse these steps, supposing that y is *some* solution. We start with

$$y' - ry = 0$$

and then multiply by e^{-rx} to get

$$e^{-rx}y' - re^{-rx}y = 0.$$

By the product rule, this last equation is

$$(e^{-rx}y)' = 0.$$

Integrating both sides with respect to x gives

$$e^{-rx}y = c,$$

where c is a constant of integration. Multiplying both sides by e^{rx} shows that y must be of the form

$$y = ce^{rx}.$$

Thus we have shown that ce^{rx} is the most general solution of $y' = ry$ in the sense that any particular solution can be obtained by specifying the value of c.

The method used in the above example consists of multiplying the expression $y' + ay$ by e^{ax} and then recognizing the result as the derivative $(e^{ax}y)' = e^{ax}y' + ae^{ax}y$. We shall use this **exponential multiplier** e^{ax} repeatedly in what follows.

Example 2. To solve the differential equation

$$y' - 3y = e^x,$$

we multiply by e^{-3x} and get

$$e^{-3x}y' - 3e^{-3x}y = e^{-2x}.$$

This is the same as

$$(e^{-3x}y)' = e^{-2x}.$$

Now we integrate both sides with respect to x, getting

$$e^{-3x}y = -\tfrac{1}{2}e^{-2x} + c,$$

where c is some constant of integration. Then multiplying by e^{3x} we obtain

$$y = -\tfrac{1}{2}e^x + ce^{3x}$$

for the most general solution. It is easy to verify directly, of course, that we have indeed found *some* solutions, one for each value of c. What we have shown additionally is that *any* solution must be of the form $-\frac{1}{2}e^x + ce^{3x}$.

Before considering more complicated examples, it will be useful to describe some notation that is often used in solving differential equations. We let D stand for differentiation with respect to some agreed-on variable, say x, and interpret $D + 2$, $D^2 - 1$, and similar expressions as linear functions acting on suitably differentiable functions y. For example:

$$\begin{aligned}(D + 2)y &= Dy + 2y \\ &= y' + 2y \\ (D^2 - 1)y &= D^2y - y \\ &= D(Dy) - y = y'' - y.\end{aligned}$$

An important observation is that D acts as a *linear function* on y; the term **linear operator** is sometimes used to avoid possible confusion over the fact that y itself is a function of x (though not necessarily a linear one). To see that D acts linearly, all we have to do is recall the familiar properties of differentiation:

$$D(y_1 + y_2) = Dy_1 + Dy_2$$
$$D(ky) = kDy, \qquad k \text{ constant.}$$

These two equations express the linearity of D. From the fact that compositions of linear functions are linear it follows that the operators D^2, D^3, and in general D^n are also linear. Because numerical multiplication is a linear operation and because the sum of linear operations is linear, the operator $(D + a)$ is linear for all constants a. Putting these facts together allows us to conclude that expressions such as

$$D^2 + a, \qquad D^2 + aD + b, \qquad (D + s)(D + t)$$

are all linear operators, with the respective interpretations

$$\begin{aligned}
(D^2 + a)y &= y'' + ay \\
(D^2 + aD + b)y &= y'' + ay' + by \\
(D + s)(D + t)y &= (D + s)(y' + ty) \\
&= D(y' + ty) + s(y' + ty) \\
&= y'' + ty' + sy' + sty \\
&= y'' + (t + s)y' + sty \\
&= \big(D^2 + (s + t)D + st\big)y.
\end{aligned}$$

The last example shows that, for constants s and t,

$$(D + s)(D + t) = D^2 + (s + t)D + st,$$

and also that

$$(D + t)(D + s) = D^2 + (s + t)D + st.$$

Thus operators of the form $D + a$ can be multiplied like polynomials in D if a is constant. It is sometimes also important to be able to factor an operator, for example $D^2 - 1$. We see immediately that for this example

$$\begin{aligned}
D^2 - 1 &= (D - 1)(D + 1) \\
&= (D + 1)(D - 1).
\end{aligned}$$

Returning to differential equations, suppose we are given one of the form

$$y'' + ay' + by = 0;$$

Equation (3) at the beginning of this section is similar to this, with $a = 3$, $b = 1$. Writing the equation using differential operators gives

$$(D^2 + aD + b)y = 0.$$

Our method of solution will be to try to factor the operator into factors of the form $(D + s)$ and $(D + t)$, and then apply the exponential multiplier method of Examples 1 and 2 repeatedly.

Example 3. Suppose we want to find all functions $y = y(x)$ that satisfy

$$y'' + 5y' + 6y = 0.$$

We write the equation in operator form as

$$(D^2 + 5D + 6)y = 0.$$

Next we try to factor the operator. We see that

$$(D^2 + 5D + 6) = (D + 3)(D + 2);$$

thus we need to solve

$$(D + 3)(D + 2)y = 0.$$

To find all solutions, we suppose that y is *some* solution. Letting

$$(D + 2)y = u$$

for the moment, we substitute u into the previous equation and arrive at

$$(D + 3)u = 0.$$

But this equation can be solved for u if we multiply through by e^{3x}. We get

$$e^{3x}Du + 3e^{3x}u = 0$$

or

$$D(e^{3x}u) = 0.$$

Therefore

$$e^{3x}u = c_1,$$

for some constant c_1, and so

$$u = c_1e^{-3x}.$$

Recall now that we have temporarily set $(D + 2)y = u$. We then have

$$(D + 2)y = c_1e^{-3x}.$$

Multiply this last equation by e^{2x} to get

$$e^{2x}Dy + 2e^{2x}y = c_1e^{-x}$$

or

$$D(e^{2x}y) = c_1e^{-x}.$$

Integrating with respect to x gives

$$e^{2x}y = -c_1e^{-x} + c_2$$

or

$$y = -c_1e^{-3x} + c_2e^{-2x}.$$

Since the constants c_1 and c_2 are arbitrary anyway, we can change the sign on the first one to get

$$y = c_1e^{-3x} + c_2e^{-2x}$$

for the form of the most general solution.

The exponential multiplier used in the previous examples is found as follows: $(D + a)y$ is multiplied by e^{ax} to produce $D(e^{ax}y)$, that is,

$$e^{ax}(D + a)y = D(e^{ax}y).$$

Repeated application of the rule leads to the following general fact.

6.1 Theorem

The differential equation

$$(D - r_1)(D - r_2) \ldots (D - r_n)y = 0,$$

with $r_1, r_2, \ldots, r_n$ *all different*, has its most general solution of the form

$$y = c_1e^{r_1x} + c_2e^{r_2x} + \ldots + c_ne^{r_nx}.$$

If some r's are equal, say $r_1 = r_2 = r_3 = \ldots = r_k$, we replace $e^{r_2x}, e^{r_3x}, \ldots, e^{r_kx}$ by $xe^{r_1x}, x^2e^{r_1x}, \ldots, x^{k-1}e^{r_1x}$, respectively, to get the general solution.

Proof. For simplicity we shall start with the case $n = 2$. Given

$$(D - r_1)(D - r_2)y = 0,$$

we set

$$(D - r_2)y = u.$$

Then the equation

$$(D - r_1)u = 0$$

is solved by multiplying through with e^{-r_1x}. We get

$$e^{-r_1x}Du - r_1e^{-r_1x}u = 0$$

or

$$D(e^{-r_1x}u) = 0.$$

Then integration with respect to x gives

$$e^{-r_1x}u = c_1$$

or

$$u = c_1e^{r_1x}.$$

Now

$$(D - r_2)y = c_1e^{r_1x}.$$

We multiply by e^{-r_2x} to get

$$e^{-r_2x}Dy - r_2e^{-r_2x}y = c_1e^{(r_1-r_2)x}.$$

Integrating both sides of

$$D(e^{-r_2x}y) = c_1e^{(r_1-r_2)x} \tag{6}$$

gives

$$e^{-r_2x}y = \frac{c_1}{r_1 - r_2}e^{(r_1-r_2)x} + c_2$$

or

$$y = \frac{c_1}{r_1 - r_2}e^{r_1x} + c_2e^{r_2x}.$$

We have assumed above that $r_1 \neq r_2$; so the last integration is correct as given. For neatness we replace the arbitrary constant c_1 by $(r_1 - r_2)c_1$, which is just as arbitrary.

In case $r_1 = r_2$, the integration just performed is not correct. We would have $r_1 - r_2 = 0$; so Equation (6) becomes

$$D(e^{-r_2x}y) = c_1.$$

Now integration gives

$$e^{-r_2x}y = c_1x + c_2$$

or

$$y = c_1xe^{r_2x} + c_2e^{r_2x},$$

as stated in the theorem.

More generally, to solve

$$(D - r_1)(D - r_2) \dots (D - r_n)y = 0,$$

we set

$$(D - r_2) \dots (D - r_n)y = u_1, \tag{7}$$

so that the previous equation becomes

$$(D - r_1)u_1 = 0.$$

Substitution of the general solution for u_1 into (7) gives a new equation which we split up by setting

$$(D - r_3) \dots (D - r_n)y = u_2.$$

Then Equation (7) becomes

$$(D - r_2)u_2 = u_1,$$

to be solved for u_2. Continuing in this way, we finally have to solve an equation

$$(D - r_n)y = u_{n-1}$$

for y. The solution of each equation reduces to the solution of a sum of equations of the form

$$(D - r)y = cx^k e^{sx}, \qquad s \neq r$$

or

$$(D - r)y = cx^k e^{rx}.$$

In the first case we multiply by e^{-rx}. Then integration by parts leads to a linear combination of powers of x times e^{sx}. In the second case, after multiplication by e^{-rx}, we only have to integrate x^k on the right. In either case we get solutions of the form stated in the theorem.

The key to the application of the theorem is the factorization of a linear differential operator into factors of the form $(D - r)$. Looked at as a polynomial in D, the expression

$$D^n + a_{n-1}D^{n-1} + \ldots + a_1 D + a_0 \tag{8}$$

is called the **characteristic polynomial** of the associated differential equation

$$y^{(n)} + a_{n-1}y^{(n-1)} + \ldots + a_1 y' + a_0 y = 0.$$

Notice that, to get the polynomial from the equation, we replace $y^{(k)}$ by D^k and that the term $a_0 y$ corresponds to $a_0 D^0 = a_0$. Finding a factorization for the polynomial depends on knowing its roots, for if the polynomial (8) has roots $r_1, \ldots, r_n$, then it has the factored form

$$(D - r_1)(D - r_2) \ldots (D - r_n).$$

Finding the roots of a polynomial exactly is impossible in general. However, when $n = 2$ (or 3 or 4 for that matter), there are well-known formulas for the roots in terms of the coefficients. For simplicity we have assumed that the leading coefficient of the characteristic polynomial is 1; otherwise we can divide through so that the leading coefficient becomes 1.

Example 4. Suppose we want to solve

$$y''' - 4y'' + 4y' = 0.$$

Writing the characteristic polynomial

$$D^3 - 4D^2 + 4D,$$

we observe that it can be written

$$D(D^2 - 4D + 4) = D(D - 2)^2$$

The roots are 0 and 2, where 2 is a repeated root. Thus the general solution to the equation is a linear combination of e^{0x}, e^{2x}, and xe^{2x}, and so

$$y = c_1 + c_2e^{2x} + c_3xe^{2x}$$

is the most general solution.

From here on the theory will be described in terms of vector spaces. We have referred to the linearity of the operators $(D - r)$ without being specific about the vector space on which they operate. There are in fact many possibilities, depending on the requirements of a particular problem. One choice is to consider $C^{(n)}$, the vector space of real-valued functions of x having continuous nth derivatives. If L is a linear differential operator of order n, that is, an operator containing differentiation of order at most n, then L acts linearly from $C^{(n)}$ to $C^{(0)}$, the space of continuous functions. Another possibility is to consider L acting on $C^{(\infty)}$, the vector space of functions having continuous derivatives of all orders. For our purposes the former choice will be more natural. Observe that the functions e^{rx} and x^ke^{rx} of Theorem 6.1 are nevertheless members of $C^{(\infty)}$, and so they are automatically also in $C^{(n)}$. Furthermore, the set $\mathcal{N}$ of solutions of

$$(D - r_1) \ldots (D - r_n)y = 0$$

is a vector subspace of $C^{(n)}$, because $\mathcal{N}$ is the null space of the linear operator

$$L = (D - r_1) \ldots (D - r_n)$$

acting on $C^{(n)}$. From Theorem 6.1 we can immediately conclude that $\mathcal{N}$ has dimension at most n, the order of L, because $\mathcal{N}$ is spanned by n distinct functions of the form e^{rx} or x^je^{rx}. (The possibility that r may be a complex number is not ruled out here but will be explained in the next section.) In fact we can prove the following theorem. The proof consists simply of showing that the basic solutions are linearly independent, but it requires some work.

6.2 Theorem

Let $\mathcal{N}$ be the subspace of $C^{(n)}$ consisting of all solutions of the linear differential equation

$$(D - r_1) \ldots (D - r_n)y = 0.$$

Then $\mathcal{N}$ has dimension exactly n.

Proof. We have already observed that $\mathcal{N}$ has dimension at most n because it is spanned by a certain set of n elements. We can show that the dimension is exactly n by reviewing the way in which the exponential multiplier method works. We solve a succession of differential equations

$$(D - r_k)y = u_{k-1}$$

where u_{k-1} has the general form

$$u_{k-1}(x) = c_1 x^{l_1} e^{r_1 x} + \ldots + c_{k-1} x^{l_{k-1}} e^{r_{k-1} x}.$$

We apply the factor $e^{-r_k x}$ to get as usual

$$D(e^{-r_k x} y) = e^{-r_k x} u_{k-1}. \tag{9}$$

We now proceed inductively to show that the set of functions of the form u_k has dimension k, assuming that the set of functions of the form u_{k-1} has dimension $k - 1$. If the set of possible functions u_{k-1} has dimension $k - 1$, then the same is true of the set of functions $e^{-r_k x} u_{k-1}$. (Why?) Thus we consider the linear operator D as acting from a domain of functions $e^{-r_k x} y$ to a range vector space of dimension $k - 1$. But the null space of D alone has dimension 1. The reason is that the solutions of

$$D(e^{-r_k x} y) = 0$$

are of the form

$$e^{-r_k x} y = c,$$

clearly a 1-dimensional subspace of the domain of D. Because

$$\dim(\text{domain}) = \dim(\text{null space}) + \dim(\text{range}),$$

we have $\dim(\text{domain}) = 1 + (k - 1) = k$. Thus the set of solutions $e^{-r_k x} y$ of Equation (9) has dimension k. It follows that the corresponding set of functions

$$y = e^{r_k x}(e^{-r_k x} y)$$

also has dimension k. Since we can prove this for $k = 1, 2, \ldots, n$, we have shown that the set $\mathcal{N}$ of all solutions to the given differential equation has dimension n.

We conclude the section by considering differential equations of the form

$$L(y) = f,$$

where f is a given continuous function of x, and L has the form

$$L = (D - r_1)(D - r_2) \ldots (D - r_n).$$

We have already treated the case in which f is identically zero.

Example 5. Given

$$y'' + 2y' + y = e^{3x},$$

we write the characteristic polynomial $D^2 + 2D + 1$ and factor it, putting the equation in the form

$$(D + 1)^2 y = e^{3x}.$$

Letting $(D + 1)y = u$, we try to solve

$$(D + 1)u = e^{3x}.$$

Multiplication by e^x gives

$$e^x Du + e^x u = e^{4x}$$

or

$$D(e^x u) = e^{4x}.$$

Then, integration gives

$$e^x u = \tfrac{1}{4}e^{4x} + c_1$$

or

$$u = \tfrac{1}{4}e^{3x} + c_1 e^{-x}.$$

Since $(D + 1)y = u$, we have

$$(D + 1)y = \tfrac{1}{4}e^{3x} + c_1 e^{-x}.$$

Again multiplying by e^x, we get

$$e^x Dy + e^x y = \tfrac{1}{4}e^{4x} + c_1$$

or

$$D(e^x y) = \tfrac{1}{4}e^{4x} + c_1.$$

Then

$$e^x y = \tfrac{1}{16}e^{4x} + c_1 x + c_2$$

or

$$y = \tfrac{1}{16}e^{3x} + c_1 x e^{-x} + c_2 e^{-x}.$$

In the above example the solution breaks naturally into a sum of two parts, y_h and y_p:

$$y_h = c_1 x e^{-x} + c_2 e^{-x},$$
$$y_p = \tfrac{1}{16}e^{3x}.$$

The function y_h is called the **homogeneous part** of the solution because it is a solution of the so-called **homogeneous equation**

$$L(y) = 0$$

associated with $L(y) = f$. The function y_p is called a **particular solution** of

$$L(y) = f$$

because it is just that: a particular solution, though not the most general one. In fact, we get y_p by setting $c_1 = c_2 = 0$ in the general solution. This breakup of the solution into two parts is an example of a general fact about linear functions, a fact that is used in solving systems of linear algebraic equations. The principle is important enough, and at the same time simple enough, that we state it here also.

6.3 Theorem

Let L be a linear function. Let f be an element of the range of L, and let y_p be any element of the domain of L such that $L(y_p) = f$. Then every solution y of

$$L(y) = f$$

can be written as a sum

$$y = y_h + y_p,$$

where y_h is an element of the null space $\mathcal{N}$ of L.

Proof. Suppose that $L(y) = f$ and that also $L(y_p) = f$. Then, since L is linear,

$$\begin{aligned} L(y - y_p) &= L(y) - L(y_p) \\ &= f - f = 0. \end{aligned}$$

It follows that $y - y_p$ is in the null space $\mathcal{N}$ of L, that is, $y - y_p = y_h$ for some y_h in $\mathcal{N}$. But then $y = y_h + y_p$, as was to be shown.

The method of Example 5 can always be used to find the most general solution to the equation $L(y) = f$ of the form

$$(D - r_1) \ldots (D - r_n)y = f.$$

However, in some examples the computations can be shortened by means of Theorem 6.1, which shows that y_h, the homogeneous part of the solution, can be written down as soon as we know the roots of the characteristic polynomial. Theorem 6.3 then says that if we find the general homogeneous part of the solution y_h (using Theorem 6.1) and can somehow find a particular solution y_p, then the general solution of the given equation is $y_h + y_p$. In finding y_p it may be convenient to take advantage of the linearity of L in case the right-hand side f is a sum of two or more terms. If we want to solve

$$L(y) = a_1 f_1 + a_2 f_2 \tag{10}$$

and we can find solutions y_1 and y_2 such that

$$L(y_1) = f_1, \qquad L(y_2) = f_2,$$

then, because L is linear, the function

$$y = a_1 y_1 + a_2 y_2$$

is a solution of Equation (10). In this context, the property of linearity is sometimes called the **superposition principle** because the desired solution is found by superposition (i.e., addition) of solutions of more than one equation.

Example 6. In Example 5 we found that the differential equation

$$(D + 1)^2y = e^{3x}$$

had the general solution

$$\tfrac{1}{16}e^{3x} + c_1xe^{-x} + c_2e^{-x}.$$

When $c_1 = c_2 = 0$ we get the particular solution $y_1 = \tfrac{1}{16}e^{3x}$. If we now wanted to solve

$$(D + 1)^2y = e^{3x} + 1, \tag{11}$$

we would not have to start all over again, but would only have to find a particular solution for

$$(D + 1)^2y = 1.$$

This could be solved, of course, by using exponential multipliers, but in this case the differential equation is so simple that we can guess a solution, namely, $y_2 = 1$. Then a particular solution of Equation (11) is $y_p = \tfrac{1}{16}e^{3x} + 1$ and the general solution is

$$y = \tfrac{1}{16}e^{3x} + 1 + c_1xe^{-x} + c_2e^{-x}.$$

On the other hand, solving

$$(D + 1)^2y = e^{3x} + e^{-x}$$

requires us to find a particular solution to

$$(D + 1)^2y = e^{-x}.$$

To do this we would return to the exponential multiplier method.

Other methods of solution, using "undetermined coefficients" and "variation of parameters," are given in Exercises 10 and 11 which follow. These methods are sometimes more efficient for finding particular solutions.

EXERCISES

1. For each of the following differential equations, find an appropriate exponential multiplier and then solve by integrating both sides of the modified equation

(a) $y' + 2y = 1$.

(b) $2\dfrac{dy}{dx} + y = x$.

(c) $y' - y = e^x$.

(d) $y' = \sin x$.

2. Show that if the expression $y' + p(x)y$ is multiplied by

$$e^{\int p(x)\,dx},$$

the resulting product can be written in the form

$$\frac{d}{dx}\left(e^{\int p(x)\,dx}\,y\right).$$

[We assume that $p(x)$ is a continuous function of x, and that $\int p(x)\,dx$ is some function with derivative $p(x)$.]

3. Use the result of Problem 2 to find an exponential multiplier for each of the following differential equations. Then solve the equation.

(a) $y' + \left(\frac{1}{x}\right)y = 1.$

(b) $\frac{dy}{dx} + xy = x.$

(c) $y' - xy = 0.$

(d) $y' + y = 0.$

4. Write each of the following differential equations in operator form, e.g., $(D^2 + 2D + 1)y = 0$, and then factor the operator into factors of order 1. Then find the general solution of the equation.

(a) $y'' + 2y' + y = 0.$

(b) $2y'' - y = 0.$

(c) $y''' + 3y'' + y' = 0.$

(d) $D(D + 3)y = 0.$

(e) $(D^2 - 1)y = 1.$

(f) $y'' - y = e^x.$

(g) $y'' - y = e^x + 1.$

5. Sketch the graph of each function of x given below. Then find a differential equation of which each one is a solution, the equation being of the form $y'' + ay' + by = 0.$

(a) $xe^{-x}.$

(b) $e^x + e^{-x}.$

(c) $1 + x.$

(d) $2e^{2x} - 3e^{3x}.$

6. Define the differential operator $D + f(x)$ to act on a function y by $(D + f(x))y = y' + f(x)y.$

(a) Show that in general $(D + f(x))(D + g(x)) \neq (D + g(x))(D + f(x)).$

(b) Show that if f and g are constant, then equality holds in part (a).

7. For $x \neq 0$, the differential equation

$$x^2y'' + (x^3 + x)y' + (x^2 - 1)y = 0$$

can be written in operator form as

$$\left(D^2 + \left(\frac{x^2 + 1}{x}\right)D + \left(\frac{x^2 - 1}{x^2}\right)\right)y = 0.$$

(a) Show that the above equation can also be written as

$$(D + x)\left(D + \frac{1}{x}\right)y = 0.$$

(b) Solve the equation in part (a) by successive application of exponential multipliers of the form given in Exercise 2.

(c) Show that

$$(D + x)\left(D + \frac{1}{x}\right) \neq \left(D + \frac{1}{x}\right)(D + x).$$

(d) Solve the differential equation

$$\left(D + \frac{1}{x}\right)(D + x)y = 0.$$

8. (a) Find the general solution of the differential equation $y'' - y = 0$.
(b) Determine the constants in the general solution $y(x)$ in part (a) so that $y(0) = 0$ and $y'(0) = 1$. Sketch the graph of the resulting particular solution.
(c) Determine the constants in the general solution $y(x)$ in part (a) so that $y(0) = 1$ and $y(1) = 0$. Sketch the graph of the resulting particular solution.

9. (a) Show that the characteristic equation of the differential equation $y'' + y = 0$ has complex roots $r_1 = i$ and $r_2 = -i$.
(b) Using the definition

$$e^{\pm ix} = \cos x \pm i \sin x,$$

show that the formal solution

$$y(x) = c_1 e^{ix} + c_2 e^{-ix}$$

to the differential equation $y'' + y = 0$ can be put in the form

$$y(x) = d_1 \cos x + d_2 \sin x.$$

(c) Verify directly that $\cos x$ and $\sin x$ are solutions of $y'' + y = 0$.

10. (Method of undetermined coefficients.) Given a *nonhomogeneous* differential equation

$$(D^2 + aD + b)y = f(x), \tag{1}$$

suppose that the function $f(x)$ can itself be recognized as a particular solution of an equation of a similar but *homogeneous* form, say,

$$(D - r_1) \ldots (D - r_n)f = 0.$$

By applying the operators $(D - r_j)$ to both sides of the given Equation (1), it follows that y must also be a solution of the higher-order homogeneous equation

$$(D - r_1) \ldots (D - r_n)(D^2 + aD + b)y = 0. \tag{2}$$

Since we have a routine for writing down the most general solution y_g of this last equation, we can find a particular solution y_p of Equation (1) by substituting the general solution of (2) into it and seeing what conditions result for the arbitrary constants, or "undetermined coefficients," in y_g. To save duplication, we can first eliminate from y_g all terms that already occur in the homogeneous part, y_h, of the solution of (1). Linear independence of the basic solutions of the homogeneous equations guarantees that when y_g is substituted into (1), then coefficients of like terms must be equal on either side. It is these equalities that are used to determine the coefficients for y_p. The general solution of (1) is then $y = y_h + y_p$. Find the general solutions of the following differential equations.

(a) $y'' - y = e^{2x}$.

(b) $y'' - y = e^{x}$.

(c) $y'' + 2y' + y = e^{x}$.

(d) $y'' - y = x$.

(e) $y'' - y = e^{x} + x$.

11. (Variation of parameters.) This method is useful for finding particular solutions of linear differential equations with nonconstant coefficients of the form

$$y'' + ay' + by = f(x). \tag{3}$$

Suppose that we can find the homogeneous solution

$$y_h = c_1u_1(x) + c_2u_2(x).$$

The constants c_1 and c_2 can now be thought of as auxiliary variables or "parameters" in which we allow "variation" as functions of x. We then try to determine $c_1(x)$ and $c_2(x)$ so that

$$y(x) = c_1(x)u_1(x) + c_2(x)u_2(x) \tag{4}$$

will be a solution of the nonhomogeneous Equation (3). We compute by the product rule

$$y' = c_1u_1' + c_2u_2' + c_1'u_1 + c_2'u_2$$

and then require for simplicity that

$$c_1'u_1 + c_2'u_2 = 0. \tag{5}$$

Thus

$$y' = c_1u' + c_2u_2'. \tag{6}$$

Next compute

$$y'' = c_1u_1'' + c_2u_2'' + c_1'u_1' + c_2'u_2'. \tag{7}$$

Substitution of the expressions for y, y', and y'' into Equation (3), under the assumption that u_1 and u_2 already satisfy the homogeneous equation

$$y'' + ay' + by = 0,$$

yields (see Problem 12)

$$c_1'u_1' + c_2'u_2' = f. \tag{8}$$

Solving Equation (5) and (8) together for $c_1'(x)$ and $c_2'(x)$ gives

$$c_1' = \frac{-u_2f}{u_1u_2' - u_2u_1'}, \qquad c_2' = \frac{u_1f}{u_1u_2' - u_2u_1'}. \tag{9}$$

We then find $c_1(x)$ and $c_2(x)$ by integration, upon which (4) becomes a particular solution y_p to be added to y_h for the general solution of (3).

Find the general solutions of the following nonhomogeneous equations.

(a) $y'' + y' - 2y = \log x$.

(b) $y'' - y = e^{2x}$.

12. Assume that u_1 and u_2 are solutions of

$$y'' + ay' + by = 0.$$

(a) Verify that substitution of Equations (4), (6), and (7) of Exercise 11 into

$$y'' + ay' + by = f$$

produces Equation (8).

(b) Verify that Equations (5) and (8) of Exercise 11 have the solution given by Equation (9).

13. A pellet fired horizontally through a viscous medium has a displacement from its initial position of the form $x = x(t)$, where t is time. Denoting derivatives with respect to time by dots, the velocity of the pellet at time t is $\dot{x}(t)$ and its acceleration is $\ddot{x}(t)$. From physical principles it can be shown that, theoretically, x satisfies the differential equation

$$m\ddot{x} + k\dot{x} = 0,$$

where m is the mass of the pellet and k is a positive constant depending on the viscosity of the medium.

(a) Show that if the initial displacement is $x(0) = 0$ and the initial velocity is $x(0) = v_0$, then, for $t \geq 0$,

$$x(t) = \frac{v_0 m}{k}(1 - e^{-kt/m}).$$

(b) Show that the displacement $x(t)$ has an upper bound equal to $v_0 m/k$. What is the effect of increasing the viscosity constant or of increasing the mass of the pellet?

(c) Show that the velocity of the pellet decreases to zero as t increases.

(d) Show that the acceleration of the pellet is negative for $t \geq 0$. What is the effect of an increase in k or an increase in m on the acceleration?

(e) Sketch the graph of $x(t)$ if $v_0 m/k = 1$.

14. Let $D = d/dt$. The first-order system of differential equations

$$(aD + \alpha)y + (bD + \beta)z = f(t)$$

$$(cD + \gamma)y + (dD + \delta)z = g(t)$$

contains the purely algebraic system

$$\alpha y + \beta z = f(t)$$

$$\gamma y + \delta z = g(t)$$

as a special case when a, b, c, and d are all zero. The method of elimination can be used to solve the differential system as well as the algebraic. To find $y(t)$ and $z(t)$, operate on both sides of the first equation by $(dD + \delta)$, and on the second equation by $(bD + \beta)$. Then subtract one equation from the other. The resulting equation can be solved for $y(t)$ since it does not contain z. Next substitute the general solution $y(t)$ into *one* of the equations and solve that for $z(t)$; to determine possible relations between the constants of integration, it *may* be necessary to substitute $y(t)$ and $z(t)$ into the *other* given equation. Sometimes simplifications in this procedure can be made.

(a) Use the method just described to find the general solution of the system

$$(D + 1)y + z = 0.$$
$$3y + (D - 1)z = 0.$$

(b) Determine the constants in the solution of part (a) so that the initial conditions $y(0) = 0$ and $z(0) = 1$ are satisfied.

(c) Find the most general solution of the system

$$(D + 1)y + Dz = 0$$
$$Dy - (D - 1)z = t.$$

(d) Determine the constants in the solution of part (c) so that the initial conditions $y(0) = 0$ and $z(0) = 0$ are satisfied.

15. Suppose that two 100-gallon tanks of salt solution have concentrations (in pounds per gallon) of salt $y(t)$ and $z(t)$, respectively. Suppose that solution is flowing from the y-tank to the z-tank at a rate of 1 gallon per minute, and from the z-tank to the y-tank at a rate of 4 gallons per minute, and that the overflow from the y-tank goes down the drain, while the z-tank is kept full by the addition of fresh water. We assume that each tank is kept thoroughly mixed at all times.

(a) Show that y and z satisfy a system of differential equations of the type discussed in Exercise 14. [*Hint*. Express Dy and Dz each as a linear combination of y and z.]

(b) Find the general solution of the system found in part (a), and then determine the constants in it to be consistent with initial concentrations $y(0) = \frac{1}{10}$ and $z(0) = \frac{1}{20}$.

(c) Draw the graphs of the particular solutions $y(t)$ and $z(t)$ found in part (*b*) and interpret the results physically.

SECTION 7

COMPLEX VECTOR SPACES

In the earlier parts of this chapter we have always understood the vector space operation of numerical multiplication to mean multiplication by a real number. However, we can replace real numbers by complex numbers, and let the definition of vector space and linear function remain otherwise the same. Then, all the theorems we have proved for real vector spaces are still true relative to the complex numbers. To prove this, all we have to do is observe that the only properties of real numbers that we used in proving theorems about an abstract vector space are properties that are shared by complex numbers. Theorems involving inner products are another matter, which we shall discuss at the end of the section. As motivation for considering complex vector spaces, we shall explain how the extension is the key to further development of the study of differential operators begun in Section 6. First we shall review the relevant facts about complex numbers and complex-valued functions.

The complex number $z = x + iy$, with **real part** $\text{Re}\, z = x$ and **imaginary part** $\text{Im}\, z = y$, can be identified with the vector (x, y) in $\mathcal{R}^2$ for the purpose of drawing pictures. Furthermore, **complex addition** is defined so that it corresponds precisely to addition of elements of $\mathcal{R}^2$:

$$(x + iy) + (x' + iy') = (x + x') + i(y + y').$$

The relation is illustrated in Fig. 7. It is also true that **numerical**

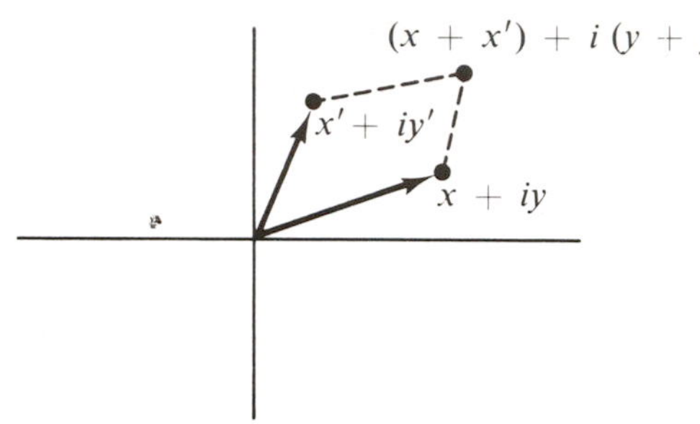

Figure 7

multiplication of a complex number by a *real* number r corresponds to numerical multiplication in $\mathcal{R}^2$:

$$r(x + iy) = rx + iry.$$

To complete the identification of the complex numbers with the vectors of $\mathcal{R}^2$ it is necessary to introduce an operation of vector multiplication in $\mathcal{R}^2$ to correspond to the multiplication of complex numbers. **Complex multiplication** is done by defining $i^2 = -1$ and computing as follows:

$$(x + iy)(x' + iy') = (xx' - yy') + i(xy' + yx').$$

The **complex conjugate** of $x + iy$, written $\overline{x + iy}$, is

$$\overline{x + iy} = x - iy.$$

Taking the conjugate $\bar{z}$ of a complex number z corresponds to reflecting it in the horizontal axis. (See Fig. 8.) Division by a nonzero complex number $x + iy$ is most easily done by multiplying both numerator and denominator by the conjugate of the denominator:

$$\frac{x' + iy'}{x + iy} = \frac{(x' + iy')(x - iy)}{(x + iy)(x - iy)} = \frac{(xx' + yy') + i(xy' - yx')}{x^2 + y^2}.$$

Finally, the **absolute value** of $x + iy$, written $|x + iy|$, is defined to be

$$|x + iy| = \sqrt{x^2 + y^2}$$

and corresponds to the length of a vector in $\mathcal{R}^2$. (See Fig. 8.) Notice that absolute value and conjugate of a complex number z

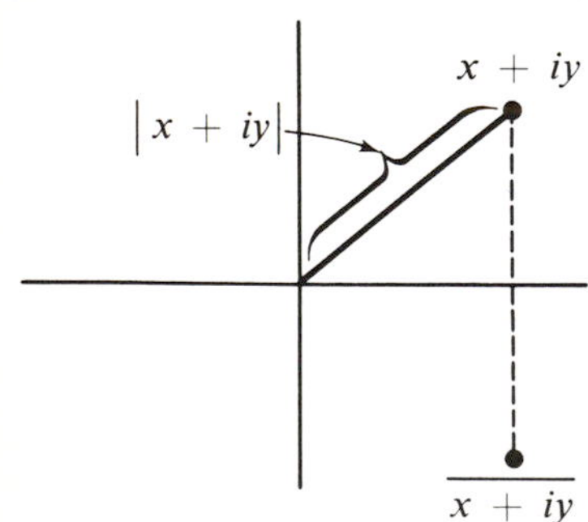

Figure 8

are related by

$$|z|^2 = z\bar{z}.$$

The geometric significance of complex multiplication is seen best by representing nonzero complex numbers in **polar form:**

$$x + iy = |x + iy|\left(\frac{x}{|x + iy|} + i\,\frac{y}{|x + iy|}\right).$$

Because $|x + iy| = \sqrt{x^2 + y^2}$, the numbers $x/\sqrt{x^2 + y^2}$ and $y/\sqrt{x^2 + y^2}$ can be written as cosine and sine, respectively, of an angle θ, called a **polar angle** of $z = x + iy$, and illustrated in Fig. 9. Of course, a complex

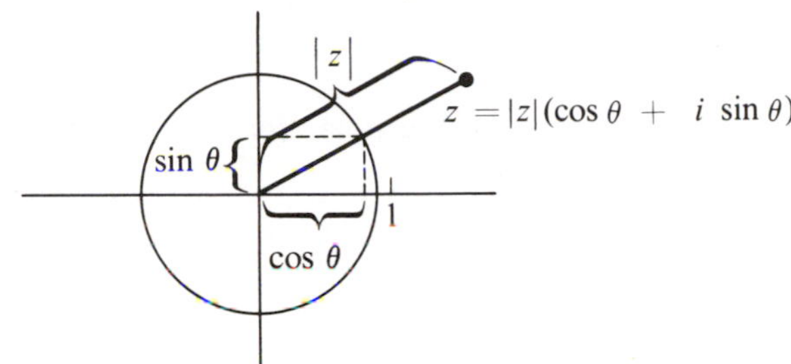

Figure 9

number has infinitely many polar angles, each pair differing by an integer multiple of 2π.

Now if z and z' are complex numbers with polar angles θ and θ', we can write their product in polar form as follows:

$$\begin{aligned}&\big(|z|\,(\cos\theta + i\sin\theta)\big)\big(|z|'\,(\cos\theta' + i\sin\theta')\big)\\ &\quad = |z|\,|z'|\,\big((\cos\theta\cos\theta' - \sin\theta\sin\theta') + i\,(\cos\theta\sin\theta' + \sin\theta\cos\theta')\big)\\ &\quad = |z|\,|z'|\,\big(\cos(\theta + \theta') + i\sin(\theta + \theta')\big).\end{aligned}$$

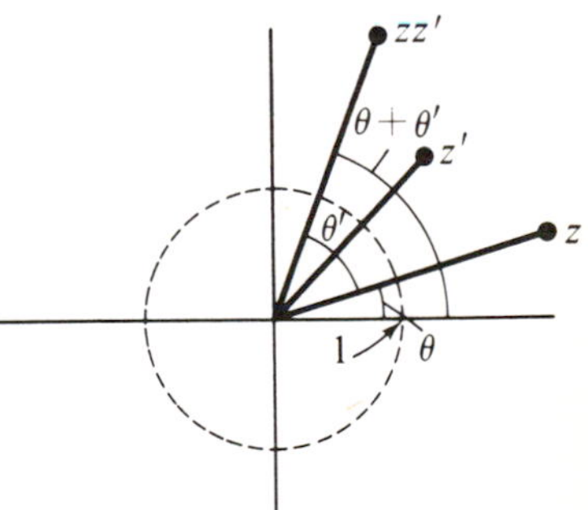

Figure 10

In the last step we have used the addition formulas for cosine and sine. The result of the computation is a number in polar form having absolute value $|z|\,|z'|$ and polar angle $\theta + \theta'$. We conclude that the absolute value of a product of complex numbers z and z' is the product of their absolute values

$$|zz'| = |z|\,|z'|,$$

and that if z and z' have polar angles $\theta(z)$ and $\theta(z')$, then zz' has a polar angle $\theta(z) + \theta(z')$. These facts are illustrated in Fig. 10.

What we have just established can be expressed most conveniently in terms of the **complex exponential** function. We define, for real numbers θ,

$$e^{i\theta} = \cos\theta + i\sin\theta.$$

Thus $e^{i\theta}$ is a complex number with $|e^{i\theta}| = \sqrt{\sin^2\theta + \cos^2\theta} = 1$, and with polar angle θ. Since polar angles are added when complex numbers are multiplied, we have

$$e^{i\theta}e^{i\theta'} = e^{i(\theta+\theta')}.$$

In particular, when $\theta' = -\theta$, we get $e^{i\theta}e^{-i\theta} = 1$; therefore

$$\frac{1}{e^{i\theta}} = e^{-i\theta} = \overline{e^{i\theta}}.$$

These last equations are justifications for using the exponential notation; the function behaves like the real exponential, for which $e^{\theta}e^{\theta'} = e^{\theta+\theta'}$ and $1/e^{\theta} = e^{-\theta}$.

In terms of the exponential, the polar form of a complex number z with polar angle $\theta(z)$ becomes

$$z = |z|e^{i\theta(z)},$$

and its conjugate is

$$\bar{z} = |z|e^{-i\theta(z)}.$$

The fact that the conjugate of zw is the product of the conjugates of z and w can be established directly, or by writing

$$\begin{aligned}\bar{z}\bar{w} &= |z|e^{-i\theta(z)}|w|e^{-i\theta(w)}\\ &= |zw|\, e^{-i(\theta(z)+\theta(w))} = \overline{zw}.\end{aligned}$$

In addition to its algebraic simplicity, another reason for using the complex exponential notation comes from the formulas for its derivative and integral. To differentiate or integrate a complex-valued function $u(x) + iv(x)$ with respect to the real variable x, we simply differentiate or integrate the real and imaginary parts. By definition,

$$\frac{d}{dx}(u(x) + iv(x)) = \frac{du}{dx}(x) + i\frac{dv}{dx}(x),$$

and

$$\int (u(x) + iv(x))\,dx = \int u(x)\,dx + i\int v(x)\,dx.$$

Then the derivative of e^{ix} with respect to x is given by

$$\begin{aligned}\frac{d}{dx}e^{ix} &= \frac{d}{dx}(\cos x + i\sin x)\\ &= -\sin x + i\cos x\\ &= i(\cos x + i\sin x) = ie^{ix}.\end{aligned}$$

In short, we have

$$\frac{d}{dx} e^{ix} = ie^{ix}.$$

Similarly

$$\int e^{ix}\, dx = \frac{1}{i} e^{ix} + c,$$

where c may be a real or complex constant. These are analogous to the formulas for the derivative and integral of e^{ax} when a is real. More generally, we can define

$$e^{(\alpha+i\beta)x} = e^{\alpha x}e^{i\beta x}$$

and compute

7.1
$$\frac{d}{dx} e^{(\alpha+i\beta)x} = (\alpha + i\beta)e^{(\alpha+i\beta)x}$$

and

7.2
$$\int e^{(\alpha+i\beta)x}\, dx = \frac{1}{\alpha + i\beta} e^{(\alpha+i\beta)x} + c, \qquad \alpha + i\beta \neq 0.$$

These computations are left as exercises.

We are now in a position to discuss the differential equation

$$(D^2 + aD + b)y = 0$$

when the factored operator

$$(D - r_1)(D - r_2) = D^2 - (r_1 + r_2)D + r_1r_2$$

contains complex numbers r_1 and r_2. We shall see that the usual techniques, as discussed for example in Section 6, still apply. In fact, the exponential multiplier method goes over formally unchanged because of Equation 7.1; we have

$$D(e^{rx}y) = e^{rx}(D + r)y,$$

whether r is real or complex.

Example 1. Consider the differential equation $y'' + y = 0$. We write the equation in operator form,

$$(D^2 + 1)y = 0,$$

and factor $D^2 + 1$ to get

$$(D - i)(D + i)y = 0.$$

Then set

$$(D + i)y = u, \tag{1}$$

and try to solve

$$(D - i)u = 0$$

for u. As in the real case, we multiply by a factor designed to make the left side the derivative of a product. The same multiplier rule suggests that the correct factor is e^{-ix}, so we write

$$e^{-ix}(D - i)u = 0$$

or, since $D(e^{-ix}u) = e^{-ix}(D - i)u$,

$$D(e^{-ix}u) = 0.$$

We integrate both sides with respect to x to get

$$e^{-ix}u = c_1 \quad \text{or} \quad u = c_1 e^{ix}.$$

Substituting this result for u into Equation (1) gives

$$(D + i)y = c_1 e^{ix},$$

which must now be solved for y. We do it by multiplying through by e^{ix} to get

$$e^{ix}(D + i)y = c_1 e^{2ix}$$

or, since $D(e^{ix}y) = e^{ix}(D + i)y$,

$$D(e^{ix}y) = c_1 e^{2ix}.$$

Integrating gives

$$e^{ix}y = \frac{1}{2i} c_1 e^{2ix} + c_2$$

or

$$y = \frac{1}{2i} c_1 e^{ix} + c_2 e^{-ix}.$$

On replacing the arbitrary constant c_1 by $2ic_1$, we have

$$\begin{aligned} y &= c_1 e^{ix} + c_2 e^{-ix} \\ &= c_1(\cos x + i \sin x) + c_2(\cos x - i \sin x) \\ &= (c_1 + c_2) \cos x + i(c_1 - c_2) \sin x. \end{aligned}$$

To make the solution simpler looking, we can set $d_1 = (c_1 + c_2)$ and $d_2 = i(c_1 - c_2)$. This involves no change in generality in the constants because, whenever d_1 and d_2 are to be real numbers, then c_1 and c_2 in general must be complex. In fact we have, solving for c_1 and c_2,

$$c_1 = \frac{d_1 + id_2}{2} \qquad \text{and} \qquad c_2 = \frac{d_1 - id_2}{2}.$$

Example 1 shows that it is important to be able to form linear combinations of elements of a vector space with complex coefficients. A **complex vector space** has the same definition as a real vector space except that numerical multiples are formed using the complex numbers $\mathcal{C}$. The definitions of linear independence and linear function are also formally the same, with complex numbers replacing real numbers. It is worth remarking that every complex vector space can be converted in a natural way into a real vector space; to obtain it we simply restrict ourselves to numerical multiplication by real numbers. As a consequence, linear independence of a set in a complex vector space automatically implies linear independence of the same set relative to the restricted real vector space. The reason is that, if

$$c_1\mathbf{x}_1 + \ldots + c_n\mathbf{x}_n = 0$$

implies that all the c's are zero whenever they are chosen from the complex numbers, then the same implication certainly holds if the c's are only chosen from the real numbers. However, the converse statement is not true, as the following example shows.

Example 2. The set $\mathcal{C}$ of complex numbers is itself a complex vector space. As a real vector space, we have seen that $\mathcal{C}$ can be identified with $\mathcal{R}^2$; hence it has dimension 2 relative to the real numbers. However, relative to the complex numbers, $\mathcal{C}$ has dimension 1 because a linear combination of more than one complex number, with complex coefficients, can always be made to be zero without all coefficients being zero. For example, in $c_1z + c_2w$, choose $c_1 = 1/z$ and $c_2 = -1/w$.

Turning to differential equations again, we observe that Theorems 6.1, 6.2, and 6.3 of the previous section are all true relative to the complex numbers. The reason is that, in the proofs of these theorems, no assumption is made about the numerical multipliers that occur other than the ordinary rules of arithmetic which apply to both real and complex numbers. In addition, the properties of the exponential function that are used hold for both real and complex exponentials. Therefore we can conclude that all solutions of the differential equation $L(y) = 0$, where

$$L = (D - r_1) \ldots (D - r_n) \tag{2}$$

are linear combinations of n functions, even when the numbers r_k are complex. These functions are e^{r_kx} or, in the case of multiple roots, $x^le^{r_kx}$. This implies, just as for real r_k:

7.3 Theorem

The set of solutions of $L(y) = 0$, with L given by Equation (2), is a vector space $\mathcal{N}$ of dimension n relative to the complex numbers.

$\mathcal{N}$ has a basis consisting of functions of the form $e^{r_k x}$ or $x^l e^{r_k x}$, where r_k may be complex.

Theorem 7.3 is not usually applied directly in the above form because operators such as

$$P(D) = D^2 + aD + b,$$

which occur in practice, most often have real numbers for the coefficients a and b. This implies that, in the factorization

$$D^2 + aD + b = (D - r_1)(D - r_2),$$

the *complex* roots r_k always occur in complex conjugate pairs. (If r is a root of $P(x)$ with *real* coefficients, then so is $\bar{r}$. Why?) As a result, solutions of the differential equation also occur in pairs of the form

$$e^{rx}, \qquad e^{\bar{r}x},$$

perhaps multiplied by some power of x. It then becomes natural to write $r = \alpha + i\beta$ and $\bar{r} = \alpha - i\beta$. We get

$$\begin{aligned} c_1 e^{rx} + c_2 e^{\bar{r}x} &= c_1 e^{(\alpha+i\beta)x} + c_2 e^{(\alpha-i\beta)x} \\ &= e^{\alpha x}(c_1(\cos\beta x + i\sin\beta x) + c_2(\cos\beta x - i\sin\beta x)) \\ &= d_1 e^{\alpha x}\cos\beta x + d_2 e^{\alpha x}\sin\beta x. \end{aligned}$$

where $d_1 = c_1 + c_2$ and $d_2 = i(c_1 - c_2)$. The functions $e^{\alpha x}\cos\beta x$ and $e^{\alpha x}\sin\beta x$ are easier to interpret geometrically than are the complex exponentials that gave rise to them. Hence the solutions are often written using the trigonometric form

$$y = d_1 e^{\alpha x}\cos\beta x + d_2 e^{\alpha x}\sin\beta x.$$

Example 3. The differential equation

$$y'' + 2y' + 2y = 0$$

has characteristic polynomial

$$D^2 + 2D + 2.$$

The roots of this polynomial are $-1 \pm i$; so in factored form, the operator equation can be written

$$(D - (-1 - i))(D - (-1 + i))y = 0.$$

The complex exponential solutions are

$$e^{(-1-i)x}, \qquad e^{(-1+i)x}.$$

Translated into trigonometric form, these give

$$e^{-x}\cos x, \qquad e^{-x}\sin x;$$

so the general solution can be written either

$$c_1e^{(-1-i)x} + c_2e^{(-1+i)x}$$

or

$$d_1e^{-x}\cos x + d_2e^{-x}\sin x.$$

The natural counterpart of Theorem 7.3 for differential equations with real coefficients is the following theorem which guarantees a basis for the space of real solutions of $L(y) = 0$.

7.4 Theorem

The set of real-valued solutions of

$$(D^n + a_{n-1}D^{n-1} + \ldots + a_1D + a_0)y = 0,$$

where the a_k are real, is a vector space $\mathcal{M}$ of dimension n relative to the real numbers. $\mathcal{M}$ has a basis consisting of functions of the form

$$x^le^{\alpha x}\cos\beta x, \qquad x^le^{\alpha x}\sin\beta x,$$

where α and β are real.

Proof. We know from Theorem 7.3 that the complex solutions of the above differential equation are linear combinations of functions of the form x^le^{rx}, where either r is real or else there is a companion solution $x^le^{\bar{r}x}$. We have seen that any solution, and in particular any real solution, can then be written as a linear combination of n functions of the form

$$x^le^{\alpha x}\cos\beta x, \qquad x^le^{\alpha x}\sin\beta x. \tag{3}$$

Since the space $\mathcal{M}$ of solutions has complex dimension n, any n functions that span it must be linearly independent over the complex numbers. But then these same n functions will automatically be linearly independent over the real numbers and so form a basis for $\mathcal{M}$. Hence n functions of the form (3) are a basis for $\mathcal{M}$.

We conclude the section with some additional remarks about complex vector spaces.

Example 4. Denote by $\mathcal{C}^n$ the set of all n-tuples of complex numbers. Then $\mathcal{C}^n$ is easily seen to be a vector space with addition and numerical multiplication defined coordinate-wise. Unless something is stated to the contrary, it is always understood that numerical multiplication in $\mathcal{C}^n$ is relative to the complex numbers. Then $\mathcal{C}^n$ has $\mathbf{e}_1 = (1, 0, \ldots, 0), \ldots,$ $\mathbf{e}_n = (0, \ldots, 0, 1)$ as a basis, and so it has complex dimension n.

Example 5. Let P_n be the vector space of polynomials in powers of x from 1 up to x^n, with complex coefficients. Then P_n has complex dimension $n + 1$ because the spanning set $1, x, \ldots, x^n$ is linearly independent. To see this, observe that

$$c_0 + c_1 x + \ldots + c_n x^n = 0 \qquad \text{for all } x$$

implies that the polynomial has more than n roots. This is possible only if all the coefficients are zero.

It is possible to define length and inner product (the analog of dot-product in $\mathcal{R}^n$) in a complex vector space. To see how this should be done, suppose that the inner product of two vectors $\mathbf{x}$ and $\mathbf{y}$ is to be a complex-valued function $\langle \mathbf{x}, \mathbf{y} \rangle$. Suppose further that $\langle \mathbf{x}, \mathbf{x} \rangle$ is to be nonnegative so that we can define the length of $\mathbf{x}$ by $|\mathbf{x}| = \sqrt{\langle \mathbf{x}, \mathbf{x} \rangle}$. If now we simply assume that $\langle \mathbf{x}, \mathbf{y} \rangle$ is complex linear in both $\mathbf{x}$ and $\mathbf{y}$, then we would have for any complex number c,

$$|c\mathbf{x}|^2 = \langle c\mathbf{x}, c\mathbf{x} \rangle = c^2 \langle \mathbf{x}, \mathbf{x} \rangle = c^2 \, |\mathbf{x}|^2.$$

But $|c\mathbf{x}|^2$ and $|\mathbf{x}|^2$ are both real numbers, while in general c^2, the square of a complex number, is not real. To get around this difficulty we require that $\langle \mathbf{x}, \mathbf{y} \rangle$ be **conjugate symmetric,**

$$\langle \mathbf{x}, \mathbf{y} \rangle = \overline{\langle \mathbf{y}, \mathbf{x} \rangle},$$

instead of symmetric. Thus we define a **complex inner product** $\langle \mathbf{x}, \mathbf{y} \rangle$ of elements $\mathbf{x}$, $\mathbf{y}$ in a complex vector space so that the following properties hold.

7.5 Positivity: $\langle \mathbf{x}, \mathbf{x} \rangle > 0$, except that $\langle 0, 0 \rangle = 0$.

Conjugate Symmetry: $\langle \mathbf{x}, \mathbf{y} \rangle = \overline{\langle \mathbf{y}, \mathbf{x} \rangle}$.

Additivity: $\langle \mathbf{x} + \mathbf{y}, \mathbf{z} \rangle = \langle \mathbf{x}, \mathbf{z} \rangle + \langle \mathbf{y}, \mathbf{z} \rangle$.

Homogeneity: $\langle c\mathbf{x}, \mathbf{y} \rangle = c \langle \mathbf{x}, \mathbf{y} \rangle$.

The conjugate symmetry implies additivity in the second vector also, so that

$$\langle \mathbf{x}, \mathbf{y} + \mathbf{z} \rangle = \langle \mathbf{x}, \mathbf{y} \rangle + \langle \mathbf{x}, \mathbf{z} \rangle.$$

However, we have **conjugate homogeneity** in the second variable:

$$\langle \mathbf{x}, c\mathbf{y} \rangle = \overline{\langle c\mathbf{y}, \mathbf{x} \rangle} = \bar{c}\overline{\langle \mathbf{y}, \mathbf{x} \rangle} = \bar{c}\langle \mathbf{x}, \mathbf{y} \rangle.$$

Example 6. In the complex vector space $\mathcal{C}^n$, let $\mathbf{z} = (z_1, \ldots, z_n)$ and $\mathbf{w} = (w_1, \ldots, w_n)$. Define

$$\langle \mathbf{z}, \mathbf{w} \rangle = z_1\bar{w}_1 + \ldots + z_n\bar{w}_n.$$

It is easy to check that this defines a complex inner product. If we define the length of $\mathbf{z}$ by

$$|\mathbf{z}| = \langle \mathbf{z}, \mathbf{z} \rangle^{1/2},$$

then $|\mathbf{z}|$ turns out to have the three properties of length listed in 5.3 of Section 5, Chapter 1.

Example 7. Let $\mathcal{E}_n$ be the vector space of exponential polynomials

$$p(x) = \sum_{k=-n}^{n} c_k e^{ikx}$$

defined for $-\pi \le x \le \pi$ with complex coefficients c_k. We can define a complex inner product on $\mathcal{E}_n$ by

$$\langle p, q \rangle = \int_{-\pi}^{\pi} p(x)\overline{q(x)}\, dx.$$

To integrate the complex function $p\bar{q}$ we simply integrate its real and imaginary parts. We define length by

$$\begin{aligned} |p| &= \langle p, p \rangle^{1/2} \\ &= \left(\int_{-\pi}^{\pi} |p(x)|^2\, dx\right)^{1/2}. \end{aligned}$$

In Examples 6 and 7, the length of a complex vector $\mathbf{z}$ was defined by $|\mathbf{z}| = \langle \mathbf{z}, \mathbf{z} \rangle^{1/2}$, using a conjugate symmetric inner product. The usual properties of length are:

7.6 Positivity: $|\mathbf{z}| > 0$, except that $|\mathbf{0}| = 0$.

Homogeneity: $|c\mathbf{z}| = |c|\,|\mathbf{z}|$.

Triangle Inequality: $|\mathbf{z} + \mathbf{w}| \le |\mathbf{z}| + |\mathbf{w}|$.

The first is obviously satisfied because of the corresponding property of the inner product. The second follows from

$$|c\mathbf{z}| = \langle c\mathbf{z}, c\mathbf{z} \rangle^{1/2} = [\langle c\bar{c}\mathbf{z}, \mathbf{z} \rangle]^{1/2} = |c|\,|\mathbf{z}|.$$

As in the case of a real inner product, and with the same proof, the triangle inequality follows from the Cauchy-Schwarz inequality: $|(\mathbf{z}, \mathbf{w})| \leq |\mathbf{z}|\ |\mathbf{w}|$. The proof of Cauchy-Schwarz is a simple modification of the real case and is left as Exercise 11 at the end of this section. The difference between the real and complex proofs here illustrates the fact that, because a complex inner product has somewhat different properties from a real inner product, we cannot expect theorems involving inner products to extend without change from real to complex vector spaces. Complex inner products are used in this chapter in the exercises following Sections 8 and 9.

EXERCISES

1. For each of the following complex numbers z, find $\bar{z}$ and $|z|$. Then write z in polar form and find $1/z$.

(a) $1 + i$.

(b) $-1 + 2i$.

(c) $2i$.

(d) $\dfrac{2 + i}{i}$.

2. Verify that, for complex numbers z_1, z_2, and z_3, the distributive law

$$z_1(z_2 + z_3) = z_1z_2 + z_1z_3,$$

the associative laws

$$z_1(z_2z_3) = (z_1z_2)z_3 \quad \text{and} \quad z_1 + (z_2 + z_3) = (z_1 + z_2) + z_3,$$

and the commutative laws

$$z_1z_2 = z_2z_1 \quad \text{and} \quad z_1 + z_2 = z_2 + z_1$$

all hold.

3. (a) Verify Equation 7.1.
(b) Verify Equation 7.2.

4. Prove directly from the definitions of conjugate and absolute values that $\overline{z_1z_2} = \bar{z}_1\bar{z}_2$ and $|z_1z_2| = |z_1|\,|z_2|$.

5. Separate the real and imaginary terms in the infinite series

$$\sum_{k=0}^{\infty} \frac{(i\theta)^k}{k!}$$

into two infinite series, and use the result to justify defining $e^{i\theta}$ by

$$e^{i\theta} = \cos\theta + i\sin\theta.$$

6. Show that if r is complex, then $(D + r)$ is linear as an operator on the complex vector space consisting of continuously differentiable functions $u(x) + iv(x)$, where x is a real variable.

7. Find the general solutions of the following differential equations.

(a) $(D^2 + 1)y = 1$.
(b) $(D^2 + D + 1)y = 0$.
(c) $y'' + y = \sin x$.
(d) $y'' + y = 1$.
(e) $y'' + y = \sin x + 1$.
(f) $y'' + y = \tan x$.

8. A spring vibrating in a viscous medium has a displacement from its initial position denoted by $x(t)$, where t is time. Denoting differentiation with respect to t by a dot, the differential equation,

$$m\ddot{x} + k\dot{x} + hx = 0,$$

can be derived from physical considerations. Here m is the mass of the spring, k is a positive constant depending on the viscosity of the medium, and h is a positive constant depending on the stiffness of the spring.

(a) By considering the roots of the characteristic polynomial, show that $x(t)$ is oscillatory or not, depending on whether $4mh > k^2$.
(b) Assuming m, h, and k all equal to 1, find the solution of the differential equation satisfying $x(0) = 0$ and $\dot{x}(0) = 3$.
(c) If in part (b) we change to $k = 2$, but leave the other conditions the same, find the corresponding solution to the differential equation.
(d) What is the maximum displacement from the initial position under the conditions of part (b)? Show that the oscillation tends to zero as t increases.
(e) Sketch the graph of the displacement function under the conditions of part (c). What is the maximum displacement and at what time does it occur?

9. (a) Show that $(a \cos cx + b \sin cx)$, where a, b, and c are real numbers, can be written in the form $r \cos (cx - \theta)$, where $r = \sqrt{a^2 + b^2}$ and θ is an angle such that $\cos \theta = a/r$ and $\sin \theta = b/r$.
(b) The result of part (a) is useful because it shows that a linear combination of $\cos cx$ and $\sin cx$ has a graph which is the graph of $\cos cx$ shifted by a suitable **phase angle,** θ, and multiplied by an **amplitude,** r. Sketch the graph of

$$\frac{1}{2}\cos 2x + \frac{1}{\sqrt{3}}\sin 2x$$

by first finding r and an appropriate θ.

10. Show directly that $e^{\alpha x} \cos \beta x$ and $e^{\alpha x} \sin \beta x$ are linearly independent relative to the real numbers by using the formulas

$$\cos \beta x = \frac{e^{i\beta x} + e^{-i\beta x}}{2}, \qquad \sin \beta x = \frac{e^{i\beta x} - e^{-i\beta x}}{2i},$$

together with the fact that e^{rx} and $e^{\bar{r}x}$ are linearly independent relative to the complex numbers when $r \neq \bar{r}$.

11. Prove the Cauchy-Schwarz inequality

$$|\langle \mathbf{z}, \mathbf{w} \rangle| \leq |\mathbf{z}|\,|\mathbf{w}|$$

for complex inner products. [Start, as in the real case, by assuming $|\mathbf{z}| = |\mathbf{w}| = 1$. Express in polar form

$$\langle \mathbf{z}, \mathbf{w} \rangle = |\langle \mathbf{z}, \mathbf{w} \rangle|\, e^{i\theta},$$

so that

$$\langle e^{-i\theta}\, \mathbf{z}, \mathbf{w} \rangle = |\langle \mathbf{z}, \mathbf{w} \rangle|.$$

Then expand $|e^{-i\theta}\, \mathbf{z} - \mathbf{w}|^2$ in terms of the complex inner product.]

12. Show that $\mathcal{C}^n$, the set of n-tuples of complex numbers, has dimension n relative to the complex numbers and dimension $2n$ relative to the real numbers.

13. Show that if a set of real-valued functions is linearly independent relative to the real numbers, then it is linearly independent relative to the complex numbers. (The somewhat simpler converse statement is true quite generally and has been treated in the text.)

SECTION 8

ORTHONORMAL BASES

If $\mathcal{V}$ is a vector space with an inner product, and $\mathcal{V}$ has a basis

$$\mathbf{u}_1, \mathbf{u}_2, \ldots, \mathbf{u}_n,$$

then it is often desirable that the basis be an **orthonormal** set. This means that distinct vectors $\mathbf{u}_i$ and $\mathbf{u}_j$ are perpendicular:

$$\mathbf{u}_i \cdot \mathbf{u}_j = 0, \qquad i \neq j, \tag{1}$$

and that each has length 1:

$$|\mathbf{u}_i| = \sqrt{\mathbf{u}_i \cdot \mathbf{u}_i} = 1, \qquad i = 1, \ldots, n. \tag{2}$$

These conditions are useful because, if a vector $\mathbf{x}$ in $\mathcal{V}$ is expressed as a linear combination of basis vectors by

$$\mathbf{x} = c_1\mathbf{u}_1 + \ldots + c_j\mathbf{u}_j + \ldots + c_n\mathbf{u}_n,$$

then the coefficients c_j can be computed easily. In fact, we have

$$\begin{aligned} \mathbf{x} \cdot \mathbf{u}_j &= c_1\mathbf{u}_1 \cdot \mathbf{u}_j + \ldots + c_j\mathbf{u}_j \cdot \mathbf{u}_j + \ldots + c_n\mathbf{u}_n \cdot \mathbf{u}_j \\ &= 0 + \ldots + c_j + \ldots + 0, \end{aligned}$$

so that

$$c_j = \mathbf{x} \cdot \mathbf{u}_j. \tag{3}$$

Example 1. The vectors (1, 0) and (0, 1) in $\mathcal{R}^2$ form an orthonormal basis. So does the pair of vectors $(1/\sqrt{2}, 1/\sqrt{2})$ and $(-1/\sqrt{2}, 1/\sqrt{2})$. In terms of the latter basis we can write

$$(x, y) = c_1\left(\frac{1}{\sqrt{2}}, \frac{1}{\sqrt{2}}\right) + c_2\left(\frac{-1}{\sqrt{2}}, \frac{1}{\sqrt{2}}\right).$$

To determine c_1 and c_2 we compute, using Equation (3),

$$c_1 = (x, y) \cdot \left(\frac{1}{\sqrt{2}}, \frac{1}{\sqrt{2}}\right) = \frac{(x + y)}{\sqrt{2}},$$

$$c_2 = (x, y) \cdot \left(\frac{-1}{\sqrt{2}}, \frac{1}{\sqrt{2}}\right) = \frac{(-x + y)}{\sqrt{2}}.$$

In particular, if $(x, y) = (1, 2)$, we have $c_1 = 3/\sqrt{2}$ and $c_2 = 1/\sqrt{2}$; therefore

$$(1, 2) = \frac{3}{\sqrt{2}}\left(\frac{1}{\sqrt{2}}, \frac{1}{\sqrt{2}}\right) + \frac{1}{\sqrt{2}}\left(\frac{-1}{\sqrt{2}}, \frac{1}{\sqrt{2}}\right).$$

Example 2. The set of functions defined for $-\pi \leq x \leq \pi$ by

$$\cos x, \sin x, \ldots, \cos nx, \sin nx$$

span the vector space $\mathscr{T}_n$ of trigonometric sums of the form

$$T(x) = \sum_{k=1}^{n} (a_n \cos kx + b_k \sin kx).$$

A natural inner product for $\mathscr{T}_n$ is

$$\langle f, g \rangle = \frac{1}{\pi} \int_{-\pi}^{\pi} f(x)g(x)\, dx,$$

and with respect to this inner product, the above set turns out to be orthonormal. In fact, computation shows that

$$\frac{1}{\pi} \int_{-\pi}^{\pi} \cos kx \cos lx\, dx = \begin{cases} 0, & k \neq l \\ 1, & k = l \end{cases}$$

$$\frac{1}{\pi} \int_{-\pi}^{\pi} \sin kx \sin lx\, dx = \begin{cases} 0, & k \neq l \\ 1, & k = l \end{cases}$$

$$\frac{1}{\pi} \int_{-\pi}^{\pi} \cos kx \sin lx\, dx = 0.$$

It follows that the coefficients a_k and b_k are given by

$$a_k = \frac{1}{\pi} \int_{-\pi}^{\pi} T(x) \cos kx\, dx,$$

$$b_k = \frac{1}{\pi} \int_{-\pi}^{\pi} T(x) \sin kx\, dx.$$

The numbers a_k and b_k are called the kth **Fourier coefficients** of T, and their applications are discussed in Section 5 of Chapter 5.

The fact that orthonormal bases are simpler to work with suggests the following question: Given a basis for a vector space with an inner product, is there some way to find an orthonormal basis? The answer is yes, and we shall describe an effective procedure. First we make the following observation.

8.1 Theorem

If $\mathbf{u}_1, \ldots, \mathbf{u}_n$ is an orthonormal set in a vector space $\mathcal{V}$, then the vectors $\mathbf{u}_i$ are linearly independent. The coefficients in the linear combination

$$\mathbf{x} = c_1\mathbf{u}_1 + \ldots + c_n\mathbf{u}_n$$

are computed by $c_j = \mathbf{x} \cdot \mathbf{u}_j$.

Proof. Suppose there are constants $c_1, \ldots, c_n$ such that

$$c_1\mathbf{u}_1 + \ldots + c_j\mathbf{u}_j + \ldots + c_n\mathbf{u}_n = 0.$$

If for some j we form the dot-product of both sides of this equation with $\mathbf{u}_j$, then the result is

$$c_1\mathbf{u}_1 \cdot \mathbf{u}_j + \ldots + c_j\mathbf{u}_j \cdot \mathbf{u}_j + \ldots + c_n\mathbf{u}_n \cdot \mathbf{u}_j = 0.$$

But $\mathbf{u}_n \cdot \mathbf{u}_j = 0$ if $k \neq j$, and $\mathbf{u}_j \cdot \mathbf{u}_j = 1$; so we get $c_j = 0$. Since j is arbitrary, the $\mathbf{u}$'s are linearly independent.

Now suppose that $\mathbf{x}_1, \ldots, \mathbf{x}_n$ is a linearly independent set in a vector space $\mathcal{V}$; we shall describe a process for constructing an orthonormal set $\mathbf{u}_1, \ldots, \mathbf{u}_n$. First we pick any of the $\mathbf{x}$'s, say $\mathbf{x}_1$, and set

$$\mathbf{u}_1 = \frac{\mathbf{x}_1}{|\mathbf{x}_1|}.$$

Then $|\mathbf{u}_1| = |\mathbf{x}_1|/|\mathbf{x}_1| = 1$. Now pick another of the $\mathbf{x}$'s, say $\mathbf{x}_2$, and form its projection on $\mathbf{u}_1$, that is, form the vector $(\mathbf{x}_2 \cdot \mathbf{u}_1)\mathbf{u}_1$. If the vectors were ordinary geometric vectors, the relationship between $\mathbf{x}_1$, $\mathbf{x}_2$, and $\mathbf{u}_1$ would be somewhat as shown in Fig. 11.

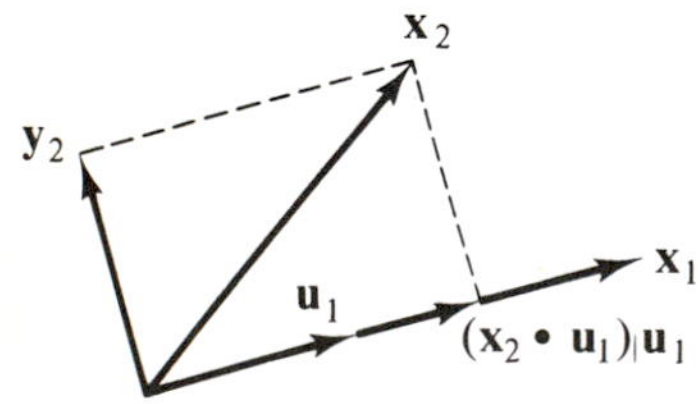

Figure 11

The vector $\mathbf{y}_2$ is defined by

$$\mathbf{y}_2 = \mathbf{x}_2 - (\mathbf{x}_2 \cdot \mathbf{u}_1)\mathbf{u}_1,$$

and we can check that $\mathbf{y}_2$ is perpendicular to $\mathbf{u}_1$, for

$$\begin{aligned}\mathbf{y}_2 \cdot \mathbf{u}_1 &= \mathbf{x}_2 \cdot \mathbf{u}_1 - (\mathbf{x}_2 \cdot \mathbf{u}_1)(\mathbf{u}_1 \cdot \mathbf{u}_1)\\ &= \mathbf{x}_2 \cdot \mathbf{u}_1 - \mathbf{x}_2 \cdot \mathbf{u}_1 = 0.\end{aligned}$$

To get a *unit* vector perpendicular to $\mathbf{u}_1$, we let

$$\mathbf{u}_2 = \frac{\mathbf{y}_2}{|\mathbf{y}_2|}.$$

The vector $\mathbf{y}_2$ cannot be zero because, by its definition, that would imply that $\mathbf{x}_2$ and $\mathbf{u}_1$ were linearly dependent.

Having found $\mathbf{u}_1$ and $\mathbf{u}_2$, we choose $\mathbf{x}_3$ and form its projection on the subspace of $\mathcal{V}$ spanned by $\mathbf{u}_1$ and $\mathbf{u}_2$. By definition, this is the vector

$$\mathbf{p} = (\mathbf{x}_3 \cdot \mathbf{u}_1)\mathbf{u}_1 + (\mathbf{x}_3 \cdot \mathbf{u}_2)\mathbf{u}_2.$$

We define $\mathbf{y}_3$ by subtracting this projection from $\mathbf{x}_3$:

$$\mathbf{y}_3 = \mathbf{x}_3 - (\mathbf{x}_3 \cdot \mathbf{u}_1)\mathbf{u}_1 - (\mathbf{x}_3 \cdot \mathbf{u}_2)\mathbf{u}_2.$$

As before we can check that $\mathbf{y}_3$ is perpendicular to $\mathbf{u}_1$ and also to $\mathbf{u}_2$. Since $\mathbf{u}_1 \cdot \mathbf{u}_1 = 1$ and $\mathbf{u}_1 \cdot \mathbf{u}_2 = 0$, we have

$$\mathbf{y}_3 \cdot \mathbf{u}_1 = (\mathbf{x}_3 \cdot \mathbf{u}_1) - (\mathbf{x}_3 \cdot \mathbf{u}_1) = 0.$$

Similarly

$$\mathbf{y}_3 \cdot \mathbf{u}_2 = (\mathbf{x}_3 \cdot \mathbf{u}_2) - (\mathbf{x}_3 \cdot \mathbf{u}_2) = 0.$$

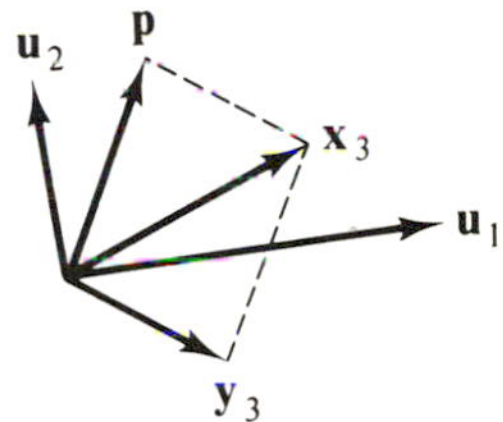

Figure 12

Figure 12 shows these vectors as geometric arrows. We define $\mathbf{u}_3$ by setting

$$\mathbf{u}_3 = \frac{\mathbf{y}_3}{|\mathbf{y}_3|}.$$

Once again $\mathbf{y}_3 = 0$ would imply linear dependence of $\mathbf{x}_3$, $\mathbf{u}_1$, and $\mathbf{u}_2$. But because the subspace spanned by $\mathbf{u}_1$ and $\mathbf{u}_2$ is the same as the one spanned by $\mathbf{x}_1$ and $\mathbf{x}_2$, this would imply linear dependence of the $\mathbf{x}$'s.

We proceed in this way, successively computing $\mathbf{u}_1, \mathbf{u}_2, \ldots, \mathbf{u}_j$. To find $\mathbf{u}_{j+1}$ we set

8.2

$$\mathbf{y}_{j+1} = \mathbf{x}_{j+1} - (\mathbf{x}_{j+1} \cdot \mathbf{u}_1)\mathbf{u}_1 - \ldots - (\mathbf{x}_{j+1} \cdot \mathbf{u}_j)\mathbf{u}_j,$$

$$\mathbf{u}_{j+1} = \frac{\mathbf{y}_{j+1}}{|\mathbf{y}_{j+1}|}.$$

As before we can verify that $\mathbf{u}_{j+1}$ is perpendicular to $\mathbf{u}_1, \mathbf{u}_2, \ldots, \mathbf{u}_j$. Equations 8.2 summarize what is known as the **Gram-Schmidt process** for finding an orthonormal set from an independent set. It can be continued until we run out of $\mathbf{x}$'s in the independent set.

The vector

$$(\mathbf{x} \cdot \mathbf{u}_1)\mathbf{u}_1 + \ldots + (\mathbf{x} \cdot \mathbf{u}_j)\mathbf{u}_j$$

is called the **projection** of $\mathbf{x}$ on the subspace spanned by $\mathbf{u}_1, \ldots, \mathbf{u}_j$. (The projection is also called the **Fourier expansion** of $\mathbf{x}$ relative to $\mathbf{u}_1, \ldots, \mathbf{u}_j$.

In Theorem 7.1 of Chapter 5 it is proved that the Fourier expansion of $\mathbf{x}$ is the linear combination of the $\mathbf{u}$'s which is nearest to $\mathbf{x}$.)

Example 3. The vectors $\mathbf{x}_1 = (1, -1, 2)$ and $\mathbf{x}_2 = (1, 0, -1)$ span a plane P in $\mathcal{R}^3$ because they are linearly independent. To find an orthonormal basis for P, we apply the Gram-Schmidt process to $\mathbf{x}_1$ and $\mathbf{x}_2$. We set

$$\mathbf{u}_1 = \frac{\mathbf{x}_1}{|\mathbf{x}_1|} = \frac{(1, -1, 2)}{\sqrt{6}} = \left(\frac{1}{\sqrt{6}}, \frac{-1}{\sqrt{6}}, \frac{2}{\sqrt{6}}\right).$$

and then compute

$$\begin{aligned}\mathbf{y}_2 &= \mathbf{x}_2 - (\mathbf{x}_2 \cdot \mathbf{u}_1)\mathbf{u}_1 \\ &= (1, 0, -1) - \left(\frac{-1}{\sqrt{6}}\right)\left(\frac{1}{\sqrt{6}}, -\frac{1}{\sqrt{6}}, \frac{2}{\sqrt{6}}\right) \\ &= (1, 0, -1) + (\tfrac{1}{6}, -\tfrac{1}{6}, \tfrac{1}{3}) = (\tfrac{7}{6}, -\tfrac{1}{6}, -\tfrac{2}{3}).\end{aligned}$$

Then

$$\begin{aligned}\mathbf{u}_2 &= \frac{\mathbf{y}_2}{|\mathbf{y}_2|} = \frac{(\frac{7}{6}, -\frac{1}{6}, -\frac{2}{3})}{\sqrt{66/36}} \\ &= \frac{1}{\sqrt{66}}(7, -1, -4).\end{aligned}$$

Thus the plane P can be represented as all linear combinations

$$u\mathbf{x}_1 + v\mathbf{x}_2$$

or all linear combinations

$$s\mathbf{u}_1 + t\mathbf{u}_2.$$

The relationship between the two pairs of vectors is shown in Fig. 13.

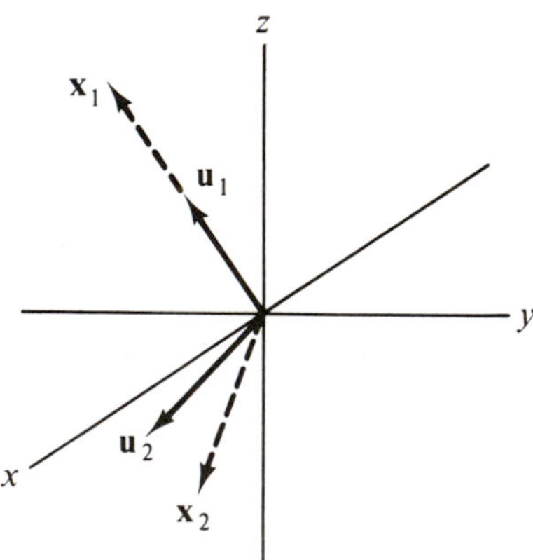

Figure 13

Example 4. Let $P_n[-1, 1]$ be the vector space of polynomials $f(x) = a_0 + a_1x + \ldots + a_nx^n$ defined for $-1 \leq x \leq 1$. We define an inner product by

$$\langle f, g\rangle = \int_{-1}^{1} f(x)g(x)\,dx.$$

We have seen in Example 5(b) of Section 4 that the functions $1, x, \ldots, x^n$ form a basis for P_n. To find an orthonormal basis, we observe that $\langle 1, 1\rangle = 2$ and therefore set $u_1(x) = 1/\sqrt{2}$. Then

$$y_2(x) = x - \left(x, \frac{1}{\sqrt{2}}\right)\frac{1}{\sqrt{2}} = x,$$

because $\langle x, 1/\sqrt{2}\rangle = \int_{-1}^{1} (x/\sqrt{2})\,dx = 0$. We then compute

$$u_2(x) = \frac{x}{\langle x, x\rangle^{1/2}} = \frac{x}{\sqrt{\int_{-1}^{1} x^2\,dx}} = \sqrt{\tfrac{3}{2}}x.$$

Next set

$$\begin{aligned} y_3(x) &= x^2 - \langle x^2, u_1(x)\rangle u_1(x) - \langle x^2, u_2(x)\rangle u_2(x) \\ &= x^2 - \left(\int_{-1}^{1} \frac{x^2}{\sqrt{2}}\,dx\right)\frac{1}{\sqrt{2}} - 0 \\ &= x^2 - \tfrac{1}{3}. \end{aligned}$$

Then

$$\begin{aligned} u_3(x) &= \frac{x^2 - \frac{1}{3}}{\langle x^2 - \frac{1}{3}, x^2 - \frac{1}{3}\rangle^{1/2}} = \frac{x^2 - \frac{1}{3}}{\sqrt{\int_{-1}^{1} (x^2 - \frac{1}{3})^2\,dx}} \\ &= \sqrt{\tfrac{45}{8}}(x^2 - \tfrac{1}{3}) = \sqrt{\tfrac{5}{8}}(3x^2 - 1). \end{aligned}$$

The process can be continued indefinitely. (The resulting polynomials are called the normalized Legendre polynomials, and another method for computing them is given in Section 7 of Chapter 5.)

Of course, if we start with two bases for the same vector space $\mathcal{V}$ and apply the Gram-Schmidt process to them, we will in general get different orthonormal bases. In particular, if one of the two given bases is already orthonormal and we apply the Gram-Schmidt process to the other one, we may get two different bases. The following theorem gives a condition under which the two resulting orthonormal bases are the same except perhaps for orientation. It is necessary to assume that $\mathcal{V}$ is a vector space with scalar multiplication by *real* numbers.

8.3 Theorem

Let $\{\mathbf{u}_1, \ldots, \mathbf{u}_n\}$ and $\{\mathbf{v}_1, \ldots, \mathbf{v}_n\}$ be two orthonormal sets in a real vector space $\mathcal{V}$. If the subspaces spanned by $\{\mathbf{u}_1, \ldots, \mathbf{u}_k\}$ and $\{\mathbf{v}_1, \ldots, \mathbf{v}_k\}$ are the same for $k = 1, 2, \ldots, n$, then $\mathbf{u}_k = \pm\mathbf{v}_k$ for $k = 1, 2, \ldots, n$.

Proof. Let $\mathbf{v}_k$ be any one of the $\mathbf{v}$'s. Since by assumption $\mathbf{v}_k$ is in the subspace spanned by $\{\mathbf{u}_1, \ldots, \mathbf{u}_k\}$, we can write

$$\mathbf{v}_k = \sum_{j=1}^{k} r_j \mathbf{u}_j$$

for some real numbers r_j. However, $\mathbf{v}_k$ is orthogonal to each of $\{\mathbf{u}_1, \ldots, \mathbf{u}_{k-1}\}$ because, by assumption, it is orthogonal to $\{\mathbf{v}_1, \ldots, \mathbf{v}_{k-1}\}$, and these two sets span the same subspace. If we form the dot product of both sides of the above equation with $\mathbf{u}_j$ for $1 \leq j \leq k$, we then get

$$0 = \mathbf{v}_k \cdot \mathbf{u}_j = r_j, \qquad j = 1, \ldots, k-1,$$
$$\mathbf{v}_k \cdot \mathbf{u}_k = r_k.$$

Thus $\mathbf{v}_k = (\mathbf{v}_k \cdot \mathbf{u}_k)\mathbf{u}_k$. Since both the $\mathbf{u}$'s and the $\mathbf{v}$'s have length 1, we must have

$$|\mathbf{v}_k \cdot \mathbf{u}_k| = 1.$$

It follows that $\mathbf{v}_k \cdot \mathbf{u}_k = \pm 1$; so $\mathbf{v}_k = \pm\mathbf{u}_k$.

Example 5. Consider the natural basis $\mathbf{e}_1 = (1, 0, 0)$, $\mathbf{e}_2 = (0, 1, 0)$, $\mathbf{e}_3 = (0, 0, 1)$ together with the basis $\mathbf{x}_1 = (3, 0, 0)$, $\mathbf{x}_2 = (1, -1, 0)$, $\mathbf{x}_3 = (1, 1, 1)$ for $\mathcal{R}^3$. Clearly, $\mathbf{e}_1$ and $\mathbf{x}_1$ span the same subspace of $\mathcal{R}^3$. Similarly, the pairs $\{\mathbf{e}_1, \mathbf{e}_2\}$ and $\{\mathbf{x}_1, \mathbf{x}_2\}$ span the xy-plane, and the complete bases both span all of $\mathcal{R}^3$. It follows from Theorem 8.3 that applying the Gram-Schmidt process to $\{\mathbf{x}_1, \mathbf{x}_2, \mathbf{x}_3\}$ will result in an orthonormal basis of the form $\{\pm\mathbf{e}_1, \pm\mathbf{e}_2, \pm\mathbf{e}_3\}$. As a matter of fact, we find that we get $\{\mathbf{e}_1, -\mathbf{e}_2, \mathbf{e}_3\}$. We leave the verification as an exercise. The vectors are shown in Fig. 14.

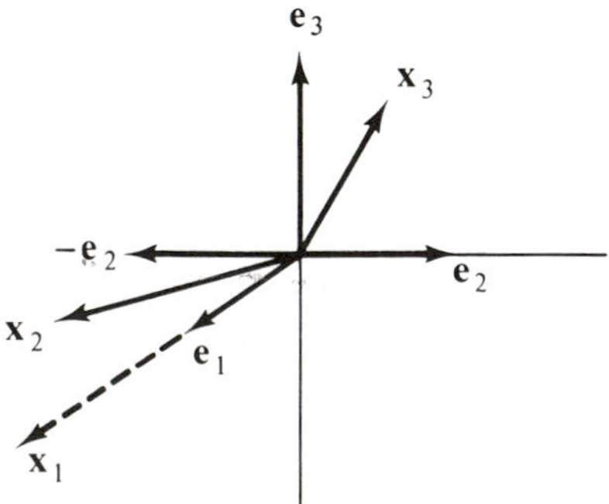

Figure 14

EXERCISES

1. (a) Find a vector (x, y, z) such that the triple of vectors $(1, 1, 1)$, $(-1, \frac{1}{2}, \frac{1}{2})$, and (x, y, z) forms an orthogonal basis for $\mathcal{R}^3$.
(b) Normalize the basis found in part (a).

2. The vectors $(1, 1, 1)$ and $(1, 2, 1)$ span a plane P in $\mathcal{R}^3$. Find an orthonormal basis for P by using the Gram-Schmidt process.

3. The three vectors $(1, 2, 1, 1)$, $(-1, 0, 1, 0)$, and $(0, 1, 0, 2)$ form a basis for a subspace $\mathcal{S}$ of $\mathcal{R}^4$. Use the Gram-Schmidt process to find an orthonormal basis for $\mathcal{S}$.

4. Let P_2 be the 3-dimensional vector space of quadratic polynomials $f(x)$ defined for $0 \leq x \leq 1$. If P_2 has the inner product

$$\langle f, g\rangle = \int_0^1 f(x)g(x)\, dx,$$

find an orthonormal basis for P_2. [*Hint*. One basis for P_2 consists of $\{1, x, x^2\}$.]

5. Show that the application of the Gram-Schmidt process to the three vectors $\mathbf{x}_1$, $\mathbf{x}_2$, $\mathbf{x}_3$ in Example 5 of the text gives the triple $\mathbf{e}_1$, $-\mathbf{e}_2$, $\mathbf{e}_3$.

6. Prove that applying the Gram-Schmidt process to an orthonormal set gives the same orthonormal set back again.

7. Let $C[-\pi, \pi]$ be the vector space of complex-valued functions $f(x)$ defined for $-\pi \leq x \leq \pi$. Let $\langle f, g\rangle$ be defined for f and g in $C[-\pi, \pi]$ by

$$\langle f, g\rangle = \frac{1}{2\pi}\int_{-\pi}^{\pi} f(x)\overline{g(x)}\, dx.$$

(a) Show that $\langle f, g\rangle$ is a complex inner product.
(b) Show that

$$\langle e^{imx}, e^{imx}\rangle = \begin{cases} 1, & m = n. \\ 0, & m \neq n. \end{cases}$$

(c) Show that the vector subspaces of $C[-\pi, \pi]$ spanned by the following two sets are the same:

$$\cos kx, \qquad \sin kx, \qquad k = 0, 1, 2, \ldots, n. \tag{1}$$

$$e^{ikx}, \qquad k = 0, \qquad \pm 1, \pm 2, \ldots, \pm n. \tag{2}$$

8. The conclusion of Theorem 8.3 is that $\mathbf{u}_k = \pm\mathbf{v}_k$. What conclusion is possible if $\mathcal{V}$ is a complex vector space? [See Problem 11, Section 7.]

SECTION 9

EIGENVECTORS

In this section we shall find a natural way to associate a basis for a vector space $\mathcal{V}$ with a given linear function f from $\mathcal{V}$ to $\mathcal{V}$. Suppose that there is a *nonzero* vector x in $\mathcal{V}$ and a number λ such that

$$f(\mathbf{x}) = \lambda \mathbf{x}.$$

Then $\mathbf{x}$ is called an **eigenvector** of the linear function f, and λ is called its associated **eigenvalue**. (The terms **characteristic vector** and **characteristic root** are sometimes used.) Since f is linear, the foregoing equation will always have the trivial solution $\mathbf{x} = 0$ for any λ; so we rule that out as being uninteresting. Of course, also by the linearity of f, if $\mathbf{x}$ is an eigenvector corresponding to λ, then so is $c\mathbf{x}$ for any number $c \neq 0$.

Example 1. Let f be the linear function from $\mathcal{R}^2$ to $\mathcal{R}^2$ defined by the matrix

$$\begin{pmatrix} 1 & 1 \\ 4 & 1 \end{pmatrix}.$$

Thus

$$f\begin{pmatrix} x \\ y \end{pmatrix} = \begin{pmatrix} 1 & 1 \\ 4 & 1 \end{pmatrix}\begin{pmatrix} x \\ y \end{pmatrix} = \begin{pmatrix} x + y \\ 4x + y \end{pmatrix}.$$

It is easy to verify that

$$\begin{pmatrix} 1 & 1 \\ 4 & 1 \end{pmatrix}\begin{pmatrix} 1 \\ 2 \end{pmatrix} = 3\begin{pmatrix} 1 \\ 2 \end{pmatrix}$$

and that

$$\begin{pmatrix} 1 & 1 \\ 4 & 1 \end{pmatrix}\begin{pmatrix} 1 \\ -2 \end{pmatrix} = (-1)\begin{pmatrix} 1 \\ -2 \end{pmatrix}.$$

That is, the vector $(1, 2)$ in $\mathcal{R}^2$ is an eigenvector corresponding to the eigenvalue 3, and the vector $(1, -2)$ is an eigenvector corresponding to the eigenvalue -1. Of course, nonzero multiples of these two vectors will be eigenvectors with the same two eigenvalues.

Before discussing how to find eigenvectors, we shall see why they are useful. Suppose that $\mathcal{V}$ is a vector space, f is a linear function from $\mathcal{V}$ to $\mathcal{V}$, and suppose that $\mathcal{V}$ has a basis $\{\mathbf{x}_1, \mathbf{x}_2, \ldots, \mathbf{x}_n\}$ consisting of eigenvectors of f, that is,

$$f(\mathbf{x}_k) = \lambda_k \mathbf{x}_k, \qquad k = 1, 2, \ldots, n,$$

for some numbers λ_k. It follows that representing an arbitrary element of $\mathcal{V}$ using this basis leads to a particularly simple form for f. Suppose

$$\mathbf{x} = c_1\mathbf{x}_1 + \ldots + c_n\mathbf{x}_n.$$

Then, using the linearity of f and the fact that the $\mathbf{x}$'s are eigenvectors, we have

$$\begin{aligned} f(\mathbf{x}) &= c_1 f(\mathbf{x}_1) + \ldots + c_n f(\mathbf{x}_n) \\ &= c_1 \lambda_1 \mathbf{x}_1 + \ldots + c_n \lambda_n \mathbf{x}_n. \end{aligned}$$

Thus, relative to the basis of eigenvectors, the action of f is simply to multiply each coefficient by the corresponding eigenvalue. In short,

9.1 $$f\left(\sum_{k=1}^{n} c_k \mathbf{x}_k\right) = \sum_{k=1}^{n} \lambda_k c_k \mathbf{x}_k,$$

where

9.2 $$f(\mathbf{x}_k) = \lambda_k \mathbf{x}_k, \qquad k = 1, 2, \ldots, n.$$

In geometric terms, using the eigenvectors for a basis allows the action of the function f to be interpreted as a succession of numerical multiplications, one on each of the basis vectors.

Example 2. Returning to the linear function of Example 1, we express an arbitrary vector in $\mathcal{R}^2$ as a linear combination of the two eigenvectors $\mathbf{x}_1 = (1, 2)$ and $\mathbf{x}_2 = (1, -2)$:

$$\mathbf{x} = u\begin{pmatrix} 1 \\ 2 \end{pmatrix} + v\begin{pmatrix} 1 \\ -2 \end{pmatrix}.$$

The corresponding eigenvalues are 3 and -1; so from

$$f(\mathbf{x}) = uf\begin{pmatrix} 1 \\ 2 \end{pmatrix} + vf\begin{pmatrix} 1 \\ -2 \end{pmatrix},$$

we get

$$f(\mathbf{x}) = 3u\begin{pmatrix} 1 \\ 2 \end{pmatrix} - v\begin{pmatrix} 1 \\ -2 \end{pmatrix}.$$

Figure 15 shows the effect of f on each of the two eigenvectors $\mathbf{x}_1$ and $\mathbf{x}_2$. It follows that the image $f(\mathbf{x})$ of any vector $\mathbf{x}$ can be constructed by

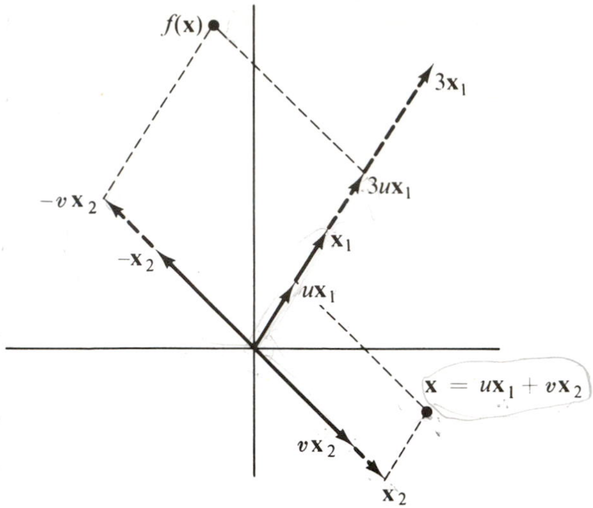

Figure 15

geometric methods. Furthermore, we can express the effect of f in words by saying that f is a combination of two transformations:

1. Stretching away from the line through $\mathbf{x}_2$ along the lines parallel to $\mathbf{x}_1$.

2. Reflection in the line through $\mathbf{x}_1$ and along the lines parallel to $\mathbf{x}_2$.

For this analysis to work, it was essential that the eigenvectors of f span all of $\mathcal{R}^2$.

Example 3. To find the eigenvectors and corresponding eigenvalues for the function f of Examples 1 and 2 (in case they had not been given), we would proceed as follows. We need to find vectors $\mathbf{x} \neq 0$, and numbers λ such that

$$f(\mathbf{x}) - \lambda\mathbf{x} = 0.$$

In matrix form, this equation is

$$\begin{pmatrix} 1 & 1 \\ 4 & 1 \end{pmatrix}\begin{pmatrix} x \\ y \end{pmatrix} - \lambda\begin{pmatrix} x \\ y \end{pmatrix} = \begin{pmatrix} 0 \\ 0 \end{pmatrix}$$

or

$$\begin{pmatrix} 1 & 1 \\ 4 & 1 \end{pmatrix}\begin{pmatrix} x \\ y \end{pmatrix} - \begin{pmatrix} \lambda & 0 \\ 0 & \lambda \end{pmatrix}\begin{pmatrix} x \\ y \end{pmatrix} = \begin{pmatrix} 0 \\ 0 \end{pmatrix}$$

or

$$\begin{pmatrix} (1-\lambda) & 1 \\ 4 & (1-\lambda) \end{pmatrix}\begin{pmatrix} x \\ y \end{pmatrix} = \begin{pmatrix} 0 \\ 0 \end{pmatrix}. \tag{1}$$

It is clear that if the foregoing 2-by-2 matrix has an inverse, then the only solutions are $x = 0$ and $y = 0$. Hence we must try to find values of λ for which the matrix *fails* to have an inverse. This will occur precisely when

$$\det \begin{pmatrix} (1-\lambda) & 1 \\ 4 & (1-\lambda) \end{pmatrix} = 0,$$

that is, when λ satisfies $(1-\lambda)^2 - 4 = 0$. This quadratic equation in λ has roots $\lambda = 3$ and $\lambda = -1$, as we can see by inspection or otherwise. To find eigenvectors corresponding to $\lambda = 3$ and $\lambda = -1$, we must find x and y, not both zero, satisfying Equation (1). Thus we consider

$$\lambda = 3: \quad \begin{pmatrix} -2 & 1 \\ 4 & -2 \end{pmatrix} \begin{pmatrix} x \\ y \end{pmatrix} = \begin{pmatrix} 0 \\ 0 \end{pmatrix}$$

and

$$\lambda = -1: \quad \begin{pmatrix} 2 & 1 \\ 4 & 2 \end{pmatrix} \begin{pmatrix} x \\ y \end{pmatrix} = \begin{pmatrix} 0 \\ 0 \end{pmatrix}.$$

These equations reduce to

$$\lambda = 3: \qquad -2x + y = 0$$
$$\lambda = -1: \qquad 2x + y = 0.$$

It follows that there are many solutions; but all we need is one for each eigenvalue. We choose for simplicity

$$\lambda = 3: \quad \begin{pmatrix} x \\ y \end{pmatrix} = \begin{pmatrix} 1 \\ 2 \end{pmatrix}$$
$$\lambda = -1: \quad \begin{pmatrix} x \\ y \end{pmatrix} = \begin{pmatrix} 1 \\ -2 \end{pmatrix},$$

though any nonzero numerical multiple of either vector would do.

The method of Example 3 can be summarized as follows. To find eigenvalues of a linear function f from $\mathcal{R}^n$ to $\mathcal{R}^n$, solve the **characteristic equation**

9.3 $$\det (A - \lambda I) = 0,$$

where A is the matrix of f. Then, for each eigenvalue $\lambda_1, \ldots, \lambda_n$ try to find nonzero vectors $\mathbf{x}_1, \ldots, \mathbf{x}_n$ that satisfy the matrix equation

9.4 $$(A - \lambda_k I)\mathbf{x}_k = 0.$$

Because Equations 9.3 and 9.4 are expressed in terms of the matrix A of f, we sometimes refer to eigenvectors and eigenvalues of the *matrix* rather than the function f. Of course Equation 9.4 is just Equation 9.2 in matrix form; but Equation 9.2 also applies to linear functions that are not representable by matrices.

Example 4. This example will show one way in which a linear function can fail to have eigenvectors that provide a suitable basis for the vector space on which f acts. Consider the linear function f from $\mathcal{R}^2$ to $\mathcal{R}^2$ defined for fixed θ, $0 \leq \theta < 2\pi$, by

$$f\begin{pmatrix} x \\ y \end{pmatrix} = \begin{pmatrix} \cos\theta & -\sin\theta \\ \sin\theta & \cos\theta \end{pmatrix}\begin{pmatrix} x \\ y \end{pmatrix}.$$

This function carries

$$\begin{pmatrix} 1 \\ 0 \end{pmatrix} \quad \text{into} \quad \begin{pmatrix} \cos\theta \\ \sin\theta \end{pmatrix}$$

and

$$\begin{pmatrix} 0 \\ 1 \end{pmatrix} \quad \text{into} \quad \begin{pmatrix} -\sin\theta \\ \cos\theta \end{pmatrix}$$

and so represents a rotation counterclockwise about the origin through an angle θ. Equation 9.3 becomes

$$\det\begin{pmatrix} \cos\theta - \lambda & -\sin\theta \\ \sin\theta & \cos\theta - \lambda \end{pmatrix} = 0$$

or

$$\lambda^2 - 2\lambda\cos\theta + 1 = 0.$$

This quadratic equation for λ has only complex conjugate roots of the form

$$\cos\theta \pm i\sin\theta = e^{\pm i\theta}.$$

Since we are looking for nonzero vectors $\mathbf{x}$ in $\mathcal{R}^2$ satisfying

$$f(\mathbf{x}) = e^{\pm i\theta}\mathbf{x},$$

we conclude that f has no eigenvectors in $\mathcal{R}^2$, unless $\theta = 0$ or $\theta = \pi$. In the first case we get the identity transformation, which clearly has every nonzero vector as an eigenvector with eigenvalue 1. In the second case we get $f(\mathbf{x}) = -\mathbf{x}$, which is a reflection in the origin and has every nonzero vector as an eigenvector with eigenvalue -1. A moment's thought shows why, in general, a rotation cannot have eigenvectors in $\mathcal{R}^2$. See Exercise 8 for an example of a linear function which fails in a different way to provide a basis of eigenvectors.

We conclude by considering the possibility that the eigenvectors of a linear function f from $\mathcal{V}$ to $\mathcal{V}$ can be chosen so that they form an orthonormal basis. Since eigenvectors are only determined to within a numerical multiple, we can always normalize them to have length 1. We can achieve the orthogonality if the function f is **symmetric** with respect to an inner product on $\mathcal{V}$, that is,

$$f(\mathbf{x}) \cdot \mathbf{y} = \mathbf{x} \cdot f(\mathbf{y})$$

for all $\mathbf{x}$ and $\mathbf{y}$ in $\mathcal{V}$. In particular, if f is a linear function from $\mathcal{R}^n$ to $\mathcal{R}^n$, it has a matrix $A = (a_{ij})$, and we can write the symmetry equation as

$$A\mathbf{x} \cdot \mathbf{y} = \mathbf{x} \cdot A\mathbf{y}.$$

Example 6. If f is a linear function from $\mathcal{R}^2$ to $\mathcal{R}^2$ with a matrix

$$A = \begin{pmatrix} a & b \\ b & c \end{pmatrix}$$

that is symmetric about its main diagonal, then f is a symmetric transformation. We can compute

$$\begin{pmatrix} a & b \\ b & c \end{pmatrix}\begin{pmatrix} x_1 \\ x_2 \end{pmatrix} \cdot \begin{pmatrix} y_1 \\ y_2 \end{pmatrix} = \begin{pmatrix} ax_1 + bx_2 \\ bx_1 + cx_2 \end{pmatrix} \cdot \begin{pmatrix} y_1 \\ y_2 \end{pmatrix}$$
$$= ax_1y_1 + b(x_2y_1 + x_1y_2) + cx_2y_2.$$

Similarly, we compute

$$\begin{pmatrix} x_1 \\ x_2 \end{pmatrix} \cdot \begin{pmatrix} a & b \\ b & c \end{pmatrix}\begin{pmatrix} y_1 \\ y_2 \end{pmatrix} = \begin{pmatrix} x_1 \\ x_2 \end{pmatrix} \cdot \begin{pmatrix} ay_1 + by_2 \\ by_1 + cy_2 \end{pmatrix}$$
$$= ax_1y_1 + b(x_1y_2 + x_2y_1) + cx_2y_2.$$

Since the two dot-products are equal, we conclude that f is symmetric.

For a symmetric transformation we have the following theorem.

9.5 Theorem

Let f be a symmetric linear function from $\mathcal{V}$ to $\mathcal{V}$, where $\mathcal{V}$ is a vector space with an inner product. If $\mathbf{x}_1$ and $\mathbf{x}_2$ are eigenvectors of f corresponding to distinct eigenvalues λ_1 and λ_2, then $\mathbf{x}_1$ and $\mathbf{x}_2$ are orthogonal.

Proof. We need to show that $\mathbf{x}_1 \cdot \mathbf{x}_2 = 0$ assuming that $f(\mathbf{x}_1) = \lambda_1\mathbf{x}_1$ and $f(\mathbf{x}_2) = \lambda_2\mathbf{x}_2$. We have

$$f(\mathbf{x}_1) \cdot \mathbf{x}_2 = (\lambda_1\mathbf{x}_1) \cdot \mathbf{x}_2 = \lambda_1(\mathbf{x}_1 \cdot \mathbf{x}_2)$$

and

$$\mathbf{x}_1 \cdot f(\mathbf{x}_2) = \mathbf{x}_1 \cdot (\lambda_2 \mathbf{x}_2) = \lambda_2(\mathbf{x}_1 \cdot \mathbf{x}_2).$$

Because f is symmetric, $f(\mathbf{x}_1) \cdot \mathbf{x}_2 = \mathbf{x}_1 \cdot f(x_2)$; so

$$\lambda_1(\mathbf{x}_1 \cdot \mathbf{x}_2) = \lambda_2(\mathbf{x}_1 \cdot \mathbf{x}_2).$$

Hence $(\lambda_1 - \lambda_2)(\mathbf{x}_1 \cdot \mathbf{x}_2) = 0$. Since λ_1 and λ_2 are assumed not equal, we must have $\mathbf{x}_1 \cdot \mathbf{x}_2 = 0$.

A more general version of Theorem 9.5 is given in Chapter 5, Section 7, where it is applied to a different class of examples.

Example 7. The function g from $\mathcal{R}^2$ to $\mathcal{R}^2$ with matrix

$$\begin{pmatrix} 1 & 2 \\ 2 & 1 \end{pmatrix}$$

has eigenvalues computed from Equation 9.3 by

$$\det \begin{pmatrix} (1-\lambda) & 2 \\ 2 & (1-\lambda) \end{pmatrix} = 0,$$

that is, $(1 - \lambda)^2 + 4 = 0$. By inspection the eigenvalues are $\lambda = 3$ and $\lambda = -1$. (Note that the transformation g is not the same as the transformation f of Example 1, even though they have the same eigenvalues.) Equation 9.4 for the eigenvectors has two interpretations, depending on which eigenvalue is used. We have

$$\lambda = 3: \quad \begin{pmatrix} -2 & 2 \\ 2 & -2 \end{pmatrix} \begin{pmatrix} x \\ y \end{pmatrix} = \begin{pmatrix} 0 \\ 0 \end{pmatrix},$$

$$\lambda = -1: \quad \begin{pmatrix} 2 & 2 \\ 2 & 2 \end{pmatrix} \begin{pmatrix} x \\ y \end{pmatrix} = \begin{pmatrix} 0 \\ 0 \end{pmatrix}.$$

These equations reduce to

$$\lambda = 3: \qquad -2x + 2y = 0,$$

$$\lambda = -1: \qquad 2x + 2y = 0.$$

We find solutions

$$\lambda = 3: \qquad (x, y) = (1, 1),$$

$$\lambda = -1: \qquad (x, y) = (1, -1).$$

The vectors $\mathbf{x}_1 = (1, 1)$, $\mathbf{x}_2 = (1, -1)$ are clearly orthogonal, and they can be normalized to give $\mathbf{u}_1 = (1/\sqrt{2}, 1/\sqrt{2})$, $\mathbf{u}_2 = (1/\sqrt{2}, -1/\sqrt{2})$. Of course the normalized vectors $\mathbf{u}_1$ and $\mathbf{u}_2$ are eigenvectors also.

We can now take advantage of the fact that $\mathbf{u}_1$ and $\mathbf{u}_2$ are eigenvectors and also that they form an orthonormal set. The latter fact enables us to express any vector in $\mathcal{R}^2$ as a linear combination of $\mathbf{u}_1$ and $\mathbf{u}_2$ very simply. From the previous section we have

$$\mathbf{x} = (\mathbf{x} \cdot \mathbf{u}_1)\mathbf{u}_1 + (\mathbf{x} \cdot \mathbf{u}_2)\mathbf{u}_2$$

for any $\mathbf{x}$ in $\mathcal{R}^2$. Now using the fact that $\mathbf{u}_1$ and $\mathbf{u}_2$ are eigenvectors of the linear function g, we can write

$$g(\mathbf{x}) = (\mathbf{x} \cdot \mathbf{u}_1)g(\mathbf{u}_1) + (\mathbf{x} \cdot \mathbf{u}_2)g(\mathbf{u}_2).$$

But $g(\mathbf{u}_1) = 3\mathbf{u}_1$ and $g(\mathbf{u}_2) = -\mathbf{u}_2$, so

$$g(\mathbf{x}) = 3(\mathbf{x} \cdot \mathbf{u}_1)\mathbf{u}_1 - (\mathbf{x} \cdot \mathbf{u}_2)\mathbf{u}_2.$$

Thus the action of g has a geometric description like that given for f in Example 2 of this section. The only difference is that the stretching and reflection takes place along different lines: perpendicular lines in the case of g, nonperpendicular lines in the case of f.

EXERCISES

1. The linear function f from $\mathcal{R}^2$ to $\mathcal{R}^2$ with matrix

$$\begin{pmatrix} 1 & 12 \\ 3 & 1 \end{pmatrix}$$

has eigenvalues 7 and -5. Which of the following vectors is an eigenvector of f? For those that are, what is the corresponding eigenvalue?

$$\begin{pmatrix} 2 \\ 1 \end{pmatrix}, \quad \begin{pmatrix} -2 \\ 1 \end{pmatrix}, \quad \begin{pmatrix} -4 \\ -2 \end{pmatrix}, \quad \begin{pmatrix} -2 \\ 2 \end{pmatrix}, \quad \begin{pmatrix} 1 \\ 1 \end{pmatrix}.$$

2. Find all the eigenvalues of each of the linear functions defined by the following matrices.

(a) $\begin{pmatrix} 1 & 4 \\ 1 & 1 \end{pmatrix}$.

(b) $\begin{pmatrix} 0 & 4 \\ 1 & 0 \end{pmatrix}$.

(c) $\begin{pmatrix} 2 & 4 \\ 1 & 2 \end{pmatrix}$.

(d) $\begin{pmatrix} 1 & 0 & 0 \\ 2 & 1 & 0 \\ 0 & 1 & 2 \end{pmatrix}$.

3. For each matrix in Exercise 2, find an eigenvector corresponding to each eigenvalue.

4. Show that 0 is an eigenvalue of a linear function f if and only if f is not one-to-one.

5. Show that if f is a one-to-one linear function having λ for an eigenvalue, then f^{-1}, the inverse of f, has $1/\lambda$ for an eigenvalue.

6. Let f be a linear function having λ for an eigenvalue.

(a) Show that λ^2 is an eigenvalue for $f \circ f$.
(b) Show that λ^n is an eigenvalue for the function got by composing f with itself n times.

7. Let $C^{(\infty)}(\mathcal{R})$ be the vector space of infinitely often differentiable functions $f(x)$ for x in $\mathcal{R}$. Then the differential operator D^2 acts linearly from $C^{(\infty)}(\mathcal{R})$ to $C^{(\infty)}(\mathcal{R})$.

(a) Show that for any real number λ the functions $\cos \lambda x$ and $\sin \lambda x$ are eigenvectors of D^2 corresponding to the eigenvalue $-\lambda^2$.
(b) Let $C^{(\infty)}[0, \pi]$ be the subspace of $C^{(\infty)}(\mathcal{R})$ consisting of functions f such that $f(0) = f(\pi) = 0$. Show that if D^2 is restricted to acting on $C^{(\infty)}[0, \pi]$, then its only eigenfunctions are of the form $\sin kx$, corresponding to $\lambda = -k^2$, where k is an integer.

8. Let h be the linear function from $\mathcal{R}^2$ to $\mathcal{R}^2$ with matrix

$$\begin{pmatrix} 1 & 1 \\ 0 & 1 \end{pmatrix}.$$

(a) Show that the eigenvectors of h span only a 1-dimensional subspace of $\mathcal{R}^2$. Thus the geometric action of h cannot be analyzed by looking only at what it does to eigenvectors.
(b) The linear function h is called a **shear transformation.** Give a geometric description of the action of h on $\mathcal{R}^2$.

9. (a) Find the eigenvalues and a corresponding pair of eigenvectors for the linear function f from $\mathcal{R}^2$ to $\mathcal{R}^2$ having matrix

$$\begin{pmatrix} 2 & 0 \\ 0 & 3 \end{pmatrix}.$$

(b) Show that the eigenvectors of the function f in part (a) form an orthogonal basis for $\mathcal{R}^2$, and use this fact to give a geometric description of the action of f on $\mathcal{R}^2$.

(c) Generalize the results you found for (a) and (b) to any linear function f from $\mathcal{R}^n$ to $\mathcal{R}^n$ having a diagonal matrix **diag** $(a_1, a_2, \ldots, a_n)$.

10. Find the eigenvalues of the function g on $\mathcal{R}^2$ with matrix

$$\begin{pmatrix} 1 & 2 \\ 1 & 1 \end{pmatrix}.$$

Show that the corresponding eigenvectors span $\mathcal{R}^2$ and describe the action of g.

11. Let $\mathcal{C}^2$ be the complex vector space of pairs $z = (z_1, z_2)$ of complex numbers, and suppose that $\mathcal{C}^2$ has the complex inner product

$$\mathbf{z} \cdot \mathbf{w} = z_1\bar{w}_1 + z_2\bar{w}_2.$$

(a) Show that the linear function f defined on $\mathcal{C}^2$ by the matrix

$$\begin{pmatrix} \cos\theta & -\sin\theta \\ \sin\theta & \cos\theta \end{pmatrix}$$

has eigenvectors and corresponding eigenvalues

$$\lambda = e^{i\theta}: \qquad \begin{pmatrix} 1 \\ -i \end{pmatrix}$$

$$\lambda = e^{-i\theta}: \qquad \begin{pmatrix} 1 \\ i \end{pmatrix}.$$

(b) Show that if $\sin\theta = 0$, then every nonzero vector in $\mathcal{C}^2$ is an eigenvector of the function f in part (a).

(c) Show that the eigenvectors in part (a) form a basis for $\mathcal{C}^2$.

(d) Show that the eigenvectors in part (a) are orthogonal with respect to the complex inner product in $\mathcal{C}^2$.

12. Explain in geometric terms why a rotation about the origin in $\mathcal{R}^2$ through an angle θ, $0 \leq \theta < 2\pi$, has no eigenvectors in $\mathcal{R}^2$ unless $\theta = 0$ or $\theta = \pi$.

13. Suppose that $\mathcal{W}$ is a complex vector space with a complex inner product, so that $\langle \mathbf{z}, \mathbf{w} \rangle = \overline{\langle \mathbf{w}, \mathbf{z} \rangle}$. Then a linear function f from $\mathcal{W}$ to $\mathcal{W}$ is called **Hermitian symmetric** if $\langle f(\mathbf{z}), \mathbf{w} \rangle = \langle \mathbf{z}, f(\mathbf{w}) \rangle$, for $\mathbf{z}$, $\mathbf{w}$ in $\mathcal{W}$.

(a) Show that if f is Hermitian symmetric and has λ for a complex eigenvalue, then λ must actually be real.

(b) Show that if $\mathcal{W}$ has complex dimension 2 and f is given by a 2-by-2 complex matrix

$$\begin{pmatrix} \alpha & \gamma \\ \beta & \delta \end{pmatrix},$$

then f is Hermitian symmetric if and only if $\beta = \bar{\gamma}$.

14. If

$$\mathbf{x}(t) = \begin{pmatrix} x(t) \\ y(t) \end{pmatrix}$$

defines a function of the real variable t taking values in $\mathcal{R}^2$, then we define the derivative of $\mathbf{x}$ by

$$\mathbf{x}'(t) = \begin{pmatrix} x'(t) \\ y'(t) \end{pmatrix}.$$

(a) Let

$$A = \begin{pmatrix} a & c \\ b & d \end{pmatrix}.$$

Show that the vector equation

$$\mathbf{x}' = A\mathbf{x}$$

is the same as the system of first-order differential equations

$$x' = ax + cy$$
$$y' = bx + dy$$

(b) Show that $\mathbf{x}(t) = e^{\lambda t}\mathbf{c}$, where $\mathbf{c}$ is a constant vector, is a solution of the vector equation $\mathbf{x}' = A\mathbf{x}$ if $\mathbf{c}$ is an eigenvector of the matrix A with corresponding eigenvalue λ.

(c) If the matrix A is

$$\begin{pmatrix} 1 & 2 \\ 2 & 1 \end{pmatrix},$$

find solutions of the vector differential equation $\mathbf{x}' = A\mathbf{x}$ in the form

$$\mathbf{x} = e^{\lambda_1 t}\mathbf{c}_1 + e^{\lambda_2 t}\mathbf{c}_2.$$

[*Note.* The eigenvectors $\mathbf{c}_1$ and $\mathbf{c}_2$ are determined only to within a nonzero numerical multiple.]

(d) Show that the second-order differential equation

$$(D - r_1)(D - r_2)y = 0$$

is equivalent to the system of first-order equations

$$x' = r_1 x$$
$$y' = x + r_2 x,$$

where $x = (D - r_2)y$. Then find the matrix A that can be used to express this system in the form of part (a).

(e) Show that the characteristic equation of the matrix A found in part (d) is the same as the characteristic equation of the operator $(D - r_1) \times (D - r_2)$ as defined in Section 6.

15. (a) Show that the functions $\cos kx$ and $\sin kx$ are eigenvectors of the differential operator D^2 acting on $C^{(\infty)}$. What are the corresponding eigenvalues?

(b) If we restrict the operator D^2 to real-valued functions defined for $-\pi \le x \le \pi$, we can define an inner product by

$$\langle f, g\rangle = \int_{-\pi}^{\pi} f(x)g(x)\,dx.$$

Show that the eigenvectors $\cos kx$ and $\sin lx$ are orthogonal with respect to this inner product for $k, l = 1, 2, 3, \ldots$. Then use Theorem 8.1 to conclude that the eigenvectors are linearly independent.

(c) Show that the functions e^{ikx} and e^{-ikx} are eigenvectors of the differential operator D^2 acting on complex-valued functions. What are the corresponding eigenvalues?

(d) Using the complex inner product

$$\langle f, g\rangle_* = \int_{-\pi}^{\pi} f(x)\overline{g(x)}\, dx,$$

show that the functions e^{ikx} are orthogonal, for $k = 0, \pm 1, \pm 2, \ldots$. Can you conclude from Theorem 8.1 that the complex exponentials are linearly independent?

SECTION 10

COORDINATES

In this section we show how any problem about finite-dimensional vector spaces and linear functions on them can be reduced to an equivalent problem about $\mathcal{R}^n$ and matrices. This is done by introducing coordinates in the vector spaces. The familiar coordinates of 2- and 3-dimensional geometry may be considered a special case of what we are about to describe. The following theorem is fundamental.

10.1 Theorem

Let $V = (\mathbf{v}_1, \ldots, \mathbf{v}_n)$ be a basis for a vector space $\mathcal{V}$, so $\dim(\mathcal{V}) = n$. For any vector $\mathbf{x}$ in $\mathcal{V}$ there is one and only one n-tuple of numbers $r_1, \ldots, r_n$ such that $\mathbf{x} = r_1\mathbf{v}_1 + \ldots + r_n\mathbf{v}_n$.

Proof. Since the $\mathbf{v}$'s span $\mathcal{V}$, $\mathbf{x}$ is a linear combination of them; so there is at least one n-tuple $(r_1, \ldots, r_n)$ satisfying the condition. If $(s_1, \ldots, s_n)$ is any other n-tuple such that

$$s_1\mathbf{v}_1 + \ldots + s_n\mathbf{v}_n = \mathbf{x} = r_1\mathbf{v}_1 + \ldots + r_n\mathbf{v}_n,$$

then

$$(r_1 - s_1)\mathbf{v}_1 + \ldots + (r_n - s_n)\mathbf{v}_n = 0.$$

Since the $\mathbf{v}$'s are independent, the numbers $r_i - s_i$ must all be zero, and so $(s_1, \ldots, s_n)$ is the same as $(r_1, \ldots, r_n)$.

Let $V = (\mathbf{v}_1, \ldots, \mathbf{v}_n)$ be a basis for a vector space $\mathcal{V}$, and let $\mathbf{x}$ be a vector in $\mathcal{V}$. The unique n-tuple $(r_1, \ldots, r_n)$ such that $\mathbf{x} = r_1\mathbf{v}_1 + \ldots + r_n\mathbf{v}_n$ is the n-tuple of **coordinates of x with respect to the basis** V. The vector in $\mathcal{R}^n$ with entries $r_1, \ldots, r_n$ is the **coordinate vector** of $\mathbf{x}$ with respect to the basis.

Example 1. (a) The vectors $\mathbf{v}_1 = (-1, 1, 0)$ and $\mathbf{v}_2 = (0, -1, 1)$ form a basis for the subspace $\mathcal{V}$ of $\mathcal{R}^3$ consisting of all vectors the sum of whose

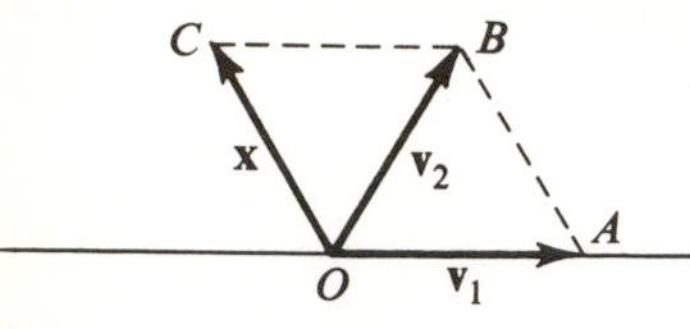

Figure 16

entries is 0. The vector $\mathbf{x} = (1, -3, 2)$ is in $\mathcal{V}$. An easy calculation shows that $\mathbf{x} = -\mathbf{v}_1 + 2\mathbf{v}_2$, so the coordinates of $\mathbf{x}$ with respect to the basis $\mathbf{v}_1, \mathbf{v}_2$ are $(-1, 2)$.

(b) In the plane, let $\mathbf{v}_1$ be the vector of unit length along the horizontal axis, and $\mathbf{v}_2$ the vector of unit length at an angle 60° counterclockwise from $\mathbf{v}_1$. Let $\mathbf{x}$ be the vector of unit length 60° beyond $\mathbf{v}_2$, shown in Fig. 16.

By geometry, CB is parallel to OA and OC is parallel to AB (why?); so $OABC$ is a parallelogram and $\mathbf{v}_2 = \mathbf{x} + \mathbf{v}_1$. Hence $\mathbf{x} = -\mathbf{v}_1 + \mathbf{v}_2$ and the coordinates of $\mathbf{x}$ with respect to this basis are $(-1, 1)$.

Coordinates provide a way of representing vectors in a concrete form suitable for calculations. They also provide a concrete way of representing linear functions. The key point is that if the value of a linear function is known for each vector of a basis, then the function is known completely. This follows from the fact that any vector $\mathbf{x}$ can be expressed as a linear combination $r_1\mathbf{v}_1 + \ldots + r_n\mathbf{v}_n$ of the vectors in a basis. Then $f(\mathbf{x})$ is the combination $r_1 f(\mathbf{v}_1) + \ldots + r_n f(\mathbf{v}_n)$ of the vectors $f(\mathbf{v}_i)$ with the same coefficients, because f is linear. Suppose $\mathcal{V} \xrightarrow{f} \mathcal{W}$ is a given linear function and that $V = (\mathbf{v}_1, \ldots, \mathbf{v}_n)$ and $W = (\mathbf{w}_1, \ldots, \mathbf{w}_m)$ are bases in $\mathcal{V}$ and $\mathcal{W}$, respectively. The possibility that $\mathcal{V}$ and $\mathcal{W}$ are the same space is not ruled out. The **matrix of f with respect to the bases V and W** is defined to be the m-by-n matrix whose jth column is the coordinate vector of $f(\mathbf{v}_j)$ with respect to the basis W. Thus if we denote by $A = (a_{ij})$ the matrix of f relative to V and W, then the entries in the jth column of A appear as the coefficients a_{ij} in the equation

$$f(\mathbf{v}_j) = a_{1j}\mathbf{w}_1 + \ldots + a_{mj}\mathbf{w}_m.$$

The matrix A contains all the information needed to evaluate $f(\mathbf{x})$ for any $\mathbf{x}$. For if $\mathbf{x}$ is represented as

$$\mathbf{x} = r_1\mathbf{v}_1 + \ldots + r_n\mathbf{v}_n$$

with coordinate vector $\mathbf{r} = (r_1, \ldots, r_n)$, then

$$\begin{aligned} f(\mathbf{x}) &= \sum_{j=1}^{n} r_j f(\mathbf{v}_j) \\ &= \sum_{j=1}^{n} r_j \sum_{i=1}^{m} a_{ij}\mathbf{w}_i \\ &= \sum_{i=1}^{m} \left(\sum_{j=1}^{n} a_{ij} r_j \right) \mathbf{w}_i. \end{aligned}$$

We can then recognize the coefficients in the representation of $f(\mathbf{x})$ relative to W as the entries in the product $A\mathbf{r}$ of the matrix A and the coordinate vector $\mathbf{r}$. We remark that when $\mathcal{V}$ and $\mathcal{W}$ are the same space it is usually appropriate, though not logically necessary, to take V and W

to be the same bases. We do this in all the examples for which $\mathcal{V}$ is the same as $\mathcal{W}$.

Example 2. (a) If f is the linear function from $\mathcal{R}^n$ to $\mathcal{R}^m$ given by multiplication by a matrix A, then the matrix of f with respect to the natural bases in $\mathcal{R}^n$ and $\mathcal{R}^m$ is A itself. This is just a rephrasing of Theorem 4.2 of Chapter 1.

(b) The function f from $\mathcal{R}^3$ to $\mathcal{R}^3$ defined by

$$f\begin{pmatrix} x_1 \\ x_2 \\ x_3 \end{pmatrix} = \begin{pmatrix} x_2 \\ x_3 \\ x_1 \end{pmatrix}$$

takes the subspace $\mathcal{V}$ spanned by $\mathbf{v}_1 = \begin{pmatrix} 1 \\ -1 \\ 0 \end{pmatrix}$ and $\mathbf{v}_2 = \begin{pmatrix} 0 \\ -1 \\ 1 \end{pmatrix}$ into itself, and it may be considered as a linear function from $\mathcal{V}$ to $\mathcal{V}$. We have

$$f(\mathbf{v}_1) = \begin{pmatrix} -1 \\ 0 \\ 1 \end{pmatrix} = -1\mathbf{v}_1 + 1\mathbf{v}_2$$

and $f(\mathbf{v}_2) = -1\mathbf{v}_1 + 0\mathbf{v}_2$, so the matrix of f with respect to the basis V is

$$\begin{pmatrix} -1 & -1 \\ 1 & 0 \end{pmatrix}.$$

(c) Let $\mathbf{v}_1$, $\mathbf{v}_2$ be the basis in $\mathcal{R}^2$ taken in Example 1(b), and let f be a rotation of 60° counterclockwise. Then $f(\mathbf{v}_1) = \mathbf{v}_2$ and $f(\mathbf{v}_2)$, the vector $\mathbf{x}$ in Fig. 16, is equal to $-\mathbf{v}_1 + \mathbf{v}_2$. The matrix of f with respect to the basis $\mathbf{v}$ is therefore

$$\begin{pmatrix} 0 & -1 \\ 1 & 1 \end{pmatrix}.$$

(d) Let P_2 be the vector space of quadratic polynomials, and let $\mathcal{W}$ be $\mathcal{R}^4$. Define $f(p)$, for p a polynomial in P_2, to be the vector in $\mathcal{R}^4$ with entries $p(1)$, $p(-1)$, $p(2)$, and $p(3)$. Take the polynomials 1, x, x^2 as a basis for P_2 and the natural basis as basis for $\mathcal{R}^4$. Then f takes the polynomial 1 into the 4-tuple $(1, 1, 1, 1)$; it takes the polynomial x into the 4-tuple $(1, -1, 2, 3)$; and it takes x^2 into $(1, 1, 4, 9)$. Its matrix with respect to the given bases is therefore

$$\begin{pmatrix} 1 & 1 & 1 \\ 1 & -1 & 1 \\ 1 & 2 & 4 \\ 1 & 3 & 9 \end{pmatrix}.$$

Given a basis $V = (\mathbf{v}_1, \ldots, \mathbf{v}_n)$ for $\mathcal{V}$, we may define a function c_V from $\mathcal{V}$ to $\mathcal{R}^n$ by setting $c_V(\mathbf{x})$ equal to the coordinate vector of $\mathbf{x}$ with respect to V for every $\mathbf{x}$ in $\mathcal{V}$. The function c_V is called the **coordinate map for the basis** V. The range of c_V is obviously all of $\mathcal{R}^n$ because, for any n-tuple $r_1, \ldots, r_n$, we can take $\mathbf{x} = r_1\mathbf{v}_1 + \ldots + r_n\mathbf{v}_n$; then $c_V(\mathbf{x})$ is the given n-tuple. By Theorem 10.1, c_V is one-to-one. Therefore there is an inverse function c_V^{-1} from $\mathcal{R}^n$ to $\mathcal{V}$. It is easily seen to be given by the formula

$$c_V^{-1}(r_1, \ldots, r_n) = r_1\mathbf{v}_1 + \ldots + r_n\mathbf{v}_n.$$

An obvious but important fact is:

10.2 Theorem

Coordinate maps and their inverses are linear functions.

Proof. To show that c_V is linear, we must verify that $c_V(r\mathbf{x} + s\mathbf{y}) = rc_V(\mathbf{x}) + sc_V(\mathbf{y})$. This follows at once from the fact that if $\mathbf{x} = \sum_{i=1}^{n} a_i\mathbf{v}_i$, and $\mathbf{y} = \sum_{i=1}^{n} b_i\mathbf{v}_i$, then $r\mathbf{x} + s\mathbf{y} = \sum_{i=1}^{n} (ra_i + sb_i)\mathbf{v}_i$. The linearity of c_V^{-1} is an immediate consequence of the same equation.

Note that if $\mathcal{V} = \mathcal{R}^n$, the coordinate map c for the natural basis is the identity function $c(\mathbf{x}) = \mathbf{x}$.

A linear function $\mathcal{V} \xrightarrow{f} \mathcal{W}$, which is one-to-one and whose range is all of $\mathcal{W}$, is called an **isomorphism,** and two spaces such that there exists an isomorphism between them are said to be **isomorphic.** We have shown that coordinate maps are isomorphisms, which proves:

10.3 Theorem

Every n-dimensional real vector space is isomorphic to $\mathcal{R}^n$.

The significance of these concepts lies in the fact that isomorphic spaces are alike in all respects, when considered as abstract vector spaces.

Any statement true for one space (provided it can be formulated entirely in terms of addition and numerical multiplication) can be carried over by an isomorphism to give a true corresponding statement for an isomorphic space. For example, if $\mathcal{V} \xrightarrow{f} \mathcal{W}$ is an isomorphism, then an equation $\mathbf{y} = \sum_{i=1}^{n} a_i\mathbf{x}_i$ is true in $\mathcal{V}$ if and only if $f(\mathbf{y}) = \sum_{i=1}^{n} a_i f(\mathbf{x}_i)$ is true in $\mathcal{W}$.

By using coordinate maps we can give an alternative description of the matrix of a linear function. Given $\mathcal{V} \xrightarrow{f} \mathcal{W}$, and bases $V = (\mathbf{v}_1, \ldots, \mathbf{v}_n)$ in $\mathcal{V}$ and $W = (\mathbf{w}_1, \ldots, \mathbf{w}_m)$ in $\mathcal{W}$, the composition $c_W \circ f \circ c_V^{-1}$ is a function from $\mathcal{R}^n$ to $\mathcal{R}^m$. It is therefore given by multiplication by some m-by-n matrix A; we claim that A is precisely the same as the matrix of f with respect to the bases V and W. To prove the assertion, note that the jth column of A is $(c_W \circ f \circ c_V^{-1})(\mathbf{e}_j) = c_W(f(\mathbf{v}_j))$, since $c_V^{-1}(\mathbf{e}_j) = \mathbf{v}_j$. Now $c_W(f(\mathbf{v}_j))$ is (by definition of c_W) the coordinate vector of $f(\mathbf{v}_j)$ with respect to the basis W, and this is just how the matrix of f with respect to bases V and W was defined. This description of the matrix of a function makes the proof of the following generalization of Theorem 4.4 of Chapter 1 very easy. It simply says that, in general, matrix multiplication corresponds to composition of functions, provided one uses bases consistently.

10.4 Theorem

Let f and g be linear functions with $\mathcal{U} \xrightarrow{f} \mathcal{V}$ and $\mathcal{V} \xrightarrow{f} \mathcal{W}$. Suppose bases U, V, and W are given in $\mathcal{U}$, $\mathcal{V}$, and $\mathcal{W}$, that A is the matrix of f with respect to the bases U and V, and that B is the matrix of g with respect to the bases V and W. Then BA is the matrix of $g \circ f$ with respect to the bases U and W.

Proof. By Theorem 4.3 of Chapter 1 and the characterization of matrices of functions by coordinate maps, this is simply the statement that

$$(c_W \circ g \circ c_V^{-1}) \circ (c_V \circ f \circ c_U^{-1}) = c_W \circ (g \circ f) \circ c_U^{-1},$$

which is clear because composition of functions is associative.

The special case in which $\mathcal{U}$, $\mathcal{V}$, and $\mathcal{W}$ are all the same space with the same basis is particularly important. If $\mathcal{V} \xrightarrow{f} \mathcal{V}$ has matrix A, then $f \circ f$ has matrix A^2, $f \circ f \circ f$ has matrix A^3, etc.

Example 3. (a) For the function f of Example 2(b), it is clear that $f \circ f \circ f$ is the identity, and it is easy to verify that

$$\begin{pmatrix} 1 & 1 \\ -1 & 0 \end{pmatrix}^3 = I.$$

(b) Differentiation is a linear function from the space of polynomials to itself. If we define $D(p)$ to be p' for elements of the space of quadratic polynomials and use the basis $(1, x, x^2)$ as in Example 2(d), then the matrix of D is

$$A = \begin{pmatrix} 0 & 1 & 0 \\ 0 & 0 & 2 \\ 0 & 0 & 0 \end{pmatrix}.$$

It is easy to compute that

$$A^2 = \begin{pmatrix} 0 & 0 & 2 \\ 0 & 0 & 0 \\ 0 & 0 & 0 \end{pmatrix}$$

and $A^3 = 0$, which corresponds to the fact that the third derivative of any quadratic polynomial is 0.

The matrix of a linear function with respect to a pair of bases, of course, depends on the bases. Since the matrix with respect to any pair completely determines the function, it should be possible to compute the matrix of a function with respect to one pair of bases from its matrix with respect to any other.

Let us look at an example. Suppose f has the matrix

$$\begin{pmatrix} 1 & 2 & 0 \\ -1 & 1 & 3 \end{pmatrix}$$

with respect to the natural bases in $\mathcal{R}^3$ and $\mathcal{R}^2$. Consider the bases

$$\mathbf{v}_1 = \begin{pmatrix} 1 \\ 2 \\ 3 \end{pmatrix}, \quad \mathbf{v}_2 = \begin{pmatrix} 0 \\ 1 \\ 2 \end{pmatrix}, \quad \mathbf{v}_3 = \begin{pmatrix} 2 \\ 0 \\ 1 \end{pmatrix}$$

in $\mathcal{R}^3$ and

$$\mathbf{w}_1 = \begin{pmatrix} 1 \\ 2 \end{pmatrix}, \quad \mathbf{w}_2 = \begin{pmatrix} 2 \\ 3 \end{pmatrix}$$

in $\mathcal{R}^2$. To find the matrix of f with respect to $(\mathbf{v}_1, \mathbf{v}_2, \mathbf{v}_3)$ and $(\mathbf{w}_1, \mathbf{w}_2)$, we compute

$$f(\mathbf{v}_1) = \begin{pmatrix} 5 \\ 10 \end{pmatrix} = 5\mathbf{w}_1,$$

$$f(\mathbf{v}_2) = \begin{pmatrix} 2 \\ 7 \end{pmatrix} = 8\mathbf{w}_1 - 3\mathbf{w}_2, \qquad f(\mathbf{v}_3) = \begin{pmatrix} 2 \\ 1 \end{pmatrix} = -4\mathbf{w}_1 + 3\mathbf{w}_2.$$

Hence the required matrix for f is

$$\begin{pmatrix} 5 & 8 & -4 \\ 0 & -3 & 3 \end{pmatrix}.$$

A basis $V = (\mathbf{v}_1, \ldots, \mathbf{v}_n)$ can be described in terms of another basis $X = (\mathbf{x}_1, \ldots, \mathbf{x}_n)$ by using the matrix whose jth column is the coordinate vector of $\mathbf{v}_j$ with respect to X. This matrix gives a function from $\mathcal{R}^n$ to $\mathcal{R}^n$, which can be recognized as $c_X \circ c_V^{-1}$ by observing its effect on the natural basis vectors $\mathbf{e}_j$. Multiplication by this matrix converts the V-coordinates of any vector $\mathbf{w}$ into the X-coordinates of $\mathbf{w}$, since

$$c_X(\mathbf{w}) = (c_X \circ c_V^{-1}) \circ c_V(\mathbf{w})$$

The inverse matrix gives the $\mathbf{x}$'s in terms of the $\mathbf{v}$'s and corresponds to $c_V \circ c_X^{-1}$. In the example in the preceding paragraph, the matrices giving V and W in terms of the natural bases in $\mathcal{R}^2$ and $\mathcal{R}^3$ are

$$\begin{pmatrix} 1 & 0 & 2 \\ 2 & 1 & 0 \\ 3 & 2 & 1 \end{pmatrix} \quad \text{and} \quad \begin{pmatrix} 1 & 2 \\ 2 & 3 \end{pmatrix}.$$

Since $\mathbf{e}_1 = -3\mathbf{w}_1 + 2\mathbf{w}_2$ and $\mathbf{e}_2 = 2\mathbf{w}_1 - \mathbf{w}_3$, the matrix giving the $\mathbf{e}$'s in terms of the $\mathbf{w}$'s is

$$\begin{pmatrix} -3 & 2 \\ 2 & -1 \end{pmatrix},$$

which is easily checked to be

$$\begin{pmatrix} 1 & 2 \\ 2 & 3 \end{pmatrix}^{-1}.$$

In the general situation we have a space $\mathcal{V}$ with bases $X = (\mathbf{x}_1, \ldots, \mathbf{x}_n)$ and $V = (\mathbf{v}_1, \ldots, \mathbf{v}_n)$, and a space $\mathcal{W}$ with bases $Y = (\mathbf{y}_1, \ldots, \mathbf{y}_m)$ and $W = (\mathbf{w}_1, \ldots, \mathbf{w}_m)$. Let P and Q be the matrices giving V in terms of X, and W in terms of Y, and let f be a linear function from $\mathcal{V}$ to $\mathcal{W}$ whose matrix with respect to X and Y is A. Then the matrix of f with respect to V and W is $Q^{-1}AP$. This is most easily seen by working with the coordinate maps. The matrix for f with respect to V and W corresponds to

$$c_W \circ f \circ c_V^{-1} = (c_W \circ c_Y^{-1}) \circ (c_Y \circ f \circ c_X^{-1}) \circ (c_X \circ c_V^{-1}),$$

and the three factors on the right correspond to Q^{-1}, A, and P, respectively.

For the previous example,

$$\begin{aligned} Q^{-1}AP &= \begin{pmatrix} -3 & 2 \\ 2 & -1 \end{pmatrix} \begin{pmatrix} 1 & 2 & 0 \\ -1 & 1 & 3 \end{pmatrix} \begin{pmatrix} 1 & 0 & 2 \\ 2 & 1 & 0 \\ 3 & 2 & 1 \end{pmatrix} \\ &= \begin{pmatrix} 5 & 8 & -4 \\ 0 & -3 & 3 \end{pmatrix}, \end{aligned}$$

which of course is the same result as before.

If $\mathcal{V}$ and $\mathcal{W}$ are the same space, and X is the same basis as Y, and V is the same basis as W, then P and Q are the same, and the new matrix for f is $P^{-1}AP$. Two matrices A and B are said to be **similar** if there exists an invertible matrix P such that $B = PAP^{-1}$. A few properties of similarity are presented in the exercises.

Example 4. The derivative function D has the matrix

$$\begin{pmatrix} 0 & 1 & 0 \\ 0 & 0 & 2 \\ 0 & 0 & 0 \end{pmatrix}$$

with respect to the basis $(1, x, x^2)$ for the space of quadratic polynomials (see Example 3(b)). The polynomials $1 + x$, $x + x^2$ and $1 + x^2$ are also a basis, given in terms of $1, x, x^2$ by the matrix

$$\begin{pmatrix} 1 & 0 & 1 \\ 1 & 1 & 0 \\ 0 & 1 & 1 \end{pmatrix}.$$

The inverse matrix is

$$\tfrac{1}{2}\begin{pmatrix} 1 & 1 & -1 \\ -1 & 1 & 1 \\ 1 & -1 & 1 \end{pmatrix}.$$

With respect to the new basis, D has the matrix

$$\tfrac{1}{2}\begin{pmatrix} 1 & 1 & -1 \\ -1 & 1 & 1 \\ 1 & -1 & 1 \end{pmatrix}\begin{pmatrix} 0 & 1 & 0 \\ 0 & 0 & 2 \\ 0 & 0 & 0 \end{pmatrix}\begin{pmatrix} 1 & 0 & 1 \\ 1 & 1 & 0 \\ 0 & 1 & 1 \end{pmatrix}$$

$$= \tfrac{1}{2}\begin{pmatrix} 1 & 1 & -1 \\ -1 & 1 & 1 \\ 1 & -1 & 1 \end{pmatrix}\begin{pmatrix} 1 & 1 & 0 \\ 0 & 2 & 2 \\ 0 & 0 & 0 \end{pmatrix}$$

$$= \tfrac{1}{2}\begin{pmatrix} 1 & 3 & 2 \\ -1 & 1 & 2 \\ 1 & -1 & -2 \end{pmatrix}$$

EXERCISES

1. Let $\mathbf{v}_1 = \begin{pmatrix} 2 \\ 0 \\ 0 \end{pmatrix}$, $\mathbf{v}_2 = \begin{pmatrix} 1 \\ -1 \\ 2 \end{pmatrix}$, and $\mathbf{v}_3 = \begin{pmatrix} 0 \\ -2 \\ 3 \end{pmatrix}$. Verify that $V = (\mathbf{v}_1, \mathbf{v}_2, \mathbf{v}_3)$ is a basis for $\mathcal{R}^3$ and find the coordinate vectors of $\mathbf{x}_1 = \begin{pmatrix} 0 \\ 0 \\ 1 \end{pmatrix}$, $\mathbf{x}_2 = \begin{pmatrix} 1 \\ 0 \\ 2 \end{pmatrix}$, and $\mathbf{x}_3 = \begin{pmatrix} 3 \\ 0 \\ 1 \end{pmatrix}$ with respect to V.

2. Find the matrix of a rotation of 45° in $\mathcal{R}^2$ with respect to:

(a) The basis $\mathbf{x}_1 = \begin{pmatrix} 1 \\ 1 \end{pmatrix}$, $\mathbf{x}_2 = \begin{pmatrix} 0 \\ -2 \end{pmatrix}$. $\left[\textit{Ans.} \begin{pmatrix} 0 & \sqrt{2} \\ -1/\sqrt{2} & \sqrt{2} \end{pmatrix}. \right]$

(b) The basis of unit vectors in the directions of $\mathbf{x}_1$ and $\mathbf{x}_2$.

3. Show that the elementary matrices E_{ij} of shape m-by-n form a basis for the vector space of m-by-n matrices. What is the dimension of the space?

4. Let $A = \begin{pmatrix} 1 & 2 \\ 0 & 1 \end{pmatrix}$. Show that each of the following functions from the space of 2-by-2 matrices to itself is linear, and find the matrix of each with respect to the basis of Exercise 3.

(a) $f(X) = AX$. (b) $g(X) = AXA^{-1}$.

5. Find bases for the null spaces of the linear functions defined on the space of 2-by-2 matrices by:

(a) $f(X) = AX - XA$, (b) $g(X) = AX - X^tA^t$,

where A is the same as in Exercise 4.

6. Let $\mathcal{V} \xrightarrow{f} \mathcal{W}$ be an isomorphism. Show that $\mathbf{v}_1, \ldots, \mathbf{v}_n$ form a basis for $\mathcal{V}$ if and only if $f(\mathbf{v}_1), \ldots, f(\mathbf{v}_n)$ form a basis for $\mathcal{W}$. This proves that isomorphic spaces have the same dimension.

7. For any two vector spaces $\mathcal{V}$, $\mathcal{W}$, let $\mathcal{L}(\mathcal{V}, \mathcal{W})$ consist of all linear functions from $\mathcal{V}$ to $\mathcal{W}$. For f and g in $\mathcal{L}(\mathcal{V}, \mathcal{W})$ and r a number, define functions

$$f + g \quad \text{by} \quad (f + g)(\mathbf{v}) = f(\mathbf{v}) + g(\mathbf{v})$$

and

$$r \cdot f \quad \text{by} \quad (r \cdot f)(\mathbf{v}) = r \cdot f(\mathbf{v}).$$

Show that $f + g$ and $r \cdot f$ are linear and so are in $\mathcal{L}(\mathcal{V}, \mathcal{W})$, and show that with these operations, $\mathcal{L}(\mathcal{V}, \mathcal{W})$ is a vector space.

8. (a) Show that $\mathcal{L}(\mathcal{R}^n, \mathcal{R}^m)$ is isomorphic to the space of all m-by-n matrices.
(b) Show that if $\dim(\mathcal{V}) = n$ and $\dim(\mathcal{W}) = m$, and bases are chosen in $\mathcal{V}$ and $\mathcal{W}$, then assigning to each f in $\mathcal{L}(\mathcal{V}, \mathcal{W})$ its matrix with respect to the given bases gives an isomorphism between $\mathcal{L}(\mathcal{V}, \mathcal{W})$ and the space of m-by-n matrices.

9. Let P_n be the space of polynomials of degree less than or equal to n. Show that the function t defined by $t(p(x)) = p(x + 1)$ (so that, for example, $t(2x + 1) = 2x + 3$) is a linear function from P_n to itself.

(a) Write the matrix of t with respect to the basis $1, x, x^2$ for the space P_2.
(b) Verify by matrix calculation that on P_2

$$t = 1 + D + \frac{D^2}{2},$$

where D is the derivative function of Example 3(b). (Here "1" is to be interpreted as the identity function, and D^2 as $D \circ D$.)
(c) Show that on P_n

$$t = 1 + D + \frac{D^2}{2} + \frac{D^3}{6} + \ldots + \frac{D^n}{n!}$$

[*Hint.* Use the definitions of the functions directly, without bringing in coordinates.]

10. Show that if A and B are similar, then

(a) $\det A = \det B$.
(b) A^{-1} is similar to B^{-1} (if the inverses exist).
(c) A^n is similar to B^n for any positive integer n.

11. The **trace** of a square matrix A is defined to be the sum of its diagonal elements and is written $\operatorname{tr}(A)$. For example $\operatorname{tr}\begin{pmatrix} 1 & 2 \\ 3 & 4 \end{pmatrix} = 1 + 4 = 5$, and if I is the n-by-n identity matrix, $\operatorname{tr}(I) = n$.

(a) Show that if A and B are square and have the same size, then $\operatorname{tr}(AB) = \operatorname{tr}(BA)$.
(b) Show that if A and B are similar, then $\operatorname{tr}(A) = \operatorname{tr}(B)$. [*Hint.* This can be proved in one line by applying part (a) to the right pair of matrices.]

12. Suppose $\mathcal{R}^3 \xrightarrow{f} \mathcal{R}^3$ is a rotation and $\mathbf{v}_1$ is a unit vector in the direction of the axis of the rotation. If $\mathbf{v}_2$ and $\mathbf{v}_3$ are chosen to make $(\mathbf{v}_1, \mathbf{v}_2, \mathbf{v}_3)$ an orthonormal basis, then $f(\mathbf{v}_1) = \mathbf{v}_1$ and $f(\mathbf{v}_2)$, $f(\mathbf{v}_3)$ will be perpendicular to $\mathbf{v}_1$.

(a) Show that the matrix of f with respect to $(\mathbf{v}_1, \mathbf{v}_2, \mathbf{v}_3)$ is

$$\begin{pmatrix} 1 & 0 & 0 \\ 0 & \cos\alpha & -\sin\alpha \\ 0 & \sin\alpha & \cos\alpha \end{pmatrix},$$

where α is the angle of the rotation. (Compare Exercise 4 of Section 4, Chapter 1.)

(b) Use the result of Exercise 11 to show that if A is the matrix of a rotation of angle α with respect to *any* basis, then $\cos \alpha = \frac{1}{2}(\text{tr}(A) - 1)$.

13. The linear function on $\mathcal{R}^3$ defined by

$$f(\mathbf{e}_1) = \mathbf{e}_2, \qquad f(\mathbf{e}_2) = \mathbf{e}_3, \qquad f(\mathbf{e}_3) = \mathbf{e}_1$$

is clearly a rotation about the line $x_1 = x_2 = x_3$. Find its angle by geometrical reasoning (what is $f \circ f \circ f$?) and check the result of Exercise 12. [*Ans.* 120°.]

14. Show that the matrix

$$A = \begin{pmatrix} \frac{2}{7} & \frac{6}{7} & \frac{3}{7} \\ \frac{3}{7} & \frac{2}{7} & -\frac{6}{7} \\ -\frac{6}{7} & \frac{3}{7} & -\frac{2}{7} \end{pmatrix}$$

represents a rotation of $\mathcal{R}^3$ and find its axis and angle. (If $\mathbf{x}$ is a unit vector in the direction of the axis, then $A\mathbf{x} = \mathbf{x}$ and so $(A - I)\mathbf{x} = 0$. To show that A represents a rotation, find its matrix with respect to an orthonormal basis that includes $\mathbf{x}$.) [*Ans.* Axis is (3, 3, −1).]

3

Derivatives

SECTION 1

VECTOR FUNCTIONS

The purpose of this section is to make the reader familiar with some specific examples of vector functions. These examples will be given by formulas that can be analyzed in terms of the elementary functions of one-variable calculus. Wherever possible we will give a pictorial description of the function, something that can often be done in more than one way. In later parts of the book we shall often return to examples like the ones here, and so develop more familiarity with them, and with the ways in which they can be applied.

Example 1. The function $\mathcal{R}^3 \xrightarrow{f} \mathcal{R}^2$ defined by

$$f(x, y, z) = \begin{pmatrix} x^2 + y^2 + z^2 \\ x + y + z \end{pmatrix}$$

has as its domain all of $\mathcal{R}^3$; but, because $x^2 + y^2 + z^2 \geq 0$, its image contains only those vectors in $\mathcal{R}^2$ for which the first coordinate is nonnegative. A different function $\mathcal{R}^3 \xrightarrow{g} \mathcal{R}^2$ is defined by

$$g(x, y, z) = \begin{pmatrix} 3x + 4y \\ 3y + 5z \end{pmatrix} = \begin{pmatrix} 3 & 4 & 0 \\ 0 & 3 & 5 \end{pmatrix} \begin{pmatrix} x \\ y \\ z \end{pmatrix},$$

and g is linear because it is given by matrix multiplication acting on the vectors of $\mathcal{R}^3$. The function $\mathcal{R}^2 \xrightarrow{h} \mathcal{R}^2$ defined by

$$h(x, y) = \begin{pmatrix} \sqrt{1 - x^2 - y^2} \\ x + y \end{pmatrix},$$

for vectors (x, y) such that $x^2 + y^2 \leq 1$, has as its domain a circular disk D in $\mathcal{R}^2$. The notation $f: D \to \mathcal{R}^m$ is often used to denote a function $\mathcal{R}^n \xrightarrow{f} \mathcal{R}^m$ with domain D contained in $\mathcal{R}^n$.

Example 2. The functions defined by the following formulas have both domain space and range space equal to the set $\mathcal{R}$ of real numbers:

$$f(x) = x^3,$$

$$g(x) = \sqrt{x}, \qquad x \geq 0,$$

$$h(x) = \sin x.$$

The domain and range of f are both equal to $\mathcal{R}$. According to the usual interpretation of the square-root symbol, the range of g is the set of non-negative real numbers, and so also is the domain of g. The domain of h is $\mathcal{R}$, while the range of h is the interval $-1 \leq y \leq 1$. To emphasize what its domain is, the function g might be denoted

$$g: [0, \infty) \to \mathcal{R},$$

where $[0, \infty)$ stands for the interval $0 \leq x < \infty$.

A vector $\mathbf{x}$ in $\mathcal{R}_n$ whose coordinates are the real numbers $x_1, \ldots, x_n$ will be written either as a horizontal n-tuple or as a column. Thus, we shall write both

$$\begin{pmatrix} x_1 \\ \cdot \\ \cdot \\ \cdot \\ x_n \end{pmatrix} \quad \text{and} \quad (x_1, \ldots, x_n)$$

for the vector $\mathbf{x}$. The practice of writing columns instead of horizontal tuples arises, of course, from the definition of matrix multiplication. If a function is determined by a matrix

$$\begin{pmatrix} a & b \\ c & d \end{pmatrix},$$

we usually write

$$\begin{pmatrix} a & b \\ c & d \end{pmatrix} \begin{pmatrix} x \\ y \end{pmatrix}$$

for the value of the function at

$$\begin{pmatrix} x \\ y \end{pmatrix} \quad \text{or} \quad (x, y).$$

A function whose range is a subset of the space $\mathcal{R}$ of all real numbers is called **real-valued.** Every function $\mathcal{R}^n \xrightarrow{f} \mathcal{R}^m$ defines a set of real-valued functions $f_1, \ldots, f_m$, called the **coordinate functions** of f; we set $f_i(\mathbf{x})$ equal to the ith coordinate of $f(\mathbf{x})$. Thus

$$f(\mathbf{x}) = (f_1(\mathbf{x}), \ldots, f_m(\mathbf{x}))$$

for every $\mathbf{x}$ in the domain of f, and f_i is called the ith coordinate function of f.

Example 3. Consider the vector function

$$f\begin{pmatrix} x \\ y \\ z \end{pmatrix} = \begin{pmatrix} x + y + z \\ xy + yz + zx \\ xyz \end{pmatrix}.$$

The coordinate functions of f are the three real-valued functions

$$f_1(x, y, z) = x + y + z,$$
$$f_2(x, y, z) = xy + yz + zx,$$
$$f_3(x, y, z) = xyz.$$

The **graph** of a function f is defined to be the set of ordered pairs $(\mathbf{x}, f(\mathbf{x}))$ where $\mathbf{x}$ is in the domain of f. In studying real-valued functions of one real variable, graphs are a considerable aid to understanding. For example, the graph of the function defined by $f(x) = x^2 - 2$ is the set of all ordered pairs (x, y) with $y = x^2 - 2$, that is, it is the subset of the xy-plane consisting of the parabola shown in Fig. 1. As another example, the

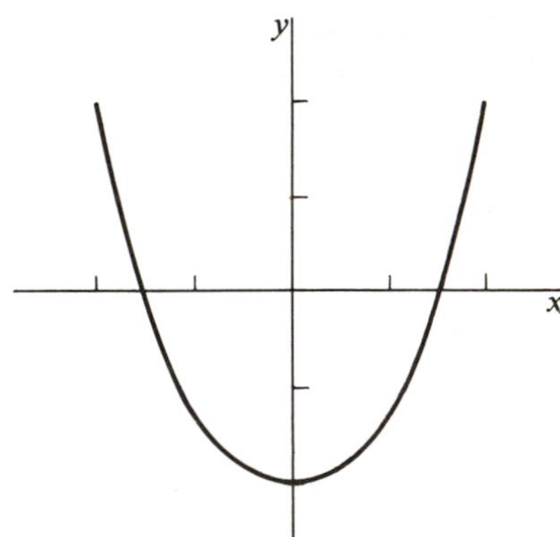

Figure 1

graph of a function $\mathcal{R}^2 \xrightarrow{f} \mathcal{R}$ is the subset of $\mathcal{R}^3$ consisting of all points

$$(x, y, z) = (x, y, f(x, y)).$$

As a means of increasing understanding by visualization, a graph is useful only for functions $\mathcal{R}^n \xrightarrow{f} \mathcal{R}^m$ for which $m + n \leq 3$.

Example 4. To sketch the graph of the function f defined by

$$f(x, y) = x^2 + y^2,$$

recall that it consists of all points in $\mathcal{R}^3$ of the form $(x, y, z) = (x, y, x^2 + y^2)$. One thing we can do is plot several of these, for example, $(1, 1, 2)$, $(0, 1, 1)$, $(1, 0, 1)$, etc. Another approach is to draw some of the curves that lie on the graph. In particular, we can draw cross sections of the graph obtained by holding either x or y fixed and letting the other one vary. The result is a curve in a plane parallel either to the yz- or xz-coordinate plane. Two of these are shown in Fig. 2. We have chosen

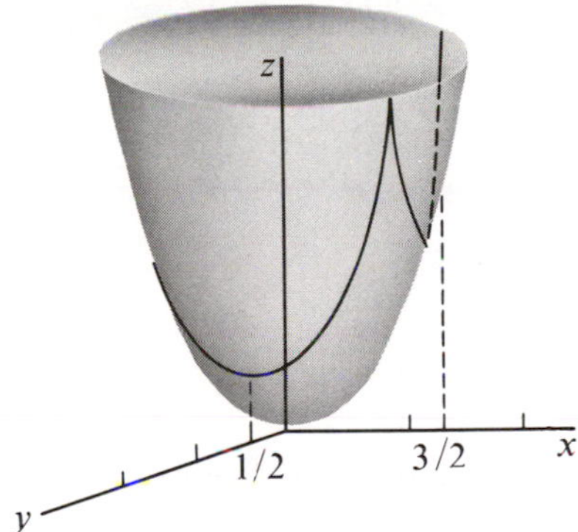

Figure 2

$x = \frac{3}{2}$ for one of them. The result is the set of points $(x, y, z) = (\frac{3}{2}, y, \frac{9}{4} + y^2)$. To sketch them we can think first of the points $(y, z) = (y, \frac{9}{4} + y^2)$ in the yz-plane. Then move the resulting curve $\frac{3}{2}$ units in the direction of the positive x-axis. Similarly, choosing $y = \frac{1}{2}$, we sketch the curve of points $(x, y, z) = (x, \frac{1}{2}, x^2 + \frac{1}{4})$. Doing this for several values of x and y gives a better picture than simply plotting individual points.

Example 5. Consider the function $\mathcal{R} \xrightarrow{f} \mathcal{R}^2$ defined by

$$f(t) = \begin{pmatrix} \cos t \\ \sin t \end{pmatrix}, \qquad \text{for every } t \text{ in } \mathcal{R}.$$

Since the length $|\mathbf{x}|$ of a vector $\mathbf{x} = (x, y)$ is given by $|\mathbf{x}| = \sqrt{x^2 + y^2}$, we have $|f(t)| = \sqrt{\cos^2 t + \sin^2 t} = 1$. Thus the range of f is a subset of the unit circle $|\mathbf{x}| = 1$ in $\mathcal{R}^2$. The number t is interpreted geometrically as the angle in radians between the vector $f(t)$ and the positive x-axis. As t runs through $\mathcal{R}$, the unit circle is covered infinitely often. It follows that the range of f is the whole unit circle. The circle is not, however, the graph of f. The latter is a subset of $\mathcal{R}^3$ and is a spiral whose axis is the t-axis. See Fig. 3(b). What we have done is sketched the points of the form $(t, x, y) = (t, \cos t, \sin t)$ that make up the graph of the function.

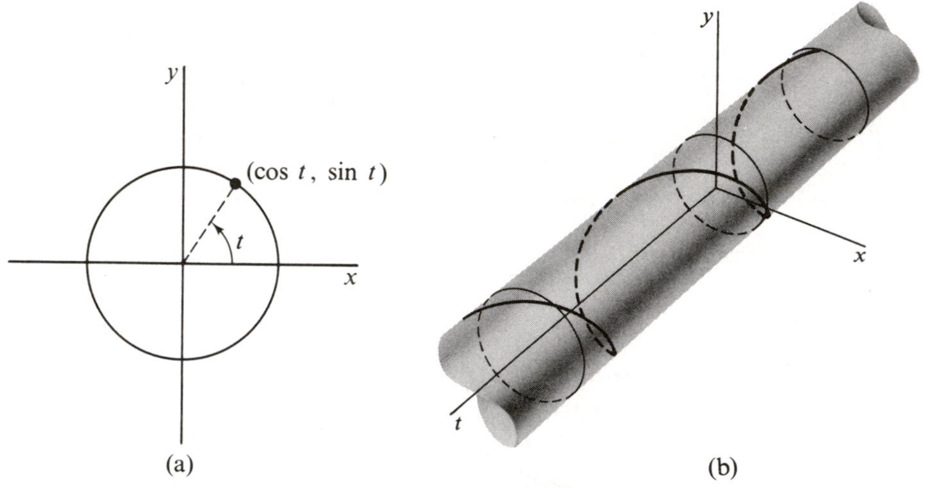

Figure 3

Example 6. Let the vector function f be defined by

$$f(t) = \begin{pmatrix} x \\ y \\ z \end{pmatrix} = \begin{pmatrix} t \\ t^2 \\ t^3 \end{pmatrix}, \qquad -\infty < t < \infty.$$

The graph of f is a subset of $\mathcal{R}^4$, so we shall not attempt to draw it. Instead we shall sketch the range. By setting $z = 0$, we obtain the equations $x = t$, $y = t^2$, which are equivalent to $y = x^2$. Thus the projected image, on the xy-plane, of the range of f is the graph of $y = x^2$. Similarly, in the yz-plane we obtain $y = z^{2/3}$, and in the xz-plane we get $z = x^3$. From this information, we have drawn in Fig. 4 that part of the range of f that lies above the first quadrant of the xy-plane.

In drawing Fig. 4 we have labeled the axes in the manner usually associated with a right-hand orientation. Of course, if we were to interchange x and y in this labeling, the picture would look different. A

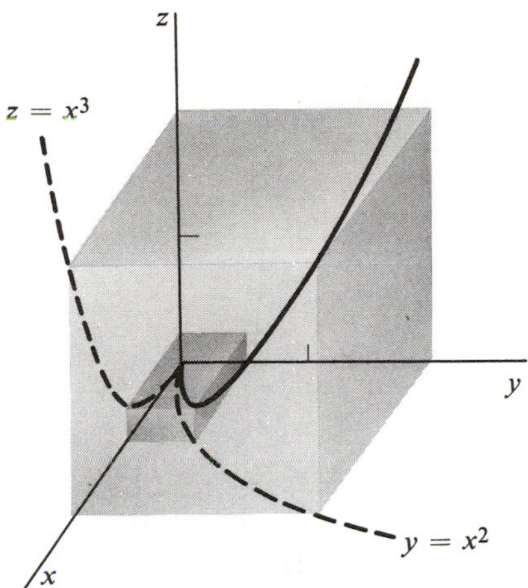

Figure 4

similar change in Fig. 5(b) would result in a ramp that spirals down, turning always to the left instead of to the right. To make it easy to see the relationship between the pictures, we have always chosen the right-hand orientation for the axes in the 3-dimensional ones. For a discussion of orientation see Section 7 of Chapter 1.

Example 7. Consider the function

$$f\begin{pmatrix} u \\ v \end{pmatrix} = \begin{pmatrix} x \\ y \\ z \end{pmatrix} = \begin{pmatrix} u \cos v \\ u \sin v \\ v \end{pmatrix} \qquad \begin{matrix} 0 \le u \le 4, \\ 0 \le v \le 2\pi. \end{matrix}$$

The domain of f is the shaded rectangle in Fig. 5(a). To sketch the range, we proceed as follows. Choose a number a in the interval $0 \le u \le 4$, and set $u = a$. Then,

$$\left.\begin{aligned} x &= a \cos v \\ y &= a \sin v \\ z &= v \end{aligned}\right\} \qquad 0 \le v \le 2\pi,$$

and $x^2 + y^2 = a^2$. We interpret v both as distance along the z-axis and as the angle between $(x, y, 0)$ and the x-axis. It follows that the image under f of the line segment $u = a$, for $0 \le v \le 2\pi$ (see Fig. 5(a)) is the spiral whose projection on the xy-plane is the circle of radius a and whose axis is the z-axis (see Fig. 5(b)). Next, choose a number b in the interval

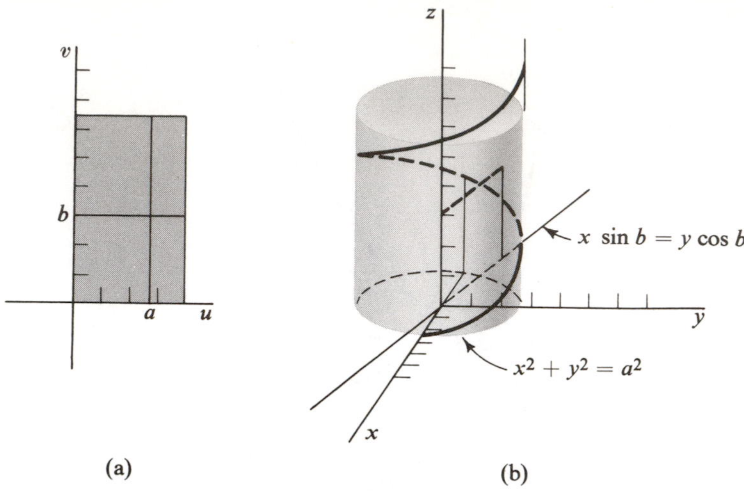

Figure 5

$0 \leq v \leq 2\pi$, and set $v = b$. Then,

$$\left.\begin{aligned} x &= u \cos b \\ y &= u \sin b \\ z &= b \end{aligned}\right\} \qquad 0 \leq u \leq 4,$$

and $x^2 + y^2 = u^2$. The image under f of the line segment $v = b, 0 \leq u \leq 4$ (see Fig. 5(a)), is the line segment $x \sin b = y \cos b$, $z = b$, of length 4, where x runs from 0 to $4 \cos b$ (see Fig. 5(b)). Letting a and b vary, we obtain the range of f as the spiral surface shown in Fig. 6.

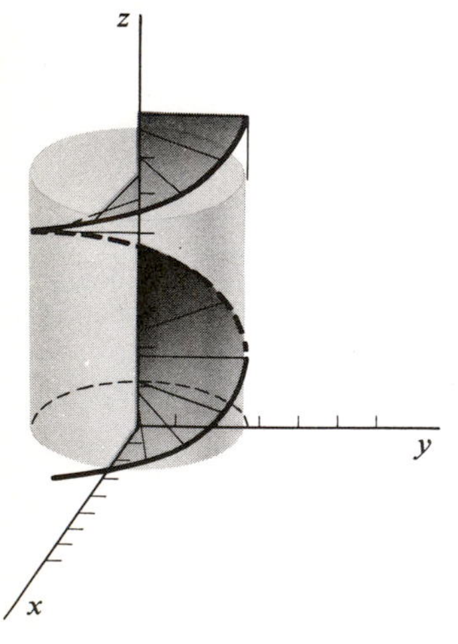

Figure 6

Notice that the range of f in Examples 5 and 6 is a curve, whereas in Example 7 it is a surface. Similarly, the graph of f in Example 4 is a surface, but in Example 5 it is a curve. The evidence suggests that, when the dimension of the domain space is 1, we get a curve, and when it is 2, we get a surface. Exceptions to this are discussed in the last section of Chapter 4.

The curves and surfaces pictured in Figs. 2, 3, 4, and 6 are related to the vector functions that define them in two essentially different ways. The bowl-shaped surface in Fig. 2 is just the graph of the function which defines it, and the same is true for the spiral curve in Fig. 3(b). We shall say that both curve and surface are defined **explicitly.** On the other hand, the curve in Fig. 4 and the surface in Fig. 6 are the ranges of their defining functions. They are said to be defined **parametrically.** This terminology is standard in discussing real-valued functions of one real variable. For example, the function

$$f(x) = \sqrt{16 - x^2}$$

explicitly defines the upper half of the circle of radius 4, and the same curve is defined parametrically by the pair of functions

$$\begin{cases} x(t) = 4\cos t, \\ y(t) = 4\sin t, \end{cases} \qquad 0 \le t \le \pi.$$

Parametric representations of lines in 3-dimensional space have been studied in Chapter 1. Let $\mathbf{x}_1$ and $\mathbf{x}_2$ be any two vectors in $\mathcal{R}^3$. If $\mathbf{x}_1 \neq 0$, the range of the function $\mathcal{R} \xrightarrow{f} \mathcal{R}^3$ defined by

$$f(t) = t\mathbf{x}_1 + \mathbf{x}_2, \qquad -\infty < t < \infty,$$

is a parametrically defined line.

A curve or a surface can also be defined **implicitly** as a level set of a function. A set S is a **level set** of f if for some point $\mathbf{k}$ in the range of f, S consists of all $\mathbf{x}$ in the domain of f such that $f(\mathbf{x}) = \mathbf{k}$. The most easily visualized examples occur for functions f from $\mathcal{R}^2$ to $\mathcal{R}$. Then the graph of f can be pictured as are the surfaces in Fig. 7. The level sets corresponding

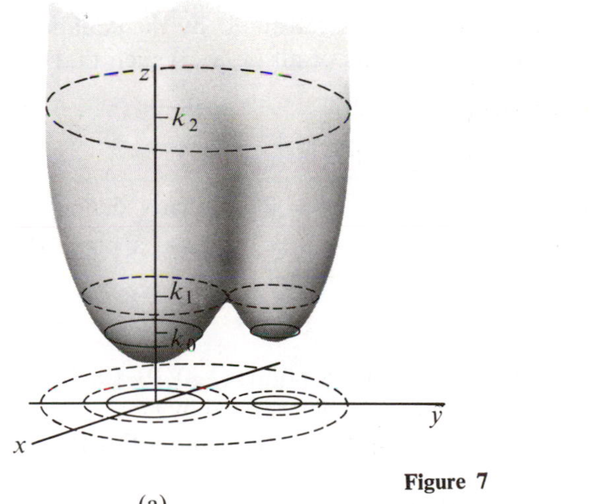

(a)

(b)

Figure 7

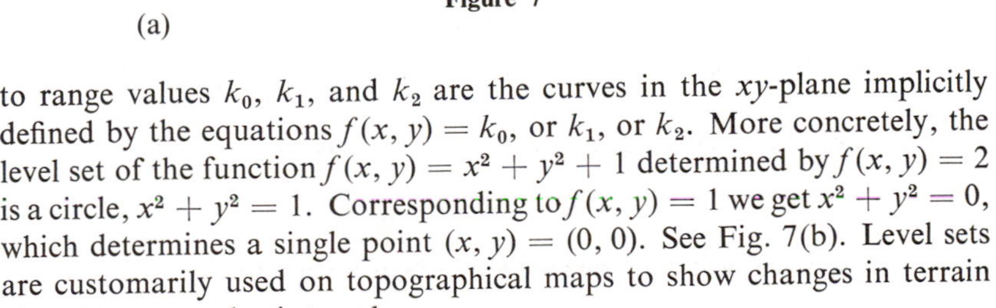

to range values k_0, k_1, and k_2 are the curves in the xy-plane implicitly defined by the equations $f(x, y) = k_0$, or k_1, or k_2. More concretely, the level set of the function $f(x, y) = x^2 + y^2 + 1$ determined by $f(x, y) = 2$ is a circle, $x^2 + y^2 = 1$. Corresponding to $f(x, y) = 1$ we get $x^2 + y^2 = 0$, which determines a single point $(x, y) = (0, 0)$. See Fig. 7(b). Level sets are customarily used on topographical maps to show changes in terrain elevation at regular intervals.

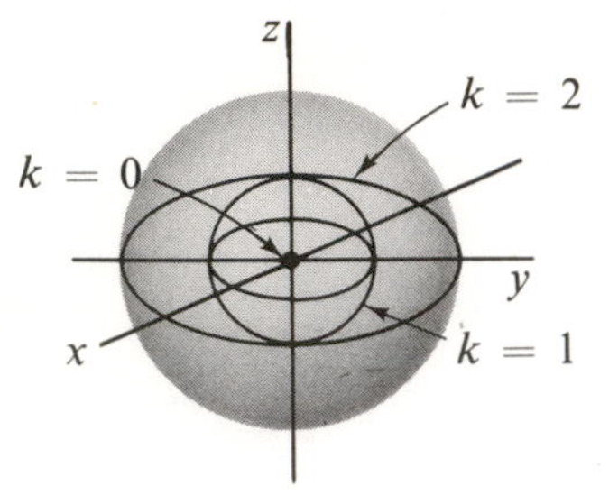

Figure 8

Example 8. Consider the function f defined by $f(x, y, z) = x^2 + y^2 + z^2$. The subset S of $\mathcal{R}^3$ consisting of all points (x, y, z) that satisfy an equation

$$x^2 + y^2 + z^2 = k \tag{1}$$

is implicitly defined by Equation (1). If $k > 0$, we get a sphere of radius $\sqrt{k}$ centered at $(0, 0, 0)$, because $x^2 + y^2 + z^2$ is the square of the distance from (x, y, z) to $(0, 0, 0)$. If $k = 0$, we get the single point $(0, 0, 0)$. Finally, if $k < 0$, the corresponding level set S is empty. In this example the graph of f is a subset of $\mathcal{R}^4$ and cannot be pictured. Hence it is desirable to at least draw the level sets and get an idea of which points in the domain are sent by f into certain fixed points in the range. Some of these are shown in Fig. 8 as a point and two concentric spheres.

Example 9. Let $\mathcal{R}^3 \xrightarrow{f} \mathcal{R}^2$ be defined by

$$f(x, y, z) = \begin{pmatrix} x^2 + y^2 + z^2 \\ x + y + z \end{pmatrix}.$$

We shall describe the level set γ implicitly defined in $\mathcal{R}^3$ by the equation $f(x, y, z) = (2, 1)$. In other words, γ consists of all (x, y, z) such that

$$\begin{aligned} x^2 + y^2 + z^2 &= 2 \\ x + y + z &= 1. \end{aligned} \tag{2}$$

We have seen in Example 8 that $x^2 + y^2 + z^2 = 2$ implicitly defines a sphere S of radius $\sqrt{2}$. The equation $x + y + z = 1$ defines a plane P, because it is satisfied by the set of (x, y, z) such that

$$(1, 1, 1) \cdot (x, y, z - 1) = 0.$$

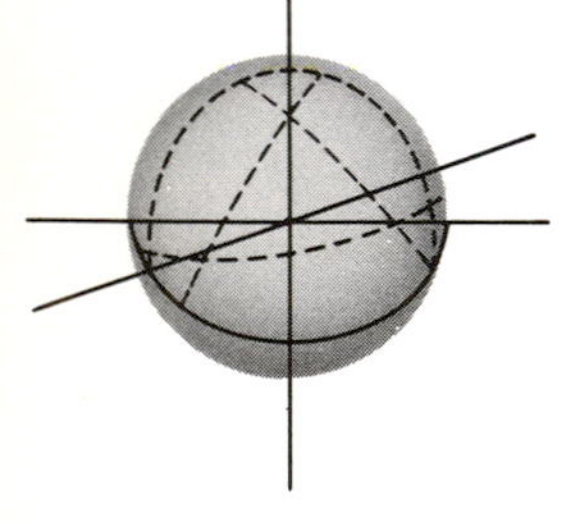

Figure 9

In fact P is evidently the plane containing the three points $(1, 0, 0)$, $(0, 1, 0)$, and $(0, 0, 1)$ and shown in Fig. 9. The level set determined by $f(x, y, z) = (2, 1)$ is then the circle C consisting of all points satisfying *both* Equations (2) and so is the intersection of S and P.

We summarize the definitions of explicit, parametric, and implicit representations. A set S is defined:

1. **explicitly** if S is the **graph** in $\mathcal{R}^{n+m}$ of a function
$$\mathcal{R}^n \xrightarrow{f} \mathcal{R}^m.$$
2. **parametrically** if S is the **range** in $\mathcal{R}^m$ of a function
$$\mathcal{R}^n \xrightarrow{f} \mathcal{R}^m.$$

3. **implicitly** if, for some function

$$\mathcal{R}^{n+m} \xrightarrow{f} \mathcal{R}^m,$$

S is a **level set** of f, that is, for some point $\mathbf{k}$ in $\mathcal{R}^m$, S is the set of all $\mathbf{x}$ in the **domain** of f such that $f(\mathbf{x}) = \mathbf{k}$.

A set S defined in some one of the above three ways will be called a curve or a surface provided that f satisfies certain smoothness conditions to be described in the last section of Chapter 4. In the meantime we shall use the terms curve and surface informally.

When the domain and range spaces of a vector function are the same, it is often helpful to picture the domain vectors $\mathbf{x}$ as points and the image vectors $f(\mathbf{x})$ as arrows. We picture $f(\mathbf{x})$ as an arrow with its tail at $\mathbf{x}$. One

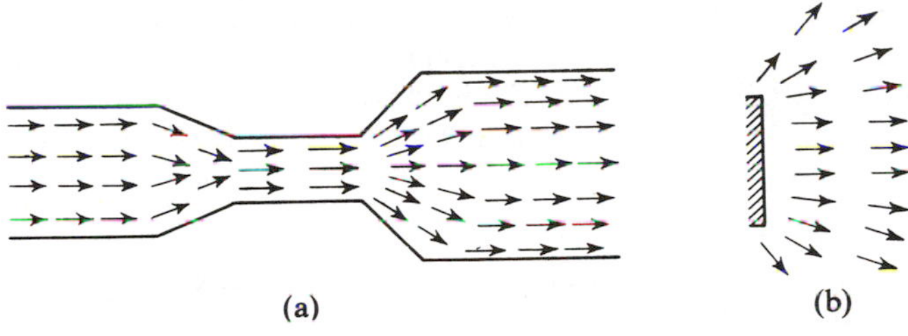

(a) (b)

Figure 10

would do this, for example, in representing a 2-dimensional fluid flow in which the image vector at each point is the velocity and direction of the flow. See Fig. 10(a). Another example is an electric field, where the value of the function at a point is the vector giving the force exerted by the field on a unit charge. See Fig. 10(b). Vector functions looked at in this way are sometimes called **vector fields.** We try to visualize a vector field by thinking of the appropriate arrow emanating from each point of the domain of the field. Hence the domain and range spaces must have the same dimension.

Example 10. A simple example of a vector field is given by the function

$$f(x, y) = \left(x, \frac{y}{2}\right).$$

To sketch it we locate some points (x, y) and some corresponding points $(x, y/2)$. Then we translate the arrow directed from $(0, 0)$ to $(x, y/2)$ parallel to itself until its tail rests at (x, y). Some of these arrows are shown in Fig. 11.

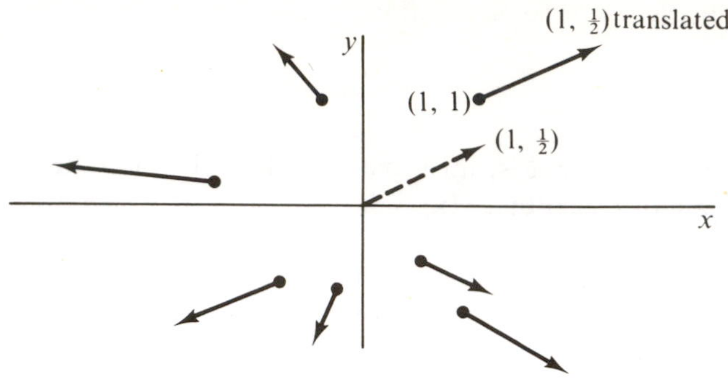

Figure 11

EXERCISES

1. Suppose that the temperature at a point (x, y, z) in space is given by $T(x, y, z) = x^2 + y^2 + z^2$. A particle moves so that at time t its location is given by $(x, y, z) = (t, t^2, t^3)$. Find the temperature at the point occupied by the particle at $t = \frac{1}{2}$. What is the rate of change of the temperature at the particle when $t = \frac{1}{2}$?

2. Let the density per unit of volume in a cubical box of side length 2 vary directly as the distance from the center and inversely as $1 + t^2$, where t is time. If the density at a corner of the box is 1 when $t = 0$, find a formula for the density at any point and at any time. What is the rate of change of the density at a point $\frac{1}{2}$ unit from the center of the box at time $t = 1$?

3. Consider the function $f(x, y) = \sqrt{4 - x^2 - y^2}$.

(a) Sketch the domain of f (take it as large as possible).
(b) Sketch the graph of f.
(c) Sketch the range of f.

4. The function $\mathcal{R} \xrightarrow{g} \mathcal{R}^2$ is defined by

$$g(t) = \begin{pmatrix} 2 \cos t \\ 3 \sin t \end{pmatrix}, \qquad 0 \leq t \leq 2\pi.$$

(a) Draw the range of g.
(b) Draw the graph of g.

5. A transformation from the xy-plane to the uv-plane is defined by

$$f\begin{pmatrix} x \\ y \end{pmatrix} = \begin{pmatrix} u \\ v \end{pmatrix} = \begin{pmatrix} x \\ y(1 + x^2) \end{pmatrix}.$$

What are the images of horizontal lines in the xy-plane? What are the coordinate functions of f?

6. For each of the following linear functions: (i) What is the domain? (ii) Describe and sketch the range. (iii) Describe and sketch the set implicitly defined by the equation $L(\mathbf{x}) = 0$. What is this set usually called?

(a) $L\begin{pmatrix} x \\ y \end{pmatrix} = \begin{pmatrix} 2 & 1 \\ 4 & 3 \end{pmatrix}\begin{pmatrix} x \\ y \end{pmatrix}$.

(b) $L\begin{pmatrix} x \\ y \end{pmatrix} = \begin{pmatrix} 2 & 1 \\ 4 & 2 \end{pmatrix}\begin{pmatrix} x \\ y \end{pmatrix}$.

(c) $L\begin{pmatrix} x \\ y \\ z \end{pmatrix} = \begin{pmatrix} 1 & 0 & 2 \\ 3 & 2 & 1 \end{pmatrix}\begin{pmatrix} x \\ y \\ z \end{pmatrix}$.

(d) $L\begin{pmatrix} x \\ y \\ z \end{pmatrix} = \begin{pmatrix} 1 & 2 & 3 \\ -1 & 4 & 2 \\ 5 & -8 & 0 \end{pmatrix}\begin{pmatrix} x \\ y \\ z \end{pmatrix}$.

7. Sketch the surfaces defined explicitly by the following functions:

(a) $f\begin{pmatrix} x \\ y \end{pmatrix} = 2 - x^2 - y^2$.

(b) $h\begin{pmatrix} x \\ y \end{pmatrix} = \dfrac{1}{x^2 + y^2}$.

(c) $g(x, y) = \sin x$.

(d) $f(x, y) = 0$.

(e) $f(x, y) = e^{x+y}$.

(f) $g(x, y) = \begin{Bmatrix} 1 & \text{if } |x| < |y|, \\ 0 & \text{if } |x| \geq |y|. \end{Bmatrix}$

8. Sketch the curves defined parametrically by the following functions:

(a) $f(t) = \begin{pmatrix} 1 \\ 2 \\ 0 \end{pmatrix} t + \begin{pmatrix} 1 \\ 1 \\ 1 \end{pmatrix}, \qquad -\infty < t < \infty.$

(b) $f(t) = \begin{pmatrix} t \\ t^2 \\ t^3 \end{pmatrix}, \qquad 0 \leq t \leq 1.$

(c) $f(t) = (2t, t), \qquad -1 \leq t \leq 1.$

(d) $h(t) = (t, t, t^2), \qquad -1 \leq t \leq 2.$

(e) $f(t) = \begin{pmatrix} 2t \\ |t| \end{pmatrix}, \qquad -1 \leq t \leq 2.$

9. Draw the surfaces defined parametrically by the following functions:

(a) $$f\begin{pmatrix} u \\ v \end{pmatrix} = \begin{pmatrix} x \\ y \\ z \end{pmatrix} = \begin{pmatrix} 1 & 0 \\ 0 & 1 \\ 1 & 0 \end{pmatrix}\begin{pmatrix} u \\ v \end{pmatrix} + \begin{pmatrix} 1 \\ 1 \\ 1 \end{pmatrix}, \qquad \begin{cases} -\infty < u < \infty, \\ -\infty < v < \infty. \end{cases}$$

(b) $$g\begin{pmatrix} u \\ v \end{pmatrix} = \begin{pmatrix} x \\ y \\ z \end{pmatrix} = \begin{pmatrix} \cos u \sin v \\ \sin u \sin v \\ \cos v \end{pmatrix}, \qquad \begin{cases} 0 \le u \le 2\pi. \\ 0 \le v \le \pi/2. \end{cases}$$

(c) $$\begin{pmatrix} x \\ y \\ z \end{pmatrix} = \begin{pmatrix} \cos u \cosh v \\ \sin u \cosh v \\ \sinh v \end{pmatrix}, \qquad \begin{cases} 0 \le u \le 2\pi, \\ -\infty < v < \infty. \end{cases}$$

10. Draw the following implicitly defined level sets:

(a) $f(x, y) = x + y = 1.$

(b) $g(x, y) = \dfrac{x^2}{a^2} + \dfrac{y^2}{b^2} = 1.$

(c) $f(x, y) = (x^2 + y^2 + 1)^2 - 4x^2 = 0.$

(d) $f(x, y, z) = x + y + z = 1.$

(e) $xyz = 0.$

(f) $g(x, y, z) = x^2 - y^2 = 2.$

(g) $$\begin{pmatrix} x - y \\ y + z \end{pmatrix} = \begin{pmatrix} 0 \\ 0 \end{pmatrix}.$$

(h) $$\begin{cases} 2x + y + z = 2, \\ x \qquad\quad - z = 3. \end{cases}$$

(i) $$\begin{pmatrix} xyz \\ x + y \end{pmatrix} = \begin{pmatrix} 0 \\ 1 \end{pmatrix}.$$

11. Suppose that the density per unit of area of a thin film, referred to plane rectangular coordinates, is given by the formula $d(x, y) = x^2 + 2y^2 - x + 1$, for $-1 \le x \le 1$ and $-1 \le y \le 1$. Sketch the set of points at which the film has density $\frac{7}{4}$.

12. Sketch the indicated vector fields.

(a) $f\begin{pmatrix} x \\ y \end{pmatrix} = \begin{pmatrix} 1 \\ x \end{pmatrix}$ for $-1 \leq x \leq 2,\ y = 0,\ y = 1.$

(b) $f\begin{pmatrix} x \\ y \end{pmatrix} = \begin{pmatrix} -x \\ y \end{pmatrix}$ for $x^2 + y^2 \leq 4.$

(c) $f\begin{pmatrix} x \\ y \end{pmatrix} = \begin{pmatrix} y \\ x \end{pmatrix}$ for $x^2 + y^2 \leq 4.$

(d) $f\begin{pmatrix} x \\ y \end{pmatrix} = \dfrac{1}{x^2 + y^2}\begin{pmatrix} 1 \\ 1 \end{pmatrix}$ for $x^2 + y^2 \leq 4.$

13. Let a transformation from the Euclidean xy-plane to itself be given by

$$f\begin{pmatrix} x \\ y \end{pmatrix} = \begin{pmatrix} x + y \\ -x + y \end{pmatrix}.$$

Show that f accomplishes an expansion out from the origin by a factor $\sqrt{2}$ and a rotation through an angle $\pi/4$.

14. The vector function f is defined by

$$f\begin{pmatrix} x \\ y \end{pmatrix} = \begin{pmatrix} x^2 - y^2 \\ 2xy \end{pmatrix}.$$

What are the coordinate functions of f? Consider the domain space to be the xy-plane and the range space to be the uv-plane.

(a) Find the image of the segment of the line $y = x$ between

$$\begin{pmatrix} 0 \\ 0 \end{pmatrix} \text{ and } \begin{pmatrix} 1 \\ 1 \end{pmatrix}.$$

(b) Find the image of the region defined by $0 < x$, $0 < y$, and $x^2 + y^2 < 1$.

(c) Find the angle between the images of the lines $y = 0$ and $y = (1/\sqrt{3})x$.
[*Ans.* $\pi/3$.]

15. A vector function f from the xy-plane to the uv-plane is defined by

$$f\begin{pmatrix} x \\ y \end{pmatrix} = \begin{pmatrix} u \\ v \end{pmatrix} = \begin{pmatrix} x \\ \dfrac{(x + y)^2}{4x} \end{pmatrix}, \qquad x \neq 0.$$

What are the coordinate functions of f? Find the image of the region bounded by the lines $x = y$, $y = x - 8$, $x = -y$, $y = 8 - x$.

SECTION 2

FUNCTIONS OF ONE VARIABLE

If a point moves in a vector space so as to occupy various positions at different times, then its position at time t can be described by a vector-valued function f with values $f(t)$. For example, a point moving on a line in $\mathcal{R}^3$ might be at the point

$$f(t) = t\mathbf{x}_1 + \mathbf{x}_0$$

at time t, where $\mathbf{x}_1$ and $\mathbf{x}_0$ are points in $\mathcal{R}^3$. More generally, a function f taking values in $\mathcal{R}^n$ might be given in the form

$$f(t) = (f_1(t), \ldots, f_n(t)),$$

where the coordinate functions $f_1, \ldots, f_n$ describe the real-valued coordinates of a point in $\mathcal{R}^n$ at different times t.

Example 1. If $\mathbf{x}_1 = (x_1, y_1, z_1)$ and $\mathbf{x}_0 = (x_0, y_0, z_0)$ are points in $\mathcal{R}^3$, then the function $\mathcal{R} \xrightarrow{f} \mathcal{R}^3$ defined by

$$\begin{aligned} f(t) &= t(x_1, y_1, z_1) + (x_0, y_0, z_0) \\ &= (tx_1 + x_0, ty_1 + y_0, tz_1 + z_0) \end{aligned}$$

gives a parametric representation of a point on a line. The function g for which

$$g(t) = (t, t^2)$$

similarly describes a curve in $\mathcal{R}^2$. In fact, because the coordinates $x = t$ and $y = t^2$ satisfy the relation $y = x^2$, the point (t, t^2) always lies on the parabola with equation $y = x^2$, shown in Fig. 12.

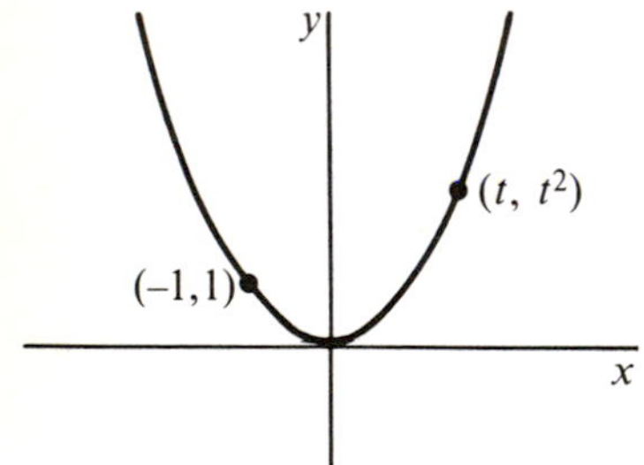

Figure 12

We can define the limit of a vector-valued function f with values in $\mathcal{R}^n$ by using limits of the real-valued coordinate functions f_k of f. Thus if

$$f(t) = (t_1(t), \ldots, f_n(t))$$

is defined for an interval $a < t < b$ containing t_0, we write

$$\lim_{t \to t_0} f(t) = \left(\lim_{t \to t_0} f_1(t), \ldots, \lim_{t \to t_0} f_n(t)\right).$$

Similarly, a function with values in $\mathcal{R}^n$ is said to be continuous if its real-valued coordinate functions are all continuous on their interval of definition.

Example 2. The function defined by $g(t) = (t, t^2)$ has limit vector $(2, 4)$ at $t = 2$ because

$$\begin{aligned} \lim_{t \to 2} (t, t^2) &= \left(\lim_{t \to 2} t, \lim_{t \to 2} t^2\right) \\ &= (2, 4). \end{aligned}$$

The function g is continuous for all real t because the coordinate functions t and t^2 are continuous.

The intuitive idea behind continuity of a vector-valued function is similar to that for a real-valued function: the values of the function should not change abruptly. These ideas are treated more fully in Section 7. At present we shall consider only continuous functions $\mathcal{R} \xrightarrow{g} \mathcal{R}^n$ with $g(t)$ defined on an interval $a < t < b$. We shall first define the derivative of g and show how it leads to a definition of tangent line to the curve γ defined as the range of g.

The function g has a **derivative** $g'(t)$ at a point t in the interval (a, b) if

$$g'(t) = \lim_{h \to 0} \frac{g(t + h) - g(t)}{h}.$$

If the limit exists for each t in (a, b), then $g'(t)$ determines a new function $\mathcal{R} \xrightarrow{g'} \mathcal{R}^n$, just as in the case $n = 1$. The derivative is often written dg/dt.

Example 3. Let $g(t) = (t^2, t^3)$. Then

$$\lim_{h \to 0} \frac{g(t + h) - g(t)}{h} = \lim_{h \to 0} \frac{1}{h} \begin{pmatrix} (t + h)^2 - t^2 \\ (t + h)^3 - t^3 \end{pmatrix} = \lim_{h \to 0} \begin{pmatrix} \dfrac{(t + h)^2 - t^2}{h} \\ \dfrac{(t + h)^3 - t^3}{h} \end{pmatrix}.$$

Since $\lim_{h \to 0} ((t + h)^2 - t^2)/h = 2t$ and $\lim_{h \to 0} ((t + h)^3 - t^3)/h = 3t^2$, the vector limit $g'(t)$ exists, and $g'(t) = (2t, 3t^2)$.

Example 3 suggests that a function $\mathcal{R} \xrightarrow{g} \mathcal{R}^n$ has a derivative at a point t if and only if each coordinate function of g has a derivative there. This is true, and in fact we have:

2.1 If $g(t) = \begin{pmatrix} g_1(t) \\ \cdot \\ \cdot \\ \cdot \\ g_n(t) \end{pmatrix}$, then $g'(t) = \begin{pmatrix} g_1'(t) \\ \cdot \\ \cdot \\ \cdot \\ g_n'(t) \end{pmatrix}$, where the derivatives $g_k'(t)$ are ordinary derivatives of real-valued functions of a real variable. The result is an immediate consequence of the relation of the limit of a vector function to the limits of its coordinate functions.

Example 4. If

$$g(t) = \begin{pmatrix} \cos t \\ \sin t \end{pmatrix}, \qquad \text{then} \quad g'(t) = \begin{pmatrix} -\sin t \\ \cos t \end{pmatrix}.$$

If

$$h(t) = \begin{pmatrix} t \\ t^2 \\ t^3 \end{pmatrix}, \qquad \text{then} \quad h'(t) = \begin{pmatrix} 1 \\ 2t \\ 3t^2 \end{pmatrix}.$$

It is clear from Fig. 13 that the vector $g(t + h) - g(t)$ has a direction which, as h tends to 0, should tend to what we would like to call the tangent direction to the curve γ at $g(t)$. However, since g is assumed continuous,

$$\lim_{h \to 0} g(t + h) - g(t) = 0,$$

and the zero vector that we get as a limit has no direction. This difficulty is overcome in most examples by dividing by h before letting h tend to zero. Observe that division by h will not change the direction of $g(t + h) - g(t)$ if h is positive; it will reverse it if h is negative. A glance at Fig. 13 shows that this reversal is desirable for our purposes. (What would be wrong with dividing by $|h|$?) The derivative $g'(t)$, if it exists and is not zero, is called the **tangent vector** to γ at $g(t)$. Of course, any nonzero multiple of $g'(t)$ is then called a tangent vector, and the line with direction vector $g'(t)$ and passing through $g(t)$ is called the **tangent line** to γ at $g(t)$. Thus, if $g(t_0)$ is a particular point on a curve, the tangent line at $g(t_0)$ will have a parametric representation of the form

$$tg'(t_0) + g(t_0).$$

The tangent vector $g'(t_0)$ is usually pictured with its tail at $g(t_0)$ as in Fig. 13.

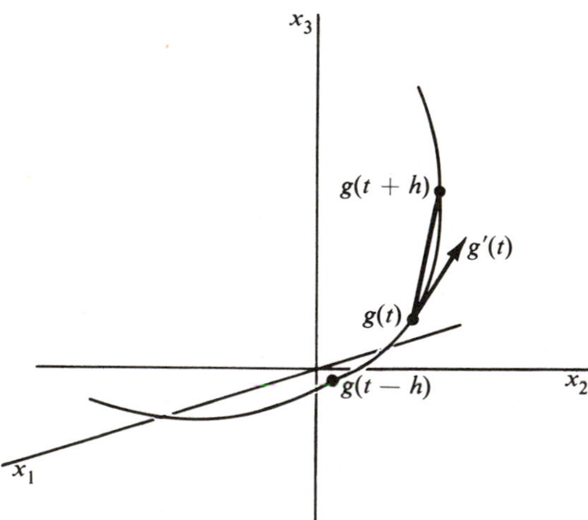

Figure 13

Example 5. The circle defined parametrically by $g(t) = (\cos t, \sin t)$ has a tangent vector at $g(t_0)$ given by $g'(t_0) = (-\sin t_0, \cos t_0)$. In particular, the tangent vector at $g(\pi/4) = (1/\sqrt{2}, 1/\sqrt{2})$ is $g'(\pi/4) = (-1/\sqrt{2}, 1/\sqrt{2})$. Hence the tangent line to the circle at $(1/\sqrt{2}, 1/\sqrt{2})$ has a parametric representation $(x, y) = t(-1\sqrt{2}, 1/\sqrt{2}) + (1/\sqrt{2}, 1/\sqrt{2})$. The line is shown in Fig. 14, together with some tangent vectors, each of which has length $|g'(t_0)| = 1$.

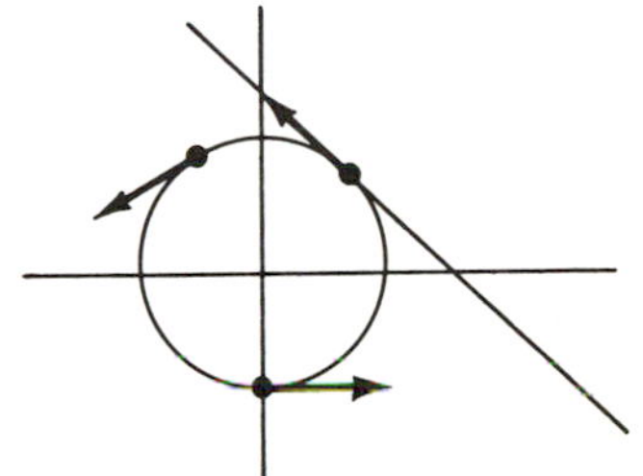

Figure 14

For the spiral curve given by $f(t) = (\cos t, \sin t, t)$, we have $f'(0) = (0, 1, 1)$. The tangent line to the spiral at $(1, 0, 0)$ can be represented by

$$\begin{pmatrix} x \\ y \\ z \end{pmatrix} = t\begin{pmatrix} 0 \\ 1 \\ 1 \end{pmatrix} + \begin{pmatrix} 1 \\ 0 \\ 0 \end{pmatrix},$$

and in this case the tangent vector $f'(0)$ has length $\sqrt{2}$.

One reason for singling out $g'(t)$ for special attention as *the* tangent vector, rather than some multiple of it, is that we often want to consider the parameter t as a time variable, and $g(t)$ as representing the path of a point moving in $\mathcal{R}^n$. Under this interpretation, the Euclidean length $|g'(t)|$ is the natural definition for the speed of motion along the path γ described by $g(t)$ as t varies. To justify the use of the term "speed," we observe that, for small h, the number $|g(t + h) - g(t)|/|h|$ is close to the average rate of traversal of γ over the interval from t to $t + h$. In addition, if $g'(t)$ exists, it is easy to show that

$$\lim_{h \to 0} \frac{|g(t + h) - g(t)|}{|h|} = |g'(t)| .$$

In fact, by the reversed triangle inequality (Exercise 12, Section 5, Chapter 1),

$$\left| \frac{|g(t + h) - g(t)|}{|h|} - |g'(t)| \right| \le \left| \frac{g(t + h) - g(t)}{h} - g'(t) \right|,$$

which tends to zero as h tends to zero. Thus $|g'(t)|$ is a limit of average rates over arbitrarily small time intervals. For this reason the real-valued function v defined by $v(t) = |g'(t)|$ is called the **speed** of g, and the vector $\mathbf{v}(t) = g'(t)$ is called its **velocity vector** at the point $g(t)$. The vector $\mathbf{v}(t)$ is, of course, the same as what we have called the tangent vector to γ at $g(t)$, provided $\mathbf{v}(t) \neq 0$.

Example 6. If a point moves in the plane so that at time t its position is $g(t) = (t^2, t^3)$, then the velocity vector is $\mathbf{v}(t) = (2t, 3t^2)$, and $v(t) = \sqrt{4t^2 + 9t^4}$. In particular, $\mathbf{v}(0) = 0$. The path traced by g is shown in Fig. 15 for $-1 \le t \le 1$, and in drawing the picture it is helpful to observe

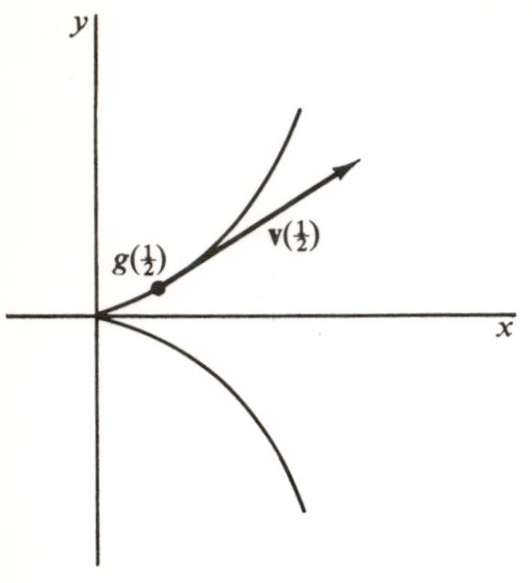

Figure 15

that the coordinates of a point on the path satisfy the equation $y^2 = x^3$.

The fact that the tangent vector shrinks to zero in this example as $g(t)$ approaches the origin is a reflection of the fact that, if the velocity vector varies continuously, the speed must be zero at an abrupt change in the direction of motion. In this way the parametrization describes the physical situation well. However, for the purpose of assigning a tangent line to the path at the origin, the given parametrization is not useful.

Example 7. Let $f(t) = (\cos t, \sin t, t)$, as in the second part of Example 5. Then $\mathbf{v}(t) = (-\sin t, \cos t, 1)$. It follows that the velocity vector is always perpendicular to the vector $(\cos t, \sin t, 0)$ that points from the axis of the spiral to $f(t)$. The speed at any time t is $v(t) = |\mathbf{v}(t)| = \sqrt{2}$.

We list here some useful formulas that hold if f and g each has a vector derivative on an interval (a, b) and φ is real-valued and differentiable there.

2.2 $(f + g)' = f' + g'$, $(cf)' = cf'$, c constant.

2.3 $(\varphi f)' = \varphi f' + \varphi' f$, φ real-valued.

2.4 $(f \cdot g)' = f \cdot g' + f' \cdot g$.

2.5 $(f(u))' = u' f'(u)$,

where u is a real-valued, differentiable function of one variable, with its range in (a, b).

These can all be proved by writing f and g in terms of their coordinate functions and then applying the corresponding differentiation formulas for real-valued function, together with Formula 2.1. For example, the proof of 2.5, a version of the chain rule for differentiation, goes like this:

$$\begin{aligned}(f(u))' &= (f_1(u), \ldots, f_n(u))' \\ &= ([f_1(u)]', \ldots, [f_n(u)]') \\ &= (f_1'(u)u', \ldots, f_n'(u)u') = u'f'(u).\end{aligned}$$

If $\mathcal{R} \xrightarrow{g} \mathcal{R}^n$ has a derivative $\mathcal{R} \xrightarrow{g'} \mathcal{R}^n$, then we can ask for the derivative of g', which we denote by g''. Thus we have $\mathcal{R} \xrightarrow{g''} \mathcal{R}^n$, though g'' may be defined at fewer points than g or even g'. We also write d^2g/dt^2 for g'', and so on for higher-order derivatives.

Example 8. Let $\mathcal{R} \xrightarrow{g} \mathcal{R}^3$ describe a path in $\mathcal{R}^3$ with velocity vector $\mathbf{v}(t)$ and speed $v(t)$ at each point $g(t)$. Then $\mathbf{t}(t) = (1/v(t))\mathbf{v}(t)$ is a tangent vector of length 1, provided $v(t) \neq 0$. In any case, we can write $g'(t) = v(t)\mathbf{t}(t)$. If we assume that g' has a derivative, we define the **acceleration vector** at $g(t)$ by $\mathbf{a}(t) = g''(t)$. The physical significance of $\mathbf{a}(t)$ is that if $g(t)$ describes the motion of a particle of constant mass m, then $m\mathbf{a}(t)$ is the **force vector** $F(t)$ acting on the particle. If we denote by $a(t)$ the length of $\mathbf{a}(t)$, then $a(t)$ is the **magnitude of the acceleration,** and $ma(t)$ is the **magnitude of the force** acting on the particle.

If $\mathbf{t}(t)$ is a unit tangent vector at $g(t)$, the equation $g'(t) = v(t)\mathbf{t}(t)$ implies that $\mathbf{a} = (v\mathbf{t})'$. Applying Formula 2.3, we get $\mathbf{a} = v'\mathbf{t} + v\mathbf{t}'$. Thus if $\mathbf{t}'(t) = 0$, the acceleration vector, and hence the force vector at $g(t)$, has either the same or else the opposite direction to the motion. On the other hand, if $\mathbf{t}'(t) \neq 0$, we can define the unit vector $\mathbf{n}$ by $\mathbf{n}(t) = \mathbf{t}'(t)/|\mathbf{t}'(t)|$, and so the acceleration vector can be written

$$\mathbf{a} = v'\mathbf{t} + v\,|\mathbf{t}'|\,\mathbf{n}.$$

This equation expresses the acceleration $\mathbf{a}(t)$ at each point $g(t)$ in terms of an orthonormal pair of vectors $\mathbf{t}(t)$ and $\mathbf{n}(t)$. We have $|\mathbf{t}| = |\mathbf{n}| = 1$ by the definition of these vectors, and application of Formula 2.4 to the equation $\mathbf{t} \cdot \mathbf{t} = 1$ gives $\mathbf{t} \cdot \mathbf{t}' = 0$. But by the definition of $\mathbf{n}$, this implies $\mathbf{t} \cdot \mathbf{n} = 0$.

The pair $\mathbf{t}(t)$, $\mathbf{n}(t)$, should be pictured at $g(t)$ as in Fig. 16. The third unit vector $\mathbf{b}(t)$ shown there is defined by $\mathbf{b} = \mathbf{t} \times \mathbf{n}$, and is called the **binormal vector** to the path, while $\mathbf{n}$ is called the **principal normal.** Thus any vector naturally associated with the point $g(t)$ on the path can be written

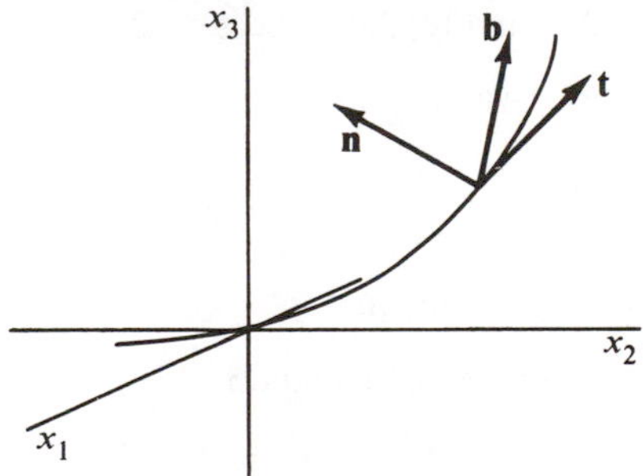

Figure 16

as a linear combination of the triple $\{\mathbf{t}(t), \mathbf{n}(t), \mathbf{b}(t)\}$, which changes as we go from point to point along the path.

EXERCISES

1. If $g(t) = (e^t, t)$ for all real t, sketch in $\mathcal{R}^2$ the curve described by g together with the tangent vectors $g'(0)$ and $g'(1)$.

2. Let $f(t) = (t, t^2, t^3)$ for $0 \leq t \leq 1$.

   (a) Sketch the curve described by f in $\mathcal{R}^3$ and the tangent line at $(\frac{1}{2}, \frac{1}{4}, \frac{1}{8})$.
   (b) Find $|f'(t)|$.
   (c) If $f(t) = (t, t^2, t^3)$ for all real t, find all points of the curve described by f at which the tangent vector is parallel to the vector $(4, 4, 3)$. Are there any points at which the tangent is perpendicular to $(4, 4, 3)$?

3. Sketch the curve represented by $(x, y) = (t^3, t^5)$, and show that the parametrization fails to assign a tangent vector at the origin. Find a parametrization of the curve that does assign a tangent at the origin.

4. Show that the curve described by $g(t) = (\sin 2t, 2\sin^2 t, 2\cos t)$ lies on a sphere centered at the origin in $\mathcal{R}^3$. Find the length of the velocity vector $\mathbf{v}(t)$ and show that the projection of this vector into the xy-plane has a constant length.

5. (a) Show that if $\mathcal{R} \xrightarrow{\mathbf{v}} \mathcal{R}^3$ is continuous for $0 \leq t$, then there is a unique path g through a given point $g(0)$ in $\mathcal{R}^3$, having $\mathbf{v}(t)$ as its velocity vector at $g(t)$.
   (b) Show that a continuous function $\mathcal{R} \xrightarrow{\mathbf{a}} \mathcal{R}^3$ defined for $0 \leq t$, determines a unique path in $\mathcal{R}^3$, having $\mathbf{a}(t)$ as its acceleration vector, provided the initial point $g(0)$ and initial velocity $\mathbf{v}(0)$ are specified.

6. Suppose a target moves with constant speed $v > 0$ on a circular path of radius r, and that a missile, also having constant speed v, pursues the target by starting from the center of the circle, always remaining between the center and the target. When does the missile hit the target?

7. Prove (a) 2.2, (b) 2.3, and (c) 2.4 of Section 2.

8. Show that if f is vector-valued, differentiable, and never zero for $a < t < b$, then

   (a) $f \cdot \dfrac{df}{dt} = |f| \dfrac{d|f|}{dt}$.

   (b) $|f|$ is constant if and only if $f \cdot f' = 0$.

9. Consider the vector differential equation

$$\mathbf{x}'' + a\mathbf{x}' + b\mathbf{x} = 0$$

   to be solved for a function $\mathbf{x}(t)$ taking values in $\mathcal{R}^n$ and defined on some interval. We assume that a and b are constants. Show that if the real equation $r^2 + ar + b = 0$ has distinct roots r_1 and r_2, then the differential equation has a solution of the form

$$\mathbf{x}(t) = \mathbf{c}_1 e^{r_1 t} + \mathbf{c}_2 e^{r_2 t},$$

where $\mathbf{c}_1$ and $\mathbf{c}_2$ are constant vectors in $\mathcal{R}^n$. What happens if $r_1 = r_2$?

10. Prove that $(f \times g)' = (f \times g') + (f' \times g)$, where f and g take values in $\mathcal{R}^3$ and are differentiable on an interval.

11. Show that if $\mathcal{R} \xrightarrow{g} \mathcal{R}^n$ has a derivative and $g'(t) = 0$ for $a < t < b$, then $g(t)$ is a constant vector on that interval. [*Hint*. Apply the mean-value theorem to each coordinate function.]

12. Let a differentiable function $g(t)$ represent the position in $\mathcal{R}^3$ at time t of a particle of possibly varying mass $m(t)$. The vector function $P(t) = m(t)\mathbf{v}(t)$ is called the **linear momentum** of the particle. The **force vector** is $F(t) = (m(t)\mathbf{v}(t))'$. The **angular momentum** about the origin is $L(t) = g(t) \times P(t)$, and the **torque** about the origin is $N(t) = g(t) \times F(t)$.

(a) Show that if F is identically zero, then P is constant. This is called the law of conservation of linear momentum.

(b) Show that $L'(t) = N(t)$, and hence that if N is identically zero, then L is constant. This is called the law of conservation of angular momentum.

13. Show that if a particle has an acceleration vector $\mathbf{a}(t)$ at time t and $v(t) \neq 0$, then $v' = \mathbf{t} \cdot \mathbf{a}$, where $\mathbf{t}$ is the unit vector $(1/v)\mathbf{v}$.

14. Let $\mathcal{R} \xrightarrow{f} \mathcal{R}^n$ be a function defined for $a \leq t \leq b$. If the coordinate functions $f_1, \ldots, f_n$ of f are integrable, we can define the integral of f over the interval $[a, b]$ by

$$\int_a^b f(t)\,dt = \left(\int_a^b f_1(t)\,dt, \ldots, \int_a^b f_n(t)\,dt\right).$$

(a) If $f(t) = (\cos t, \sin t)$ for $0 \leq t \leq \pi/2$, compute $\displaystyle\int_0^{\pi/2} f(t)\,dt$.

(b) If $g(t) = (t, t^2, t^3)$ for $0 \leq t \leq 1$, compute $\displaystyle\int_0^1 g(t)\,dt$.

15. If $\mathcal{R} \xrightarrow{f} \mathcal{R}^n$ and $\mathcal{R} \xrightarrow{g} \mathcal{R}^n$ are both integrable over $[a, b]$, show by using the corresponding properties of integrals of real-valued functions that:

$$\int_a^b kf(t)\,dt = k\int_a^b f(t)\,dt, \qquad k \text{ any real number},$$

$$\int_a^b (f(t) + g(t))\,dt = \int_a^b f(t)\,dt + \int_a^b g(t)\,dt,$$

where the integrals are as defined in the previous exercise.

16. If $\mathcal{R} \xrightarrow{f} \mathcal{R}^n$ is defined for $a \leq t \leq b$, and f' is continuous there, prove the following extension of the fundamental theorem of calculus:

$$\int_a^b f'(t)\, dt = f(b) - f(a).$$

17. If $\mathcal{R} \xrightarrow{f} \mathcal{R}^n$ is continuous over $[a, b]$,

(a) Show that

$$\int_a^b \mathbf{k} \cdot f(t)\, dt = \mathbf{k} \cdot \int_a^b f(t)\, dt,$$

where $\mathbf{k}$ is a constant vector.

(b) Show that

$$\left| \int_a^b f(t)\, dt \right| \leq \int_a^b |f(t)|\, dt.$$

[*Hint.* By the Cauchy-Schwarz inequality

$$f(u) \cdot \int_a^b f(t)\, dt \leq |f(u)| \left| \int_a^b f(t)\, dt \right|, \qquad \text{for each } u.$$

Integrate with respect to u, and apply the result of part (a).]

SECTION 3

ARC LENGTH

The definition of length for vectors can be used to define the length of a parametrized curve γ. We assume that γ is described by a continuous function $\mathcal{R} \xrightarrow{g} \mathcal{R}^n$, where the domain of g is a closed interval $a \leq t \leq b$. Thus γ is the image of $[a, b]$ under g. Corresponding to any finite set P of numbers $a = t_0 < t_1 < \cdots < t_K = b$, there are points $g(t_k)$, $k = 0, \ldots, K$, on γ. We join these points in order by a polygonal path as shown in Fig. 17. The length of the kth segment of the polygonal approximation to γ is $|g(t_k) - g(t_{k-1})|$, and the total length of the polygon is

$$l(P) = \sum_{k=1}^{K} |g(t_k) - g(t_{k-1})| \,.$$

Let $l(\gamma)$ denote the least upper bound of the numbers $l(P)$. This will, of course, be infinite if the set of numbers $l(P)$ is unbounded. If $l(\gamma)$ is finite, then γ is said to be **rectifiable,** and $l(\gamma)$ is called its **length.** It is clear from the definition that $l(\gamma)$ depends on the function g that describes γ and not just on γ itself. This is reasonable if we want to take into account the fact that some part of γ may be traced more than once by g. In practice this is very often what is wanted. If it should happen that g is not one-to-one, then we may write $l(g)$ instead of $l(\gamma)$ to emphasize the dependence on g.

The length of a path is usually awkward to compute directly from the definition. However, if γ is parametrized by a function g such that the tangent vector $g'(t)$ varies continuously with t, then $l(\gamma)$ is finite and equal to $\int_a^b |g'(t)|\, dt$. In fact, it is enough to assume that g is **piecewise smooth,** that is, g' is continuously extendable to the endpoints of finitely many intervals which placed end to end form the interval $a \leq t \leq b$. This allows us to find the length of some curves for which the tangent has an abrupt change, as in Fig. 17.

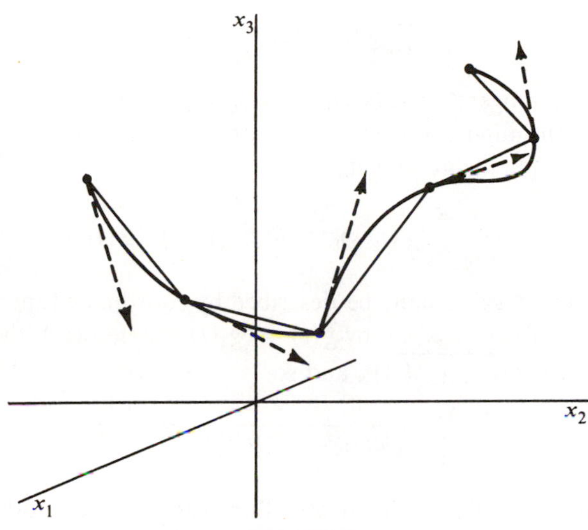

Figure 17

3.1 Theorem

Let a curve γ be parametrized by a piecewise smooth function $\mathcal{R} \xrightarrow{g} \mathcal{R}^n$, defined for $a \leq t \leq b$. Then $l(\gamma)$ is finite and

$$l(\gamma) = \int_a^b |g'(t)|\, dt.$$

The proof of this theorem is given in Section 3 of the Appendix, and we shall give here only an argument that makes the integral formula plausible. Since, by definition, the existence of $g'(t_{k-1})$ means

$$\lim_{t_k \to t_{k-1}} \frac{g(t_k) - g(t_{k-1})}{t_k - t_{k-1}} = g'(t_{k-1}),$$

we have

$$g(t_k) - g(t_{k-1}) - (t_k - t_{k-1})g'(t_{k-1}) = (t_k - t_{k-1})Z(t_k - t_{k-1}),$$

where $Z(t_k - t_{k-1})$ satisfies

$$\lim_{t_k \to t_{k-1}} Z(t_k - t_{k-1}) = 0.$$

Thus $(t_k - t_{k-1})g'(t_{k-1})$ becomes a better approximation to $g(t_k) - g(t_{k-1})$ as $(t_k - t_{k-1})$ is made small. We are led to approximate

$$l(P) = \sum_{k=1}^{K} |g(t_k) - g(t_{k-1})|$$

by

$$\sum_{k=1}^{K} |g'(t_{k-1})|\,(t_k - t_{k-1}).$$

(Some of the tangent vectors $(t_k - t_{k-1})g'(t_{k-1})$ are shown in Fig. 17.) But if g' is continuous, so is $|g'|$, and, letting $m(P) = \max_{1 \le k \le K} (t_k - t_{k-1},)$ we have, to conclude the argument,

$$\lim_{m(P)\to 0} \sum_{k=1}^{K} |g'(t_{k-1})|\,(t_k - t_{k-1}) = \int_a^b |g'(t)|\, dt.$$

A curve in $\mathcal{R}^n$ will usually be described by coordinate functions. Thus, if γ is defined for $a \le t \le b$ by $g(t) = (g_1(t), g_2(t), g_3(t))$, then $|g'(t)| = \sqrt{(g_1'(t))^2 + (g_2'(t))^2 + (g_3'(t))^2}$, and so

$$l(g) = \int_a^b \sqrt{(g_1'(t))^2 + (g_2'(t))^2 + (g_3'(t))^2}\, dt,$$

with a similar formula holding in $\mathcal{R}^n$. For example, the spiral curve in $\mathcal{R}^3$ defined by $g(t) = (\cos t, \sin t, t)$ for $0 \le t \le 1$ has length

$$\begin{aligned} l(g) &= \int_0^1 \sqrt{(-\sin t)^2 + (\cos t)^2 + 1}\, dt \\ &= \sqrt{2}. \end{aligned}$$

Example 1. The plane curve defined by $g(t) = (t\,|t|, |t|)$ for $-1 \le t \le 1$ is shown in Fig. 18. Since g is piecewise smooth, $l(g)$ is finite. We have

$$g'(t) = \begin{cases} (-2t, -1), & -1 \le t < 0 \\ (2t, 1), & 0 < t \le 1, \end{cases}$$

so $|g'(t)| = \sqrt{4t^2 + 1}$ for $-1 \le t \le 1$, $t \ne 0$. Then

$$\begin{aligned} l(g) &= \int_{-1}^1 \sqrt{4t^2 + 1}\, dt \\ &= \sqrt{5} + (\tfrac{1}{2}) \log (2 + \sqrt{5}). \end{aligned}$$

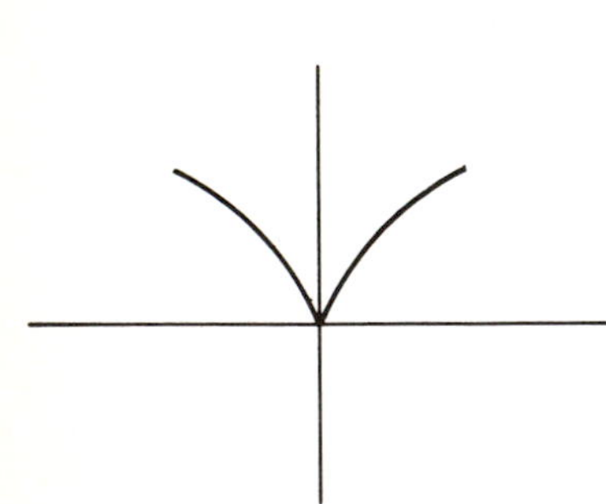

Figure 18

Example 2. The graph γ of a real-valued continuous function f, defined for $a \le x \le b$, is a curve in $\mathcal{R}^2$ that can be described parametrically by the one-to-one function

$$g(x) = \begin{pmatrix} x \\ f(x) \end{pmatrix}, \qquad a \le x \le b.$$

If f' is continuous, then so is $|g'(x)| = \sqrt{1 + (f'(x))^2}$, and the formula for finding $l(\gamma)$ becomes in this case

$$l(\gamma) = \int_a^b \sqrt{1 + (f'(x))^2}\, dx.$$

If γ is a piecewise smooth curve described by a one-to-one function $g(t)$, $a \leq t \leq b$, we can think of $l(\gamma)$ as representing the total mass of γ, assuming γ has a uniform density equal to 1 at each point. More generally, if p is a real-valued function defined on γ, we can form the integral

$$\int_a^b p(g(t))\,|g'(t)|\, dt,$$

and, if it exists, call it the integral of p over γ. In particular, if p is a non-negative function that can be interpreted as the density per unit of length of a mass distribution over γ, the integral becomes the definition of the **total mass** m of the distribution.

Example 3. Consider a full turn of the spiral curve described by $g(t) = (\cos t, \sin t, t)$ for $0 \leq t \leq 2\pi$. Suppose that the density of the curve at a point is equal to the square of the distance of that point from the midpoint $\mathbf{q}$ of the axis of the spiral. Relative to our description of the curve, this midpoint has coordinates $(0, 0, \pi)$. Thus the density at $g(t)$ is equal to

$$\begin{aligned} |g(t) - \mathbf{q}|^2 &= \cos^2 t + \sin^2 t + (t - \pi)^2 \\ &= 1 + (t - \pi)^2. \end{aligned}$$

Also, we have seen that $|g'(t)| = \sqrt{2}$. Hence, the total mass of the distribution is given by

$$\begin{aligned} m &= \int_0^{2\pi} [1 + (t - \pi)^2]\sqrt{2}\, dt \\ &= \sqrt{2}\int_{-\pi}^{\pi} (1 + t^2)\, dt = (\tfrac{2}{3})\sqrt{2}\,(3\pi + \pi^3). \end{aligned}$$

If $g(t)$, with $a \leq t \leq b$, defines a smooth curve γ having a nonzero tangent at every point, then the function $s(t)$ defined by

3.2 $$s(t) = \int_a^t |g'(u)|\, du$$

is the length of the part of γ corresponding to the interval from a to t. Since $|g'(t)|$ is positive, the function $s(t)$ is strictly increasing. Thus $s(t)$ is a one-to-one function from the interval $a \leq t \leq b$ to the interval

$0 \leq s \leq l(\gamma)$, and so has an inverse function. We denote this inverse simply by $r(s)$. We now form the vector function $h(s) = g(r(s))$, which describes the same curve γ that $g(t)$ does, but with a new parametrization in which the variable s, with $0 \leq s \leq l(\gamma)$, represents the length of the path along γ from $h(0)$ to $h(s)$. The curve γ is then said to be parametrized by arc length.

Example 4. Let $g(t) = (t, t^2)$ for $0 \leq t \leq 1$. Then

$$\begin{aligned} s(t) &= \int_0^t \sqrt{1 + 4u^2}\, du \\ &= \tfrac{1}{2}t\sqrt{1 + 4t^2} + \tfrac{1}{4}\log\left(2t + \sqrt{1 + 4t^2}\right) \end{aligned}$$

Since $s'(t) = \sqrt{1 + 4t^2} > 0$, $s(t)$ is strictly increasing and so has an inverse.

Example 4 shows that $r(s)$, the inverse of $s(t)$, may be awkward to compute explicitly. However, its use has several theoretical advantages. For example, if a curve γ is parametrized by arc length, then the integral of a real-valued function p over γ takes a simpler form. Suppose γ is given originally by a function $g(t)$, with $a \leq t \leq b$. Then the new parametrization is given by a function $h(s)$ defined for $0 \leq s \leq l(\gamma)$, where $h(s) = g(r(s))$. Changing variable in the integral of p over γ we get, because $s'(t) = |g'(t)|$,

$$\begin{aligned} \int_a^b p(g(t))\,|g'(t)|\, dt &= \int_a^b p(g(t))s'(t)\, dt \\ &= \int_0^{l(\gamma)} p(h(s))\, ds. \end{aligned}$$

The expression $|g'(t)|\, dt$, or its simpler counterpart ds, is sometimes called the **element of arc length** on the curve γ. From the Equation 3.2 that relates s and t, we can derive another in terms of derivatives:

3.3
$$\frac{ds}{dt}(t) = |g'(t)| = v(t).$$

To prove this, simply differentiate both sides of 3.2 with respect to t, and use the fact that the speed $v(t)$ has been *defined* to be $|g'(t)|$. Equation 3.3 gives further justification for the definition of speed: the derivative of arc length with respect to time turns out to equal speed.

Example 5. Recall that the equation

$$\mathbf{a}(t) = \mathbf{v}'(t)\mathbf{t}(t) + v(t)\,|\mathbf{t}'(t)|\,\mathbf{n}(t), \tag{1}$$

derived in Example 8 of the previous section, expresses the acceleration vector of a curve at each point of the curve as a linear combination of two unit vectors, one tangent and one perpendicular to the curve. The coefficient of $\mathbf{n}(t)$ can be written more meaningfully. For, denoting by $r(s)$ the inverse of $s(t)$, we have by Equation 2.5 that

$$\frac{d}{ds}\mathbf{t}\big(r(s)\big) = \mathbf{t}'\big(r(s)\big)\frac{dr}{ds}(s). \tag{2}$$

For inverse functions r and s we have, in general, $r\big(s(t)\big) = t$; therefore, by the chain rule,

$$\frac{dr}{ds}\big(s(t)\big)\frac{ds}{dt}(t) = 1.$$

It then follows from Equation 3.3 that

$$\frac{dr}{ds}(s) = \frac{1}{v(t)}. \tag{3}$$

Then Equations (2) and (3) give

$$\mathbf{t}'(t) = v(t)\frac{d}{ds}\mathbf{t}(t)$$

Since $v(t) \geq 0$, when we take the length of the vectors on both sides, we can take $v(t)$ outside getting

$$|\mathbf{t}'(t)| = v(t)\left|\frac{d}{ds}\mathbf{t}(t)\right|.$$

The factor

$$\kappa(t) = \left|\frac{d}{ds}\mathbf{t}(t)\right| = \frac{|\mathbf{t}'(t)|}{v(t)}$$

is called the **curvature** of the curve at the point corresponding to t. Then Equation (1) becomes, in terms of curvature,

3.4 $$\mathbf{a}(t) = v'(t)\mathbf{t}(t) + v^2(t)\kappa(t)\mathbf{n}(t).$$

The terms in the sum are called the **tangential** and **normal** (or **centripetal**) **components** of the acceleration, respectively, and the numerical factors are sometimes denoted $a_t = v'$ and $a_n = v^2\kappa$.

From Equation 3.4 we can immediately conclude several things about $\mathbf{a}$, the acceleration vector. If the speed v is a constant v_0, then $v' = 0$, and $\mathbf{a}(t) = v_0^2\kappa(t)\mathbf{n}(t)$. This means that $\mathbf{a}$ is perpendicular to the curve and that its length, $v_0^2\kappa(t)$, varies only with the curvature κ. At the other extreme, if the path of motion is a straight line, then the unit tangent vector **t** is a constant vector so

$$\frac{d\mathbf{t}}{ds}(r(s)) = 0\,.$$

Thus $\kappa = 0$, and we have $\mathbf{a}(t) = v'(t)\mathbf{t}(t)$, which shows that the acceleration vector has either the same direction as the tangent or the opposite direction, depending on the sign of v'. Exercise 10 shows that κ, the curvature of the path, is a measure of how rapidly the tangent vector **t** is turning.

EXERCISES

1. Find the length of the following curves.

(a) $(x, y) = (t, \log\cos t)$, $0 \le t \le 1$. [*Ans.* log (sec 1 + tan 1.]

(b) $(x, y) = (t^2, \frac{2}{3}t^3 - \frac{1}{2}t)$, $0 \le t \le 2$. [*Ans.* $\frac{19}{3}$.]

(c) $y = x^{3/2}$, $0 \le x \le 5$.

(d) $g(t) = (6t^2, 4\sqrt{2}\,t^3, 3t^4)$, $-1 \le t \le 2$. [*Ans.* 81.]

2. (a) Set up the integral for the arc length of the ellipse

$$(x, y) = (a\cos t, b\sin t),\ 0 \le t \le 2\pi.$$

(b) Show that the computation of the integral in part (a) can be reduced to the computation of a standard elliptic integral of the form

$$\int_0^{\pi/2} \sqrt{1 - k^2\sin^2\theta}\, d\theta.$$

(c) By using a table of elliptic integrals or by direct numerical calculation, find an approximate value for the arc length of an ellipse with $a = 1$ and $b = 2$. [*Ans.* 9.689.]

3. Suppose a curve γ is parametrically defined by two continuously differentiable functions

$$f(t), \qquad a \le t \le b,$$

$$g(u), \qquad \alpha \le u \le \beta.$$

These functions are called **equivalent** parametrizations of γ if there is a continuously differentiable function φ such that

$$a = \varphi(\alpha) \quad \text{and} \quad b = \varphi(\beta),$$

$$f(\varphi(u)) = g(u), \qquad \alpha \le u \le \beta,$$

$$\varphi'(u) > 0, \qquad \alpha < u < \beta.$$

(a) Show that equivalent parametrizations of γ assign the same length to γ.

(b) Show that

$$(x, y) = (\cos t, \sin t), \qquad 0 \le t \le \frac{\pi}{2}$$

and

$$(x, y) = \left(\frac{1 - u^2}{1 + u^2}, \frac{2u}{1 + u^2}\right), \qquad 0 \le u \le 1$$

are equivalent parametrizations of a quarter circle.

(c) Find a pair of nonequivalent parametrizations of some curve.

4. Show that the curve

$$(x, y) = (\cos s, \sin s), \qquad 0 \le s \le 2\pi$$

is parametrized by arc length, and sketch the velocity and acceleration vectors, together with the curve, at $s = \pi/2$.

5. Let γ be a continuously differentiable curve with endpoints $\mathbf{p}_1$ and $\mathbf{p}_2$. Let λ be the line segment $\mathbf{p}_1 + t(\mathbf{p}_2 - \mathbf{p}_1)$, $0 \le t \le 1$. Prove that $l(\lambda) \le l(\gamma)$.

6. Consider the spiral curve

$$\begin{pmatrix} x \\ y \\ z \end{pmatrix} = \begin{pmatrix} a \cos \omega t \\ a \sin \omega t \\ bt \end{pmatrix}, \qquad 0 \le t.$$

(a) Find explicitly the arc length parametrization of the curve.
(b) Find the unit tangent and principal normal vectors at an arbitrary point.
(c) Find the curvature at an arbitrary point.

7. (a) Show that for a line given by $g(t) = t\mathbf{x}_1 + \mathbf{x}_0$, the curvature is identically zero.

(b) Show that if a curve γ, parametrized by arc length and given by a function $f(s)$, has a tangent at every point and has curvature identically zero, then γ is a straight line.

8. Find the total mass of the spiral given by $g(t) = (a \cos t, a \sin t, bt)$, $0 \le t \le 2\pi$, if its density per unit of length at (x, y, z) is equal to $x^2 + y^2 + z^2$.

9. Show that if γ is the *graph* of a function $\mathcal{R} \xrightarrow{f} \mathcal{R}^2$, defined for $a \le x \le b$. then

$$l(\gamma) = \int_a^b \sqrt{1 + (f_1'(x))^2 + (f_2'(x))^2}\, dx,$$

where f_1 and f_2 are the coordinate functions of γ, assumed continuously differentiable.

10. If a curve is parametrized by arc length, its curvature is $\kappa(s) = |(d/ds)\mathbf{t}(s)|$, Show that if $\theta(s, h)$ is the angle between $\mathbf{t}(s)$ and $\mathbf{t}(s + h)$, which tends to zero as h tends to zero, then

$$\kappa(s) = \lim_{h \to 0} \left| \frac{\theta(s, h)}{h} \right| .$$

[*Hint*. Show that $|\mathbf{t}(s + h) - \mathbf{t}(s)| = \sqrt{2 - 2 \cos \theta(s, h)}$.]

11. Show that if a curve is given parametrically by a function $g(t)$, then in terms of derivatives with respect to t, the curvature at $g(t)$ is

$$\frac{\sqrt{|g'(t)|^2\,|g''(t)|^2 - (g'(t)\cdot g''(t))^2}}{|g'(t)|^3}.$$

[*Hint.* Express $|\mathbf{t}'|$ in terms of g' and g''.]

12. (a) Find the curvature function $\kappa(t)$ of the parabola $\mathbf{x}(t) = (t, t^2)$ for $-\infty < t < \infty$.
(b) Show that, for a circle of radius r, the curvature is given by $\kappa(t) = 1/r$.

13. A piece of wire is coiled in a uniform spiral 3 inches in diameter and 2 feet long. Find the length of the wire if it contains 6 complete turns.

14. Let γ be a continuously differentiable curve having a mass distribution of density $p(\mathbf{x})$ at each point $\mathbf{x}$ of γ. Let m be the total mass of γ. If γ is given by $\mathcal{R} \xrightarrow{g} \mathcal{R}^n$, $a \le t \le b$, the **center of mass** of the distribution is the vector

$$\mathbf{z} = \frac{1}{m}\int_a^b g(t)p(g(t))\,|g'(t)|\,dt.$$

(See Exercise 14 of the previous section for the definition of the vector integral.)

(a) Find the center of mass of the spiral $g(t) = (a\cos t, a\sin t, bt)$, $0 \le t \le 2\pi$, with density at (x, y, z) equal to $x^2 + y^2 + z^2$.
(b) Show that if γ has uniform density 1 and is parametrized by arc length, then

$$\mathbf{z} = \frac{1}{m}\int_0^{l(\gamma)} g(s)\,ds.$$

(c) Use the results of Problem 17(b) of the previous section to show that the center of mass then satisfies

$$|\mathbf{z}| \le \frac{1}{m}\int_0^{l(\gamma)} |g(s)|\,ds.$$

SECTION 4

LINE INTEGRALS

The integral $\int_a^b f(x)\,dx$ of a real-valued function of a real variable can be generalized in several ways. One generalization that has applications in physics is the line integral, which we describe here. Let $\mathcal{R}^3 \xrightarrow{F} \mathcal{R}^3$ be a continuous vector field defined in a region D of $\mathcal{R}^3$. Let γ be a curve lying in D and parametrized by a function $\mathcal{R} \xrightarrow{g} \mathcal{R}^3$ with $g(t)$ defined and continuously differentiable for $a \le t \le b$. To say that γ lies in D means simply that the range of g lies in D. A typical situation is shown in Fig. 19. At each point $g(t)$ of γ we picture the tangent vector $g'(t)$ as an arrow with its initial point at $g(t)$. Also at $g(t)$ we locate the arrow describing the vector

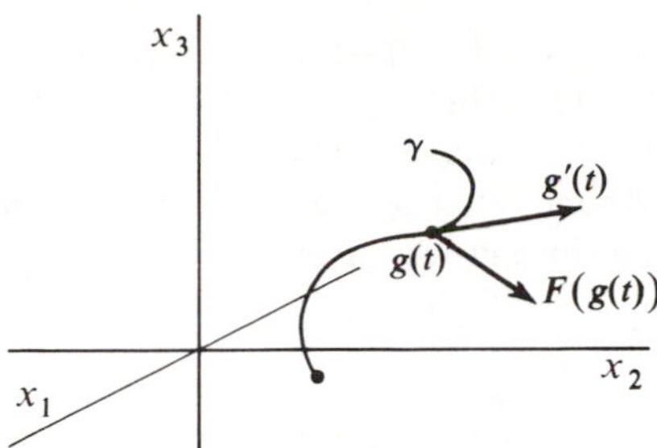

Figure 19

field F at $g(t)$. The dot-product $F(g(t)) \cdot g'(t)$ is then a continuous real-valued function of t for $a \leq t \leq b$, and the integral

$$\int_a^b F(g(t)) \cdot g'(t)\, dt$$

is called the **line integral** of F over γ.

Example 1. If a vector field is given in $\mathcal{R}^3$ by $F(x, y, z) = (x^2, y^2, z^2)$ and γ is given by $g(t) = (t, t^2, t^3)$ for $0 \leq t \leq 1$, then the integral of F over γ is

$$\int_0^1 (t^2, t^4, t^6) \cdot (1, 2t, 3t^2)\, dt = \int_0^1 (t^2 + 2t^5 + 3t^8)\, dt = 1.$$

The line integral can be interpreted in qualitative terms as follows. The dot-product

$$F(g(t)) \cdot \frac{g'(t)}{|g'(t)|}$$

is the coordinate of $F(g(t))$ in the direction of the unit tangent vector to γ at $g(t)$. Then $F(g(t)) \cdot g'(t)$, the integrand in Formula (1), is the tangential coordinate of $F(g(t))$ times $|g'(t)|$, the speed of traversal of γ at $g(t)$. In particular, if $F(g(t))$ is always perpendicular to γ at $g(t)$, the integrand, and hence the integral will be zero. At the other extreme, for a given field F, if the speed $|g'(t)|$ is prescribed at each point of the curve, then the integrand will be maximized by choosing a curve γ that at each point has the same direction as the field there. Thus the integrand in the line integral can be thought of as a measure of the effect of the vector field along γ.

Formula (1) can be generalized to any number of dimensions. Thus if $\mathcal{R}^n \xrightarrow{F} \mathcal{R}^n$, and $\mathcal{R} \xrightarrow{g} \mathcal{R}^n$ describes for $a \leq t \leq b$ a smooth curve, γ, lying in D, the line integral of F over γ is still defined by Formula (1), in which the dot-product is now formed in $\mathcal{R}^n$.

Example 2. Let $F(x, y) = (x, y)$ define a vector field in $\mathcal{R}^2$. The curve given by $g(t) = (\cos t, \sin t)$ for $0 \leq t \leq \pi/2$ is a quarter circle shown in Fig. 20 together with some tangent vectors and some vectors of the field. Because the field is perpendicular to the curve at each point, we expect the integral to be zero, and in fact we have

$$\int_0^{\pi/2} F(g(t)) \cdot g'(t)\, dt = \int_0^{\pi/2} (\cos t, \sin t) \cdot (-\sin t, \cos t)\, dt$$
$$= \int_0^{\pi/2} (-\cos t \sin t + \sin t \cos t)\, dt = 0.$$

Example 3. An important physical interpretation of the line integral arises as follows. Suppose that the function $\mathcal{R}^3 \xrightarrow{F} \mathcal{R}^3$ determines a continuous force field in a region D in $\mathcal{R}^3$. To define W, the work done in

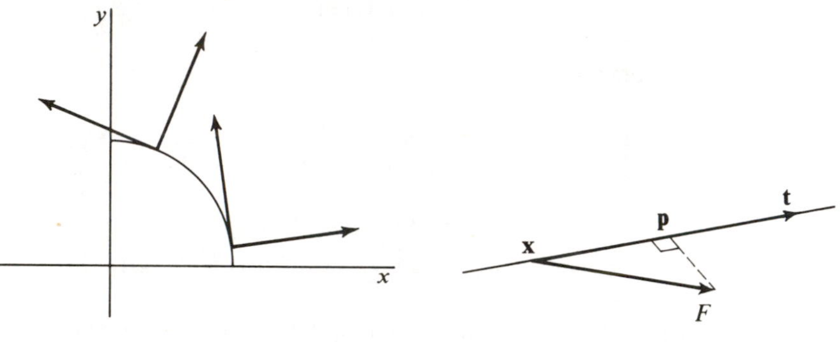

Figure 20 **Figure 21**

moving a particle along a curve γ in D, we use the definition that, for linear motion in a constant field, work is force times distance. That is,

$$W = (F_{\mathbf{t}})(s),$$

where s is the distance traversed and $F_{\mathbf{t}}$ is the coordinate of the force in the direction of motion. In Fig. 21 a particle moves along a line having direction vector $\mathbf{t}$ with $|\mathbf{t}| = 1$, and it is subject at each point $\mathbf{x}$ to the same force vector $F(\mathbf{x})$. The coordinate of F in the direction of motion is $F_{\mathbf{t}} = F \cdot \mathbf{t}$. Then $W = (F \cdot \mathbf{t})s$.

For motion along a continuously differentiable curve, we begin by approximating the curve by tangent vectors, as is done in defining arc length. If the curve γ is given parametrically by $\mathcal{R} \xrightarrow{g} \mathcal{R}^3$ with $g(t)$ defined for $a \leq t \leq b$, then the arrows representing the tangent vectors $g'(t_{k-1})(t_k - t_{k-1})$, $t_0 < t_1 < \ldots < t_K$, will approximate γ as shown in Fig. 22. Let us fix a point $\mathbf{x}_k = g(t_k)$ on γ, and near $\mathbf{x}_k$ approximate F by the constant field $F(\mathbf{x}_k)$. That is, near $\mathbf{x}_k$ we approximate $F(\mathbf{x})$ by the vector

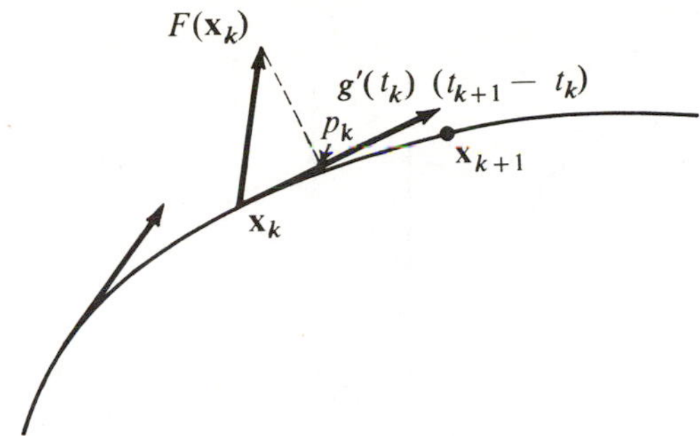

Figure 22

field that assigns the constant vector $F(\mathbf{x}_k)$ to every point. The tangential coordinate of $F(\mathbf{x}_k)$ is $F(\mathbf{x}_k) \cdot \mathbf{t}(t_k)$, where $\mathbf{t}(t) = g'(t)/|g'(t)|$.

Thus the work done in moving a particle along γ from $\mathbf{x}_k$ to $\mathbf{x}_{k+1}$ is approximately

$$\begin{aligned} W_k &= \big(F(\mathbf{x}_k) \cdot \mathbf{t}(t_k)\big)\, |g'(t_k)|\ (t_{k+1} - t_k) \\ &= F\big(g(t_k)\big) \cdot g'(t_k)(t_{k+1} - t_k). \end{aligned}$$

Letting $m(P) = \max\limits_{1 \le k \le K} (t_k - t_{k-1})$, we get

$$\lim_{m(P) \to 0} \sum_{k=0}^{K-1} W_k = \int_a^b F\big(g(t)\big) \cdot g'(t)\, dt,$$

which we define to be the **work** done by the field F in moving the particle through the domain of F along γ.

The assumptions made in Example 3 that F be continuous and that g' be continuous assured that the integrand $F\big(g(t)\big) \cdot g'(t)$ would be continuous and hence that the line integral would exist. However these conditions are stronger than necessary. It is enough to assume that the path of integration is piecewise smooth and then that the vector field F is sufficiently regular so that the integral in Formula (1) exists. Thus the derivative g' may be discontinuous at finitely many points, so that γ has sharp corners as shown in Fig. 23.

Example 4. Let a vector field be defined in $\mathcal{R}^3$ by $F(x, y, z) = (x, y, z)$. Let $g(t) = (\cos t, \sin t, |t - \pi/2|)$ describe a curve γ in $\mathcal{R}^3$ for $0 \le t \le \pi$. Then γ has a corner at $(0, 1, 0)$, where $t = \pi/2$. Indeed g is not differentiable there, and in fact $\lim\limits_{t \to \pi/2-} g'(t) = (-1, 0, -1)$ and $\lim\limits_{t \to \pi/2+} g'(t) = (-1, 0, 1)$, showing that the direction of the tangent jumps abruptly at $t = \pi/2$. Nevertheless, the integral of F over γ exists. To compute it, the interval of integration would ordinarily be broken at $t = \pi/2$. However, in this particular case $F\big(g(t)\big) \cdot g'(t) = t - \pi/2$ unless $t = \pi/2$, and

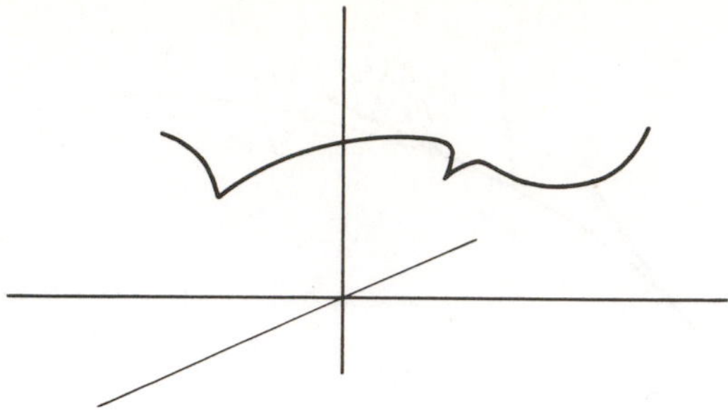

Figure 23

the line integral is easily seen to have the value zero over the interval $0 \leq t \leq \pi$.

A convenient notation for line integrals arises if we denote the parametrization of γ by $g(t) = \big(x(t), y(t), z(t)\big)$ for $a \leq t \leq b$. Denoting the coordinate functions of F by F_1, F_2, and F_3, and suppressing the variable t in the integrand, we get

$$\int_a^b F\big(g(t)\big) \cdot g'(t)\, dt$$

$$= \int_a^b \left[F_1(x, y, z) \frac{dx}{dt} + F_2(x, y, z) \frac{dy}{dt} + F_3(x, y, z) \frac{dz}{dt} \right] dt.$$

The last integral can be abbreviated

$$\int_\gamma F_1\, dx + F_2\, dy + F_3\, dz$$

and can be still further shortened by writing $d\mathbf{x} = (dx, dy, dz)$. Then the formula becomes

$$\int_\gamma F \cdot d\mathbf{x}.$$

The meaning of this notation is developed further in Chapter 7, Section 7. In the meantime we shall simply use it as a shorthand.

Example 5. Let $\mathcal{R}^4 \xrightarrow{F} \mathcal{R}^4$ be given by

$$F(x, y, z, w) = (x - y, y - z, z - w, w - x).$$

The curve γ, given by $g(t) = (t, -t, t^2, -t^2)$ for $0 \leq t \leq 1$, passes through the field. The integral of F over γ is

$$\int_\gamma F \cdot d\mathbf{x} = \int_0^1 [(2t)(1) + (-t - t^2)(-1) + (2t^2)(2t) + (-t^2 - t)(-2t)]\, dt$$

$$= 4.$$

Let γ be a differentiable curve in $\mathcal{R}^n$ given by $\mathcal{R} \xrightarrow{g} \mathcal{R}^n$ for $a \leq t \leq b$. The integral over γ of a real-valued function p, with its domain containing γ, is defined to be

$$\int_a^b p(g(t))\,|g'(t)|\,dt. \tag{2}$$

The function p can be thought of as the density per unit of length of a mass distribution over γ, in which case the integral represents the total mass distributed along γ. If we write the line integral of a vector field over γ in the form

$$\int_a^b F(g(t)) \cdot \frac{g'(t)}{|g'(t)|}\,|g'(t)|\,dt, \tag{3}$$

the relationship between Formulas (2) and (3) becomes clear. The integral of the vector field over γ depends on the direction of the tangent to γ at each point and not just the length of the tangent as in the case of the integral of a real-valued function over γ. If the curve γ is parametrized by arc length, then the two integrals take the respective forms

$$\int_0^{l(\gamma)} p(g(s))\,ds \quad \text{and} \quad \int_0^{l(\gamma)} F(g(s)) \cdot \mathbf{t}(s)\,ds,$$

where $\mathbf{t}(s) = (dg/ds)(s)$ is a tangent vector to γ of length 1.

It is clear from the definition of the line integral that, in general, the value depends on the parametrization of the curve γ. The extent to which the value is independent of parametrization is taken up in the exercises.

EXERCISES

1. Compute the following line integrals.

(a) $\int_L x\,dx + x^2\,dy + y\,dz$, where L is given by $g(t) = (t, t, t)$, for $0 \leq t \leq 1$.

(b) $\int_P (x + y)\,dx + dy$, where P is given by $g(t) = (t, t^2)$, $0 \leq t \leq 1$.

(c) $\int_{\gamma_1} x\,dy$ and $\int_{\gamma_2} x\,dy$, where γ_1 is given by $g(t) = (\cos t, \sin t)$ for $0 \leq t \leq 2\pi$, and where γ_2 is given by $h(t) = (\cos t, \sin t)$ for $0 \leq t \leq 4\pi$.

(d) $\int_{\gamma_1} (dx + dy)$, where γ_1 is given by $(x, y) = (\cos t, \sin t)$, $0 \leq t \leq 2\pi$.

(e) $\displaystyle\int_{\gamma_1} \frac{dx + dy}{x^2 + y^2}$, where γ_1 is the curve in part (d).

(f) $\int_\gamma (e^x\,dx + z\,dy + \sin z\,dz)$, where γ is given by $(x, y, z) = (t, t^2, t^3)$, $0 \leq t \leq 1$.

(g) $\int_\gamma F \cdot d\mathbf{x}$, where $F(x, y, z, w) = (x, x, y, xw)$ and γ is given by $(x, y, z, w) = (t, 1, t, t)$, $0 \leq t \leq 2$.

2. Let γ_1 be given by $(x, y) = (\cos t, \sin t)$, $0 \le t \le \pi/2$, and γ_2 be given by $(x, y) = (1 - u, u)$, $0 \le u \le 1$. Compute $\int_{\gamma_1} (f\,dx + g\,dy)$ and $\int_{\gamma_2} (f\,dx + g\,dy)$ for the choices of f and g given below.

(a) $f(x, y) = x, g(x, y) = x + 1$.
(b) $f(x, y) = x + y, g(x, y) = 1$.

(c) $f(x, y) = \dfrac{1}{x^2 + y^2}, g(x, y) = \dfrac{1}{x^2 + y^2}$.

3. Find the work done in moving a particle along the curve $(x, y, z) = (t, t, t^2)$, $0 \le t \le 2$, under the influence of the field $F(x, y, z) = (x + y, y, y)$.

4. Prove that if γ is a curve given parametrically by a function $\mathcal{R} \xrightarrow{f} \mathcal{R}^n$ with

$$\mathbf{x} = f(t), \qquad 0 \le t \le 1,$$

and if $-\gamma$ is the curve described by

$$\mathbf{x} = f(1 - t), \qquad 0 \le t \le 1,$$

then

$$\int_{-\gamma} F \cdot d\mathbf{x} = -\int_{\gamma} F \cdot d\mathbf{x}.$$

5. Let $F(x, y) = (y, x)$ describe a vector field in $\mathcal{R}^2$. Find a curve γ_1 passing through the field and starting at the point (2,1) such that γ_1 has length 1 and the integral of F over γ_1 is zero.

6. Prove that if γ is given by $\mathcal{R} \xrightarrow{g} \mathcal{R}^n$ for $\alpha \le t \le b$ and γ is then reparametrized by arc length s so that $t = t(s)$, then

$$\int_a^b F(g(t)) \cdot g'(t)\,dt = \int_0^{l(\gamma)} F(h(s)) \cdot \mathbf{t}(s)\,ds,$$

where $h(s) = g(t(s))$ and $\mathbf{t}(s) = (dh/ds)(s)$. [*Hint*. Use the change of variable theorem for integrals.]

7. Let $\mathcal{R}^n \xrightarrow{F} \mathcal{R}^n$ be a vector field and γ a curve such that the line integral $\int_\gamma F \cdot d\mathbf{x}$ exists. Prove that if $|F(\mathbf{x})| \le M$, a constant, on γ, then $|\int_\gamma F \cdot d\mathbf{x}| \le Ml(\gamma)$.

8. (a) If $g(t)$ and $h(u)$ are equivalent parametrizations of γ as defined in Exercise 3 of the previous section on arc length, show that $\int_\gamma F \cdot d\mathbf{x}$ has the same value when computed with either parametrization.
(b) Find nonequivalent parametrizations of the circle $x^2 + y^2 = 1$ in $\mathcal{R}^2$ and a vector field F such that the integrals of F with respect to the two parametrizations are different.

9. Find the total mass of the wire $\begin{pmatrix} x \\ y \\ z \end{pmatrix} = \begin{pmatrix} 6t^2 \\ 4\sqrt{2}\,t^3 \\ 3t^4 \end{pmatrix}, 0 \le t \le 1$:

(a) If the density at the point corresponding to t is t^2.

(b) If the density s units from the origin measured along the curve is $(s + 1)$.
(c) If the density at a point is equal to its distance from the origin measured in $\mathcal{R}^3$.

10. Show that if $\int_\gamma F \cdot d\mathbf{x}$ and $\int_\gamma G \cdot d\mathbf{x}$ exist, then $\int_\gamma (aF + bG) \cdot d\mathbf{x} = a \int_\gamma F \cdot d\mathbf{x} + b \int_\gamma G \cdot d\mathbf{x}$, where a and b are any constants.

11. Show that if γ and η are smooth curves described by functions $\mathcal{R} \xrightarrow{g} \mathcal{R}^n$ defined on $[a, b]$, and $\mathcal{R} \xrightarrow{h} \mathcal{R}^n$ defined on $[b, c]$, with $g(b) = h(b)$, then

$$\int_\delta F \cdot d\mathbf{x} = \int_\gamma F \cdot d\mathbf{x} + \int_\eta F \cdot d\mathbf{x},$$

where δ is the curve given by

$$f(t) = \begin{cases} g(t), & a \leq t \leq b \\ h(t), & b \leq t \leq c, \end{cases}$$

and F is a continuous vector field on $\gamma \cup \eta$.

12. Let a function $g(t)$ represent the position of a particle of varying mass $m(t)$ in $\mathcal{R}^3$ at time t. Then the velocity vector of the particle is $\mathbf{v}(t) = g'(t)$, and the force vector acting on the particle at $g(t)$ is $F(g(t)) = [m(t)\mathbf{v}(t)]'$.

(a) Show that $F(g(t)) \cdot g'(t) = m'(t)v^2(t) + m(t)v(t)v'(t)$, where v is the speed of the particle.
(b) Show that if $m(t)$ is constant, then the work done in moving the particle over its path between times $t = a$ and $t = b$ is $w = (m/2)(v^2(b) - v^2(a))$. (The function $(\frac{1}{2})mv^2(t)$ is the kinetic energy of the particle.)

SECTION 5

PARTIAL DERIVATIVES

To extend the techniques of calculus to functions defined in $\mathcal{R}^n$ we need partial derivatives. Let f be a real-valued function with domain space $\mathcal{R}^n$. For each $i = 1, \ldots, n$, we define a new real-valued function called the **partial derivative of f with respect to the ith variable** and denoted by $\partial f/\partial x_i$. For each $\mathbf{x} = (x_1, \ldots, x_n)$ in the domain of f, the number $(\partial f/\partial x_i)(\mathbf{x})$ is by definition

5.1 $$\frac{\partial f}{\partial x_i}(\mathbf{x}) = \lim_{t \to 0} \frac{f(x_1, \ldots, x_i + t, \ldots, x_n) - f(x_1, \ldots, x_i, \ldots, x_n)}{t}.$$

The domain space of $\partial f/\partial x_i$ is $\mathcal{R}^n$, and the domain of $\partial f/\partial x_i$ is the subset of the domain of f consisting of all $\mathbf{x}$ for which the above limit exists. Thus the domain of $\partial f/\partial x_i$ could conceivably be the empty set. The number $(\partial f/\partial x_i)(\mathbf{x})$ is simply the derivative at x_i of the function of one variable obtained by holding $x_1, \ldots, x_{i-1}, x_{i+1}, \ldots, x_n$ fixed and by considering f to be a function of the ith variable only. As a result, the differentiation formulas of one-variable calculus apply directly.

Example 1. Let $f(x, y, z) = x^2y + y^2z + z^2x$. Then

$$\frac{\partial f}{\partial x}(x, y, z) = 2xy + z^2,$$

$$\frac{\partial f}{\partial y}(x, y, z) = x^2 + 2yz,$$

$$\frac{\partial f}{\partial z}(x, y, z) = y^2 + 2zx.$$

The partial derivatives at $\mathbf{x} = (1, 2, 3)$ are

$$\frac{\partial f}{\partial x}(1, 2, 3) = 4 + 9 = 13,$$

$$\frac{\partial f}{\partial y}(1, 2, 3) = 1 + 12 = 13,$$

$$\frac{\partial f}{\partial z}(1, 2, 3) = 4 + 6 = 10.$$

Example 2. Let $f(u, v) = \sin u \cos v$. Then

$$\frac{\partial f}{\partial u} = \frac{\partial \sin u \cos v}{\partial u} = \cos u \cos v,$$

$$\frac{\partial f}{\partial v} = \frac{\partial \sin u \cos v}{\partial v} = -\sin u \sin v,$$

$$\frac{\partial f}{\partial v}\left(\frac{\pi}{2}, \frac{\pi}{2}\right) = \frac{\partial \sin u \cos v}{\partial v}\left(\frac{\pi}{2}, \frac{\pi}{2}\right) = -\sin \frac{\pi}{2} \sin \frac{\pi}{2} = -1.$$

We can repeat the operation of taking partial derivatives. The partial derivative of $\partial f/\partial x_i$ with respect to the jth variable is $\partial/\partial x_j\ (\partial f/\partial x_i)$ and is denoted by $\partial^2 f/\partial x_j\, \partial x_i$. This can be repeated indefinitely, provided the derivatives exist. An alternative notation for higher-order partial derivatives is illustrated below, in which the variable of differentiation is denoted by a subscript.

$$\frac{\partial f}{\partial x_i} = f_{x_i}$$

$$\frac{\partial}{\partial x_j}\left(\frac{\partial f}{\partial x_i}\right) = \frac{\partial^2 f}{\partial x_j\, \partial x_i} = f_{x_i x_j}$$

$$\frac{\partial}{\partial x_i}\left(\frac{\partial f}{\partial x_i}\right) = \frac{\partial^2 f}{\partial x_i^2} = f_{x_i x_i}$$

$$\frac{\partial}{\partial x_i}\left(\frac{\partial^2 f}{\partial x_j\, \partial x_i}\right) = \frac{\partial^3 f}{\partial x_i\, \partial x_j\, \partial x_i} = f_{x_i x_j x_i}.$$

Example 3. Consider $f(x, y) = xy - x^2$.

$$f_x = \frac{\partial f}{\partial x} = y - 2x$$

$$f_{xy} = \frac{\partial^2 f}{\partial y\, \partial x} = 1$$

$$f_{xx} = \frac{\partial^2 f}{\partial x^2} = -2$$

$$f_{yxx} = \frac{\partial^3 f}{\partial x^2\, \partial y} = 0.$$

To interpret partial derivatives geometrically we can use the fact that, for a real-valued function of a single real variable, the value of the derivative at a point is the slope of the tangent line to the graph of the function at that point. For illustrative purposes it will be enough to consider the graph of a function $\mathcal{R}^2 \xrightarrow{f} \mathcal{R}$, namely, the set of points $(x, y, f(x, y))$ in $\mathcal{R}^3$ where (x, y) is in the domain of f. Such a graph is shown in Fig. 24 as a surface lying over a rectangle in the xy-plane. The intersection of the surface with the vertical plane determined by the condition $y = b$ is a curve satisfying the conditions

$$z = f(x, y), \qquad y = b.$$

Consider as a subset of 2-dimensional space the curve defined by the

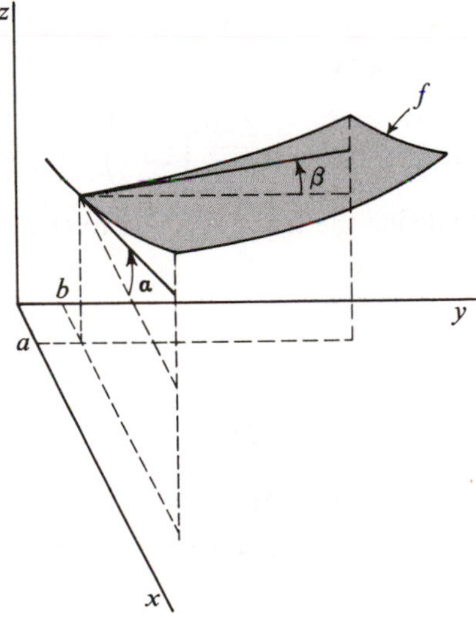

Figure 24

function $g(x) = f(x, b)$. Its slope at $x = a$ is

$$g'(a) = \frac{\partial f}{\partial x}(a, b).$$

Similarly, at $y = b$ the curve defined by $h(y) = f(a, y)$ has slope equal to

$$h'(b) = \frac{\partial f}{\partial y}(a, b).$$

The angles α and β shown in Fig. 24 therefore satisfy

$$\tan \alpha = \frac{\partial f}{\partial x}(a, b), \qquad \tan \beta = \frac{\partial f}{\partial y}(a, b).$$

The numbers $\tan \alpha$ and $\tan \beta$ are slopes of tangent lines to two curves contained in the graph of the function f. For this reason it is natural to try to define a tangent plane to the graph of f to be the plane containing these two lines. (If f satisfies the condition of differentiability defined in Section 8, this is what is done.) We see easily that the set of points (x, y, z) satisfying

$$z = f(a, b) + (x - a)\frac{\partial f}{\partial x}(a, b) + (y - b)\frac{\partial f}{\partial y}(a, b) \tag{1}$$

is a plane containing the tangent lines found above. In fact, specifying $y = b$ and $x = a$ determines the two lines.

Example 4. The part of the graph of

$$f(x, y) = 1 - 2x^2 - y^2$$

corresponding to $x \geq 0$, $y \geq 0$ is shown in Fig. 25. The function f has partial derivatives at $(\frac{1}{2}, \frac{1}{2})$ given by

$$\frac{\partial f}{\partial x}\left(\frac{1}{2}, \frac{1}{2}\right) = -2, \qquad \frac{\partial f}{\partial y}\left(\frac{1}{2}, \frac{1}{2}\right) = -1.$$

Using Equation (1) to define the tangent plane to the graph of f at $(\frac{1}{2}, \frac{1}{2})$,

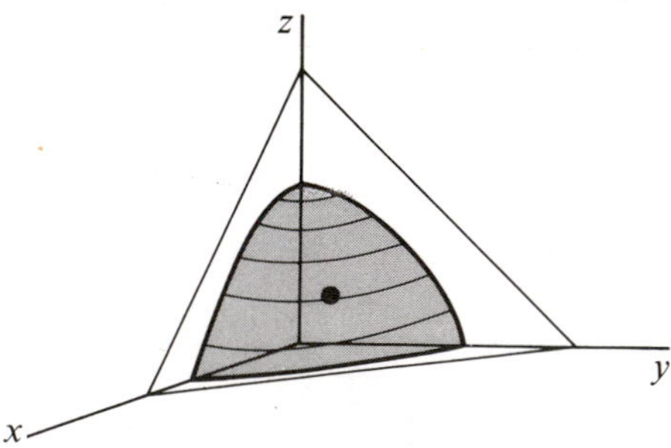

Figure 25

we find also $f(\frac{1}{2}, \frac{1}{2}) = \frac{1}{4}$. Thus the equation of the tangent is

$$\begin{aligned} z &= \tfrac{1}{4} - 2(x - \tfrac{1}{2}) - (y - \tfrac{1}{2}) \\ &= -2x - y + \tfrac{7}{4}. \end{aligned}$$

We can sketch the tangent plane by drawing the two tangent lines in it determined by $x = \frac{1}{2}$ and $y = \frac{1}{2}$. It is somewhat easier to locate three points on the plane, for simplicity

$$(\tfrac{7}{8}, 0, 0), \qquad (0, \tfrac{7}{4}, 0), \qquad (0, 0, \tfrac{7}{4}).$$

The point of tangency is $(\frac{1}{2}, \frac{1}{2}, \frac{1}{4})$.

Continuity for functions of more than one variable is discussed extensively in Section 7. At this point we shall consider briefly the case $\mathcal{R}^2 \xrightarrow{f} \mathcal{R}$. For convenience we assume, for each $\mathbf{x} = (x, y)$ in the domain of f, that $f(\mathbf{z})$ is defined for all vectors $\mathbf{z} = (z, w)$ satisfying $|\mathbf{x} - \mathbf{z}| < \delta$, where δ is some positive number. We then say that f is continuous if, for every point $\mathbf{x}$ in the domain of f,

$$\lim_{|\mathbf{x}-\mathbf{z}|\to 0} f(\mathbf{z}) = f(\mathbf{x}).$$

The limit relation means that $f(\mathbf{z})$ can be made arbitrarily close to $f(\mathbf{x})$ if $|\mathbf{x} - \mathbf{z}|$, the distance from $\mathbf{x}$ to $\mathbf{z}$, is made small enough. As usual, the intuitive idea of continuity is that the values of the function f should not change abruptly, resulting, for example, in breaks in the graph of f. The graphs shown in Figs. 24 and 25 are those of continuous functions, while Fig. 26 shows a simple example of the graph of a discontinuous function. The condition that the domain D of f contain all points $\mathbf{z}$ sufficiently near every $\mathbf{x}$ in D is expressed by saying that D is an **open set.**

It is a consequence of certain continuity conditions on f that the higher-order partial derivatives of $\mathcal{R}^2 \xrightarrow{f} \mathcal{R}$ are independent of the order of differentiation. The precise statement follows, though we remark that a slightly stronger theorem can be proved. (See Exercise 8 of Chapter 6, Section 2.)

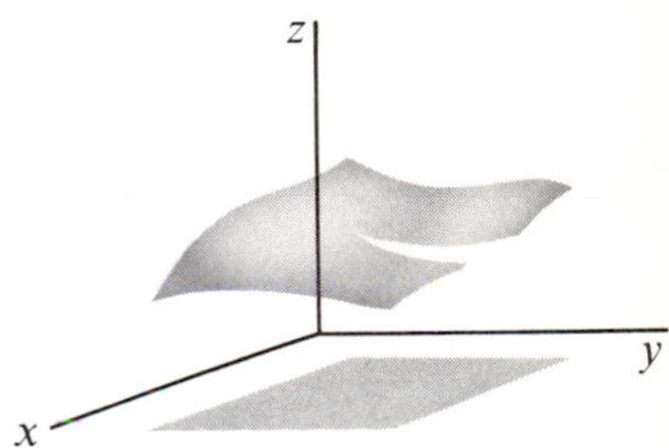

Figure 26

5.2 Theorem

Let $\mathcal{R}^2 \xrightarrow{f} \mathcal{R}$ be a continuous function such that f_x, f_y, f_{xy}, and f_{yx} are also continuous on the same domain as f. Then $f_{xy} = f_{yx}$.

Proof. Choose x, y, and $\delta > 0$ so that the function

$$\begin{aligned} F(h, k) = {} & [f(x + h, y + k) - f(x + h, y)] \\ & - [f(x, y + k) - f(x, y)] \end{aligned}$$

is defined whenever $\sqrt{h^2 + k^2} < \delta$. We now assume that neither h nor k is zero, and apply the mean-value theorem to the function

$$G(u) = f(u, y + k) - f(u, y)$$

on the interval with endpoints x and $x + h$. We find

$$G(x + h) - G(x) = hG'(x_1),$$

where x_1 is between x and $x + h$. In terms of F and f, this last equation is

$$F(h, k) = h[f_x(x_1, y + k) - f_x(x_1, y)].$$

Now apply the mean-value theorem again, this time to the function $H(v) = f_x(x_1, v)$ on the interval with endpoints y and $y + k$. We find

$$F(h, k) = hkf_{xy}(x_1, y_1),$$

where y_1 is between y and $y + k$.

Rewriting F in the form

$$F(h, k) = [f(x + h, y + k) - f(x, y + k)] - [f(x + h, y) - f(x, y)]$$

allows us to follow the same general procedure, this time differentiating with respect to y, then x. We find

$$F(h, k) = hkf_{yx}(x_2, y_2),$$

where x_2 and y_2 lie between $(x, x + h)$ and $(y, y + k)$, respectively. Equating the two expressions found for $F(h, k)$, and canceling the factor hk, gives

$$f_{xy}(x_1, y_1) = f_{yx}(x_2, y_2).$$

Now let both h and k tend to zero. It follows that the distances

$$\sqrt{(x_1 - x)^2 + (y_1 - y)^2} \quad \text{and} \quad \sqrt{(x_2 - x)^2 + (y_2 - y)^2}$$

both tend to zero; therefore, by the continuity of f_{xy} and f_{yx}, we get $f_{xy}(x, y) = f_{yx}(x, y)$. But the point (x, y) was chosen arbitrarily, so $f_{xy} = f_{yx}$ on the domain of f.

Theorem 5.2 can be applied successively to still higher-order partial derivatives, provided the analogous differentiability and continuity requirements are satisfied. Moreover, by considering only two variables at a time, we can apply it to functions $\mathcal{R}^n \xrightarrow{f} \mathcal{R}$ where $n > 2$. Thus, for the commonly encountered functions, which have partial derivatives of arbitrarily high order, we have typically

$$\frac{\partial^2 f}{\partial x\,\partial y} = \frac{\partial^2 f}{\partial y\,\partial x}$$

$$\frac{\partial^3 g}{\partial x\,\partial y\,\partial x} = \frac{\partial^3 g}{\partial x^2\,\partial y}$$

$$\frac{\partial^4 h}{\partial z\,\partial x\,\partial y\,\partial z} = \frac{\partial^4 h}{\partial x\,\partial y\,\partial z^2}, \quad \text{etc.}$$

The last two formulas follow from repeated application of the two-variable formula by interchanging two differentiations at a time.

EXERCISES

1. Find $\dfrac{\partial f}{\partial x}$ and $\dfrac{\partial f}{\partial y}$, where $f(x, y)$ is:

(a) $x^2 + x \sin (x + y)$.
(b) $\sin x \cos (x + y)$.
(c) e^{x+y+1}.
(d) $\arctan (y/x)$.
(e) x^y.
(f) $\log_x y$.

$$\left[\textit{Ans.}\ \frac{\partial f}{\partial x} = -\frac{\ln y}{x(\ln x)^2}. \right]$$

2. Find $\dfrac{\partial^2 f}{\partial y\,\partial x}$ and $\dfrac{\partial^2 f}{\partial x\,\partial y}$, where f is

(a) $xy + x^2y^3$. (b) $\sin (x^2 + y^2)$. (c) $\dfrac{1}{x^2 + y^2}$.

3. Find the first-order partial derivatives of the following functions:

(a) $f(x, y, z) = x^2 e^{x+y+z} \cos y$.

(b) $f(x, y, z, w) = \dfrac{x^2 - y^2}{z^2 + w^2}$.

(c) $f(x, y, z) = x^{(y^z)}$.

4. Find $\dfrac{\partial^3 f(x, y)}{\partial x^2\,\partial y}$ if $f(x, y) = \log (x + y)$.

5. Show that $\dfrac{\partial^2 f}{\partial x^2} + \dfrac{\partial^2 f}{\partial y^2} = 0$ is satisfied by

(a) $\log (x^2 + y^2)$. (b) $x^3 - 3xy^2$.

6. If $f(x, y, z) = 1/(x^2 + y^2 + z^2)^{1/2}$, show that

$$f_{xx} + f_{yy} + f_{zz} = 0.$$

7. If $f(x_1, x_2, \ldots, x_n) = 1/(x_1^2 + x_2^2 + \ldots + x_n^2)^{(n-2)/2}$, show that

$$f_{x_1x_1} + f_{x_2x_2} + \ldots + f_{x_nx_n} = 0.$$

8. Prove directly that if $f(x, y)$ is a polynomial, then

$$\frac{\partial^2 f}{\partial x\, \partial y} = \frac{\partial^2 f}{\partial y\, \partial x}.$$

9. If

$$f(x, y) = \begin{cases} 2xy\dfrac{x^2 - y^2}{x^2 + y^2}, & \text{for } x^2 + y^2 \neq 0 \\ 0, & \text{for } x = y = 0, \end{cases}$$

show that $f_{xy}(0, 0) = -2$ and $f_{yx}(0, 0) = 2$.

10. For each of the following functions find an explicit representation for the tangent plane to the graph at the indicated point. Also, sketch the graph of the given function together with the tangent plane.

(a) $\sqrt{1 - x^2 - y^2}$ at $(\frac{1}{2}, \frac{1}{2}, 1/\sqrt{2})$.
(b) e^{x+y} at $(1, 2, e^3)$.
(c) $\sin(x^2 + y^2)$ at $(0, 0, 0)$.

11. Find a parametric representation for the line perpendicular to the tangent plane found in Exercise 10(a) and passing through the point of tangency.

SECTION 6

VECTOR PARTIAL DERIVATIVES

In the previous section we considered partial derivatives of real-valued functions only. If $\mathcal{R}^n \xrightarrow{f} \mathcal{R}^m$ is a vector-valued function, then $\partial f/\partial x_i$ is still defined by Equation 5.1 as a limit in one real variable. The difference is that, in Equation 5.1, the quotient and, hence, the limit are now vectors. Since, as described in Section 2, vector limits are computed by taking the limit of each coordinate function, it follows immediately that

6.1 If

$$f(\mathbf{x}) = \begin{pmatrix} f_1(\mathbf{x}) \\ \cdot \\ \cdot \\ \cdot \\ f_m(\mathbf{x}) \end{pmatrix},$$

then

$$\frac{\partial f}{\partial x_i}(\mathbf{x}) = \begin{pmatrix} \dfrac{\partial f_1}{\partial x_i}(\mathbf{x}) \\ \cdot \\ \cdot \\ \cdot \\ \dfrac{\partial f_m}{\partial x_i}(\mathbf{x}) \end{pmatrix}.$$

The geometric significance of the vector partial derivative is as follows. If all the coordinates but one, say x_i, are held fixed, and x_i alone is allowed to vary, then $f(\mathbf{x})$ traces a curve in $\mathcal{R}^m$. The vector $\partial f/\partial x_i(\mathbf{x})$ is a tangent vector to the curve as defined in Section 2.

Example 1. Let

$$f(u, v) = \begin{pmatrix} u \cos v \\ u \sin v \\ v \end{pmatrix}.$$

Then

$$\frac{\partial f}{\partial u}(u_0, v_0) = \begin{pmatrix} \cos v_0 \\ \sin v_0 \\ 0 \end{pmatrix}, \quad \frac{\partial f}{\partial v}(u_0, v_0) = \begin{pmatrix} -u_0 \sin v_0 \\ u_0 \cos v_0 \\ 1 \end{pmatrix}.$$

In Example 1, we can restrict the vector variable (u, v) so that $v = v_0$ is held fixed. Then $f(u, v_0)$ describes a curve as u varies. Similarly $f(u_0, v)$ describes another curve as v varies. Such curves are called parameter curves in the range of f. We shall assume that the coordinate functions of f are continuous functions. According to the definition of Section 2, the vectors

$$\frac{\partial f}{\partial u}(u_0, v_0) \quad \text{and} \quad \frac{\partial f}{\partial v}(u_0, v_0)$$

are tangent vectors to the two parameter curves obtained by varying u and v, respectively, at the point where they intersect, namely, $f(u_0, v_0)$. An example is shown in Fig. 27, where the two curves are singled out by the

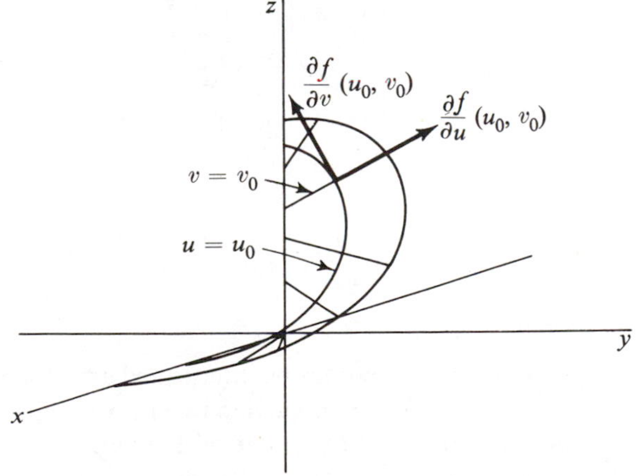

Figure 27

relations $v = v_0$ and $u = u_0$ that determine them; the varying parameter in case $v = v_0$ is u, and in case $u = u_0$ it is v.

Under certain circumstances, and in particular in the most commonly met examples, the parameter curves of a function $\mathcal{R}^2 \xrightarrow{f} \mathcal{R}^3$ fit together to form a surface in the range space $\mathcal{R}^3$. The technical conditions are discussed in the last section of Chapter 4; at this point we shall simply consider some examples.

Example 2. Returning to the function f of Example 1, the curves determined by conditions of the form $u = u_0$ are spirals around the vertical axis in $\mathcal{R}^3$ at a distance u_0 from the axis. Taken together, these curves form a surface with the shape of a spiral ramp, as discussed in Example 7 of Section 1. The ramp can also be thought of as formed by the straight lines determined by the conditions $v = v_0$ and radiating out, perpendicular to the vertical axis. The point $(u, v) = (1, \pi/2)$ in the domain of f corresponds to

$$f(1, \pi/2) = (0, 1, \pi/2)$$

on the surface. At this point on the surface, the tangent vectors

$$\frac{\partial f}{\partial u}(1, \pi/2) = \begin{pmatrix} 0 \\ 1 \\ 0 \end{pmatrix}, \qquad \frac{\partial f}{\partial v}(1, \pi/2) = \begin{pmatrix} -1 \\ 0 \\ 1 \end{pmatrix}$$

determine a plane which, when translated parallel to itself so that it passes through $f(1, \pi/2)$, it is natural to call a tangent plane. A parametric representation for this tangent plane would be given by

$$u\begin{pmatrix} 0 \\ 1 \\ 0 \end{pmatrix} + v\begin{pmatrix} -1 \\ 0 \\ 1 \end{pmatrix} + \begin{pmatrix} 0 \\ 1 \\ \frac{\pi}{2} \end{pmatrix}.$$

In general we shall say that if the range of a function $\mathcal{R}^2 \xrightarrow{f} \mathcal{R}^3$ is a surface with a tangent plane at $f(u_0, v_0)$, then the tangent plane is represented parametrically by

$$u\frac{\partial f}{\partial u}(u_0, v_0) + v\frac{\partial f}{\partial v}(u_0, v_0) + f(u_0, v_0),$$

with u and v as parameters.

A function $\mathcal{R}^3 \xrightarrow{f} \mathcal{R}^2$ can sometimes be interpreted as a 2-dimensional **flow** as follows. A point (x, y, t) in $\mathcal{R}^3$ is carried by f into a point $(x', y') = f(x, y, t)$ in $\mathcal{R}^2$, where (x, y) is to be thought of as the position of a particle in $\mathcal{R}^2$ at time $t = 0$, and $f(x, y, t)$ is to be the position of that same

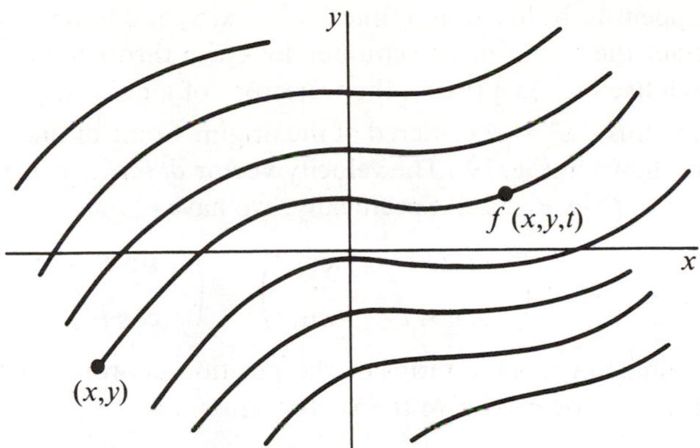

Figure 28

particle after time t has elapsed. Figure 28 shows a picture of a 2-dimensional flow. In the picture, the path of a particle that happens to be at position (x, y) at time $t = 0$ is shown as a curve containing also the position $f(x, y, t)$ of the same particle t time units later. Thus, by holding (x, y) fixed and allowing t to vary, we get a curve called a **trajectory** of the flow. The family of all trajectories determined by f gives the geometric picture of the flow. The tangent vectors $f_t(x, y, t)$, or

$$\frac{\partial f}{\partial t}(x, y, t),$$

for each fixed t form a vector field with the tail of the arrow $f_t(x, y, t)$ now located at the point $(x', y') = f(x, y, t)$, initially at (x, y). This vector field is called the velocity field of the flow at time t. If the velocity field of a flow is independent of t, then the flow is called **steady.** As the next example shows, a steady flow is not necessarily one for which the function $f_t(x, y, t)$ is itself independent of t. On the other hand, it should not be assumed that for a nonsteady flow every particle on a given trajectory follows the same path as the particle that initiates the trajectory.

Example 3. The function

$$\begin{aligned} f(x, y, t) &= \begin{pmatrix} x \cos t - y \sin t \\ x \sin t + y \cos t \end{pmatrix} \\ &= \begin{pmatrix} \cos t & -\sin t \\ \sin t & \cos t \end{pmatrix} \begin{pmatrix} x \\ y \end{pmatrix} \end{aligned}$$

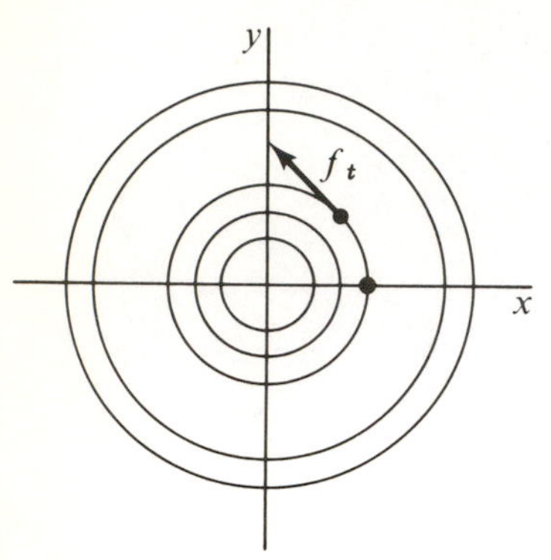

Figure 29

happens to be linear as a function of (x, y) and is in fact a rotation about the origin in $\mathcal{R}^2$, counterclockwise through an angle t, for each fixed t. As a result, the trajectory of a point (x, y) is a circle of radius $\sqrt{x^2 + y^2}$ centered at the origin. Some of the trajectories are shown in Fig. 29. The velocity vector $\partial f/\partial t(x, y, t)$ is denoted $f_t(x, y, t)$ in Fig. 29. Specifically, we have

$$f_t(x, y, t) = \begin{pmatrix} -x \sin t - y \cos t \\ x \cos t - y \sin t \end{pmatrix} = \begin{pmatrix} -\sin t & -\cos t \\ \cos t & -\sin t \end{pmatrix} \begin{pmatrix} x \\ y \end{pmatrix}.$$

To find $f_t(x, y, t)$ in terms of the position coordinates $(x', y') = f(x, y, t)$, we can solve the vector equation

$$\begin{pmatrix} x' \\ y' \end{pmatrix} = \begin{pmatrix} \cos t & -\sin t \\ \sin t & \cos t \end{pmatrix} \begin{pmatrix} x \\ y \end{pmatrix}$$

for (x, y) and find

$$\begin{pmatrix} x \\ y \end{pmatrix} = \begin{pmatrix} \cos t & \sin t \\ -\sin t & \cos t \end{pmatrix} \begin{pmatrix} x' \\ y' \end{pmatrix}.$$

It follows that $f_t(x, y, t)$ is given by the product

$$\begin{pmatrix} -\sin t & -\cos t \\ \cos t & -\sin t \end{pmatrix} \begin{pmatrix} \cos t & \sin t \\ \sin t & \cos t \end{pmatrix} \begin{pmatrix} x' \\ y' \end{pmatrix} = \begin{pmatrix} 0 & -1 \\ 1 & 0 \end{pmatrix} \begin{pmatrix} x' \\ y' \end{pmatrix}.$$

Thus the velocity field v is given by

$$v(x', y') = \begin{pmatrix} -y' \\ x' \end{pmatrix}.$$

Since v is independent of time, the flow is steady.

Example 3 illustrates a basic assumption that is always made about a flow: two particles cannot occupy the same position at the same time; in other words, if

$$f(\mathbf{x}_0, t) = f(\mathbf{x}_1, t),$$

then $\mathbf{x}_0 = \mathbf{x}_1$. The assumption does not imply that trajectories cannot cross one another (or themselves), but only that, for t fixed, the equation

$$f(\mathbf{x}, t) = \mathbf{y}$$

has at most one solution $\mathbf{x}$ for any given vector $\mathbf{y}$. Of course a function $\mathcal{R}^4 \xrightarrow{f} \mathcal{R}^3$ of the form $f(\mathbf{x}, t) = f(x, y, z, t)$ would be used to describe a 3-dimensional flow. Flows of higher dimension are also important in theoretical mechanics.

EXERCISES

1. Find formulas for the vector partial derivatives $\partial f/\partial x(x, y)$ and $\partial f/\partial y(x, y)$ if

(a) $f(x, y) = \begin{pmatrix} x + y \\ x - y \\ x^2 + y^2 \end{pmatrix}.$

(b) $f(x, y) = (e^x \cos y, e^x \sin y)$.

2. For each of the following functions, find the vector partial derivatives at the indicated point. Sketch the coordinate curves for which these vectors are tangents at the given point, and sketch the tangent plane to the range of f at that point.

(a) $f(u, v) = (u, v, u^2 + v^2)$ at $(1, 1, 2)$.

(b) $f(u, v) = (\cos u \sin v, \sin u \sin v, \cos v)$ for $(u, v) = (\pi/4, \pi/4)$.

[*Hint.* The range of f in part (b) lies on a sphere.]

3. Find a parametric representation for the tangent plane of Exercise 2(a) and also a representation for the line perpendicular to the tangent plane and passing through the point of tangency.

4. Consider the 2-dimensional flow

$$f(x, y, t) = \begin{pmatrix} x + t \\ y + t^2 \end{pmatrix}, \qquad \text{for} \quad t \leq 0.$$

(a) Sketch the trajectories of the flow that start at $(x, y) = (0, 0)$, $(0, 1)$, and $(1, 1)$.

(b) For $t = 1$ sketch the velocity vectors at the points $f(x, y, 1)$, with (x, y) chosen as in part (a).

(c) Show that the flow determined by f is not steady. [*Hint.* It is not enough to show that $\partial f/\partial t$ depends on t. Consider $f(0, 0, 1)$ and $\partial f/\partial t(0, 0, 1)$ as compared with $f(1, 1, 0)$ and $\partial f/\partial t(1, 1, 0)$.]

5. Consider the 2-dimensional flow

$$f(x, y, t) = (xe^t, ye^t), \qquad \text{for} \quad t \geq 0.$$

(a) Sketch the trajectories of the flow that start at $(x, y) = (0, 1)$ and $(1, 1)$.

(b) For $t = 1$ sketch the velocity vectors at the points $f(x, y, 1)$, with (x, y) chosen as in part (a).

(c) Solve the equation $(x', y') = f(x, y, t)$ for (x, y) in terms of (x', y'), and substitute the result into $\partial f/\partial t(x, y, t)$ to show that the flow determined by f is a steady flow.

6. The flow

$$g(x, y, z, t) = \begin{pmatrix} x \cos t - y \sin t \\ x \sin t + y \cos t \\ z + t \end{pmatrix}$$

is a 3-dimensional extension of the flow considered in Example 3 of the text. Find $\partial g/\partial t$ and show that the flow determined by g is a steady flow.

SECTION 7

LIMITS AND CONTINUITY

Limits and continuity have been introduced for functions of one variable in Section 2 and for real-valued functions in Section 5. Here we shall unify and extend the definitions and show how to construct continuous vector functions from continuous real functions. To begin, the definition of limit is based on the idea of nearness. To say for example that

$$\lim_{x\to 0} \frac{\sin x}{x} = 1$$

is to say that $(\sin x)/x$ is arbitrarily close to 1 provided x is sufficiently close to 0. Nearness on the real-number line can be expressed by inequalities. For example, $|x - 3| < 0.4$ says that the distance between the number x and the number 3 is less than 0.4 or, equivalently, that x lies in the interior of the interval with center 3 and half-length 0.4.

The statement "$(\sin x)/x$ is arbitrarily close to 1 provided x is sufficiently close to 0" is translated in terms of inequalities as: For any positive number ϵ, there exists a positive number δ such that if

$$0 < |x - 0| = |x| < \delta,$$

then

$$\left| \frac{\sin x}{x} - 1 \right| < \epsilon.$$

We shall extend these ideas to $\mathcal{R}^n$.

In $\mathcal{R}^n$ a definition of limit also requires the means of asserting that one point is close to another. Distance will be defined with respect to Euclidean length. For any $\epsilon > 0$ and point $\mathbf{x}_0$ in $\mathcal{R}^n$, the set of all vectors $\mathbf{x}$ in $\mathcal{R}^n$ that satisfy the inequality

$$|\mathbf{x} - \mathbf{x}_0| < \epsilon$$

is a **spherical ball** with radius ϵ and center $\mathbf{x}_0$. For example, if $\mathbf{x}_0 = (1, 2, 1)$, the set of all $\mathbf{x}$ in $\mathcal{R}^3$ such that

$$|\mathbf{x} - \mathbf{x}_0| = \sqrt{(x - 1)^2 + (y - 2)^2 + (z - 1)^2} < 0.5$$

is the ball shown in Fig. 30.

Let S be a subset of $\mathcal{R}^n$ and $\mathbf{x}$ a point in $\mathcal{R}^n$. Then $\mathbf{x}$ is a **limit point** of S if, for any $\epsilon > 0$, there exists a point $\mathbf{y}$ in S such that $0 < |\mathbf{x} - \mathbf{y}| < \epsilon$. Translated into English, the definition says that $\mathbf{x}$ is a limit point of S if

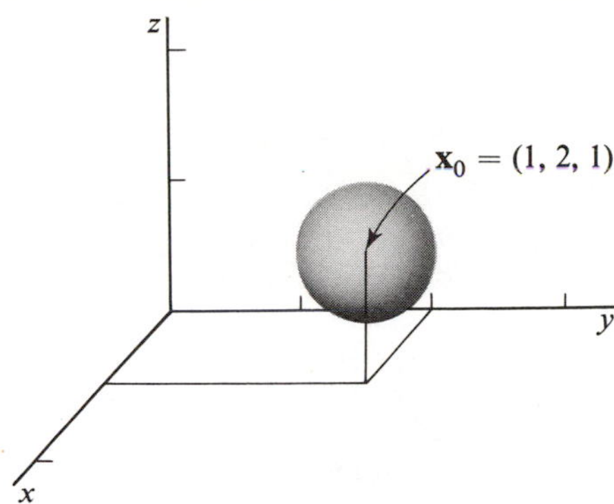

Figure 30

there are points in S other than $\mathbf{x}$ that are contained in a ball of arbitrarily small radius with center at $\mathbf{x}$. If, for example, S is the disk defined by $x^2 + y^2 < 1$, together with the single point $(2, 0)$, then the set of limit points of S consists of S together with the circle $x^2 + y^2 = 1$. See Fig. 31. However, the point $(2, 0)$ is not a limit point of S even though it is a point of S.

We come now to the definition of limit for a function $\mathcal{R}^n \xrightarrow{f} \mathcal{R}^m$. Let $\mathbf{y}_0$ be a point in $\mathcal{R}^m$ and $\mathbf{x}_0$ a limit point of the domain of f. Typical points of the domain of f will be denoted $\mathbf{x}$. Then $\mathbf{y}_0$ is the limit of f at $\mathbf{x}_0$ if, for any $\epsilon > 0$, there is a $\delta > 0$ such that $|f(\mathbf{x}) - \mathbf{y}_0| < \epsilon$ whenever $\mathbf{x}$ satisfies $0 < |\mathbf{x} - \mathbf{x}_0| < \delta$. The relation is written

$$\lim_{\mathbf{x} \to \mathbf{x}_0} f(\mathbf{x}) = \mathbf{y}_0.$$

To put it a little less formally, the definition says that $f(\mathbf{x})$ is arbitrarily close to $\mathbf{y}_0$ provided $\mathbf{x}$ is sufficiently close to $\mathbf{x}_0$ and $\mathbf{x} \neq \mathbf{x}_0$. Geometrically the idea is this: Given any ϵ-ball B_ϵ centered at $\mathbf{y}_0$, there exists a δ-ball B_δ centered at $\mathbf{x}_0$ whose intersection with the domain of f, except possibly for

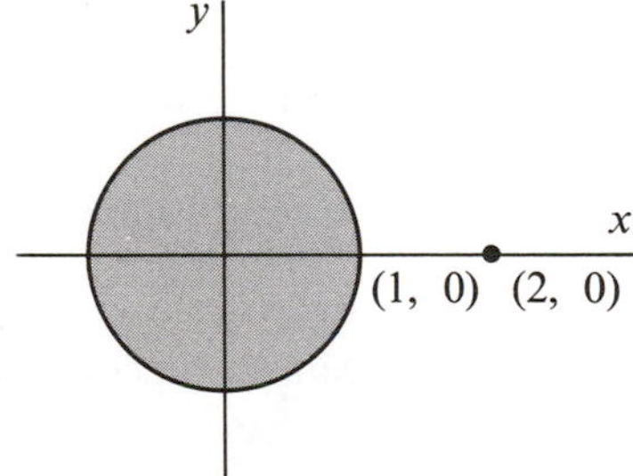

Figure 31

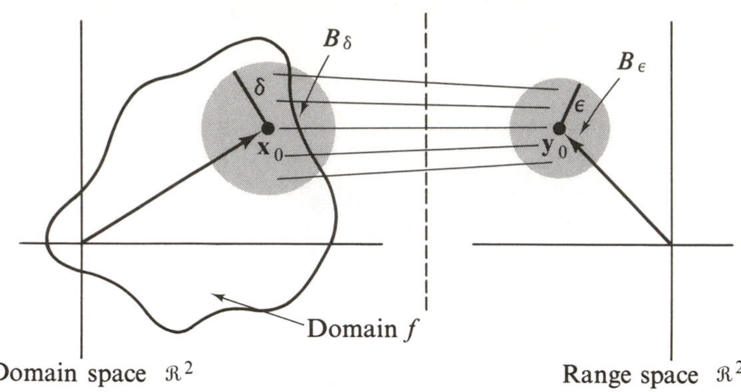

Figure 32

$\mathbf{x}_0$ itself, is sent by f into B_ϵ. A 2-dimensional example is pictured in Fig. 32. The statement

$$\lim_{\mathbf{x}\to\mathbf{x}_0} f(\mathbf{x}) = \mathbf{y}_0$$

is also commonly read "The limit of $f(\mathbf{x})$, as $\mathbf{x}$ approaches $\mathbf{x}_0$, is $\mathbf{y}_0$."

Example 1. Consider the function defined by

$$f(t) = (\cos t, \sin t).$$

The domain of f is all of $\mathcal{R}$, and at an arbitrary point t_0 of $\mathcal{R}$, f has limit $f(t_0)$. To see this we use known facts about $\cos t$ and $\sin t$ and consider

$$\begin{aligned} |f(t) - f(t_0)| &= \sqrt{(\cos t - \cos t_0)^2 + (\sin t - \sin t_0)^2} \\ &\leq |\cos t - \cos t_0| + |\sin t - \sin t_0|. \end{aligned} \tag{1}$$

This holds because $\sqrt{a^2 + b^2} \leq |a| + |b|$. (See Problem 15.) Since, using the fact that $\sin t$ and $\cos t$ are continuous,

$$\lim_{t\to t_0} \cos t = \cos t_0, \quad \text{and} \quad \lim_{t\to t_0} \sin t = \sin t_0,$$

we can choose a $\delta > 0$ such that for any preassigned $\epsilon > 0$

$$|\cos t - \cos t_0| < \frac{\epsilon}{2} \quad \text{and} \quad |\sin t - \sin t_0| < \frac{\epsilon}{2},$$

whenever $|t - t_0| < \delta$. Hence, Equation (1) shows that $|f(t) - f(t_0)| < \epsilon$ whenever $|t - t_0| < \delta$.

Example 2. Consider the real-valued function defined in all of $\mathcal{R}^2$ except $(x, y) = (0, 0)$ by

$$f(x, y) = \frac{1}{x^2 + y^2}.$$

There is no limit as $(x, y) \to (0, 0)$, for example along the line $y = x$. We can write

$$\lim_{\mathbf{x} \to 0} f(\mathbf{x}) = \infty$$

to describe what happens.

Example 3. The range space and the domain are the same as in the preceding example and

$$f(x, y) = \frac{x^2 - y^2}{x^2 + y^2}.$$

There is no limit as $(x, y) \to (0, 0)$. If (x, y) approaches $(0, 0)$ along the line $y = \alpha x$, we obtain

$$\lim_{\mathbf{x} \to 0} \frac{x^2 - y^2}{x^2 + y^2} = \lim_{x \to 0} \frac{x^2(1 - \alpha^2)}{x^2(1 + \alpha^2)} = \frac{1 - \alpha^2}{1 + \alpha^2}.$$

The limit is obviously not independent of α; it equals 0 if $\alpha = 1$, and 1 if $\alpha = 0$.

The functions in the last two examples are both real-valued. The following theorem shows that the problem of the existence and evaluation of a limit for any function $\mathcal{R}^n \xrightarrow{f} \mathcal{R}^m$ reduces to the same problem for the coordinate functions. The latter are, of course, real-valued.

7.1 Theorem

Given $\mathcal{R}^n \xrightarrow{f} \mathcal{R}^m$, with coordinate functions $f_1, \ldots, f_m$, and a point $\mathbf{y}_0 = (y_1, \ldots, y_m)$ in $\mathcal{R}^m$, then

$$\lim_{\mathbf{x} \to \mathbf{x}_0} f(\mathbf{x}) = \mathbf{y}_0 \tag{2}$$

if and only if

$$\lim_{\mathbf{x} \to \mathbf{x}_0} f_i(\mathbf{x}) = y_i, \qquad i = 1, \ldots, m. \tag{3}$$

Proof. To say that Equations (2) and (3) are equivalent is to say that the distance

$$|f(\mathbf{x}) - \mathbf{y}_0| = \sqrt{(f_1(\mathbf{x}) - y_1)^2 + \ldots + (f_m(\mathbf{x}) - y_m)^2}$$

can be made arbitrarily small if and only if all the absolute values

$$|f_1(\mathbf{x}) - y_1|, \ldots, |f_m(\mathbf{x}) - y_m|$$

can be made arbitrarily small. But the equivalence of these last two statements follows at once from the elementary inequalities

$$|f(\mathbf{x}) - \mathbf{y}_0| \geq |f_i(\mathbf{x}) - y_i|, \qquad i = 1, \ldots, m$$

and

$$|f(\mathbf{x}) - \mathbf{y}_0| \leq \sqrt{m} \max_{1 \leq i \leq m} \{|f_i(\mathbf{x}) - y_i|\}.$$

We leave these as exercises.

Example 4. Vector functions f_1 and f_2 are defined by

$$f_1(t) = \begin{pmatrix} t \\ t^2 \\ \sin t \end{pmatrix}, \qquad f_2(t) = \begin{pmatrix} t \\ t^2 \\ \sin \dfrac{1}{t} \end{pmatrix}.$$

Then

$$\lim_{t \to 0} f_1(t) = \begin{pmatrix} 0 \\ 0 \\ 0 \end{pmatrix};$$

but $\lim f_2(t)$ does not exist because the coordinate function $\sin(1/t)$ has no limit at $t = 0$.

The concept of continuity is fundamental to calculus. Roughly speaking, a continuous function f is one whose values do not change abruptly. That is, if $\mathbf{x}$ is close to $\mathbf{x}_0$, then $f(\mathbf{x})$ must be close to $f(\mathbf{x}_0)$. This idea is related to the notion of limit, and the definition of continuity is as follows: **A function f is continuous at $\mathbf{x}_0$** if $\mathbf{x}_0$ is in the domain of f and $\lim_{\mathbf{x} \to \mathbf{x}_0} f(\mathbf{x}) = f(\mathbf{x}_0)$. At a nonlimit or **isolated** point of the domain of f, we cannot ask for a limit; instead we simply define f to be automatically continuous at such a point. It is an immediate corollary of Theorem 7.1 that:

7.2 Theorem

A vector function is continuous at a point if and only if its coordinate functions are continuous there.

Example 5. The function

$$f_1(t) = \begin{pmatrix} t \\ t^2 \\ \sin t \end{pmatrix}$$

is continuous at every value of t. On the other hand, the function

$$f_2(t) = \begin{pmatrix} t \\ t^2 \\ \sin \dfrac{1}{t} \end{pmatrix}$$

is continuous except at $t = 0$.

7.3 Theorem

Every linear function $\mathcal{R}^n \xrightarrow{L} \mathcal{R}^m$ is continuous, and for such an L there is a number k such that

$$|L(\mathbf{x})| \leq k\,|\mathbf{x}|, \qquad \text{for every } \mathbf{x} \text{ in } \mathcal{R}^n.$$

Proof. We prove the inequality first. Let $\mathbf{e}_1, \ldots, \mathbf{e}_n$ be the natural basis for $\mathcal{R}^n$. If $\mathbf{x} = (x_1, \ldots, x_n)$, then $\mathbf{x} = x_1\mathbf{e}_1 + \ldots + x_n\mathbf{e}_n$. Since L is linear,

$$L(\mathbf{x}) = x_1 L(\mathbf{e}_1) + \ldots + x_n L(\mathbf{e}_n).$$

By the homogeneity and triangle properties of the norm, we have

$$|L(\mathbf{x})| \leq |x_1|\,|L(\mathbf{e}_1)| + \ldots + |x_n|\,|L(\mathbf{e}_n)|.$$

Setting $k = |L(\mathbf{e}_1)| + \ldots + |L(\mathbf{e}_n)|$, and using the fact that $|x_i| \leq |\mathbf{x}|$ for $i = 1, \ldots, n$, gives the desired inequality.

We use the inequality to show that L is continuous. If $\mathbf{x}$ and $\mathbf{x}_0$ are vectors in $\mathcal{R}^n$, then

$$\begin{aligned} |L(\mathbf{x}) - L(\mathbf{x}_0)| &= |L(\mathbf{x} - \mathbf{x}_0)| \\ &\leq k\,|\mathbf{x} - \mathbf{x}_0|. \end{aligned}$$

This shows that, as $\mathbf{x}$ tends to $\mathbf{x}_0$, $L(\mathbf{x})$ tends to $L(\mathbf{x}_0)$.

A function is simply called **continuous** if it is continuous at every point of its domain. From Theorem 7.2 we conclude that a continuous vector-valued function of a single variable, $\mathcal{R} \xrightarrow{f} \mathcal{R}^n$, is precisely one for which the coordinate functions $f_1, \ldots, f_n$ are continuous real-valued functions of a real variable. The latter of course include most of the functions of ordinary calculus, such as x^2, $\sin x$, and, for $x > 0$, $\log x$. These same functions can be used to construct examples of the continuous coordinate functions that go to make up a vector-valued function of a vector variable,

$\mathcal{R}^n \xrightarrow{f} \mathcal{R}^m$. For example, the coordinate functions of

$$f(x, y) = \left(\frac{\sin xy}{e^{x+y}}, \frac{\cos xy}{e^{x+y}}\right)$$

are continuous. The continuity of these and other examples can be deduced from repeated application of the following three theorems, together with Theorem 7.2.

7.4 The functions $\mathcal{R}^n \xrightarrow{P_k} \mathcal{R}$, where $P_k(x_1, \ldots, x_n) = x_k$, are continuous for $k = 1, 2, \ldots, n$.

7.5 The functions $\mathcal{R}^2 \xrightarrow{S} \mathcal{R}$ and $\mathcal{R}^2 \xrightarrow{M} \mathcal{R}$, defined by $S(x, y) = x + y$ and $M(x, y) = xy$, are continuous.

7.6 If $\mathcal{R}^n \xrightarrow{f} \mathcal{R}^m$ and $\mathcal{R}^m \xrightarrow{g} \mathcal{R}^p$ are continuous, then the composition $g \circ f$ given by $g \circ f(\mathbf{x}) = g(f(\mathbf{x}))$ is continuous wherever it is defined.

Proving the continuity of P_k, S, and M is left as an exercise.

Proof of 7.6. We assume that $\mathbf{x}_0$ is a limit point of the domain of $g \circ f$ and show that $\lim_{\mathbf{x} \to \mathbf{x}_0} g \circ f(\mathbf{x}) = g \circ f(\mathbf{x}_0)$. If $\epsilon > 0$, we can, by the continuity of g, find a $\delta > 0$ such that $|g(\mathbf{y}) - g(f(\mathbf{x}_0))| < \epsilon$ whenever $|\mathbf{y} - f(\mathbf{x}_0)| < \delta$ and $\mathbf{y}$ is in the domain of g. But since f is also continuous, we can find a $\delta' > 0$ such that $|f(\mathbf{x}) - f(\mathbf{x}_0)| < \delta$ if $|\mathbf{x} - \mathbf{x}_0| < \delta'$ and $\mathbf{x}$ is in the domain of f. It follows that

$$|g \circ f(\mathbf{x}) - g \circ f(\mathbf{x}_0)| = |g(f(\mathbf{x})) - g(f(\mathbf{x}_0))| < \epsilon$$

for these same vectors $\mathbf{x}$, and hence that $g \circ f$ is continuous at $\mathbf{x}_0$. Since $\mathbf{x}_0$ was an arbitrary limit point of the domain of $g \circ f$, the proof is complete.

Example 6. The function $f(x, y) = \sqrt{1 - x^2 - y^2}$, defined for $|(x, y)| \leq 1$, is continuous, because it can be written

$$f(x, y) = \sqrt{1 - (P_1(x, y))^2 - (P_2(x, y))^2},$$

and so is a composition of continuous functions. Similarly, $g(x, y) = \log(x + y)$, defined for $x + y > 0$, is continuous. The product of f and g, given by

$$h(x, y) = \sqrt{1 - x^2 - y^2} \log(x + y),$$

is defined on the half-disk which is the intersection of the domains of f and g, and which is shown in Figure 33. The product is a continuous function because it is the composition of the continuous vector function

$$F(x, y) = (f(x, y), g(x, y))$$

with the function M of 7.5.

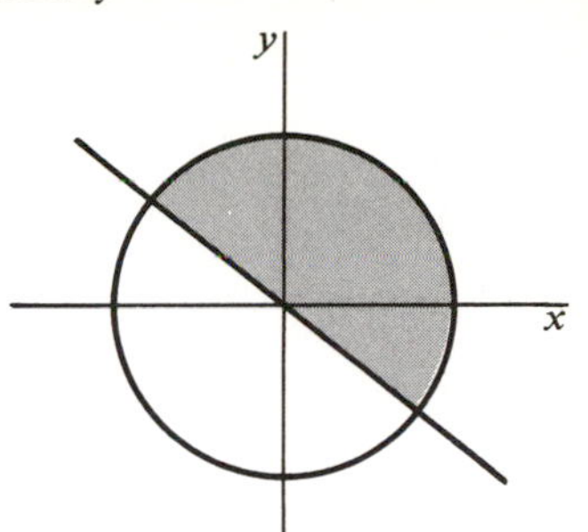

Figure 33

A vector $\mathbf{x}_0$ is an interior point of a subset of a vector space if all points sufficiently close to $\mathbf{x}_0$ are also in the subset. Consider, for example, the subset S of $\mathcal{R}^2$ consisting of all points (x, y) such that $0 < x \leq 2$ and $-1 \leq y < 1$ (cf. Fig. 34(a)). The points $(1, 0)$, $(\frac{1}{2}, \frac{1}{2})$, $(1, -1)$, $(2, 0)$ all belong to S. The first two are interior points and the last two are not. More generally, the interior points of S are precisely those (x, y) that satisfy the inequalities $0 < x < 2$ and $-1 < y < 1$. The formal definition is as follows: $\mathbf{x}_0$ is an **interior point** of a subset S of $\mathcal{R}^n$ if there exists a positive real number δ such that $\mathbf{x}$ belongs to S whenever $|\mathbf{x} - \mathbf{x}_0| < \delta$.

A subset of $\mathcal{R}^n$, all of whose points are interior, is called **open**. Notice that according to this definition the whole space $\mathcal{R}^n$ is an open set. So also is the empty subset $\varnothing$ of $\mathcal{R}^n$. Since $\varnothing$ contains no points, the condition for openness is vacuously satisfied. An open set containing a particular point is often called a **neighborhood** of that point.

Example 7. For any $\epsilon > 0$ and any $\mathbf{x}_0$ in $\mathcal{R}^n$, we can show that the set B_ϵ, of all vectors $\mathbf{x}$ such that $|\mathbf{x} - \mathbf{x}_0| < \epsilon$ is open. In 3-dimensional space, for example, B_ϵ is the ϵ-ball pictured in Fig. 30 for $\epsilon = 0.5$. Let $\mathbf{x}_1$ be an arbitrary point in B_ϵ. Then $|\mathbf{x}_1 - \mathbf{x}_0| < \epsilon$. We must show that every vector sufficiently close to $\mathbf{x}_1$ is in B_ϵ. Set, as in Fig. 34(b),

$$\delta = \epsilon - |\mathbf{x}_1 - \mathbf{x}_0|.$$

Then δ is positive. Suppose $\mathbf{x}$ is any vector such that $|\mathbf{x} - \mathbf{x}_1| < \delta$. By

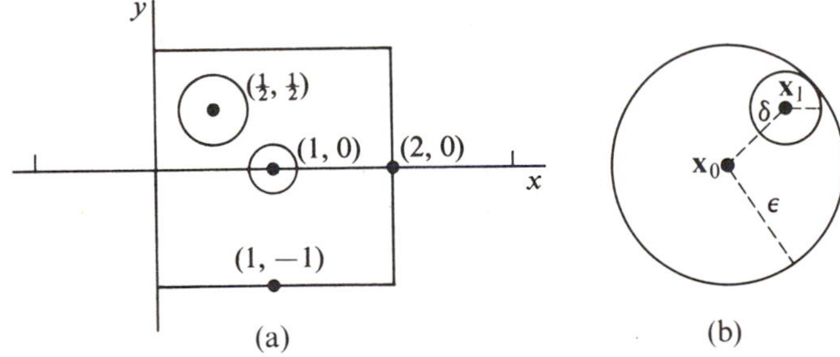

Figure 34

the triangle inequality,

$$\begin{aligned}|\mathbf{x} - \mathbf{x}_0| &= |(\mathbf{x} - \mathbf{x}_1) + (\mathbf{x}_1 - \mathbf{x}_0)| \\ &\leq |\mathbf{x} - \mathbf{x}_1| + |\mathbf{x}_1 - \mathbf{x}_0| \\ &< \delta + (\epsilon - \delta) = \epsilon.\end{aligned}$$

Hence, $\mathbf{x}$ is in B_ϵ, and the proof is complete.

Example 8. Let I be a finite set of points in $\mathcal{R}^n$. Then the set consisting of all points in $\mathcal{R}^n$ that are not in I is open. Thus a vector function that is defined at all points of $\mathcal{R}^n$ except for some finite set has for its domain an open subset of $\mathcal{R}^n$.

A set S is **closed** if it contains all its limit points, and the **closure** of S is the set S together with its limit points. The **boundary** of S is the closure of S with the interior of S deleted. Thus an interval $a \leq x \leq b$, denoted $[a, b]$, is called a closed interval and is in fact a closed set in the above sense. An open interval $a < x < b$, denoted (a, b), is on the other hand an open set. The diagram on the left shows an open set S, then the closure of S, and then the boundary of S.

S

EXERCISES

In Exercises 1 and 2 take the domains of the functions to be as large as possible.

1. At which points do the following functions fail to have limits?

(a) $f\begin{pmatrix} x \\ y \end{pmatrix} = \begin{pmatrix} y + \tan x \\ \ln (x + y) \end{pmatrix}$.

(b) $f\begin{pmatrix} x \\ y \end{pmatrix} = \begin{pmatrix} \dfrac{y}{x^2 + 1} \\ \dfrac{x}{y^2 - 1} \end{pmatrix}$.

(c) $f(x, y) = \dfrac{x}{\sin x} + y$.

(d) $f(x, y) = \begin{cases} \dfrac{x}{\sin x} + y, & \text{if } x \neq 0. \\ 2 + y, & \text{if } x = 0. \end{cases}$

(e) $f(t) = \begin{pmatrix} \sin t \\ \cos t \\ \dfrac{1}{\sin t^2} \end{pmatrix}$.

2. At which points do the following functions fail to be continuous?

(a) $f\begin{pmatrix} x \\ y \end{pmatrix} = \begin{pmatrix} \dfrac{1}{x^2} + \dfrac{1}{y^2} \\ x^2 + y^2 \end{pmatrix}.$

(b) $f\begin{pmatrix} u \\ v \end{pmatrix} = \begin{pmatrix} 3u - 4v \\ u + 8v \end{pmatrix}.$

(c) $f(x, y) = \begin{cases} \dfrac{x}{\sin x} + y, & \text{if } x \neq 0. \\ 1 + y, & \text{if } x = 0. \end{cases}$

(d) $f\begin{pmatrix} x \\ y \end{pmatrix} = \begin{cases} \dfrac{x^2 - y^2}{x^2 + y^2}, & \text{if } \ x^2 + y^2 \neq 0. \\ 0, & \text{if } \ x^2 + y^2 = 0. \end{cases}$

(e) $f\begin{pmatrix} u \\ v \end{pmatrix} = \begin{pmatrix} v \tan u \\ u \sec v \\ v \end{pmatrix}.$

(f) $f(\mathbf{x}) = \dfrac{|\mathbf{x}|}{1 - |\mathbf{x}|^2}.$

3. When $\mathbf{x}_0 = (1, 2)$, draw the set of all vectors $\mathbf{x}$ in $\mathcal{R}^2$ such that

(a) $|\mathbf{x} - \mathbf{x}_0| \leq 3.$
(b) $|\mathbf{x} - \mathbf{x}_0| = 3.$
(c) $|\mathbf{x} - \mathbf{x}_0| < 3.$

4. Identify as open, closed, both, or neither, the subset of $\mathcal{R}^2$ consisting of all vectors $\mathbf{x} = (x, y)$ such that

(a) $|\mathbf{x} - (1, 2)| \leq 0.5.$
(b) $|\mathbf{x} - (1, 2)| < 0.5.$
(c) $|\mathbf{x} - (1, 2)| < -0.5.$
(d) $0 < x < 3$ and $0 < y < 2.$
(e) $2 \leq x < 3$ and $0 < y < 2.$
(f) $\dfrac{x^2}{a^2} + \dfrac{y^2}{b^2} < 1.$
(g) $\mathbf{x} \neq (0, 2)$ or $(1, 2).$
(h) $x^2 + y^2 > 0.$
(i) $x > 0.$
(j) $x > y.$

5. Let the set S consist of the points (x, y) in $\mathcal{R}^2$ satisfying $0 < x^2 + y^2 < 1$, together with the interval $1 \leq x < 2$ of the x-axis.

(a) Describe the boundary of S.
(b) What are the interior points of S?
(c) Describe the closure of S.

6. (a) Prove that the union of an arbitrary collection of open subsets of $\mathcal{R}^n$ is open.
   (b) Prove that the intersection of a *finite* collection of open subsets of $\mathcal{R}^n$ is open.
   (c) Give an example to show that a nonempty intersection of *infinitely* many open subsets of $\mathcal{R}^n$ may fail to be open.

7. Prove that $\mathbf{x}$ is a boundary point of a set S if and only if every neighborhood of $\mathbf{x}$ contains a point of S and a point of the complement of S.

8. A vector function f is said to have a **removable discontinuity** at $\mathbf{x}_0$ if (a) f is not continuous at $\mathbf{x}_0$, and (b) there is a vector $\mathbf{y}_0$ such that $\lim_{\mathbf{x}\to\mathbf{x}_0} f(\mathbf{x}) = \mathbf{y}_0$. Give an example of a function f and a point $\mathbf{x}_0$ such that f is not continuous at $\mathbf{x}_0$ and (1) f has a removable discontinuity at $\mathbf{x}_0$, (2) f does not have a removable discontinuity at $\mathbf{x}_0$.

9. Prove that every translation is a continuous vector function. A vector function $\mathcal{R}^n \xrightarrow{t} \mathcal{R}^n$ is a **translation** if there exists a vector $\mathbf{y}_0$ in $\mathcal{R}^n$ such that $t(\mathbf{x}) = \mathbf{x} + \mathbf{y}_0$ for all $\mathbf{x}$ in $\mathcal{R}^n$.

10. Prove 7.4 and 7.5 of the text.

11. If f and g are vector functions with the same domain and same range space, prove
$$\lim_{\mathbf{x}\to\mathbf{x}_0} (f(\mathbf{x}) + g(\mathbf{x})) = \lim_{\mathbf{x}\to\mathbf{x}_0} f(\mathbf{x}) + \lim_{\mathbf{x}\to\mathbf{x}_0} g(\mathbf{x}),$$
provided that $\lim_{\mathbf{x}\to\mathbf{x}_0} f(\mathbf{x})$ and $\lim_{\mathbf{x}\to\mathbf{x}_0} g(\mathbf{x})$ exist.

12. Let L be a line and P a plane in $\mathcal{R}^3$. Is either P or L an open subset?

13. Let S be a closed subset of $\mathcal{R}^n$. Prove that the complement of S in $\mathcal{R}^n$ is open.

14. Converse of Exercise 13: If S is an open subset of $\mathcal{R}^n$, show that the complement of S in $\mathcal{R}^n$ is closed.

15. Show that $\sqrt{a^2 + b^2} \leq |a| + |b|$.

16. Prove the inequalities
$$|(\mathbf{z}_1, \ldots, \mathbf{z}_m)| \geq |\mathbf{z}_i|, \qquad i = 1, \ldots, m$$
and
$$|(\mathbf{z}_1, \ldots, \mathbf{z}_m)| \leq \sqrt{m} \max \{|\mathbf{z}_i|\}.$$

17. Show that Theorem 7.3 can also be proved by using Theorems 7.4 and 7.5.

SECTION 8

DIFFERENTIALS

Many of the techniques of calculus have as their foundation the idea of approximating a vector function by a linear function or by an affine function. Recall that a function $\mathcal{R}^n \xrightarrow{A} \mathcal{R}^m$ is **affine** if there exists a linear function $\mathcal{R}^n \xrightarrow{L} \mathcal{R}^m$ and a vector $\mathbf{y}_0$ in $\mathcal{R}^m$ such that
$$A(\mathbf{x}) = L(\mathbf{x}) + \mathbf{y}_0, \qquad \text{for every } \mathbf{x} \text{ in } \mathcal{R}^n.$$

We shall see that affine functions form the basis of the differential calculus of vector functions.

Example 1. In terms of coordinate variables an affine function is defined by linear equations. For example, consider the point

$$\mathbf{y}_0 = \begin{pmatrix} 1 \\ 2 \end{pmatrix}$$

and the linear function $\mathcal{R}^3 \xrightarrow{L} \mathcal{R}^2$ defined by the matrix

$$\begin{pmatrix} 2 & 3 & 0 \\ 1 & 0 & 5 \end{pmatrix}.$$

The affine function $A(\mathbf{x}) = L(\mathbf{x}) + \mathbf{y}_0$, where

$$\mathbf{x} = \begin{pmatrix} x \\ y \\ z \end{pmatrix}$$

is described by the equations

$$u = 2x + 3y + 1$$
$$v = x + 5z + 2$$

Since any system of linear equations can be systematically solved, any affine function can be similarly analyzed.

We shall now study the possibility of approximating an arbitrary vector function f near a point $\mathbf{x}_0$ of its domain by an affine function A. The general idea is the possibility of replacing near $\mathbf{x}_0$ what may be a very complicated function by a simple one. Before trying to decide whether or not an approximation exists, we first have to say what we shall mean by an approximation. We begin by requiring that $f(\mathbf{x}_0) = A(\mathbf{x}_0)$. Since $A(\mathbf{x}) = L(\mathbf{x}) + \mathbf{y}_0$, where L is linear, we obtain $f(\mathbf{x}_0) = L(\mathbf{x}_0) + \mathbf{y}_0$, and so

$$A(\mathbf{x}) = L(\mathbf{x} - \mathbf{x}_0) + f(\mathbf{x}_0). \tag{1}$$

An apparently natural requirement is that

$$\lim_{\mathbf{x} \to \mathbf{x}_0} (f(\mathbf{x}) - A(\mathbf{x})) = 0. \tag{2}$$

However, Equation (2) may appear to say more than it really does. To see what it really says, we observe that from (1) we get

$$f(\mathbf{x}) - A(\mathbf{x}) = f(\mathbf{x}) - f(\mathbf{x}_0) - L(\mathbf{x} - \mathbf{x}_0).$$

Now every linear function is continuous, so $\lim_{\mathbf{x}\to\mathbf{x}_0} L(\mathbf{x}-\mathbf{x}_0) = L(0) = 0$. Hence,

$$\lim_{\mathbf{x}\to\mathbf{x}_0}(f(\mathbf{x}) - A(\mathbf{x})) = \lim_{\mathbf{x}\to\mathbf{x}_0}(f(\mathbf{x}) - f(\mathbf{x}_0)).$$

It follows that Equation (2) is precisely the statement that the vector function f is continuous at $\mathbf{x}_0$. This is significant, but it says nothing about L. Thus, in order for our notion of approximation to distinguish one affine function from another or to measure in any way how well A approximates f, some additional requirement is necessary. A natural condition, and the one we shall require, is that $f(\mathbf{x}) - A(\mathbf{x})$ approach 0 faster than $\mathbf{x}$ approaches $\mathbf{x}_0$. That is, we demand that

$$\lim_{\mathbf{x}\to\mathbf{x}_0}\frac{f(\mathbf{x}) - f(\mathbf{x}_0) - L(\mathbf{x}-\mathbf{x}_0)}{|\mathbf{x}-\mathbf{x}_0|} = 0.$$

Equivalently, we can ask that f be representable in the form

$$f(\mathbf{x}) = f(\mathbf{x}_0) + L(\mathbf{x}-\mathbf{x}_0) + |\mathbf{x}-\mathbf{x}_0|\, Z(\mathbf{x}-\mathbf{x}_0),$$

where $Z(\mathbf{y})$ is some function that tends to zero as $\mathbf{y}$ tends to zero.

A function $\mathcal{R}^n \xrightarrow{f} \mathcal{R}^m$ will be called **differentiable** at $\mathbf{x}_0$ if

1. $\mathbf{x}_0$ is an interior point of the domain of f.

2. There is an affine function that approximates f near $\mathbf{x}_0$. That is, there exists a linear function $\mathcal{R}^n \xrightarrow{L} \mathcal{R}^m$ such that

$$\lim_{\mathbf{x}\to\mathbf{x}_0}\frac{f(\mathbf{x}) - f(\mathbf{x}_0) - L(\mathbf{x}-\mathbf{x}_0)}{|\mathbf{x}-\mathbf{x}_0|} = 0.$$

The linear function L is called the **differential** of f at $\mathbf{x}_0$. The function f is said simply to be **differentiable** if it is differentiable at every point of its domain.

According to the definition, the domain of a differentiable function is an open set. It is, however, convenient to extend the definition sufficiently to speak of a differentiable function f defined on an arbitrary subset S of the domain space. By such an f we shall mean the restriction to S of a differentiable function whose domain is open.

Example 2. The function f defined by $f(x, y) = \sqrt{1 - x^2 - y^2}$ has for its domain the disk $x^2 + y^2 \leq 1$. Its graph is shown in Fig. 35. The interior points of the domain are those (x, y) such that $x^2 + y^2 < 1$. We shall see that this function is differentiable at all these points.

In dimension 1, an affine function has the form $ax + b$. Hence a real-valued function $f(x)$ of a real variable x that is differentiable at x_0 can be

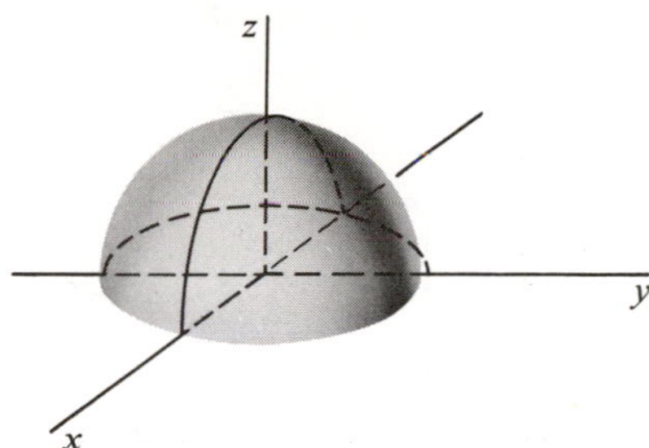

Figure 35

approximated near x_0 by a function $A(x) = ax + b$. Since $f(x_0) = A(x_0) = ax_0 + b$, we obtain

$$A(x) = ax + b = a(x - x_0) + f(x_0).$$

The linear part of A (denoted earlier by L) is in this case just multiplication by the real number a. The Euclidean norm of a real number is its absolute value, so condition 2 of the definition of differentiability becomes

$$\lim_{x \to x_0} \frac{f(x) - f(x_0) - a(x - x_0)}{|x - x_0|} = 0.$$

This is equivalent to

$$\lim_{x \to x_0} \frac{f(x) - f(x_0)}{x - x_0} = a.$$

The number a is commonly denoted by $f'(x_0)$ and is called the **derivative of** f at x_0. The affine function A is therefore given by

$$A(x) = f(x_0) + f'(x_0)(x - x_0).$$

Its graph is the **tangent line** to the graph of f at x_0, and a typical example is drawn in Fig. 36. Thus we have seen that the general definition of differentiability for vector functions reduces in dimension 1 to the definition usually encountered in a one-variable calculus course.

A linear function $\mathcal{R}^n \xrightarrow{L} \mathcal{R}^m$ must be representable by an m-by-n matrix. We shall see below that the matrix of any L satisfying conditions 1 and 2 of the definition of differentiability can be computed in terms of partial derivatives of f. It follows that L is *uniquely* determined by f at each interior point of the domain of f. Thus we can speak of *the* differential of f and $\mathbf{x}_0$ and denote it by $d_{\mathbf{x}_0}f$.

To find the matrix of $d_{\mathbf{x}_0}f$ if $\mathcal{R}^n \xrightarrow{f} \mathcal{R}^m$, we consider the natural basis $(\mathbf{e}_1, \ldots, \mathbf{e}_n)$ for the domain space $\mathcal{R}^n$. If $\mathbf{x}_0$ is an interior point of the domain of f, the vectors

$$\mathbf{x}_j = \mathbf{x}_0 + t\mathbf{e}_j, \qquad j = 1, \ldots, n,$$

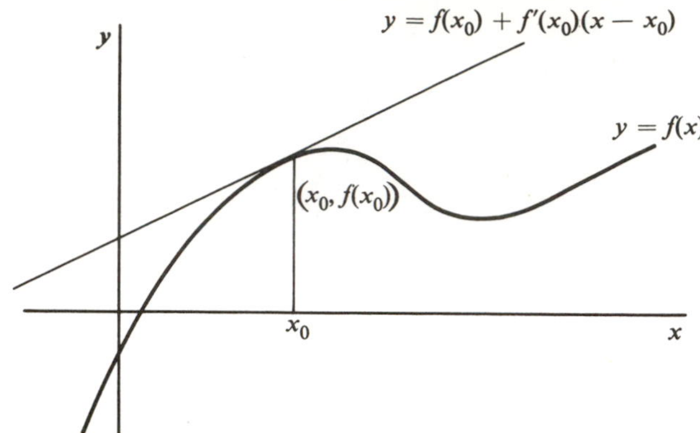

Figure 36

are all in the domain of f for sufficiently small t. (Why?) By condition 2 of the definition of the differential, we have

$$\lim_{t\to 0} \frac{f(\mathbf{x}_j) - f(\mathbf{x}_0) - d_{\mathbf{x}_0}f(t\mathbf{e}_j)}{t} = 0,$$

for $j = 1, \ldots, n$. Since $d_{\mathbf{x}_0}f$ is a linear function, this means that

$$\lim_{t\to 0} \frac{f(\mathbf{x}_j) - f(\mathbf{x}_0)}{t} = d_{\mathbf{x}_0}f(\mathbf{e}_j),$$

for $j = 1, \ldots, n$. But, by Theorem 4.2 of Chapter 1, $d_{\mathbf{x}_0}f(\mathbf{e}_j)$ is the jth column of the matrix of $d_{\mathbf{x}_0}f$. On the other hand, the vector $\mathbf{x}_j$ differs from $\mathbf{x}_0$ only in the jth coordinate, and in that coordinate the difference is just the number t. Therefore, the left side of the last equation is precisely the partial derivative $(\partial f/\partial x_j)(\mathbf{x}_0)$. Thus it is this vector which is the jth column of the matrix of $d_{\mathbf{x}_0}f$. If the coordinate functions of f are $f_1, \ldots, f_m$, then

$$\frac{\partial f}{\partial x_j}(\mathbf{x}_0) = \begin{pmatrix} \dfrac{\partial f_1}{\partial x_j}(\mathbf{x}_0) \\ \cdot \\ \cdot \\ \cdot \\ \dfrac{\partial f_m}{\partial x_j}(\mathbf{x}_0) \end{pmatrix},$$

and the entire matrix of $d_{\mathbf{x}_0}f$ has the form

$$\begin{pmatrix} \frac{\partial f_1}{\partial x_1}(\mathbf{x}_0) & \frac{\partial f_1}{\partial x_2}(\mathbf{x}_0) & \dots & \frac{\partial f_1}{\partial x_n}(\mathbf{x}_0) \\ \frac{\partial f_2}{\partial x_1}(\mathbf{x}_0) & \frac{\partial f_2}{\partial x_2}(\mathbf{x}_0) & \dots & \frac{\partial f_2}{\partial x_n}(\mathbf{x}_0) \\ \cdot & \cdot & & \cdot \\ \cdot & \cdot & & \cdot \\ \cdot & \cdot & & \cdot \\ \frac{\partial f_m}{\partial x_1}(\mathbf{x}_0) & \frac{\partial f_m}{\partial x_2}(\mathbf{x}_0) & \dots & \frac{\partial f_m}{\partial x_n}(\mathbf{x}_0) \end{pmatrix}$$

This matrix is called the **Jacobian matrix** or **derivative** of f at $\mathbf{x}_0$, and is denoted $f'(\mathbf{x}_0)$. We can summarize what we have just proved as follows.

8.1 Theorem

If the function $\mathcal{R}^n \xrightarrow{f} \mathcal{R}^m$ is differentiable at $\mathbf{x}_0$, then the differential $d_{\mathbf{x}_0}f$ is uniquely determined, and its matrix is the Jacobian matrix of f. That is, for all vectors $\mathbf{y}$ in $\mathcal{R}^n$,

$$d_{\mathbf{x}_0}f(\mathbf{y}) = f'(\mathbf{x}_0)\mathbf{y}.$$

While the linear transformation $d_{\mathbf{x}_0}f$ and it matrix $f'(\mathbf{x}_0)$ are logically distinct, the last equation shows that they can be identified in practice, provided it is understood that the matrix of $d_{\mathbf{x}_0}f$ is taken with respect to the natural bases in $\mathcal{R}^n$ and $\mathcal{R}^m$.

Example 3. The function

$$f(x, y, z) = \begin{pmatrix} x^2 + e^y \\ x + y \sin z \end{pmatrix}$$

has coordinate functions $f_1(x, y, z) = x^2 + e^y$ and $f_2(x, y, z) = x + y \sin z$. The Jacobian matrix at (x, y, z) is by definition

$$f'(x, y, z) = \left(\frac{\partial f}{\partial x}(x, y, z) \frac{\partial f}{\partial y}(x, y, z) \frac{\partial f}{\partial z}(x, y, z)\right),$$

so for this example

$$f'(x, y, z) = \begin{pmatrix} 2x & e^y & 0 \\ 1 & \sin z & y \cos z \end{pmatrix}.$$

Thus the differential of f at $(1, 1, \pi)$ is the linear function whose matrix is

$$f'(1, 1, \pi) = \begin{pmatrix} 2 & e & 0 \\ 1 & 0 & -1 \end{pmatrix}.$$

Example 4. The function f defined by

$$f(x, y) = (x^2 + 2xy + y^2, xy^2 + x^2y)$$

has differential $d_{\mathbf{x}}f$ at $\mathbf{x} = (x, y)$ represented by the Jacobian matrix

$$f'(x, y) = \begin{pmatrix} 2x + 2y & 2x + 2y \\ y^2 + 2xy & 2xy + x^2 \end{pmatrix}.$$

How can one tell whether or not a vector function is differentiable? Theorem 8.1 says only that if f is differentiable, then the differential is represented by the Jacobian matrix. It does not go the other way. Thus Examples 3 and 4 are inconclusive to the extent that we have simply assumed that the functions appearing in them are differentiable. Just as the derivative of a real-valued function of one variable may fail to exist, so in general a vector function need not be differentiable at every point. The next theorem is a convenient criterion for differentiability.

8.2 Theorem

If the domain of $\mathcal{R}^n \xrightarrow{f} \mathcal{R}^m$ is an open set D on which all partial derivatives $\partial f_i/\partial x_j$ of the coordinate functions of f are continuous, then f is differentiable at every point of D.

Proof. Let L be the linear function defined by the Jacobian matrix of f. The theorem will have been proved if it can be shown that L satisfies

$$\lim_{\mathbf{x} \to \mathbf{x}_0} \frac{f(\mathbf{x}) - f(\mathbf{x}_0) - L(\mathbf{x} - \mathbf{x}_0)}{|\mathbf{x} - \mathbf{x}_0|} = 0. \tag{3}$$

Since by Theorem 7.1 a vector function approaches a limit if and only if the coordinate functions approach the coordinates of the limit, it is enough to prove the theorem for the coordinate functions of f, or, what is notationally simpler, to prove it under the assumption that f is real-valued. If

$$\mathbf{x} = (x_1, \ldots, x_n)$$

and

$$\mathbf{x}_0 = (a_1, \ldots, a_n),$$

set

$$\mathbf{y}_k = (x_1, \ldots, x_k, a_{k+1}, \ldots, a_n), \qquad k = 0, 1, \ldots, n,$$

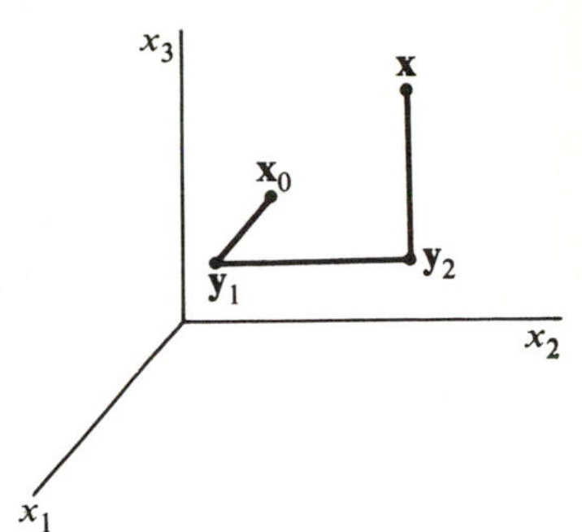

Figure 37

so that $\mathbf{y}_0 = \mathbf{x}_0$ and $\mathbf{y}_n = \mathbf{x}$. These vectors are illustrated for three dimensions in Fig. 37. Then, because of cancellation between successive terms in the sum below, only the first and last terms survive, and we have

$$f(\mathbf{x}) - f(\mathbf{x}_0) = \sum_{k=1}^{n} \big(f(\mathbf{y}_k) - f(\mathbf{y}_{k-1})\big).$$

Because $\mathbf{y}_k$ and $\mathbf{y}_{k-1}$ differ only in their kth coordinates, we can apply the mean-value theorem for real functions of a real variable to get

$$f(\mathbf{y}_k) - f(\mathbf{y}_{k-1}) = (x_k - a_k)\frac{\partial f}{\partial x_k}(\mathbf{z}_k),$$

where $\mathbf{z}_k$ is a point on the segment joining $\mathbf{y}_k$ and $\mathbf{y}_{k-1}$. Then

$$f(\mathbf{x}) - f(\mathbf{x}_0) = \sum_{k=1}^{n} (x_k - a_k)\frac{\partial f}{\partial x_k}(\mathbf{z}_k).$$

We also have, by the definition of L,

$$L(\mathbf{x} - \mathbf{x}_0) = \left(\frac{\partial f}{\partial x_1}(\mathbf{x}_0) \ldots \frac{\partial f}{\partial x_n}(\mathbf{x}_0)\right)\begin{pmatrix} x_1 - a_1 \\ \cdot \\ \cdot \\ \cdot \\ x_n - a_n \end{pmatrix}$$

$$= \sum_{k=1}^{n} (x_k - a_k)\frac{\partial f}{\partial x_k}(\mathbf{x}_0).$$

Hence

$$|f(\mathbf{x}) - f(\mathbf{x}_0) - L(\mathbf{x} - \mathbf{x}_0)| = \left|\sum_{k=1}^{n}\left(\frac{\partial f}{\partial x_k}(\mathbf{z}_k) - \frac{\partial f}{\partial x_k}(\mathbf{x}_0)\right)(x_k - a_k)\right|$$

$$\leq \sum_{k=1}^{n}\left|\frac{\partial f}{\partial x_k}(\mathbf{z}_k) - \frac{\partial f}{\partial x_k}(\mathbf{x}_0)\right| |\mathbf{x} - \mathbf{x}_0|,$$

where we have used the triangle inequality and the fact that

$$|x_k - a_k| \leq |\mathbf{x} - \mathbf{x}_0| \quad \text{for} \quad k = 1, 2, \ldots, n.$$

Since the partial derivatives are assumed continuous at $\mathbf{x}_0$, and the $\mathbf{z}_k$ tend to $\mathbf{x}_0$ as $\mathbf{x}$ does, the limit Equation (3) follows immediately.

The entries in the Jacobian matrices that appear in Examples 3 and 4 are continuous functions. As a result of the theorem just proved, we conclude that the two functions in those examples are differentiable.

Example 5. The function

$$f(x, y) = \sqrt{1 - x^2 - y^2}$$

defined for all (x, y) such that $x^2 + y^2 < 1$ is the same as in Example 2 except that we have removed the boundary of the disk so that the domain is an open set. The Jacobian matrix is

$$(f_x \quad f_y) = \left(\frac{-x}{\sqrt{1 - x^2 - y^2}} \quad \frac{-y}{\sqrt{1 - x^2 - y^2}}\right).$$

The entries are continuous on the open disk, and we conclude, by Theorem 8.2, that f is differentiable there.

Example 6. Consider the function

$$f(t) = \begin{pmatrix} \cos t \\ \sin t \end{pmatrix}, \qquad -\infty < t < \infty.$$

The derivative $f'(t_0)$ is the 2-by-1 matrix

$$\begin{pmatrix} -\sin t_0 \\ \cos t_0 \end{pmatrix}.$$

It is instructive to consider the matrix as a vector in the range space of f and to draw it with its tail at the image point $f(t_0)$. For $t_0 = 0, \pi/4, \pi/3, \pi/2$, and π, the respective matrices of the differential $d_{t_0}f$ are

$$\begin{pmatrix} 0 \\ 1 \end{pmatrix}, \quad \begin{pmatrix} -\sqrt{2}/2 \\ \sqrt{2}/2 \end{pmatrix}, \quad \begin{pmatrix} -\sqrt{3}/2 \\ 1/2 \end{pmatrix}, \quad \begin{pmatrix} -1 \\ 0 \end{pmatrix}, \quad \text{and} \quad \begin{pmatrix} 0 \\ -1 \end{pmatrix}.$$

These vectors, drawn with their tails at their corresponding image points under f, are shown in Fig. 38. Evidently, for curves at least, the differential is related to the notion of a tangent vector. The affine function that best approximates $f(t)$ in some neighborhood of t_0 is the vector function of t given by

$$f'(t_0)(t - t_0) + f(t_0),$$

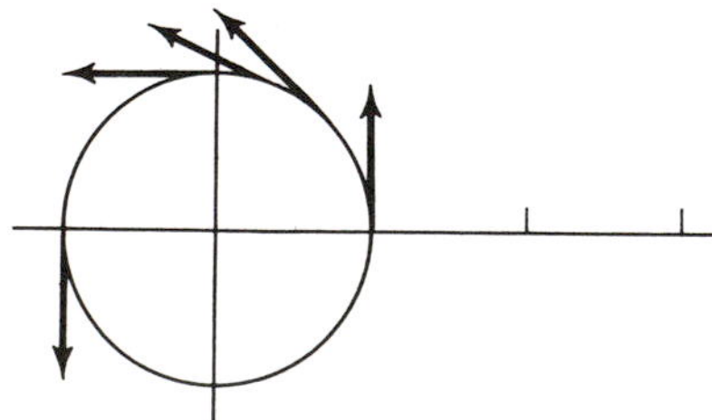

Figure 38

which in terms of matrices becomes

$$t\begin{pmatrix} -\sin t_0 \\ \cos t_0 \end{pmatrix} + \begin{pmatrix} \cos t_0 + t_0 \sin t_0 \\ \sin t_0 - t_0 \cos t_0 \end{pmatrix}.$$

This is the equation of the line tangent to the range of f at $f(t_0)$.

A good geometric picture of the differential of a real-valued function $f(x, y)$ at $\mathbf{x}_0 = (x_0, y_0)$ is obtained by looking at the surface defined explicitly by f, together with the tangent plane to the surface at the point

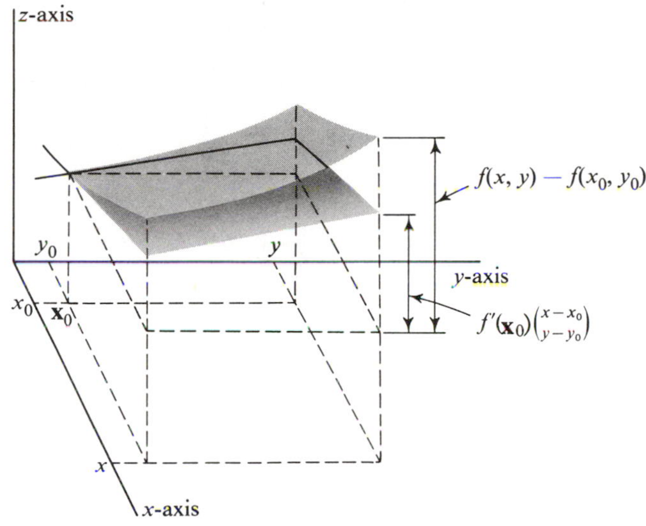

Figure 39

$(x_0, y_0, f(\mathbf{x}_0))$. An example is shown in Fig. 39. The tangent plane is the graph of the affine function defined by

$$A(x, y) = f(x_0, y_0) + f'(x_0, y_0)\begin{pmatrix} x - x_0 \\ y - y_0 \end{pmatrix}, \qquad \text{all } (x, y) \text{ in } \mathcal{R}^2.$$

The difference $A(x, y) - f(x_0, y_0)$ is a good approximation to the increment $f(x, y) - f(x_0, y_0)$, provided (x, y) is close to (x_0, y_0). Figure 39 is the analog of a similar picture, included in many one-variable calculus texts, that exhibits the differential for real-valued functions of one variable.

The example given in Problem 19 below shows that continuity of the derivative f' is not necessary for differentiability of f. However, the hypotheses of Theorem 8.2 are important enough to be given a special name. A vector function f is **continuously differentiable** on an open set D if the entries in the Jacobian matrix of f are continuous on D. Thus each of the functions of Examples 2–6 is not only differentiable but even continuously differentiable in the interior of its domain.

We finally remark that *if a function f is differentiable, then f is necessarily continuous*, (even though the entries in f' may not be continuous). The proof is very simple and we leave it as an exercise. (See Problem 11.)

EXERCISES

1. A linear function $\mathcal{R}^n \xrightarrow{L} \mathcal{R}^m$ is defined by a matrix (a_{ij}), and

$$\mathbf{y}_0 = \begin{pmatrix} b_1 \\ \cdot \\ \cdot \\ \cdot \\ b_m \end{pmatrix}$$

is a vector in $\mathcal{R}^m$. Construct a specific example by choosing m and n, a matrix (a_{ij}), and a vector

$$\begin{pmatrix} b_1 \\ \cdot \\ \cdot \\ \cdot \\ b_m \end{pmatrix}$$

so that the affine function

$$A(\mathbf{x}) = L(\mathbf{x}) + \mathbf{y}_0, \qquad \text{all } \mathbf{x} \text{ in } \mathcal{R}^n,$$

(a) Explicitly defines a line in $\mathcal{R}^2$.
(b) Explicitly defines a line in $\mathcal{R}^3$.
(c) Explicitly defines a plane in $\mathcal{R}^3$.
(d) Parametrically defines a line in $\mathcal{R}^3$.
(e) Parametrically defines a plane in $\mathcal{R}^3$.

What condition must the matrix (a_{ij}) satisfy in order to give an example for (e)?

2. If f is the vector function defined by

$$f\begin{pmatrix} x \\ y \end{pmatrix} = \begin{pmatrix} x^2 - y^2 \\ 2xy \end{pmatrix},$$

find the derivative of f at the following points:

(a) $\begin{pmatrix} x \\ y \end{pmatrix}$, (b) $\begin{pmatrix} a \\ b \end{pmatrix}$, (c) $\begin{pmatrix} 1 \\ 0 \end{pmatrix}$, (d) $\begin{pmatrix} 1/\sqrt{2} \\ 1/\sqrt{2} \end{pmatrix}$.

[*Ans.* (c) $\begin{pmatrix} 2 & 0 \\ 0 & 2 \end{pmatrix}$.]

3. Find the derivative of each of the following functions at the indicated points.

(a) $f\begin{pmatrix} x \\ y \end{pmatrix} = x^2 + y^2$ at $\begin{pmatrix} x \\ y \end{pmatrix} = \begin{pmatrix} 1 \\ 1 \end{pmatrix}$.

(b) $g(x, y, z) = xyz$ at $(x, y, z) = (1, 0, 0)$.

(c) $f(t) = \begin{pmatrix} \sin t \\ \cos t \end{pmatrix}$ at $t = \dfrac{\pi}{4}$ [*Ans.* $\begin{pmatrix} 1/\sqrt{2} \\ -1/\sqrt{2} \end{pmatrix}$.]

(d) $f(t) = \begin{pmatrix} e^t \\ t \\ t^2 \end{pmatrix}$ at $t = 1$.

(e) $g(x, y) = \begin{pmatrix} x + y \\ x^2 + y^2 \end{pmatrix}$ at $(x, y) = (1, 2)$.

(f) $A\begin{pmatrix} u \\ v \end{pmatrix} = \begin{pmatrix} u + v \\ u - v \\ 1 \end{pmatrix}$ at $\begin{pmatrix} u \\ v \end{pmatrix} = \begin{pmatrix} 1 \\ 0 \end{pmatrix}$.

(g) $T\begin{pmatrix} u \\ v \end{pmatrix} = \begin{pmatrix} u \cos v \\ u \sin v \\ v \end{pmatrix}$ at $\begin{pmatrix} u \\ v \end{pmatrix} = \begin{pmatrix} 1 \\ \pi \end{pmatrix}$. [*Ans.* $\begin{pmatrix} -1 & 0 \\ 0 & -1 \\ 0 & 1 \end{pmatrix}$.]

(h) $f(x, y, z) = (x + y + z, xy + yz + zx, xyz)$ at (x, y, z).

4. Let P be a function from 3-dimensional to 2-dimensional Euclidean space defined by $P(x, y, z) = (x, y)$.

(a) What is the geometric interpretation of this transformation?

(b) Show that P is differentiable at all points and find the matrix of the differential of P at $(1, 1, 1)$. [*Ans.* $\begin{pmatrix} 1 & 0 & 0 \\ 0 & 1 & 0 \end{pmatrix}$.]

5. (a) Draw the curve in $\mathcal{R}^2$ defined parametrically by the function

$$g(t) = (t - 1, t^2 - 3t + 2), \qquad -\infty < t < \infty.$$

(b) Find the affine function that approximates g
(1) near $t = 0$. (2) near $t = 2$. [*Ans.* $A(t) = (t - 1, t - 2)$.]

(c) Draw the curve defined parametrically by the affine function.

6. Let f be the function given in Exercise 2, and let

$$\mathbf{x}_0 = \begin{pmatrix} 1 \\ 0 \end{pmatrix}, \quad \mathbf{y}_1 = \begin{pmatrix} 0.1 \\ 0 \end{pmatrix}, \quad \mathbf{y}_2 = \begin{pmatrix} 0 \\ 1.0 \end{pmatrix}, \quad \mathbf{y}_3 = \begin{pmatrix} 0.1 \\ 0.1 \end{pmatrix}.$$

(a) Compute $f(\mathbf{x}_0 + \mathbf{y}_i)$ for $i = 1, 2, 3$.
(b) Find the affine function A that approximates f near $\mathbf{x}_0$.
(c) Use A to find approximations to the vectors

$$f(\mathbf{x}_0 + \mathbf{y}_i), \qquad i = 1, 2, 3.$$

7. (a) Sketch the surface in $\mathcal{R}^3$ defined explicitly by the function

$$f(x, y) = 4 - x^2 - y^2.$$

(b) Find the affine function that approximates f
(1) near $(0, 0)$. (2) near $(2, 0)$. [*Ans.* $A(x, y) = 8 - 4x$.]
(c) Draw the graphs of the affine functions in (b).

8. What is the derivative of the affine function

$$\begin{pmatrix} a_1 & a_2 & a_3 \\ b_1 & b_2 & b_3 \\ c_1 & c_2 & c_3 \end{pmatrix} \begin{pmatrix} x \\ y \\ z \end{pmatrix} + \begin{pmatrix} a_0 \\ b_0 \\ c_0 \end{pmatrix} ?$$

9. Prove that every linear function is its own differential.

10. Prove that if the vector function f is differentiable at $\mathbf{x}_0$, then

$$f'(\mathbf{x}_0)\mathbf{x} = \lim_{t \to 0} \frac{f(\mathbf{x}_0 + t\mathbf{x}) - f(\mathbf{x}_0)}{t}.$$

11. Prove that if a vector function is differentiable at a point, then it is continuous there. [*Hint.* Multiply the quotient in the definition of differential by $|\mathbf{x} - \mathbf{x}_0|$.]

12. At which points do the following functions fail to be differentiable? Why?

(a) $f\begin{pmatrix} x \\ y \end{pmatrix} = \begin{pmatrix} \dfrac{1}{x^2} + \dfrac{1}{y^2} \\ x^2 + y^2 \end{pmatrix}.$

(b) $g\begin{pmatrix} x \\ y \end{pmatrix} = \begin{pmatrix} \sqrt{x^2 - y^2} \\ x + y \end{pmatrix}.$

(c) $f(u, v) = |u + v|$.

(d) $h(x, y) = \begin{cases} (x \sin(1/x), x^2 + y^2) & \text{if } x \neq 0. \\ (0, x^2 + y^2), & \text{if } x = 0. \end{cases}$

13. Prove that every translation is differentiable. What is the differential?

14. Consider the function $\mathcal{R}^n \xrightarrow{g} \mathcal{R}$ defined by $f(\mathbf{x}) = |\mathbf{x}|^2 = \mathbf{x} \cdot \mathbf{x}$. Prove that $f'(\mathbf{x})\mathbf{y} = 2\mathbf{x} \cdot \mathbf{y}$, for any $\mathbf{x}$ and $\mathbf{y}$ in $\mathcal{R}^n$.

15. Is the function $\mathcal{R}^n \xrightarrow{g} \mathcal{R}$ defined by $g(\mathbf{x}) = |\mathbf{x}|$ differentiable at every point of its domain?

16. Consider the function

$$N(\mathbf{x}) = \max \{|x_1|, \ldots, |x_n|\},$$

for $\mathbf{x} = (x_1, \ldots, x_n)$ in $\mathcal{R}^n$. For what points does the function fail to be differentiable? Answer for (a) $n = 1$, (b) $n = 2$, and (c) arbitrary n.

17. Show that if f and g are differentiable at $\mathbf{x}_0$ and a is a real number, then $f + g$ and af are differentiable at $\mathbf{x}_0$ and

(a) $d_{\mathbf{x}_0}(f + g) = d_{\mathbf{x}_0}f + d_{\mathbf{x}_0}g$.

(b) $d_{\mathbf{x}_0}(af) = a(d_{\mathbf{x}_0}f)$.

The domain of $f + g$ is the intersection of the domains of f and g. [*Hint.* Use the uniqueness of the differential.]

18. Verify that the function

$$f(x, y) = \begin{cases} \dfrac{xy}{x^2 - y^2}, & x \neq \pm y, \\ 0, & x = \pm y, \end{cases}$$

has a Jacobian matrix at (0, 0), but that it is not differentiable there.

19. Show that the function defined by

$$f(x) = \begin{cases} x^2 \sin \dfrac{1}{x}, & x \neq 0 \\ 0, & x = 0 \end{cases}$$

is differentiable for all x but is not continuously differentiable at $x = 0$.

SECTION 9

NEWTON'S METHOD

In this section we treat Newton's method for approximating a solution of an equation $f(\mathbf{x}) = 0$, where $\mathcal{R}^n \xrightarrow{f} \mathcal{R}^n$ is a nonlinear function. If f is linear or affine, the discussion in Chapter 2 holds. We begin by looking at the idea of approximating a vector in $\mathcal{R}^n$ by the entries in a sequence of vectors in $\mathcal{R}^n$. We are used to thinking of a real number like $\sqrt{2}$ as being approximated by a sequence of rational numbers, say, 1, 1.4, 1.41, 1.414, ... The idea extends immediately to vectors.

Example 1. Consider the vector $(\sqrt{2}, \pi)$ in $\mathcal{R}^2$. Suppose that $\sqrt{2}$ is approximated by a sequence 1, 1.4, 1.41, 1.414, . . . and that π is approximated by 3, 3.1, 3.14, 3.141, ... Then we can form the sequence of vectors (1, 3), (1.4, 3.1), (1.41, 3.14), (1.414, 3.141), . . . to approximate the vector $(\sqrt{2}, \pi)$. It is natural to pair the numbers as we have, and not in some other order, because it so happens that the entries in each pair are accurate approximations to $\sqrt{2}$ and π to as many decimal places as are given.

To make the ideas precise we define the limit of a sequence in $\mathcal{R}^n$. Let $\mathbf{x}_1, \mathbf{x}_2, \mathbf{x}_3, \ldots$ be an infinite sequence of vectors in $\mathcal{R}^n$, just one vector corresponding to each positive integer. Suppose there is a vector $\mathbf{x}$ in $\mathcal{R}^n$ such that, for any $\epsilon > 0$, there is an integer N for which

$$|\mathbf{x}_k - \mathbf{x}| < \epsilon$$

whenever $k \geq N$. Then we say that the given sequence **converges** to the **limit x**, and we write

$$\lim_{k \to \infty} \mathbf{x}_k = \mathbf{x}.$$

We can summarize by saying that the sequence $\mathbf{x}_1, \mathbf{x}_2, \mathbf{x}_3, \ldots$ converges to $\mathbf{x}$ if $|\mathbf{x}_k - \mathbf{x}|$ is arbitrarily small for all sufficiently large k. Figure 40

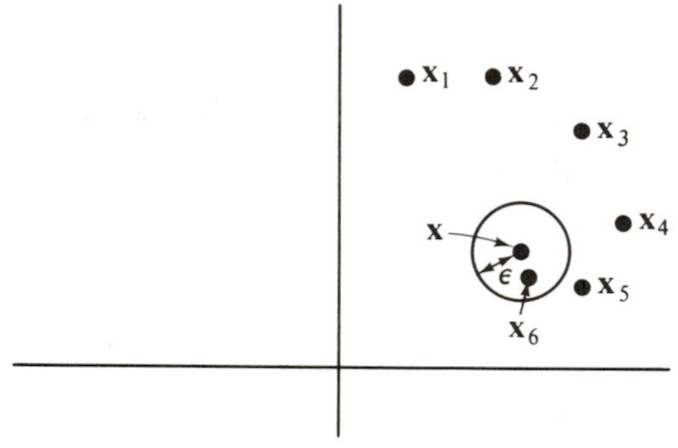

Figure 40

shows a sequence with entries lying within ϵ of $\mathbf{x}$ whenever $k \geq 6$. We leave as an exercise showing that if $x_1, x_2, x_3, \ldots$ and $y_1, y_2, y_3, \ldots$ are the sequences of Example 1, approximating $\sqrt{2}$ and π respectively, then the fact that $\lim_{k \to \infty} x_k = \sqrt{2}$ and $\lim_{k \to \infty} y_k = \pi$ implies that $\lim_{k \to \infty} (x_k, y_k) = (\sqrt{2}, \pi)$.

We turn now to Newton's method for approximating a solution of an equation $f(\mathbf{x}) = 0$, in the case where f is real-valued and $\mathbf{x}$ is a real variable. We assume that f is continuously differentiable. If the graph of f should happen to be convex as shown in Fig. 41, then it is geometrically clear that the tangent line to the graph at $(x_0, f(x_0))$ crosses the x-axis at a point x_1

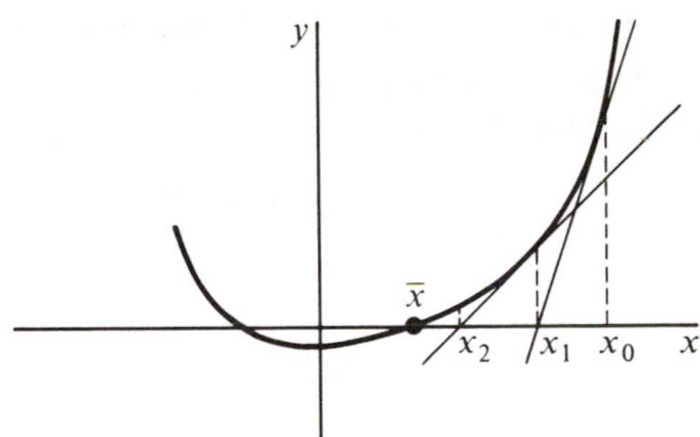

Figure 41

which is a better approximation to the solution $\bar{x}$ than x_0 is. Having chosen x_0 somewhat arbitrarily, and having found x_1, we can repeat the process. This time we use the tangent line at $(x_1, f(x_1))$ and call its intersection with the x-axis x_2. Thus we can generate a sequence of numbers $x_0, x_1, x_2, \ldots$ approximating $\bar{x}$.

In practice, we need a formula for computing the sequence $x_1, x_2, \ldots$. We observe first that the tangent line at $(x_0, f(x_0))$ has the equation

$$y = f'(x_0)(x - x_0) + f(x_0).$$

Since the approximation x_1 is found by intersecting the tangent with the x-axis, we set $y = 0$ in the above equation and solve for x_1. The result is

$$0 = f'(x_0)(x_1 - x_0) + f(x_0),$$

whence

$$f'(x_0)(x_1 - x_0) = -f(x_0).$$

If $f'(x_0) \neq 0$,

$$x_1 - x_0 = -\frac{f(x_0)}{f'(x_0)}$$

or

$$x_1 = x_0 - \frac{f(x_0)}{f'(x_0)}.$$

Having found x_1, to find x_2 we need only replace x_0 by x_1 in the last formula to get

$$x_2 = x_1 - \frac{f(x_1)}{f'(x_1)}.$$

In general, we compute x_{k+1} by

$$x_{k+1} = x_k - \frac{f(x_k)}{f'(x_k)}.$$

Example 2. The equation $x^2 - 3 = 0$ has two solutions, $\sqrt{3}$ and $-\sqrt{3}$. To approximate $\sqrt{3}$ we choose $x_0 = 2$ and compute x_{k+1} from x_k by the Formula (1) which in this case is

$$x_{k+1} = x_k - \frac{(x_k^2 - 3)}{2x_k}$$
$$= \frac{(x_k^2 + 3)}{2x_k}.$$

Thus we get $x_1 = \frac{7}{4} = 1.75$. Substituting this value in the above formula for $k = 1$ gives $x_2 = \frac{97}{56} \approx 1.7321$. This approximation to $\sqrt{3}$ happens to be correct to three decimal places, though we have not proved that.

We can follow a similar procedure to the one just described if f is a function from $\mathcal{R}^n$ to $\mathcal{R}^n$. The difference is that, in this case, $\mathbf{x}_1$, $\mathbf{x}_0$, and $f(\mathbf{x}_0)$ are vectors in $\mathcal{R}^n$ and $f'(\mathbf{x}_0)$ is an n-by-n Jacobian matrix. To approximate a solution of a vector equation $f(\mathbf{x}) = 0$, we consider the equation that defines the value of the best affine approximation to f near $\mathbf{x}_0$, that is,

$$\mathbf{y} = f'(\mathbf{x}_0)(\mathbf{x} - \mathbf{x}_0) + f(\mathbf{x}_0), \tag{2}$$

where $\mathbf{x}_0$ is chosen as an initial approximation to the desired solution $\bar{\mathbf{x}}$. In Fig. 41, Equation (2) can be interpreted as the equation of the tangent to the graph of f at $(\mathbf{x}_0, f(\mathbf{x}_0))$. As before, we set $\mathbf{y} = 0$ in Equation (2) to get

$$0 = f'(\mathbf{x}_0)(\mathbf{x}_1 - \mathbf{x}_0) + f(\mathbf{x}_0),$$

whence

$$f'(\mathbf{x}_0)(\mathbf{x}_1 - \mathbf{x}_0) = -f(\mathbf{x}_0).$$

Now if $f'(\mathbf{x}_0)$ has an inverse matrix $[f'(\mathbf{x}_0)]^{-1}$, we can apply the inverse to both sides, getting

$$\mathbf{x}_1 - \mathbf{x}_0 = -[f'(\mathbf{x}_0)]^{-1}f(\mathbf{x}_0).$$

Finally

$$\mathbf{x}_1 = \mathbf{x}_0 - [f'(\mathbf{x}_0)]^{-1}f(\mathbf{x}_0).$$

In this equation $[f'(\mathbf{x}_0)]^{-1}f(\mathbf{x}_0)$ is the vector obtained by applying the inverse of the matrix $f'(\mathbf{x}_0)$ to the vector $f(\mathbf{x}_0)$. The vector $\mathbf{x}_1$ is the first improvement on the initial approximation $\mathbf{x}_0$ to the solution $\bar{\mathbf{x}}$.

What we have just done can be repeated, replacing $\mathbf{x}_0$ by $\mathbf{x}_1$ to get

$$\mathbf{x}_2 = \mathbf{x}_1 - [f'(\mathbf{x}_1)]^{-1}f(\mathbf{x}_0).$$

After $k + 1$ steps we have

9.1 $$\mathbf{x}_{k+1} = \mathbf{x}_k - [f'(\mathbf{x}_k)]^{-1}f(\mathbf{x}_k).$$

Example 3. Consider the pair of equations

$$x^2 + y^2 = 2$$
$$x^2 - y^2 = 1.$$

To find approximate solutions by Newton's method we define

$$f(x, y) = \begin{pmatrix} x^2 + y^2 - 2 \\ x^2 - y^2 - 1 \end{pmatrix}$$

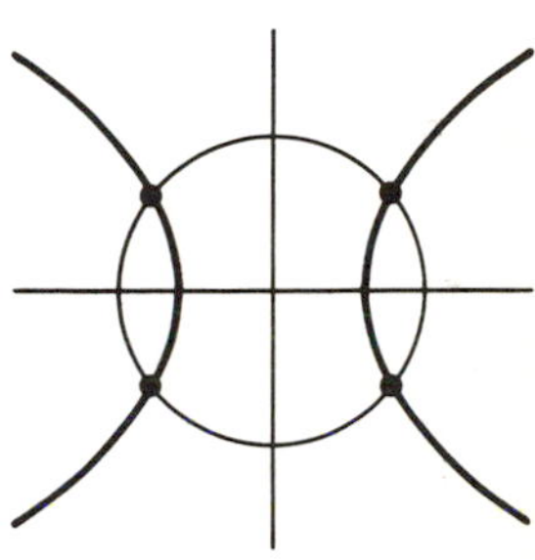

Figure 42

and solve the equation $f(x, y) = 0$. Clearly f is a function from $\mathfrak{R}^2$ to $\mathfrak{R}^2$. Since we require both $x^2 + y^2 - 2 = 0$ and $x^2 - y^2 - 1 = 0$, it will be helpful to sketch the curves defined by these two equations. The exact solutions are represented by the four points of intersection of the circle $x^2 + y^2 - 2 = 0$ and the hyperbola $x^2 - y^2 - 1 = 0$ shown in Fig. 42. The choice of an initial approximation depends on which solution we want to approximate. Looking for the solution in the first quadrant, we try $\mathbf{x}_0 = (1, 1)$. Since

$$f(x, y) = \begin{pmatrix} x^2 + y^2 - 2 \\ x^2 - y^2 - 1 \end{pmatrix},$$

we have

$$f'(x, y) = \begin{pmatrix} 2x & 2y \\ 2x & -2y \end{pmatrix}$$

and

$$[f'(x, y)]^{-1} = \begin{pmatrix} \frac{1}{4}x^{-1} & \frac{1}{4}x^{-1} \\ \frac{1}{4}y^{-1} & -\frac{1}{4}y^{-1} \end{pmatrix}.$$

Therefore,

$$\begin{pmatrix} x \\ y \end{pmatrix} - [f'(x, y)]^{-1}f(x, y) = \begin{pmatrix} x \\ y \end{pmatrix} - \begin{pmatrix} \frac{1}{4}x^{-1} & \frac{1}{4}x^{-1} \\ \frac{1}{4}y^{-1} & -\frac{1}{4}y^{-1} \end{pmatrix} \begin{pmatrix} x^2 + y^2 - 2 \\ x^2 - y^2 - 1 \end{pmatrix}$$

$$= \begin{pmatrix} x \\ y \end{pmatrix} - \begin{pmatrix} \dfrac{2x^2 - 3}{4x} \\ \dfrac{2y^2 - 1}{4y} \end{pmatrix} = \begin{pmatrix} \dfrac{2x^2 + 3}{4x} \\ \dfrac{2y^2 + 1}{4y} \end{pmatrix}. \quad (3)$$

This vector is the analog of the expression $(x^2 + 3)/2x$ in the previous example and is the formula by which the sequence of approximations is actually computed. Setting $\mathbf{x}_0 = (x_0, y_0) = (1, 1)$, we get

$$\mathbf{x}_1 = \begin{pmatrix} \dfrac{2x_0^2 + 3}{4x_0} \\ \dfrac{2y_0^2 + 1}{4y_0} \end{pmatrix} = \begin{pmatrix} \dfrac{5}{4} \\ \dfrac{3}{4} \end{pmatrix} = \begin{pmatrix} 1.25 \\ 0.75 \end{pmatrix}.$$

Substituting $\mathbf{x}_1$ into (3) gives

$$\mathbf{x}_2 = \begin{pmatrix} \dfrac{2(1.25)^2 + 3}{4\ (1.25)} \\ \dfrac{2(0.75)^2 + 3}{4(0.75)} \end{pmatrix} \approx \begin{pmatrix} 1.225 \\ 0.70833 \end{pmatrix}.$$

Substituting our approximate value for x_2 gives

$$\mathbf{x}_3 = \begin{pmatrix} \dfrac{2(1.225)^2 + 3}{4(1.225)} \\ \dfrac{2(1.70833)^2 + 1}{4(1.70833)} \end{pmatrix} \approx \begin{pmatrix} 1.22574 \\ 0.707108 \end{pmatrix}.$$

Similarly we get

$$\mathbf{x}_4 \approx \begin{pmatrix} 1.22474 \\ 0.707107 \end{pmatrix}$$

and

$$\mathbf{x}_5 \approx \begin{pmatrix} 1.22474 \\ 0.707107 \end{pmatrix}.$$

As in the previous example, further iteration using only five places after the decimal point can't produce any change.

In this example, the two simultaneous equations can actually be solved by elimination to yield $\mathbf{x} = (\sqrt{1.5}, \sqrt{0.5})$. The approximation $x_4 = (1.22474, 0.707107)$ happens to be correct to that many decimal places. The other three vector solutions can be obtained by symmetry. Referring to Fig. 42, we get them by changing one or both signs of the coordinates to minus. The numerical procedure could have been applied by taking as initial estimate $\mathbf{x}_0$ one of the vectors $(-1, -1)$, $(-1, 1)$, or $(1, -1)$.

In choosing an initial approximation $\mathbf{x}_0$, some care must be used in getting a sufficiently close approximation. For instance, if in Example 3 we wanted the solution in the first quadrant, then it is clear that too gross an error in choosing $\mathbf{x}_0$ could lead to approximating the wrong solution. In many examples a sketch or similar geometric analysis of the function f will show how $\mathbf{x}_0$ should be chosen.

In using Newton's method for large dimension, it may be very time-consuming to invert the matrix $f'(\mathbf{x}_n)$ at each step of the iteration. In such cases we can use the modified Newton formula:

9.2 $$\mathbf{x}_{k+1} = \mathbf{x}_k - [f'(\mathbf{x}_0)]^{-1} f(\mathbf{x}_k)$$

to derive a sequence of approximations to a solution of $f(\mathbf{x}) = 0$. For $k = 0$, the formula defining $\mathbf{x}_1$ is the same as the Newton formula. For $k \geq 1$, x_{k+1} as defined by Equation 9.2 will in general be different from the corresponding value determined by the Newton formula, because the matrix $[f'(x_0)]^{-1}$ remains the same at each step in Equation 9.2.

Example 4. Returning to the equation of Example 3, namely,

$$\begin{pmatrix} x^2 + y^2 - 2 \\ x^2 - y^2 - 1 \end{pmatrix} = \begin{pmatrix} 0 \\ 0 \end{pmatrix},$$

we apply Equation 9.2 with $\mathbf{x}_0 = (1, 1)$. Then

$$[f'(\mathbf{x}_0)]^{-1} = \begin{pmatrix} \frac{1}{4} & \frac{1}{4} \\ \frac{1}{4} & -\frac{1}{4} \end{pmatrix},$$

so

$$\mathbf{x} - [f'(\mathbf{x}_0)]^{-1} f(\mathbf{x}) = \begin{pmatrix} x \\ y \end{pmatrix} - \begin{pmatrix} \frac{1}{4} & \frac{1}{4} \\ \frac{1}{4} & -\frac{1}{4} \end{pmatrix} \begin{pmatrix} x^2 + y^2 - 2 \\ x^2 - y^2 - 1 \end{pmatrix}$$

$$= \begin{pmatrix} x \\ y \end{pmatrix} - \begin{pmatrix} \dfrac{2x^2 - 3}{4} \\ \dfrac{2y^2 - 1}{4} \end{pmatrix} = \begin{pmatrix} \dfrac{-2x^2 + 4x + 3}{4} \\ \dfrac{-2y^2 + 4y + 1}{4} \end{pmatrix}.$$

Then $\mathbf{x}_1$, defined by $\mathbf{x}_1 = \mathbf{x}_0 - [r'(\mathbf{x}_0)]^{-1} f(\mathbf{x}_0)$, is

$$\mathbf{x}_1 = \begin{pmatrix} \dfrac{-2x_0^2 + 4x_0 + 3}{4} \\ \dfrac{2y_0^2 + 4y_0 + 1}{4} \end{pmatrix} = \begin{pmatrix} 1.25 \\ 0.75 \end{pmatrix}.$$

In the next step

$$\mathbf{x}_2 = \begin{pmatrix} \dfrac{-2(1.25)^2 + 4(1.25) + 3}{4} \\ \dfrac{-2(0.75)^2 + 4(0.75) + 1}{4} \end{pmatrix} = \begin{pmatrix} 1.21875 \\ 0.71875 \end{pmatrix}$$

Continuing, we arrive at

$$\mathbf{x}_{10} = \begin{pmatrix} 1.22474 \\ 0.707107 \end{pmatrix},$$

which agrees with the result obtained in Example 3, though it takes more steps.

In deciding whether to use the Newton formula 9.1 or its modification 9.2, two facts should be remembered. In general, Formula 9.1 produces faster convergence than 9.2, that is, it achieves a smaller error in a given number of steps. On the other hand, Formula 9.2 has the advantage that it requires calculation of the inverse of the derivative matrix f' at only one point. Thus if computing the inverse matrices $[f'(\mathbf{x}_k)]^{-1}$ is going to be particularly time-consuming, it may be worth taking the extra iteration steps that 9.2 may require to achieve the desired accuracy.

In Section 5 of the Appendix we prove Theorem 5.1 which guarantees convergence of the sequence in 9.2 under certain conditions. Theorem 5.1 is also used to prove the inverse-function theorem, taken up in Section 6, Chapter 4.

EXERCISES

1. (a) Sketch the graph of $f(x) = \sqrt[3]{x} - x$ for $-2 < x < 2$.
(b) Sketch the tangent lines to the graph of f at $x_0 = \frac{3}{4}$, $x_0 = \frac{3}{4}$, and $x_0 = -\frac{1}{4}$.
(c) For each of the three choices for x_0 in part (b), what solution of $f(x) = 0$ can the Newton iteration be expected to converge to?
(d) Discuss the choice $x_0 = \sqrt[3]{3}/9$ for an initial approximation to a solution of $f(x) = 0$.

2. Show that Newton's method will not produce convergence to the unique solution of $\sqrt[3]{x} = 0$ unless the initial choice $x_0 = 0$ is made. [*Hint.* Show that $x_k = (-2)^k x_0$.]

3. (a) To approximate the solution of $\cos x - x = 0$ by Newton's method, show that when x_0 is chosen, then for $k \geq 0$,

$$x_{k+1} = \frac{x_k \sin x_k + \cos x_k}{1 + \sin x_k}.$$

(b) Use a computer to find x_{10}.

4. Find approximate solutions to the pair of equations

$$x^2 + y - 1 = 0$$
$$x + y^2 - 2 = 0$$

by following the steps below:

(a) Sketch the curves satisfying each of the two equations.
(b) Defining f by

$$f(x, y) = \begin{pmatrix} x^2 + y - 1 \\ x + y^2 - 2 \end{pmatrix},$$

find $f'(x, y)$, $[f'(x, y)]^{-1}$, and $(x, y) - [f'(x, y)]^{-1} f(x, y)$.
(c) Using the sketch in part (a), choose an initial approximation $\mathbf{x}_0 = (x_0, y_0)$ to the solution of $f(x, y) = 0$ that lies in the fourth quadrant of the xy-plane.

(d) Compute $x_1 = (x_1, y_1)$ by Formula 9.1.
(e) Use a computer to find $\mathbf{x}_5 = (x_5, y_5)$.

5. Let

$$f(x, y, z) = \begin{pmatrix} x + y + z \\ x^2 + y^2 + z^2 \\ x^3 + y^3 + z^3 \end{pmatrix}$$

and find $f'(1, 2, -1)$. Taking $\mathbf{x}_0 = (1, 2, -1)$, apply the modified Newton Formula 9.2 to approximate a solution to $f(x, y, z) = (2, 6, 8)$.

6. Let

$$g(u, v) = \begin{pmatrix} u^2 + uv^2 \\ u + v^3 \end{pmatrix}.$$

Noting that $g(1, 1) = (2, 2)$, use Newton's method 9.1 or its modification 9.2 to approximate a solution to $g(u, v) = (1.9, 2.1)$.

4

Vector Calculus

SECTION 1

DIRECTIONAL DERIVATIVES

A partial derivative of a real-valued function measures the rate of change of the function in a particular coordinate direction. To measure the rate of change in an arbitrary direction we use the directional derivative. Recall that a **unit vector u** is a vector of length 1, that is, $|\mathbf{u}| = 1$. Let $\mathcal{R}^n \xrightarrow{f} \mathcal{R}$ be a real-valued function, and let $\mathbf{u}$ be a unit vector in the domain space $\mathcal{R}^n$. The **directional derivative of f with respect to u**, denoted by $\partial f/\partial \mathbf{u}$, is the real-valued function defined by

$$\frac{\partial f}{\partial \mathbf{u}}(\mathbf{x}) = \lim_{t \to 0} \frac{f(\mathbf{x} + t\mathbf{u}) - f(\mathbf{x})}{t}.$$

The domain of $\partial f/\partial \mathbf{u}$ is the subset of the domain of f for which the above limit exists.

The connection between the derivative with respect to a unit vector and the differential is provided in the following theorem.

1.1 Theorem

If f is differentiable at $\mathbf{x}$, then

$$\frac{\partial f}{\partial \mathbf{u}}(\mathbf{x}) = f'(\mathbf{x})\mathbf{u}$$

for every unit vector $\mathbf{u}$ in $\mathcal{R}^n$.

Proof. The existence of the derivative $f'(\mathbf{x})$ implies that

$$\lim_{t\to 0} \frac{f(\mathbf{x} + t\mathbf{u}) - f(\mathbf{x}) - f'(\mathbf{x})(t\mathbf{u})}{|t\mathbf{u}|} = 0,$$

which is equivalent to

$$\lim_{t\to 0} \frac{1}{|\mathbf{u}|}\left|\frac{f(\mathbf{x} + t\mathbf{u}) - f(\mathbf{x})}{t} - f'(\mathbf{x})\mathbf{u}\right| = 0.$$

This in turn is equivalent to

$$\lim_{t\to 0} \frac{f(\mathbf{x} + t\mathbf{u}) - f(\mathbf{x})}{t} = f'(\mathbf{x})\mathbf{u},$$

and the proof is finished.

The equation

$$\frac{\partial f}{\partial \mathbf{u}}(\mathbf{x}) = f'(\mathbf{x})\mathbf{u}$$

shows that, for a differentiable function, the directional derivative involves nothing very new. However, it does give an important interpretation of the derivative $f'(\mathbf{x})$ applied to a unit vector $\mathbf{u}$.

For each vector $\mathbf{u}$ in $\mathcal{R}^n$ of length $|\mathbf{u}| = 1$, we have defined the **directional derivative of f in the direction of $\mathbf{u}$** to be the function $\partial f/\partial \mathbf{u}$. The reason for the name "directional derivative" is that in a Euclidean space there is a natural way to associate a vector to each direction, namely, take the unit vector in that direction. The number $(\partial f/\partial \mathbf{u})(\mathbf{x})$ is then regarded as a standard measure of the rate of change of the values of f in the direction of $\mathbf{u}$.

Example 1. The domain space of the function

$$f(x, y, z) = xyz + e^{2x+y}$$

is assumed to be Euclidean 3-dimensional space. We find the directional derivative of f in the direction of $\mathbf{u} = (\frac{1}{2}, \frac{1}{2}, 1/\sqrt{2})$. Setting $\mathbf{x} = (x, y, z)$ and using Theorem 1.1, we obtain

$$\frac{\partial f}{\partial \mathbf{u}}(\mathbf{x}) = f'(\mathbf{x})\mathbf{u} = (yz + 2e^{2x+y} \quad xz + e^{2x+y} \quad xy)\begin{pmatrix} 1/2 \\ 1/2 \\ 1/\sqrt{2} \end{pmatrix}$$

$$= \frac{yz + xz + \sqrt{2}\,xy}{2} + \frac{3e^{2x+y}}{2}.$$

It follows that the directional derivative of f in the direction of $\mathbf{u}$ has at the origin the value $\partial f/\partial \mathbf{u}\,(0, 0, 0) = \frac{3}{2}$.

Let $\mathcal{R}^2 \xrightarrow{f} \mathcal{R}$ be a function whose graph is a surface in 3-dimensional Euclidean space, and let $\mathbf{u}$ be a unit vector in $\mathcal{R}^2$, i.e., $|\mathbf{u}| = 1$. An example

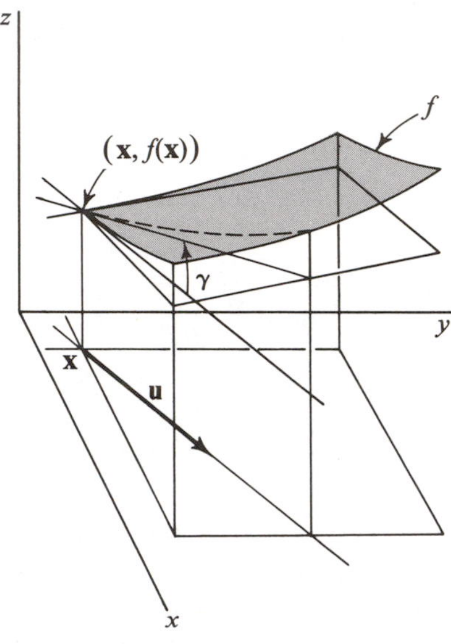

Figure 1

is shown in Fig. 1. The value of the directional derivative $\partial f/\partial \mathbf{u}$ at $\mathbf{x} = (x, y)$ is by definition

$$\frac{\partial f}{\partial \mathbf{u}}(\mathbf{x}) = \lim_{t \to 0} \frac{f(\mathbf{x} + t\mathbf{u}) - f(\mathbf{x})}{t}.$$

The distance between the points $\mathbf{x} + t\mathbf{u}$ and $\mathbf{x}$ is given by

$$|(\mathbf{x} + t\mathbf{u}) - \mathbf{x}| = |t\mathbf{u}| = |t|.$$

Hence, the ratio

$$\frac{f(\mathbf{x} + t\mathbf{u}) - f(\mathbf{x})}{t}$$

is the slope of the line through the points $f(\mathbf{x} + t\mathbf{u})$ and $f(\mathbf{x})$. It follows that the limit, $(\partial f/\partial \mathbf{u})(\mathbf{x})$, of the ratio is the slope of the tangent line at $(\mathbf{x}, f(\mathbf{x}))$ to the curve formed by the intersection of the graph of f with the plane that contains $\mathbf{x}$ and $\mathbf{x} + \mathbf{u}$, and is parallel to the z-axis. This curve is drawn with a dashed line in the figure. The angle γ shown in Fig. 1

therefore satisfies the equation

$$\tan \gamma = \frac{\partial f}{\partial \mathbf{u}}(\mathbf{x}).$$

The situation here is a generalization of that illustrated in Fig. 24 of Chapter 3, Section 5. If we choose $\mathbf{u} = (1, 0)$, the angle γ becomes the angle α in the earlier figure and

$$\frac{\partial f}{\partial \mathbf{u}} = \frac{\partial f}{\partial x}.$$

On the other hand, if we take $\mathbf{v} = (0, 1)$, then γ is the angle β in Figure 24, and

$$\frac{\partial f}{\partial \mathbf{v}} = \frac{\partial f}{\partial y}.$$

The mean-value theorem for real-valued functions of a real variable can be extended to *real-valued* functions of a vector variable as follows.

1.2 Theorem

Let $\mathcal{R}^n \xrightarrow{f} \mathcal{R}$ be differentiable on an open set containing the segment S joining two vectors $\mathbf{x}$ and $\mathbf{y}$ in $\mathcal{R}^n$. Then there is a point $\mathbf{x}_0$ on S such that

$$f(\mathbf{y}) - f(\mathbf{x}) = f'(\mathbf{x}_0)(\mathbf{y} - \mathbf{x}).$$

Proof. Consider the function $g(t) = f(t(\mathbf{y} - \mathbf{x}) + \mathbf{x})$, defined for $0 \le t \le 1$. Let us set $m(t) = t(\mathbf{y} - \mathbf{x}) + \mathbf{x}$. Then if h is a real number,

$$\begin{aligned} g(t + h) - g(t) &= f((t + h)(\mathbf{y} - \mathbf{x}) + \mathbf{x}) - f(t(\mathbf{y} - \mathbf{x}) + \mathbf{x}) \\ &= f'(m(t))h(\mathbf{y} - \mathbf{x}) + |h(\mathbf{y} - \mathbf{x})|\, Z(h(\mathbf{y} - \mathbf{x})), \end{aligned}$$

where $\lim_{h \to 0} Z(h(\mathbf{y} - \mathbf{x})) = 0$. This last relation simply says that f is differentiable at $m(t)$. Dividing by h, we get

$$\frac{g(t + h) - g(t)}{h} = f(m(t))(\mathbf{y} - \mathbf{x}) \pm |\mathbf{y} - \mathbf{x}|\, Z(h(\mathbf{y} - \mathbf{x})),$$

whence

$$g'(t) = f'(m(t))(\mathbf{y} - \mathbf{x}). \tag{1}$$

But by the mean-value theorem for functions of one variable,

$$\begin{aligned} \frac{g(1) - g(0)}{1 - 0} &= f(\mathbf{y}) - f(\mathbf{x}) \\ &= g'(t_0), \end{aligned}$$

for some t_0 satisfying $0 < t_0 < 1$. Setting $t = t_0$ and $m(t_0) = \mathbf{x}_0$ in Equation (1) gives the required formula.

One of the most important conclusions to be drawn from the mean-value theorem for functions of one variable is that a function with zero derivative on an interval is constant. For a function f of a vector variable, we shall replace the domain interval by an open set D in $\mathcal{R}^n$ that we assume to be **polygonally connected.** A polygonally connected set S is one such that any two points in it can be joined by a polygon in S, that is, by a finite sequence of line segments lying in S.

1.3 Theorem

If $\mathcal{R}^n \xrightarrow{f} \mathcal{R}^m$ is differentiable on a polygonally connected open set D and $f'(\mathbf{x}) = 0$ for every $\mathbf{x}$ in D, then f is constant.

Proof. We need only prove that each coordinate function of f is constant, and so we can assume that f is real-valued. If $\mathbf{x}_1$ and $\mathbf{x}_2$ are points of D joined by a single line segment, then Theorem 1.2 and the assumption that $f'(\mathbf{x}) = 0$ in D together imply that $f(\mathbf{x}_1) = f(\mathbf{x}_2)$. Obviously the same conclusion holds for two points, $\mathbf{x}_1$ and $\mathbf{x}_2$, joined by a finite sequence of segments. So f is constant.

EXERCISES

1. For each of the following functions defined on 3-dimensional Euclidean space, find the directional derivative in the direction of the unit vector $\mathbf{u}$ at the point $\mathbf{x}$.

   (a) $f(x, y, z) = x^2 + y^2 + z^2$, $\mathbf{u} = (1/\sqrt{3}, 1/\sqrt{3}, 1/\sqrt{3})$, $\mathbf{x} = (1, 0, 1)$.
   (b) $h(x, y, z) = xyz$, $\mathbf{u} = (\cos\alpha \sin\beta, \sin\alpha \sin\beta, \cos\beta)$, $\mathbf{x} = (1, 0, 0)$.
   (c) $f(x, y)$, $\mathbf{x} = (x, y)$, and $\mathbf{y} = (\cos\alpha, \sin\alpha)$. (Assume that f is real-valued and differentiable.)

2. For each of the following real-valued functions defined on Euclidean space, find the directional derivative at $\mathbf{x}$ in the direction indicated.

   (a) $f(x, y) = x^2 - y^2$ at $\mathbf{x} = (1, 1)$ and in the direction
   $$\left(\frac{1}{\sqrt{5}}, \frac{2}{\sqrt{5}}\right).$$
   $\left[\textit{Ans. } \frac{-2}{\sqrt{5}}.\right]$

   (b) $f(x, y) = e^x \sin y$ at $\mathbf{x} = (1, 0)$ and in the direction $(\cos\alpha, \sin\alpha)$.

   (c) $f(x, y) = e^{x+y}$ at $\mathbf{x} = (1, 1)$ in the direction of the curve defined by $g(t) = (t^2, t^3)$ at $g(2)$ for t increasing.

3. Find the absolute value of the directional derivative at $(1, 1, 0)$ of the function $f(x, y, z) = x^2 + ye^z$ in the direction of the tangent line at $g(0)$ to the curve in 3-dimensional Euclidean space defined parametrically by $g(t) = (3t^2 + t + 1, 2t, t^2)$.

4. Find the directional derivative at $(1, 0, 0)$ of the function $f(x, y, z) = x^2 + ye^z$ in the direction of increasing t along the curve in Euclidean $\mathcal{R}^3$ defined by $g(t) = (t^2 - t + 2, t, t + 2)$ at $g(0)$. [*Ans.* $-1/\sqrt{3}$.]

5. Find the absolute value of the directional derivative at $(1, 0, 1)$ of the function $f(x, y, z) = 4x^2y + y^2z$ in the direction of the perpendicular at $(1, 1, 1)$ to the surface in Euclidean 3-space defined implicitly by $x^2 + 2y^2 + z^2 = 4$. [*Ans.* $8/\sqrt{6}$.]

6. (a) Show that the vector $(y_1z_2 - y_2z_1, z_1x_2 - z_2x_1, x_1y_2 - x_2y_1)$ is perpendicular to (x_1, y_1, z_1) and (x_2, y_2, z_2).
   (b) Find the absolute value of the directional derivative at $(1, 2, 1)$ of the function $f(x, y, z) = x^3 + y^2 + z$ in the direction of the perpendicular at $(1, 2, 1)$ to the surface defined parametrically by

$$(x, y, z) = (u^2v, u + v, u).$$ [*Ans.* $2/\sqrt{3}$.]

7. If the temperature at a point (x, y, z) of a solid ball of radius 3 centered at $(0, 0, 0)$ is given by $T(x, y, z) = yz + zx + xy$, find the direction in which T is increasing most rapidly at $(1, 1, 2)$.

8. Show that the mean-value formula of Theorem 1.2 can be written in the form

$$\frac{f(\mathbf{y}) - f(\mathbf{x})}{|\mathbf{y} - \mathbf{x}|} = \frac{\partial f}{\partial \mathbf{u}}(\mathbf{x}_0),$$

where $\mathbf{u} = (\mathbf{y} - \mathbf{x})/|\mathbf{y} - \mathbf{x}|$.

9. Show that the function f defined by

$$f(x, y) = \begin{cases} \dfrac{x\,|y|}{\sqrt{x^2 + y^2}}, & (x, y) \neq (0, 0) \\ 0, & (x, y) = (0, 0) \end{cases}$$

has a directional derivative in every direction at $(0, 0)$, but that f is not differentiable at $(0, 0)$.

SECTION 2
THE GRADIENT

In the first section of Chapter 3 we looked at some examples of vector fields. Many of the most important ones arise as gradient fields, which are described below.

If f is a differentiable real-valued function, $\mathcal{R}^n \xrightarrow{f} \mathcal{R}$, then the function ∇f defined by

$$\nabla f(\mathbf{x}) = \left(\frac{\partial f}{\partial x_1}(\mathbf{x}), \ldots, \frac{\partial f}{\partial x_n}(\mathbf{x})\right)$$

is called the **gradient** of f. The gradient is evidently a function from $\mathcal{R}^n$ to $\mathcal{R}^n$, and it is most often pictured as a vector field.

Example 1. The function $f(x, y) = \frac{1}{6}(x^2 + y^3)$ is differentiable in all of $\mathcal{R}^2$, and so $\nabla f(x, y) = (\frac{1}{3}x, \frac{1}{2}y^2)$ is also defined in $\mathcal{R}^2$. The field is shown in Fig. 2 at several points.

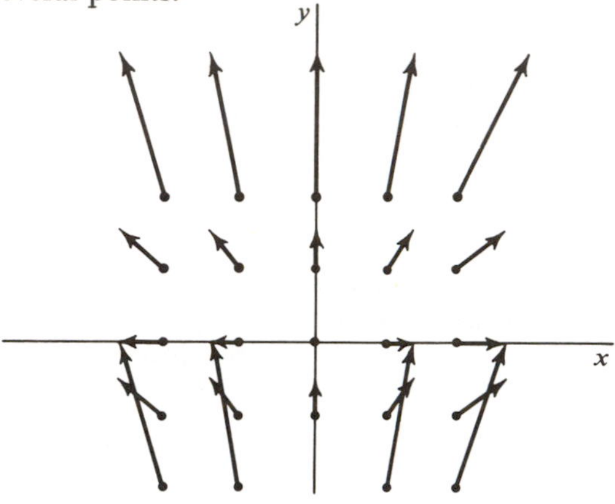

Figure 2

The function $g(x, y, z) = x^2 + y^2 + z^2$ has gradient $\nabla g(x, y, z) = (2x, 2y, 2z)$, and the direction of the field is directly away from the origin at each point.

The gradient of a function is important for several reasons. To begin, we remark that the directional derivative of a real-valued function f with respect to a unit vector $\mathbf{u}$ can be written in terms of the gradient of f as

2.1
$$\frac{\partial f}{\partial \mathbf{u}}(\mathbf{x}) = \nabla f(\mathbf{x}) \cdot \mathbf{u}.$$

The reason is that $(\partial f/\partial \mathbf{u})(\mathbf{x}) = f'(\mathbf{x})\mathbf{u}$, and the application of the matrix $f'(\mathbf{x})$ to $\mathbf{u}$ is the same as the dot product $\nabla f(\mathbf{x}) \cdot \mathbf{u}$. Using Equation 2.1, we can easily prove the following theorem. (This is the origin of the use of the term gradient, usually meaning incline or slope.)

2.2 Theorem

Let $\mathcal{R}^n \xrightarrow{f} \mathcal{R}$ be differentiable in an open set D in $\mathcal{R}^n$. Then at each point $\mathbf{x}$ in D for which $\nabla f(\mathbf{x}) \neq 0$, the vector $\nabla f(\mathbf{x})$ points in the direction of maximum increase for f. The number $|\nabla f(\mathbf{x})|$ is the maximum rate of increase.

Proof. Given a unit vector $\mathbf{u}$, we have, by Equation 2.1 and the Cauchy-Schwarz inequality,

$$\begin{aligned}\frac{\partial f}{\partial \mathbf{u}}(\mathbf{x}) &= \nabla f(\mathbf{x}) \cdot \mathbf{u} \leq |\nabla f(\mathbf{x})|\,|\mathbf{u}| \\ &= |\nabla f(\mathbf{x})|\end{aligned}$$

But when $\mathbf{u} = \nabla f(\mathbf{x})/|\nabla f(\mathbf{x})|$, and only then, we have

$$\frac{\partial f}{\partial \mathbf{u}}(\mathbf{x}) = |\nabla f(\mathbf{x})|.$$

Thus the rate of increase $\partial f/\partial \mathbf{u}(\mathbf{x})$ is never greater than $|\nabla f(\mathbf{x})|$ and is equal to it in the direction of the gradient.

Example 2. Let $f(x, y) = e^{xy}$. Then $\nabla f(x, y) = (ye^{xy}, xe^{xy})$; thus at $(1, 2)$ the function f increases most rapidly in the direction $\nabla f(1, 2) = (2e^2, e^2)$, which has the same direction as the unit vector $(2/\sqrt{5}, 1/\sqrt{5})$. The rate of increase in that direction is $|\nabla f(1, 2)| = \sqrt{5}\, e^2$. Similarly $\nabla f(-1, 2) = (2e^{-2}, -e^{-2})$ and has direction $(2/\sqrt{5}, -1/\sqrt{5})$, with maximum rate of increase at $(-1, 2)$ equal to $\sqrt{5}\, e^{-2}$. The maximum rate of decrease occurs in the opposite direction, namely, $(-2/\sqrt{5}, 1/\sqrt{5})$.

Next we shall prove a chain rule for differentiating the composition $g(f(t))$, of a function $\mathcal{R} \xrightarrow{f} \mathcal{R}^n$ and a function $\mathcal{R}^n \xrightarrow{g} \mathcal{R}$. Thus if $\mathcal{R} \xrightarrow{f} \mathcal{R}^3$ is given by $f(t) = (t, t^2, t)$ and $\mathcal{R}^3 \xrightarrow{g} \mathcal{R}$ by $g(x, y, z) = x \cos (y + z)$, then $g(f(t)) = t \cos (t^2 + t)$. This defines a new function from $\mathcal{R}$ to $\mathcal{R}$. For example, if g is defined in a region D of $\mathcal{R}^3$ and f describes the motion of a point along a path lying in D, we may be interested in finding the rate of change of the composite function with respect to t. The theorem gives a formula for doing this in terms of the gradient of g.

2.3 Theorem

Let g be real-valued and continuously differentiable on an open set D in $\mathcal{R}^n$ and let $f(t)$ be defined and differentiable for $a < t < b$, taking its values in D. Then the composite function $F(t) = g(f(t))$ is differentiable for $a < t < b$ and

$$F'(t) = \nabla g(f(t)) \cdot f'(t).$$

Proof. By definition,

$$F'(t) = \lim_{h \to 0} \frac{F(t+h) - F(t)}{h}$$
$$= \lim_{h \to 0} \frac{g(f(t+h)) - g(f(t))}{h},$$

if the limit exists. Since f is differentiable, it is continuous. Then we can choose $\delta > 0$ such that, whenever $|h| < \delta$, $f(t+h)$ is always inside an open ball centered at $f(t)$ and contained in D. We now apply the mean-value theorem of the previous section to g, getting

$$g(\mathbf{y}) - g(\mathbf{x}) = g'(\mathbf{x}_0)(\mathbf{y} - \mathbf{x})$$
$$= \nabla g(\mathbf{x}_0) \cdot (\mathbf{y} - \mathbf{x}),$$

where $\mathbf{x}_0$ is some point on the segment joining $\mathbf{y}$ and $\mathbf{x}$. Letting $\mathbf{x} = f(t)$ and $\mathbf{y} = f(t+h)$, with $|h| < \delta$, we have

$$\frac{F(t+h) - F(t)}{h} = \nabla g(\mathbf{x}_0) \cdot \frac{f(t+h) - f(t)}{h}.$$

The vector $\mathbf{x}_0$ is now some point on the segment joining $f(t)$ and $f(t+h)$. (Why is $\mathbf{x}_0$ in the domain of f?) Since g was assumed continuously differentiable, $\nabla g(\mathbf{x})$ is continuous, and so $\nabla g(\mathbf{x}_0)$ tends to $\nabla g(f(t))$ as h tends to zero. The dot product is continuous, so $F'(t)$ exists, with

$$F'(t) = \lim_{h \to 0} \nabla g(\mathbf{x}_0) \cdot \frac{f(t+h) - f(t)}{h}.$$
$$= \nabla g(f(t)) \cdot f'(t).$$

Example 3. Let $g(x, y) = x^2y + xy^3$ for (x, y) in $\mathcal{R}^2$. Let $f(t)$ be differentiable in some neighborhood of $t = t_0$ and take its values in $\mathcal{R}^2$. If it is known only that $f(t_0) = (-1, 1)$ and $f'(t_0) = (2, 3)$, then the composition, $F(t) = g(f(t))$, is known only at $t = t_0$, and $F'(t_0)$ cannot be computed by direct differentiation. However, by the previous theorem we have

$$F'(t_0) = \nabla g(f(t_0)) \cdot f'(t_0).$$

We find that $\nabla g(x, y) = (2xy + y^3, x^2 + 3xy^2)$, so $\nabla g(f(t_0)) = (-1, -2)$. Then $F'(t_0) = (-1, -2) \cdot (2, 3) = -8$.

An extension of Theorem 2.3 to the case in which f is vector-valued is proved in the next section.

The gradient is particularly useful in analyzing the level sets of a real-valued function. Recall that a level set S of a function f is a set of points $\mathbf{x}$ satisfying $f(\mathbf{x}) = k$ for some constant k. For $\mathcal{R}^2 \xrightarrow{f} \mathcal{R}$ we are usually interested in S when it is a curve, and for $\mathcal{R}^3 \xrightarrow{f} \mathcal{R}$, the sets S most often considered are surfaces. Examples are shown in Fig. 3.

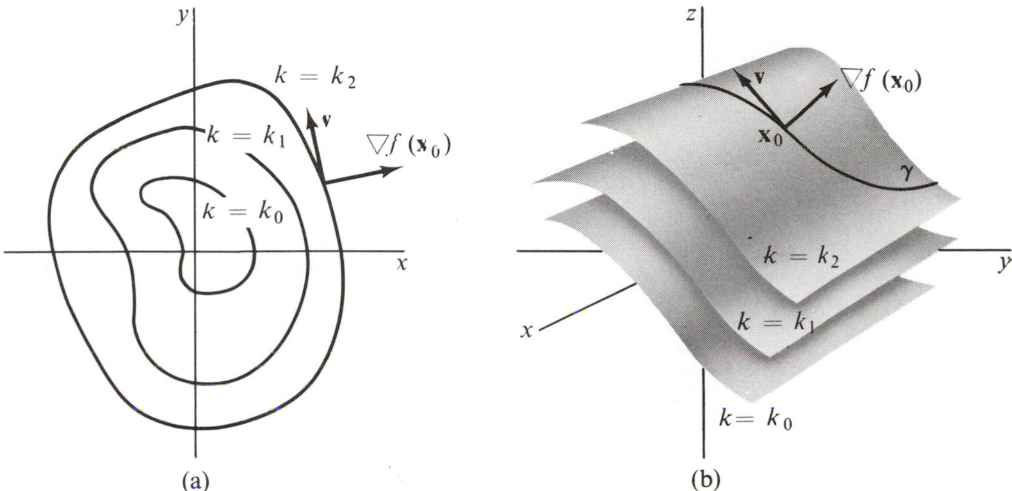

Figure 3

To say that a point $\mathbf{x}_0$ is on the level set S corresponding to level k is to say that $f(\mathbf{x}_0) = k$. Now suppose that there is a curve γ lying in S and parametrized by a continuously differentiable function $\mathcal{R} \xrightarrow{g} \mathcal{R}^n$. Suppose also that $g(t_0) = \mathbf{x}_0$ and $g'(t_0) = \mathbf{v} \neq 0$, so that $\mathbf{v}$ is a tangent vector to γ at $\mathbf{x}_0$, as shown in Fig. 3. Applying the chain rule to the function $h(t) = f(g(t))$ at t_0 gives

$$\begin{aligned} h'(t_0) &= \nabla f(g(t_0)) \cdot g'(t_0) \\ &= \nabla f(\mathbf{x}_0) \cdot \mathbf{v}. \end{aligned}$$

But since γ lies on S, we have $h(t) = f(g(t)) = k$, that is, h is constant. Thus $h'(t_0) = 0$, and

2.4 $$\nabla f(\mathbf{x}_0) \cdot \mathbf{v} = 0.$$

This says that $\nabla f(\mathbf{x}_0)$, if it is not zero, is perpendicular to every tangent vector to an arbitrary smooth curve lying on S and passing through $\mathbf{x}_0$. For this reason it is natural to say that $\nabla f(\mathbf{x}_0)$ is perpendicular or **normal** to S at $\mathbf{x}_0$ and to take as the **tangent** plane (or line) to S at $\mathbf{x}_0$ the set of all points $\mathbf{x}$ satisfying

$$\nabla f(\mathbf{x}_0) \cdot (\mathbf{x} - \mathbf{x}_0) = 0, \qquad \text{if} \quad \nabla f(\mathbf{x}_0) \neq 0.$$

Example 4. The function $f(x, y, z) = x^2 + y^2 - z^2$ has for one of its level surfaces a cone C consisting of all points satisfying $x^2 + y^2 - z^2 = 0$. The point $\mathbf{x}_0 = (1, 1, \sqrt{2})$ lies on C, and to find the tangent plane to C at $\mathbf{x}_0$ we compute $\nabla f(\mathbf{x}_0) = (2, 2, -2\sqrt{2})$. Then

$$\nabla f(\mathbf{x}_0) \cdot (\mathbf{x} - \mathbf{x}_0) = (2, 2, -2\sqrt{2}) \cdot (x - 1, y - 1, z - \sqrt{2}),$$

and the tangent plane is given by $(x - 1) + (y - 1) - \sqrt{2}(z - \sqrt{2}) = 0$, or $x + y - \sqrt{2}\, z = 0$. This plane is shown in Fig. 4. Notice that both C

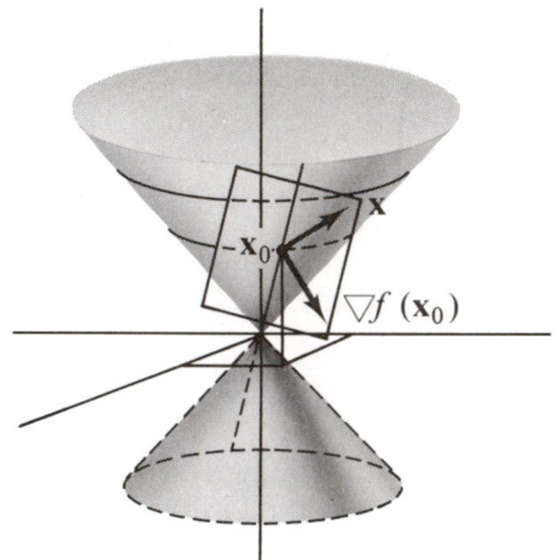

Figure 4

and its tangent contain a common line with direction $(1, 1, \sqrt{2})$, and that the normal vector to the tangent is perpendicular to that line.

Putting together Theorem 2.2 with 2.4, we see that the direction of maximum increase of a real-valued differentiable function at a point is perpendicular to the level set of the function through that point.

Example 5. The function $f(x, y) = xy$ has level curves $xy = k$. If $k \neq 0$, these curves are all hyperbolas, and each one of them is intersected perpendicularly by every member of the family of hyperbolas $g(x, y) = x^2 - y^2 = k$. (These are shown in Fig. 5.) To see this, observe that $\nabla f(x, y) = (y, x)$ and $\nabla g(x, y) = (2x, -2y)$. Hence, for each (x, y) we have $\nabla f(x, y) \cdot \nabla g(x, y) = 0$. Thus the normal vectors, and hence the tangents, are perpendicular. It also follows that a tangent to a curve from one family points in a direction of maximum increase for the defining function of the other family. The argument fails at $(0, 0)$. See Exercise 6 following this section.

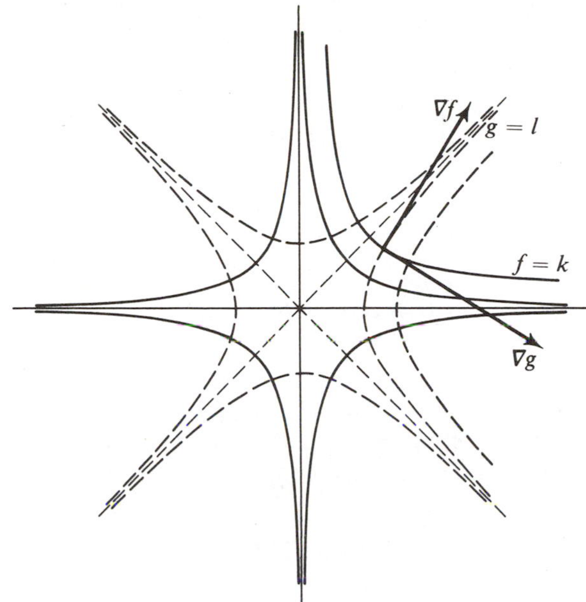

Figure 5

As another example of the use of the gradient, we shall prove the following theorem, which is an extension of the familiar formula

$$\int_a^b f'(t)\,dt = f(b) - f(a), \tag{1}$$

the fundamental theorem of calculus.

2.5 Theorem

Let f be a continuously differentiable real-valued function defined in an open set D of $\mathcal{R}^n$. (Thus ∇f is a continuous vector field in D.) If γ is a piecewise smooth curve in D with initial and terminal points $\mathbf{a}$ and $\mathbf{b}$, then

$$\int_\gamma \nabla f \cdot d\mathbf{x} = f(\mathbf{b}) - f(\mathbf{a}).$$

In particular, the value of the line integral of a *gradient* field over a curve depends only on the endpoints of the curve; thus, in this case, the notation

$$\int_{\mathbf{a}}^{\mathbf{b}} \nabla f \cdot d\mathbf{x} = f(\mathbf{b}) - f(\mathbf{a})$$

is justified.

Proof. Suppose γ is parametrized by $G(t)$ with $a \leq t \leq b$, and $G(a) = \mathbf{a}$, $G(b) = \mathbf{b}$. Using first the definition of the line integral, and then Theorem 2.3, we have

$$\begin{aligned}\int_\gamma \nabla f \cdot d\mathbf{x} &= \int_a^b \nabla f\big(G(t)\big) \cdot G'(t)\, dt \\ &= \int_a^b \frac{d}{dt} f\big(G(t)\big)\, dt.\end{aligned}$$

But by Equation (1), the fundamental theorem, the last integral is equal to $f\big(G(b)\big) - f\big(G(a)\big) = f(\mathbf{b}) - f(\mathbf{a})$.

Example 6. Consider the field $\nabla f(x, y)$ in $\mathcal{R}^2$, where $f(x, y) = \frac{1}{2}(x^2 + y^2)$. Then $\nabla f(x, y) = (x, y)$. If γ is any continuously differentiable curve with initial and final endpoints (x_1, y_1), and (x_2, y_2) then

$$\begin{aligned}\int_\gamma x\, dx + y\, dy &= f(x_2, y_2) - f(x_1, y_1) \\ &= \tfrac{1}{2}(x_2^2 - x_1^2) + \tfrac{1}{2}(y_2^2 - y_1^2).\end{aligned}$$

This is what we would expect formally from the fundamental theorem. If, on the other hand, we let $g(x, y) = xy$, we have $\nabla g(x, y) = (y, x)$, and for any curve of the kind previously considered we have

$$\int_\gamma y\, dx + x\, dy = x_2 y_2 - x_1 y_1.$$

EXERCISES

1. Find the gradient, ∇f, of each of the following functions at the indicated points.

(a) $f(x, y, z) = (x - y)e^{xz}$; $(1, 2, -1)$.
(b) $f(x, y) = x^2 - y^2 - \sin y$, for arbitrary (x, y) in $\mathcal{R}^2$.
(c) $f(x, y) = x + y$; $(2, 3)$.
(d) $f(\mathbf{x}) = |\mathbf{x}|^2$, for arbitrary $\mathbf{x}$ in $\mathcal{R}^n$.
(e) $f(\mathbf{x}) = |\mathbf{x}|^\alpha$, for arbitrary nonzero $\mathbf{x}$ in $\mathcal{R}^n$.
(f) $f(x, y) = 0$ identically in $\mathcal{R}^2$; $(1, 1)$.

2. For the functions in the previous problem find the direction and rate of maximum increase at the indicated point.

3. (a) The notation grad f is often used for the gradient ∇f. Show that

$$\frac{\partial f}{\partial \mathbf{y}}(\mathbf{x}) = \text{grad}\, f(\mathbf{x}) \cdot \mathbf{y}.$$

(b) The notation $\nabla_{\mathbf{y}} f$ is often used for the derivative $\partial f/\partial \mathbf{y}$. Show that

$$\nabla_{\mathbf{y}} f(\mathbf{x}) = \nabla f(\mathbf{x}) \cdot \mathbf{y}.$$

4. Find, if possible, a normal vector and the tangent plane to each of the following level curves or surfaces at the indicated points.

(a) $x^2 + y^2 - z^2 = 2$ at $(x, y, z) = (1, 1, 0)$ and at $(x, y, z) = (0, 0, 0)$.
(b) $x \sin y = 0$ at $(x, y) = (0, \pi/2)$ and at $(x, y) = (0, 0)$.
(c) $|\mathbf{x}| = 1$ at $\mathbf{x} = \mathbf{e}_1$, the first natural basis vector in $\mathcal{R}^n$.
(d) $x^2y + yz + w = 3$ at $(x, y, z, w) = (1, 1, 1, 1)$.
(e) $xyz = 1$ at $(x, y, z) = (1, 1, 1)$.
(f) $xyz = 0$ at $(x, y, z) = (1, 2, 0)$.

5. If $\mathcal{R}^2 \xrightarrow{f} \mathcal{R}$ is continuously differentiable, its graph can be defined implicitly in $\mathcal{R}^3$ as the level surface S of the function $F(x, y, z) = z - f(x, y)$ given by $F(x, y, z) = 0$.

(a) Show that $\nabla F = (-\partial f/\partial x, -\partial f/\partial y, 1)$, which is never the zero vector.
(b) Find a normal vector and the tangent plane to the graph of $f(x, y) = xy + ye^x$ at $(x, y) = (1, 1)$.

6. (a) The example $f(x, y) = x^2 + y^2$ has $\nabla f(0, 0) = 0$, which fails to indicate that there is a direction of maximum increase for f at $(x, y) = (0, 0)$. Is this reasonable? What happens at $(0, 0)$?
(b) In Example 5 of Section 2, the point $(x, y) = (0, 0)$ has been avoided. What are the directions of maximum increase for $f(x, y) = xy$ and $g(x, y) = x^2 - y^2$ at $(x, y) = (0, 0)$?

7. If $g(x, y) = e^{x+y}$ and $f'(0) = (1, 2)$, find $F'(0)$, where $F(t) = g(f(t))$ and $f(0) = (1, -1)$.

8. Let γ be a curve in $\mathcal{R}^3$ being traversed at time $t = 1$ with speed 2 and in the direction of $(1, -1, 2)$. If $t = 1$ corresponds to the point $(1, 1, 1)$ on γ find the rate of change of the function $x + y + xy$ along γ at $t = 1$.

9. If $f(x, y, z) = \sin x$ and $F(t) = (\cos t, \sin t, t)$, find $g'(\pi)$, where $g(t) = f(F(t))$.

10. Let $\mathcal{R} \xrightarrow{F} \mathcal{R}^n$ be differentiable. Let $\mathcal{R}^n \xrightarrow{f} \mathcal{R}$ be continuously differentiable and such that the composition $g(t) = f(F(t))$ exists. If $F'(t_0)$ is tangent to a level surface of f at $F(t_0)$, show that $g'(t_0) = 0$.

11. Given a vector field $\mathcal{R}^n \xrightarrow{F} \mathcal{R}^n$, the problem of finding a function $\mathcal{R}^n \xrightarrow{f} \mathcal{R}$ such that $\nabla f = F$ is equivalent to solving the following system of equations for f, where $F_1, \ldots, F_n$ are the coordinate functions of F:

$$\frac{\partial f}{\partial x_1} = F_1, \ldots, \frac{\partial f}{\partial x_n} = F_n.$$

(a) For the case $n = 2$, show that if the system

$$\frac{\partial f}{\partial x}(x, y) = F_1(x, y), \qquad \frac{\partial f}{\partial y}(x, y) = F_2(x, y)$$

has a solution f, then f must have both of the forms

$$f(x, y) = \int F_1(x, y)\, dx + C_1(y)$$

and

$$f(x, y) = \int F_2(x, y)\, dy + C_2(x),$$

where each indefinite integration is performed with the other variable held fixed.

(b) Find f, if $\nabla f(x, y) = (y^2 + 2xy, 2xy + x^2)$, by using part (a).
(c) Find f, if $\nabla f(x, y, z) = (y + z, z + x, x + y)$, by using an appropriate extension of part (a).
(d) Find f, if $\nabla f(x, y, z) = (yz + z, xz, xy + x)$.

12. (a) Find the function f of Problem 11(b) by direct computation of a line integral of ∇f from $(0, 0)$ to (x, y).
(b) Find the function f of Problem 11(c) by direct computation of a line integral from $(0, 0, 0)$ to (x, y, z).

13. (a) Show that if $F_1, \ldots, F_n$ are continuously differentiable, then a necessary condition for the system of equations

$$\frac{\partial f}{\partial x_1} = F_1, \ldots, \frac{\partial f}{\partial x_n} = F_n$$

to have a solution f is that

$$\frac{\partial F_i}{\partial x_j} = \frac{\partial F_j}{\partial x_i}, \qquad i, j = 1, \ldots, n.$$

(b) Show that the functions $\mathcal{R}^2 \xrightarrow{F} \mathcal{R}^2$ and $\mathcal{R}^3 \xrightarrow{G} \mathcal{R}^3$, given by $F(x, y) = (xy, x^2)$ and $G(x, y, z) = (y, x, -zx)$, are not the gradients of real-valued functions.

14. (a) Compute the line integral $\int_\gamma y\, dx + x\, dy$ along an arbitrary continuously differentiable curve from $(0, 0)$ to $(2, 3)$. [*Hint*. Guess a function $\mathcal{R}^2 \xrightarrow{f} \mathcal{R}$ such that $\nabla f(x, y) = (y, x)$.]
(b) Compute the line integral $\int_\gamma x\, dx + y\, dy + z\, dz$ along an arbitrary continuously differentiable curve from $(0, 0, 0)$ to $(1, 2, 3)$ and from $(0, 0, 0)$ to (x, y, z).

15. Let $F(x, y) = \left(\dfrac{-y}{x^2 + y^2}, \dfrac{x}{x^2 + y^2}\right)$ for $(x, y) \neq (0, 0)$.

(a) If F_1 and F_2 are the coordinate functions of F, show that

$$\frac{\partial F_1}{\partial y} = \frac{\partial F_2}{\partial x}.$$

(b) Show that in any region not meeting the y-axis, F is the gradient of $f(x, y) = \arctan(y/x)$, the principal branch.

(c) Show that F is not the gradient of any function f differentiable in a region completely surrounding the origin—for example, the region defined by $x^2 + y^2 > 0$.

16. Show that if a force field in a region D of $\mathcal{R}^3$ has the form ∇f for some continuously differentiable function $\mathcal{R}^3 \xrightarrow{f} \mathcal{R}$, then the work done in moving a particle through the field depends only on the level surfaces of f on which the particle starts and finishes.

17. The level surfaces of a function $\mathcal{R}^3 \xrightarrow{f} \mathcal{R}$ are called the **equipotential surfaces** of the vector field ∇f, and f is a called the **potential function** of the field.

(a) Show that the equipotential surfaces are orthogonal to the field.
(b) Find the equipotential surfaces of the field $\nabla f(x, y, z) = (x, y, z)$.
(c) Find the field of which $f(x, y, z) = (x^2 + y^2 + z^2)^{-1/2}$ (the **Newtonian potential**) is the potential function.
(d) Find the field of which $f(x, y) = -\frac{1}{2} \log (x^2 + y^2)$ (the **logarithmic potential**) is the potential function.
(e) Show that if $f(\mathbf{x}) = |\mathbf{x}|^{2-n}$ (the generalized Newtonian potential in $\mathcal{R}^n$, $n \geq 3$), then $\nabla f(\mathbf{x}) = (2 - n)\, |\mathbf{x}|^{-n}\mathbf{x}$.

18. Let $\mathcal{R} \xrightarrow{F} \mathcal{R}^3$ define a smooth curve γ for $\alpha \leq t \leq \beta$, with γ lying in a region D of $\mathcal{R}^3$ in which $\mathcal{R}^3 \xrightarrow{f} \mathcal{R}$ is continuously differentiable. Show that

$$\frac{d}{dt} \int_{F(a)}^{F(t)} \nabla f \cdot d\mathbf{x} = \nabla f(F(t)) \cdot F'(t).$$

19. Suppose that $\int_{x_0}^{x} F(\mathbf{x}) \cdot d\mathbf{x}$ is independent of the piecewise smooth curve γ joining $\mathbf{x}_0$ to $\mathbf{x}$ for all $\mathbf{x}$ in some open set D in $\mathcal{R}^n$. Show that $\nabla f = F$, where $f(\mathbf{x}) = \int_{x_0}^{x} F \cdot d\mathbf{x}$. [*Hint.* Look at the integral over a curve γ approaching $\mathbf{x}$ in an arbitrary direction, and apply the fundamental theorem of calculus.]

20. If $T(x, y, z)$ represents the temperature at a point (x, y, z) of a region R in $\mathcal{R}^3$, the vector field ∇T is called the **temperature gradient.** Under certain physical assumptions $\nabla T(x, y, z)$ is proportional to the vector that represents the direction and rate per unit of area of heat flow at (x, y, z). The sets on which T is constant are called isotherms. If the isotherms of a temperature function are concentric spheres, prove that the temperature gradient points either toward or away from the center of the spheres.

SECTION 3

THE CHAIN RULE

One of the most useful formulas of one-variable calculus is the chain rule, used to compute the derivative of the composition of one function with another:

$$[g(f(x))]' = g'(f(x))f'(x).$$

The generalization to several variables is just as valuable and, properly formulated, is just as easy to state.

If two functions f and g are so related that the range space of f is the same as the domain space of g, we may form the **composite function** $g \circ f$ by first applying f and then g. Thus,

$$g \circ f(\mathbf{x}) = g(f(\mathbf{x}))$$

for every vector $\mathbf{x}$ such that $\mathbf{x}$ is in the domain of f and $f(\mathbf{x})$ is in the domain of g. The domain of $g \circ f$ consists of those vectors $\mathbf{x}$ that are carried by f into the domain of g. An abstract picture of the composition of two functions is shown in Fig. 6.

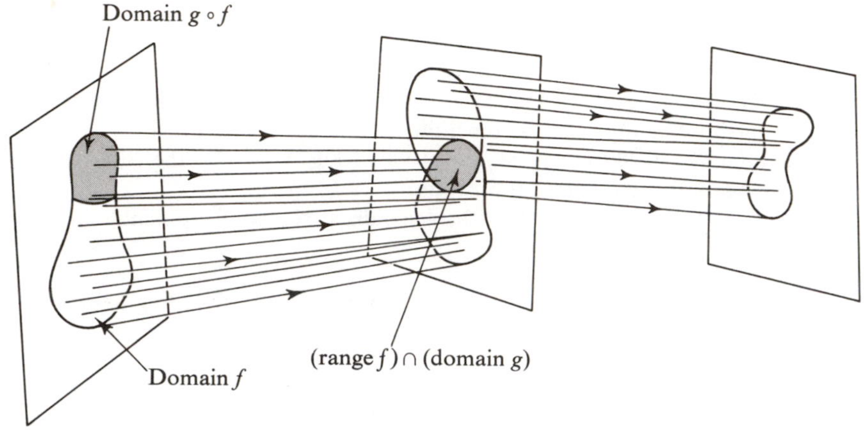

Figure 6

Example 1. Suppose that we are given a 2-dimensional region in which the points move about according to some specified law. It may be known that, for a given initial position with coordinates (u, v), a point is always to be found at some definite later time in a position (x, y). Then (x, y) and (u, v) may be related by equations of the form

$$x = g_1(u, v)$$
$$y = g_2(u, v).$$

In vector notation these equations might be written

$$\mathbf{x} = g(\mathbf{u}),$$

where $\mathbf{x} = (x, y)$, $\mathbf{u} = (u, v)$, and g has coordinate functions g_1, g_2. Now suppose that the initial position $\mathbf{u} = (u, v)$ of a point is itself determined as a function of other variables (s, t) by equations

$$u = f_1(s, t)$$
$$v = f_2(s, t).$$

These may be written in vector form as

$$\mathbf{u} = f(\mathbf{s}),$$

where $\mathbf{s} = (s, t)$ and f has coordinate functions f_1, f_2. Then (x, y) and (s, t) are related by

$$x = g_1(f_1(s, t), f_2(s, t))$$
$$y = g_2(f_1(s, t), f_2(s, t)),$$

or

$$\mathbf{x} = g(f(\mathbf{s})).$$

Using the notation $g \circ f$ for the composition of g and f, we can also write

$$\mathbf{x} = g \circ f(\mathbf{s}).$$

To determine the derivative of $g \circ f$ in terms of the derivatives of g and f, suppose that $\mathcal{R}^n \xrightarrow{f} \mathcal{R}^m$ is differentiable at $\mathbf{x}_0$ and that $\mathcal{R}^m \xrightarrow{g} \mathcal{R}^p$ is differentiable at $\mathbf{y}_0 = f(\mathbf{x}_0)$. Then $g'(\mathbf{y}_0)$ is a p-by-m matrix and $f'(\mathbf{x}_0)$ is an m-by-n matrix. It follows that the product $g'(\mathbf{y}_0)f'(\mathbf{x}_0)$ is defined and is a p-by-n matrix. The chain rule says that this product matrix is the derivative of $g \circ f$ at $\mathbf{x}_0$. Because the matrix product corresponds to composition of linear functions, the result can be stated in terms of differentials: the differential of a composition is a composition of differentials.

Example 2. Consider the special case in which f is a function of a single real variable and g is real-valued. Then $g \circ f$ is a real function of a real variable. Theorem 2.3 shows that if f and g are continuously differentiable, then

$$(g \circ f)'(t) = \nabla g(f(t)) \cdot f'(t).$$

That is, in terms of coordinate functions,

$$(g \circ f)'(t) = \left(\frac{\partial g}{\partial y_1}(f(t)), \ldots, \frac{\partial g}{\partial y_m}(f(t))\right) \cdot (f_1'(t), \ldots, f_m'(t)).$$

The right side of this last equation can be written as a matrix product in terms of the derivative matrices

$$g'(f(t)) = \left(\frac{\partial g}{\partial y_1}(f(t)), \ldots, \frac{\partial g}{\partial y_m}(f(t))\right),$$

and

$$f'(t) = \begin{pmatrix} f_1'(t) \\ \cdot \\ \cdot \\ \cdot \\ f_m'(t) \end{pmatrix}$$

as $(g \circ f)'(t) = g'(f(t))f'(t)$. The product of $g'(f(t))$ and $f'(t)$ is defined by matrix multiplication, in this case 1-by-m times m-by-1, and is equivalent to the dot-product of the two matrices looked at as vectors in $\mathcal{R}^m$. Thus, for the case in which the domain of f and the range of g are both 1-dimensional, the formulas

$$\nabla g(f(t)) \cdot f'(t) \quad \text{and} \quad g'(f(t))f'(t)$$

are practically the same.

The following theorem gives an extension to any dimension for the domain and range of g and f.

3.1 The Chain Rule

Let $\mathcal{R}^n \xrightarrow{f} \mathcal{R}^m$ be continuously differentiable at $\mathbf{x}$ and let

$$\mathcal{R}^m \xrightarrow{g} \mathcal{R}^p$$

be continuously differentiable at $f(\mathbf{x})$. If $g \circ f$ is defined on an open set containing $\mathbf{x}$, then $g \circ f$ is continuously differentiable at $\mathbf{x}$, and

$$(g \circ f)'(\mathbf{x}) = g'(f(\mathbf{x}))f'(\mathbf{x}).$$

Proof. We need only show that the derivative matrix of $g \circ f$ at $\mathbf{x}$ has continuous entries given by the entries in the product of $g'(f(\mathbf{x}))$ and $f'(\mathbf{x})$. These matrices have the form

$$\begin{pmatrix} \dfrac{\partial g_1}{\partial y_1}(f(\mathbf{x})) & \cdots & \dfrac{\partial g_1}{\partial y_m}(f(\mathbf{x})) \\ \cdot & & \cdot \\ \cdot & & \cdot \\ \cdot & & \cdot \\ \dfrac{\partial g_p}{\partial y_1}(f(\mathbf{x})) & \cdots & \dfrac{\partial g_p}{\partial y_m}(f(\mathbf{x})) \end{pmatrix} \text{and} \begin{pmatrix} \dfrac{\partial f_1}{\partial x_1}(\mathbf{x}) & \cdots & \dfrac{\partial f_1}{\partial x_n}(\mathbf{x}) \\ \cdot & & \cdot \\ \cdot & & \cdot \\ \cdot & & \cdot \\ \dfrac{\partial f_m}{\partial x_1}(\mathbf{x}) & \cdots & \dfrac{\partial f_m}{\partial x_n}(\mathbf{x}) \end{pmatrix}.$$

The product of the matrices has for its ijth entry the sum of products

$$\sum_{k=1}^{m} \frac{\partial g_i}{\partial y_k}(f(\mathbf{x})) \frac{\partial f_k}{\partial x_j}(\mathbf{x}). \tag{1}$$

But this expression is just the dot-product of two vectors $\nabla g_i(f(\mathbf{x}))$ and $\partial f/\partial x_j(\mathbf{x})$. It follows from Theorem 2.3 that

$$\nabla g_i(f(\mathbf{x})) \cdot \frac{\partial f}{\partial x_j}(\mathbf{x}) = \frac{\partial (g_i \circ f)}{\partial x_j}(\mathbf{x}), \tag{2}$$

because we are differentiating with respect to the single variable x_j. But this establishes the matrix relation, because the entries in $(g \circ f)'(\mathbf{x})$ are by definition given by the right side of Equation (2). Since g and f are continuously differentiable Formula (1) represents a continuous function of $\mathbf{x}$ for each i and j. Hence $g \circ f$ is continuously differentiable.

In Section 2 of the Appendix we give a version of the chain rule in which we assume only differentiability of f and g, and we conclude only differentiability of $g \circ f$ rather than continuous differentiability. Otherwise the theorem is the same as 3.1.

Example 3. Let $f(x, y) = (x^2 + y^2, x^2 - y^2)$ and let $g(u, v) = (uv, u + v)$. We find

$$g'(u, v) = \begin{pmatrix} v & u \\ 1 & 1 \end{pmatrix} \quad \text{and} \quad f'(x, y) = \begin{pmatrix} 2x & 2y \\ 2x & -2y \end{pmatrix}.$$

To find $(g \circ f)'(2, 1)$, we note that $f(2, 1) = (5, 3)$ and compute

$$g'(5, 3) = \begin{pmatrix} 3 & 5 \\ 1 & 1 \end{pmatrix} \quad \text{and} \quad f'(2, 1) = \begin{pmatrix} 4 & 2 \\ 4 & -2 \end{pmatrix}.$$

Then the product of these last two matrices gives

$$(g \circ f)'(2, 1) = \begin{pmatrix} 32 & -4 \\ 8 & 0 \end{pmatrix}.$$

It is common practice in calculus to denote a function by the same symbol as a typical element of its range. Thus the derivative of a function $\mathcal{R} \xrightarrow{f} \mathcal{R}$ is often denoted, in conjunction with the equation $y = f(x)$, by dy/dx. Similarly, the partial derivatives of a function $\mathcal{R}^3 \xrightarrow{f} \mathcal{R}$ are commonly written as

$$\frac{\partial w}{\partial x}, \quad \frac{\partial w}{\partial y}, \quad \text{and} \quad \frac{\partial w}{\partial z}$$

along with the explanatory equation $w = f(x, y, z)$. For example, if $w = xy^2e^{x+3z}$, then

$$\frac{\partial w}{\partial x} = y^2e^{x+3z} + xy^2e^{x+3z},$$

$$\frac{\partial w}{\partial y} = 2xye^{x+3z},$$

$$\frac{\partial w}{\partial z} = 3xy^2e^{x+3z}.$$

This notation has the disadvantage that it does not contain specific reference to the function being differentiated. On the other hand, it is notationally convenient and is, moreover, the traditional language of calculus. To illustrate its convenience, suppose that the functions g and f are given by

$$w = g(x, y, z), \qquad x = f_1(s, t), \qquad y = f_2(s, t), \qquad z = f_3(s, t).$$

Then, by the chain rule,

$$\begin{pmatrix} \frac{\partial w}{\partial s} & \frac{\partial w}{\partial t} \end{pmatrix} = \begin{pmatrix} \frac{\partial g}{\partial x} & \frac{\partial g}{\partial y} & \frac{\partial g}{\partial z} \end{pmatrix} \begin{pmatrix} \frac{\partial x}{\partial s} & \frac{\partial x}{\partial t} \\ \frac{\partial y}{\partial s} & \frac{\partial y}{\partial t} \\ \frac{\partial z}{\partial s} & \frac{\partial z}{\partial t} \end{pmatrix}.$$

Matrix multiplication yields

$$\left.\begin{aligned} \frac{\partial w}{\partial s} &= \frac{\partial g}{\partial x}\frac{\partial x}{\partial s} + \frac{\partial g}{\partial y}\frac{\partial y}{\partial s} + \frac{\partial g}{\partial z}\frac{\partial z}{\partial s} \\ \frac{\partial w}{\partial t} &= \frac{\partial g}{\partial x}\frac{\partial x}{\partial t} + \frac{\partial g}{\partial y}\frac{\partial y}{\partial t} + \frac{\partial g}{\partial \partial}\frac{\partial z}{\partial t} \end{aligned}\right\}. \tag{3}$$

A slightly different-looking application of the chain rule is obtained if the domain space of f is 1-dimensional, that is, if f is a function of one variable. Consider, for example,

$$w = g(u, v), \qquad \begin{pmatrix} u \\ v \end{pmatrix} = f(t) = \begin{pmatrix} f_1(t) \\ f_2(t) \end{pmatrix}.$$

The composition $g \circ f$ is in this case a real-valued function of one variable. Its differential is defined by the 1-by-1 matrix whose entry is the derivative

$$\frac{d(g \circ f)}{dt} = \frac{dw}{dt}.$$

The derivatives of g and f are defined, respectively, by the Jacobian matrices

$$\begin{pmatrix} \frac{\partial w}{\partial u} & \frac{\partial w}{\partial v} \end{pmatrix} \quad \text{and} \quad \begin{pmatrix} \frac{du}{dt} \\ \frac{dv}{dt} \end{pmatrix}.$$

Hence, the chain rule implies that

$$\frac{dw}{dt} = \begin{pmatrix} \frac{\partial w}{\partial u} & \frac{\partial w}{\partial v} \end{pmatrix} \begin{pmatrix} \frac{du}{dt} \\ \frac{dv}{dt} \end{pmatrix} = \frac{\partial w}{\partial u}\frac{du}{dt} + \frac{\partial w}{\partial v}\frac{dv}{dt}. \tag{4}$$

Finally, let us suppose that both f and g are real-valued functions of one variable. This is the situation encountered in one-variable calculus. The derivatives of f at t, of g at $s = f(t)$, and of $g \circ f$ at t are represented by the three 1-by-1 Jacobian matrices $f'(t)$, $g'(s)$, and $(g \circ f)'(t)$, respectively. The chain rule implies that

3.2 $$(g \circ f)'(t) = g'(s)f'(t).$$

If the functions are presented in the form

$$x = g(s), \qquad s = f(t),$$

the more explicit Formula 3.2 can be written as the famous equation

$$\frac{dx}{dt} = \frac{dx}{ds}\frac{ds}{dt}. \tag{5}$$

Example 4. Given that

$$\begin{cases} x = u^2 + v^3 \\ y = e^{uv}, \end{cases} \qquad \text{and} \qquad \begin{cases} u = t + 1 \\ v = e^t, \end{cases}$$

find dx/dt and dy/dt at $t = 0$. Let $\mathcal{R} \xrightarrow{f} \mathcal{R}^2$ and $\mathcal{R}^2 \xrightarrow{g} \mathcal{R}^2$ be the functions defined by

$$f(t) = \begin{pmatrix} t+1 \\ e^t \end{pmatrix} = \begin{pmatrix} u \\ v \end{pmatrix}, \qquad -\infty < t < \infty,$$

$$g\begin{pmatrix} u \\ v \end{pmatrix} = \begin{pmatrix} u^2 + v^3 \\ e^{uv} \end{pmatrix} = \begin{pmatrix} x \\ y \end{pmatrix}, \qquad \begin{cases} -\infty < u < \infty. \\ -\infty < v < \infty. \end{cases}$$

The differential of f at t is defined by the 2-by-1 Jacobian matrix

$$\begin{pmatrix} \frac{du}{dt} \\ \frac{dv}{dt} \end{pmatrix} = \begin{pmatrix} 1 \\ e^t \end{pmatrix}.$$

The matrix of the differential of g at $\begin{pmatrix} u \\ v \end{pmatrix}$ is

$$\begin{pmatrix} \frac{\partial x}{\partial u} & \frac{\partial x}{\partial v} \\ \frac{\partial y}{\partial u} & \frac{\partial y}{\partial v} \end{pmatrix} = \begin{pmatrix} 2u & 3v^2 \\ ve^{uv} & ue^{uv} \end{pmatrix}.$$

The dependence of x and y on t is given by

$$\begin{pmatrix} x \\ y \end{pmatrix} = (g \circ f)(t), \qquad -\infty < t < \infty.$$

Hence, the two derivatives dx/dt and dy/dt are the entries in the Jacobian matrix that defines the differential of the composite function $g \circ f$. The chain rule therefore implies that

$$\begin{pmatrix} \frac{dx}{dt} \\ \frac{dy}{dt} \end{pmatrix} = \begin{pmatrix} \frac{\partial x}{\partial u} & \frac{\partial x}{\partial v} \\ \frac{\partial y}{\partial u} & \frac{\partial y}{\partial v} \end{pmatrix} \begin{pmatrix} \frac{du}{dt} \\ \frac{dv}{dt} \end{pmatrix}.$$

That is,

$$\left.\begin{aligned} \frac{dx}{dt} &= \frac{\partial x}{\partial u}\frac{du}{dt} + \frac{\partial x}{\partial v}\frac{dv}{dt} = 2u + 3v^2e^t \\ \frac{dy}{dt} &= \frac{\partial y}{\partial u}\frac{du}{dt} + \frac{\partial y}{\partial v}\frac{dv}{dt} = ve^{uv} + ue^{uv+t} \end{aligned}\right\}. \tag{6}$$

If $t = 0$, then $\begin{pmatrix} u \\ v \end{pmatrix} = f(0) = \begin{pmatrix} 1 \\ 1 \end{pmatrix}$, and we get $u = v = 1$. It follows that

$$\frac{dx}{dt}(0) = 2 + 3 = 5,$$

$$\frac{dy}{dt}(0) = e + e + 2e.$$

The definition of matrix multiplication gives the derivative formulas that result from applications of the chain rule a formal pattern that is easy to memorize. The pattern is particularly in evidence when the coordinate functions are denoted by real variables as in Formulas (3), (4), (5), and (6). All formulas of the general form

$$\cdots + \frac{\partial z}{\partial x}\frac{\partial x}{\partial t} + \frac{\partial z}{\partial y}\frac{\partial y}{\partial t} + \cdots$$

have the disadvantage, however, of not containing explicit reference to the points at which the various derivatives are evaluated. It is, of course, essential to know this information. It can be found by going to the formula

$$(g \circ f)'(\mathbf{x}) = g'(f(\mathbf{x}))f'(\mathbf{x}).$$

It follows that derivatives appearing in the matrix $f'(\mathbf{x})$ are evaluated at $\mathbf{x}$, and those in the matrix $g'(f(\mathbf{x}))$ are evaluated at $f(\mathbf{x})$. This is the reason for setting $t = 0$ and $u = v = 1$ in Formula 6 to obtain the final answers in Example 4.

Example 5. Let

$$z = xy \quad \text{and} \quad \begin{cases} x = f(u, v). \\ y = g(u, v). \end{cases}$$

Suppose that when $u = 1$ and $v = 2$, we have

$$\frac{\partial x}{\partial u} = -1, \qquad \frac{\partial x}{\partial v} = 3, \qquad \frac{\partial y}{\partial u} = 5, \qquad \frac{\partial y}{\partial v} = 0.$$

Suppose also that $f(1, 2) = 2$ and $g(1, 2) = -2$. What is $\partial z/\partial u(1, 2)$? The chain rule implies that

$$\frac{\partial z}{\partial u} = \frac{\partial z}{\partial x}\frac{\partial x}{\partial u} + \frac{\partial z}{\partial y}\frac{\partial y}{\partial u}. \tag{7}$$

When $u = 1$ and $v = 2$, we are given that $x = f(1, 2) = 2$ and $y = g(1, 2) = -2$. Hence

$$\frac{\partial z}{\partial x}(2, -2) = y|_{x=2,y=-2} = -2$$

$$\frac{\partial z}{\partial y}(2, -2) = x|_{x=2,y=-2} = 2.$$

To obtain $\partial z/\partial u$ at $(u, v) = (1, 2)$, it is necessary to know at what points to evaluate the partial derivatives that appear in Equation (7). In greater detail, the chain rule implies that

$$\frac{\partial z}{\partial u}(1, 2) = \frac{\partial z}{\partial x}(2, -2)\frac{\partial x}{\partial u}(1, 2) + \frac{\partial z}{\partial y}(2, -2)\frac{\partial y}{\partial u}(1, 2).$$

Hence

$$\frac{\partial z}{\partial u}(1, 2) = (-2)(-1) + (2)(5) = 12.$$

Example 6. If $w = f(ax^2 + bxy + cy^2)$ and $y = x^2 + x + 1$, find $dw/dx(-1)$. The solution relies on formulas that follow from the chain rule [like (3), (4), (5), and (7)]. Set

$$z = ax^2 + bxy + cy^2.$$

Then, $w = f(z)$ and

$$\frac{dz}{dx} = \frac{\partial z}{\partial x} + \frac{\partial z}{\partial y}\frac{dy}{dx}.$$

Hence,

$$\begin{aligned}\frac{dw}{dx} &= \frac{df}{dz}\frac{dz}{dx} = \frac{df}{dz}\left(\frac{\partial z}{\partial x} + \frac{\partial z}{\partial y}\frac{dy}{dx}\right)\\ &= f'(z)\big(2ax + by + (bx + 2cy)(2x + 1)\big).\end{aligned}$$

If $x = -1$, then $y = 1$, and so $z = a - b + c$. Thus,

$$\frac{dw}{dx}(-1) = f'(a - b + c)(-2a + 2b - 2c).$$

The Jacobian matrix, or derivative, of a function f from $\mathcal{R}^n$ to $\mathcal{R}^n$ is a square matrix and so has a determinant. This determinant, $\det f'(\mathbf{x})$, is a real-valued function of $\mathbf{x}$ called the Jacobian determinant of f; it plays a particularly important role in the change-of-variable theorem for integrals taken up in Chapter 6. At this point we remark on a simple corollary of the chain rule and the product rule for determinants:

3.3 Theorem

If $\mathcal{R}^n \xrightarrow{f} \mathcal{R}^n$ is differentiable at $\mathbf{x}_0$ and $\mathcal{R}^n \xrightarrow{g} \mathcal{R}^n$ is differentiable at $\mathbf{y}_0 = f(\mathbf{x}_0)$, then the Jacobian determinant of $g \circ f$ at $\mathbf{x}_0$ is the product of the Jacobian determinant of f at $\mathbf{x}_0$ and that of g at $\mathbf{y}_0$.

If f is defined by

$$f\begin{pmatrix} x_1 \\ \cdot \\ \cdot \\ \cdot \\ x_n \end{pmatrix} = \begin{pmatrix} f_1(x_1, \dots, x_n) \\ \cdot \\ \cdot \\ \cdot \\ f_n(x_1, \dots, \;\; {}_n) \end{pmatrix} = \begin{pmatrix} y_1 \\ \cdot \\ \cdot \\ \cdot \\ y_n \end{pmatrix},$$

then the Jacobian determinant $\det f'$ is often denoted by

$$\frac{\partial(f_1, \dots, f_n)}{\partial(x_1, \dots, x_n)},$$

or equivalently

$$\frac{\partial(y_1, \dots, y_n)}{\partial(x_1, \dots, x_n)}.$$

Example 7. Let

$$f\begin{pmatrix} r \\ \theta \end{pmatrix} = \begin{pmatrix} r\cos\theta \\ r\sin\theta \end{pmatrix} = \begin{pmatrix} x \\ y \end{pmatrix} \quad \text{and} \quad g\begin{pmatrix} x \\ y \end{pmatrix} = \begin{pmatrix} x^2 - y^2 \\ 2xy \end{pmatrix} = \begin{pmatrix} w \\ z \end{pmatrix}.$$

Then,

$$\frac{\partial(x, y)}{\partial(r, \theta)} = \det\begin{pmatrix} \cos\theta & -r\sin\theta \\ \sin\theta & r\cos\theta \end{pmatrix} = r(\cos^2\theta + \sin^2\theta) = r,$$

and

$$\frac{\partial(w, z)}{\partial(x, y)} = \det\begin{pmatrix} 2x & -2y \\ 2y & 2x \end{pmatrix} = 4(x^2 + y^2).$$

The Jacobian determinant of the composite function $g \circ f$ is denoted in this case by $\partial(w, z)/\partial(r, \theta)$. If

$$\begin{pmatrix} x_0 \\ y_0 \end{pmatrix} = \begin{pmatrix} r_0\cos\theta_0 \\ r_0\sin\theta_0 \end{pmatrix},$$

Theorem 3.3 implies that

$$\begin{aligned} \frac{\partial(w, z)}{\partial(r, \theta)}(r_0, \theta_0) &= \frac{\partial(w, z)}{\partial(x, y)}(x_0, y_0)\frac{\partial(x, y)}{\partial(r, \theta)}(r_0, \theta_0) \\ &= 4(x_0^2 + y_0^2)r_0 = 4r_0^3. \end{aligned}$$

EXERCISES

1. Given that

$$f\begin{pmatrix} x \\ y \end{pmatrix} = \begin{pmatrix} x^2 + xy + 1 \\ y^2 + 2 \end{pmatrix}, \quad g\begin{pmatrix} u \\ v \end{pmatrix} = \begin{pmatrix} u + v \\ 2u \\ v^2 \end{pmatrix},$$

find the matrix of the differential of the composite function $g \circ f$ at

$$\mathbf{x}_0 = \begin{pmatrix} 1 \\ 1 \end{pmatrix}. \qquad \left[Ans. \begin{pmatrix} 3 & 3 \\ 6 & 2 \\ 0 & 12 \end{pmatrix}. \right]$$

2. Let

$$f(t) = \begin{pmatrix} t \\ t + 1 \\ t^2 \end{pmatrix} = \begin{pmatrix} x \\ y \\ z \end{pmatrix}$$

and

$$g\begin{pmatrix} x \\ y \\ z \end{pmatrix} = \begin{pmatrix} x + 2y + z^2 \\ x^2 - y \end{pmatrix} = \begin{pmatrix} u \\ v \end{pmatrix}.$$

(a) Find the Jacobian matrix of $g \circ f$ at $t = a$. $\left[\textit{Ans. } \begin{pmatrix} 3 + 4a^3 \\ 2a - 1 \end{pmatrix}.\right]$

(b) Find du/dt in terms of the derivatives of x, y, z, and the partial derivatives of u.

3. Consider the curve defined parametrically by

$$f(t) = \begin{pmatrix} t \\ t^2 - 4 \\ e^{t-2} \end{pmatrix}, \qquad -\infty < t < \infty.$$

Let g be a real-valued differentiable function with domain $\mathcal{R}^3$. If

$$\mathbf{x}_0 = \begin{pmatrix} 2 \\ 0 \\ 1 \end{pmatrix}$$

and

$$\frac{\partial g}{\partial x}(\mathbf{x}_0) = 4, \qquad \frac{\partial g}{\partial y}(\mathbf{x}_0) = 2, \qquad \frac{\partial g}{\partial z}(\mathbf{x}_0) = 2,$$

find $d(g \circ f)/dt$ at $t = 2$. [*Ans.* 14.]

4. Consider the functions

$$f\begin{pmatrix} u \\ v \end{pmatrix} = \begin{pmatrix} u + v \\ u - v \\ u^2 - v^2 \end{pmatrix} = \begin{pmatrix} x \\ y \\ z \end{pmatrix}$$

and

$$F(x, y, z) = x^2 + y^2 + z^2 = w.$$

(a) Find the matrix that defines the differential of $F \circ f$ at $\begin{pmatrix} a \\ b \end{pmatrix}$.

(b) Find $\partial w/\partial u$ and $\partial w/\partial v$.

5. Let $u = f(x, y)$. Make the change of variables $x = r\cos\theta$, $y = r\sin\theta$. Given that

$$\frac{\partial f}{\partial x} = x^2 + 2xy - y^2 \quad \text{and} \quad \frac{\partial f}{\partial y} = x^2 - 2xy + 2,$$

find $\partial f/\partial\theta$, when $r = 2$ and $\theta = \pi/2$. [*Ans.* 8.]

6. If $w = \sqrt{x^2 + y^2 + z^2}$ and

$$\begin{pmatrix} x \\ y \\ z \end{pmatrix} = \begin{pmatrix} r \cos \theta \\ r \sin \theta \\ r \end{pmatrix},$$

find $\partial w/\partial r$ and $\partial w/\partial \theta$ using the chain rule. Check the result by direct substitution.

7. Vector functions f and g are defined by

$$f\begin{pmatrix} u \\ v \end{pmatrix} = \begin{pmatrix} u \cos v \\ u \sin v \end{pmatrix}, \qquad \begin{cases} 0 < u < \infty, \\ -\pi/2 < v < \pi/2, \end{cases}$$

$$g\begin{pmatrix} x \\ y \end{pmatrix} = \begin{pmatrix} \sqrt{x^2 + y^2} \\ \arctan \dfrac{y}{x} \end{pmatrix}, \qquad 0 < x < \infty.$$

(a) Find the Jacobian matrix of $g \circ f$ at $\begin{pmatrix} u \\ v \end{pmatrix}$.

(b) Find the Jacobian matrix of $f \circ g$ at $\begin{pmatrix} x \\ y \end{pmatrix}$.

(c) Are the following statements true or false?
1. domain of f = domain of $g \circ f$.
2. domain of g = domain of $f \circ g$.

8. A function I is called an **identity function** if $I(\mathbf{x}) = \mathbf{x}$ for all $\mathbf{x}$ in the domain of I.

(a) Show that if differentiable vector functions f and g are so related that the composite function $g \circ f$ is an identity function, then the transformation $(d_{f(\mathbf{x})} g) \circ (d_{\mathbf{x}} f)$ is also an identity function for $\mathbf{x}$ in the domain of $g \circ f$.
(b) How does this exercise apply to the preceding one?

9. Let $\mathbf{x}_1$ be a tangent vector at $\mathbf{x}_0$ to a curve defined parametrically by a differentiable vector function f. If $\mathbf{x}_0$ is in the domain of a differentiable vector function F, prove that $F'(\mathbf{x}_0)\mathbf{x}_1$, if not zero, is a tangent vector at $F(\mathbf{x}_0)$ to the curve defined parametrically by $F \circ f$.

10. The convention of denoting coordinate functions by real variables has its pitfalls. Resolve the following paradox: Let $w = f(x, y, z)$ and $z = g(x, y)$. By the chain rule

$$\frac{\partial w}{\partial x} = \frac{\partial w}{\partial x}\frac{\partial x}{\partial x} + \frac{\partial w}{\partial y}\frac{\partial y}{\partial x} + \frac{\partial w}{\partial z}\frac{\partial z}{\partial x}.$$

The quantities x and y are unrelated, so that $\partial y/\partial x = 0$. Clearly $\partial x/\partial x = 1$.

Hence,

$$\frac{\partial w}{\partial x} = \frac{\partial w}{\partial x} + \frac{\partial w}{\partial z}\frac{\partial z}{\partial x},$$

and so

$$0 = \frac{\partial w}{\partial z}\frac{\partial z}{\partial x}.$$

In particular, take $w = 2x + y + 3z$ and $z = 5x + 18$. Then

$$\frac{\partial w}{\partial z} = 3 \quad \text{and} \quad \frac{\partial z}{\partial x} = 5.$$

It follows that $0 = 15$, which of course is false.

11. If $y = f(x - at) + g(x + at)$, where a is constant and f and g are twice differentiable, show that

$$a^2 \frac{\partial^2 y}{\partial x^2} = \frac{\partial^2 y}{\partial t^2}. \qquad \text{(Wave equation)}$$

12. If $z = f(x, y)$ is differentiable and $\begin{pmatrix} x \\ y \end{pmatrix} = \begin{pmatrix} r\cos\theta \\ r\sin\theta \end{pmatrix}$, show that

$$\left(\frac{\partial z}{\partial x}\right)^2 + \left(\frac{\partial z}{\partial y}\right)^2 = \left(\frac{\partial z}{\partial r}\right)^2 + \frac{1}{r^2}\left(\frac{\partial z}{\partial \theta}\right)^2.$$

13. If $f(tx, ty) = t^n f(x, y)$ for some integer n, and for all x, y, and t, show that

$$x\frac{\partial f}{\partial x} + y\frac{\partial f}{\partial y} = nf(x, y).$$

14. Show that for a differentiable real-valued function $g(x, y)$,

$$\frac{dg(x, x)}{dx} = \frac{\partial g}{\partial x}(x, x) + \frac{\partial g}{\partial y}(x, x).$$

Using the function $f(x) = (x, x)$ show that this result is equivalent to the statement $(g \circ f)'(x) = g'(f(x))f'(x)$.
Apply the equation to the function $g(x, y) = x^y$.

15. (a) If

$$w = f(x, y, z, t), \qquad x = g(u, z, t), \qquad \text{and} \quad z = h(u, t),$$

write a formula for dw/dt, where by this symbol is meant the rate of change of w with respect to t, and where all the interrelations of w, x, z, t are taken into account.

(b) If

$$f(x, y, z, t) = 2xy + 3z + t^2,$$

$$g(u, z, t) = ut\sin z,$$

$$h(u, t) = 2u + t,$$

evaluate the above dw/dt at the point $u = 1, t = 2, y = 3$, by using the formula you derived in part (a) and also by substituting in the functions for x and z and then differentiating.

16. Consider a real-valued function $f(x, y)$ such that

$$f_x(2, 1) = 3, \qquad f_y(2, 1) = -2, \qquad f_{xx}(2, 1) = 0,$$
$$f_{xy}(2, 1) = f_{yx}(2, 1) = 1, \qquad f_{yy}(2, 1) = 2.$$

Let $\mathcal{R}^2 \xrightarrow{g} \mathcal{R}^2$ be defined by

$$g(u, v) = (u + v, uv).$$

Find $\partial^2(f \circ g)/\partial v\, \partial u$ at $(1, 1)$. [*Ans.* 2.]

17. Calculate the Jacobian determinants of the following functions at the points indicated.

(a) $f\begin{pmatrix} u \\ v \end{pmatrix} = \begin{pmatrix} u^2 + 2uv + 3v \\ u - v \end{pmatrix} = \begin{pmatrix} x \\ y \end{pmatrix}$, at $\mathbf{u}_0 = \begin{pmatrix} 0 \\ 2 \end{pmatrix}$. [*Ans.* -7.]

(b) $g\begin{pmatrix} x \\ y \end{pmatrix} = \begin{pmatrix} x^2 - y^2 \\ 2xy \end{pmatrix} = \begin{pmatrix} z \\ w \end{pmatrix}$, at $\mathbf{x}_0 = \begin{pmatrix} 6 \\ -2 \end{pmatrix}$. [*Ans.* 160.]

(c) $A\begin{pmatrix} x \\ y \end{pmatrix} = \begin{pmatrix} a & b \\ c & d \end{pmatrix}\begin{pmatrix} x \\ y \end{pmatrix}$, at an arbitrary $\begin{pmatrix} x \\ y \end{pmatrix}$.

(d) An affine transformation $\mathcal{R}^n \xrightarrow{A} \mathcal{R}^n$, $A(\mathbf{x}) = L(\mathbf{x}) + \mathbf{y}_0$, at an arbitrary $\mathbf{x}_0$.

(e) $T\begin{pmatrix} r \\ \phi \\ \theta \end{pmatrix} = \begin{pmatrix} r\cos\theta\sin\phi \\ r\sin\theta\sin\phi \\ r\cos\phi \end{pmatrix}$, at $\begin{pmatrix} r \\ \phi \\ \theta \end{pmatrix}$.

18. Using the functions f and g in Exercises 17(a) and (b), compute the Jacobian determinant of the composite function $g \circ f$ at $\begin{pmatrix} 0 \\ 2 \end{pmatrix}$.

[*Ans.* -1120.]

19. In terms of functions $\mathcal{R}^2 \xrightarrow{f} \mathcal{R}^2$ and $\mathcal{R}^2 \xrightarrow{g} \mathcal{R}^2$, what do the following equations say and how do they follow from Theorem 3.3?

(a) $\dfrac{\partial(x, y)}{\partial(u, v)}\dfrac{\partial(u, v)}{\partial(r, \theta)} = \dfrac{\partial(x, y)}{\partial(r, \theta)}$. (*b*) $\dfrac{\partial(x, y)}{\partial(u, v)}\dfrac{\partial(u, v)}{\partial(x, y)} = 1$.

SECTION 4

IMPLICITLY DEFINED FUNCTIONS

It may happen that two vectors are related by a formula that doesn't express either one directly as a function of the other. For example the formula

$$\frac{pv}{t} = k_0$$

may express the relationship between pressure p on the one hand, and volume and temperature (v, t) on the other hand, of the gas in some container. Or the equations

$$x^2 + y^2 + z^2 = 1$$
$$x + y + z = 0$$

may be interpreted as a relation between the three coordinates of a point on a certain sphere of radius 1 centered at $(0, 0, 0)$ in $\mathcal{R}^3$. In neither example do the equations give an explicit formula for any of the coordinates in terms of others. In this section we study the application of calculus to such relations.

For any two functions $\mathcal{R}^2 \xrightarrow{F} \mathcal{R}$ and $\mathcal{R} \xrightarrow{f} \mathcal{R}$, **the equation**

$$F(x, y) = 0 \tag{1}$$

defines f implicitly if $F(x, f(x)) = 0$ for every x in the domain of f. The zero on the right side of Equation (1) can be replaced by any constant c. But since $F(x, y) = c$ is equivalent to $G(x, y) = F(x, y) - c = 0$, it is customary to absorb such a constant into the function F.

Example 1. Let $F(x, y) = x^2 + y^2 - 1$. Then the condition that $F(x, f(x)) = x^2 + (f(x))^2 - 1 = 0$, for every x in the domain of f, is satisfied by each of the following choices for f.

$$f_1(x) = \sqrt{1 - x^2}, \qquad -1 \leq x \leq 1.$$
$$f_2(x) = -\sqrt{1 - x^2}, \qquad -1 \leq x \leq 1.$$
$$f_3(x) = \begin{cases} \sqrt{1 - x^2}, & -\frac{1}{2} \leq x < 0. \\ -\sqrt{1 - x^2}, & 0 \leq x \leq 1. \end{cases}$$

Their graphs are shown in Fig. 7. It follows from the definition of an implicitly defined function that all three functions f_1, f_2, f_3 are defined implicitly by the equation $x^2 + y^2 - 1 = 0$.

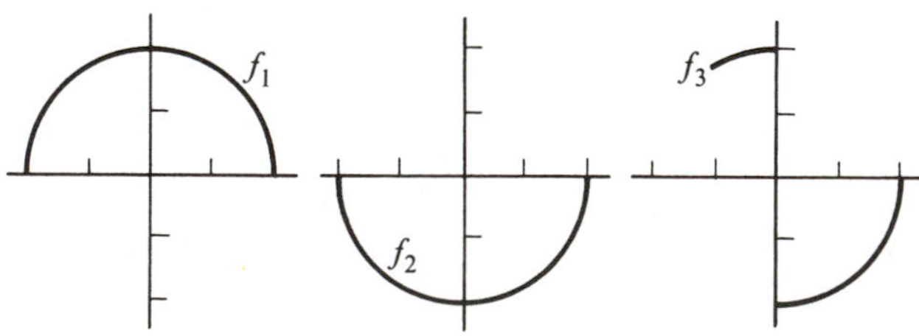

Figure 7

Consider a function $\mathcal{R}^{n+m} \xrightarrow{F} \mathcal{R}^m$. An arbitrary element in $\mathcal{R}^{n+m}$ can be written as $(x_1, \ldots, x_n, y_1, \ldots, y_m)$ or as a pair $(\mathbf{x}, \mathbf{y})$, where $\mathbf{x} = (x_1, \ldots, x_n)$ and $\mathbf{y} = (y_1, \ldots, y_m)$. In this way F can be thought of either as a function of the two vector variables, $\mathbf{x}$ in $\mathcal{R}^n$ and $\mathbf{y}$ in $\mathcal{R}^m$, or else as a function of the single vector variable $(\mathbf{x}, \mathbf{y})$ in $\mathcal{R}^{n+m}$. **The function $\mathcal{R}^n \xrightarrow{f} \mathcal{R}^m$ is defined implicitly by the equation**

$$F(\mathbf{x}, \mathbf{y}) = 0$$

if $F(\mathbf{x}, f(x)) = 0$ for every $\mathbf{x}$ in the domain of f.

Example 2. The equations

$$\begin{aligned} x + y + z - 1 &= 0 \\ 2x \qquad + z + 2 &= 0 \end{aligned} \tag{2}$$

determine y and z as functions of x. We get

$$y = x + 3, \qquad z = -2x - 2.$$

In terms of a function $\mathcal{R}^3 \xrightarrow{F} \mathcal{R}^2$, Equations (2) can be written

$$\begin{aligned} F\left(x, \begin{pmatrix} y \\ z \end{pmatrix}\right) &= \begin{pmatrix} x + y + z - 1 \\ 2x \qquad + z + 2 \end{pmatrix} = \begin{pmatrix} 0 \\ 0 \end{pmatrix} \\ &= \begin{pmatrix} 1 \\ 2 \end{pmatrix} x + \begin{pmatrix} 1 & 1 \\ 0 & 1 \end{pmatrix} \begin{pmatrix} y \\ z \end{pmatrix} + \begin{pmatrix} -1 \\ 2 \end{pmatrix} = \begin{pmatrix} 0 \\ 0 \end{pmatrix}. \end{aligned}$$

The implicitly defined function $\mathcal{R} \xrightarrow{f} \mathcal{R}^2$ is

$$f(x) = \begin{pmatrix} y \\ z \end{pmatrix} = \begin{pmatrix} x + 3 \\ -2x - 2 \end{pmatrix}.$$

Although Example 1 shows that an implicitly defined function need not be continuous, we shall be primarily concerned in this section with functions that are not only continuous but also differentiable. The *implicit function theorem* appearing at the end of Section 6 gives conditions for the existence of a differentiable f defined by an equation $F(\mathbf{x}, f(\mathbf{x})) = 0$. Before discussing this theorem, however, we consider the problem of finding the differential of f when it is known to exist. Suppose the functions $\mathcal{R}^2 \xrightarrow{F} \mathcal{R}$ and $\mathcal{R} \xrightarrow{f} \mathcal{R}$ are differentiable and that

$$F(x, f(x)) = 0$$

for every x in the domain of f. Then the chain rule applied to $F(x, f(x))$ yields

$$F_x(x, f(x)) + F_y(x, f(x))f'(x) = 0.$$

Hence,

4.1 $$f'(x) = -\frac{F_x(x, f(x))}{F_y(x, f(x))}, \qquad \text{if} \quad F_y(x, f(x)) \neq 0.$$

For vector-valued functions a similar computation is possible.

Example 3. Given the equations

$$x^2 + y^2 + z^2 - 5 = 0, \qquad xyz + 2 = 0, \tag{3}$$

suppose that x and y are differentiable functions of z, that is, the function defined implicitly by Equations (3) is of the form $(x, y) = f(z)$. To compute dx/dz and dy/dz, we apply the chain rule to the given equations to get

$$2x\frac{dx}{dz} + 2y\frac{dy}{dz} + 2z = 0,$$

$$yz\frac{dx}{dz} + xz\frac{dy}{dz} + xy = 0.$$

These new equations can be solved for dx/dz and dy/dz. The solution is

$$\begin{pmatrix} \dfrac{dx}{dz} \\ \\ \dfrac{dy}{dz} \end{pmatrix} = \begin{pmatrix} \dfrac{x(y^2 - z^2)}{z(x^2 - y^2)} \\ \\ \dfrac{y(z^2 - x^2)}{z(x^2 - y^2)} \end{pmatrix},$$

which is the matrix $f'(z)$. Notice that the corresponding values for x and y have to be known to make the formula completely explicit. That is, from the information given so far, there is no possible way of evaluating $(dx/dz)(1)$. On the other hand, given the point $(x, y, z) = (1, -2, 1)$, we have $(dx/dz)(1) = -1$. The reason is that, just as in Example 1, there is more than one function f defined implicitly by Equations (3). By specifying a particular point on its graph, we determine f uniquely in the vicinity of the point.

Example 4. Consider

$$xu + yv + zw = 1,$$

$$x + y + z + u + v + w = 0,$$

$$xy + zuv + w = 1.$$

Suppose that each of x, y, and z is a function of u, v, and w. To find the partial derivatives of x, y, and z with respect to w, we differentiate the three equations using the chain rule.

$$u\frac{\partial x}{\partial w} + v\frac{\partial y}{\partial w} + w\frac{\partial z}{\partial w} + z = 0,$$

$$\frac{\partial x}{\partial w} + \frac{\partial y}{\partial w} + \frac{\partial z}{\partial w} + 1 = 0,$$

$$y\frac{\partial x}{\partial w} + x\frac{\partial y}{\partial w} + uv\frac{\partial z}{\partial w} + 1 = 0.$$

Then, solving for $\partial x/\partial w$ gives

$$\frac{\partial x}{\partial w} = \frac{uv^2 + xz + w - zuv - xw - v}{u^2v + vy + wx - yw - ux - uv^2}.$$

Similarly, we could solve for $\partial y/\partial w$ and $\partial z/\partial w$. To find partials with respect to u, differentiate the original equations with respect to u and solve for $\partial x/\partial u$, $\partial y/\partial u$, and $\partial z/\partial u$. Partials with respect to v are found by the same method.

The computation indicated in Example 4 leads to the nine entries in the matrix of the differential of an implicitly defined vector function. In order for the computation to work, it is necessary to have the number of given equations equal the number of implicitly defined coordinate functions. To get some insight into the reason for this requirement, suppose we are given a differentiable vector function

$$F(u, v, x, y) = \begin{pmatrix} F_1(u, v, x, y) \\ F_2(u, v, x, y) \end{pmatrix}$$

and that the equations

$$F_1(u, v, x, y) = 0, \qquad F_2(u, v, x, y) = 0 \tag{4}$$

implicitly define a differentiable function $(x, y) = f(u, v)$. Differentiating Equations (4) with respect to u and v by means of the chain rule, we get

$$\frac{\partial F_1}{\partial u} + \frac{\partial F_1}{\partial x}\frac{\partial x}{\partial u} + \frac{\partial F_1}{\partial y}\frac{\partial y}{\partial u} = 0, \qquad \frac{\partial F_1}{\partial v} + \frac{\partial F_1}{\partial x}\frac{\partial x}{\partial v} + \frac{\partial F_1}{\partial y}\frac{\partial y}{\partial v} = 0,$$

$$\frac{\partial F_2}{\partial u} + \frac{\partial F_2}{\partial x}\frac{\partial x}{\partial u} + \frac{\partial F_2}{\partial y}\frac{\partial y}{\partial u} = 0, \qquad \frac{\partial F_2}{\partial v} + \frac{\partial F_2}{\partial x}\frac{\partial x}{\partial v} + \frac{\partial F_2}{\partial y}\frac{\partial y}{\partial v} = 0.$$

These equations can be written in matrix form as follows:

$$\begin{pmatrix} \dfrac{\partial F_1}{\partial u} & \dfrac{\partial F_1}{\partial v} \\ \dfrac{\partial F_2}{\partial u} & \dfrac{\partial F_2}{\partial v} \end{pmatrix} + \begin{pmatrix} \dfrac{\partial F_1}{\partial x} & \dfrac{\partial F_1}{\partial y} \\ \dfrac{\partial F_2}{\partial x} & \dfrac{\partial F_2}{\partial y} \end{pmatrix} \begin{pmatrix} \dfrac{\partial x}{\partial u} & \dfrac{\partial x}{\partial v} \\ \dfrac{\partial y}{\partial u} & \dfrac{\partial y}{\partial v} \end{pmatrix} = 0. \tag{5}$$

The last matrix on the right is the matrix of the differential of f at (u, v). Solving for it, we get

$$\begin{pmatrix} \dfrac{\partial x}{\partial u} & \dfrac{\partial x}{\partial v} \\ \dfrac{\partial y}{\partial u} & \dfrac{\partial y}{\partial v} \end{pmatrix} = - \begin{pmatrix} \dfrac{\partial F_1}{\partial x} & \dfrac{\partial F_1}{\partial y} \\ \dfrac{\partial F_2}{\partial x} & \dfrac{\partial F_2}{\partial y} \end{pmatrix}^{-1} \begin{pmatrix} \dfrac{\partial F_1}{\partial u} & \dfrac{\partial F_1}{\partial v} \\ \dfrac{\partial F_2}{\partial u} & \dfrac{\partial F_2}{\partial v} \end{pmatrix}. \tag{6}$$

To be able to solve Equation (5) uniquely for the matrix $f'(u, v)$, it is essential that the inverse matrix that appears in Equation (6) exist. This implies, in particular, that the number of equations originally given equals the number of variables implicitly determined or, equivalently, that the range spaces of F and f must have the same dimension.

The analog of Equation (6) holds for an arbitrary number of coordinate functions F_i and is proved in exactly the same way. We can summarize the result in the following generalization of 4.1.

4.2 Theorem

If $\mathcal{R}^{n+m} \xrightarrow{F} \mathcal{R}^m$ and $\mathcal{R}^n \xrightarrow{f} \mathcal{R}^m$ are differentiable, and if $\mathbf{y} = f(\mathbf{x})$ satisfies $F(\mathbf{x}, \mathbf{y}) = 0$, then

$$f'(\mathbf{x}) = -F_{\mathbf{y}}^{-1}(\mathbf{x}, f(\mathbf{x}))F_{\mathbf{x}}(\mathbf{x}, f(\mathbf{x})),$$

provided $F_{\mathbf{y}}$ has an inverse. The derivative $F_{\mathbf{y}}$ is computed with $\mathbf{x}$ held fixed, and $F_{\mathbf{x}}$ is computed with $\mathbf{y}$ held fixed.

The notation used above is illustrated in the next example.

Example 5. Suppose that

$$F(x, y, z) = \begin{pmatrix} x^2y + xz \\ xz + yz \end{pmatrix}$$

and that we choose $\mathbf{x} = x, \quad \mathbf{y} = (y, z)$. Then

$$F_x(x, y, z) = \begin{pmatrix} 2xy + z \\ z \end{pmatrix}$$

and

$$F_{(y,z)}(x, y, z) = \begin{pmatrix} x^2 & x \\ z & x + y \end{pmatrix}.$$

Newton's method provides, in many cases, an effective way of computing an approximate value for an implicitly defined function. Suppose that $f(\mathbf{x})$ satisfies the equation $F(\mathbf{x}, f(\mathbf{x})) = 0$ in some neighborhood of $\mathbf{x} = \mathbf{x}_0$. To compute $f(\mathbf{x}_0)$ we need to solve the equation $F(\mathbf{x}_0, \mathbf{y}) = 0$ for $\mathbf{y}$. We apply Newton's method to the function $g(\mathbf{y}) = F(x_0, \mathbf{y})$. The iteration equation $\mathbf{y}_{k+1} = \mathbf{y}_k - [g'(\mathbf{y}_k)]^{-1}g(\mathbf{y}_k)$ becomes:

4.3 $$\mathbf{y}_{k+1} = \mathbf{y}_k - [F_{\mathbf{y}}(\mathbf{x}_0, \mathbf{y}_k)]^{-1}F(\mathbf{x}_0, \mathbf{y}_k).$$

Of course, some choice for $\mathbf{y}_0$ has to be made using more detailed information about F.

Example 6. If $F(x, y) = x^2y + xy^2$, then $F_y(x, y) = x^2 + 2xy$, and Equation 4.3 becomes

$$y_{k+1} = y_k - \frac{x_0^2 y_k + x_0 y_k^2}{x_0^2 + 2x_0 y_k} = \frac{y_k^2}{x_0 + 2y_k}. \tag{7}$$

Suppose that $x_0 = 1$ and that we choose $y_0 = 1$ as an initial approximation to the number $f(x_0)$ satisfying $f(x_0, f(x_0)) = 0$. Then

$$y_1 = \tfrac{1}{3},$$

$$y_2 = \frac{(\frac{1}{3})^2}{1 + 2(\frac{1}{3})} = \frac{1}{15},$$

$$y_3 = \frac{(\frac{1}{15})^2}{1 + 2(\frac{1}{15})} = \frac{1}{255}.$$

These values for y suggest that $y = 0$ may be a solution, and indeed we can check that $F(1, 0) = 0$.

On the other hand, if we try $y_0 = -2$ together with $x_0 = 1$, we get

$$y_1 = \frac{(-2)^2}{1 + 2(-2)} = -\frac{4}{3},$$

$$y_2 = \frac{(-\frac{4}{3})^2}{1 + 2(-\frac{4}{3})} = -\frac{16}{15},$$

$$y_3 = \frac{(-\frac{16}{15})^2}{1 + 2(-\frac{16}{15})} = -\frac{256}{255}.$$

These values of y seem to be converging to $y = -1$, and indeed $F(1, -1) = 0$. To solve $F(1.1, y) = 0$ approximately, we would set $x_0 = 1.1$ in Equation (7) and try either $y_0 = 0$ or $y_0 = 1$.

Example 7. Suppose we are given

$$F(x, y, z) = \begin{pmatrix} x^2y + z \\ x + y^2z \end{pmatrix}$$

and are asked to find approximate values for y and z satisfying $F(1, y, z) = 0$. The derivative matrix of F with respect to (y, z) is

$$F_{(y,z)}(x, y, z) = \begin{pmatrix} x^2 & 1 \\ 2yz & y^2 \end{pmatrix};$$

so

$$F_{(y,z)}(1, y, z) = \begin{pmatrix} 1 & 1 \\ 2yz & y^2 \end{pmatrix}.$$

Computing the inverse matrix by the formula

$$\begin{pmatrix} a & b \\ c & d \end{pmatrix}^{-1} = \frac{1}{ad - bc}\begin{pmatrix} d & -b \\ -c & a \end{pmatrix}$$

gives

$$[F_{(y,z)}(1, y, z)]^{-1} = \frac{1}{y^2 - 2yz}\begin{pmatrix} y^2 & -1 \\ -2yz & 1 \end{pmatrix}.$$

Then the iteration Formula 4.3 is

$$\begin{aligned} \begin{pmatrix} y_{k+1} \\ z_{k+1} \end{pmatrix} &= \begin{pmatrix} y_k \\ z_k \end{pmatrix} - \frac{1}{y_k^2 - 2y_kz_k}\begin{pmatrix} y_k^2 & -1 \\ -2y_kz_k & 1 \end{pmatrix}\begin{pmatrix} y_k + z_k \\ 1 + y_k^2z_k \end{pmatrix} \\ &= \frac{1}{y_k^2 - 2y_kz_k}\begin{pmatrix} 1 - 2y_k^2z_k \\ -1 + 2y_k^2z_k \end{pmatrix}. \end{aligned}$$

This formula shows that, no matter how we choose y_0 and z_0 so that $y_0^2 - 2y_0z_0 \neq 0$, we must always have $y_{k+1} = -z_{k+1}$ for $k \geq 0$. Thus we might try $y_0 = 1$, $z_0 = -1$. Substitution into the iteration formula shows that $y_{k+1} = 1$, $z_{k+1} = -1$, for every $k \geq 0$, and in fact that $F(1, 1, -1) = 0$. Thus no further numerical computation is necessary in this example because we made a lucky choice for y_0 and z_0. In the event that we had not been so lucky, we would compute successive approximations (x_k, y_k) to the solution.

EXERCISES

1. If

$$x^2y + yz = 0, \qquad xyz + 1 = 0,$$

find dx/dz and dy/dz at $(x, y, z) = (1, 1, -1)$.

$$\left[Ans. \ \frac{dx}{dz} = -\frac{1}{2}, \frac{dy}{dz} = \frac{3}{2}.\right]$$

2. If Exercise 1 is expressed in the general vector notation of Theorem 4.2, what are F, $\mathbf{x}$, $\mathbf{y}$, $F_{\mathbf{y}}$, and $F_{\mathbf{x}}$?

3. If

$$x + y - u - v = 0,$$
$$x - y + 2u + v = 0,$$

find $\partial x/\partial u$ and $\partial y/\partial u$ by:

(a) Solving for x and y in terms of u and v.
(b) Implicit differentiation.

4. If Exercise 3 is expressed in the vector notation of Theorem 4.2, what is the matrix $f'(\mathbf{x})$?

5. If $x^2 + yu + xv + w = 0$, $x + y + uvw + 1 = 0$, then, regarding x and y as functions of u, v, and w, find

$$\frac{\partial x}{\partial u} \quad \text{and} \quad \frac{\partial y}{\partial u} \quad \text{at} \quad (x, y, u, v, w) = (1, -1, 1, 1, -1).$$

$$\left[Ans. \ \frac{\partial x}{\partial u} = 0, \frac{\partial y}{\partial u} = 1.\right]$$

6. (a) The equations $2x^3y + yx^2 + t^2 = 0$, $x + y + t - 1 = 0$ implicitly define a curve

$$f(t) = \begin{pmatrix} x(t) \\ y(t) \end{pmatrix}$$

which satisfies

$$f(1) = \begin{pmatrix} -1 \\ 1 \end{pmatrix}.$$

Find the tangent line to f at $t = 1$. $\left[Ans. \ t\begin{pmatrix} -\frac{3}{5} \\ -\frac{2}{5} \end{pmatrix} + \begin{pmatrix} -1 \\ 1 \end{pmatrix}.\right]$

(b) Apply Newton's method to approximate $f(1.1)$.

7. Let the equation $x^2/4 + y^2 + z^2/9 - 1 = 0$ define z implicitly as a function $z = f(x, y)$ near the point $x = 1$, $y = \sqrt{11}/6$, $z = 2$. The graph of the function f is a surface. Find its tangent plane at $(1, \sqrt{11}/6, 2)$.

8. Suppose the equation $F(x, y, z) = 0$ implicitly defines $z = f(x, y)$ and that $z_0 = f(x_0, y_0)$. Suppose further that the surface that is the graph of $z = f(x, y)$ has a tangent plane at (x_0, y_0). Show that

$$(x - x_0)\frac{\partial F}{\partial x}(x_0, y_0, z_0) + (y - y_0)\frac{\partial F}{\partial y}(x_0, y_0, z_0) + (z - z_0)\frac{\partial F}{\partial z}(x_0, y_0, z_0) = 0$$

is the equation for this tangent plane.

9. The equations

$$\begin{aligned} 2x + y + 2z + u - v - 1 &= 0 \\ xy + z - u + 2v - 1 &= 0 \\ yz + xz + u^2 + v &= 0 \end{aligned}$$

near $(x, y, z, u, v) = (1, 1, -1, 1, 1)$ define x, y, and z as functions of u and v.

(a) Find the matrix of the differential of the implicitly defined function

$$\begin{pmatrix} x \\ y \\ z \end{pmatrix} = \begin{pmatrix} x(u, v) \\ y(u, v) \\ z(u, v) \end{pmatrix} = f(u, v) \quad \text{at} \quad (u, v) = (1, 1).$$

(b) The function f parametrically defines a surface in the (x, y, z) space. Find the tangent plane to it at the point $(1, 1, -1)$.

(c) Apply Newton's method to approximate $f(1.1, 1.1)$.

SECTION 5

CURVILINEAR COORDINATES

Formulas that occur in mathematics and its applications can often be simplified by good descriptions of the quantities to be singled out for special attention. Since in practice these quantities are usually represented by vectors whose entries are real-number coordinates, the problem can be viewed as one of choosing the most useful system of coordinates. Thus we consider introducing coordinates in $\mathcal{R}^n$ different from the natural coordinates x_i that appear in the designation of a typical point $(x_1, \ldots, x_n)$. Specifically, to each point $(x_1, \ldots, x_n)$ there will be assigned a new n-tuple $(u_1, \ldots, u_n)$. Clearly, if we are to be able to switch back and forth from one set of coordinates to the other, the assignment described above must be one-to-one, that is, for each $(x_1, \ldots, x_n)$ there should be just one n-tuple $(u_1, \ldots, u_n)$ and vice versa. In practice, the new coordinate assignment is often made for some specific subregion of $\mathcal{R}^n$ rather than for the whole space. The vector space of new coordinates $(u_1, \ldots, u_n)$ will be denoted by $\mathcal{U}^n$ to avoid confusion with $\mathcal{R}^n$, whose points $(x_1, \ldots, x_n)$ are being assigned the new coordinates.

Example 1. (*Polar coordinates in the plane.*) Consider two copies of 2-dimensional space: the xy-plane, denoted by $\mathcal{R}^2$, and the $r\theta$-plane, denoted by $\mathcal{U}^2$. The function $\mathcal{U}^2 \xrightarrow{T} \mathcal{R}^2$ defined by

$$\begin{pmatrix} x \\ y \end{pmatrix} = T\begin{pmatrix} r \\ \theta \end{pmatrix} = \begin{pmatrix} r\cos\theta \\ r\sin\theta \end{pmatrix}, \qquad \begin{cases} 0 < r < \infty \\ -\infty < \theta < \infty \end{cases} \tag{1}$$

has a simple geometric description. The image under T of a point $\begin{pmatrix} r \\ \theta \end{pmatrix}$ is the point

$$\mathbf{x} = \begin{pmatrix} x \\ y \end{pmatrix}$$

whose distance from the origin is r and such that the angle from the positive x-axis to $\mathbf{x}$ in the counterclockwise direction is θ. See Fig. 8.

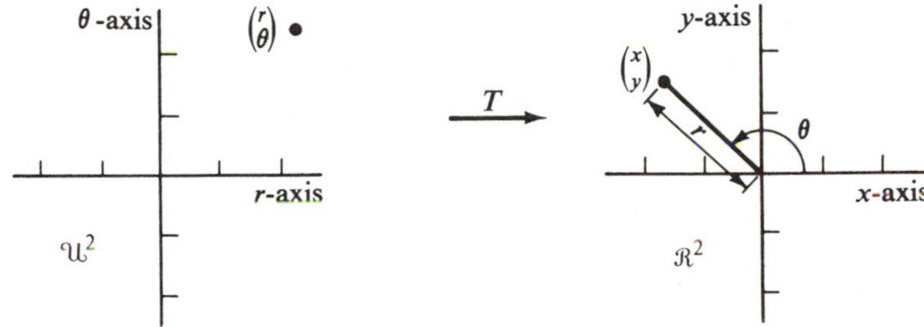

Figure 8

For any two points

$$\begin{pmatrix} r_1 \\ \theta_1 \end{pmatrix} \quad \text{and} \quad \begin{pmatrix} r_2 \\ \theta_2 \end{pmatrix}$$

in the domain of T, it is easy to prove that

$$T\begin{pmatrix} r_1 \\ \theta_1 \end{pmatrix} = T\begin{pmatrix} r_2 \\ \theta_2 \end{pmatrix}$$

if and only if $r_1 = r_2$ and $\theta_1 = \theta_2 + 2\pi m$ for some integer m. The range of T consists of all of $\mathcal{R}^2$ except for the origin. It follows that, for any point

$$\mathbf{x} = \begin{pmatrix} x \\ y \end{pmatrix} \quad \text{in } \mathcal{R}^2 \text{ except} \quad \begin{pmatrix} 0 \\ 0 \end{pmatrix},$$

there exist numbers r and θ, called **polar coordinates** of $\mathbf{x}$, such that

$$T\begin{pmatrix} r \\ \theta \end{pmatrix} = \mathbf{x}.$$

Furthermore, the polar coordinates of $\mathbf{x}$ are uniquely specified up to an integer multiple of 2π in the second coordinate.

From a slightly different point of view, the preceding paragraph says that T is not one-to-one, but that it becomes so if its domain is restricted to be a subset of a rectangular half-strip in the $r\theta$-plane defined by inequalities

$$0 < r < \infty, \qquad \theta_0 \leq \theta < \theta_0 + 2\pi.$$

So restricted, T has an inverse function. Solving the equations $x = r\cos\theta$, $y = r\sin\theta$ for r and θ, we obtain, for $x \neq 0$,

$$r = \sqrt{x^2 + y^2}, \qquad \theta = \arctan\frac{y}{x} + k\pi.$$

We have used the common convention that any inverse trigonometric function is the principal branch of the corresponding multiple-valued function. Hence, the range of the function arctan is the interval $-\pi/2 < \theta < \pi/2$. It follows that the function defined by

$$\begin{pmatrix} r \\ \theta \end{pmatrix} = \begin{pmatrix} \sqrt{x^2 + y^2} \\ \arctan\frac{y}{x} \end{pmatrix}, \qquad x > 0,$$

is the inverse of the restriction of T to the region $0 < r < \infty$, $-\pi/2 < \theta < \pi/2$. Similarly, the function defined by

$$\begin{pmatrix} r \\ \theta \end{pmatrix} = \begin{pmatrix} \sqrt{x^2 + y^2} \\ \operatorname{arccot}\frac{x}{y} \end{pmatrix}, \qquad y > 0,$$

is the inverse of the restriction of T to $0 < r < \infty$, $0 < \theta < \pi$.

We have not defined polar coordinates for the origin of the xy-plane simply because

$$\begin{pmatrix} 0\cos\theta \\ 0\sin\theta \end{pmatrix} = \begin{pmatrix} 0 \\ 0 \end{pmatrix} \quad \text{for all } \theta,$$

and so the one-to-one requirement fails at the origin. This fact causes no real difficulty. For example, the equation in rectangular coordinates of the lemniscate,

$$(x^2 + y^2)^2 = 2(x^2 - y^2), \tag{2}$$

becomes, upon introduction of polar coordinates,

$$r^2 = 2\cos 2\theta, \qquad r > 0. \tag{3}$$

The image under T of the set of pairs (r, θ) that satisfy Equation (3) is precisely the set of pairs (x, y) that satisfy Equation (2), except for the origin. We may simply fill in this one point. See Fig. 9.

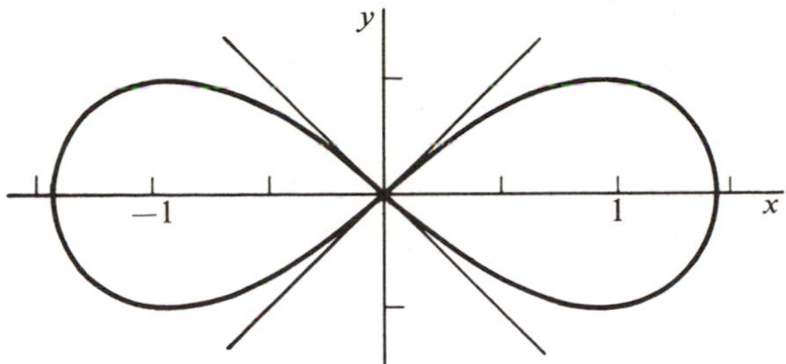

Figure 9

Example 2. (*Spherical coordinates in* 3-*dimensional space.*) Consider the function $\mathcal{U}^3 \xrightarrow{T} \mathcal{R}^3$, defined by

$$T\begin{pmatrix} r \\ \phi \\ \theta \end{pmatrix} = \begin{pmatrix} r \sin \phi \cos \theta \\ r \sin \phi \sin \theta \\ r \cos \phi \end{pmatrix}, \qquad \begin{cases} 0 < r < \infty \\ 0 < \phi < \pi \\ 0 \leq \theta < 2\pi. \end{cases} \tag{4}$$

Here for simplicity we have restricted the domain of T from the outset so that it is one-to-one. Its range is all of $\mathcal{R}^3$ with the exception of the z-axis. Hence, it assigns **spherical coordinates** (r, ϕ, θ) to every point of $\mathcal{R}^3$ except those on the z-axis. As with polar coordinates in the plane, the spherical coordinates (r, ϕ, θ) of a point $\mathbf{x} = (x, y, z)$ have a simple geometric interpretation (see Fig. 10): The number r is the distance from $\mathbf{x}$ to the

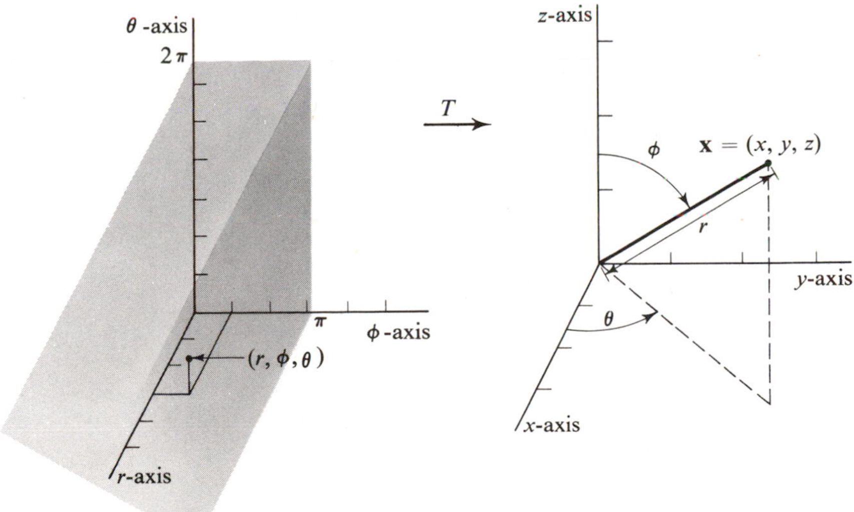

Figure 10

origin. The coordinate ϕ is the angle in radians between the vector $\mathbf{x}$ and the positive z-axis. Finally, θ is the angle in radians from the positive x-axis to the projected image $(x, y, 0)$ of $\mathbf{x}$ on the xy-plane. The symbols ϕ and θ are sometimes interchanged, particularly in physical applications.

We can compute an explicit expression for the inverse function, which we denote by T^{-1}, by solving the equations

$$\begin{aligned} x &= r \sin \phi \cos \theta, \\ y &= r \sin \phi \sin \theta, \\ z &= r \cos \phi, \end{aligned}$$

for r, θ, and ϕ. We get, for $y \geq 0$,

$$\begin{pmatrix} r \\ \phi \\ \theta \end{pmatrix} = T^{-1}\begin{pmatrix} x \\ y \\ z \end{pmatrix} = \begin{pmatrix} \sqrt{x^2 + y^2 + z^2} \\ \arccos \dfrac{z}{\sqrt{x^2 + y^2 + z^2}} \\ \arccos \dfrac{x}{\sqrt{x^2 + y^2}} \end{pmatrix}, \qquad x^2 + y^2 > 0.$$

Since the range of *arccos* (the principal branch) is the interval $0 \leq \theta \leq \pi$, this function is actually the inverse of the function obtained by restricting the domain of T by the further condition $0 \leq \theta \leq \pi$. To get values of θ in the interval $\pi < \theta < 2\pi$, corresponding to $y < 0$, we add π to the third coordinate in the formula above.

Three surfaces in $\mathcal{R}^3$ implicitly defined by spherical coordinate equations $r = 1$, $\phi = \pi/4$, and $\theta = \pi/3$, respectively, are shown in Fig. 11.

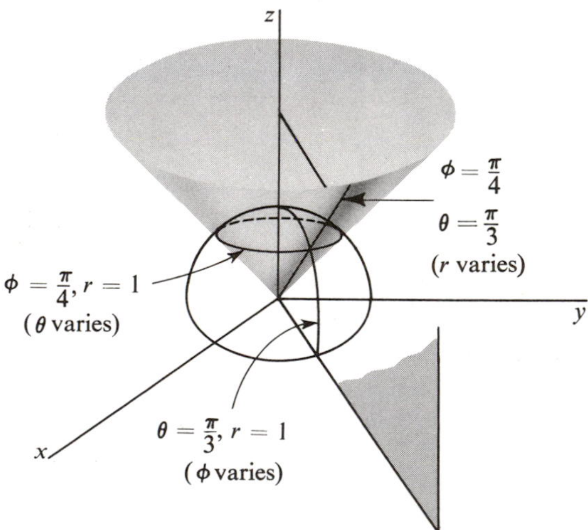

Figure 11

The corresponding rectangular coordinate equations derived from the above expressions for T^{-1} are

$$x^2 + y^2 + z^2 = 1, \qquad x^2 + y^2 > 0$$

$$z = \frac{\sqrt{2}}{2}\sqrt{x^2 + y^2 + z^2}, \qquad z > 0$$

$$y = \sqrt{3}\,x, \qquad x > 0,$$

respectively.

The name "curvilinear" is applied to coordinates for the reason that, if all but one of the nonrectangular coordinates are held fixed, and the remaining one is varied, the coordinate transformation defines a curve in $\mathcal{R}^n$. Thus in plane polar coordinates the coordinate curves are circles and straight lines as shown in Fig. 12. For spherical coordinates, typical

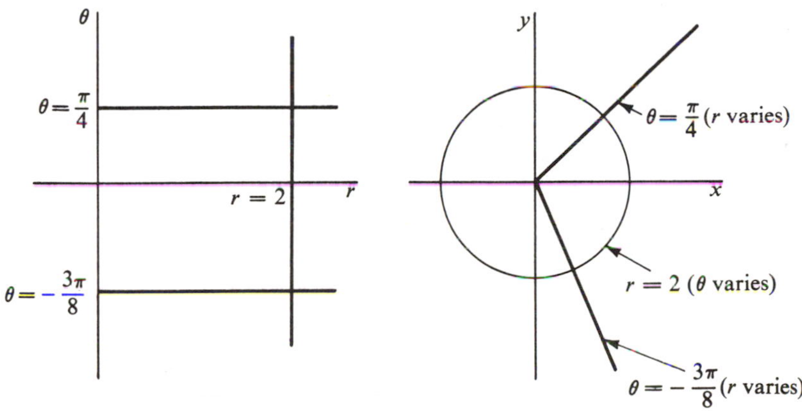

Figure 12

coordinate curves are the circle, semicircle, and half-line obtained as intersections of the pairs of surfaces shown in Fig. 11. The curves and surfaces got by varying one or more curvilinear coordinate variables play the same role that the natural coordinate lines and planes of $\mathcal{R}^n$ do. For example, to say that a point in $\mathcal{R}^3$ has rectangular coordinates $(x, y, z) = (1, 2, 1)$ is to say that it lies at the intersection of the coordinate planes $x = 1, y = 2$, and $z = 1$. Similarly, to say that a point in $\mathcal{R}^3$ has spherical coordinates $(r, \phi, \theta) = (1, \pi/4, \pi/3)$ is to say that it lies at the intersection of the surfaces shown in Fig. 11.

Generalizing from the preceding examples, we see that a system of curvilinear coordinates in $\mathcal{R}^n$ is determined by a function $\mathcal{U}^n \xrightarrow{T} \mathcal{R}^n$. It is assumed that for some open subset N in the domain of T, the restriction of T to N is one-to-one and therefore has an inverse T^{-1}. The **curvilinear**

coordinates, determined by T and N, of a point $\mathbf{x}$ lying in the image set $T(N)$ are

$$\begin{pmatrix} u_1 \\ \cdot \\ \cdot \\ \cdot \\ u_n \end{pmatrix} = T^{-1}\begin{pmatrix} x_1 \\ \cdot \\ \cdot \\ \cdot \\ x_n \end{pmatrix}.$$

It is convenient to impose fairly stringent regularity conditions on a coordinate transformation. Specifically, we shall assume that, at every point $\mathbf{u}$ of N, the function T is continuously differentiable and that $T'(\mathbf{u})$ is invertible.

The polar and spherical coordinate changes represented by

$$\begin{pmatrix} x \\ y \end{pmatrix} = \begin{pmatrix} r\cos\theta \\ r\sin\theta \end{pmatrix}$$

and

$$\begin{pmatrix} x \\ y \\ z \end{pmatrix} = \begin{pmatrix} r\sin\phi\cos\theta \\ r\sin\phi\sin\theta \\ r\cos\phi \end{pmatrix}$$

have Jacobian matrices

$$\begin{pmatrix} \cos\theta & -r\sin\theta \\ \sin\theta & r\cos\theta \end{pmatrix} \tag{5}$$

and

$$\begin{pmatrix} \sin\phi\cos\theta & r\cos\phi\cos\theta & -r\sin\phi\sin\theta \\ \sin\phi\sin\theta & r\cos\phi\sin\theta & r\sin\phi\cos\theta \\ \cos\phi & -r\sin\phi & 0 \end{pmatrix}, \tag{6}$$

respectively. The matrices (5) and (6), and more generally the Jacobian matrices of differentiable coordinate transformations, have a simple geometric interpretation. Each column of the Jacobian is obtained by differentiation of the coordinate functions with respect to a single variable, while holding the other variables fixed. This means that the jth column of the matrix represents a tangent vector to the curvilinear coordinate curve for which the jth coordinate is allowed to vary. That is, let the coordinate transformation be given by $\mathcal{U}^n \xrightarrow{T} \mathcal{R}^n$. Then the jth column of the matrix of the derivative $T'(\mathbf{u}_0)$ is a tangent vector, which we shall denote by $\mathbf{c}_j$, at $\mathbf{x}_0 = T(\mathbf{u}_0)$, to the curvilinear coordinate curve formed by allowing only the jth coordinate of $\mathbf{u}_0$ to vary. Tangent vectors are shown (with their

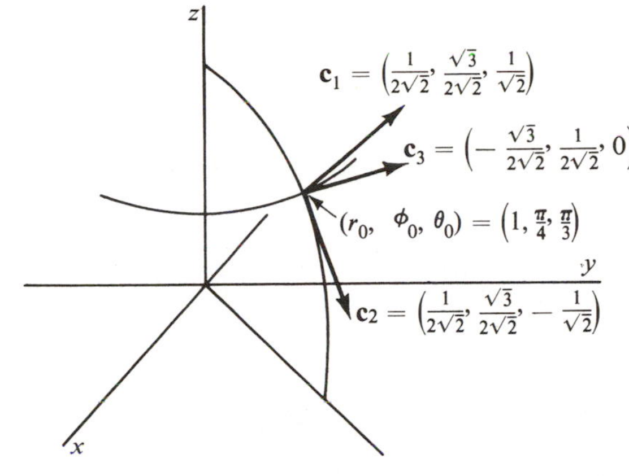

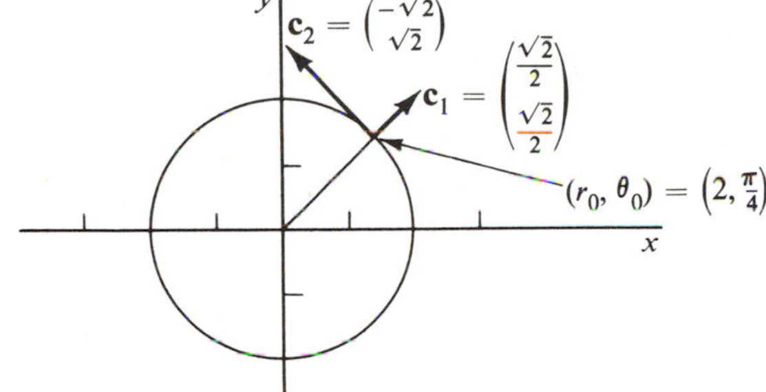

Figure 13

initial points translated to the point $\mathbf{x}_0$) in Fig. 13 for some polar and spherical coordinate curves. Since the coordinate curves lie in the cartesian space, the coordinates of the tangent vectors $\mathbf{c}_1, \ldots, \mathbf{c}_n$ are rectangular coordinates, not curvilinear coordinates.

We shall show that the Jacobian matrix itself of a coordinate transformation can be interpreted as the matrix of a certain linear change of coordinates. To see this, consider curvilinear coordinates in $\mathcal{R}^n$ given by $\mathbf{x} = T(\mathbf{u})$, where $\mathbf{u}$ is the curvilinear coordinate variable. Fix a point $\mathbf{x}_0$ having curvilinear coordinates $\mathbf{u}_0$. At $\mathbf{x}_0$ we can introduce a new origin and new unit vectors $\mathbf{e}_1, \ldots, \mathbf{e}_n$ with the same directions as the natural basis vectors for $\mathcal{R}^n$. Then multiplication by the matrix $T'(\mathbf{u}_0)$ transforms the vectors $\mathbf{e}_1, \ldots, \mathbf{e}_n$ into the vectors $\mathbf{c}_1, \ldots, \mathbf{c}_n$ that are the tangent vectors to the curvilinear coordinate curves. Figure 14 illustrates the relation between the $\mathbf{e}_i$ and the $\mathbf{c}_i$. Notice that the vectors $\mathbf{c}_1, \ldots, \mathbf{c}_n$ will be linearly independent if and only if $T'(\mathbf{u}_0)$ is invertible. This is one reason for requiring not only that a coordinate transformation be one-to-one in a neighborhood of a point, but also that its derivative be invertible.

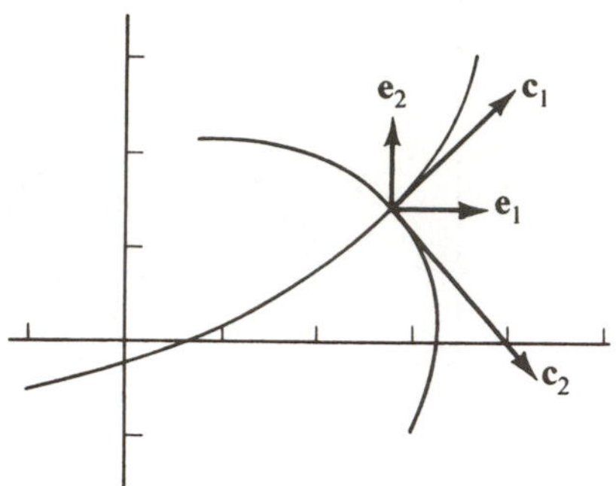

Figure 14

Example 3. (*Cylindrical coordinates in* $\mathcal{R}^3$.) The coordinate transformation is defined by

$$\begin{pmatrix} x \\ y \\ z \end{pmatrix} = \begin{pmatrix} r\cos\theta \\ r\sin\theta \\ z \end{pmatrix}, \qquad \begin{cases} 0 < r < \infty \\ -\pi < \theta \leq \pi \\ -\infty < z < \infty. \end{cases}$$

The Jacobian matrix is

$$\begin{pmatrix} \cos\theta & -r\sin\theta & 0 \\ \sin\theta & r\cos\theta & 0 \\ 0 & 0 & 1 \end{pmatrix}.$$

Curvilinear coordinate surfaces and tangent vectors to curvilinear coordinate curves are shown in Fig. 15. Notice that the Jacobian determinant

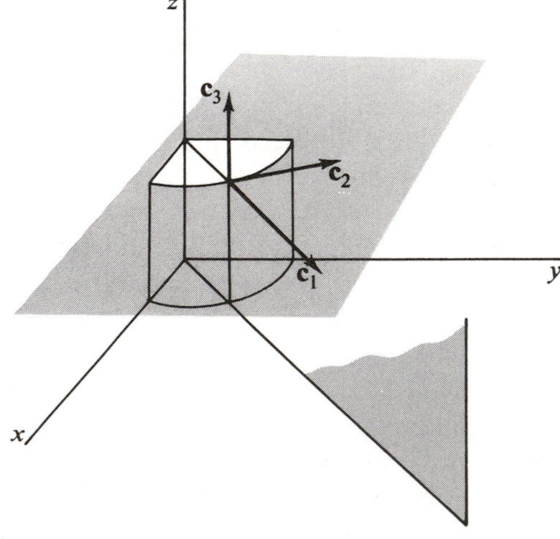

Figure 15

is

$$\frac{\partial(x, y, z)}{\partial(r, \theta, z)} = r.$$

Computations involving curvilinear coordinates are much simpler if the coordinate curves, and the vectors $\mathbf{c}_k$, are orthogonal. (This is so in the examples we have considered.) In this case it is customary to replace the $\mathbf{c}_k$ by unit vectors having the same direction. Thus, letting $h_1 = |\mathbf{c}_1|, \ldots,$ $h_n = |\mathbf{c}_n|$, we have a local orthonormal basis $(1/h_1\mathbf{c}_1, \ldots, 1/h_n\mathbf{c}_n)$. The result is that the matrix H, whose columns are the rectangular coordinates of $1/h_1\mathbf{c}_1, \ldots, 1/h_n\mathbf{c}_n$, is an orthogonal matrix whose inverse is equal to its transpose.

Example 4. For spherical coordinates in $\mathcal{R}^3$ we have $h_1 = 1$, $h_2 = r$, $h_3 = r \sin \phi$, so that the matrix H is given by

$$H = \begin{pmatrix} \sin\phi\cos\theta & \cos\phi\cos\theta & -\sin\theta \\ \sin\phi\sin\theta & \cos\phi\sin\theta & \cos\theta \\ \cos\phi & -\sin\phi & 0 \end{pmatrix}.$$

We have assumed $0 < \phi < \pi$, so that $\sin \phi > 0$. Then, because the matrix H is orthogonal, $H^{-1} = H^t$ and

$$H^{-1} = \begin{pmatrix} \sin\phi\cos\theta & \sin\phi\sin\theta & \cos\phi \\ \cos\phi\cos\theta & \cos\phi\sin\theta & -\sin\phi \\ -\sin\theta & \cos\theta & 0 \end{pmatrix}.$$

We have seen that the vectors $\mathbf{c}_1, \ldots, \mathbf{c}_n$, tangent to coordinate curves, describe approximately the nature of a curvilinear coordinate system. The inner products of these vectors among themselves occur sufficiently often that there is a special notation for them:

$$\mathbf{c}_i \cdot \mathbf{c}_j = g_{ij}, \qquad i, j = 1, \ldots, n.$$

Since $\mathbf{c}_i \cdot \mathbf{c}_j = \mathbf{c}_j \cdot \mathbf{c}_i$, we have $g_{ij} = g_{ji}$. If the vectors $\mathbf{c}_1, \ldots, \mathbf{c}_n$ are orthogonal, as is often the case in practice, then only the inner products $\mathbf{c}_i \cdot \mathbf{c}_i = g_{ii}$ will be different from zero. A number of important formulas can be expressed in curvilinear coordinates entirely in terms of the functions g_{ij} and without explicit reference to the particular curvilinear coordinate functions being used.

In the xy-plane suppose curvilinear coordinates are given by

$$\begin{pmatrix} x \\ y \end{pmatrix} = \begin{pmatrix} x(u, v) \\ y(u, v) \end{pmatrix}.$$

The Jacobian of the coordinate transformation is

$$\begin{pmatrix} \dfrac{\partial x}{\partial u} & \dfrac{\partial x}{\partial v} \\[2ex] \dfrac{\partial y}{\partial u} & \dfrac{\partial y}{\partial v} \end{pmatrix},$$

whence

$$g_{11} = \left(\frac{\partial x}{\partial u}\right)^2 + \left(\frac{\partial y}{\partial u}\right)^2,$$

$$g_{12} = g_{21} = \frac{\partial x}{\partial u}\frac{\partial x}{\partial v} + \frac{\partial y}{\partial u}\frac{\partial y}{\partial v},$$

$$g_{22} = \left(\frac{\partial x}{\partial v}\right)^2 + \left(\frac{\partial y}{\partial v}\right)^2.$$

Now suppose that

$$\begin{pmatrix} u(t) \\ v(t) \end{pmatrix}$$

is a differentiable function from an interval $[a, b]$ to $\mathcal{U}^2$. Then the equation

$$f(t) = \begin{pmatrix} x(u(t), v(t)) \\ y(u(t), v(t)) \end{pmatrix}, \qquad a \le t \le b,$$

is the parametric representation of a curve in $\mathcal{R}^2$. The tangent vector to the curve can be computed by the chain rule to be

$$\frac{df}{dt} = \begin{pmatrix} \dfrac{dx}{dt} \\[2ex] \dfrac{dy}{dt} \end{pmatrix} = \begin{pmatrix} \dfrac{\partial x}{\partial u}\dfrac{du}{dt} + \dfrac{\partial x}{\partial v}\dfrac{dv}{\partial t} \\[2ex] \dfrac{\partial y}{\partial u}\dfrac{du}{dt} + \dfrac{\partial y}{\partial v}\dfrac{dv}{dt} \end{pmatrix}.$$

The length of the tangent vector df/dt is then

$$\begin{aligned} \left|\frac{df}{dt}\right| &= \left(\left(\frac{dx}{dt}\right)^2 + \left(\frac{dy}{dt}\right)^2\right)^{1/2} \\ &= \left(\left[\left(\frac{\partial x}{\partial u}\right)^2 + \left(\frac{\partial y}{\partial u}\right)^2\right]\left(\frac{du}{dt}\right)^2 + 2\left[\frac{\partial x}{\partial u}\frac{\partial x}{\partial v} + \frac{\partial y}{\partial u}\frac{\partial y}{\partial v}\right]\frac{du}{dt}\frac{dv}{dt}\right. \\ &\qquad\qquad \left. + \left[\left(\frac{\partial x}{\partial v}\right)^2 + \left(\frac{\partial y}{\partial v}\right)^2\right]\left(\frac{dv}{dt}\right)^2\right)^{1/2} \\ &= \left(g_{11}\left(\frac{du}{dt}\right)^2 + 2g_{12}\frac{du}{dt}\frac{dv}{dt} + g_{22}\left(\frac{dv}{dt}\right)^2\right)^{1/2}. \end{aligned}$$

A similar computation in $\mathcal{R}^n$ leads to the formula

$$\left|\frac{df}{dt}\right| = \left(\sum_{i,j=1}^{n} g_{ij} \frac{du_i}{dt}\frac{du_j}{dt}\right)^{1/2}, \tag{7}$$

where $f(t) = \big(x_1(u_1(t), \ldots, u_n(t)), \ldots, x_n(u_1(t), \ldots, u_n(t))\big)$.

Once the g_{ij} have been computed for the particular coordinate system, Equation (7) can be used for any differentiable curve by substituting in the components of the vector

$$\begin{pmatrix} \dfrac{du}{dt} \\ \\ \dfrac{dv}{dt} \end{pmatrix}.$$

To be more specific, suppose we are given plane polar coordinates

$$\begin{pmatrix} x \\ y \end{pmatrix} = \begin{pmatrix} r\cos\theta \\ r\sin\theta \end{pmatrix}.$$

The Jacobian matrix is

$$\begin{pmatrix} \cos\theta & -r\sin\theta \\ \sin\theta & r\cos\theta \end{pmatrix}.$$

Hence

$$g_{11} = 1, \qquad g_{12} = g_{21} = 0, \qquad g_{22} = r^2.$$

A curve $f(\theta) = \begin{pmatrix} r(\theta)\cos\theta \\ r(\theta)\sin\theta \end{pmatrix}$ has a tangent vector $df/d\theta$ of length

$$\left|\frac{df}{d\theta}\right| = \left(\left(\frac{dr}{d\theta}\right)^2 + r^2\right)^{1/2}.$$

Example 5. If (r, ϕ, θ) are spherical coordinates in $\mathcal{R}^3$, the length of a curve γ defined by an equation

$$\begin{pmatrix} r \\ \phi \\ \theta \end{pmatrix} = \begin{pmatrix} r(t) \\ \phi(t) \\ \theta(t) \end{pmatrix}, \qquad a \le t \le b,$$

can be computed from Equation (7). The derivative of the coordinate transformation

$$\begin{pmatrix} x \\ y \\ z \end{pmatrix} = \begin{pmatrix} r\sin\phi\cos\theta \\ r\sin\phi\sin\theta \\ r\cos\phi \end{pmatrix}$$

is the matrix

$$\begin{pmatrix} \sin\phi\cos\theta & r\cos\phi\cos\theta & -r\sin\phi\sin\theta \\ \sin\phi\sin\theta & r\cos\phi\sin\theta & r\sin\phi\cos\theta \\ \cos\phi & -r\sin\phi & 0 \end{pmatrix}.$$

The function g_{ij} is the inner product of the ith and the jth column of the preceding matrix. Hence

$$\begin{pmatrix} g_{11} & g_{12} & g_{13} \\ g_{21} & g_{22} & g_{23} \\ g_{31} & g_{32} & g_{33} \end{pmatrix} = \begin{pmatrix} 1 & 0 & 0 \\ 0 & r^2 & 0 \\ 0 & 0 & r^2\sin^2\phi \end{pmatrix}.$$

And so Equation (7) yields

$$l(\gamma) = \int_a^b \sqrt{\left(\frac{dr}{dt}\right)^2 + r^2\left(\frac{d\phi}{dt}\right)^2 + r^2\sin^2\phi\left(\frac{d\theta}{dt}\right)^2}\,dt.$$

In particular, the curve λ defined by

$$\begin{pmatrix} r \\ \phi \\ \theta \end{pmatrix} = \begin{pmatrix} 1 \\ t \\ t \end{pmatrix}, \qquad 0 \le t \le \frac{\pi}{2}$$

has length

$$l(\lambda) = \int_0^{\pi/2} \sqrt{1 + \sin^2 t}\,dt.$$

This integral cannot be computed by means of an elementary indefinite integral. It can, however, be approximated by simple methods. Or it can be reduced to a standard elliptic integral as follows. Replacing t by $\pi/2 - \theta$, and using $\cos^2\theta = 1 - \sin^2\theta$, we get

$$\int_0^{\pi/2} \sqrt{1 + \sin^2 t}\,dt = \sqrt{2}\int_0^{\pi/2} \sqrt{1 - \tfrac{1}{2}\sin^2\theta}\,d\theta.$$

The latter integral has been tabulated,† and we estimate

$$l(\lambda) \approx \sqrt{2}\,(1.35) \approx 1.91.$$

EXERCISES

1. Sketch the three curves given below in polar coordinates.

(a) $r = \theta$, $0 \le \theta \le \pi/2$.
(b) $r(\sin\theta - \cos\theta) = \pi/2$, $\pi/2 \le \theta \le \pi$.
(c) $r = \pi/2\cos\theta$, $\pi \le \theta \le 3\pi/2$.

† See, for example, R. S. Burington, *Handbook of Mathematical Tables*, Handbook Publishers, Inc., Sandusky, Ohio, 1965.

2. In $\mathcal{R}^3$, sketch the curves and surfaces given below in spherical coordinates.

(a) $r = 2$, $0 \leq \theta \leq \pi/4$, $\pi/4 \leq \phi \leq \pi/2$.
(b) $1 \leq r \leq 2$, $\theta = \pi/2$, $\phi = \pi/4$.
(c) $0 \leq r \leq 1$, $0 \leq \theta \leq \pi/2$, $\phi = \pi/4$.
(d) $0 \leq r \leq 1$, $\theta = \pi/4$, $0 \leq \phi \leq \pi/4$.

3. Use cylindrical coordinates in $\mathcal{R}^3$ to describe the region defined in rectangular coordinates by $0 \leq x$, $x^2 + y^2 \leq 1$.

4. Let (r, θ) be polar coordinates in $\mathcal{R}^2$. The equation

$$\begin{pmatrix} r \\ \theta \end{pmatrix} = \begin{pmatrix} \sin t \\ t \end{pmatrix}, \qquad 0 \leq t \leq \frac{\pi}{2},$$

describes a curve in $\mathcal{U}^2$. Sketch this curve, and sketch its image in $\mathcal{R}^2$ under the polar coordinate transformation.

5. Let (r, ϕ, θ) be spherical coordinates in $\mathcal{R}^3$. The equation

$$\begin{pmatrix} r \\ \phi \\ \theta \end{pmatrix} = \begin{pmatrix} 1 \\ t \\ t \end{pmatrix}$$

determines a curve in $\mathcal{R}^3$ (as well as in the $r\phi\theta$-space $\mathcal{U}^3$). Sketch the curve in $\mathcal{R}^3$. [*Suggestion.* The curve lies on a sphere.]

6. Prove that the Jacobian matrices (5) and (6) of the polar and spherical coordinate transformations given in Examples 2 and 3 have inverses.

7. The equations

$$\left.\begin{aligned} x &= ar \sin \phi \cos \theta \\ y &= br \sin \phi \sin \theta \\ z &= cr \cos \phi \end{aligned}\right\}, \qquad a, b, c > 0,$$

define ellipsoidal coordinates in $\mathcal{R}^3$. For $a = 1$, $b = c = 2$, sketch a typical example of each of the three kinds of coordinate surface.

8. Compute the cartesian components of the tangent vectors to the coordinate curves for the general ellipsoidal coordinates given in Exercise 7, when $a = b = 1$, $c = 2$, and $r = \frac{1}{2}$, $\phi = \theta = \pi/2$.

9. Let r, ϕ, and θ be spherical coordinates in $\mathcal{R}^3$. The equation

$$\begin{pmatrix} r \\ \phi \\ \theta \end{pmatrix} = \begin{pmatrix} 1 \\ t \\ t^2 \end{pmatrix}$$

determines a curve in $\mathcal{R}^3$. Compute the cartesian components of the tangent vector to the curve.

10. Prove that in the 3-dimensional spherical coordinates of Example 2 in the text, the sphere $x_1^2 + x_2^2 + x_3^2 = 1$ has the equation $r = 1$.

11. Let "elliptic" coordinates in the plane be determined by

$$\begin{pmatrix} x \\ y \end{pmatrix} = \begin{pmatrix} ar\cos\theta \\ br\sin\theta \end{pmatrix}, \qquad a > 0, \quad b > 0.$$

(a) Compute the coefficients g_{ij} for this coordinate system.
(b) For what choices of a and b will it always be true that $g_{ij} = 0$ for $i \neq j$?

12. Verify the assertion made in the text that the jth column of the Jacobian matrix at $\mathbf{u}_0$ of a coordinate transformation $\mathcal{U}^n \xrightarrow{T} \mathcal{R}^n$ is a tangent vector at $\mathbf{x}_0 = T(\mathbf{u}_0)$ to the curvilinear coordinate curve obtained by letting the jth coordinate of $\mathbf{u}_0$ vary.

13. Show that if $f(x, y) = \bar{f}(r, \theta)$, where $(x, y) = (r\cos\theta, r\sin\theta)$, then

$$\frac{\partial^2 f}{\partial x^2} + \frac{\partial^2 f}{\partial y^2} = \frac{\partial^2 \bar{f}}{\partial r^2} + \frac{1}{r^2}\frac{\partial^2 \bar{f}}{\partial \theta^2} + \frac{1}{r}\frac{\partial \bar{f}}{\partial r}.$$

14. (a) Find the formula for the arc length of a curve determined in plane polar coordinates by an equation of the form

$$\begin{pmatrix} r \\ \theta \end{pmatrix} = \begin{pmatrix} r(t) \\ \theta(t) \end{pmatrix}, \qquad a \leq t \leq b.$$

(b) Compute the length of the curve $\begin{pmatrix} r \\ \theta \end{pmatrix} = \begin{pmatrix} 2t \\ t \end{pmatrix}, \qquad 0 \leq t \leq 2.$

[*Ans.* $2\sqrt{5} + \log(2 + \sqrt{5})$.]

(c) Sketch the curve.

15. An equation

$$\begin{pmatrix} u \\ v \end{pmatrix} = \begin{pmatrix} u(t) \\ v(t) \end{pmatrix}, \qquad a \leq t \leq b,$$

determines a curve γ on the conical surface in $\mathcal{R}^3$

$$\begin{pmatrix} x \\ y \\ z \end{pmatrix} = \begin{pmatrix} u\cos v\sin\alpha \\ u\sin v\sin\alpha \\ u\cos\alpha \end{pmatrix}, \qquad \begin{cases} 0 \leq u \leq \infty, \\ 0 \leq v \leq 2\pi, \end{cases}$$

where α is fixed, $0 < \alpha < \pi/2$. Find the general formula for the arc length of γ.

16. Let $T(u_1, \ldots, u_n) = (x_1, \ldots, x_n)$ define a curvilinear coordinate system in a region D of $\mathcal{R}^n$.

(a) Show that $g_{ij} = \sum_{k=1}^{n} \frac{\partial x_k}{\partial u_i}\frac{\partial x_k}{\partial u_j}$, for $i, j = 1, \ldots, n$.

(b) If (g^{ij}) is the matrix inverse to (g_{ij}), show that

$$g^{ij} = \sum_{k=1}^{n} \frac{\partial u_i}{\partial x_k}\frac{\partial u_j}{\partial x_k}, \qquad i, j = 1, \ldots, n.$$

(c) Show that if $f(u_1, \ldots, u_n)$ represents a function differentiable in a region D of $\mathcal{R}^n$, then

$$\nabla f = \sum_{j=1}^{n}\left[\sum_{i=1}^{n} \frac{\partial f}{\partial u_i} g^{ij}\right]\mathbf{c}_j.$$

(d) Show that if T defines an orthogonal coordinate system in D, then

$$\nabla f = \sum_{i=1}^{n} \frac{1}{h^2}\frac{\partial f}{\partial u_i}\mathbf{c}_i.$$

SECTION 6

INVERSE AND IMPLICIT FUNCTION THEOREMS

If a function f is thought of as sending vectors $\mathbf{x}$ into vectors $\mathbf{y}$ in the range of f, then it is natural to start with $\mathbf{y}$ and ask what vector or vectors $\mathbf{x}$ are sent by f into $\mathbf{y}$. More particularly, we may ask if there is a function that reverses the action of f. If there is a function f^{-1} with the property

$$f^{-1}(\mathbf{y}) = \mathbf{x} \quad \text{if and only if} \quad f(\mathbf{x}) = \mathbf{y},$$

then f^{-1} is called the **inverse function** of f. It follows that the domain of f^{-1} is the range of f and that the range of f^{-1} is the domain of f. Some familiar examples of functions and their inverses are

$$\begin{cases} f(x) = x^2, & x \geq 0. \\ f^{-1}(y) = \sqrt{y}, & y \geq 0. \end{cases}$$

$$\begin{cases} f(x) = e^x, & -\infty < x < \infty, \\ f^{-1}(y) = \ln y, & y > 0. \end{cases}$$

$$\begin{cases} f(x) = \sin x, & -\dfrac{\pi}{2} \leq x \leq \dfrac{\pi}{2}. \\ f^{-1}(y) = \arcsin y, & -1 \leq y \leq 1. \end{cases}$$

The inverse function f^{-1} should not be confused with the reciprocal $1/f$. For example, if $f(x) = x^2$, then $f^{-1}(2) = \sqrt{2}$, whereas $(f(2))^{-1} = 1/f(2) = \frac{1}{4}$.

We recall that a function is one-to-one if each element in the range is the image of precisely one element in the domain. A fact that is used repeatedly is that a function f has an inverse if and only if it is one-to-one.

The inverse function L^{-1} of every invertible linear function $\mathcal{R}^n \xrightarrow{L} \mathcal{R}^m$ is itself linear. That is, by applying L^{-1} to both sides, we see that

$$L^{-1}(a\mathbf{y}_1 + b\mathbf{y}_2) = aL^{-1}(\mathbf{y}_1) + bL^{-1}(\mathbf{y}_2),$$

whenever $\mathbf{y}_1$ and $\mathbf{y}_2$ are in the range of L. If the dimension of $\mathcal{R}^n$ is less than that of $\mathcal{R}^m$, the range of L is a proper subspace of $\mathcal{R}^m$. In this case, L^{-1} is not defined on all of $\mathcal{R}^m$. On the other hand, if $\mathcal{R}^n$ and $\mathcal{R}^m$ have the same dimension, the domain of L^{-1} is all of $\mathcal{R}^m$. Thus the inverse function of every one-to-one linear function $\mathcal{R}^n \xrightarrow{L} \mathcal{R}^n$ is a linear function $\mathcal{R}^n \xrightarrow{L^{-1}} \mathcal{R}^n$.

Example 1. Consider the affine function $\mathcal{R}^3 \xrightarrow{A} \mathcal{R}^3$ defined by

$$A\begin{pmatrix} x \\ y \\ z \end{pmatrix} = \begin{pmatrix} 4 & 0 & 5 \\ 0 & 1 & -6 \\ 3 & 0 & 4 \end{pmatrix}\begin{pmatrix} x-1 \\ y-0 \\ z-1 \end{pmatrix} + \begin{pmatrix} 1 \\ 5 \\ 2 \end{pmatrix}.$$

It is obvious that any affine function $A(\mathbf{x}) = L(\mathbf{x} - \mathbf{x}_0) + \mathbf{y}_0$ is one-to-one if and only if the linear function L is one-to-one also. In this example,

$$\mathbf{x}_0 = \begin{pmatrix} 1 \\ 0 \\ 1 \end{pmatrix} \quad \text{and} \quad L(\mathbf{x}) = \begin{pmatrix} 4 & 0 & 5 \\ 0 & 1 & -6 \\ 3 & 0 & 4 \end{pmatrix}\begin{pmatrix} x \\ y \\ z \end{pmatrix}.$$

The inverse matrix of

$$\begin{pmatrix} 4 & 0 & 5 \\ 0 & 1 & -6 \\ 3 & 0 & 4 \end{pmatrix}$$

can be computed to be

$$\begin{pmatrix} 4 & 0 & -5 \\ -18 & 1 & 24 \\ -3 & 0 & 4 \end{pmatrix}.$$

It follows that L, and therefore A, has an inverse. In fact, if $A(\mathbf{x}) = \mathbf{y}$, then

$$A(\mathbf{x}) = L(\mathbf{x} - \mathbf{x}_0) + \mathbf{y}_0$$

and

$$A^{-1}(\mathbf{y}) = L^{-1}(\mathbf{y} - \mathbf{y}_0) + \mathbf{x}_0. \tag{1}$$

That this is the correct expression for A^{-1} may be checked by substituting $A(\mathbf{x})$ for $\mathbf{y}$. We get

$$A^{-1}(A(\mathbf{x})) = L^{-1}(L(\mathbf{x} - \mathbf{x}_0)) + \mathbf{x}_0 = \mathbf{x}.$$

Hence

$$A^{-1}\begin{pmatrix} u \\ v \\ w \end{pmatrix} = \begin{pmatrix} 4 & 0 & -5 \\ -18 & 1 & 24 \\ -3 & 0 & 4 \end{pmatrix}\begin{pmatrix} u-1 \\ v-5 \\ w-2 \end{pmatrix} + \begin{pmatrix} 1 \\ 0 \\ 1 \end{pmatrix}.$$

Obviously this method will enable us to find the inverse of any affine transformation $\mathcal{R}^n \xrightarrow{A} \mathcal{R}^n$ if the inverse exists.

We have the following criteria for deciding whether a linear function $\mathcal{R}^n \xrightarrow{L} \mathcal{R}^m$ has an inverse. If M is the matrix of L, then by Theorem 4.2 of Chapter 1 the columns of M are the vectors $L(\mathbf{e}_j)$ and so span the range of L. Hence L is one-to-one, and has an inverse, if and only if the columns of M are linearly independent. Alternatively, *if M is a square matrix*, then L has an inverse if and only if the inverse matrix M^{-1} exists. We recall that M^{-1} exists if and only if $\det M \neq 0$.

The principal purpose of this section is the study of inverses of non-linear vector functions. Given a function $\mathcal{R}^n \xrightarrow{f} \mathcal{R}^n$ one may ask: (1) Does it have an inverse? and (2) If it does, what are its properties? In general it is not easy to answer these questions just by looking at the function. On the other hand, we do know how to tell whether or not an affine transformation has an inverse and, what is more, how to compute it explicitly when it does exist. Furthermore, if f is differentiable at a point $\mathbf{x}_0$, it can be approximated near that point by an affine transformation A. For this reason, one might conjecture that if the domain of f is restricted to points close to $\mathbf{x}_0$, then f will have an inverse if A does. In addition, one might guess that A^{-1} is the approximating affine transformation to f^{-1} near $f(\mathbf{x}_0)$. Except for details, these statements are correct and are the content of the inverse function theorem.

6.1 The Inverse Function Theorem

Let $\mathcal{R}^n \xrightarrow{f} \mathcal{R}^n$ be a continuously differentiable function such that $f'(\mathbf{x}_0)$ has an inverse. Then there is an open set N containing $\mathbf{x}_0$ such that f, when restricted to N, has a continuously differentiable inverse f^{-1}. The image set $f(N)$ is open. In addition,

$$[f^{-1}]'(\mathbf{y}_0) = [f'(\mathbf{x}_0)]^{-1},$$

where $\mathbf{y}_0 = f(\mathbf{x}_0)$. That is, the differential of the inverse function at $\mathbf{y}_0$ is the inverse of the differential of f at $\mathbf{x}_0$.

The existence of f^{-1} is proved in the Appendix. Once the existence has been established, we can write $f^{-1} \circ f = I$, where $\mathcal{R}^n \xrightarrow{I} \mathcal{R}^n$ is the identity transformation on the neighborhood N. Then by the chain rule we have, since the identity transformation is its own differential,

$$[f^{-1}]'(\mathbf{y}_0)f'(\mathbf{x}_0) = I \quad \text{or} \quad [f^{-1}]'(\mathbf{y}_0) = [f'(\mathbf{x}_0)]^{-1}.$$

For real-valued functions of one variable, the existence of an inverse function is not hard to prove. Let $\mathcal{R} \xrightarrow{f} \mathcal{R}$ satisfy the differentiability condition of the theorem, and suppose that $f'(x_0)$ has a matrix inverse. Since the inverse matrix exists whenever $f'(x_0) \neq 0$, the geometric meaning of the condition that $f'(x_0)$ have an inverse is that the graph of f should not have a horizontal tangent. To be specific, suppose that $f'(x_0) > 0$. Since f' is continuous, we have $f'(x) > 0$ for every x in some interval $a < x < b$ that contains x_0, as shown in Fig. 16. We contend that f restricted to this

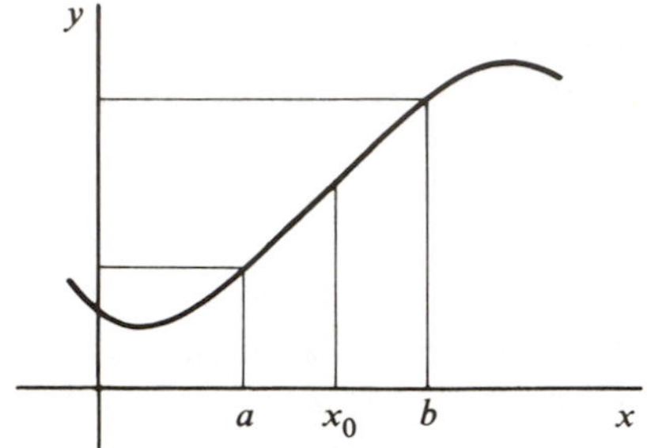

Figure 16

interval is one-one. For suppose x_1 and x_2 are any two points in the interval such that $x_1 < x_2$. By the mean-value theorem it follows that

$$\frac{f(x_2) - f(x_1)}{x_2 - x_1} = f'(x_0),$$

for some x_0 in the interval $x_1 < x < x_2$. Since $f'(x_0) > 0$, and $x_2 - x_1 > 0$, we obtain

$$f(x_2) - f(x_1) > 0.$$

Thus, f is strictly increasing in the interval $a < x < b$, and our contention is proved. It follows that f restricted to this interval has an inverse. The other conclusions of the inverse function theorem can also be obtained in a straightforward way for this special case.

Example 2. Consider the function f defined by

$$f\begin{pmatrix} x \\ y \end{pmatrix} = \begin{pmatrix} x^3 - 2xy^2 \\ x + y \end{pmatrix}, \qquad \begin{cases} -\infty < x < \infty, \\ -\infty < y < \infty. \end{cases}$$

At the point

$$\mathbf{x}_0 = \begin{pmatrix} 1 \\ -1 \end{pmatrix}$$

the differential $d_{\mathbf{x}_0}f$ is defined by the Jacobian matrix

$$\begin{pmatrix} 3x^2 - 2y^2 & -4xy \\ 1 & 1 \end{pmatrix}_{x=1,\, y=-1} = \begin{pmatrix} 1 & 4 \\ 1 & 1 \end{pmatrix}.$$

The inverse of this matrix is

$$\begin{pmatrix} -\frac{1}{3} & \frac{4}{3} \\ \frac{1}{3} & -\frac{1}{3} \end{pmatrix}.$$

Since f is obviously differentiable, we conclude from the inverse function theorem that in some open set containing $\mathbf{x}_0$ the function f has an inverse f^{-1}. Moreover, if

$$\mathbf{y}_0 = f(\mathbf{x}_0) = \begin{pmatrix} -1 \\ 0 \end{pmatrix},$$

the matrix of the differential $d_{\mathbf{y}_0} f^{-1}$ is

$$\begin{pmatrix} -\frac{1}{3} & \frac{4}{3} \\ \frac{1}{3} & -\frac{1}{3} \end{pmatrix}.$$

Although it would be difficult to evaluate f^{-1} explicitly, it is easy to write down the affine transformation that approximates f^{-1} in the vicinity of the point $\mathbf{y}_0$. It is the inverse A^{-1} of the affine transformation A that approximates f near $\mathbf{x}_0$. We have, either by the inverse function theorem or by Formula (1) of Example 1,

$$\begin{aligned} A(\mathbf{x}) &= f(\mathbf{x}_0) + f'(\mathbf{x}_0)(\mathbf{x} - \mathbf{x}_0) \\ &= \mathbf{y}_0 + f'(\mathbf{x}_0)(\mathbf{x} - \mathbf{x}_0). \\ A^{-1}(\mathbf{y}) &= f^{-1}(\mathbf{y}_0) + [f^{-1}]'(\mathbf{y}_0)(\mathbf{y} - \mathbf{y}_0) \\ &= \mathbf{x}_0 + [f'(\mathbf{x}_0)]^{-1}(\mathbf{y} - \mathbf{y}_0). \end{aligned}$$

Hence, if we set $\mathbf{y} = \begin{pmatrix} u \\ v \end{pmatrix}$,

$$\begin{aligned} A^{-1}\begin{pmatrix} u \\ v \end{pmatrix} &= \begin{pmatrix} 1 \\ -1 \end{pmatrix} + \begin{pmatrix} -\frac{1}{3} & \frac{4}{3} \\ \frac{1}{3} & -\frac{1}{3} \end{pmatrix}\begin{pmatrix} u + 1 \\ v - 0 \end{pmatrix} \\ &= \begin{pmatrix} -\frac{1}{3} & \frac{4}{3} \\ \frac{1}{3} & -\frac{1}{3} \end{pmatrix}\begin{pmatrix} u \\ v \end{pmatrix} + \begin{pmatrix} \frac{2}{3} \\ -\frac{2}{3} \end{pmatrix}. \end{aligned}$$

Example 3. The equations

$$u = x^4 y + x, \qquad v = x + y^3$$

define a transformation from $\mathcal{R}^2$ to $\mathcal{R}^2$. The matrix of the differential of the transformation at $(x, y) = (1, 1)$ is

$$\begin{pmatrix} 4x^3 y + 1 & x^4 \\ 1 & 3y^2 \end{pmatrix}_{(x,y)=(1,1)} = \begin{pmatrix} 5 & 1 \\ 1 & 3 \end{pmatrix}.$$

Since the columns of this matrix are independent, the differential has an inverse, and according to the inverse function theorem the transformation has an inverse also, in an open neighborhood of $(x, y) = (1, 1)$. The inverse transformation would be given by equations of the form

$$x = F(u, v), \qquad y = G(u, v).$$

The actual computation of F and G is difficult, but we can easily compute the partial derivatives of F and G with respect to u and v at the point $(u, v) = (2, 2)$ that corresponds to $(x, y) = (1, 1)$. These partial derivatives occur in the Jacobian matrix of F and G or, equivalently, in the inverse matrix of the differential of the given functions. We have

$$\begin{pmatrix} \dfrac{\partial F}{\partial u}(2, 2) & \dfrac{\partial F}{\partial v}(2, 2) \\ \dfrac{\partial G}{\partial u}(2, 2) & \dfrac{\partial G}{\partial v}(2, 2) \end{pmatrix} = \begin{pmatrix} 5 & 1 \\ 1 & 3 \end{pmatrix}^{-1} = \begin{pmatrix} \dfrac{3}{14} & -\dfrac{1}{14} \\ -\dfrac{1}{14} & \dfrac{5}{14} \end{pmatrix}.$$

Suppose $\mathcal{R}^n \xrightarrow{f} \mathcal{R}^n$ is a function for which the hypotheses of the inverse function theorem are satisfied at some point $\mathbf{x}_0$. It is important to realize that the theorem does not settle the question of the existence of an inverse for the whole function f, but only for f restricted to some open set containing $\mathbf{x}_0$. For example, the transformation

$$\begin{pmatrix} x \\ y \end{pmatrix} = \begin{pmatrix} u \cos v \\ u \sin v \end{pmatrix}, \qquad 0 < u,$$

has Jacobian matrix

$$\begin{pmatrix} \cos v & -u \sin v \\ \sin v & u \cos v \end{pmatrix}$$

with inverse matrix

$$\begin{pmatrix} \cos v & \sin v \\ -\dfrac{1}{u} \sin v & \dfrac{1}{u} \cos v \end{pmatrix}.$$

The inverse matrix exists for all (u, v) satisfying $u > 0$. However, the otherwise unrestricted transformation clearly has no inverse, for the same image point is obtained whenever v increases by 2π. Two corresponding regions are shown in Fig. 17. If the transformation is restricted so that, for instance, $0 < v < 2\pi$, then it becomes one-to-one and has an inverse.

In Section 4 we considered the problem of finding derivatives of an implicitly defined function f under the assumption that f satisfied an equation $F(\mathbf{x}, f(\mathbf{x})) = 0$, with both f and F differentiable. We saw that to solve for $f'(\mathbf{x}_0)$ by matrix methods it was necessary for $F_{\mathbf{y}}(\mathbf{x}_0, f(\mathbf{x}_0))$ to have an inverse. It is natural that the same condition occur in the next

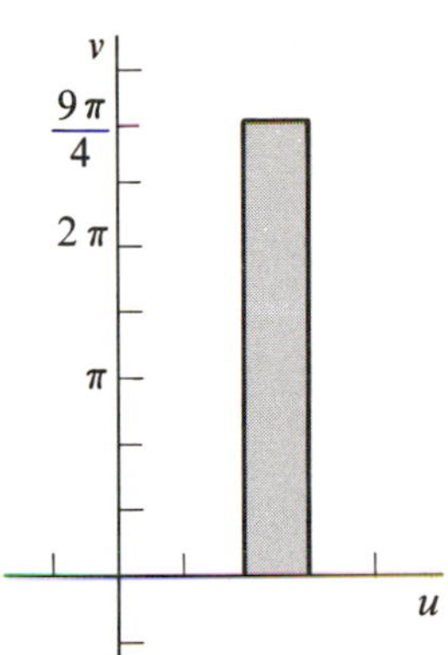

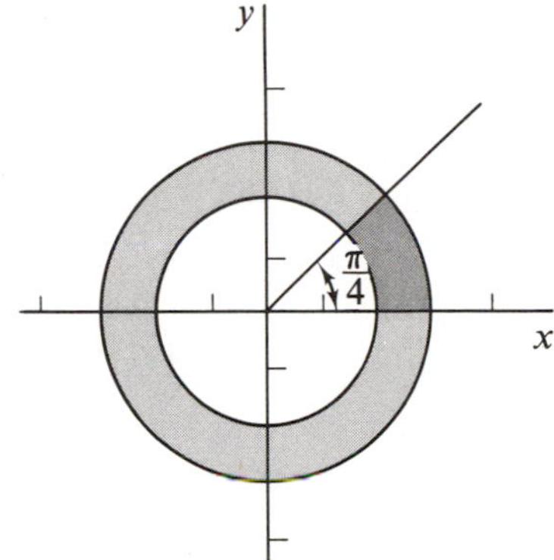

Figure 17

theorem, which treats the question whether there exists a differentiable f defined implicitly by F. The proof can be made to depend on the inverse function theorem, and we give both proofs together in Section 5 of the Appendix.

6.2 Implicit Function Theorem

Let $\mathcal{R}^{n+m} \xrightarrow{F} \mathcal{R}^m$ be a continuously differentiable function. Suppose that for some $\mathbf{x}_0$ in $\mathcal{R}_n$ and $\mathbf{y}_0$ in $\mathcal{R}^m$

1. $F(\mathbf{x}_0, \mathbf{y}_0) = 0$.
2. $F_{\mathbf{y}}(\mathbf{x}_0, \mathbf{y}_0)$ has an inverse.

Then there exists a continuously differentiable function $\mathcal{R}^n \xrightarrow{f} \mathcal{R}^m$ defined on some neighborhood N of $\mathbf{x}_0$ such that $f(\mathbf{x}_0) = \mathbf{y}_0$ and $F(\mathbf{x}, f(\mathbf{x})) = 0$, for all $\mathbf{x}$ in N.

As we showed in Theorem 4.2, the derivative of f is then given by

$$f'(\mathbf{x}) = -\,F_{\mathbf{y}}^{-1}(\mathbf{x}, f(\mathbf{x}))F_{\mathbf{x}}(\mathbf{x}, f(\mathbf{x})).$$

Example 4. The equation $x^3y + y^3x - 2 = 0$ defines $y = f(x)$ implicitly in a neighborhood of $x = 1$ if $f(1) = 1$. As a function of y, $x^3y + y^3x - 2$ has Jacobian $(1 + 3y^2)$ at $x = 1$, and the latter is invertible at $y = 1$, that is,

$$1 + 3y^2\big|_{y=1} = 4 \neq 0.$$

The solution can be computed by standard methods to be

$$y = \sqrt[3]{\frac{1}{x} + \frac{1}{x}\sqrt{\frac{x^{10} + 27}{27}}} + \sqrt[3]{\frac{1}{x} - \frac{1}{x}\sqrt{\frac{x^{10} + 27}{27}}}.\dagger$$

Example 5. The equations

$$z^3x + w^2y^3 + 2xy = 0, \qquad xyzw - 1 = 0 \tag{2}$$

can be written in the form $F(\mathbf{x}, \mathbf{y}) = 0$, where $\mathbf{x} = \begin{pmatrix} x \\ y \end{pmatrix}$, $\mathbf{y} = \begin{pmatrix} z \\ w \end{pmatrix}$, and

$$F(\mathbf{x}, \mathbf{y}) = \begin{pmatrix} z^3x + w^2y^3 + 2xy \\ xyzw - 1 \end{pmatrix}.$$

Let $\mathbf{x}_0 = \begin{pmatrix} -1 \\ -1 \end{pmatrix}$ and $\mathbf{y}_0 = \begin{pmatrix} 1 \\ 1 \end{pmatrix}$. Then

$$F(\mathbf{x}_0, \mathbf{y}) = \begin{pmatrix} -z^3 - w^2 + 2 \\ zw - 1 \end{pmatrix}$$

and the matrix $F_{\mathbf{y}}(1, 1)$ is

$$\begin{pmatrix} -3z^2 & 2w \\ w & z \end{pmatrix}_{\binom{z}{w}=\binom{1}{1}} = \begin{pmatrix} -3 & -2 \\ 1 & 1 \end{pmatrix}.$$

The inverse exists and is the matrix

$$\begin{pmatrix} -1 & -2 \\ 1 & 3 \end{pmatrix}.$$

It is then a consequence of the implicit function theorem that Equations (2) implicitly define a function f in an open set about $\mathbf{x}_0$ such that $f(\mathbf{x}_0) = \mathbf{y}_0$. That is, we have

$$\begin{pmatrix} z \\ w \end{pmatrix} = f\begin{pmatrix} x \\ y \end{pmatrix},$$

and so each of z and w is a function of x and y near

$$\begin{pmatrix} -1 \\ -1 \end{pmatrix}.$$

EXERCISES

1. Can a function have two different inverses?

2. Show that $(f^{-1})^{-1} = f$.

† See R. S. Burington, *Mathematical Tables and Formulas*, Handbook Publishers, Inc., Sandusky, Ohio, 1965.

3. Which of the following functions have inverses?

(a) $y = \cosh x, \ -\infty < x < \infty$.
(b) $y = \cosh x, 0 \leq x < \infty$.
(c) $f(x) = \tan x, \begin{cases} \frac{7}{4}\pi \leq x < \frac{11}{4}\pi, \\ x \neq \frac{10}{4}\pi. \end{cases}$
(d) $f(x) = \tan x, 0 \leq x \leq \pi/4$.
(e) $y = x^2 - 2x + 1, 1 \leq x < \infty$.
(f) $y = x^2 - 3x + 2, 0 \leq x < \infty$.

4. Compute A^{-1} for the following affine functions:
(a) $A(x) = 7x + 2$.

(b) $A\begin{pmatrix} u \\ v \end{pmatrix} = \begin{pmatrix} 1 & 3 \\ 2 & 4 \end{pmatrix}\begin{pmatrix} u - 1 \\ v - 2 \end{pmatrix} + \begin{pmatrix} 3 \\ 4 \end{pmatrix}$.

$$\left[\textit{Ans. } A^{-1}\begin{pmatrix} x \\ y \end{pmatrix} = \begin{pmatrix} -2 & \frac{3}{2} \\ 1 & -\frac{1}{2} \end{pmatrix}\begin{pmatrix} x - 3 \\ y - 4 \end{pmatrix} + \begin{pmatrix} 1 \\ 2 \end{pmatrix}.\right]$$

5. Using the inverse function theorem, show that the following functions have inverses when restricted to some open set containing x_0.

(a) $f(x) = \tan x, x_0 = \pi/6$.
(b) $y = x^2 - 3x + 2, x_0 = 4$.
(c) $y = x^3 - 7x + 6, x_0 = 4$.

(d) $f(x) = \int_{-\infty}^{x} e^{-t^2}\,dt, x_0 = 0$.

6. Let $f\begin{pmatrix} x \\ y \end{pmatrix} = \begin{pmatrix} x^2 - y^2 \\ 2xy \end{pmatrix}$.

(a) Show that, for every point $\mathbf{x}_0$ except

$$\mathbf{x}_0 = \begin{pmatrix} 0 \\ 0 \end{pmatrix},$$

the restriction of f to some open set containing $\mathbf{x}_0$ has an inverse.

(b) Show that, with domain unrestricted, f has no inverse.

(c) If f^{-1} is the inverse of f in a neighborhood of the point $\begin{pmatrix} 1 \\ 2 \end{pmatrix}$, compute the affine transformation that approximates f^{-1} close to

$$f\begin{pmatrix} 1 \\ 2 \end{pmatrix} = \begin{pmatrix} -3 \\ 4 \end{pmatrix}.$$

$$\left[\textit{Ans. } \begin{pmatrix} \frac{1}{10} & \frac{1}{5} \\ -\frac{1}{5} & \frac{1}{10} \end{pmatrix}\begin{pmatrix} u \\ v \end{pmatrix} + \begin{pmatrix} \frac{1}{2} \\ 1 \end{pmatrix}.\right]$$

7. Find the affine function that best approximates the inverse of the function

$$f\begin{pmatrix} x \\ y \end{pmatrix} = \begin{pmatrix} x^3 + 2xy + y^2 \\ x^2 + y \end{pmatrix}$$

near the point $f\begin{pmatrix} 1 \\ 1 \end{pmatrix}$. Notice that to find the precise inverse would be difficult.

$$\left[\textit{Ans.} \begin{pmatrix} -\frac{1}{3} & \frac{4}{3} \\ \frac{2}{3} & -\frac{5}{3} \end{pmatrix} \begin{pmatrix} u \\ v \end{pmatrix} + \begin{pmatrix} -\frac{1}{3} \\ \frac{5}{3} \end{pmatrix}. \right]$$

8. (a) Let T be defined by

$$\begin{pmatrix} x \\ y \end{pmatrix} = T\begin{pmatrix} r \\ \theta \end{pmatrix} = \begin{pmatrix} r\cos\theta \\ r\sin \end{pmatrix}, \qquad \begin{cases} 0 < r, \\ 0 \le \theta < 2\pi. \end{cases}$$

Find $T'(\mathbf{u})$ and its inverse for those points

$$\mathbf{u} = \begin{pmatrix} r \\ \theta \end{pmatrix}$$

for which they exist.

(b) Let S be defined by

$$\begin{pmatrix} x \\ y \\ z \end{pmatrix} = S\begin{pmatrix} r \\ \phi \\ \theta \end{pmatrix} = \begin{pmatrix} r\sin\phi\cos\theta \\ r\sin\phi\sin\theta \\ r\cos\phi \end{pmatrix}, \qquad \begin{cases} 0 < r, \\ 0 < \phi < \pi/2, \\ 0 < \theta < 2\pi. \end{cases}$$

Find $S'(\mathbf{u})$ and its inverse for those points

$$\mathbf{u} = \begin{pmatrix} r \\ \phi \\ \theta \end{pmatrix}$$

for which they exist.

(c) Compute an explicit representation for S^{-1}.

9. Suppose that the function T defined by

$$\begin{pmatrix} u \\ v \end{pmatrix} = T\begin{pmatrix} x \\ y \end{pmatrix} = \begin{pmatrix} f(x, y) \\ g(x, y) \end{pmatrix}$$

has a differentiable inverse function S defined by

$$\begin{pmatrix} x \\ y \end{pmatrix} = S\begin{pmatrix} u \\ v \end{pmatrix} = \begin{pmatrix} h(u, v) \\ k(u, v) \end{pmatrix}.$$

If $f(1, 2) = 3$, $g(1, 2) = 4$, and $T'(1, 2)$ equals

$$\begin{pmatrix} 3 & 5 \\ 4 & 7 \end{pmatrix}, \qquad \text{find } \frac{\partial h}{\partial v}(3, 4). \qquad [\textit{Ans.} -5.]$$

10. If

$$\begin{cases} x = u + v + w, \\ y = u^2 + v^2 + w^2, \\ z = u^3 + v^3 + w^3, \end{cases}$$

compute $\partial v / \partial y$ at the image of $(u, v, w) = (1, 2, -1)$, namely, $(x, y, z) = (2, 6, 8)$. [*Ans.* 0.]

11. Let

$$f\begin{pmatrix} u \\ v \end{pmatrix} = \begin{pmatrix} u^2 + u^2 v + 10v \\ u + v^3 \end{pmatrix}.$$

(a) Show that f has an inverse f^{-1} in the vicinity of the point $\begin{pmatrix} 1 \\ 1 \end{pmatrix}$.

(b) Find an approximate value of $f^{-1}\begin{pmatrix} 11.8 \\ 2.2 \end{pmatrix}$.

12. Does $f(t) = \begin{pmatrix} t \\ t \\ t \end{pmatrix}$ have an inverse?

13. Show that the differentiable function

$$F(x, y, z) = \begin{pmatrix} f(x, y, z) \\ g(x, y, z) \\ f(x, y, z) + g(x, y, z) \end{pmatrix}$$

can never have a differentiable inverse.

14. Although the condition that the differential $d_{\mathbf{x}_0} f$ have an inverse is needed for the proof of the inverse function theorem, it is perfectly possible for this condition to fail even though an inverse exists. Verify this fact with the example $f(x) = x^3$.

15. The inverse function theorem is the correct modification of the simple but false assertion that if $d_{\mathbf{x}} f$ has an inverse, then f has an inverse. The converse—namely, if f has an inverse, then $d_{\mathbf{x}} f$ has an inverse—is also false (see Exercise 14). It too, however, is almost true. Using the chain rule, prove the corrected form: *If f is differentiable and has a differentiable inverse, then $d_{\mathbf{x}} f$ is one-to-one.*

16. Consider the function $\mathcal{R} \xrightarrow{f} \mathcal{R}$ defined by

$$f(x) = \begin{cases} \dfrac{x}{2} + x^2 \sin \dfrac{1}{x}, & \text{if } x \neq 0. \\ 0, & \text{if } x = 0. \end{cases}$$

Show that $d_0 f$ is one-to-one but that f has no inverse in the vicinity of $x = 0$. What does the example show about the inverse function theorem?

17. What is the inverse function of the linear function

$$L\begin{pmatrix} u \\ v \end{pmatrix} = \begin{pmatrix} u + v \\ u - v \\ u \end{pmatrix}?$$

18. The inverse function theorem can be generalized as follows:

Let $\mathcal{R}^n \xrightarrow{f} \mathcal{R}^m$, *where* $n < m$, *be continuously differentiable. If* $d_{\mathbf{x}_0} f$ *is one-to-one, there is an open set* N *containing* $\mathbf{x}_0$ *such that* f, *restricted to* N, *has an inverse* f^{-1}.

An important difference between this theorem and the inverse function theorem as we have stated it is that here the image $f(N)$ is not an open subset of $\mathcal{R}^m$. One consequence of this is that f^{-1} is not differentiable at $f(\mathbf{x}_0)$.

(a) If

$$f\begin{pmatrix} u \\ v \end{pmatrix} = \begin{pmatrix} u + v \\ (u + v)^2 \\ (u + v)^3 \end{pmatrix},$$

for what points (u_0, v_0) does f have an inverse in a neighborhood of $f(u_0, v_0)$?

(b) Prove the generalized inverse function theorem. [*Hint.* Let the vectors $\mathbf{y}_1, \ldots, \mathbf{y}_n$ be a basis for the range of $d_{\mathbf{x}_0} f$. Extend to a basis $\mathbf{y}_1, \ldots, \mathbf{y}_n, \mathbf{y}_{n+1}, \ldots, \mathbf{y}_m$ for all of $\mathcal{R}^m$, and define $\mathcal{R}^m \xrightarrow{g} \mathcal{R}^n$ by

$$g\left(\sum_{i=1}^{m} a_i \mathbf{y}_i\right) = (a_1, \ldots, a_n).$$

Show that $(g \circ f)$ and $d_{\mathbf{x}_0}(g \circ f) = (d_{f(\mathbf{x}_0)} g) \circ (d_{\mathbf{x}_0} f)$ satisfy the condition of the inverse function theorem.]

19. Consider the equation $(x - 2)^3 y + x e^{y-1} = 0$.

(a) Is y defined implicitly as a function of x in a neighborhood of $(x, y) = (1, 1)$?
(b) In a neighborhood of $(0, 0)$?
(c) In a neighborhood of $(2, 1)$?

20. The point $(x, y, t) = (0, 1, -1)$ satisfies the equations

$$xyt + \sin xyt = 0, \qquad x + y + t = 0.$$

Are x and y defined implicitly as functions of t in a neighborhood of $(0, 1, -1)$?

21. Requirement 2 in the implicit function theorem that $F_{\mathbf{y}}(\mathbf{x}_0, \mathbf{y}_0)$ have an inverse is not a necessary condition for the equation $F(\mathbf{x}, \mathbf{y}) = 0$ to define a unique differentiable function f such that $f(\mathbf{x}_0) = \mathbf{y}_0$. Show this by taking $F(x, y) = x^9 - y^3$ and $(x_0, y_0) = (0, 0)$.

22. Show that if N' is an open subset of $\mathcal{R}^{n+m}$ containing the point $(\mathbf{x}_0, \mathbf{y}_0)$, then the subset N of all $\mathbf{x}$ in $\mathcal{R}^n$ such that $(\mathbf{x}, \mathbf{y}_0)$ is in N' is an open subset of $\mathcal{R}^n$.

23. Prove that under the assumptions of Theorem 6.2 there is only one function f defined by $F(\mathbf{x}, f(\mathbf{x})) = 0$ in a neighborhood of $\mathbf{x}_0$ and satisfying $f(\mathbf{x}_0) = \mathbf{y}_0$. [*Hint.* Use the function $H(\mathbf{x}, \mathbf{y}) = (\mathbf{x}, F(\mathbf{x}\ \ \mathbf{y}))$.]

SECTION 7

SURFACES AND TANGENTS

While explicit, implicit, and parametric representations of surfaces have so far been used for illustrative purposes, a precise definition of the term "surface" has not been given. In this section we shall define a smooth surface in terms of each of the three representations, give a unified definition of tangent for each mode of representation, and show how the three are related. In particular, we shall see that for each of the three ways of representing a surface—as an image, as a graph, or as a level set—a representation for a tangent is obtained by taking the image, graph, or level set of the affine approximation to the given function.

An n-dimensional **plane** in $\mathcal{R}^m$ is either an n-dimensional linear subspace (that is, the set spanned by n linearly independent vectors) or else an affine subspace (that is, the translation of a linear subspace by a fixed vector $\mathbf{y}_0$). When $n = 1$ or $n = 2$ we get a line or an ordinary plane. A parametric representation of an n-dimensional plane in $\mathcal{R}^m$ is obtained by looking at the range of an affine function $\mathcal{R}^n \xrightarrow{A} \mathcal{R}^m$, where $A(\mathbf{x}) = L(\mathbf{x}) + \mathbf{y}_0$, and L is a linear function defined on $\mathcal{R}^n$. To ensure that the range of A is n-dimensional, we can require that the matrix of L have n linearly independent columns, since the columns $L(\mathbf{e}_j)$ span the range of L.

Parametrically Defined Surfaces

Let a set S be defined parametrically by a function $\mathcal{R}^n \xrightarrow{f} \mathcal{R}^m$. According to the definition of Section 1 of Chapter 3 this means that S is the image under f of the domain of f. Next we restrict f to a neighborhood of some point $\mathbf{x}_0$ on which f is one-to-one. If f is differentiable at $\mathbf{x}_0$, then the affine function A that approximates f near $\mathbf{x}_0$ is given by $A(\mathbf{x}) = f(\mathbf{x}_0) + f'(\mathbf{x}_0)(\mathbf{x} - \mathbf{x}_0)$, for all $\mathbf{x}$ in $\mathcal{R}^n$. Then, if A defines parametrically an n-dimensional plane, this plane is called the **tangent plane** to S at $f(\mathbf{x}_0)$. Notice that n, the dimension of the plane, is required to be the same as the dimension of the domain space of f. If, in addition, f is continuously differentiable on its domain, then the set S is called a **smooth surface** (or **smooth curve** if $n = 1$) at every point at which there is a tangent.

Example 1. Consider the surface S in 3-dimensional space $\mathcal{R}^3$ defined parametrically by

$$f\begin{pmatrix} u \\ v \end{pmatrix} = \begin{pmatrix} u \cos v \\ u \sin v \\ v \end{pmatrix}, \qquad \begin{cases} 0 \le u \le 4, \\ 0 \le v \le 2\pi. \end{cases}$$

This function is discussed in Example 7, Section 1 Chapter 3, and its range, which is the surface S, is pictured in Fig. 6. At $(u_0, v_0) = (2, \pi/2)$ the matrix of the differential is

$$\begin{pmatrix} \cos v_0 & -u_0 \sin v_0 \\ \sin v_0 & u_0 \cos v_0 \\ 0 & 1 \end{pmatrix} = \begin{pmatrix} 0 & -2 \\ 1 & 0 \\ 0 & 1 \end{pmatrix}.$$

The affine function $A(\mathbf{x}) = f(\mathbf{x}_0) + f'(\mathbf{x}_0)(\mathbf{x} - \mathbf{x}_0)$ that approximates f near $\mathbf{x}_0$ is therefore given by

$$A\begin{pmatrix} u \\ v \end{pmatrix} = \begin{pmatrix} 0 \\ 2 \\ \frac{\pi}{2} \end{pmatrix} + \begin{pmatrix} 0 & -2 \\ 1 & 0 \\ 0 & 1 \end{pmatrix} \begin{pmatrix} u - 2 \\ v - \frac{\pi}{2} \end{pmatrix}$$

$$= u\begin{pmatrix} 0 \\ 1 \\ 0 \end{pmatrix} + v\begin{pmatrix} -2 \\ 0 \\ 1 \end{pmatrix} + \begin{pmatrix} \pi \\ 0 \\ 0 \end{pmatrix},$$

for all u and v. Since the vectors

$$\begin{pmatrix} 0 \\ 1 \\ 0 \end{pmatrix} \quad \text{and} \quad \begin{pmatrix} -2 \\ 0 \\ 1 \end{pmatrix}$$

are linearly independent, we conclude that the range of A is a plane. Hence, the surface S has a tangent plane at $(0, 2, \pi/2)$. Eliminating u and v from the equations

$$x = -2v + \pi$$
$$y = u$$
$$z = v,$$

we obtain

$$x = -2z + \pi$$

as the equation that implicitly defines the tangent plane to S at $(0, 2, \pi/2)$. See Fig. 18.

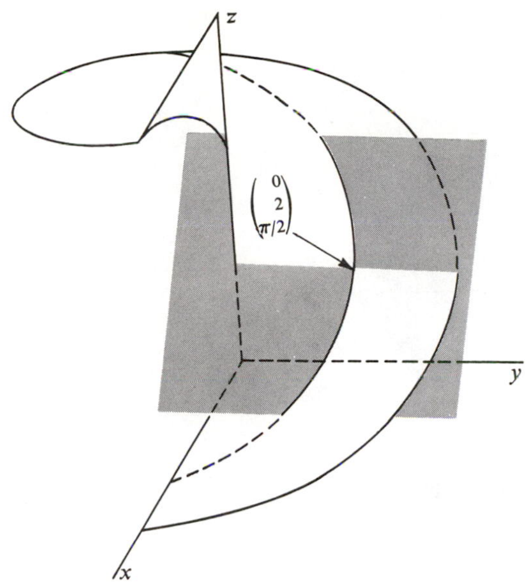

Figure 18

Example 2. The function of t defined by

$$f(t) = \begin{pmatrix} x \\ y \\ z \end{pmatrix} = \begin{pmatrix} t \\ t^2 \\ t^3 \end{pmatrix}, \qquad -\infty < t < \infty,$$

and discussed in Example 6, Section 1 of Chapter 3, parametrically defines the curve shown in Fig. 19. The differential of f at t_0 is defined by the Jacobian matrix

$$\begin{pmatrix} 1 \\ 2t_0 \\ 3t_0^2 \end{pmatrix},$$

and the affine function

$$A(t) = f(t_0) + f'(t_0)(t - t_0)$$

that approximates f near t_0 is given by

$$A(t) = \begin{pmatrix} t_0 \\ t_0^2 \\ t_0^3 \end{pmatrix} + (t - t_0)\begin{pmatrix} 1 \\ 2t_0 \\ 3t_0^2 \end{pmatrix}$$

$$= t\begin{pmatrix} 1 \\ 2t_0 \\ 3t_0^2 \end{pmatrix} - \begin{pmatrix} 0 \\ t_0^2 \\ 2t_0^2 \end{pmatrix}, \qquad -\infty < t < \infty.$$

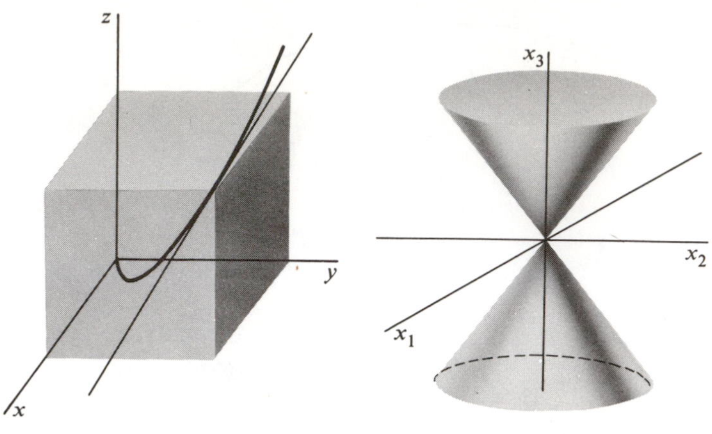

Figure 19

Figure 20

Since $(1, 2t_0, 3t_0^2) \neq 0$, it follows that the range of A is the tangent line to the curve at $f(t_0)$. Figure 19 shows the curve and the tangent line to it at $f(1) = (1, 1, 1)$.

The condition that the affine approximation to $\mathcal{R}^n \xrightarrow{f} \mathcal{R}^m$ defines an n-dimensional plane is important both because of the restriction it places on the tangent and because of the smoothness that it requires of the surface if f is continuously differentiable. As far as the tangent goes, it is clear that its dimension at $f(\mathbf{x}_0)$ is the same as the dimension of the range of the differential of f at $\mathbf{x}_0$. But because the columns of the m-by-n Jacobian matrix $f'(\mathbf{x}_0)$ span the range of the differential, it is enough to require this matrix to have n linearly independent columns. That is,

7.1 Theorem

Let $\mathcal{R}^n \xrightarrow{f} \mathcal{R}^m$ be differentiable. Then the tangent to the range of f at $f(\mathbf{x}_0)$ exists (and has dimension n) if and only if the matrix $f'(\mathbf{x}_0)$ has n linearly independent columns.

The requirement that f be continuously differentiable signifies for a smooth surface S that the tangent varies continuously from point to point on S. To see the effect of the dimension requirement for a smooth surface, we consider the following example.

Example 3. The function $\mathcal{R}^2 \xrightarrow{f} \mathcal{R}^3$ defined by

$$f(u, v) = \begin{pmatrix} u \cos v \\ u \sin v \\ u \end{pmatrix}, \qquad \text{for } (u, v) \text{ in } \mathcal{R}^2,$$

is continuously differentiable because the Jacobian matrix

$$f'(u, v) = \begin{pmatrix} \cos v & -u \sin v \\ \sin v & u \cos v \\ 1 & 0 \end{pmatrix}$$

has continuous entries. The range of f is a cone shown in Fig. 20. Points of the cone not at the vertex correspond to values of $u \neq 0$, and it is easy to check that for $u \neq 0$ the columns of $f'(u, v)$ are linearly independent. Thus the tangent plane at such a point has the expected dimension, namely 2. However, at the vertex, $u = 0$. Therefore, $f'(0, v)$ has only one nonzero column. Thus any attempt to use the affine approximation to f to define a tangent at the vertex leads to a 1-dimensional tangent. Indeed, it seems natural to say that the cone has no tangent at its vertex. However, the cone satisfies the definition of smooth surface at every other point. The lack of smoothness at the vertex is not associated with a lack of differentiability in f, but rather with the failure of the tangent to exist.

Explicitly Defined Surfaces

Suppose a set S is defined explicitly by a function $\mathcal{R}^n \xrightarrow{f} \mathcal{R}^m$. This means that S is the graph of f in $\mathcal{R}^{n+m}$, consisting of the points of the form $(\mathbf{x}, f(\mathbf{x}))$, for all $\mathbf{x}$ in the domain of f. If f is differentiable at $\mathbf{x}_0$, the affine function

$$A(\mathbf{x}) = f(\mathbf{x}_0) + f'(\mathbf{x}_0)(\mathbf{x} - \mathbf{x}_0)$$

explicitly defines an n-dimensional plane, and this plane is called the tangent to S at $\mathbf{x}_0$. If in addition f is continuously differentiable, then S is a smooth surface.

The complication in the parametric theory that requires checking that the differential has n-dimensional range does not occur here. If $f'(\mathbf{x}_0)$ exists, then A always has as its graph an n-dimensional plane.

7.2 Theorem

The graph of every affine function $\mathcal{R}^n \xrightarrow{A} \mathcal{R}^m$ is an n-dimensional plane.

Proof. By the definition of an affine function, there exists a linear function $\mathcal{R}^n \xrightarrow{L} \mathcal{R}^m$ and a vector $\mathbf{y}_0$ in $\mathcal{R}^m$ such that

$$A(\mathbf{x}) = L(\mathbf{x}) + \mathbf{y}_0, \qquad \text{for all } \mathbf{x} \text{ in } \mathcal{R}^n.$$

The graph of A is the set of all points

$$(\mathbf{x}, A(\mathbf{x})) = (\mathbf{x}, L(\mathbf{x})) + (0, \mathbf{y}_0), \qquad \mathbf{x} \text{ in } \mathcal{R}^n.$$

It is therefore the image under translation by $(0, \mathbf{y}_0)$ of the graph of L. Hence, the graph of A is an n-dimensional plane if and only if the graph of L, which is a subspace of $\mathcal{R}^{n+m}$, has dimension n. But if $(\mathbf{e}_1, \ldots, \mathbf{e}_n)$ is the natural basis for $\mathcal{R}^n$, then any point $(\mathbf{x}, L(\mathbf{x}))$ on the graph of L can be written

$$(\mathbf{x}, L(\mathbf{x})) = x_1(\mathbf{e}_1, L(\mathbf{e}_1)) + \ldots + x_n(\mathbf{e}_n, L(\mathbf{e}_n)).$$

Clearly, the n vectors $(\mathbf{e}_j, L(\mathbf{e}_j))$ are linearly independent, and since they span the graph of L, that graph has dimension n.

Example 4. The hemisphere shown in Fig. 21 is defined explicitly by

$$g\begin{pmatrix} x \\ y \end{pmatrix} = \sqrt{9 - x^2 - y^2}.$$

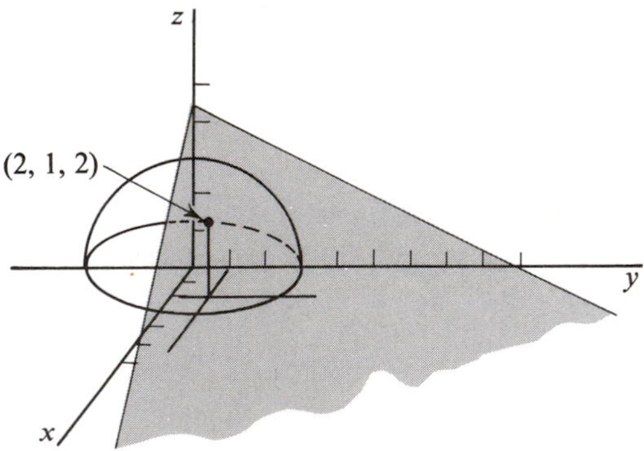

Figure 21

The differential of g at $\mathbf{x}_0 = \begin{pmatrix} 2 \\ 1 \end{pmatrix}$ is defined by the Jacobian matrix

$$\left(\frac{-x}{\sqrt{9 - x^2 - y^2}} \quad \frac{-y}{\sqrt{9 - x^2 - y^2}}\right)_{x=2, y=1} = \begin{pmatrix} -1 & -\frac{1}{2} \end{pmatrix}.$$

The tangent plane to the hemisphere at (2, 1, 2) is the graph of the approximating affine function

$$\begin{aligned} A(\mathbf{x}) &= g(\mathbf{x}_0) + g'(\mathbf{x}_0)(\mathbf{x} - \mathbf{x}_0) \\ &= 2 + \begin{pmatrix} -1 & -\frac{1}{2} \end{pmatrix} \begin{pmatrix} x - 2 \\ y - 1 \end{pmatrix} \\ &= \frac{9}{2} - x - \frac{1}{2} y. \end{aligned}$$

The plane is implicitly defined by the equation $z = A(\mathbf{x})$, that is, by

$$2x + y + 2z = 9.$$

The graph of any function $\mathcal{R}^n \xrightarrow{f} \mathcal{R}^m$ can always be represented parametrically by a function $\mathcal{R}^n \xrightarrow{g} \mathcal{R}^{n+m}$ of the form $g(\mathbf{x}) = (\mathbf{x}, f(\mathbf{x}))$. This raises the question of whether the sets S that can be represented in both ways have the same tangents and of whether the notion of smooth surface is the same in both representations. It is clear that g is continuously differentiable if and only if f is, because $g'(\mathbf{x}_0)$ has the form

$$g'(\mathbf{x}_0) = \begin{pmatrix} I \\ f'(\mathbf{x}_0) \end{pmatrix},$$

where I is the n-by-n identity matrix. Thus

$$\begin{aligned} g(\mathbf{x}_0) + g'(\mathbf{x}_0)(\mathbf{x} - \mathbf{x}_0) &= (\mathbf{x}_0, f(\mathbf{x}_0)) + (\mathbf{x} - \mathbf{x}_0, f'(\mathbf{x}_0)(\mathbf{x} - \mathbf{x}_0)) \\ &= (\mathbf{x}, f(\mathbf{x}_0) + f'(\mathbf{x}_0)(\mathbf{x} - \mathbf{x}_0)), \end{aligned}$$

and this shows that the graph of the affine approximation to f is the same as the range of the affine approximation to g. Hence the two definitions of tangent and of smooth surface are the same where they overlap.

We recall that the null space of a linear function $\mathcal{R}^{n+m} \xrightarrow{L} \mathcal{R}^m$ is a subspace of $\mathcal{R}^{n+m}$. Then, for a fixed vector $\mathbf{x}_0$, the set of all $\mathbf{x}$ such that $\mathbf{x} - \mathbf{x}_0$ is in the null space of L is a plane in $\mathcal{R}^{n+m}$. Clearly, the plane passes through $\mathbf{x}_0$. For nonlinear functions we have the following.

Implicitly Defined Surfaces

Consider a function $\mathcal{R}^{n+m} \xrightarrow{F} \mathcal{R}^m$ and a fixed vector $\mathbf{z}_0$ in $\mathcal{R}^m$. Let S be the level set defined by the equation $F(\mathbf{x}) = \mathbf{z}_0$. If F is differentiable at a point $\mathbf{x}_0$ in $\mathcal{R}^{n+m}$ and the affine approximation $A(\mathbf{x}) = F(\mathbf{x}_0) + F'(\mathbf{x}_0)(\mathbf{x} - \mathbf{x}_0)$ determines an n-dimensional plane implicitly by $A(\mathbf{x}) = \mathbf{z}_0$, then this plane is called the tangent to S at $\mathbf{x}_0$. Since $F(\mathbf{x}_0) = \mathbf{z}_0$, the defining equation of the plane reduces to

$$F'(\mathbf{x}_0)(\mathbf{x} - \mathbf{x}_0) = 0. \tag{1}$$

If in addition F is continuously differentiable on its domain, then S is representable as a smooth surface near every point at which there is a tangent.

Example 5. The equation $x^2 + y^2 + z^2 = 9$ implicitly defines a sphere of radius 3 with center at the origin in 3-dimensional Euclidean space. An equation of the tangent plane to the sphere at

$$\mathbf{x}_0 = (x_0, y_0, z_0) = (2, 1, 2)$$

is determined as follows. If $F(x, y, z) = x^2 + y^2 + z^2$, the Jacobian matrix $F'(\mathbf{x}_0)$ is

$$(2x_0 \quad 2y_0 \quad 2z_0) = (4 \quad 2 \quad 4).$$

Equation (1), which implicitly defines the tangent plane, is therefore

$$(4 \quad 2 \quad 4)\begin{pmatrix} x - 2 \\ y - 1 \\ z - 2 \end{pmatrix} = 0.$$

This is equivalent to $4x + 2y + 4z = 18$ and thence to $2x + y + 2z = 9$. Notice that we have found the same equation as that obtained for the tangent plane in Example 4.

If the plane determined by the equation $F'(\mathbf{x}_0)(\mathbf{x} - \mathbf{x}_0) = 0$ is not n-dimensional, then, according to the definition, the function $\mathcal{R}^{n+m} \xrightarrow{F} \mathcal{R}^m$ does not assign a tangent to the level set $F(\mathbf{x}) = \mathbf{z}_0$ at $\mathbf{x} = \mathbf{x}_0$. This is similar to the complication that occurs in the parametric theory and that gave rise to Theorem 7.1. In the present case we need to know that n is the dimension of the null space of the linear transformation with matrix $F'(\mathbf{x}_0)$. Since the dimension of the null space is equal to the dimension of the domain minus the dimension of the range, we want the dimension of the range to be m. (Theorem 4.7, Chapter 2.) Hence

7.3 Theorem

Let $\mathcal{R}^{n+m} \xrightarrow{F} \mathcal{R}^m$ be differentiable. Then the tangent to the level set $F(\mathbf{x}) = \mathbf{z}_0$ at $\mathbf{x}_0$ exists (and has dimension n) if and only if $F'(\mathbf{x}_0)$ has m linearly independent columns.

Example 3 shows that in the parametric case a surface may fail to be smoothly represented because the dimension of the tangent is too small. The following example is fairly typical of the way in which the standard method of assigning tangents in the implicit case may fail.

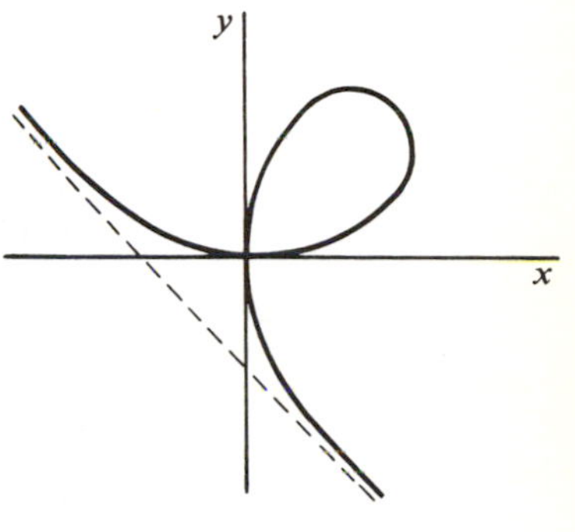

Figure 22

Example 6. Figure 22 shows the folium of Descartes, which is the level set determined by $x^3 + y^3 - 3xy = 0$. The function $F(x, y) = x^3 + y^3 - 3xy$ is continuously differentiable with derivative

$$F'(x, y) = (3x^2 - 3y, 3y^2 - 3x).$$

The criterion of Theorem 7.3 requires that one or another of the two entries be different from zero, in which case the tangent at a point (x_0, y_0) satisfies

$$(x_0^2 - y_0)(x - x_0) + (y_0^2 - x_0)(y - y_0) = 0.$$

The one point at which the differential fails to assign a tangent is $(x, y) = (0, 0)$. There the null space of the differential has dimension 2.

The function $\mathcal{R} \xrightarrow{f} \mathcal{R}^2$ given by

$$f(t) = \left(\frac{3t}{1 + t^3}, \frac{3t^2}{1 + t^3}\right), \qquad -1 < t < \infty$$

is a parametrization of the part of the curve that lies in the first and second quadrants, and it assigns the curve a horizontal tangent at the origin. Interchanging the coordinate functions of f gives a parametrization of the part of the curve in the first and fourth quadrants.

If $\mathcal{R}^{n+m} \xrightarrow{F} \mathcal{R}^m$ is continuously differentiable and its level set $F(\mathbf{x}) = 0$ has an implicitly determined tangent at a point $\mathbf{x}_0$, then according to Theorem 7.3 the m-by-$(n + m)$ matrix $F'(\mathbf{x}_0)$ has m linearly independent columns. Denote the variables corresponding to these columns by the vector $\mathbf{y}$ and observe that the implicit-function theorem applies. The conclusion is that, writing $\mathbf{x} = (\mathbf{v}, \mathbf{y})$, there is a continuously differentiable function $\mathcal{R}^n \xrightarrow{f} \mathcal{R}^m$ satisfying $F(\mathbf{v}, f(\mathbf{v})) = 0$ in some neighborhood of $\mathbf{x}_0$. The significance of this result is that, restricted to a neighborhood of $\mathbf{x}_0$, the level set $F(\mathbf{x}) = 0$ can be represented as the graph of the function f. It is routine to show that we get the same tangent by using the explicit or the implicit representation, and we leave the computation as Exercise 12.

EXERCISES

1. Find a parametric representation $t\mathbf{x}_1 + \mathbf{x}_2$ for the tangent line to each of the curves defined parametrically by the following functions at the points indicated. Sketch the curve and the tangent line.

(a) $f(t) = \begin{pmatrix} t \\ e^t \end{pmatrix}$, at $f(0)$. $\left[\textit{Ans. } t\begin{pmatrix} 1 \\ 1 \end{pmatrix} + \begin{pmatrix} 0 \\ 1 \end{pmatrix}.\right]$

(b) $g(t) = \begin{pmatrix} x \\ y \\ z \end{pmatrix} = \begin{pmatrix} t^2 + 1 \\ t - 1 \\ t^2 \end{pmatrix}$, at $\begin{pmatrix} 2 \\ 0 \\ 1 \end{pmatrix}$.

2. Find the tangent plane to each of the surfaces defined parametrically by the following functions at the points indicated. Sketch the surface and the tangent plane in (a) and (b).

(a) $f\begin{pmatrix} u \\ v \end{pmatrix} = \begin{pmatrix} u + v \\ u - v \\ u^2 - v^2 \end{pmatrix}$, $\begin{pmatrix} 0 \le u \le 2 \\ 0 \le v \le 2 \end{pmatrix}$ at $f\begin{pmatrix} 1 \\ 1 \end{pmatrix}$.

$\left[\textit{Ans. } \begin{pmatrix} u + v + 2 \\ u - v \\ 2u - 2v \end{pmatrix}.\right]$

(b) $g\begin{pmatrix} u \\ v \end{pmatrix} = \begin{pmatrix} x \\ y \\ z \end{pmatrix} = \begin{pmatrix} \cos u \sin v \\ \sin u \sin v \\ \cos v \end{pmatrix}$, $\begin{Bmatrix} 0 \le u \le 2\pi \\ 0 \le v \le \pi/2 \end{Bmatrix}$ at $g\begin{pmatrix} \pi \\ \pi/4 \end{pmatrix}$.

(c) $\begin{pmatrix} x \\ y \\ z \end{pmatrix} = \begin{pmatrix} u + \ v + 1 \\ 2u + 3v \quad 1 \\ u + 2v - 2 \end{pmatrix}$, at $\begin{pmatrix} 3 \\ 5 \\ 1 \end{pmatrix}$.

3. Find the tangent plane or tangent line to each of the following explicitly defined curves and surfaces at the points indicated.

(a) $f(x) = (x - 1)(x - 2)(x - 3)$, at $(0, -6)$.

(b) $f(x, y) = \dfrac{1}{x^2 + y^2}$, at $(x_0, y_0, f(x_0, y_0)) = (0, 2, \frac{1}{4})$.

(c) $g(t) = \begin{pmatrix} t \\ e^t \end{pmatrix}$, at $\begin{pmatrix} 1 \\ 1 \\ e \end{pmatrix}$. $\left[\textit{Ans. } L(t) = t\begin{pmatrix} 1 \\ e \end{pmatrix}.\right]$

(d) $g(x, y) = \cosh(x^2 + y^2)$, at $(x_0, y_0, g(x_0, y_0)) = (1, 2, \cosh 5)$.

4. Find the tangent line or tangent plane to each of the following implicitly defined curves and surfaces at the points indicated. Sketch the curve or surface and the tangent in (b), (d), and (e).

(a) $xy + yz + zx = 1$, at $\mathbf{x}_0 = (2, -1, 3)$.

(b) $\dfrac{x^2}{4} + \dfrac{y^2}{9} + z^2 = 1$, at $\mathbf{x}_0 = \left(1, 0, \dfrac{\sqrt{3}}{2}\right)$.

$$\left[\textit{Ans. } \frac{x}{2} + \sqrt{3}\, z = 2.\right]$$

(c) $5x + 5y + 2z = 8$ at $(1, 1, -1)$.

(d) $\dfrac{x^2}{a^2} + \dfrac{y^2}{b^2} = 1$, at $(x_0, y_0) = \left(\dfrac{\sqrt{3}\, a}{2}, \dfrac{b}{2}\right)$.

(e) $F\begin{pmatrix} x \\ y \\ z \end{pmatrix} = \begin{pmatrix} x^2 + y^2 + z^2 \\ x + y \end{pmatrix} = \begin{pmatrix} 9 \\ 3 \end{pmatrix}$, at $\mathbf{x}_0 = \begin{pmatrix} 2 \\ 1 \\ 2 \end{pmatrix}$.

$$\left[\textit{Ans. } \begin{pmatrix} 4 & 2 & 4 \\ 1 & 1 & 0 \end{pmatrix}\begin{pmatrix} x \\ y \\ z \end{pmatrix} = \begin{pmatrix} 18 \\ 3 \end{pmatrix}.\right]$$

5. In each of Exercises 1(a), 2(a), 2(b), and 4(b), find a normal to the given curve or surface at the point indicated.

$$\left[\textit{Ans. } 1(a) \begin{pmatrix} 1 \\ -1 \end{pmatrix}.\right]$$

6. Each of the following curves and surfaces fails, according to our definitions, to have a tangent line or plane at the indicated point. Why?

(a) $f(t) = \begin{pmatrix} t \\ |t| \\ t^2 \end{pmatrix}$, at $f(0)$.

(b) $g(t) = \begin{pmatrix} t^2 - 2t \\ t^3 - 3t \\ t^4 - t^3 - t \end{pmatrix}$, at $\begin{pmatrix} -1 \\ -2 \\ -1 \end{pmatrix}$.

(c) $f\begin{pmatrix} u \\ v \end{pmatrix} = \begin{pmatrix} u^2v^4 \\ uv^2 \\ u^2 + v^4 \end{pmatrix}$, at $f\begin{pmatrix} 1 \\ 1 \end{pmatrix}$.

(d) $f(x, y) = \sqrt{1 - x^2 - y^2}$, at $\left(\frac{\sqrt{2}}{2}, \frac{\sqrt{2}}{2}, 0\right)$.

(e) $F\begin{pmatrix} x \\ y \\ z \end{pmatrix} = \begin{pmatrix} z^2 e^{x+y} \\ 2xyz^2 \end{pmatrix} = \begin{pmatrix} 4e^2 \\ 8 \end{pmatrix}$, at $\mathbf{x}_0 = \begin{pmatrix} 1 \\ 1 \\ 2 \end{pmatrix}$.

7. Find all points at which the surface defined parametrically by the function

$$f\begin{pmatrix} u \\ v \end{pmatrix} = \begin{pmatrix} u^2v^2 \\ uv \\ uv + 1 \end{pmatrix}$$

fails to have a tangent plane.

8. Different vector functions can define the same curve or surface. Show that the functions

$$f_1(t) = (\cos t, \sin t), \qquad 0 < t < 2\pi,$$

$$f_2(s) = \left(\frac{s^2 - 1}{s^2 + 1}, \frac{2s}{s^2 + 1}\right), \qquad -\infty < s < \infty,$$

parametrically define the same curve in 2-dimensional Euclidean space.

9. Consider the vector functions

$$f(t) = \begin{pmatrix} t^3 \\ t^3 \\ t^3 \end{pmatrix}, \qquad -\infty < t < \infty,$$

$$g\begin{pmatrix} u \\ v \end{pmatrix} = \begin{pmatrix} u^3 \\ v^3 \\ 0 \end{pmatrix}, \qquad \begin{cases} -\infty < u < \infty, \\ -\infty < v < \infty. \end{cases}$$

(a) What curve and what surface are parametrically defined by f and g, respectively?

(b) Show that according to our definition the curve does not have a tangent line and the surface does not have a tangent plane at (0, 0, 0).

10. Let

$$f(t) = \begin{pmatrix} t^3 \\ t^3 \end{pmatrix}, \qquad -\infty < t < \infty.$$

(a) Show that the curve in $\mathcal{R}^3$ defined explicitly by f has a tangent line at every point.

(b) Show that the curve in $\mathcal{R}^2$ defined parametrically by f fails to have a tangent line at one point.

(c) Interpret (a) and (b) geometrically. What is the relation between the tangent line in (a) and in (b)?

11. Let $\mathbf{y}_0 = 0$ lie in the range of a function $\mathcal{R}^n \xrightarrow{F} \mathcal{R}^m$. The surface S defined implicitly by the equation $F(\mathbf{x}) = 0$ is assumed to have a tangent $\mathcal{T}$ at $\mathbf{x}_0$.

(a) Check that the surface S' in $\mathcal{R}^{n+m}$ defined explicitly by F has a tangent $\mathcal{T}'$ at $(\mathbf{x}_0, \mathbf{y}_0)$.

(b) Let P be the plane in $\mathcal{R}^{n+m}$ defined by the equation $\mathbf{y} = 0$, and show that $S = S' \cap P$.

(c) Prove that $\mathcal{T} = \mathcal{T}' \cap P$.

(d) Using the equation $F(x, y) = x^2 + y^2 - 2 = 0$ and the point $\mathbf{x}_0 = (1, 1)$, draw a picture illustrating S, $\mathcal{T}$, S', $\mathcal{T}'$, and P.

(e) Using the equation $F(x, y) = 4x^2 - 4xy + y^2 = 0$ and the point $\mathbf{x}_0 = (1, 2)$, draw a picture illustrating S, S', $\mathcal{T}'$, and P. What happens to $\mathcal{T}$?

12. Show that if $\mathcal{R}^{n+m} \xrightarrow{F} \mathcal{R}^m$ determines a smooth surface S, then the tangent to S at $\mathbf{x}_0$ determined by $F'(\mathbf{x}_0)(\mathbf{x} - \mathbf{x}_0) = 0$ is the same as the tangent to the graph of the function $\mathcal{R}^n \xrightarrow{f} \mathcal{R}^m$ which satisfies $F(\mathbf{v}, f(\mathbf{v})) = 0$, and whose existence is guaranteed by the implicit function theorem.

13. Verify that the two parametrically defined tangents at the origin that are described in Example 6 of the text are horizontal and vertical, respectively.

14. Show that if P is an n-dimensional plane through the origin in $\mathcal{R}^{m+n}$, then P is precisely the null space of some linear function taking values in $\mathcal{R}^m$.

5

Real-Valued Functions

SECTION 1

EXTREME VALUES

The problem of finding the maximum and minimum values of a real-valued function of several variables is important in many branches of applied mathematics, as well as in pure mathematics. Familiar examples are extremes of temperature, speed, or economic profit, each of which may be a function of more than one variable in a practical problem.

A real-valued function f has an **absolute maximum value at $\mathbf{x}_0$** if, for all $\mathbf{x}$ in the domain of f,

$$f(\mathbf{x}) \leq f(\mathbf{x}_0),$$

and an **absolute minimum value** if instead

$$f(\mathbf{x}_0) \leq f(\mathbf{x}).$$

The number $f(\mathbf{x}_0)$ is called a **local maximum value** or a **local minimum value** if there is a neighborhood N of $\mathbf{x}_0$ such that, respectively,

$$f(\mathbf{x}) \leq f(\mathbf{x}_0) \quad \text{or} \quad f(\mathbf{x}_0) \leq f(\mathbf{x}),$$

for all $\mathbf{x}$ in N. A maximum or minimum value of f is called an **extreme value.** A point $\mathbf{x}_0$ at which an extreme value occurs is called an **extreme point.**

Example 1. Consider the function $f(x, y) = x^2 + y^2$ whose domain is the set of points (x, y) that lie inside or on the ellipse $x^2 + 2y^2 = 1$. The graph of f is shown in Fig. 1. Suppose that f has an extreme value (i.e., maximum or minimum) at a point (x_0, y_0) in the interior of the ellipse. Then obviously, both functions f_1 and f_2 defined by

$$f_1(x) = f(x, y_0), \qquad f_2(y) = f(x_0, y)$$

must also have extreme values at x_0 and y_0, respectively. Applying the familiar criterion for differentiable functions of one variable, we have

$$f_1'(x_0) = f_2'(y_0) = 0.$$

Since

$$f_1'(x_0) = \frac{\partial f}{\partial x}(x_0, y_0)$$

and

$$f_2'(y_0) = \frac{\partial f}{\partial y}(x_0, y_0),$$

a necessary condition for f to have an extreme value at (x_0, y_0) is

$$\frac{\partial f}{\partial x}(x_0, y_0) = \frac{\partial f}{\partial y}(x_0, y_0) = 0.$$

In this example,

$$\frac{\partial f}{\partial x}(x, y) = 2x \quad \text{and} \quad \frac{\partial f}{\partial y}(x, y) = 2y,$$

and so the only extreme value of f in the interior of the ellipse occurs at $(x_0, y_0) = (0, 0)$. It is obvious from the graph of f, shown in Fig. 1, that

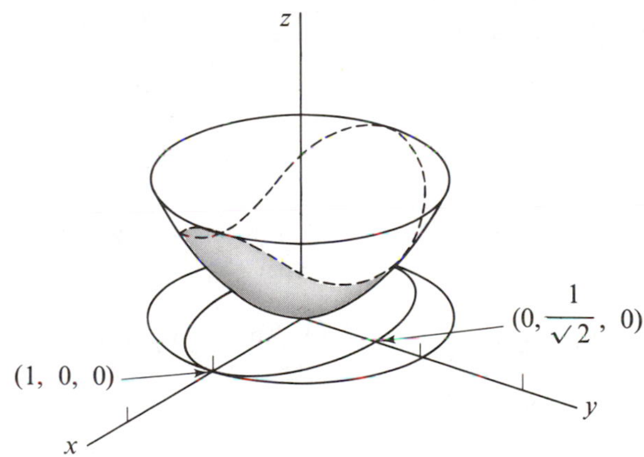

Figure 1

this value is a minimum. We next consider the values of f on the boundary curve itself. The ellipse can be defined parametrically by the function

$$g(t) = (x, y) = \left(\cos t, \frac{1}{\sqrt{2}} \sin t\right), \qquad 0 \le t < 2\pi.$$

Thus, the values of f on the ellipse are given as the values of the composition $f \circ g$. Any extreme values of f on the ellipse will be extreme for $f \circ g$. The latter is a real-valued function of one variable, and we treat it in the usual way, that is, by setting its derivative equal to zero. By the chain rule, we obtain

$$\frac{d}{dt}(f \circ g) = \left(2\cos t \;\; \frac{2}{\sqrt{2}}\sin t\right)\begin{pmatrix} -\sin t \\ \frac{1}{\sqrt{2}}\cos t \end{pmatrix}$$

$$= -2\cos t \sin t + \sin t \cos t$$

$$= -\tfrac{1}{2}\sin 2t.$$

Extreme values therefore may occur at $t = 0, \pi/2, \pi$, and $3\pi/2$. The corresponding values of (x, y) are $(1, 0)$, $(0, 1/\sqrt{2})$, $(-1, 0)$, and $(0, -1/\sqrt{2})$, and those of f are $1, \frac{1}{2}, 1$, and $\frac{1}{2}$, respectively. We see that the absolute minimum of f is 0 at $(0, 0)$ and that the absolute maximum of f occurs at the two points $(1, 0)$ and $(-1, 0)$. Notice that the two extreme values of $f \circ g$ that occur at $t = \pi/2$ and $3\pi/2$ are not extreme for f, as can be seen by looking at Fig. 1.

The methods used in the preceding example are valid in any number of dimensions. The next theorem is the principal criterion used in this extension, and while it can be proved by reducing it to the one-variable method, we give a proof that contains the one-variable situation as a special case.

1.1 Theorem

If a differentiable function $\mathcal{R}^n \xrightarrow{f} \mathcal{R}$ has a local extreme value at a point $\mathbf{x}_0$ interior to its domain, then $f'(\mathbf{x}_0) = 0$.

Proof. Suppose f has a local minimum at $\mathbf{x}_0$. For any unit vector $\mathbf{u}$ in $\mathcal{R}^n$, there is an $\epsilon > 0$ such that if $-\epsilon < t < \epsilon$, then $f(\mathbf{x}_0) \le f(\mathbf{x}_0 + t\mathbf{u})$. Hence, for $0 < t < \epsilon$,

$$0 \le \frac{f(\mathbf{x}_0 + t\mathbf{u}) - f(\mathbf{x}_0)}{t},$$

$$0 \le \frac{f(\mathbf{x}_0 - t\mathbf{u}) - f(\mathbf{x}_0)}{t}.$$

It follows by Theorem 1.1 of Chapter 4 that

$$\frac{\partial f}{\partial \mathbf{u}}(\mathbf{x}_0) = f'(\mathbf{x}_0)\mathbf{u}.$$

Therefore,

$$0 \leq \lim_{t \to 0+} \frac{f(\mathbf{x}_0 + t\mathbf{u}) - f(\mathbf{x}_0)}{t} = f'(\mathbf{x}_0)\mathbf{u},$$

$$0 \leq \lim_{t \to 0+} \frac{f(\mathbf{x}_0 - t\mathbf{u}) - f(\mathbf{x}_0)}{t} = f'(\mathbf{x}_0)(-\mathbf{u}) = -f'(\mathbf{x}_0)\mathbf{u}.$$

We conclude that $f'(\mathbf{x}_0)\mathbf{u} = 0$. Because $\mathbf{u}$ is arbitrary, $f'(\mathbf{x}_0) = 0$. The argument for a maximum value is analogous.

This result is what we should expect. Recall that

$$\frac{\partial f}{\partial \mathbf{u}}(\mathbf{x}_0) = f'(\mathbf{x}_0)\mathbf{u},$$

and that the derivative with respect to $\mathbf{u}$ measures the rate of change of f in the direction of $\mathbf{u}$. At an extreme point in the interior of the domain of f, this rate should be zero in every direction. The importance of the theorem is that of all the interior points $\mathbf{x}$ of the domain of f we need to look for extreme points only among those for which $f'(\mathbf{x}) = 0$. Points $\mathbf{x}$ for which $f'(\mathbf{x}) = 0$ are called **critical points** of f.

In practice we are often given a function f that is differentiable on an open set and want to find the extreme points of f when it is restricted to some subset S of the domain of f. In the next example the following two remarks are illustrated: (1) *a point* $\mathbf{x}$ *such that* $f'(\mathbf{x}) = 0$ *is not necessarily an extreme point for* f; (2) f *may have an extreme point* $\mathbf{x}$ *when restricted to a set* S *without having* $f'(\mathbf{x}) = 0$.

Example 2. Let $f(x, y, z) = xyz$ in the region defined by $|x| \leq 1$, $|y| \leq 1$, $|z| \leq 1$. Thus the domain of f is the cube with each edge of length 2 illustrated in Fig. 2. The condition for critical points, $f'(\mathbf{x}) = 0$, is equivalent to $(yz, xz, xy) = (0, 0, 0)$. The solutions of this equation are the points satisfying $x = y = 0$, or $x = z = 0$, or $y = z = 0$; in other words, the coordinate axes. Since f has the value zero at any one of its critical points, and since f has both positive and negative values in the neighborhood of any one of these points, no critical point can be an extreme point. Furthermore, a little thought shows that f has maximum value 1 and minimum value -1. These values occur at the corners of the cube, none of which is a critical point.

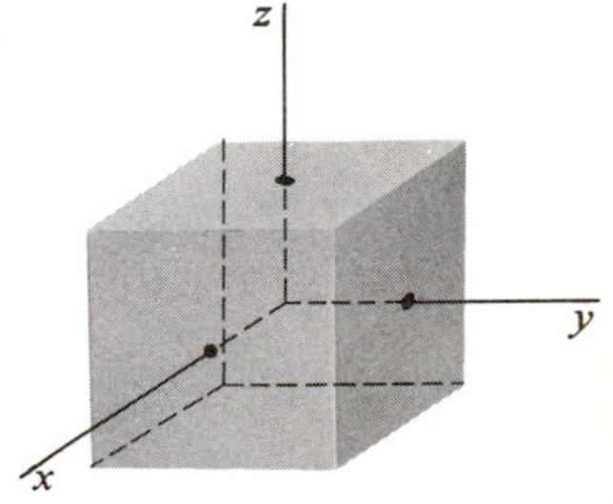

Figure 2

The problem of finding the extreme values of a function f on the boundary of a subregion R of $\mathcal{R}^n$ is one in which f has been restricted to a set S of lower dimension than that of R. Then, as we have seen in Example 2, it is not sufficient just to examine the critical points of f as a function on R. More generally, we may be interested in f when it is restricted to a lower-dimensional set S that is not necessarily the boundary of any region at all.

Example 3. The function $f(x, y, z) = y^2 - z - x$ has a differential defined by the matrix

$$(-1 \quad 2y \quad -1),$$

so f has no critical points as a function defined on $\mathcal{R}^3$. Suppose, however, that f is restricted to the curve γ defined parametrically by

$$\begin{pmatrix} x \\ y \\ z \end{pmatrix} = \begin{pmatrix} t \\ t^2 \\ t^3 \end{pmatrix}, \qquad -\infty < t < \infty.$$

On γ, f takes the values $F(t) = f(t, t^2, t^3) = t^4 - t^3 - t$ while t varies over $(-\infty, \infty)$. We have

$$F'(t) = 4t^3 - 3t^2 - 1 = (t - 1)(4t^2 + t + 1).$$

Then $F'(t)$ is zero only at $t = 1$. Furthermore, since $F''(t) = 12t^2 - 6t$, we have $F''(1) > 0$. It follows that the point $(1, 1, 1)$ is a relative minimum for f restricted to the curve γ. The minimum value of f on γ is -1, and there are no other extreme values.

Example 4. Suppose the function $f(x, y, z) = x + y + z$ is restricted to the intersection of the two surfaces

$$x^2 + y^2 = 1, \qquad z = 2$$

shown in Fig. 3. The curve C of intersection can be parametrized by

$$\begin{pmatrix} x \\ y \\ z \end{pmatrix} = \begin{pmatrix} \cos t \\ \sin t \\ 2 \end{pmatrix}, \qquad 0 \leq t < 2\pi.$$

The function f on C takes the value $F(t) = \cos t + \sin t + 2$. We have $F'(t) = -\sin t + \cos t$, so $F'(t) = 0$ at $t = \pi/4$ and $t = 5\pi/4$. Since $F''(\pi/4) < 0$ and $F''(5\pi/4) > 0$,

$$f(\sqrt{2}/2, \sqrt{2}/2, 2) = \sqrt{2} + 2$$

is the maximum and

$$f(-\sqrt{2}/2, -\sqrt{2}/2, 2) = -\sqrt{2} + 2$$

is the minimum value for f on C.

The solution of the previous problem depended on our being able to find a concrete parametric representation for the curve of intersection of the

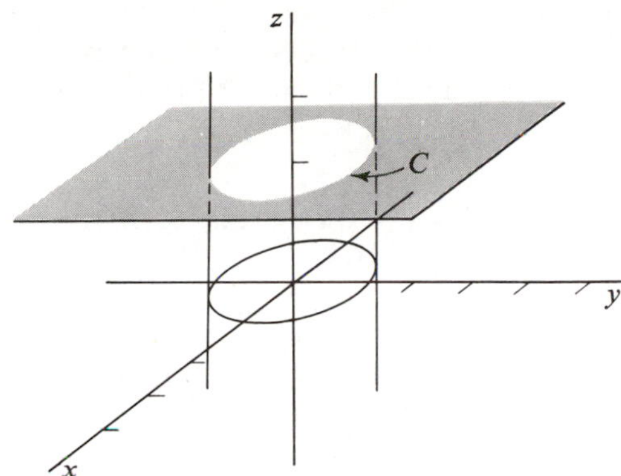

Figure 3

cylinder $x^2 + y^2 - 1 = 0$ and the plane $z - 2 = 0$. When a specific parametrization is not readily available, we can still sometimes apply the method of **Lagrange multipliers** to be described below. The method consists in verifying the pure existence of a parametric representation and then deriving necessary conditions for there to be an extreme point for a function f when restricted to the parametrized curve or surface.

1.2 Theorem. Lagrange Multiplier Method

Let the function $\mathcal{R}^n \xrightarrow{G} \mathcal{R}^m$, $n > m$, be continuously differentiable and have coordinate functions $G_1, G_2, \ldots, G_m$. Suppose the equations

$$\begin{aligned} G_1(x_1, \ldots, x_n) &= 0 \\ G_2(x_1, \ldots, x_n) &= 0 \\ \vdots \quad & \quad \vdots \\ G_m(x_1, \ldots, x_n) &= 0 \end{aligned}$$

implicitly define a surface S in $\mathcal{R}^n$, and that at a point $\mathbf{x}_0$ of S the matrix $G'(\mathbf{x}_0)$ has some m columns linearly independent.

If $\mathbf{x}_0$ is an extreme point of a differentiable function $\mathcal{R}^n \xrightarrow{f} \mathcal{R}$, when restricted to S, then $\mathbf{x}_0$ is a critical point of the function

$$f + \lambda_1 G_1 + \ldots + \lambda_m G_m$$

for some constants $\lambda_1 \ldots, \lambda_m$.

The complete proof of the theorem is given in the Appendix. We give here a geometric argument that makes the result plausible. Recall that the gradient of a differentiable, real-valued function defined in $\mathcal{R}^n$ is the vector valued function ∇f defined by

$$\nabla f(\mathbf{x}) = \left(\frac{\partial f}{\partial x_1}(\mathbf{x}), \ldots, \frac{\partial f}{\partial x_n}(\mathbf{x})\right).$$

In terms of the gradient, the Lagrange condition

$$f'(\mathbf{x}_0) + \lambda_1 G_1'(\mathbf{x}_0) + \ldots + \lambda_m G_m'(\mathbf{x}_0) = 0$$

can be expressed as

$$\nabla f(\mathbf{x}_0) + \lambda_1 \nabla G_1(\mathbf{x}_0) + \ldots + \lambda_m \nabla G_m(\mathbf{x}_0) = 0. \tag{1}$$

Equation 2.4 of Chapter 4 says that the vector $\nabla G_i(\mathbf{x}_0)$ is perpendicular to the surface S_i defined implicitly by $G_i(\mathbf{x}) = 0$ at x_0. But then each vector $\nabla G_i(\mathbf{x}_0)$ is also perpendicular, at the same point, to the intersection S of all the surfaces S_i. Since by Equation (1), $\nabla f(\mathbf{x}_0)$ is a linear combination of the vectors $\nabla G_i(\mathbf{x}_0)$, the gradient of f itself is perpendicular to S at $\mathbf{x}_0$. Now recall that by Theorem 2.2 of Chapter 4, the gradient of f points in the direction of greatest increase for f. That this direction should be perpendicular to S at an extreme point $\mathbf{x}_0$ for f restricted to S is reasonable, because otherwise we would expect to find a larger or smaller value for f by moving along S in the direction of $\nabla f(\mathbf{x}_0)$ projected onto S.

In applying the theorem it is important to verify that some m columns of $G'(\mathbf{x})$ are independent for $\mathbf{x}$ in S. Points for which this condition fails must be examined separately in looking for extreme points. All extreme points $\mathbf{x}_0$ for which the condition is satisfied are such that there are constants $\lambda_1, \ldots, \lambda_m$ for which

$$f + \lambda_1 G_1 + \ldots + \lambda_m G_m$$

has $\mathbf{x}_0$ as a critical point, or in other words,

$$f'(\mathbf{x}_0) + \lambda_1 G_1'(\mathbf{x}_0) + \ldots + \lambda_m G_m'(\mathbf{x}_0) = 0. \tag{2}$$

Example 5. The problem of Example 4 is that of finding the extreme points of $f(x, y, z) = x + y + z$ subject to the conditions

$$x^2 + y^2 - 1 = 0, \qquad z - 2 = 0. \tag{3}$$

We write down

$$(x + y + z) + \lambda_1(x^2 + y^2 - 1) + \lambda_2(z - 2).$$

The critical points of this function occur when

$$1 + 2\lambda_1 x = 0, \qquad 1 + 2\lambda_1 y = 0, \qquad 1 + \lambda_2 = 0.$$

In addition, we must satisfy Equations (3). Solving for λ_1 and λ_2, as well as x, y, and z, we get

$$\lambda_2 = -1, \qquad \lambda_1 = \pm\frac{1}{\sqrt{2}}, \qquad x = y = \pm\frac{1}{\sqrt{2}}, \qquad z = 2.$$

That is, the critical points are

$$\left(\frac{1}{\sqrt{2}}, \frac{1}{\sqrt{2}}, 2\right) \quad \text{and} \quad \left(-\frac{1}{\sqrt{2}}, \frac{-1}{\sqrt{2}}, 2\right).$$

As in Example 4, we easily see that f has its maximum value, $\sqrt{2} + 2$, at the first of these points and its minimum value, $-\sqrt{2} + 2$, at the other. Notice that, while the λ's are not needed in the final answer, it is necessary to consider all values of the λ's for which the equations can be satisfied.

Example 6. Find the maximum value of $f(x, y, z) = x - y + z$, subject to the condition $x^2 + y^2 + z^2 = 1$. The function

$$x - y + z + \lambda(x^2 + y^2 + z^2 - 1)$$

has critical points satisfying

$$1 + 2\lambda x = 0, \qquad -1 + 2\lambda y = 0, \qquad 1 + 2\lambda z = 0,$$

and

$$x^2 + y^2 + z^2 = 1.$$

The solutions of these equations are

$$\lambda = \pm\frac{\sqrt{3}}{2}, \qquad x = -y = z = \mp\frac{1}{\sqrt{3}}.$$

The maximum of f occurs at $(1/\sqrt{3}, -1/\sqrt{3}, 1/\sqrt{3})$. The maximum value is $\sqrt{3}$.

Example 7. Let $g(x_1, x_2, \ldots, x_n) = 0$ implicitly define a surface S in $\mathcal{R}^n$ and let $\mathbf{a} = (a_1, a_2, \ldots, a_n)$ be a fixed point. Minimizing the distance from S to $\mathbf{a}$ is the same thing as minimizing the square of the distance. Thus, $\mathbf{p}$, the nearest point to $\mathbf{a}$ on S, must be among the critical points of

$$\sum_{k=1}^{n}(x_k - a_k)^2 + \lambda g(x_1, \ldots, x_n)$$

for some λ. The critical points satisfy, in addition to $g(x_1, \ldots, x_n) = 0$, the equations

$$2(x_1 - a_1) + \lambda \frac{\partial g}{\partial x_1}(x_1, \ldots, x_n) = 0$$
$$\vdots$$
$$2(x_n - a_n) + \lambda \frac{\partial g}{\partial x_n}(x_1, \ldots, x_n) = 0.$$

In vector form these equations reduce at the critical point $\mathbf{p}$ to

$$\begin{pmatrix} p_1 - a_1 \\ \cdot \\ \cdot \\ \cdot \\ p_n - a_n \end{pmatrix} = -\frac{\lambda}{2} \begin{pmatrix} \frac{\partial g}{\partial x_1}(\mathbf{p}) \\ \cdot \\ \cdot \\ \cdot \\ \frac{\partial g}{\partial x_n}(\mathbf{p}) \end{pmatrix},$$

where $\mathbf{p} = (p_1, \ldots, p_n)$. The vector $\mathbf{p} - \mathbf{a}$ on the left is then parallel to the normal vector to S at $\mathbf{p}$, which appears on the right side of the equation. In other words, $\mathbf{p} - \mathbf{a}$ is perpendicular to S. A 2-dimensional example is illustrated in Fig. 4.

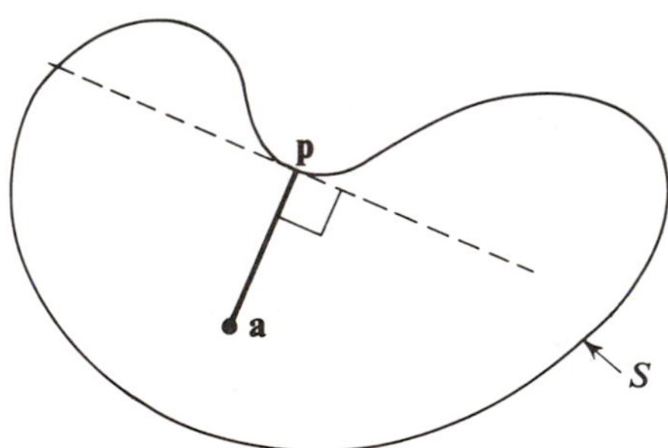

Figure 4

Example 8. Suppose that a cylindrical can is to contain a fixed volume V and that its surface area, with top and bottom, is to be as small as possible. If the radius of the can is x, and its height is y, then $V = \pi x^2 y$. We want to minimize the total area $2\pi x^2 + 2\pi xy$ of the top, bottom, and sides. We write

$$f(x, y) = 2\pi x^2 + 2\pi xy + \lambda(\pi x^2 y - V)$$

and look for critical points of f. We find that $f_x = 0, f_y = 0$ reduce to

$$2x + y + \lambda xy = 0, \qquad 2x + \lambda x^2 = 0.$$

The second equation is satisfied if $x = 0$ or if $\lambda x = -2$. But $x = 0$ would require $V = 0$. So we substitute $\lambda x = -2$ into the first equation to get $2x = y$. Thus height y must equal diameter $2x$. The value of x for a given volume V can be then determined from the equation $2\pi x^3 = \pi x^2 y = V$.

Example 9. The planes

$$x + y + z - 1 = 0 \quad \text{and} \quad x + y - z = 0$$

intersect in a line S as shown in Fig. 5. Let $f(x, y) = xy$, and restrict f to the line S. Using the Lagrange method, we consider

$$xy + \lambda(x + y + z - 1) + \mu(x + y - z).$$

Its critical points occur when

$$y + \lambda + \mu = 0, \qquad x + \lambda + \mu = 0, \qquad \lambda - \mu = 0.$$

The only point that satisfies these conditions, together with the condition that it lie on S, is $\mathbf{x}_0 = (\frac{1}{4}, \frac{1}{4}, \frac{1}{2})$. We have $\nabla f(\mathbf{x}_0) = (\frac{1}{4}, \frac{1}{4}, 0)$, which is perpendicular to S. The unit vector $\mathbf{u}$ in the direction of $\nabla f(\mathbf{x}_0)$ is shown in Fig. 5 with its initial point moved to $\mathbf{x}_0$.

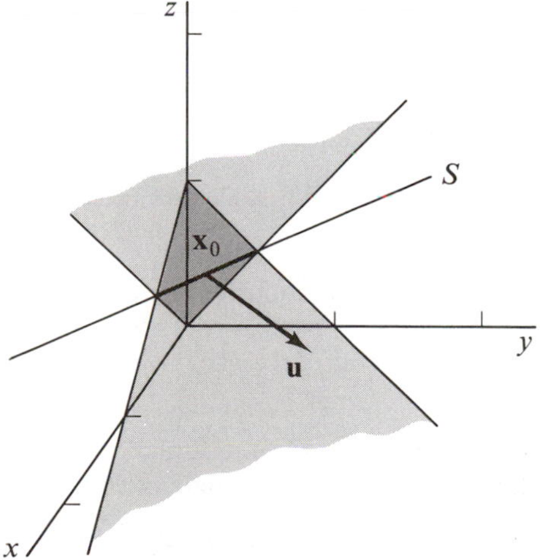

Figure 5

EXERCISES

1. Find the critical points of $x^2 + 4xy - y^2 - 8x - 6y$. [*Ans.* (2, 1).]

2. Find the points at which the largest and smallest values are attained by the following functions.

(a) $x + y$ in the square with corners $(\pm 1, \pm 1)$.
[*Ans.* Max. (1, 1), min $(-1, -1)$.]

(b) $x + y + z$ in the region $x^2 + y^2 + z^2 \leq 1$.
[*Ans.* Max. $(1/\sqrt{3}, 1/\sqrt{3}, 1/\sqrt{3})$, min. $(-1/\sqrt{3}, -1/\sqrt{3}, -1/\sqrt{3})$.]

(c) $x^2 + 24xy + 8y^2$ in the region $x^2 + y^2 \leq 25$.
[*Ans.* Max. $\pm(3, 4)$, min. $\pm(4, -3)$.]

(d) $1/(x^2 + y^2)$ in the region $(x - 2)^2 + y^2 \leq 1$.
[*Ans.* Max. (1, 0), min. (3, 0).]

(e) $x^2 + y^2 + (2\sqrt{2}/3)xy$ in the ellipse $x^2 + 2y^2 \leq 1$.
[*Ans.* Max. $(\pm 2/\sqrt{5}, \pm 1/\sqrt{10})$, min. (0, 0).]

3. Find the point on the curve

$$\begin{pmatrix} x \\ y \\ z \end{pmatrix} = \begin{pmatrix} \cos t \\ \sin t \\ \sin t/2 \end{pmatrix}$$

that is farthest from the origin. [*Ans.* $(-1, 0, 1)$.]

4. Find the critical points of the following functions.

(a) $x + y \sin x$. (b) $xy + xz$. (c) $x^2 + y^2 + z^2 - 1$.

5. Find the maximum value of the function $x(y + z)$, given that $x^2 + y^2 = 1$ and $xz = 1$. [*Ans.* $\frac{3}{2}$.]

6. Find the minimum value of $x + y^2$, subject to the condition $2x^2 + y^2 = 1$.
[*Ans.* $-1/\sqrt{2}$.]

7. Let $f(x, y)$ and $g(x, y)$ be continuously differentiable, and suppose that, subject to the condition $g(x, y) = 0$, $f(x, y)$ attains its maximum value M at (x_0, y_0). Show that the level curve $f(x, y) = M$ is tangent to the curve $g(x, y) = 0$ at (x_0, y_0).

8. A rectangular box with no top is to have surface area 32 square units. Find the dimensions that will give it the maximum volume.

9. Find the minimum distance in $\mathcal{R}^2$ from the ellipse $x^2 + 2y^2 = 1$ to the line $x + y = 4$. [*Hint.* Treat the square of the distance as a function of four variables.]

10. (a) Find the maximum value of $x^2 + xy + y^2 + yz + z^2$, subject to the condition $x^2 + y^2 + z^2 = 1$. [*Ans.* $1 + 1/\sqrt{2}$.]

(b) Find the maximum value of the same function subject to the conditions $x^2 + y^2 + z^2 = 1$ and $ax + by + cz = 0$, where (a, b, c) is a point at which the maximum is attained in (a). [*Ans.* 1.]

11. Consider a differentiable function $\mathcal{R}^n \xrightarrow{f} \mathcal{R}$ and a continuously differentiable function $\mathcal{R}^n \xrightarrow{G} \mathcal{R}^m$, $m < n$. Suppose the surface S defined by $G(\mathbf{x}) = 0$ has a tangent $\mathcal{T}$ of dimension $n - m$ at $\mathbf{x}_0$, and that the function f restricted to S has an extreme value at $\mathbf{x}_0$. Show that $\mathcal{T}$ is parallel to the tangent to the surface defined explicitly by f at the point $(\mathbf{x}_0, f(\mathbf{x}_0))$.

12. (a) Find the points $\mathbf{x}_0$ at which $f(x, y) = x^2 - y^2 - y$ attains its maximum on the circle $x^2 + y^2 = 1$. [*Ans.* $(\pm\sqrt{15}/4, -\frac{1}{4})$.]

(b) Find the directions in which f increases most rapidly at $\mathbf{x}_0$. [*Ans.* $(\pm\sqrt{15}/4, -\frac{1}{4})$.]

13. The planes $x + y - z - 2w = 1$ and $x - y + z + w = 2$ intersect in a set $\mathcal{F}$ in $\mathcal{R}^4$. Find the point on $\mathcal{F}$ that is nearest to the origin. [*Ans.* $(\frac{27}{19}, -\frac{7}{19}, \frac{7}{19}, -\frac{3}{19})$.]

14. Let $\mathbf{x}_1, \ldots, \mathbf{x}_N$ be points in $\mathcal{R}^n$, and let

$$f(\mathbf{x}) = \sum_{k=1}^{N} |\mathbf{x} - \mathbf{x}_k|^2.$$

Find the point at which f attains its minimum and find the minimum value.

15. Prove by solving an appropriate minimum problem that if $a_k > 0$, $k = 1, \ldots, n$, then

$$(a_1 a_2 \ldots a_n)^{1/n} \leq \frac{a_1 + a_2 + \ldots + a_n}{n}$$

SECTION 2 QUADRATIC POLYNOMIALS

Let $F(\mathbf{x}, \mathbf{y}) = \mathbf{x} \cdot \mathbf{y}$, be the Euclidean dot-product of two vectors $\mathbf{x}$ and $\mathbf{y}$ in $\mathcal{R}^n$. In addition to having the property $F(\mathbf{x}, \mathbf{x}) \geq 0$, the function F satisfies

$$F(\mathbf{x}, \mathbf{y}) = F(\mathbf{y}, \mathbf{x}) \tag{1}$$

$$F(\mathbf{x} + \mathbf{x}', \mathbf{y}) = F(\mathbf{x}, \mathbf{y}) + F(\mathbf{x}', \mathbf{y}) \tag{2}$$

$$F(a\mathbf{x}, \mathbf{y}) = aF(\mathbf{x}, \mathbf{y}), \tag{3}$$

where a is any real number. As a consequence of the definition of F, or of the above three properties, F is linear in the second variable also. Because of the symmetry property (1) and the linearity in both variables, a real-valued function F satisfying (1)–(3) for all pairs of vectors $\mathbf{x}$ and $\mathbf{y}$ in $\mathcal{R}^n$ is called a **symmetric bilinear function.** Such a function can be written in terms of coordinates as follows. Let $\mathbf{x} = (x_1, \ldots, x_n)$ and $\mathbf{y} = (y_1, \ldots, y_n)$. Then,

$$\mathbf{x} = \sum_{i=1}^{n} x_i \mathbf{e}_i, \qquad \mathbf{y} = \sum_{j=1}^{n} y_j \mathbf{e}_j,$$

where $\mathbf{e}_k$, $k = 1, 2, \ldots, n$, are the natural basis vectors

$$\begin{pmatrix} 1 \\ 0 \\ \cdot \\ \cdot \\ \cdot \\ 0 \end{pmatrix}, \qquad \begin{pmatrix} 0 \\ 1 \\ \cdot \\ \cdot \\ \cdot \\ 0 \end{pmatrix}, \qquad \ldots, \qquad \begin{pmatrix} 0 \\ 0 \\ \cdot \\ \cdot \\ \cdot \\ 1 \end{pmatrix},$$

of $\mathcal{R}^n$. We have from (2) and (3)

$$F(\mathbf{x}, \mathbf{y}) = F\left(\sum_{1}^{n} x_i \mathbf{e}_i, \sum_{1}^{n} y_j \mathbf{e}_j\right) = \sum_{i=1}^{n} \sum_{j=1}^{n} F(\mathbf{e}_i, \mathbf{e}_j) x_i y_j,$$

where, by (1) $F(\mathbf{e}_i, \mathbf{e}_j) = F(\mathbf{e}_j, \mathbf{e}_i)$. Conversely, an arbitrary choice of the numbers $F(\mathbf{e}_i, \mathbf{e}_j) = a_{ij}$, consistent with $a_{ij} = a_{ji}$, determines the most general symmetric bilinear function. In summary, symmetric bilinear functions are just those which, in terms of coordinates, have the form

$$F(\mathbf{x}, \mathbf{y}) = \sum_{i,j=1}^{n} a_{ij} x_i y_j, \qquad a_{ij} = a_{ji}. \tag{4}$$

In particular, if $a_{ii} = 1$ and $a_{ij} = 0$ for $i \neq j$, we get our original example

$$\mathbf{x} \cdot \mathbf{y} = \sum_{i=1}^{n} x_i y_i.$$

If F is a symmetric bilinear function on $\mathcal{R}^n$, the real-valued function of a single vector defined by

$$Q(\mathbf{x}) = F(\mathbf{x}, \mathbf{x}) \quad \text{for all } \mathbf{x} \text{ in } \mathcal{R}^n$$

is called a **homogeneous quadratic polynomial** or sometimes a **quadratic form.** Thus, by definition, every Q is associated with some bilinear F, and vice versa. From (4) it follows that, in coordinate form,

$$Q(\mathbf{x}) = \sum_{i,j=1}^{n} a_{ij} x_i y_j, \qquad a_{ij} = a_{ji}. \tag{5}$$

For example, if F is the Euclidean dot-product, we have associated with it the quadratic polynomial

$$\mathbf{x} \cdot \mathbf{x} = \sum_{i=1}^{n} x_i^2.$$

The word **homogeneous** applied to a polynomial, means that all terms have the same degree in the coordinate variables x_i. Throughout this section the phrase "quadratic polynomial" will be understood to mean "homogeneous quadratic polynomial."

Equation (4) can be written as a matrix product as follows:

$$F(\mathbf{x}, \mathbf{y}) = (x_1 \quad x_2 \quad \dots \quad x_n) \begin{pmatrix} a_{11} & a_{12} & \dots & a_{1n} \\ a_{21} & & & \cdot \\ \cdot & & & \cdot \\ \cdot & & & \cdot \\ \cdot & & & \cdot \\ a_{n1} & \dots & & a_{nn} \end{pmatrix} \begin{pmatrix} y_1 \\ y_2 \\ \cdot \\ \cdot \\ \cdot \\ y_n \end{pmatrix}$$

or

$$F(\mathbf{x}, \mathbf{y}) = \mathbf{x}^t A \mathbf{y} = \mathbf{x} \cdot A\mathbf{y}.$$

This follows immediately from the definition of matrix multiplication. The condition $a_{ji} = a_{ij}$ means that the matrix $A = (a_{ij})$ is symmetric about its principal diagonal.

In the matrix notation, (5) becomes

$$Q(\mathbf{x}) = \mathbf{x}^t A \mathbf{x},$$

and we have as a familiar special case

$$Q(\mathbf{x}) = \mathbf{x}^t I \mathbf{x} = \mathbf{x} \cdot \mathbf{x}.$$

In case $a_{ij} = 0$ for $i \neq j$, A is a diagonal matrix and Q is said to be represented in **diagonal form**

Example 1. We give some examples of quadratic polynomials.

$$(x \quad y) \begin{pmatrix} 1 & 2 \\ 2 & 4 \end{pmatrix} \begin{pmatrix} x \\ y \end{pmatrix} = x^2 + 4xy + y^2,$$

$$(x \quad y \quad z) \begin{pmatrix} 1 & 0 & 1 \\ 0 & 1 & 0 \\ 1 & 0 & 1 \end{pmatrix} \begin{pmatrix} x \\ y \\ z \end{pmatrix} = x^2 + y^2 + z^2 + 2xz,$$

$$(x_1 \quad x_2 \quad x_3 \quad x_4) \begin{pmatrix} 1 & 0 & 0 & 0 \\ 0 & 2 & 0 & 0 \\ 0 & 0 & 3 & 0 \\ 0 & 0 & 0 & 4 \end{pmatrix} \begin{pmatrix} x_1 \\ x_2 \\ x_3 \\ x_4 \end{pmatrix} = x_1^2 + 2x_2^2 + 3x_3^2 + 4x_4^2.$$

A quadratic polynomial Q is called **positive definite** if $Q(\mathbf{x}) > 0$ except for $\mathbf{x} = 0$. We remark that if Q is positive definite and F is its associated bilinear function, then F is an inner product in $\mathcal{R}^n$. This is so because (1)–(3) together with the positive definiteness condition are the characteristic properties of an inner product.

Example 2. The graphs of two quadratic polynomials are shown in Fig. 6. The one on the left is positive definite. The other one is not; its

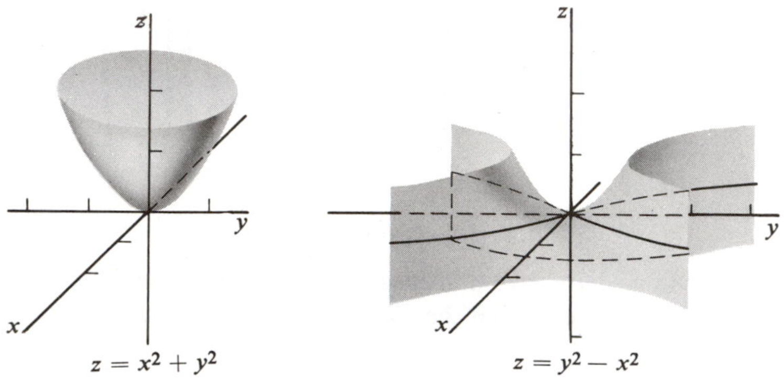

Figure 6

graph is called a **hyperbolic paraboloid.**

Example 3. The quadratic polynomial

$$Q_1(x, y, z) = (x - y - z)^2 = x^2 + y^2 + z^2 - 2xy - 2xz + 2yz$$

is nonnegative. However, it is not positive definite because it is zero on the plane $x - y - z = 0$.

The polynomial

$$Q_2(x, y, z) = (x + y + z)^2 - (x - y - z)^2 = 4xy + 4xz$$

changes sign. In fact, Q_2 is negative on the plane $x + y + z = 0$ and positive on the plane $x - y - z = 0$, except along the line of intersection of the two planes, where Q_2 is zero.

The polynomial

$$Q_3(x, y, z) = x^2 + y^2$$

is nonnegative, but not positive definite, because $Q_3(0, 0, z) = 0$ for arbitrary z.

In the examples just given, we have seen illustrations of the fact that if a quadratic polynomial can be written, say, in the diagonal form

$$Q(x, y, z) = a_1x^2 + a_2y^2 + a_3z^2,$$

then Q is positive definite if and only if all the coefficients a_i are positive. Furthermore, if some coefficients are negative or zero, it is possible to determine regions for which Q is positive or negative. In the following examples we consider one way in which a polynomial $Q(x, y)$ can be written in diagonal form.

Example 4. In $\mathcal{R}^2$ we get the most general symmetric bilinear function by choosing a, b, and c arbitrarily in

$$F((x, y), (x', y')) = (x \quad y)\begin{pmatrix} a & b \\ b & c \end{pmatrix}\begin{pmatrix} x' \\ y' \end{pmatrix}.$$

The general quadratic polynomial is then

$$Q(x, y) = ax^2 + 2bxy + cy^2.$$

To determine conditions under which Q is positive definite, notice first that we could not have both $a = 0$ and $c = 0$. For then $Q(x, y) = 2bxy$, and, if $b \neq 0$, this polynomial assumes both positive and negative values. Suppose then that $a \neq 0$. Completing the square, we have

$$Q(x, y) = \frac{1}{a}\left[a^2\left(x + \frac{b}{a}y\right)^2 + (ac - b^2)y^2\right]. \tag{6}$$

Similarly, if $c \neq 0$,

$$Q(x, y) = \frac{1}{c}\left[c^2\left(y + \frac{b}{c}x\right)^2 + (ac - b^2)x^2\right]. \tag{7}$$

We see directly that Q is positive definite if and only if $ac - b^2 > 0$ and either $a > 0$ or $c > 0$.

Example 5. Having written Q in one of the two forms (6) or (7), an obvious change of variable can be used to simplify the polynomial. To be specific, suppose $a \neq 0$ and that (6) holds. Letting

$$u = x + \frac{b}{a}y, \qquad v = 0x + y,$$

we can write Q in the form

$$au^2 + \frac{1}{a}(ac - b^2)v^2.$$

This transformation of coordinates corresponds to a change of basis in which the natural basis of $\mathcal{R}^2$ is replaced by the basis

$$\mathbf{x}_1 = (1, 0), \qquad \mathbf{x}_2 = \left(-\frac{b}{a}, 1\right).$$

The coordinate relations between x, y and u, v can be written in matrix form as

$$\begin{pmatrix} u \\ v \end{pmatrix} = \begin{pmatrix} 1 & \frac{b}{a} \\ 0 & 1 \end{pmatrix} \begin{pmatrix} x \\ y \end{pmatrix}, \quad \begin{pmatrix} x \\ y \end{pmatrix} = \begin{pmatrix} 1 & -\frac{b}{a} \\ 0 & 1 \end{pmatrix} \begin{pmatrix} u \\ v \end{pmatrix}.$$

To see concretely the geometric significance of the choice of new basis, we consider a numerical example. Let

$$Q(\mathbf{x}) = Q(x, y) = x^2 + 2xy + 3y^2.$$

Then $a = b = 1$ and $c = 3$. The new basis consists of the vectors $\mathbf{x}_1 = (1, 0)$ and $\mathbf{x}_2 = (-1, 1)$. With respect to the new coordinates we have

$$Q(\mathbf{x}) = u^2 + 2v^2,$$

where

$$\begin{pmatrix} u \\ v \end{pmatrix} = \begin{pmatrix} 1 & 1 \\ 0 & 1 \end{pmatrix} \begin{pmatrix} x \\ y \end{pmatrix}, \quad \begin{pmatrix} x \\ y \end{pmatrix} = \begin{pmatrix} 1 & -1 \\ 0 & 1 \end{pmatrix} \begin{pmatrix} u \\ v \end{pmatrix}.$$

Clearly Q is positive definite. The vectors $\mathbf{x}_1$ and $\mathbf{x}_2$, together with the level curve $Q(\mathbf{x}) = 1$, are shown in Fig. 7.

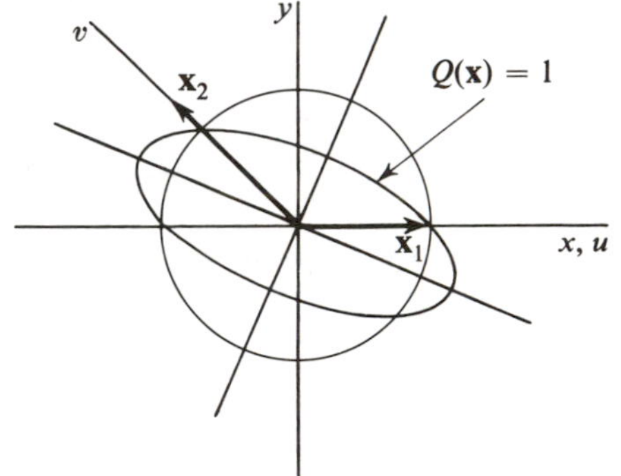

Figure 7

We have seen in Example 5 that a quadratic polynomial can be reduced to diagonal form if it is written in terms of the coordinates of an appropriately chosen basis. However, this was done with basis vectors which were not necessarily perpendicular. The next theorem shows that we can always find a diagonalizing basis consisting of perpendicular vectors of length 1.

2.1 Theorem

Let Q be a quadratic polynomial on $\mathcal{R}^n$. There exists an orthonormal basis $\mathbf{x}_1, \ldots, \mathbf{x}_n$ such that if $y_1, \ldots, y_n$ are the coordinates of a vector $\mathbf{x}$ with respect to this basis, then

$$Q(\mathbf{x}) = \lambda_1 y_1^2 + \ldots + \lambda_n y_n^2.$$

As a result,

$$Q(\mathbf{x}_k) = \lambda_k, \qquad k = 1, \ldots, n.$$

Proof. The basis vectors $\mathbf{x}_1, \ldots, \mathbf{x}_n$ will be chosen successively as follows. Let S^{n-1} be the set of unit vectors in $\mathcal{R}^n$, that is, the set of all $\mathbf{x}$ such that $|\mathbf{x}| = 1$.† Let $\mathbf{x}_1$ be a maximum point on S^{n-1} for the function Q restricted to S^{n-1}. (See the introduction to the Appendix.) By its choice, $\mathbf{x}_1$ is a unit vector. Let $\mathcal{V}_{n-1}$ be the $(n-1)$-dimensional-subspace of $\mathcal{R}^n$ consisting of the vectors $\mathbf{x}$ of $\mathcal{R}^n$ that are perpendicular to $\mathbf{x}_1$, and let S^{n-2} be the unit sphere in $\mathcal{V}_{n-1}$. Restrict $\mathbf{x}$ to S^{n-2}, and let $\mathbf{x}_2$ be a vector on S^{n-2} such that $Q(\mathbf{x})$ assumes its maximum, for $\mathbf{x}$ on S^{n-2}, at $\mathbf{x}_2$. By its choice, $\mathbf{x}_2$ is a unit vector perpendicular to $\mathbf{x}_1$. Assuming that $\mathbf{x}_1, \ldots, \mathbf{x}_k$, $k < n$, have been chosen in this way, let $\mathcal{V}_{n-k}$ be the subspace of $\mathcal{R}^n$ consisting of all vectors perpendicular to $\mathbf{x}_1, \ldots, \mathbf{x}_k$. Let S^{n-k-1} be the unit sphere in $\mathcal{V}_{n-k}$, and let Q assume its maximum on S^{n-k-1} at the point $\mathbf{x}_{k+1}$. Continue the process until n unit vectors have been chosen in this way, each perpendicular to those already chosen.

The vectors $\mathbf{x}_1, \ldots, \mathbf{x}_n$ clearly form an orthonormal basis for $\mathcal{R}^n$. We now show that this basis is a diagonalizing basis for Q. Since Q has a maximum at $\mathbf{x}_1$ when restricted to the unit sphere $|\mathbf{x}|^2 - 1 = 0$, by Lagrange's theorem, Theorem 1.2, the function f defined by

$$f(\mathbf{x}) = Q(\mathbf{x}) - \lambda(|\mathbf{x}|^2 - 1)$$

must have a critical point at $\mathbf{x}_1$ for some λ. That is $f'(\mathbf{x}_1) = 0$. Direct computation shows that every quadratic polynomial Q on $\mathcal{R}^n$ and its associated bilinear function F satisfy the equation

$$Q'(\mathbf{x})\mathbf{y} = 2F(\mathbf{x}, \mathbf{y})$$

for any $\mathbf{x}$ and $\mathbf{y}$ in $\mathcal{R}^n$. (See Exercise 10.) Hence, at the critical point $\mathbf{x}_1$,

$$0 = f'(\mathbf{x}_1)\mathbf{y} = 2F(\mathbf{x}_1, \mathbf{y}) - 2\lambda \mathbf{x}_1 \cdot \mathbf{y},$$

and we conclude that

$$F(\mathbf{x}_1, \mathbf{y}) = \lambda \mathbf{x}_1 \cdot \mathbf{y}$$

for any $\mathbf{y}$ in $\mathcal{R}^n$. It follows that

$$F(\mathbf{x}_1, \mathbf{x}_k) = 0, \qquad k = 2, \ldots, n, \qquad F(\mathbf{x}_1, \mathbf{x}_1) = \lambda = Q(\mathbf{x}_1).$$

By restricting Q to $\mathcal{V}_{n-1}$ the subspace of $\mathcal{R}^n$ perpendicular to $\mathbf{x}_1$, we

† The set of all unit vectors in $\mathcal{R}^n$ is an $(n-1)$-dimensional surface implicitly defined by the equation $|\mathbf{x}| = 1$. For this reason, we write the index $n - 1$ on S^{n-1}.

can repeat the same argument and obtain

$$F(\mathbf{x}_2, \mathbf{x}_k) = 0, \qquad k = 3, \ldots, n.$$

Continuing in this way, we obtain finally

$$F(\mathbf{x}_i, \mathbf{x}_k) = 0, \qquad \text{if} \quad i \neq k.$$

If an arbitrary vector $\mathbf{x}$ is written in terms of the basis $\mathbf{x}_1, \ldots, \mathbf{x}_n$ as $\mathbf{x} = y_1\mathbf{x}_1 + \ldots + y_n\mathbf{x}_n$, we obtain

$$Q(\mathbf{x}) = \sum_{j,k=1}^{n} F(\mathbf{x}_j, \mathbf{x}_k) y_j y_k$$

$$= \sum_{k=1}^{n} F(\mathbf{x}_k, \mathbf{x}_k) y_k^2 = \sum_{k=1}^{n} Q(\mathbf{x}_k) y_k^2.$$

This completes the proof.

A further consequence of the proof just given can be stated as follows.

2.2 Theorem

The basis vectors $\mathbf{x}_1, \ldots, \mathbf{x}_n$ with respect to which a quadratic polynomial Q has the form

$$Q(\mathbf{x}) = \sum_{k=1}^{n} \lambda_k y_k^2$$

can be chosen by requiring that $Q(\mathbf{x}_k)$ be the maximum value of Q restricted to the unit sphere of the subspace of $\mathcal{R}^n$ perpendicular to $\mathbf{x}_1, \mathbf{x}_2, \ldots, \mathbf{x}_{k-1}$.

The maximum value property of the basis vectors $\mathbf{x}_k$ can be used to compute them, as in the next example.

Example 6. Suppose the quadratic polynomial

$$Q(x, y) = 3x^2 + 2xy + 3y^2$$

is expressed using the coordinates of the natural basis for $\mathcal{R}^2$. We restrict Q to the unit circle

$$x^2 + y^2 - 1 = 0.$$

By Lagrange's theorem, Theorem 1.2, Q will have its maximum at the critical points of

$$3x^2 + 2xy + 3y^2 - \lambda(x^2 + y^2 - 1),$$

for some λ. That is, for some λ, the vector (x, y) must satisfy

$$\begin{aligned}(3-\lambda)x + \quad\quad y &= 0\\ x + (3-\lambda)y &= 0\end{aligned}$$

in addition to $x^2 + y^2 = 1$. (λ has been replaced by $-\lambda$.) Nonzero solutions to these equations will exist only if the columns of the matrix

$$\begin{pmatrix} 3-\lambda & 1 \\ 1 & 3-\lambda \end{pmatrix}$$

are dependent. Since dependence is equivalent to

$$\begin{vmatrix} 3-\lambda & 1 \\ 1 & 3-\lambda \end{vmatrix} = 0,$$

we must have $(3-\lambda)^2 - 1 = 0$ or $\lambda = 2, 4$. The corresponding solutions for (x, y) are

$$\lambda = 2: \quad (x, y) = \left(\pm\frac{1}{\sqrt{2}}, \mp\frac{1}{\sqrt{2}}\right).$$

$$\lambda = 4: \quad (x, y) = \left(\pm\frac{1}{\sqrt{2}}, \pm\frac{1}{\sqrt{2}}\right).$$

The maximum of Q occurs at $(\pm 1/\sqrt{2}, \pm 1/\sqrt{2})$, so we can choose $\mathbf{x}_1 = (1/\sqrt{2}, 1/\sqrt{2})$. For $\mathbf{x}_2$ we can take either the vector $(-1/\sqrt{2}, 1/\sqrt{2})$ or its negative.

Let $\mathbf{x}_2 = (-1/\sqrt{2}, 1/\sqrt{2})$. The change of coordinate equation is then

$$\begin{pmatrix} x \\ y \end{pmatrix} = \begin{pmatrix} \frac{1}{\sqrt{2}} & \frac{-1}{\sqrt{2}} \\ \frac{1}{\sqrt{2}} & \frac{1}{\sqrt{2}} \end{pmatrix} \begin{pmatrix} u \\ v \end{pmatrix}.$$

In terms of the new variables we have

$$Q(\mathbf{x}) = 4u^2 + 2v^2.$$

In Fig. 8, level curves of Q are shown in their relation to the new and to the original basis vectors.

For some purposes it is unnecessary to compute the orthonormal basis vectors $\mathbf{x}$ of Theorem 2.1 provided that the numbers λ can be found. For example, it is clear just from knowing the λ_k whether Q is positive definite or not. The following theorem enables us to compute, or estimate, the λ_k.

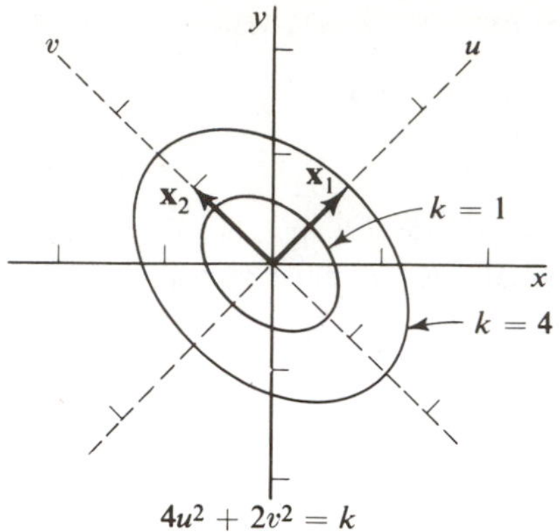

Figure 8

2.3 Theorem

Let Q be a quadratic polynomial in $\mathcal{R}^n$ given by

$$Q(\mathbf{x}) = \mathbf{x}^t A \mathbf{x},$$

where A is a symmetric matrix. Suppose that, with respect to the coordinates of the orthonormal basis $\mathbf{x}_1, \ldots, \mathbf{x}_n$, Q has the form

$$Q(\mathbf{x}) = \sum_{k=1}^{n} \lambda_k y_k^2, \tag{8}$$

where $Q(\mathbf{x}_k) = \lambda_k$. Then the numbers λ_k are the roots of the equation

$$\det\,(A - \lambda I) = 0. \tag{9}$$

Although the existence of the basis vectors $\mathbf{x}_1, \ldots, \mathbf{x}_k$ was proved in Theorem 2.1, it is not necessary to know what they are in order to find the λ_k. The λ_k can be computed by solving Equation (9). Equation (9) is called the **characteristic equation** of Q, and the roots λ_k are called **characteristic roots** or **eigenvalues**. The next theorem provides another method for computing the basis vectors $\mathbf{x}_1, \ldots, \mathbf{x}_n$.

2.4 Theorem

Let $\mathbf{z}_1, \ldots, \mathbf{z}_n$ be any orthonormal basis such that, for each $k = 1, \ldots, n$, the vector $\mathbf{z}_k$ satisfies the matrix equation

$$(A - \lambda_k I)\mathbf{z}_k = 0. \tag{10}$$

Then, with respect to this basis, Q has the diagonal form (8).

Vectors $\mathbf{z}_k$ that satisfy Equation (10) are called **characteristic vectors** or **eigenvectors corresponding to** λ_k.

Proof (of Theorem 2.3). Suppose that the orthonormal basis vectors $\mathbf{x}_1, \ldots, \mathbf{x}_n$ that diagonalize Q are

$$\mathbf{x}_1 = \begin{pmatrix} b_{11} \\ \cdot \\ \cdot \\ \cdot \\ b_{n1} \end{pmatrix}, \quad \mathbf{x}_2 = \begin{pmatrix} b_{12} \\ \cdot \\ \cdot \\ \cdot \\ b_{n2} \end{pmatrix}, \quad \ldots, \quad \mathbf{x}_n = \begin{pmatrix} b_{1n} \\ \cdot \\ \cdot \\ \cdot \\ b_{nn} \end{pmatrix}.$$

Let B be the n by n matrix with columns $\mathbf{x}_1, \ldots, \mathbf{x}_n$. According to Chapter 2, Section 10, coordinates of the same point are related by the equation $\mathbf{x} = B\mathbf{y}$, where

$$\mathbf{x} = \begin{pmatrix} x_1 \\ \cdot \\ \cdot \\ \cdot \\ x_n \end{pmatrix} \quad \text{and} \quad \mathbf{y} = \begin{pmatrix} y_1 \\ \cdot \\ \cdot \\ \cdot \\ y_n \end{pmatrix},$$

and where $y_1, \ldots, y_n$ are the coordinates with respect to $\mathbf{x}_1, \ldots, \mathbf{x}_n$. Then substituting $B\mathbf{y}$ for $\mathbf{x}$ gives

$$\begin{aligned} Q(\mathbf{x}) &= (B\mathbf{y})^t A(B\mathbf{y}) \\ &= \mathbf{y}^t(B^t AB)\mathbf{y} \\ &= \mathbf{y}^t \Lambda \mathbf{y}. \end{aligned}$$

By the choice of the columns of B, the matrix $\Lambda = B^t AB$ is a diagonal matrix with diagonal entries $\lambda_1, \ldots, \lambda_n$. Furthermore, since the columns of B are the coordinates of perpendicular unit vectors with respect to an orthonormal basis, we have directly, by matrix multiplication, $B^t B = I$. In other words, $B^t = B^{-1}$. Then $\Lambda = B^{-1}AB$. Subtracting λI from both sides of this equation and factoring the right-hand member, we get

$$\begin{aligned} \Lambda - \lambda I &= B^{-1}AB - \lambda I \\ &= B^{-1}(A - \lambda I)B. \end{aligned}$$

But $\Lambda - \lambda I$ is a diagonal matrix with diagonal entries $\lambda_k - \lambda$, so

$$\begin{aligned}(\lambda_1 - \lambda)(\lambda_2 - \lambda) \ldots (\lambda_n - \lambda) &= \det(\Lambda - \lambda I) \\ &= \det B^{-1} \det(A - \lambda I) \det B \\ &= \det(A - \lambda I).\end{aligned}$$

This shows that the roots of $\det(A - \lambda I) = 0$ are $\lambda_1, \ldots, \lambda_n$.

Proof (of Theorem 2.4). Let $\mathbf{z}_1, \ldots, \mathbf{z}_n$ be an orthonormal basis satisfying $A\mathbf{z}_k = \lambda_k \mathbf{z}_k$, for $k = 1, 2, \ldots, n$. Let C be the matrix with columns $\mathbf{z}_1, \mathbf{z}_2, \ldots, \mathbf{z}_n$. The equation

$$\mathbf{x} = C\mathbf{z}$$

gives the relation between the coordinates of the basis $\mathbf{z}_1, \ldots, \mathbf{z}_n$ in $\mathcal{R}^n$ and the natural coordinates. Then

$$Q(\mathbf{x}) = \mathbf{x}^t A \mathbf{x} = (C\mathbf{z})^t A (C\mathbf{z}) = \mathbf{z}^t (C^t A C)\mathbf{z}.$$

All we have to do is verify that the matrix C^tAC is diagonal with diagonal entries $\lambda_1, \ldots, \lambda_n$. Schematically, we write

$$\begin{aligned}C^tAC &= \begin{pmatrix} \mathbf{z}_1^t \\ \cdot \\ \cdot \\ \cdot \\ \mathbf{z}_n^t \end{pmatrix} A(\mathbf{z}_1 \quad \ldots \quad \mathbf{z}_n) \\ &= \begin{pmatrix} \mathbf{z}_1^t \\ \cdot \\ \cdot \\ \cdot \\ \mathbf{z}_n^t \end{pmatrix} (A\mathbf{z}_1 \quad \ldots \quad A\mathbf{z}_n) \\ &= \begin{pmatrix} \mathbf{z}_1^t \\ \cdot \\ \cdot \\ \cdot \\ \mathbf{z}_n^t \end{pmatrix} (\lambda_1\mathbf{z}_1 \quad \ldots \quad \lambda_n\mathbf{z}_n).\end{aligned}$$

Using the fact that

$$\mathbf{z}_i^t \mathbf{z}_j = \mathbf{z}_i \cdot \mathbf{z}_j = \begin{cases} 1, & \text{if} \quad i = j, \\ 0, & \text{if} \quad i \neq j, \end{cases}$$

we get

$$C^tAC = \begin{pmatrix} \lambda_1 & 0 & \dots & 0 \\ 0 & \lambda_2 & & \\ \cdot & & & \cdot \\ \cdot & & & \cdot \\ \cdot & & & \cdot \\ 0 & & & \lambda_n \end{pmatrix}.$$

This completes the proof.

Example 7. Let $Q(x, y, z) = xy + yz + zx$. The matrix of Q is

$$\begin{pmatrix} 0 & \frac{1}{2} & \frac{1}{2} \\ \frac{1}{2} & 0 & \frac{1}{2} \\ \frac{1}{2} & \frac{1}{2} & 0 \end{pmatrix}$$

and the characteristic equation is

$$\begin{pmatrix} -\lambda & \frac{1}{2} & \frac{1}{2} \\ \frac{1}{2} & -\lambda & \frac{1}{2} \\ \frac{1}{2} & \frac{1}{2} & -\lambda \end{pmatrix} = 0$$

or

$$-\lambda^3 + \tfrac{3}{4}\lambda + \tfrac{1}{4} = 0.$$

The characteristic roots are $\lambda = 1, -\frac{1}{2}, -\frac{1}{2}$. So there is an orthonormal system of coordinates (u, v, w) with respect to which Q has the form

$$u^2 - \tfrac{1}{2}v^2 - \tfrac{1}{2}w^2.$$

To find the related basis vectors we look for the unit vector solutions of the equations

$$\begin{pmatrix} -\lambda & \frac{1}{2} & \frac{1}{2} \\ \frac{1}{2} & -\lambda & \frac{1}{2} \\ \frac{1}{2} & \frac{1}{2} & -\lambda \end{pmatrix} \begin{pmatrix} x \\ y \\ z \end{pmatrix} = 0,$$

with $\lambda = 1$ and $\lambda = -\frac{1}{2}$. With $\lambda = 1$ we get $x = y = z$ for a solution, so we can choose $\mathbf{x}_1 = (1/\sqrt{3}, 1/\sqrt{3}, 1/\sqrt{3})$. When $\lambda = -\frac{1}{2}$, the matrix equation simply requires that the two remaining basis vectors lie in the plane $x + y + z = 0$, perpendicular to $\mathbf{x}_1$. Then $\mathbf{x}_2$ and $\mathbf{x}_3$ can be chosen to be arbitrary perpendicular vectors in that plane, for example,

$$\left(\frac{1}{\sqrt{2}}, -\frac{1}{\sqrt{2}}, 0\right) \quad \text{and} \quad \left(\frac{1}{\sqrt{6}}, \frac{1}{\sqrt{6}}, -\frac{2}{\sqrt{6}}\right).$$

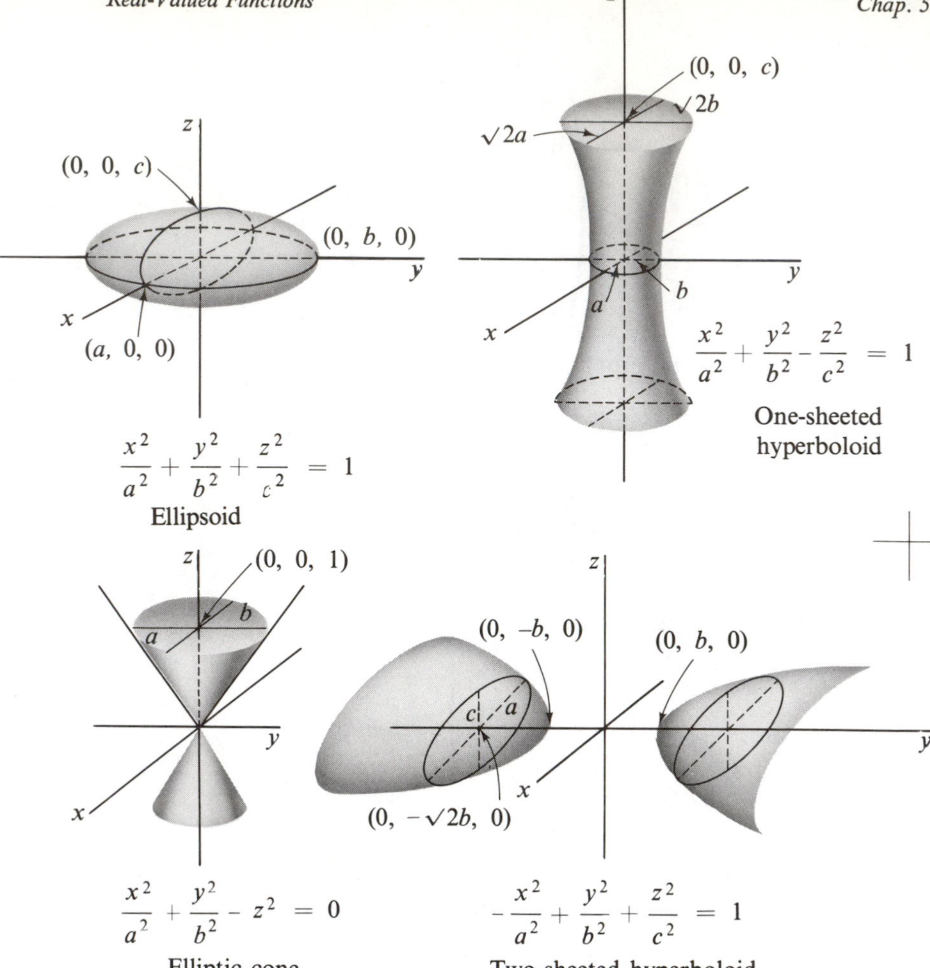

Figure 9

A level surface of a quadratic polynomial is called a **quadratic surface**. Since every quadratic polynomial can be diagonalized with respect to some orthonormal basis, it is sufficient to be able to picture the level surfaces of some standard quadratic polynomials in order to be able to picture a more general quadratic surface. Figure 9 shows some illustrations of quadratic surfaces in $\mathcal{R}^3$.

Example 8. The quadratic polynomial $xy + yz$ can be represented by a symmetric matrix as

$$xy + yz = (x \quad y \quad z)\begin{pmatrix} 0 & \frac{1}{2} & 0 \\ \frac{1}{2} & 0 & \frac{1}{2} \\ 0 & \frac{1}{2} & 0 \end{pmatrix}\begin{pmatrix} x \\ y \\ z \end{pmatrix}.$$

The characteristic equation is $-\lambda^3 + \frac{1}{2}\lambda = 0$, and this equation has roots $\lambda = 1/\sqrt{2}, 0, -1/\sqrt{2}$. The corresponding characteristic vector equations, together with their unit vector solutions, are as follows:

$$\lambda = \frac{1}{\sqrt{2}}: \quad \begin{pmatrix} -\frac{1}{\sqrt{2}} & \frac{1}{2} & 0 \\ \frac{1}{2} & -\frac{1}{\sqrt{2}} & \frac{1}{2} \\ 0 & \frac{1}{2} & -\frac{1}{\sqrt{2}} \end{pmatrix} \begin{pmatrix} x \\ y \\ z \end{pmatrix} \quad 0,$$

$$(x, y, z) = \pm\left(\frac{1}{2}, \frac{1}{\sqrt{2}}, \frac{1}{2}\right)$$

$$\lambda = 0: \quad \begin{pmatrix} 0 & \frac{1}{2} & 0 \\ \frac{1}{2} & 0 & \frac{1}{2} \\ 0 & \frac{1}{2} & 0 \end{pmatrix} \begin{pmatrix} x \\ y \\ z \end{pmatrix} = 0, \qquad (x, y, z) = \pm\left(\frac{1}{\sqrt{2}}, 0, -\frac{1}{\sqrt{2}}\right)$$

$$\lambda = -\frac{1}{\sqrt{2}}: \quad \begin{pmatrix} \frac{1}{\sqrt{2}} & \frac{1}{2} & 0 \\ \frac{1}{2} & \frac{1}{\sqrt{2}} & \frac{1}{2} \\ 0 & \frac{1}{2} & \frac{1}{\sqrt{2}} \end{pmatrix} \begin{pmatrix} x \\ y \\ z \end{pmatrix} = 0,$$

$$(x, y, z) = \pm\left(\frac{1}{2}, -\frac{1}{\sqrt{2}}, \frac{1}{2}\right).$$

We choose the diagonalizing basis

$$\mathbf{x}_1 = \left(\frac{1}{2}, \frac{1}{\sqrt{2}}, \frac{1}{2}\right), \quad \mathbf{x}_2 = \left(\frac{1}{\sqrt{2}}, 0, -\frac{1}{\sqrt{2}}\right), \quad \mathbf{x}_3 = \left(\frac{1}{2}, -\frac{1}{\sqrt{2}}, \frac{1}{2}\right).$$

If u, v, and w are the coordinates with respect to the new basis, then the polynomial will have the form $(1/\sqrt{2})u^2 - (1/\sqrt{2})w^2$. The level surface

$$\frac{1}{\sqrt{2}}u^2 - \frac{1}{\sqrt{2}}w^2 = 1$$

is the hyperbolic cylinder one sheet of which is shown in Fig. 10.

Homogeneous polynomials of degree $N > 2$ can be defined in very much the same way as quadratic polynomials. We can start with a real-valued function $F(\mathbf{x}_1, \ldots, \mathbf{x}_N)$ that is symmetric in its N vector variables, and linear in each of them, and defined for every N vectors $\mathbf{x}_1, \ldots, \mathbf{x}_N$.

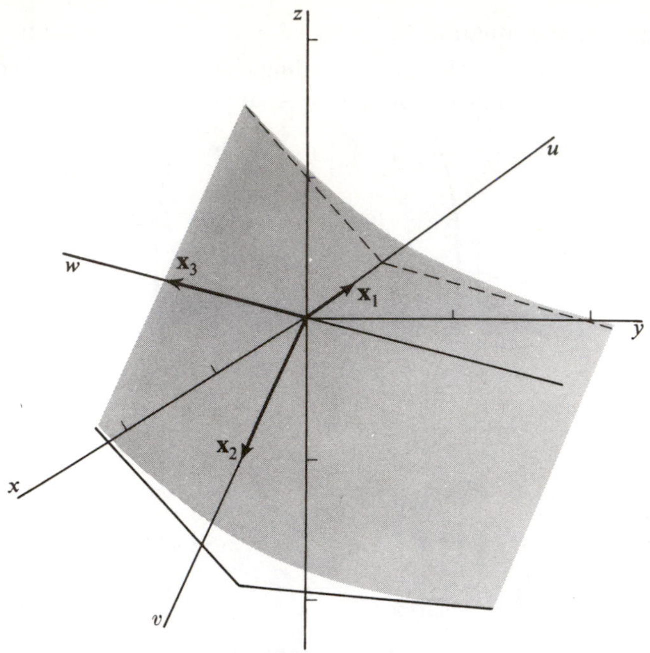

Figure 10

Then

$$P(\mathbf{x}) = F(\mathbf{x}, \mathbf{x}, \ldots, \mathbf{x})$$

will be a polynomial of degree N. Alternatively, we can consider functions of $\mathbf{x} = (x_1, x_2, \ldots, x_n)$ having the form

$$P(\mathbf{x}) = \sum_{i_1,\ldots,i_N=1}^{n} a_{i_1\ldots i_N} x_{i_1} \ldots x_{i_N},$$

where the coefficients $a_{i_1\ldots i_N}$ are symmetric in the subscripts. In both cases we get the same class of functions. The symmetry condition on the coefficients of F is a convenience, and if it were not assumed, it could be obtained simply by averaging the nonsymmetric coefficients. The details are left as an exercise (Exercise 12).

EXERCISES

1. By changing coordinates, write each of the following quadratic polynomials as a sum of squares. In each problem exhibit an orthonormal basis that does the job, and write the coordinate transformation.

(a) $3x^2 + 2\sqrt{2}\,xy + 4y^2$.

$$\left[\textit{Ans. } Q(\mathbf{x}) = 2u^2 + 5v^2;\ \mathbf{x}_1 = \left(\frac{\sqrt{2}}{\sqrt{3}}, -\frac{1}{\sqrt{3}}\right), \mathbf{x}_2 = \left(\frac{1}{\sqrt{3}}, \frac{\sqrt{2}}{\sqrt{3}}\right).\right]$$

(b) $3x^2 + 2\sqrt{3}\,xy + 5y^2$.

$$\left[\textit{Ans. } Q(\mathbf{x}) = 2u^2 + 6v^2;\ \mathbf{x}_1 = \left(-\frac{\sqrt{3}}{2}, \frac{1}{2}\right),\ \mathbf{x}_2 = \left(\frac{1}{2}, \frac{\sqrt{3}}{2}\right).\right]$$

(c) $(x \quad y)\begin{pmatrix} 2 & 2 \\ 2 & 5 \end{pmatrix}\begin{pmatrix} x \\ y \end{pmatrix}$.

(d) $2x^2 - 5xy + 2y^2 - 2xz + 4z^2 - 2yz$.

$$\left[\begin{aligned} &\textit{Ans. } Q(\mathbf{x}) = \frac{9}{2}u^2 - \frac{9}{2}v^2 + 0w^2; \qquad \mathbf{x}_1 = \left(\frac{1}{\sqrt{2}}, -\frac{1}{\sqrt{2}}, 0\right), \\ &\mathbf{x}_2 = \left(\frac{1}{3\sqrt{2}}, \frac{1}{3\sqrt{2}}, \frac{4}{3\sqrt{2}}\right), \qquad \mathbf{x}_3 = \left(\frac{3}{2}, \frac{2}{3}, -\frac{1}{3}\right).\end{aligned}\right]$$

(e) $(x \quad y \quad z)\begin{pmatrix} -1 & 2 & 0 \\ 2 & 0 & 2 \\ 0 & 2 & 1 \end{pmatrix}\begin{pmatrix} x \\ y \\ z \end{pmatrix}$.

2. (a) For each polynomial Q in Exercise 1, find the maximum of Q when Q is restricted to the unit sphere, $|\mathbf{x}| = 1$, of the Euclidean space on which Q is defined. [*Hint.* See Theorem 2.2.]

(b) Find the maximum of the polynomial in 1(a), restricted to the circle $x^2 + y^2 = 3$. [*Ans.* 15.]

3. Classify the following quadratic polynomials as positive definite, negative definite, or neither. Give reasons. (Q is **negative definite** if $Q < 0$ except for $Q(0) = 0$.)

(a) $2x^2 - 7xy + 5y^2$. [*Ans.* Neither.]
(b) $2x^2 - 3xy + 5y^2$. [*Ans.* Positive definite.]
(c) $-x^2 + 2xy - 6y^2$. [*Ans.* Negative definite.]
(d) $3x^2 + xy + 3y^2 + 5z^2$. [*Ans.* Positive definite.]

(e) $(x \quad y \quad z)\begin{pmatrix} 1 & 3 & 0 \\ 3 & 1 & 1 \\ 0 & 1 & 3 \end{pmatrix}\begin{pmatrix} x \\ y \\ z \end{pmatrix}$. [*Ans.* Neither.]

4. Prove that $x^2 + y^2 + z^2 - xy - xz - yz$ is not positive definite, but becomes so when restricted to the plane $x + y + z = 0$.

5. Sketch the level curves $Q(\mathbf{x}) = 1$ and $Q(\mathbf{x}) = 0$ for each of the following polynomials in $\mathcal{R}^2$.

(a) xy. (b) $x^2 + xy + y^2$. (c) $x^2 + xy - 2y^2$.

6. Sketch the level surfaces $Q(\mathbf{x}) = 1$ and $Q(\mathbf{x}) = 0$ for the following polynomials in $\mathcal{R}^3$.

   (a) $x^2 - xy + y^2 + z^2$. (b) $x^2 + xy$. (c) $x^2 - 2xy + y^2 - z^2$.

7. Show that every quadratic polynomial Q satisfies
$$Q(a\mathbf{x}) = a^2Q(\mathbf{x}),$$
for every real number a.

8. Let Q be an arbitrary quadratic polynomial on $\mathcal{R}^n$, and let F be its associated symmetric bilinear function. Prove that
$$F(\mathbf{x}, \mathbf{y}) = \tfrac{1}{2}[Q(\mathbf{x} + \mathbf{y}) - Q(\mathbf{x}) - Q(\mathbf{y})],$$
for all vectors $\mathbf{x}$ and $\mathbf{y}$ in $\mathcal{R}^n$. This equation proves that F is uniquely determined by Q.

9. Prove that every quadratic polynomial Q on $\mathcal{R}^n$ is a continuous function.

10. Let Q be an arbitrary quadratic polynomial on $\mathcal{R}^n$, and let F be its associated symmetric bilinear function, that is, $Q(\mathbf{x}) = F(\mathbf{x}, \mathbf{x})$. Prove that Q is a differentiable vector function or, more explicitly, that
$$Q'(\mathbf{x})\mathbf{y} = 2F(\mathbf{x}, \mathbf{y}).$$

11. Prove that every quadratic polynomial Q is a continuously differentiable function.

12. What follows illustrates the fact that the condition of symmetry on a bilinear function can be obtained by averaging out the nonsymmetry. Let G be a real-valued function defined for all pairs of vectors $\mathbf{x}$ and $\mathbf{y}$ and linear in each variable (we do *not* assume symmetry). Show that the function F defined by
$$F(\mathbf{x}, \mathbf{y}) = \tfrac{1}{2}(G(\mathbf{x}, \mathbf{y}) + G(\mathbf{y}, \mathbf{x}))$$
is a symmetric bilinear function. Show that $G(\mathbf{x}, \mathbf{x}) = F(\mathbf{x}, \mathbf{x})$ and, hence, that G and F define the same quadratic polynomial.

13. Let Q be a quadratic polynomial on $\mathcal{R}^n$. Prove that there exists a basis $(\mathbf{x}_1, \ldots, \mathbf{x}_n)$ for $\mathcal{R}^n$ such that, for any vector $\mathbf{x} = y_1\mathbf{x}_1 + \ldots + y_n\mathbf{x}_n$,
$$Q(\mathbf{x}) = \sum_{i=1}^{n} \lambda_i y_i^2, \qquad \text{with} \quad \lambda_i = 0, 1, \text{ or } -1.$$

14. Prove that if Q is a positive-definite quadratic polynomial on $\mathcal{R}^n$, there exists a positive real number m such that
$$Q(\mathbf{x}) \geq m\,|\mathbf{x}|^2, \qquad \text{for all } \mathbf{x} \text{ in } \mathcal{R}^n.$$
[*Suggestion*. Diagonalize Q.] A corollary is that the values of Q on the unit sphere $|\mathbf{x}| = 1$ are bounded away from zero.

15. Let Q be a positive-definite quadratic polynomial in $\mathcal{R}^2$, and let λ_1 and λ_2 be its characteristic roots. Show that $\lambda_1^{-1/2}$ and $\lambda_2^{-1/2}$ are the lengths of the principal axes of the ellipse $Q(\mathbf{x}) = 1$.

16. Verify that if $a \neq 0$ and $ab - f^2 \neq 0$, then

$$(x \quad y \quad z)\begin{pmatrix} a & f & e \\ f & b & d \\ e & d & c \end{pmatrix}\begin{pmatrix} x \\ y \\ z \end{pmatrix} = a\left(x + \frac{f}{a}y + \frac{e}{a}z\right)^2$$

$$+ \frac{\begin{vmatrix} a & f \\ f & b \end{vmatrix}}{a}\left(y + \frac{\begin{vmatrix} a & f \\ e & d \end{vmatrix}}{\begin{vmatrix} a & f \\ f & b \end{vmatrix}}z\right)^2 + \frac{\begin{vmatrix} a & f & e \\ f & b & d \\ e & d & c \end{vmatrix}}{\begin{vmatrix} a & f \\ f & b \end{vmatrix}}z^2.$$

Conclude that the above polynomial is positive definite if and only if the three determinants are positive:

$$a, \quad \begin{vmatrix} a & f \\ f & b \end{vmatrix}, \quad \begin{vmatrix} a & f & e \\ f & b & d \\ e & d & c \end{vmatrix}$$

What is the condition for negative definiteness? The criterion can be extended to quadratic polynomials in $\mathcal{R}^n$. See R. M. Thrall and L. Tornheim, *Vector Spaces and Matrices*, Wiley, 1957, p. 170.

17. Show that the linear transformation $\mathcal{R}^n \xrightarrow{L} \mathcal{R}^n$ determined by a symmetric matrix A has the property that

$$L(\mathbf{x}) \cdot \mathbf{y} = \mathbf{x} \cdot L(\mathbf{y})$$

for all $\mathbf{x}$ and $\mathbf{y}$ in $\mathcal{R}^n$. Conversely, show that if this equation is satisfied, then A is a symmetric matrix.

18. A transformation $\mathcal{R}^n \xrightarrow{L} \mathcal{R}^n$ for which the equation in Exercise 17 holds is called a symmetric transformation. Show that for a given symmetric transformation there is an orthonormal basis such that, with respect to this basis, the matrix of the transformation is diagonal.

19. (a) Give a geometric description of the action on $\mathcal{R}^3$ of the symmetric transformations with matrices

$$\begin{pmatrix} \lambda_1 & 0 & 0 \\ 0 & 1 & 0 \\ 0 & 0 & 1 \end{pmatrix}, \quad \begin{pmatrix} 1 & 0 & 0 \\ 0 & \lambda_2 & 0 \\ 0 & 0 & 1 \end{pmatrix}, \quad \text{and} \quad \begin{pmatrix} 1 & 0 & 0 \\ 0 & 1 & 0 \\ 0 & 0 & \lambda_3 \end{pmatrix}.$$

(b) Use the result of Exercise 18 to give a geometric description of the action of a symmetric transformation in $\mathcal{R}^3$.

20. The proof of Theorem 2.1 works just as well when the Euclidean inner product is replaced by an arbitrary one. Use this fact to prove the following theorem: If Q_1 and Q_2 are quadratic polynomials in $\mathcal{R}^n$, and Q_1 is positive definite, then there is an orthonormal basis for $\mathcal{R}^n$ with respect to which Q_1 and Q_2 both have diagonal form.

21. Prove the converse of Theorem 2.4, namely, that if $\mathbf{x}_1, \ldots, \mathbf{x}_n$ is an orthonormal diagonalizing basis for $Q(\mathbf{x}) = \mathbf{x}^t A\mathbf{x}$, then $(A - \lambda_k I)\mathbf{x}_k = 0$, where $\lambda_k = Q(\mathbf{x}_k)$.

SECTION 3

TAYLOR EXPANSIONS

We begin by reviewing the definition and the simplest properties of the Taylor expansion for functions of one variable. If $f(x)$ has an Nth derivative at x_0, its **Taylor expansion of degree N about x_0** is the polynomial

$$f(x_0) + \frac{1}{1!}f'(x_0)(x - x_0) + \frac{1}{2!}f''(x_0)(x - x_0)^2 + \ldots + \frac{1}{N!}f^{(N)}(x_0)(x - x_0)^N.$$

The relation between f and its Taylor expansion can be expressed conveniently by the following **integral remainder formula.**

3.1 Theorem

If f has a continuous Nth derivative in a neighborhood of x_0, then in that neighborhood

$$f(x) = f(x_0) + \frac{1}{1!}f'(x_0)(x - x_0) + \ldots + \frac{1}{N!}f^{(N)}(x_0)(x - x_0)^N + R_N, \quad (1)$$

where

$$R_N = \frac{1}{(N-1)!}\int_{x_0}^{x}(x - t)^{N-1}[f^{(N)}(t) - f^{(N)}(x_0)]\, dt.$$

Proof. The remainder can be written as the difference

$$R_N = \frac{1}{(N-1)!}\int_{x_0}^{x}(x - t)^{N-1}f^{(N)}(t)\, dt - \frac{f^{(N)}(x_0)}{(N-1)!}\int_{x_0}^{x}(x - t)^{N-1}\, dt.$$

The second of these integrals is directly computed to be

$$\frac{1}{N!}f^{(N)}(x_0)(x - x_0)^N,$$

which is just the last term of the Taylor expansion. The first integral can be integrated by parts to give

$$\frac{1}{(N-2)!}\int_{x_0}^{x}(x - t)^{N-2}[f^{(N-1)}(t) - f^{(N-1)}(x_0)]\, dt = R_{N-1}.$$

We therefore obtain

$$R_N = -\frac{1}{N!} f^{(N)}(x_0)(x - x_0)^N + R_{N-1}.$$

If we substitute the preceding equation into (1), we get (1) back again with N replaced by $N - 1$. The induction is completed by noticing that, for $N = 1$, Equation (1) is just

$$f(x) = f(x_0) + f'(x_0)(x - x_0) + \int_{x_0}^{x} [f'(t) - f'(x_0)]\, dt,$$

and that this is a valid equation.

It follows from the remainder formula that a polynomial of degree N is equal to its Taylor expansion of degree N. For if f is of degree N, $f^{(N)}$ is a constant function and so the remainder is identically zero. We list some common examples. It is only in the first one that we have equality. The expansions are all about $x = 0$.

$$(1 + x)^N = \sum_{k=0}^{N} \binom{N}{k} x^k,$$

$$\frac{1}{(1 - x)^M}: \quad \sum_{k=0}^{N} \binom{M + k - 1}{M - 1} x^k,$$

$$e^x: \quad 1 + \frac{1}{1!}x + \frac{1}{2!}x^2 + \ldots + \frac{1}{N!}x^N,$$

$$\log(1 - x): \quad -x - \tfrac{1}{2}x^2 - \tfrac{1}{3}x^3 - \ldots - \frac{1}{N}x^N, \tag{2}$$

$$\cos x: \quad 1 - \frac{x^2}{2!} + \frac{x^4}{4!} - \ldots + (-1)^k \frac{x^{2k}}{(2k)!},$$

$$\sin x: \quad x - \frac{x^3}{3!} + \frac{x^5}{5!} - \ldots + (-1)^k \frac{x^{2k+1}}{(2k + 1)!}.$$

Figure 11 shows the graphical relationship between the functions e^x and $\cos x$, and their second-degree Taylor expansions.

To see how to generalize Taylor expansions to functions of more than one variable, suppose first we are given a function $f(x, y)$ having partial derivatives of order N at a point (x_0, y_0). For the moment, consider a function F of one variable defined by

$$F(t) = f(tu + x_0, tv + y_0),$$

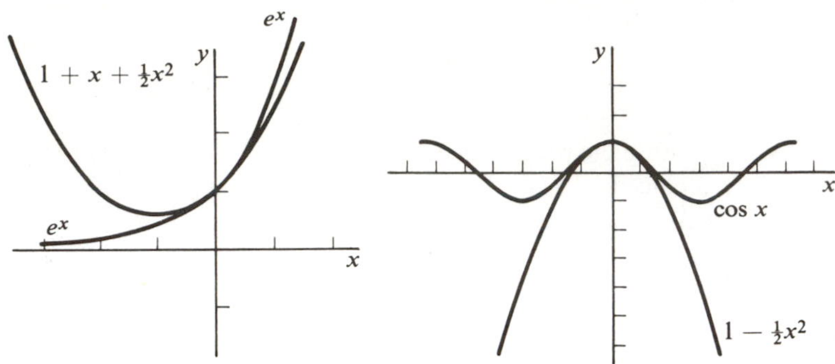

Figure 11

where u and v are held fixed. We attempt to find a Taylor expansion for F about $t = 0$ and, therefore, compute the successive derivatives of F at 0. We find, by the chain rule,

$$F'(t) = u\frac{\partial f}{\partial x}(tu + x_0, ty + y_0) + v\frac{\partial f}{\partial y}(tv + x_0, ty + y_0),$$

and hence that

$$F'(0) = u\frac{\partial f}{\partial x}(x_0, y_0) + v\frac{\partial f}{\partial y}(x_0, y_0).$$

Similarly, we compute $F''(t)$, set $t = 0$, and find

$$F''(0) = u^2\frac{\partial^2 f}{\partial x^2}(x_0, y_0) + 2uv\frac{\partial^2 f}{\partial x\,\partial y}(x_0, y_0) + v^2\frac{\partial^2 f}{\partial y^2}(x_0, y_0).$$

Further calculation shows that, in general,

$$F^{(N)}(0) = \sum_{k=0}^{N}\binom{N}{k}u^k v^{N-k}\frac{\partial^k f}{\partial x^k}(x_0, y_0)\frac{\partial^{N-k} f}{\partial y^{N-k}}(x_0, y_0),$$

where $\binom{N}{k}$ is the binomial coefficient $N!/k!(N-k)!$. Now replace u by $(x - x_0)$ and v by $(y - y_0)$. The Taylor expansion

$$F(t) \sim F(0) + \frac{1}{1!}F'(0)t + \frac{1}{2!}F''(0)t^2 + \ldots$$

becomes, for $t = 1$,

$$f(x, y) \sim f(x_0, y_0) + \frac{1}{1!}\left(\frac{\partial f}{\partial x}(x_0, y_0)(x - x_0) + \frac{\partial f}{\partial y}(x_0, y_0)(y - y_0)\right)$$
$$+ \frac{1}{2!}\left(\frac{\partial^2 f}{\partial x^2}(x_0, y_0)(x - x_0)^2 + 2\frac{\partial^2 f}{\partial x\,\partial y}(x - x_0)(y - y_0)\right.$$
$$\left. + \frac{\partial^2 f}{\partial y^2}(x_0, y_0)(y - y_0)^2\right) + \cdots.$$

Thus we arrive at the following definition. For a function $f(x, y)$ of two variables having continuous Nth-order partial derivatives in a neighborhood of (x_0, y_0) the **Taylor expansion of degree** N **about** (x_0, y_0) is the polynomial

$$f(x_0, y_0) + \frac{1}{1!}((x - x_0)f_x(x_0, y_0) + (y - y_0)f_y(x_0, y_0))$$
$$+ \frac{1}{2!}((x - x_0)^2 f_{xx}(x_0, y_0) + 2(x - x_0)(y - y_0)f_{xy}(x_0, y_0)$$
$$+ (y - y_0)^2 f_{yy}(x_0, y_0)) \quad (3)$$
$$\vdots$$
$$+ \frac{1}{N!}\sum_{k=0}^{N}\binom{N}{k}(x - x_0)^k(y - y_0)^{N-k} f_{x^k y^{N-k}}(x_0, y_0).$$

Example 1. Let $f(x, y) = \sqrt{1 + x^2 + y^2}$. To expand about $(0, 0)$ through the second degree, we compute

$$f_x(0, 0) = f_y(0, 0) = 0,$$
$$f_{xx}(0, 0) = f_{yy}(0, 0) = 1 \quad \text{and} \quad f_{xy}(0, 0) = 0.$$

Then Formula (3) reduces to the second-degree polynomial $1 + \frac{1}{2}(x^2 + y^2)$. The graphs of f and its second-degree Taylor expansion are shown in Fig. 12.

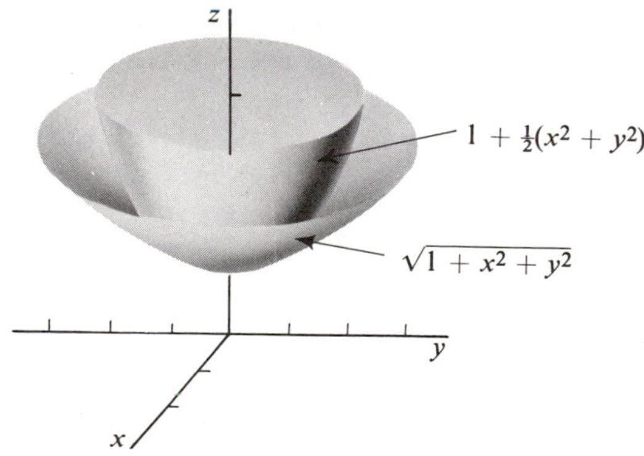

Figure 12

To simplify the writing of the terms of the Taylor expansion, we can use the following notation. The differential operator

$$\left(x\frac{\partial}{\partial x}+y\frac{\partial}{\partial y}\right)$$

applied to f and evaluated at $\mathbf{x}_0 = (x_0, y_0)$ is by definition the first-degree polynomial

$$d_{\mathbf{x}_0}f(x, y) = \left(x\frac{\partial}{\partial x}+y\frac{\partial}{\partial y}\right)_{\mathbf{x}_0} f = x\frac{\partial f}{\partial x}(\mathbf{x}_0) + y\frac{\partial f}{\partial y}(\mathbf{x}_0).$$

Differentials of order k, $k > 1$, can also be defined. They are homogeneous polynomials of degree k. If f has the required derivatives, the definition is

$$\begin{aligned} d^k_{\mathbf{x}_0} f(x, y) &= \left(x\frac{\partial}{\partial x}+y\frac{\partial}{\partial y}\right)^k_{\mathbf{x}_0} f \\ &= \sum_{j=0}^{k}\binom{k}{j}x^j y^{k-j}\frac{\partial^k f}{\partial x^j\,\partial y^{k-j}}(\mathbf{x}_0). \end{aligned}$$

Here the operator

$$\left(x\frac{\partial}{\partial x}+y\frac{\partial}{\partial y}\right)^k$$

has been multiplied out according to the binomial expansion. The operator is applied to f, and the partial derivatives are then evaluated at $\mathbf{x}_0 = (x_0, y_0)$. Notice that x and y are the only variables that appear in the preceding equation, since

$$\binom{k}{j}\frac{\partial^k f}{\partial x^j\,\partial y^{k-j}}(\mathbf{x}_0)$$

is a constant for a fixed $\mathbf{x}_0$. The Nth-degree Taylor expansion of f at $\mathbf{x}_0$ can now be written

$$f(x_0, y_0) + \frac{1}{1!}d_{\mathbf{x}_0} f(x - x_0, y - y_0) + \ldots + \frac{1}{N!}d^N_{\mathbf{x}_0} f(x - x_0, y - y_0).$$

Example 2. If $f(x, y)$ has the required derivatives then the differential of order 3 at $(1, 1)$ evaluated at $(x - 1, y - 1)$ is the polynomial

$$\begin{aligned} d^3_{(1,1)}f(x-1, y-1) &= \left((x-1)\frac{\partial}{\partial x}+(y-1)\frac{\partial}{\partial y}\right)^3_{(1,1)} f \\ &= (x-1)^3\frac{\partial^3 f}{\partial x^3}(1, 1) \\ &\quad + 3(x-1)^2(y-1)\frac{\partial^3 f}{\partial x^2\,\partial y}(1, 1) \\ &\quad + 3(x-1)(y-1)^2\frac{\partial^3 f}{\partial x\,\partial y^2}(1, 1) \\ &\quad + (y-1)^3\frac{\partial^3 f}{\partial y^3}(1, 1). \end{aligned}$$

Example 3. When the polynomial $(x_1 + x_2 + \dots + x_n)^N$ is multiplied out, each term will consist of a constant times a factor of the form $x_1^{k_1}x_2^{k_2} \dots x_n^{k_n}$ where the nonnegative integers k_i satisfy $k_1 + \dots + k_n = N$. The **multinomial expansion** has the form

$$(x_1 + \dots + x_n)^N = \sum_{k + \dots + k_n = N} \binom{N}{k_1 \dots k_n} x_1^{k_1} \dots x_n^{k_n}.$$

The **multinomial coefficients** can be computed to be

$$\binom{N}{k_1 \dots k_n} = \frac{N!}{k_1! \dots k_n!}.$$

This computation will be done later using Taylor's theorem (Theorem 3.2). The coefficients can also be computed by counting. See Kemeny, Snell, Thompson, *Finite Mathematics*, Prentice-Hall, Inc., 1966, p. 109.

For a function of n variables, the **kth-order differential at $\mathbf{x}_0 =$** $(a_1, \dots, a_n)$ is defined to be the following polynomial in $\mathbf{x} = (x_1, \dots, x_n)$:

$$\begin{aligned} d_{\mathbf{x}_0}^k f(\mathbf{x}) &= \left(x_1 \frac{\partial}{\partial x_1} + \dots + x_n \frac{\partial}{\partial x_n}\right)_{\mathbf{x}_0}^k f \\ &= \sum_{k_1 + \dots + k_n = k} \binom{k}{k_1 \dots k_n} x_1^{k_1} \dots x_n^{k_n} \frac{\partial^k f}{\partial x_1^{k_1} \dots \partial x_n^{k_n}} (a_1, \dots, a_n). \end{aligned}$$

In terms of differentials the Taylor expansion about $\mathbf{x}_0$ is defined to be

$$f(\mathbf{x}_0) + \frac{1}{1!} d_{\mathbf{x}_0} f(\mathbf{x} - \mathbf{x}_0) + \frac{1}{2!} d_{\mathbf{x}_0}^2 f(\mathbf{x} - \mathbf{x}_0) + \dots - \frac{1}{N!} d_{\mathbf{x}_0}^N f(\mathbf{x} - \mathbf{x}_0).$$

The function $d_{\mathbf{x}_0} f$ is exactly the same as the differential defined in Section 8 of Chapter 3. For completeness we can also define the 0th differential by

$$d_{\mathbf{x}_0}^0 f(\mathbf{x}) = f(\mathbf{x}_0).$$

Example 4. The second-degree Taylor expansion of $e^{x_1 + \dots x_n}$ about $\mathbf{x} = 0$ is

$$\begin{aligned} 1 &+ \frac{1}{1!}\left(x_1 \frac{\partial}{\partial x_1} + \dots + x_n \frac{\partial}{\partial x_n}\right)_0 f + \frac{1}{2!}\left(x_1 \frac{\partial}{\partial x_1} + \dots + x_n \frac{\partial}{\partial x_n}\right)_0^2 f \\ &= 1 + \frac{1}{1!}(x_1 + \dots + x_n) + \frac{1}{2!}(x_1^2 + 2x_1x_2 + \dots + x_2^2 + \dots + x_n^2). \end{aligned}$$

According to the preceding paragraphs, the Taylor expansion of a function f is defined in such a way that the coefficients of the polynomial can be computed in a routine manner from the derivatives of f. The Taylor expansion is important because it provides a polynomial approximation to f near $\mathbf{x}_0$ that exhibits in a simple way many of the characteristics of f near

$\mathbf{x}_0$. Furthermore, as higher-degree terms are included in the expansion, the approximation gets better. Consider first the one-variable case. The expansion

$$f(x_0) + \frac{1}{1!} f'(x_0)(x - x_0) = f(x_0) + d_{x_0}f(x - x_0)$$

is the affine approximation of f provided by $d_{x_0}f$. In other words, as x approaches x_0,

$$f(x) - f(x_0) - f'(x_0)(x - x_0)$$

tends to zero faster than $x - x_0$.

Having found a first-degree approximation, we now ask for one of the second-degree. Indeed, the next theorem shows that the desired approximation is the second-degree Taylor expansion and that as x approaches x_0,

$$f(x) - f(x_0) - \frac{1}{1!} f'(x_0)(x - x_0) - \frac{1}{2!} f''(x_0)(x - x_0)^2$$

tends to zero faster than $(x - x_0)^2$. For a function of two variables, the first-degree Taylor expansion can be written

$$f(x_0, y_0) + \frac{1}{1!}\left(\frac{\partial f}{\partial x}(x_0, y_0)(x - x_0) + \frac{\partial f}{\partial y}(x_0, y_0)(y - y_0)\right)$$
$$= f(x_0, y_0) + (d_{\binom{x_0}{y_0}} f)\begin{pmatrix} x - x_0 \\ y - y_0 \end{pmatrix}$$

and so is just the affine approximation to f provided by the differential. We shall see that to find a second-degree approximation of a similar kind we need to take the second-degree Taylor expansion. The complete statement follows, and the proof is given in the Appendix.

3.2 Taylor's Theorem

Let $\mathcal{R}^n \xrightarrow{f} \mathcal{R}$ have all derivatives of order N continuous in a neighborhood of $\mathbf{x}_0$. Let $T_N(\mathbf{x} - \mathbf{x}_0)$ be the Nth-degree Taylor expansion of f about $\mathbf{x}_0$. That is,

$$T_N(\mathbf{x} - \mathbf{x}_0) = f(\mathbf{x}_0) + d_{\mathbf{x}_0}f(\mathbf{x} - \mathbf{x}_0) + \ldots + \frac{1}{N!} d^N_{\mathbf{x}_0} f(\mathbf{x} - \mathbf{x}_0).$$

Then

$$\lim_{\mathbf{x} \to \mathbf{x}_0} \frac{(f(\mathbf{x}) - T_N(\mathbf{x} - \mathbf{x}_0))}{|\mathbf{x} - \mathbf{x}_0|^N} = 0,$$

and T_N is the only Nth-degree polynomial having this property.

The uniqueness statement at the end of Taylor's theorem shows that a polynomial of degree N is equal to its Nth-degree Taylor expansion about an arbitrary point $\mathbf{x}_0$.

Example 5. The polynomial $x^2y + x^3 + y^3$ can be written as a polynomial in $(x - 1)$ and $(y + 1)$ by computing its Taylor expansion about $(1, -1)$. The result is

$$\begin{aligned} x^2y + x^3 + y^3 &= -1 + \frac{1}{1!}((x-1) + 4(y+1)) \\ &\quad + \frac{1}{2!}(4(x-1)^2 + 4(x-1)(y+1) - 6(y+1)^2) \\ &\quad + \frac{1}{3!}(6(x-1)^3 + 6(x-1)^2(y+1) + 6(y+1)^3) \\ &= -1 + (x-1) + 4(y+1) + 2(x-1)^2 \\ &\quad + 2(x-1)(y+1) - 3(y+1)^2 + (x-1)^3 \\ &\quad + (x-1)^2(y+1) + (y+1)^3. \end{aligned}$$

Example 6. The infinite series expansion $e^t = 1 + t + (1/2!)t^2 + \ldots$ is valid for all t. Letting $t = x + y$ we get

$$e^{x+y} = 1 + \frac{1}{1!}(x+y) + \frac{1}{2!}(x^2 + 2xy + y^2) + \ldots, \quad \text{for all } x \text{ and } y.$$

It follows that

$$1 + \frac{1}{1!}(x+y) + \frac{1}{2!}(x^2 + 2xy + y^2)$$

is the 2nd-degree Taylor expansion of e^{x+y}. The remainder, of the form $(x+y)^3/3! + \ldots$, tends to zero when it is divided by $(\sqrt{x^2 + y^2})^2$, as (x, y) tends to $(0, 0)$. According to Taylor's theorem, there is only one polynomial of degree two having this property.

Example 7. Let $f(x, y) = e^{xy} \sin(x + y)$. Since

$$e^{xy} = 1 + xy + \frac{1}{2!}x^2y^2 + R_1$$

and

$$\sin(x+y) = (x+y) - \frac{1}{3!}(x+y)^3 + R_2,$$

we can multiply the expansions together, putting into the remainder all terms of degree greater than three. The result is

$$f(x, y) = e^{xy} \sin (x + y) = (x + y) + x^2y + xy^2 - \frac{1}{3!}(x + y)^3 + R,$$

where $R/|(x, y)|^3$ tends to zero as (x, y) tends to $(0, 0)$. In other words, we have found the third-degree Taylor expansion of $e^{xy} \sin (x + y)$ about $(0, 0)$. In standard form the expansion looks like

$$\begin{aligned} f(x, y) &= e^{xy} \sin (x + y) \\ &= \frac{1}{1!}(x + y) + \frac{1}{3!}(-x^3 + 3x^2y + 3xy^2 - y^3) + R. \end{aligned}$$

We can conclude that

$$\frac{\partial^3 f}{\partial y^3}(0, 0) = \frac{\partial^3 f}{\partial x^3}(0, 0) = -1,$$

$$\frac{\partial^3 f}{\partial x\, \partial y^2}(0, 0) = \frac{\partial^3 f}{\partial x^2\, \partial y}(0, 0) = 1.$$

Example 8. The functions e^x and $\cos x$ have second-degree expansions about $x = 0$

$$e^x = 1 + x + \frac{x^2}{2} + R(x), \qquad \cos x = 1 - \frac{x^2}{2} + R'(x).$$

Then

$$\begin{aligned} e^{\cos x} = 1 + \left(1 - \frac{x^2}{2} + R'(x)\right) + \frac{1}{2}\left(1 - \frac{x^2}{2} + R'(x)\right)^2 \\ + R\left(1 - \frac{x^2}{2} + R'(x)\right). \end{aligned}$$

Since $R(1 - x^2/2 + R'(x))$ does not even tend to zero as x tends to zero, we must proceed differently to find a Taylor expansion of $e^{\cos x}$. We have

$$\begin{aligned} e^{\cos x} &= e(e^{-1+\cos x}) \\ &= e\left[1 + \left(-\frac{x^2}{2} + R'(x)\right) + \frac{1}{2}\left(-\frac{x^2}{2} + R'(x)\right)^2 \right. \\ &\qquad \left. + R\left(-\frac{x^2}{2} + R'(x)\right)\right] \\ &= e\left(1 - \frac{x^2}{2}\right) + R''(x), \end{aligned}$$

where $R''(x)/x^2$ tends to zero as x tends to zero. The coefficients can also be found by direct computation of the derivatives of $e^{\cos x}$.

Example 9. Since $1/(1 + x^2) = 1 - x^2 + R(x)$,

$$\prod_{k=1}^{n} \frac{1}{1 + x_k^2} = \prod_{k=1}^{n} (1 - x_k^2 + R(x_k))$$
$$= 1 - (x_1^2 + x_2^2 + \ldots + x_n^2) + R'(x_1, \ldots, x_n),$$

where $R'/|(x_1, \ldots, x_n)|^2$ tends to zero as $(x_1, \ldots, x_n)$ tends to zero.

Example 10. The Taylor expansion of $f(\mathbf{x}) = (x_1 + \ldots + x_n)^N$ about $\mathbf{x}_0 = 0$ is

$$(x_1 + \ldots + x_n)^N = \frac{1}{N!}\left(x_1 \frac{\partial}{\partial x_1} + \ldots + x_n \frac{\partial}{\partial x_n}\right)_0^N f$$
$$= \frac{1}{N!} \sum_{k_1 + \ldots + k_n = N} \binom{N}{k_1 \ldots k_n} x_1^{k_1} \ldots x_n^{k_n} \frac{\partial^N f}{\partial x_1^{k_1} \ldots \partial x_n^{k_n}}$$
$$= \sum_{k_1 + \ldots + k_n = N} \binom{N}{k_1 \ldots k_n} x_1^{k_1} \ldots x_n^{k_n}.$$

Only the Nth-order differential is different from zero at $\mathbf{x}_0 = 0$. To compute the multinomial coefficient, differentiate both sides by $\partial^N/(\partial x_1^{k_1} \ldots \partial x_n^{k_n})$. Then, setting $\mathbf{x} = 0$, we get

$$N! = \binom{N}{k_1 \ldots k_n} k_1! \ldots k_n!,$$

from which the formula of Example 3 follows.

EXERCISES

1. Find the third-degree Taylor expansion of $(u + v)^3$
   (a) about the point $(u_0, v_0) = (0, 0)$.
   (b) about the point $(u_0, v_0) = (1, 2)$.

2. Find the best second-degree approximation to the function $f(x, y) = xe^y$ near the point $(x_0, y_0) = (2, 0)$.
   [*Ans.* $2 + (x - 2) + 2y + y^2 + (x - 2)y$.]

3. Find the best second-degree approximation to the function $f(x, y) = x^{y+1}$ near the point $(x_0, y_0) = (2, 0)$.

4. Find the quadratic terms of the Taylor expansion of xe^{x+y} about $(0, 0)$
   (a) by computing derivatives.
   (b) by substitution.

5. Find the quadratic terms of the Taylor expansion of $e^{\sin(x+y)}$ about $(0, 0)$.

6. If $f(x, y) = (x^2 + y^2)e^{x^2+y^2}$, use a Taylor expansion of f to compute

$$\frac{\partial^3 f}{\partial x^2 \partial y}(0, 0).$$

[*Ans.* 0.]

7. Compute the second-degree Taylor expansion of $\sqrt{1 + x^2 + y^2}$ about the point $(x_0, y_0) = (-1, 1)$.

8. Compute the second-degree Taylor expansion of

(a) $f(x, y, z) = (x^2 + 2xy + y^2)e^z$, about $(x_0, y_0, z_0) = (1, 2, 0)$.
(b) $g(x, y, z) = xy^2z^3$, about $(x_0, y_0, z_0) = (1, 2, -1)$.

9. Write the polynomial xy^2z^3 as a polynomial in $(x - 1)$, y, and $z + 1$.

10. Compute the second-degree Taylor expansion of $\log_2 x$ at $x = 1$. Sketch the graph of the expansion near $x = 1$.

11. Compute the second-degree Taylor expansion of $\log \cos (x + y)$ at $(x_0, y_0) = (0, 0)$

(a) by computing derivatives
(b) by substitution.

12. Compute the second-degree Taylor expansion of $\exp(-x_1^2 - x_2^2 - \ldots - x_n^2)$ about $(x_1, x_2, \ldots, x_n) = (0, 0 \ldots 0)$.

13. Prove that the Taylor expansion of degree N of a polynomial of degree K, $K \geq N$, consists of the terms of the polynomial that are of degree less than or equal to N.

14. Compute the differentials $d^k_{\mathbf{x}_0} f(\mathbf{y})$ for arbitrary $\mathbf{y}$.

(a) $k = 2$, $\mathbf{x}_0 = (1, 2)$, $f(x, y) = x^3y + 3x^2 + 2xy^3$.
[*Ans.* $18x^2 + 54xy + 24y^2$.]
(b) $k = 1$, $\mathbf{x}_0 = (a, b, c)$, $f(x, y, z) = 1/(x + y + z + 1)$.
(c) $k = 2$, $\mathbf{x}_0 = (0, 0, 0)$, $f(x, y, z) = 1/(x + y + z + 1)$.
[*Ans.* $2(x + y + z)^2$.]
(d) $k = 4$, $\mathbf{x}_0 = (0, 0, 0)$, $f(x, y, z) = x^3 + 3xy^2 + 4xy^3 + 6x^2y^3 + 7y^5$.
[*Ans.* $96xy^3$.]

15. Find the second differential of $\sin(x_1 + x_2 + \ldots + x_n)$ at

$$(x_1, x_2, \ldots, x_n) = (0, 0, \ldots 0).$$

[*Ans.* 0.]

SECTION 4

TAYLOR EXPANSIONS AND EXTREME VALUES

The tangent $\mathcal{T}$ to the graph of a function $\mathcal{R}^n \xrightarrow{f} \mathcal{R}$ at a point $(\mathbf{x}_0, f(\mathbf{x}_0))$ is found by computing the first-degree Taylor expansion of f about $\mathbf{x}_0$. We now consider the question of whether or not the graph of f crosses $\mathcal{T}$ at $(\mathbf{x}_0, f(\mathbf{x}_0))$. The possibilities for a function of one variable are shown in Fig. 13.

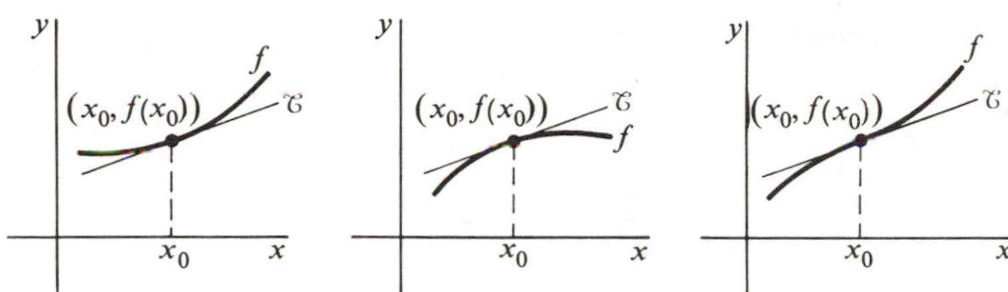

Figure 13

Consider first a very much simplified situation in which f is equal to its second-degree Taylor expansion. That is, assume that

$$f(\mathbf{x}) = f(\mathbf{x}_0) + d_{\mathbf{x}_0}f(\mathbf{x} - \mathbf{x}_0) + \tfrac{1}{2}\, d^2_{\mathbf{x}_0}f(\mathbf{x} - \mathbf{x}_0).$$

The first two terms of the expansion constitute the best affine approximation to f near $\mathbf{x}_0$, and the graph of $f(\mathbf{x}_0) + d_{\mathbf{x}_0}f(\mathbf{x} - \mathbf{x}_0)$ is the tangent $\mathcal{T}$ to the graph of f at $\mathbf{p}_0 = (\mathbf{x}_0, f(\mathbf{x}_0))$. It is clear that if $d^2_{\mathbf{x}_0}f(\mathbf{x} - \mathbf{x}_0)$ is positive for all $\mathbf{x}$ in $\mathcal{R}^n$ except $\mathbf{x} = \mathbf{x}_0$, then

$$f(\mathbf{x}) > f(\mathbf{x}_0) + d_{\mathbf{x}_0}f(\mathbf{x} - \mathbf{x}_0), \qquad \text{for} \quad \mathbf{x} \neq \mathbf{x}_0,$$

which implies that the graph of f lies above the tangent $\mathcal{T}$. Similarly, if $d^2_{\mathbf{x}_0}f(\mathbf{x} - \mathbf{x}_0)$ is negative except for $\mathbf{x} = \mathbf{x}_0$, the graph of f lies below $\mathcal{T}$. On the other hand, if $d^2_{\mathbf{x}_0}f(\mathbf{x} - \mathbf{x}_0)$ changes sign at $\mathbf{x}_0$, then the graph of f will cross $\mathcal{T}$ at $\mathbf{p}_0$.

To say that the quadratic polynomial $Q(\mathbf{x} - \mathbf{x}_0) = d^2_{\mathbf{x}_0}f(\mathbf{x} - \mathbf{x}_0)$ changes sign at $\mathbf{x}_0$ means that there are points $\mathbf{x}_1$ and $\mathbf{x}_2$ arbitrarily close to $\mathbf{x}_0$ such that

$$Q(\mathbf{x}_1 - \mathbf{x}_0) > 0 \quad \text{and} \quad Q(\mathbf{x}_2 - \mathbf{x}_0) < 0. \tag{1}$$

The phrase "arbitrarily close" can really be omitted, because every homogeneous quadratic polynomial Q has the property

$$Q(t\mathbf{x}) = t^2 Q(\mathbf{x}). \tag{2}$$

It follows that if (1) holds for two vectors $\mathbf{x}_1$ and $\mathbf{x}_2$, not necessarily close to $\mathbf{x}_0$, then (1) also holds for $t(\mathbf{x}_1 - \mathbf{x}_0) + \mathbf{x}_0$ and $t(\mathbf{x}_2 - \mathbf{x}_0) + \mathbf{x}_0$, for any $t \neq 0$. By choosing t small enough, we can bring the latter vectors as close to $\mathbf{x}_0$ as we like.

Example 1. The function

$$f(x, y) = 2x^2 - xy - 3y^2 - 3x + 7y$$

equals its second-degree Taylor expansion. It has one critical point at $\mathbf{x}_0 = (1, 1)$, and the tangent plane $\mathcal{T}$ at $(1, 1, 2)$ is therefore horizontal.

The second differential is given by

$$d^2_{\mathbf{x}_0}f\begin{pmatrix} x-1 \\ y-1 \end{pmatrix} = 4(x-1)^2 - 2(x-1)(y-1) - 6(y-1)^2.$$

Trying $\mathbf{x}_1 = (2, 1)$ and $\mathbf{x}_2 = (1, 2)$, we obtain

$$d^2_{\mathbf{x}_0}f(\mathbf{x}_1 - \mathbf{x}_0) = 4 > 0,$$
$$d^2_{\mathbf{x}_0}f(\mathbf{x}_2 - \mathbf{x}_0) = -6 < 0.$$

We conclude that the graph of f crosses the tangent plane $\mathcal{T}$ at (1, 1, 2) and, consequently, that f has neither a local maximum nor minimum at $\mathbf{x}_0$.

The assumption that f equals its second-degree Taylor expansion is too strong to be of much practical value. However, the next theorem shows that under more general hypothesis, the sign of the second differential still determines whether or not f crosses its tangent. The quadratic polynomial $d^2_{\mathbf{x}_0}f(\mathbf{x} - \mathbf{x}_0)$ is always zero at $\mathbf{x} = \mathbf{x}_0$. If it is positive except at that one point, it is said to be **positive definite,** and if it is negative except at that one point then it is said to be **negative definite.**

4.1 Theorem

Let $\mathcal{R}^n \xrightarrow{f} \mathcal{R}$ have all its second partial derivatives continuous in a neighborhood of $\mathbf{x}_0$, and denote the tangent to the graph of f at $\mathbf{p}_0 = (\mathbf{x}_0, f(\mathbf{x}_0))$ by $\mathcal{T}$.

(a) If $d^2_{\mathbf{x}_0}f(\mathbf{x} - \mathbf{x}_0)$ is positive definite, then f lies above $\mathcal{T}$ in some neighborhood of $\mathbf{x}_0$.

(b) If $d^2_{\mathbf{x}_0}f(\mathbf{x} - \mathbf{x}_0)$ is negative definite, then f lies below $\mathcal{T}$ in some neighborhood of $\mathbf{x}_0$.

(c) If $d^2_{\mathbf{x}_0}f(\mathbf{x} - \mathbf{x}_0)$ assumes both positive and negative values, then f crosses the tangent $\mathcal{T}$ at $\mathbf{p}_0$.

Notice that not all possible cases are covered by parts (a), (b), and (c). It may happen, for example, that the second differential is zero somewhere other than at $\mathbf{x}_0$, but that it still does not change sign.

Proof (of 4.1). By Taylor's theorem we have

$$f(\mathbf{x}) - f(\mathbf{x}_0) - d_{\mathbf{x}_0}f(\mathbf{x} - \mathbf{x}_0) = \tfrac{1}{2}d^2_{\mathbf{x}_0}f(\mathbf{x} - \mathbf{x}_0) + R, \qquad (3)$$

where

$$\lim_{\mathbf{x}\to\mathbf{x}_0} \frac{R}{|\mathbf{x}-\mathbf{x}_0|^2} = 0.$$

Under assumption (a), we must show that

$$\tfrac{1}{2}d^2_{\mathbf{x}_0}f(\mathbf{x}-\mathbf{x}_0) + R > 0$$

in some neighborhood of $\mathbf{x}_0$, excluding $\mathbf{x}_0$ itself. The homogeneity of the quadratic polynomial $d^2_{\mathbf{x}_0}f$ (see Equation (2)) implies that

$$\frac{d^2_{\mathbf{x}_0}f(\mathbf{x}-\mathbf{x}_0)}{|\mathbf{x}-\mathbf{x}_0|^2} = d^2_{\mathbf{x}_0}f\left(\frac{\mathbf{x}-\mathbf{x}_0}{|\mathbf{x}-\mathbf{x}_0|}\right).$$

Since $d^2_{\mathbf{x}_0}f$ is positive definite, its values for unit vectors are bounded away from zero by a constant $m > 0$. (See Exercise 14, Section 2). Now choose $\delta > 0$ so that, for $0 < |\mathbf{x}-\mathbf{x}_0| < \delta$,

$$\frac{|R|}{|\mathbf{x}-\mathbf{x}_0|^2} \le \frac{m}{4}.$$

It follows that

$$\tfrac{1}{2}d^2_{\mathbf{x}_0}f(\mathbf{x}-\mathbf{x}_0) + R \ge \frac{m}{2}|\mathbf{x}-\mathbf{x}_0|^2 - \frac{m}{4}|\mathbf{x}-\mathbf{x}_0|^2 > 0$$

which, according to Equation (3), is what we wanted to show.

The proof of part (b) is practically the same as the proof just given. To prove part (c), suppose that $d^2_{\mathbf{x}_0}f(\mathbf{x}_1-\mathbf{x}_0) > 0$ and $d^2_{\mathbf{x}_0}f(\mathbf{x}_2-\mathbf{x}_0) < 0$. Set

$$\mathbf{x}_i(t) = t(\mathbf{x}_i - \mathbf{x}_0) + \mathbf{x}_0, \qquad i = 1, 2, \qquad -\infty < t < \infty.$$

Using the homogeneity property of the polynomial $d^2_{\mathbf{x}_0}f$, and also of the norm, we obtain, for any $t \neq 0$,

$$f(\mathbf{x}_i(t)) - f(\mathbf{x}_0) - d_{\mathbf{x}_0}f(\mathbf{x}_i(t)-\mathbf{x}_0)$$
$$= t^2\left[\tfrac{1}{2}d^2_{\mathbf{x}_0}f(\mathbf{x}_i-\mathbf{x}_0) + |\mathbf{x}_i-\mathbf{x}_0|^2 \frac{R}{|\mathbf{x}_i(t)-\mathbf{x}_0|^2}\right]. \quad (4)$$

Since

$$\lim_{t\to 0} \frac{R}{|\mathbf{x}_i(t)-\mathbf{x}_0|^2} = 0,$$

it follows that, for any nonzero t sufficiently small, the left side of Equation (4) is positive if $i = 1$ and negative if $i = 2$. In other words, the graph of f lies both above and below the tangent $\mathcal{T}$ for some values of $\mathbf{x}$ arbitrarily close to $\mathbf{x}_0$. This completes the proof.

Example 2. The function

$$f(x, y) = (x^2 + y^2)e^{x^2-y^2}$$

has its critical points at (0, 0), (0, 1), and (0, −1). This implies that the tangents at these points are horizontal planes. The second-degree Taylor expansions at the three points are, respectively,

$$f(0, 0) + \tfrac{1}{2}d^2_{(0,0)}f(x, y) = x^2 + y^2,$$

$$f(0, 1) + \tfrac{1}{2}d^2_{(0,1)}f(x, y - 1) = \frac{1}{e}\,[1 + 2x^2 - 2(y - 1)^2],$$

$$f(0, -1) + \tfrac{1}{2}d^2_{(0,-1)}f(x, y + 1) = \frac{1}{e}\,[1 + 2x^2 - 2(y + 1)^2].$$

Clearly, $d^2_{(0,0)}f$ is positive definite, while $d^2_{(0,1)}f$ and $d^2_{(0,-1)}f$ assume both positive and negative values arbitrarily close to their respective critical points. We conclude that f has a local minimum value at (0, 0) and neither a maximum nor a minimum at the points (0, 1) and (0, −1). The graph of f and the horizontal tangent planes at (0, 0, 0) and (0, 1, $1/e$) are shown in Fig. 14.

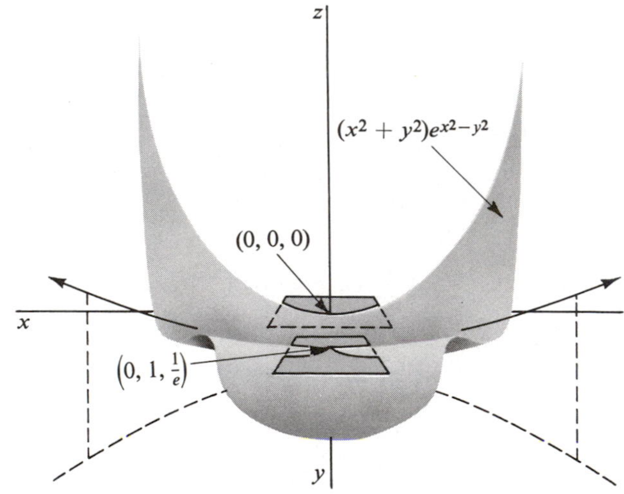

Figure 14

Example 3. The function

$$f(x, y, z) = x \sin z + z \sin y$$

has a critical point at $\mathbf{x}_0 = (-1, \pi/2, 0)$. The second differential at $\mathbf{x}_0$ is

$$d^2_{\mathbf{x}_0}f\left(x + 1, y - \frac{\pi}{2}, z\right) = 2z(x + 1).$$

Since $d^2_{\mathbf{x}_0}f$ has both positive and negative values, the function f has neither a local maximum nor minimum at $\mathbf{x}_0$.

Example 4. If $f(x) = x \sin^3 x$, the first four terms of the Taylor expansion at $x = 0$ are identically zero, while

$$d_0^4 fx = 24x^4.$$

The criteria of Theorem 4.1 do not cover this example. However, a similar proof would show that f behaves like its Taylor expansion in the matter of crossing its tangent. The conclusion is that $x \sin^3 x$ has a local minimum at $x = 0$.

For distinguishing among the critical points of a function those that are maximum points or minimum points, a detailed examination of the polynomial $d_{\mathbf{x}_0}^2 f(\mathbf{x} - \mathbf{x}_0)$ is not necessary. It is enough to know that this quadratic approximation is positive or negative definite. In addition to the criteria of Section 2, we list here for reference a simple test for quadratic polynomials in two or three variables. (See Section 2, Example 4 and Exercise 16.)

The polynomial

$$ax^2 + 2bxy + cy^2 = (x \quad y)\begin{pmatrix} a & b \\ b & c \end{pmatrix}\begin{pmatrix} x \\ y \end{pmatrix}$$

is positive definite if and only if

$$a > 0 \quad \text{and} \quad \begin{vmatrix} a & b \\ b & c \end{vmatrix} > 0$$

and is negative definite if and only if

$$a < 0 \quad \textit{and} \quad \begin{vmatrix} a & b \\ b & c \end{vmatrix} > 0.$$

The character of the polynomial

$$ax^2 + by^2 + cz^2 + 2dyz + 2exz + 2fxy = (x \quad y \quad z)\begin{pmatrix} a & f & e \\ f & b & d \\ e & d & c \end{pmatrix}\begin{pmatrix} x \\ y \\ z \end{pmatrix}$$

depends on the sign of the three determinants

$$a, \qquad \begin{vmatrix} a & f \\ f & b \end{vmatrix}, \qquad \begin{vmatrix} a & f & e \\ f & b & d \\ e & d & c \end{vmatrix}.$$

The polynomial is positive definite if and only if all three determinants are positive; it is negative definite if and only if the middle one is positive and the other two are negative.

Example 5. The function $g(x, y, z) = x^2 + y^2 + z^2 + xy$ has critical points only when

$$\begin{aligned} g_x &= 2x + y = 0, \\ g_y &= 2y + x = 0, \\ g_z &= 2z \qquad = 0; \end{aligned}$$

so the only critical point occurs at (0, 0, 0). Since $d_0^2 g = 2g$, we can test g itself for positive definiteness. We have

$$a = 1, \qquad \begin{vmatrix} a & f \\ f & b \end{vmatrix} = \begin{vmatrix} 1 & \frac{1}{2} \\ \frac{1}{2} & 1 \end{vmatrix} = \tfrac{3}{4}, \qquad \begin{vmatrix} a & f & e \\ f & b & d \\ e & d & c \end{vmatrix} = \begin{vmatrix} 1 & \frac{1}{2} & 0 \\ \frac{1}{2} & 1 & 0 \\ 0 & 0 & 1 \end{vmatrix} = \tfrac{3}{4}.$$

Thus g is positive definite, and, as a result, g has minimum value 0 at (0, 0, 0).

EXERCISES

1. Find all the critical points of the following functions:

(a) $(x + y)e^{-xy}$. [*Ans.* $(\pm 1/\sqrt{2}, \pm 1/\sqrt{2})$.]

(b) $xy + xz$. [*Ans.* $(0, y, -y)$, any y.]

(c) $(x^2 + y^2) \ln (x^2 + y^2)$.

(d) $\cos (x^2 + y^2 + z^2)$.

(e) $x^2 + y^2 + z^2$.

2. Compute the second-degree Taylor expansion of the function in Exercise 1(a) at each of its critical points.

3. Classify the critical points in 1(a), 1(b), and 1(e) as maximum, minimum, or neither.

4. In each of the following, consider the tangent to the graph of the function at the point indicated. Decide whether the function lies above or below the tangent near the indicated point, or whether it crosses there.

(a) $x^2 \sin x$ at $x = 1$. [*Ans.* Lies above.]

(b) $1/(x - y)$ at $(x, y) = (2, 1)$.

(c) $x^4 + y^4$ at (0, 0).

(d) $e^{z+w} - x^2 - y^2$ at (0, 0, 0, 0). [*Ans.* Crosses.]

5. Locate all the critical points $\mathbf{x}_0$ of each of the following functions and, by looking at $d^2_{\mathbf{x}_0} f$, decide whether the function has a local maximum, or a local minimum, or neither, at $\mathbf{x}_0$. If examination of the second differential fails to give any information, consider the next highest term of the Taylor expansion that does give information.

(a) $\sin x \cos x$.
(b) x^2y^2. [*Ans.* $x = 0$ or $y = 0$, min.]
(c) $x^2 + 4xy - y^2 - 8x - 6y$.
(d) $x^2 - xy - y^2 + 5y - 1$. [*Ans.* $(1, 2)$, neither.]
(e) $x^2 + 2y^2 - x$.
(f) $x \sin y$.
(g) $x^4 + y^4$. [*Ans.* $(0, 0)$, min.]
(h) $(x - y)^4$.
(i) $\exp(-x_1^2 - x_2^2 - \ldots - x_n^2)$.

SECTION 5

FOURIER SERIES

The Nth degree Taylor expansion of a function f requires the existence of at least N derivatives at a point x_0. Furthermore, because the derivatives $f^{(k)}(x_0)$ that occur in the Taylor expansion are determined by the properties of f in a neighborhood of x_0, the expansion in general approximates f near x_0 only. Another approach to the problem of approximating a function is by using a **trigonometric polynomial** of degree N,

$$s_N(x) = \frac{a_0}{2} + \sum_{k=1}^{N}(a_k \cos kx + b_k \sin kx). \tag{1}$$

Instead of using the values of the function f to be approximated near only a single point, the Fourier method is to compute the coefficients a_k and b_k by taking certain weighted averages of f over the interval on which it is defined. We shall assume at first that f is continuous on the interval $-\pi \leq x \leq \pi$ and define the **Fourier coefficients** of f by the formulas

5.1 $$a_k = \frac{1}{\pi}\int_{-\pi}^{\pi} f(x) \cos kx\, dx, \qquad b_k = \frac{1}{\pi}\int_{-\pi}^{\pi} f(x) \sin kx\, dx.$$

The trigonometric polynomial obtained by substituting these particular values of a_k and b_k into Equation (1) can often be used to approximate f. (In fact, Exercise 2(b) shows that if f itself is a trigonometric polynomial, then substitution into Equation (1) gives f back again exactly.) The results of Examples 1 and 2 below give some idea of how well this choice of coefficients works. (See also Theorem 5.2.)

The trigonometric polynomial (1), with its coefficients determined by Formula 5.1, is called the Nth **Fourier approximation** to f on the interval $[-\pi, \pi]$. The polynomial is evidently a periodic function of period 2π, that is, it satisfies the equation $s_N(x + 2\pi) = s_N(x)$ for all real x. From the periodicity it follows that

$$s_N(x + 2k\pi) = s_N(x)$$

for all x and all integers k. For this reason it is possible to restrict attention to those x's lying in some fixed interval of length 2π, say, $-\pi \leq x \leq \pi$. We shall compute some examples of Fourier approximations to get some idea of how they work.

Example 1. Let $f(x) = |x|$ for $-\pi \leq x \leq \pi$.
Then

$$a_k = \frac{1}{\pi}\int_{-\pi}^{\pi} |x| \cos kx \, dx, \qquad b_k = \frac{1}{\pi}\int_{-\pi}^{\pi} |x| \sin kx \, dx.$$

Clearly, $|x| \sin kx$ has integral zero over $[-\pi, \pi]$, because the integrals over $[-\pi, 0]$ and $[0, \pi]$ are negatives of one another. Hence $b_k = 0$ for $k = 1, 2, \ldots$. On the other hand, the graph of $|x| \cos kx$ is symmetric about the y-axis. For $k \neq 0$ we integrate by parts, getting

$$\begin{aligned} a_k &= \frac{2}{\pi}\int_0^{\pi} x \cos kx \, dx \\ &= \frac{2}{\pi}\left[\frac{x \sin kx}{k}\right]_0^{\pi} - \frac{2}{\pi k}\int_0^{\pi} \sin kx \, dx \\ &= \left[\frac{2}{\pi k^2} \cos kx\right]_0^{\pi} = \frac{2}{\pi k^2}(\cos k\pi - 1) \\ &= \begin{cases} 0, & k = 2, 4, 6, \ldots, \\ -\dfrac{4}{\pi k^2}, & k = 1, 3, 5, \ldots. \end{cases} \end{aligned}$$

When $k = 0$, we have

$$a_0 = \frac{2}{\pi}\int_0^{\pi} x \, dx = \pi.$$

To summarize,

$$a_0 = \pi, \qquad a_k = \begin{cases} 0, & k = 2, 4, 6, \ldots, \\ -\dfrac{4}{\pi k^2}, & k = 1, 3, 5, \ldots, \end{cases}$$

$$b_k = 0, \qquad k = 1, 2, 3, \ldots.$$

Hence, the Nth Fourier approximation is given for $N = 1, 3, 5, \ldots$ by the trigonometric polynomial

$$s_N(x) = \frac{\pi}{2} - \frac{4}{\pi}\cos x - \frac{4}{\pi}\frac{\cos 3x}{3^2} - \cdots - \frac{4}{\pi}\frac{\cos Nx}{N^2}.$$

If N is even, we have $s_N(x) = s_{N-1}(x)$. Figure 15 shows how the graphs of s_0, s_1, and s_3 approximate that of $|x|$ on $[-\pi, \pi]$.

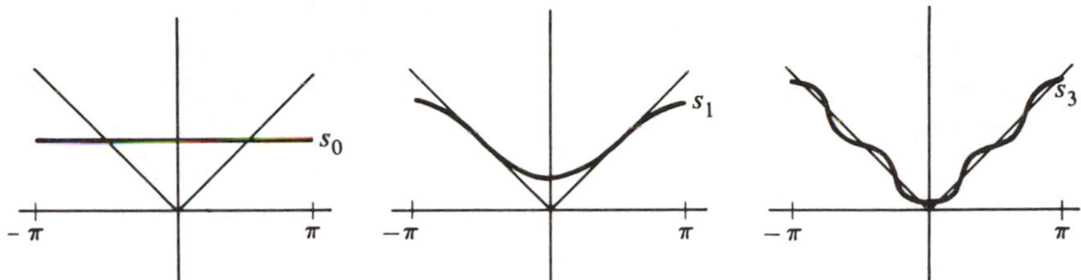

Figure 15

Example 2. Let

$$g(x) = \begin{cases} 1, & 0 \leq x \leq \pi \\ -1, & -\pi \leq x < 0. \end{cases}$$

Then the Fourier coefficients of g are given by integration as

$$a_k = 0, \qquad k = 0, 1, 2, \ldots$$

$$b_k = \begin{cases} 0, & k = 2, 4, 6, \ldots \\ \dfrac{4}{\pi k}, & k = 1, 3, 5, \ldots. \end{cases}$$

Hence, for N odd, the Nth Fourier approximation to g is given by

$$s_N(x) = \frac{4}{\pi} \sin x + \frac{4}{\pi} \frac{\sin 3x}{3} + \ldots + \frac{4}{\pi} \frac{\sin Nx}{N}.$$

The graphs of s_1, s_3, and s_5 are shown in Fig. 16, together with that of $g(x)$.

An important question is whether, for specific values of x, a Fourier approximation $s_N(x)$ converges as $N \to \infty$ to $f(x)$, where f is the function from which the Fourier coefficients are computed. We define the **Fourier series** of f to be the infinite series

$$\frac{a_0}{2} + \sum_{k=1}^{\infty} (a_k \cos kx + b_k \sin kx), \tag{2}$$

where a_k and b_k are given by Formula 5.1. Theorem 5.2 below gives some conditions on f under which the Fourier series can be used to represent f.

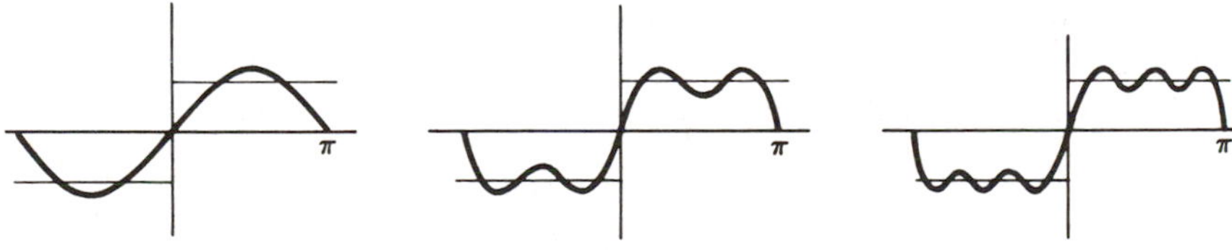

Figure 16

Indeed, suppose that the graph of f is **piecewise smooth.** This means that the interval $[-\pi, \pi]$ can be broken into finitely many subintervals, with endpoints $-\pi < x_1 < x_2 < \ldots < x_k < \pi$, such that f can be extended continuously from each open interval (x_k, x_{k+1}) to the closed interval $[x_k, x_{k+1}]$ so that f' is continuous on $[x_k, x_{k+1}]$. Then we can show that the Fourier series of f converges to $f(x)$ wherever f is continuous, and, at a possible discontinuity at x_k, will converge to the average value

$$(\tfrac{1}{2})[f(x_k-) + f(x_k+)].$$

Here $f(x-)$ stands for the left-hand limit of f at x, and $f(x+)$ represents the right-hand limit. The graph of a typical piecewise smooth function is shown in Fig. 17, with the average value indicated by a dot at each jump.

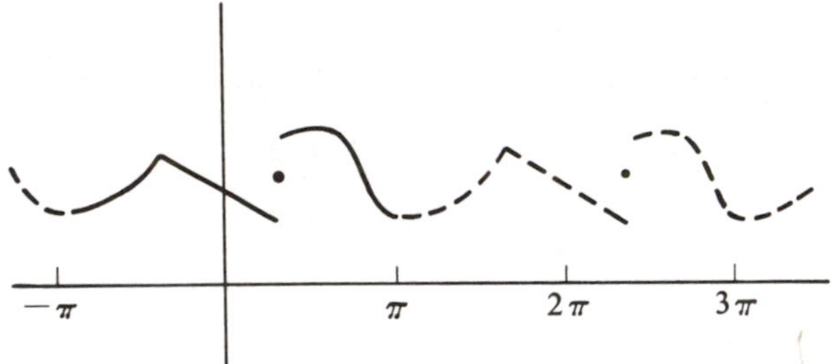

Figure 17

5.2 Theorem

Let f be piecewise smooth on $(-\pi, \pi)$. Then the Fourier series of f converges at every point x of the interval to $(\frac{1}{2})[f(x-) + f(x+)]$. In particular, if f is continuous at x, then the series converges to $f(x)$. At $x = \pm\pi$ the series converges to $(\frac{1}{2})[f(\pi-) + f(-\pi+)]$.

The proof is given in the Appendix.

Examples 1 and 2 gave an indication of the way in which the partial sums of a Fourier series converge. In each of those two examples, the function satisfies the condition of piecewise smoothness; hence the series converges to the appropriate value of the function for $-\pi < x < \pi$.

Example 3. The function g defined in Example 2 is rather arbitrarily defined to have the value 1 at $x = 0$. In spite of this arbitrariness, Theorem 5.2 allows us to conclude that the Fourier series of g converges as follows:

$$\sum_{k=0}^{\infty} \frac{4}{\pi} \frac{\sin(2k+1)x}{2k+1} = \begin{cases} 1, & 0 < x < \pi, \\ 0, & x = 0, \\ -1, & -\pi < x < 0. \end{cases}$$

To be very specific, we can set $x = \pi/2$ and arrive at the alternating series expansion

$$\sum_{k=0}^{\infty} \frac{(-1)^k}{2k+1} = \frac{\pi}{4}.$$

Theorem 5.2 shows to some extent the reason for choosing the coefficients in a trigonometric polynomial according to Formula 5.1. The reason is that, under favorable circumstances, the resulting sequence of trigonometric polynomials converges, as the sequence of partial sums of a Fourier series, to the function f. We shall now explain another reason. Consider the vector space $\mathcal{C}_N$ of trigonometric polynomials of degree N, restricted to the interval $-\pi \le x \le \pi$. Clearly, the $2N - 1$ functions

$$\frac{1}{\sqrt{2}}, \cos x, \sin x, \ldots, \cos Nx, \sin Nx \tag{3}$$

can be formed into linear combinations so as to span the vector space $\mathcal{C}_N$. Thus $\mathcal{C}_N$ has dimension at least $2N - 1$. If we now introduce the inner product

$$\langle f, g \rangle = \frac{1}{\pi} \int_{-\pi}^{\pi} f(x)g(x)\, dx \tag{4}$$

in $\mathcal{C}_N$, we can show that the set of functions (3) is an orthonormal set and, therefore, is linearly independent according to Theorem 8.1 of Chapter 2. In fact we have the next theorem.

5.3 Theorem

The set of functions (3) satisfies the orthogonality relations

$$\langle \cos km, \cos nx \rangle = \begin{cases} 1, & m = n \neq 0, \\ 0, & m \neq n, \end{cases}$$

$$\langle \sin mx, \sin nx \rangle = \begin{cases} 1, & m = n, \\ 0, & m \neq n, \end{cases}$$

$$\langle \cos mx, \sin nx \rangle = 0, \qquad \text{all } m, n,$$

$$\left\langle \frac{1}{\sqrt{2}}, \cos mx \right\rangle = \left\langle \frac{1}{\sqrt{2}}, \sin mx \right\rangle = 0, \qquad \text{all } m,$$

$$\left\langle \frac{1}{\sqrt{2}}, \frac{1}{\sqrt{2}} \right\rangle = 1.$$

The proof of these relations is a routine calculation of some integrals and is left as an exercise. (For example, to say that

$$\langle \cos mx, \sin nx \rangle = 0$$

is to say that

$$\frac{1}{\pi}\int_{-\pi}^{\pi} \cos mx \sin nx \, dx = 0.)$$

From Theorem 5.3 it follows that the trigonometric functions (3) form an orthonormal set in $\mathcal{C}_N$ and, hence, by Theorem 8.1 of Chapter 2, an orthonormal basis. It follows further, from Theorem 8.1 of Chapter 2, that the coefficients in a basis expansion

$$f(x) = \frac{a_0}{\sqrt{2}} + \sum_{k=1}^{N}(a_k \cos kx + b_k \sin kx)$$

can be computed by the formulas

$$a_0 = \left\langle \frac{1}{\sqrt{2}}, f(x) \right\rangle,$$

$$a_k = \langle \cos kx, f(x) \rangle, \qquad k > 0,$$

$$b_k = \langle \sin kx, f(x) \rangle, \qquad k > 0.$$

Because of the definition of the inner product, these formulas are the same as the formulas in 5.1 except that in 5.1 the formula for a_0 has the factor $1/\sqrt{2}$ taken out and included with the trigonometric polynomial. Thus the Fourier coefficient formulas are the correct formulas for computing coefficients relative to the orthonormal basis (3) and inner product (4). This fact explains why it is not necessary to recompute the previously found coefficients if the space $\mathcal{C}_N$ is extended to $\mathcal{C}_{N+1}$ by including $\cos (N + 1)x$ and $\sin (N + 1)x$.

We conclude this section by showing how the use of complex exponential functions can simplify some Fourier series formulas. Recall from Section 7 of Chapter 2 that, by definition,

$$e^{\pm ikx} = \cos kx \pm i \sin kx.$$

The identities

$$\cos kx = \frac{e^{ikx} + e^{-ikx}}{2}, \qquad \sin kx = \frac{e^{ikx} - e^{-ikx}}{2i} \tag{5}$$

follow immediately by substitution. As a result, the trigonometric polynomial (1) can be written in the form

$$s_N(x) = \sum_{k=-N}^{N} c_k e^{ikx}, \tag{6}$$

where

$$c_k = \frac{a_k - ib_k}{2}, \qquad k > 0,$$

$$c_0 = \frac{a_0}{2}, \tag{7}$$

$$c_k = \frac{a_{-k} + ib_{-k}}{2}, \qquad k < 0.$$

It follows, again by direct substitution, that for all integers k:

5.4
$$c_k = \frac{1}{2\pi}\int_{-\pi}^{\pi} e^{-ikx} f(x)\, dx.$$

Formula 5.4 is the complex counterpart of Formula 5.1. In general, of course, the coefficients c_k are complex numbers. But if f is real-valued, the trigonometric polynomial (6) is still real-valued also and, indeed, is necessarily equal to the polynomial (1), provided the coefficients are related by Equations (6). The advantages of the complex form are exemplified by the simpler orthogonality relation obtained by using the complex inner product

$$\langle f, g\rangle_c = \frac{1}{2\pi}\int_{-\pi}^{\pi} f(x)\bar{g}(x)\, dx.$$

In terms of this inner product we find

$$\begin{aligned}\langle e^{ikx}, e^{ilx}\rangle &= \frac{1}{2\pi}\int_{-\pi}^{\pi} e^{ikx}e^{-ilx}\, dx \\ &= \frac{1}{2\pi}\int_{-\pi}^{\pi} e^{i(k-l)x}\, dx = \begin{cases} 1, & k = l, \\ 0, & k \neq l. \end{cases}\end{aligned}$$

The details of this last computation are left as an exercise.

EXERCISES

1. Compute the Fourier coefficients of each of the following functions. Sketch the graph of each function, together with the first three Fourier approximations $s_N(x)$ that differ from one another

   (a) $f_1(x) = x, \qquad -\pi \le x \le \pi.$ [*Ans.* $b_k = (2(-1)^{k+1})/k$.]

   (b) $f_2(x) = \begin{cases} (-\pi - x), & -\pi \le x \le 0. \\ (\pi - x), & 0 < x \le \pi. \end{cases}$

   (c) $f_3(x) = x^2, \qquad -\pi \le x \le \pi.$

2. (a) With respect to the inner product $\langle f, g\rangle$ of two continuous functions defined by

$$\langle f, g\rangle = \frac{1}{\pi}\int_{-\pi}^{\pi} f(x)g(x)\,dx,$$

prove the orthogonality relations

$$\langle \cos nx, \cos mx\rangle = \begin{cases} 1, & m = n \neq 0 \\ 0, & m \neq n \end{cases}$$

$$\langle \sin nx, \sin mx\rangle = \begin{cases} 1, & m = n \neq 0 \\ 0, & m \neq n \end{cases}$$

$$\langle \sin nx, \cos mx\rangle = 0, \quad \text{all } n, m,$$

where n and m are integers.

(b) Use the result of part (a) to show that if

$$f(x) = \frac{c_0}{2} + \sum_{k=1}^{N} (c_k \cos kx + d_k \sin kx),$$

then the Fourier coefficients of f are given by

$$\langle f(x), \cos kx\rangle = c_k, \qquad k = 0, 1, 2, \ldots, N$$

and

$$\langle f(x), \sin kx\rangle = d_k, \qquad k = 1, 2, \ldots, N,$$

and that $\langle f(x), \cos kx\rangle = \langle f(x), \sin kx\rangle = 0$ if $k > N$.

3. Using the identities in Equation (5) of the text, express the following trigonometric polynomials as polynomials (i.e., linear combinations of powers) in e^{ikx} and e^{-ikx}.

(a) $1 + \cos x$.
(b) $\cos x - 1 \sin 2x$.

4. The result of Exercise 2(b) shows that if a function f can be represented as a trigonometric polynomial, then that polynomial is the Fourier expansion of f. For example, the identity

$$\cos^2 x = \tfrac{1}{2} + \tfrac{1}{2}\cos 2x$$

expresses $\cos^2 x$ as a trigonometric polynomial and so provides the Fourier expansion of $\cos^2 x$. Find the Fourier expansion of each of the following functions.

(a) $\sin^2 x$.
(b) $\cos^3 x$.
(c) $\sin 2x \cos x$.

5. (a) Prove the identity

$$\frac{1}{2} + \sum_{k=1}^{N} \cos ku = \frac{\sin (N + 1/2)u}{2 \sin (u/2)}.$$

[*Hint.* Sum the identity $2 \sin (u/2) \cos ku = \sin (k + \frac{1}{2})u - \sin (k - \frac{1}{2})u$ for k from 1 to N.]

(b) Is the sum on the left the Fourier expansion of the function on the right?

6. Let f be a complex-valued continuous function defined on $[-\pi, \pi]$, and define $c_k = \dfrac{1}{2\pi}\displaystyle\int_{-\pi}^{\pi} e^{-ikx}f(x)\,dx$, $k = 0, \pm 1, \pm 2, \ldots$. Then $\displaystyle\sum_{k=-N}^{N} c_k e^{ikx}$ is called the Nth complex Fourier approximation to f.

(a) Show that if

$$f(x) = \sum_{k=-N}^{N} d_k e^{-ikx},$$

then the constants d_k are the complex Fourier coefficients of f.

(b) Show that if f is real-valued, then

$$2c_0 = a_0, \qquad 2c_k = a_k - ib_k,$$

and

$$2c_{-k} = a_k + ib_k \quad \text{for} \quad k = 1, 2, 3, \ldots.$$

7. Prove that if s_N and t_N are the Nth Fourier approximations to f and g respectively, then $as_N + bt_N$ is the Nth Fourier approximation to $af + bg$, where a and b are constants.

SECTION 6

MODIFIED FOURIER EXPANSIONS

The direct application of Fourier methods to practical problems usually requires some modification of the standard formulation presented in the previous section. In the present section we describe some of these modifications and calculate some examples.

While the interval $[-\pi, \pi]$ is a natural one for Fourier expansions because it is a period interval for the trigonometric functions, it may be that a function encountered in an application needs to be approximated on some other interval.

If the function f to be approximated is defined not on the interval $[-\pi, \pi]$ but on $[-p, p]$, a suitable change in the computation of the approximation can be made as follows. With f defined on $[-p, p]$, we define

$$f_p(x) = f\left(\frac{px}{\pi}\right), \qquad -\pi \le x \le \pi.$$

Then we can compute the Fourier coefficients of f_p by Formula 5.1. The resulting trigonometric polynomials s_N will approximate f_p on $[-\pi, \pi]$. To approximate f on $[-p, p]$, we consider

$$s_N\left(\frac{\pi x}{p}\right) = \frac{a_0}{2} + \sum_{k=1}^{N}\left(a_k \cos\frac{k\pi x}{p} + b_k \sin\frac{k\pi x}{p}\right), \qquad -p \le x \le p$$

The coefficients a_k and b_k can be computed directly in terms of f by making a change of variable. We have

$$a_k = \frac{1}{\pi}\int_{-\pi}^{\pi} f_p(x)\cos kx\,dx = \frac{1}{\pi}\int_{-\pi}^{\pi} f\left(\frac{px}{\pi}\right)\cos kx\,dx$$
$$= \frac{1}{p}\int_{-p}^{p} f(x)\cos\left(\frac{k\pi x}{p}\right)dx.$$

A similar computation holds for b_k, and we have

6.1 $$a_k = \frac{1}{p}\int_{-p}^{p} f(x)\cos\frac{k\pi x}{p}\,dx, \qquad b_k = \frac{1}{p}\int_{-p}^{p} f(x)\sin\frac{k\pi x}{p}\,dx$$

for the coefficients in the Fourier approximation

$$\frac{a_0}{2} + \sum_{k=1}^{N}\left(a_k\cos\frac{k\pi x}{p} + b_k\sin\frac{k\pi x}{p}\right)$$

to the function f defined on $[-p, p]$.

Example 1. If

$$h(x) = \begin{cases} 1, 0 \le x \le p \\ -1, -p \le x \le 0, \end{cases}$$

then

$$a_k = 0, \qquad k = 0, 1, 2, \ldots,$$
$$b_k = \frac{2}{p}\int_0^p \sin\frac{k\pi x}{p}\,dx$$
$$= \frac{2}{\pi}\int_0^{\pi}\sin kx\,dx = \begin{cases} 0, & k = 2, 4, 6, \ldots, \\ \dfrac{4}{\pi k}, & k = 1, 3, 5, \ldots. \end{cases}$$

Hence, the Nth Fourier approximation to h is given, for odd N, by

$$s_N(x) = \frac{4}{\pi}\sin\frac{\pi x}{p} + \frac{4}{3\pi}\sin\frac{3\pi x}{p} + \ldots + \frac{4}{N\pi}\sin\frac{N\pi x}{p}, \qquad -p \le x \le p.$$

For a function f defined on an arbitrary interval $a \le x \le b$, it is helpful to think of a periodic extension f_E of f having period $b - a$ and defined for all real numbers x. Such an extension is illustrated in Fig. 18. We set $2p = b - a$ so that $p = (b - a)/2$ and $-p = -(b - a)/2$. We then compute the Fourier coefficients of f_E over the interval $[-p, p]$ according to Formula 6.1. Furthermore, because the integrands in Formula 6.1 have period $2p$, we can use the fact, geometrically obvious, that

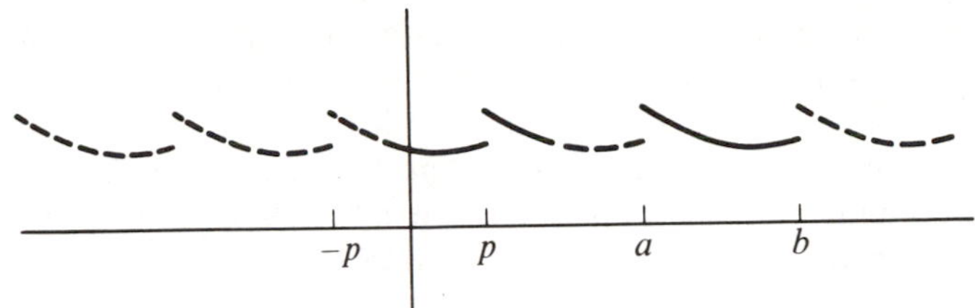

Figure 18

the integration can be performed over any interval of length $2p = b - a$, in particular, over $[a, b]$. Thus Formula 6.1 can be rewritten:

6.2

$$a_k = \frac{2}{b-a}\int_a^b f(x)\cos\frac{2k\pi x}{b-a}\,dx,$$

$$b_k = \frac{2}{b-a}\int_a^b f(x)\sin\frac{2k\pi x}{b-a}\,dx.$$

The associated trigonometric polynomials are

$$s_N(x) = \frac{a_0}{2} + \sum_{k=0}^{N}\left(a_k\cos\frac{2k\pi x}{b-a} + b_k\sin\frac{2k\pi x}{b-a}\right).$$

Example 2. Let $f(x) = x$, for $0 \le x \le 1$. We find, integrating by parts,

$$\begin{aligned} a_k &= 2\int_0^1 x\cos 2k\pi x\,dx \\ &= 2\left[x\frac{\sin 2k\pi x}{2k\pi}\right]_0^1 - \frac{2}{2k\pi}\int_0^1 \sin 2k\pi x\,dx = 0, \\ b_k &= 2\int_0^1 x\sin 2k\pi x\,dx \\ &= 2\left[-x\frac{\cos 2k\pi x}{2k\pi}\right]_0^1 + \frac{2}{2k\pi}\int_0^1 \cos 2k\pi x\,dx \\ &= -\frac{\cos 2k\pi}{k\pi} = -\frac{1}{k\pi}. \end{aligned}$$

Then the Fourier series is

$$\frac{1}{2} - \frac{1}{\pi}\left(\sin 2\pi x + \frac{\sin 4\pi x}{2} + \frac{\sin 6\pi x}{3} + \ldots\right).$$

It sometimes happens that an expansion in terms only of cosines or only of sines is more convenient to use than a general Fourier expansion. We begin with the observation that cosine is an even function, i.e., $\cos(-x) = \cos x$, and sine is an odd function, i.e., $\sin(-x) = -\sin x$. In general, a function f is said to be **even** if $f(-x) = f(x)$ for all x in the domain of f and, alternatively, to be **odd** if $f(-x) = -f(x)$ always holds. The graphs of some even and odd functions are shown in Fig. 19.

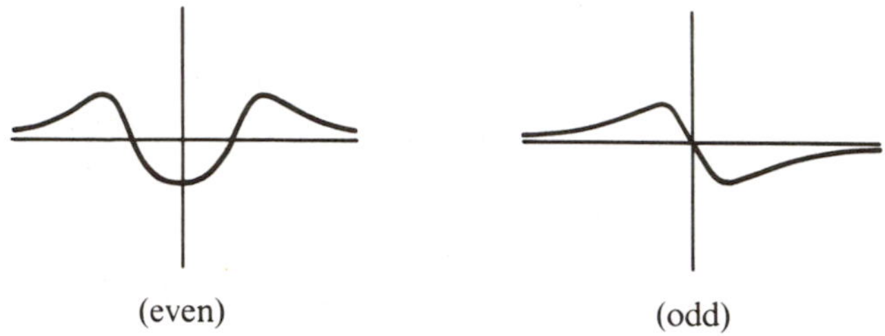

Figure 19

Geometrically, a function is even if its graph is symmetric with respect to the y-axis and odd if its graph is symmetric with respect to the origin. It follows that

$$\int_{-p}^{p} f(x)\,dx = 2\int_{0}^{p} f(x)\,dx, \qquad \text{for even } f \tag{1}$$

and

$$\int_{-p}^{p} f(x)\,dx = 0, \qquad \text{for odd } f. \tag{2}$$

Thus, if f is an even periodic function, the product

$$f(x)\sin\frac{\pi x}{p}$$

is odd. (Why?) Therefore, for the Fourier sine coefficient b_k, we have by 6.1

$$b_k = \frac{1}{p}\int_{-p}^{p} f(x)\sin\frac{\pi x}{p}\,dx = 0. \tag{3}$$

It follows that *an even function has only cosine terms in its Fourier expansion.* Similarly, if f is an odd periodic function, the product

$$f(x)\cos\frac{\pi x}{p}$$

is also odd; so for the Fourier cosine coefficient we have

$$a_k = \frac{1}{p}\int_{-p}^{p} f(x)\cos\frac{\pi x}{p}\,dx = 0. \tag{4}$$

Thus *an odd function has only sine terms in its Fourier expansion.*

The facts in the preceding paragraph are the key to solving the following problem: given a function $f(x)$ defined just on the interval $0 \le x \le p$, find a trigonometric series expansion for f consisting only of cosine terms or, alternatively, only of sine terms. The trick is to extend the definition of f from the interval $0 \le x \le p$ to all real x in such a way that the extension is periodic of period $2p$ and either is even or is odd. We then compute the Fourier series of the extension. If f_e is an even periodic extension of f, then f_e will have only cosine terms in its Fourier series but will still agree with f on $0 \le x \le p$. Similarly, if f_o is an odd periodic extension of f, then f_o has only sine terms in its expansion but will agree with f for $0 \le x \le p$. We illustrate the procedure with two examples.

Example 3. We shall compute the cosine expansion for the function defined by $f(x) = 1 - x$ for $0 \le x \le 2$. We consider the even periodic extension shown in Fig. 20. To find the extension we define f_e by $f_e(x) =$

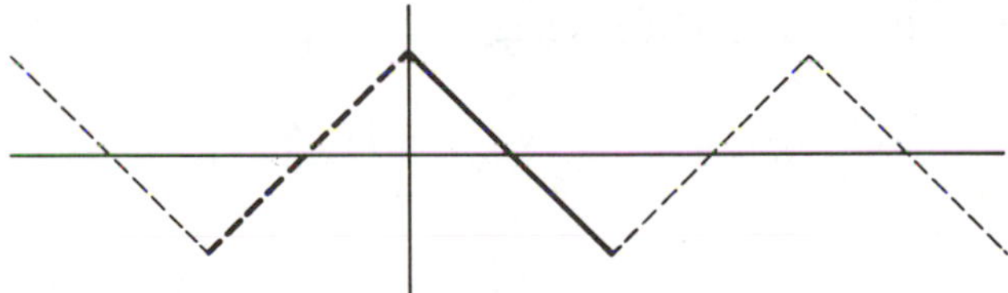

Figure 20

$f(-x)$ for $-2 \le x \le 0$, and then extend periodically, with period 4, to the whole x-axis. We can use Formula 6.1 to compute the Fourier coefficients of f_e. Since f_e is even, Equation (3) shows that $b_k = 0$ for all k. Also, Equation (1) allows us to write

$$\begin{aligned} a_k &= \frac{1}{2}\int_{-2}^{2} f_e(x)\cos\frac{k\pi x}{2}\,dx \\ &= \int_{0}^{2} f_e(x)\cos\frac{k\pi x}{2}\,dx. \end{aligned}$$

Since, for $0 \le x \le 2$, the function f_e is the same as the given function

$f(x) = 1 - x$, we have, for $k > 0$,

$$\begin{aligned} a_k &= \int_0^2 (1 - x) \cos \frac{k\pi x}{2}\, dx \\ &= \left[\frac{2}{k\pi}(1 - x) \sin \frac{k\pi x}{2}\right]_0^2 + \frac{2}{k\pi}\int_0^2 \sin \frac{k\pi x}{2}\, dx \\ &= \frac{4}{\pi^2 k^2}[1 - \cos k\pi] \\ &= \begin{cases} 0, & k \text{ odd}, \\ \dfrac{8}{\pi^2 k^2}, & k \text{ even}. \end{cases} \end{aligned}$$

Finally,

$$a_0 = \int_0^2 (1 - x)\, dx = 0.$$

Thus the cosine expansion of f on $0 \leq x \leq 2$ has for its general nonzero term

$$\frac{8}{\pi^2 k^2} \cos \frac{k\pi x}{2}, \qquad k \text{ even}.$$

Written out, the expansion looks like

$$\frac{2}{\pi^2}\left(\cos \pi x + \frac{\cos 2\pi x}{4} + \frac{\cos 3\pi x}{9} + \ldots\right).$$

Example 4. Starting with the same function as in Example 2, $f(x) = 1 - x$ for $0 \leq x \leq 2$, we compute a sine expansion by considering the odd periodic extension shown in Fig. 21. We first define $f_o(x) = -f(-x)$ for $-2 \leq x \leq 0$, and then extend periodically with period 4. Using Formula 6.1 and Equation (4) we find, as we intended, that $a_k = 0$ for all k. Also,

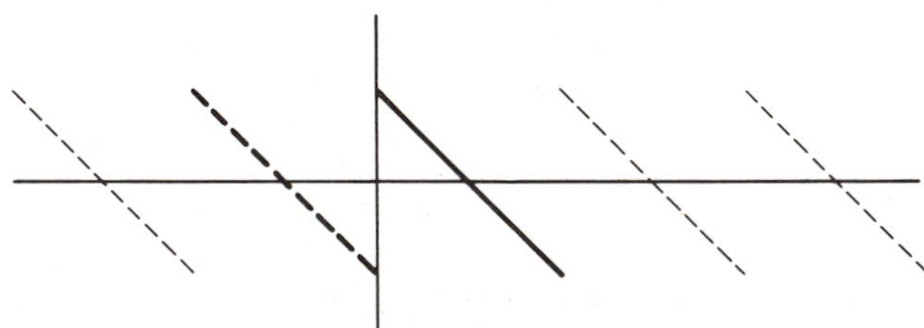

Figure 21

by Equation (1),

$$b_k = \frac{1}{2}\int_{-2}^{2} f_o(x)\sin\frac{k\pi x}{2}\,dx$$
$$= \int_0^2 f_o(x)\sin\frac{k\pi x}{2}\,dx.$$

But $f_o(x) = 1 - x$ for $0 \le x \le 2$, so

$$b_k = \int_0^2 (1-x)\sin\frac{k\pi x}{2}\,dx$$
$$= \left[-\frac{2}{k\pi}(1-x)\cos\frac{k\pi x}{2}\right]_0^2 - \frac{1}{k\pi}\int_0^2 \cos\frac{k\pi x}{2}\,dx$$
$$= \left[\frac{2}{k\pi}(-1)^k + \frac{2}{k\pi}\right] - \frac{1}{k\pi}\left[\frac{2}{k\pi}\sin\frac{k\pi x}{2}\right]_0^2$$
$$= \begin{cases} 0, & k \text{ odd}, \\ \dfrac{4}{k\pi}, & k \text{ even}. \end{cases}$$

Thus the general nonzero term in the sine expansion is

$$\frac{4}{k\pi}\sin\frac{k\pi x}{2}, \qquad \text{even } k.$$

The sine expansion is then

$$\frac{2}{\pi}\left(\sin \pi x + \frac{\sin 2\pi x}{2} + \frac{\sin 3\pi x}{3} + \ldots\right).$$

EXERCISES

1. Find the Fourier series for the function

$$f(x) = -x, \qquad -2 < x < 2.$$

To what values will the series converge at $x = 2$ and $x = -2$?

2. Find the Fourier series for the function

$$f(x) = 1 + x, \qquad 1 < x < 2.$$

To what values will the series converge at $x = 1$ and $x = 2$?

3. Find (a) the Fourier cosine expansion and (b) the Fourier sine expansion of the function

$$f(x) = x, \qquad 0 < x < \pi.$$

(c) Compare the results of (a) and (b) with the complete Fourier expansion of

$$g(x) = x, \qquad -\pi < x < \pi.$$

4. Show that every real-valued function f defined on a symmetric interval $[-a, a]$ can be written as the sum of an even function f_e and an odd function f_o. [*Hint*. Let $f_e(x) = (f(x) + f(-x))/2$.]

5. Show that, if the appropriate combinations are defined,

(a) a product of even functions is even.
(b) a product of odd functions is even.
(c) the product of an even function and an odd function is odd.
(d) a linear combination of even functions is even, and that of odd functions is odd.

6. Using elementary properties of integrals, prove Equations (1) and (2) of the text.

7. Let f be an odd function on $[-\pi, \pi]$ (i.e., $f(-x) = -f(x)$), and let g be an even function (i.e., $g(-x) = g(x)$). Let a_k, b_k and a'_k, b'_k be the Fourier coefficients of f and g, respectively. Show that

$$a_k = 0, \qquad b_k = \frac{2}{\pi}\int_0^{\pi} f(x) \sin kx\, dx,$$

$$a'_k = \frac{2}{\pi}\int_0^{\pi} g(x) \cos kx\, dx, \qquad b'_k = 0.$$

SECTION 7

HEAT AND WAVE EQUATIONS

In this section we show how a Fourier series can be used to solve some problems in heat conduction and wave motion. We first find a differential equation that is satisfied by the physical quantity being studied and then apply Fourier series to solve the equation.

Suppose we are given a thin wire of uniform density and length p. Let $u(x, t)$ be the temperature, at time t, at a point x units from one end. Thus $0 \le x \le p$, and we assume $t \ge 0$. We shall assume that the only heat transfer is along the direction of the wire and that the temperature at the two ends is held fixed. For this reason we can, without loss of generality, represent the wire as a straight segment along an x-axis and picture the temperature as the graph of a function $u = u(x, t)$, an example of which is shown in Fig. 22.

The basic physical principle of heat conduction is that heat flow is proportional to, and in the direction opposite to, the temperature gradient ∇u. Recall that ∇u is the direction in which the temperature is increasing most rapidly, so it is reasonable that heat should flow in the opposite direction, from hotter to colder. Since the medium is 1-dimensional and is represented by a segment of the x-axis, the gradient is represented by

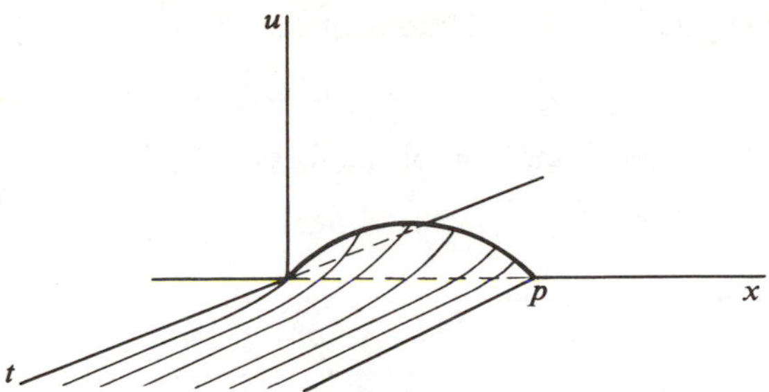

Figure 22

$\partial u/\partial x$. Thus the rate of change of total heat in a segment $[x_1, x_2]$ is proportional to

$$-\frac{\partial u}{\partial x}(x_1, t) + \frac{\partial u}{\partial x}(x_2, t). \tag{1}$$

But the rate of change of heat in the segment is also proportional to the total change in temperature,

$$\int_{x_1}^{x_2} \frac{\partial u}{\partial t}(x, t)\, dx, \tag{2}$$

the total being taken over the interval. By the fundamental theorem of calculus, the expression (1) can be written as

$$\int_{x_1}^{x_2} \frac{\partial^2 u}{\partial x^2}(x, t)\, dx. \tag{3}$$

Hence the two proportional expressions (2) and (3) for rate of change of total heat can be combined to give

$$a^2 \int_{x_1}^{x_2} \frac{\partial^2 u}{\partial x^2}(x, t)\, dx = \int_{x_1}^{x_2} \frac{\partial u}{\partial t}(x, t)\, dx,$$

where a^2 is a positive proportionality constant. Allowing x_2 to vary, we can differentiate both sides of this last equation with respect to x_2, getting

7.1
$$a^2 \frac{\partial^2 u}{\partial x^2}(x, t) = \frac{\partial u}{\partial t}(x, t).$$

This is the 1-dimensional **heat** or **diffusion equation.**

Equation 7.1 is linear in the sense that if u_1 and u_2 are solutions, then so are linear combinations $b_1 u_1 + b_2 u_2$. To single out particular solutions, we impose linear boundary conditions of the form

$$u(0, t) = 0 \quad \text{and} \quad u(p, t) = 0, \tag{4}$$

together with an initial condition of the form

$$y(x, 0) = h(x).$$

The standard method of solution is by **separation of variables,** in which we start by trying to find product solutions of the form

$$u(x, t) = G(x)H(t)$$

with boundary condition $G(0) = G(p) = 0$. If such exist, substitution into $a^2u_{xx} = u_t$ gives

$$a^2G''(x)H(t) = G(x)H'(t),$$

for $0 \leq x \leq p$, $0 < t$. Dividing through by $G(x)H(t)$ gives

$$a^2\frac{G''(x)}{G(x)} = \frac{H'(t)}{H(t)}. \tag{5}$$

For this equation to be satisfied for varying x and t, both sides must be equal to a constant, which we denote by $-\lambda^2$. This procedure is the origin of the term separation of variables.

Setting both sides of (5) equal to $-\lambda^2$ gives two equations:

$$a^2G'' + \lambda^2G = 0, \tag{6}$$

$$H' + \lambda^2H = 0. \tag{7}$$

Equation (6) has solutions

$$G(x) = c_1 \cos\left(\frac{\lambda}{a}\right)x + c_2 \sin\left(\frac{\lambda}{a}\right)x.$$

But $G(0) = 0$ requires $c_1 = 0$, and $G(p) = 0$ then requires $c_2 \sin(\lambda/a)p = 0$. This condition can be achieved, without making $c_2 = 0$, only by choosing λ so that $(\lambda/a)p = k\pi$, where k is an integer. That is, we must take $\lambda = (ka\pi)/p$, with the result that G has the form

$$G(x) = c_2 \sin\left(\frac{k\pi}{p}\right)x, \qquad k = 1, 2, \ldots.$$

Equation (7) has solution

$$H(t) = e^{-\lambda^2 t}$$

which, because $\lambda^2 = (k^2a^2\pi^2)/p^2$, becomes

$$H(t) = e^{-k^2(a^2\pi^2/p^2)t}.$$

The product solution $u(x, t)$ is thus given, except for a constant factor, by

$$u_k(x, t) = e^{-k^2(a^2\pi^2/p^2)t} \sin\left(\frac{k\pi}{p}\right)x, \qquad k = 1, 2, \ldots.$$

Thus a limit of linear combinations

$$\sum_{k=1}^{N} b_k u_k(x, t)$$

looks like

$$u(x, t) = \sum_{k=1}^{\infty} b_k e^{-k^2(a^2\pi^2/p^2)t} \sin\left(\frac{k\pi}{p}\right)x. \tag{8}$$

To satisfy the initial condition $u(x, 0) = h(x)$, we require

$$\sum_{h=1}^{\infty} b_k \sin\left(\frac{k\pi}{p}\right)x = h(x), \tag{9}$$

for some N. If $h(x)$ can be expressed in this form, we can expect that a solution to the problem is given by Equation (8). The boundary conditions, $u(0, t) = u(p, t) = 0$, require that the temperature remain zero at the ends, and the initial condition $u(x, 0) = h(x)$ specifies the initial temperature at each point x of the wire between 0 and p. Thus we are naturally led to the problem of finding a Fourier sine series representation for $h(x)$. The matter of conditions under which the infinite series (8) actually represents a solution of Equation 7.1 is taken up in Section 8.

Example 1. To be more specific about solving the heat equation, we assume for simplicity that $p = \pi$ and recall that, to solve $a^2u_{xx} = u_t$ with boundary condition $u(0, t) = u(\pi, t) = 0$, and initial condition $u(x, 0) = h(x)$, we want in general to be able to represent h by an infinite series of the form

$$h(x) = \sum_{k=1}^{\infty} b_k \sin kx. \tag{10}$$

Suppose, for example, that h is given in $[0, \pi]$ by

$$h(x) = \begin{cases} x, & 0 \leq x \leq \pi/2 \\ \pi - x, & \pi/2 \leq x \leq \pi. \end{cases}$$

The graph of h is shown in Fig. 23.

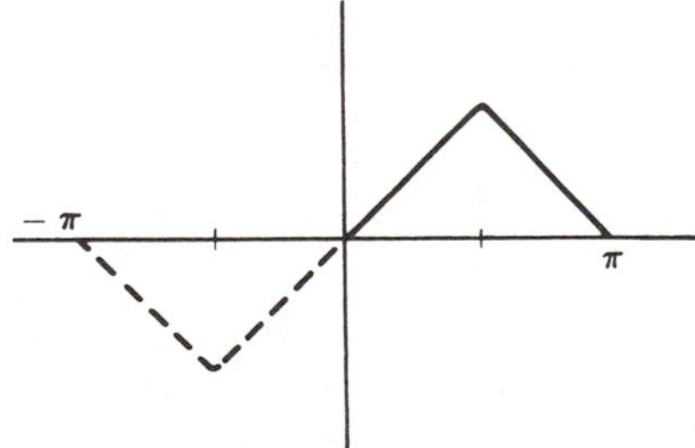

Figure 23

To make Equation (10) represent the Fourier expansion of h on $[0, \pi]$, we extend h to the interval $[-\pi, \pi]$ in such a way that the cosine terms in the expansion of h will all be zero, leaving only the sine terms to be computed. We do this by extending the graph of h symmetrically about the origin, as shown in Fig. 23. Then

$$a_k = \frac{1}{\pi}\int_{-\pi}^{\pi} h(x)\cos kx\, dx = 0,$$

because $h(-x)\cos k(-x) = -h(x)\cos kx$; therefore, the integrals over $[-\pi, 0]$ and $[0, \pi]$ are negatives of one another. To compute b_k we use the fact that $h(-x)\sin k(-x) = h(x)\sin kx$, so that the graph of this function is symmetric about the y-axis. Then

$$\begin{aligned} b_k &= \frac{1}{\pi}\int_{-\pi}^{\pi} h(x)\sin kx\, dx \\ &= \frac{2}{\pi}\int_{0}^{\pi} h(x)\sin kx\, dx \\ &= \frac{2}{\pi}\int_{0}^{\pi/2} x\sin kx\, dx + \frac{2}{\pi}\int_{\pi/2}^{\pi}(\pi - x)\sin kx\, dx \\ &= \frac{4}{\pi k^2}\sin\left(\frac{k\pi}{2}\right). \end{aligned}$$

Hence,

$$b_k = \begin{cases} 0, & k \text{ even}, \\ \dfrac{4}{\pi k^2}, & k = 1, 5, 9, \ldots, \\ \dfrac{-4}{\pi k^2}, & k = 3, 7, 11, \ldots. \end{cases}$$

Theorem 5.2 then implies that

$$h(x) = \frac{4}{\pi}\left(\frac{\sin x}{1^2} - \frac{\sin 3x}{3^2} + \frac{\sin 5x}{5^2} - \frac{\sin 7x}{7^2} + - \ldots\right)$$

for each x in $[0, \pi]$. Finally, from Equation 8 we expect the solution to the equation $a^2u_{xx} = u_t$ to be given by

$$u(x, t) = \frac{4}{\pi}\left(e^{-a^2t}\sin x - \frac{1}{3^2}e^{-3^2a^2t}\sin 3x + \frac{1}{5^2}e^{-5^2a^2t}\sin 5x - \frac{1}{7^2}e^{-7^2a^2t}\sin 7x + - \ldots\right)$$

Verification that $u(0, t) = u(\pi, t) = 0$ follows immediately from setting $x = 0$ and $x = \pi$. By setting $t = 0$, we get the representation of h by its

Fourier series, which is guaranteed by Theorem 5.2. That $u(x, t)$ satisfies the equation $a^2 u_{xx} = u_t$ depends on term-by-term differentiation of the series for u. This will be justified in the next section, though the formal verification is left as Exercise 7 at the end of this section.

Next we take up the 1-dimensional wave equation. Consider for physical motivation a stretched elastic string of length p and uniform density ρ placed along the x-axis in $\mathcal{R}^3$. Suppose that the ends of the string are held fixed at $x = 0$ and $x = p$ by opposite forces of magnitude F. If the string is somehow made to vibrate, our problem is to determine in $\mathcal{R}^3$ the position $\mathbf{x}(s, t)$ at time t of a point on the string a distance s from the end fixed at $x = 0$. Figure 24 shows a possible configuration. We imagine the string subdivided into short pieces of length Δs and then derive two different expressions for the total force acting on a typical segment of the subdivision. If $\mathbf{t}(s)$ is the unit tangent vector to the string at $\mathbf{x}(s)$, then the opposing forces at $\mathbf{x}(s_0)$ and $\mathbf{x}(s_0 + \Delta s)$ are

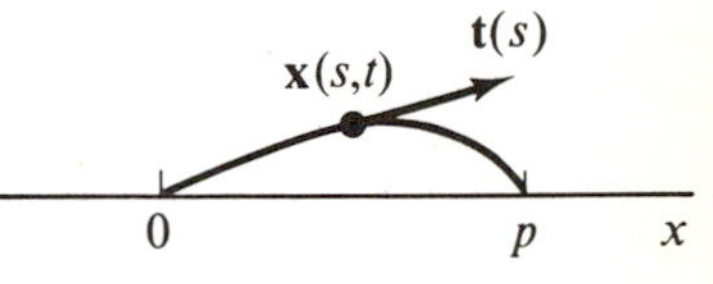

Figure 24

$$F\mathbf{t}(s_0 + \Delta s) \quad \text{and} \quad -F\mathbf{t}(s_0).$$

Hence the total force is

$$F[\mathbf{t}(s_0 + \Delta s) - \mathbf{t}(s_0)].$$

On the other hand, by Newton's law, the force equals mass, $\rho\,\Delta s$, times acceleration $\mathbf{a}(s_0)$. Hence

$$\rho\mathbf{a} = F\left[\frac{\mathbf{t}(s_0 + \Delta s) - \mathbf{t}(s_0)}{\Delta s}\right].$$

But

$$\mathbf{a} = \frac{\partial^2 \mathbf{x}}{\partial t^2}(s, t) \quad \text{and} \quad \mathbf{t}(s) = \frac{\partial \mathbf{x}}{\partial s}(s, t);$$

so, letting $\Delta s \to 0$, we get

$$\rho \frac{\partial^2 \mathbf{x}}{\partial t^2}(s, t) = F \frac{\partial^2 \mathbf{x}}{\partial s^2}(s, t). \tag{11}$$

This vector differential equation is of course equivalent to a system of three scalar equations

$$\frac{\partial^2 x}{\partial t^2} = a^2 \frac{\partial^2 x}{\partial s^2},$$
$$\frac{\partial^2 y}{\partial t^2} = a^2 \frac{\partial^2 y}{\partial s^2},$$
$$\frac{\partial^2 z}{\partial t^2} = a^2 \frac{\partial^2 z}{\partial s^2},$$

where we have set $a^2 = F/\rho > 0$. In a formal sense, the problem of finding solution functions $x(s, t)$, $y(s, t)$, and $z(s, t)$ is the same for all three equations. In practice, however, the equation for $x(s, t)$ is usually set aside when the longitudinal motion (along the x-axis) is slight; we make this assumption. Between the other two equations there is little difference in physical significance unless some other special assumption is made. We suppose to be specific that the string has been plucked in such a way that its motion takes place entirely in the xy-plane, with $z(s, t) \equiv 0$. Finally we assume that the displacements are small enough that we can replace $\partial^2 y/\partial s^2$ by $\partial^2 y/\partial x^2$. This last assumption is one that requires experimental justification in actual practice. Thus, with some loss of generality, we consider instead of the vector Equation (11) a single scalar equation for $y(x, t)$:

7.2
$$\frac{\partial^2 y}{\partial t^2} = a^2 \frac{\partial^2 y}{\partial x^2}.$$

This differential equation does not completely specify the vibration of a string unless we impose some initial conditions:

$$y(x, 0) = f(x), \tag{12}$$

$$\frac{\partial y}{\partial t}(x, 0) = g(x). \tag{13}$$

The first condition specifies the initial ($t = 0$) displacement in the y-direction as a function of x, rather than s, and the second equation specifies the initial velocity in the y-direction also as a function of x. In addition, we also impose boundary conditions

$$y(0, t) = 0 \quad \text{and} \quad y(p, t) = 0. \tag{14}$$

As in the case of the heat equation, we use separation of variables and rely on the linearity of Equation 7.2 and the conditions (14) for constructing solutions that satisfy the conditions (12) and (13) also. We try

$$y(x, t) = G(x)H(t),$$

upon which 7.2 becomes

$$G(x)H''(t) = a^2 G''(x)H(t)$$

or

$$\frac{H''(t)}{H(t)} = a^2 \frac{G''(x)}{G(x)}.$$

Since the left side is independent of t, both sides are constant; so we write

$$\frac{G''(x)}{G(x)} = \lambda \quad \text{and} \quad \frac{H''(t)}{H(t)} = a^2\lambda.$$

The first equation, $G''(x) = \lambda G(x)$, has solutions

$$G(x) = c_1 e^{\sqrt{\lambda}\, x} + c_2 e^{-\sqrt{\lambda}\, x}.$$

The boundary conditions (14) require

$$G(0) = 0 \quad \text{and} \quad G(p) = 0;$$

so

$$c_1 + c_2 = 0 \quad \text{and} \quad c_1 e^{\sqrt{\lambda}p} + c_2 e^{-\sqrt{\lambda}p} = 0.$$

Solving for c_1 and c_2 we find that, for nonzero solutions to exist, we must have $c_1 = -c_2$ and $e^{-\sqrt{\lambda}p} = e^{\sqrt{\lambda}p}$, or $e^{2\sqrt{\lambda}p} = 1$. Thus, allowing complex exponents, we conclude that $2\sqrt{\lambda}p = 2\pi ki$ for some integer k, so that

$$\sqrt{\lambda} = \frac{\pi ki}{p}.$$

Solutions $G(x)$ are then of the form

$$\begin{aligned} G(x) &= c_1 e^{(\pi ki/p)x} - c_1 e^{(-\pi ki/p)x} \\ &= 2c_1 i \sin \frac{\pi k}{p} x. \end{aligned}$$

We write

$$G_k(x) = b_k \sin \frac{\pi k}{p} x.$$

The differential equation for H, namely, $H''(t) = a^2\lambda H(t)$, now takes the form

$$H''(t) - \frac{\pi^2 k^2 a^2}{p^2} H(t) = 0,$$

since we have determined that $\lambda = \pi^2 k^2/p^2$. Solutions of this equation are of the form

$$H_k(t) = C_k \cos \frac{\pi ka}{p} t + D_k \sin \frac{\pi ka}{p} t;$$

therefore, product solutions $y_k = G_k H_k$ take the form

$$y_k(x, t) = \left[A_k \cos \frac{\pi ka}{p} t + B_k \sin \frac{\pi ka}{p} t\right] \sin \frac{\pi k}{p} x.$$

To satisfy the initial conditions (12) and (13), we form finite or infinite sums of the type

$$y(x, t) = \sum_{k=0}^{\infty} \left[A_k \cos \frac{\pi ka}{p} t + B_k \sin \frac{\pi ka}{p} t\right] \sin \frac{\pi k}{p} x.$$

Thus the initial conditions become formally

$$y(x, 0) = \sum_{k=0}^{\infty} A_k \sin \frac{\pi k}{p} x = f(x) \tag{15}$$

and

$$y_t(x, 0) = \sum_{k=0}^{\infty} \frac{\pi ka}{p} B_k \sin \frac{\pi k}{p} x = g(x). \tag{16}$$

The coefficients A_k and $(\pi ka/p)B_k$ are then determined so that they are the Fourier sine coefficients of f and g, respectively.

Example 2. For a simple example we consider a string which is initially stationary; hence, $y_t(x, 0) = 0$. We assume the initial displacement is given by

$$g(x, 0) = b \sin \frac{\pi x}{p}, \tag{16}$$

and $y(0, t) = y(p, t) = 0$. We choose $B_k = 0$ in Equation (16) and $A_1 = b$ in Equation (15), with $A_k = 0$ for $k \neq 1$. (It happens that f and g are so simple in this case that we can find their Fourier sine coefficients by inspection.) The solution to 7.2 then takes the form

$$y(x, t) = b \cos \frac{\pi a}{p} t \sin \frac{\pi}{p} x.$$

Recall that for Equation 7.2 to be physically realistic the vibrations of the string should be fairly small—in other words, the coefficient b should be small. In Fig. 25, the graph of our solution is shown with $a = 1$, $p = \pi$,

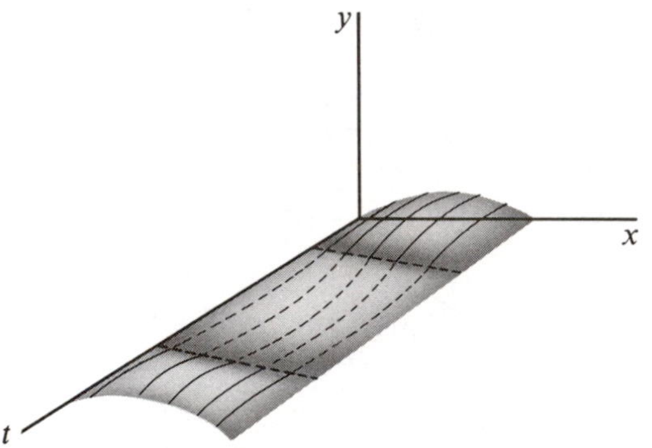

Figure 25

and b chosen unrealistically large in order to bring out some qualitative features of the picture.

EXERCISES

1. Use the method of separation of variables to solve the 1-dimensional heat equation $a^2 u_{xx} = u_t$, subject to each of the following boundary and initial conditions.

   (a) $u(0, t) = u(p, t) = 0$, $u(x, 0) = \sin(\pi x/p)$.
   (b) $u(0, t) = u(\pi, t) = 0$, $u(x, 0) = x(\pi - x)$.
   (c) $u_x(0, t) = u_x(\pi, t) = 0$, $u(x, 0) = \sin x$.

2. Use the method of separation of variables to solve the 1-dimensional wave equation $a^2u_{xx} = u_{tt}$, subject to each of the following boundary and initial conditions.

(a) $u(0, t) = u(\pi, t) = 0$, $u(x, 0) = \sin x$, $u_t(x, 0) = 0$.

(b) $u(0, t) = u(\pi, t) = 0$, $u(x, 0) = \begin{cases} x, & 0 \le x \le \pi/2 \\ \pi - x, & \pi/2 \le x \le \pi \end{cases}$, $u_t(x, 0) = 0$.

(c) $u(0, t) = u(\pi, t) = 0$, $u(x, 0) = 0$, for $0 \le x \le \pi$,

$$u_t(x, 0) = \begin{cases} 0, & 0 \le x < \pi/2 \\ 1, & \pi/2 \le x \le \pi. \end{cases}$$

3. The 2-dimensional heat equation is $a^2(u_{xx} + u_{yy}) = u_t$, and if u is independent of t, $u_t \equiv 0$. This results in the Laplace equation $u_{xx} + u_{yy} = 0$ for the temperature in a 2-dimensional steady-state heat flow problem. Using the method of separation of variables, solve the equation $u_{xx} + u_{yy} = 0$ in the rectangle $0 \le x \le \pi$, $0 \le y \le \pi$, subject to the boundary conditions $u(0, y) = u(\pi, y) = 0$, $u(x, 0) = 0$, $u(\pi, x) = \sin x$.

4. Let L be defined as an operator by

(a) $L(u) = u_t - a^2u_{xx}$.
(b) $L(y) = y_{tt} - a^2y_{xx}$.

Show that L is a linear operator, and conclude that linear combinations of solutions of (a) the heat equation and (b) the wave equation are also solutions.

5. (a) Show that a boundary condition of the form $u(a, t) = u(b, t) = 0$ is "linear" in the sense that, if two functions satisfy it, then so does *any* linear combination of the two functions.
(b) Show that a boundary condition of the form $u(a, t) = 1$ is not linear in the sense of part (a).
(c) Show that the initial condition $u(x, t) = f(x)$ is not linear in the sense of part (a) unless f is identically zero.

6. The 2-dimensional Laplace equation in polar coordinates is, by Chapter 4, Section 5, Problem 13,

$$r^2 \frac{\partial^2 u}{\partial r^2} + r\frac{\partial u}{\partial r} + \frac{\partial^2 u}{\partial \theta^2} = 0.$$

(a) By letting $u(r, \theta) = G(r)H(\theta)$, show that the method of separation of variables leads to the two differential equations

$$H'' - \lambda H = 0,$$

$$r^2G'' + rG' + \lambda G = 0.$$

(b) Show that $H'' - \lambda H = 0$ has solutions satisfying $H(0) = H(2\pi)$ if and only if $\lambda = -k^2$, where k is an integer, and that the solutions can then be written $\cos k\theta$ and $\sin k\theta$.

(c) Show that $r^2G'' + rG' - k^2G = 0$ has solutions r^k and r^{-k} for $k = 0, 1, 2, \ldots$, but that negative exponents are ruled out if $u(r, \theta) = G(r)H(\theta)$ is to be finite for $r = 0$.

(d) Show that if

$$f(\theta) = \frac{a_0}{2} + \sum_{k=1}^{N} (a_k \cos k\theta + b_k \sin k\theta),$$

then

$$u(r, \theta) = \frac{a_0}{2} + \sum_{k=1}^{N} (a_k r^k \cos k\theta + b_k r^k \sin k\theta)$$

satisfies the Laplace equation in polar coordinates together with the boundary condition $u(1, \theta) = f(\theta)$.

7. Verify that the series expansion for $u(x, t)$ given in Equation (8) of the text is a formal solution of $a^2u_{xx} = u_t$. Use term-by-term differentiation of the series. (The method is justified by Theorem 8.5 of the next section.)

SECTION 8

UNIFORM CONVERGENCE

Let $f_k(\mathbf{x})$, $k = 1, 2, 3, \ldots$, be a sequence of real-valued functions defined for all $\mathbf{x}$ in some set S. Then for each $\mathbf{x}$, we consider the series $\sum_{k=1}^{\infty} f_k(\mathbf{x})$. If it converges for each $\mathbf{x}$ in S, we say that the series **converges pointwise** on S. Calling the limit $f(\mathbf{x})$ for each $\mathbf{x}$ in S, we write

$$\begin{aligned} f(\mathbf{x}) &= \sum_{k=1}^{\infty} f_k(\mathbf{x}) \\ &= \lim_{N\to\infty} \sum_{k=1}^{N} f_k(\mathbf{x}). \end{aligned}$$

Recall that this means that, for each $\mathbf{x}$ in S, there is a number $f(\mathbf{x})$ such that, given $\epsilon > 0$, there is an integer K sufficiently large that

$$\left| \sum_{k=1}^{N} f_k(\mathbf{x}) - f(\mathbf{x}) \right| < \epsilon,$$

whenever $N \geq K$.

Example 1. The series $\sum_{k=0}^{\infty} x^k$ has as $(N + 1)$th partial sum the finite sum

$$\sum_{k=0}^{N} x_k = \begin{cases} \dfrac{1 - x^{N+1}}{1 - x}, & x \neq 1, \\ N + 1, & x = 1. \end{cases}$$

Then

$$\sum_{k=0}^{\infty} x^k = \lim_{N\to\infty} \sum_{k=0}^{N} x^k = \frac{1}{1 - x}, \qquad \text{for} \quad -1 < x < 1.$$

For real values of x outside the interval $(-1, 1)$, the series fails to converge.

The trigonometric series $\sum_{k=1}^{\infty} (\sin kx)/k^2$ converges pointwise for all real x. The reason is that its terms can be compared with those of the convergent series $\sum_{k=1}^{\infty} 1/k^2$, by observing that

$$\left| \frac{\sin kx}{k^2} \right| \leq \frac{1}{k^2}, \qquad k = 1, 2, \ldots.$$

The result is that the given series even converges absolutely.

An infinite series $\sum_{k=1}^{\infty} f_k(\mathbf{x})$ that converges for each $\mathbf{x}$ in a set S to a number $f(\mathbf{x})$ defines a function f on S. However, in general, very little can be concluded about the properties of f from pointwise convergence alone. For this reason it is sometimes helpful to consider a stronger form of convergence on S. We say that $\sum_{k=1}^{\infty} f_k$ **converges uniformly** to a function f on a set S, if, given $\epsilon > 0$, there is an integer K such that for all $\mathbf{x}$ in S and for all $N \geq K$,

$$\left| \sum_{k=1}^{N} f_k(\mathbf{x}) - f(\mathbf{x}) \right| < \epsilon.$$

The definition just given should be compared carefully with that of pointwise convergence. Notice that uniform convergence implies pointwise convergence, but not conversely. Roughly speaking, uniform convergence of a series of functions defined on a set S means that the series converges with at least a certain minimum rate for all points in S. A pointwise con-

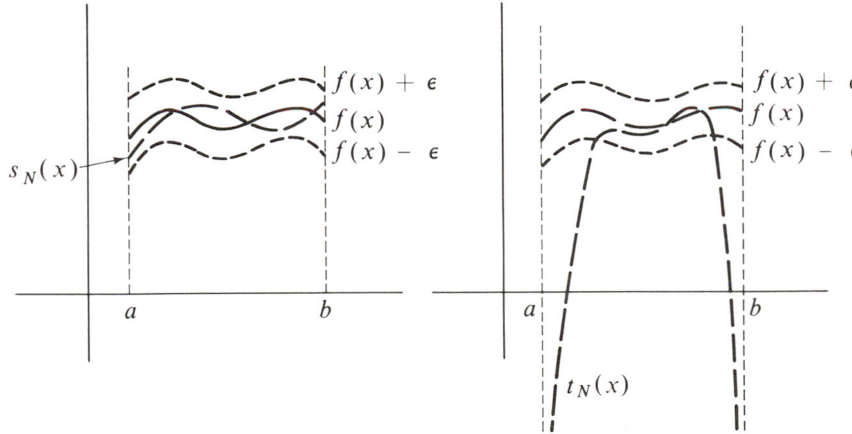

Figure 26

vergent series may have points at which the convergence is arbitrarily slow. Figure 26 is a picture of uniform and nonuniform convergence to the same function f; $s_N(x)$ and $t_N(x)$ are Nth partial sums of two series.

To determine that a series converges uniformly we have the following.

8.1 Weierstrass Test

Let $\sum_{k=1}^{\infty} f_k$ be a series of real-valued functions defined on a set S. If there is a constant series $\sum_{k=1}^{\infty} p_k$, such that

1. $|f_k(\mathbf{x})| \leq p_k$ for all $\mathbf{x}$ in S and for $k = 1, 2, \ldots,$
2. $\sum_{k=1}^{\infty} p_k$ converges,

then $\sum_{k=1}^{\infty} f_k$ converges uniformly to a function f defined on S.

Proof. The comparison test for series shows that $\sum_{k=1}^{\infty} f_k(\mathbf{x})$ converges (even absolutely) for each $\mathbf{x}$ in S to a number that we shall write $f(\mathbf{x})$. Hence we can write

$$f(\mathbf{x}) - \sum_{k=1}^{N} f_k(\mathbf{x}) = \sum_{k=1}^{\infty} f_k(\mathbf{x}) - \sum_{k=1}^{N} f_k(\mathbf{x})$$
$$= \sum_{k=N+1}^{\infty} f_k(\mathbf{x}).$$

It follows that

$$\left| f(\mathbf{x}) - \sum_{k=1}^{N} f_k(\mathbf{x}) \right| \leq \sum_{k=N+1}^{\infty} |f_k(\mathbf{x})| \leq \sum_{k=N+1}^{\infty} p_k.$$

Since $\sum_{k=1}^{\infty} p_k$ converges, we can, given $\epsilon > 0$, find a K such that $\sum_{k=N}^{\infty} p_k < \epsilon$ if $N > K$. This completes the proof, because the number K depends only on ϵ and not on $\mathbf{x}$.

Example 2. The trigonometric series $\sum_{k=1}^{\infty} (\sin kx)/k^2$ converges uniformly for all real x, because

$$\left| \frac{\sin kx}{k^2} \right| \leq \frac{1}{k^2},$$

and $\sum_{k=1}^{\infty} 1/k^2$ converges. However, the power series $\sum_{k=0}^{\infty} x^k$, while it converges pointwise for $-1 < x < 1$, fails to converge uniformly on $(-1, 1)$. See

Problem 6. The Weierstrass test can be applied on any closed subinterval $[-r, r]$ by observing that $|x^k| \leq r^k$ for x on $[-r, r]$ and that $\sum_{k=0}^{\infty} r^k$ converges if $0 \leq r \leq 1$. Hence, the power series converges pointwise on $(-1, 1)$ and uniformly on $[-r, r]$ for any $r < 1$.

The next four theorems are about uniformly convergent series of functions. They all assert that certain limit operations can be interchanged with the summing of a series, provided that some series converges uniformly. If uniform convergence is replaced by pointwise convergence, then the resulting statements are false. See Problem 9.

8.2 Theorem

Let $f_1, f_2, f_3, \ldots$ be a sequence of functions defined on a set S in $\mathcal{R}^n$. Suppose $\mathbf{x}_0$ is a limit point of S, and suppose that the limit

$$\lim_{\mathbf{x}\to\mathbf{x}_0} f_k(\mathbf{x})$$

exists for $k = 1, 2, \ldots$. Then

$$\lim_{\mathbf{x}\to\mathbf{x}_0} \sum_{k=1}^{\infty} f_k(\mathbf{x}) = \sum_{k=1}^{\infty} \lim_{\mathbf{x}\to\mathbf{x}_0} f_k(\mathbf{x}) \ ,$$

provided the series of numbers on the right converges, and the series on the left converges uniformly on S.

Proof. Let $\lim_{\mathbf{x}\to\mathbf{x}_0} f_k(\mathbf{x}) = a_k$. Then, adding and subtracting $\sum_{k=1}^{N} f_k(\mathbf{x})$ and $\sum_{k-1}^{N} a_k$, we get

$$\left| \sum_{k=1}^{\infty} f_k(\mathbf{x}) - \sum_{k=1}^{\infty} a_k \right| \leq \left| \sum_{k=1}^{\infty} f_k(\mathbf{x}) - \sum_{k=1}^{N} f_k(\mathbf{x}) \right| + \left| \sum_{k=1}^{N} f_k(\mathbf{x}) - \sum_{k=1}^{N} a_k \right| + \left| \sum_{k=1}^{N} a_k - \sum_{k=1}^{\infty} a_k \right| . \tag{1}$$

Now let $\epsilon > 0$. Since $\sum_{k=1}^{\infty} f_k$ converges uniformly we can choose K such that $N > K$ implies

$$\left| \sum_{k=1}^{\infty} f_k(\mathbf{x}) - \sum_{k=1}^{N} f_k(\mathbf{x}) \right| < \frac{\epsilon}{3}, \qquad \text{for all } \mathbf{x} \text{ in } S.$$

Then choose an $N > K$ such that

$$\left| \sum_{k=1}^{N} a_k - \sum_{k=1}^{\infty} a_k \right| < \frac{\epsilon}{3} .$$

Finally, pick $\delta > 0$ so that $|\mathbf{x} - \mathbf{x}_0| < \delta$ implies, by the relation $\lim_{\mathbf{x}\to\mathbf{x}_0} \sum_{k=1}^{N} f_k(\mathbf{x}) = \sum_{k=1}^{N} a_k$, that

$$\left| \sum_{k=1}^{N} f_k(\mathbf{x}) - \sum_{k=1}^{N} a_k \right| < \frac{\epsilon}{3}.$$

Then for $\mathbf{x}$ satisfying $|\mathbf{x} - \mathbf{x}_0| < \delta$, the left side of (1) is less than ϵ.

8.3 Corollary

If $\sum_{k=1}^{\infty} f_k$ is a uniformly convergent series of continuous functions f_k defined on a set S in $\mathcal{R}_n$, then the function f defined by $f(\mathbf{x}) = \sum_{k=1}^{\infty} f_k(\mathbf{x})$ is continuous on S.

In the next two theorems we restrict ourselves to functions of one variable, though by treating one variable at a time, we can apply them to functions of several variables.

8.4 Theorem

If the series $\sum_{k=1}^{\infty} f_k$ converges uniformly on the interval $[a, b]$, and the functions f_k are continuous on $[a, b]$, then

$$\sum_{k=1}^{\infty} \int_a^b f_k(x)\, dx = \int_a^b \left[\sum_{k=1}^{\infty} f_k(x) \right] dx.$$

Proof. By Theorem 8.3 the function $\sum_{k=1}^{\infty} f_k(x)$ is continuous on $[a, b]$ and so is integrable there. We have

$$\int_a^b \left[\sum_{k=1}^{\infty} f_k(x) \right] dx - \sum_{k=1}^{N} \int_a^b f_k(x)\, dx = \int_a^b \sum_{k=N+1}^{\infty} f_k(x)\, dx. \qquad (2)$$

Let $\epsilon > 0$, and choose K so large that if $N > K$, then $\left| \sum_{k=N+1}^{\infty} f_k(x) \right| < \epsilon(b - a)^{-1}$, for all x in $[a, b]$. Then using the fact that, for continuous g,

$$\left| \int_a^b g(x)\, dx \right| \leq (b - a) \max_{a \leq x \leq b} |g(x)|,$$

we have

$$\left| \int_a^b \sum_{k=N+1}^{\infty} f_k(x)\, dx \right| \leq (b-a) \cdot \epsilon \cdot (b-a)^{-1} = \epsilon, \qquad \text{for} \quad N > K.$$

Thus the left side of Equation (2) is less than ϵ in absolute value for $N > K$, which was to be shown.

The interchange of differentiation with the summing of a series requires somewhat more in the way of hypotheses than did the previous theorem on integration.

8.5 Theorem

Let $f_1, f_2, f_3, \ldots$ be a sequence of continuously differentiable functions defined on an interval $[a, b]$. If $\sum_{k=1}^{\infty} f_k(x) = f(x)$ for all x in $[a, b]$ (pointwise convergence), and if $\sum_{k=1}^{\infty} df_k/dx$ converges uniformly on $[a, b]$, then f is continuously differentiable, and

$$\frac{d}{dx} \sum_{k=1}^{\infty} f_k(x) = \sum_{k=1}^{\infty} \frac{df_k}{dx}(x).$$

Proof. By the fundamental theorem of calculus,

$$\begin{aligned} \sum_{k=1}^{N} [f_k(x) - f_k(a)] &= \sum_{k=1}^{N} \int_a^x f_k'(t)\, dt \\ &= \int_a^x \left[\sum_{k=1}^{N} f_k'(t) \right] dt. \qquad (3) \end{aligned}$$

Letting N tend to infinity, we get $\sum_{k=1}^{\infty} f_k(x) = f(x)$; so

$$f(x) - f(a) = \int_a^x \left[\sum_{k=1}^{\infty} f_k'(t) \right] dt,$$

where we have used pointwise convergence on the left side of Equation (3) and, on the right, have used uniform convergence, together with Theorem 8.4. Differentiation of both sides of the last equation gives

$$f'(x) = \sum_{k=1}^{\infty} f_k'(x),$$

which is the conclusion of the theorem.

Example 3. Consider the trigonometric series

$$\sum_{k=1}^{\infty} \frac{\sin kx}{k^4}.$$

Clearly the series converges for all real x. Furthermore the series of derivatives of the terms of the given series is

$$\sum_{k=1}^{\infty} \frac{\cos kx}{k^3}.$$

This series converges uniformly for all x by the Weierstrass test, because

$$\left|\frac{\cos kx}{k^3}\right| \le \frac{1}{k^3},$$

and $\sum_{1}^{\infty} (1/k^3)$ converges. Hence, by Theorem 8.5,

$$\frac{d}{dx}\sum_{k=1}^{\infty} \frac{\sin kx}{k^4} = \sum_{k=1}^{\infty} \frac{\cos kx}{k^3}.$$

The same argument can be applied to give

$$\frac{d^2}{dx^2}\sum_{k=1}^{\infty} \frac{\sin kx}{k^4} = -\sum_{k=1}^{\infty} \frac{\sin kx}{k^2}.$$

EXERCISES

1. Show that the series $\sum_{k=0}^{\infty} x^k$ converges uniformly for $-d \le x \le d$ if $0 < d < 1$.

2. (a) Show that the trigonometric series $\sum_{k=1}^{\infty} \frac{\cos kx}{k^2}$ converges uniformly for all real x.

(b) Prove that the series of part (a) defines a continuous function for all real x.

3. (a) Show that if a trigonometric series

$$\frac{a_0}{2} + \sum_{k=1}^{\infty} a_k \cos kx + b_k \sin kx$$

converges uniformly on $[-\pi, \pi]$ then it converges uniformly for *all* real x.

(b) Prove that the uniformly convergent series of part (a) is necessarily the Fourier series of the function it represents. [*Hint.* Use Theorem 8.4.]

4. (a) Show that if $|c_k| \le B$ for some fixed number B, then the series

$$u(x, t) = \sum_{k=1}^{\infty} c_k e^{-k^2 t} \sin kx$$

is a solution of the differential equation

$$u_{xx} = u_t \quad \text{for} \quad t > 0 \quad \text{and } x \text{ in } [0, \pi],$$

satisfying $u(0, t) = u(\pi, t) = 0$. [*Hint.* Use Theorem 8.5.]

(b) Show that if $u(x, t)$ in part (a) is defined for $t = 0$ by a uniformly convergent series, then $u(x, t)$ is continuous on the set S in $\mathcal{R}^2$ defined by $0 \le t$, $0 \le x \le \pi$.

(c) Show that the function $u(x, t)$ is infinitely often differentiable with respect to both x and t, for $t > 0$.

5. Show that if a trigonometric series of the form shown in Problem 3(a) satisfies $|a_n| \le A/n^2$, $|b_n| \le B/n^2$ for $n = 1, 2, 3, \ldots$, and some constants A and B, then the trigonometric series is a Fourier series.

6. By considering the partial sums of the power series $\sum_{k=0}^{\infty} x^k$ for $-1 < x < 1$, show that the series fails to converge uniformly on $(-1, 1)$.

7. Show that $\sum_{k=1}^{\infty} (-1)^k (1 - x)x^k$ converges uniformly on $[0, 1]$, but that $\sum_{k=1}^{\infty} (1 - x)x^k$ only converges pointwise on $[0, 1]$.

8. (a) Assume that the series $\sum_{k=1}^{\infty} k^2 a_n$ and $\sum_{k=1}^{\infty} k^2 b_n$ both converge absolutely. Show that

$$w(x, t) = \sum_{k=1}^{\infty} \sin kx(a_k \cos kat + b_k \sin kat)$$

is a solution of the 1-dimensional wave equation $a^2 w_{xx} = w_{tt}$. [*Hint.* Use the Weierstrass test and Theorem 8.5.]

(b) Show that the solution $w(x, t)$ of part (a) satisfies the boundary conditions $w(0, t) = w(\pi, t) = 0$ for $t \ge 0$ and an initial condition $w(x, 0) = h(x)$, where h is twice continuously differentiable.

9. Show that, with uniform convergence replaced by pointwise convergence, the statements of Theorems (a) 8.3, (b) 8.4, and (c) 8.5 become false.

10. Can Theorems 8.4 and 8.5 be proved under the more general assumption that the functions involved are vector-valued, with values in $\mathcal{R}^n$? [*Hint.* See Problems 16 and 17 of Chapter 3, Section 2.]

SECTION 9

ORTHOGONAL FUNCTIONS

Some of the properties of Fourier expansions are shared by a large class of similar expansions in which the functions $\sin kx$ and $\cos kx$ are replaced by some sequence $\{\varphi_k\}_{k=1,2,\ldots}$ of functions, all of which are mutually orthogonal. The orthogonality is measured in terms of an inner product on a vector space of functions. For example, if f and g are elements in the space $C[-\pi, \pi]$ of continuous functions on the interval $[-\pi, \pi]$, we can define an inner product of f and g by

$$\langle f, g\rangle = \int_{-\pi}^{\pi} f(x)g(x)\,dx. \tag{1}$$

It follows by direct computation (compare Theorem 5.3) that if we set $\varphi_1(x) = 1/\sqrt{2\pi}$, $\varphi_{2n}(x) = (\cos nx)/\sqrt{\pi}$, and $\varphi_{2n+1}(x) = (\sin nx_n)/\sqrt{\pi}$, then

$$\langle \varphi_k, \varphi_l\rangle = \begin{cases} 1, & k = l \\ 0, & k \neq l. \end{cases}$$

The sequence $\{\varphi_k\}_{k=1,2,\ldots}$ is **orthonormal** with respect to the inner product given by Equation (1). The term "normal" comes from the fact the functions have been normalized by requiring $\|\varphi_k\| = \langle \varphi_k, \varphi_k\rangle^{1/2} = 1$. In the trigonometric case, the normalization is achieved by dividing the sines and cosines by $\sqrt{\pi}$.

To see the importance of orthonormal sequences in general, we consider the following problem: Let $\langle f, g\rangle$ be an inner product on a vector space, and let $\{\varphi_k\}_{k=1,2,\ldots}$ be a sequence of elements, orthonormal with respect to the inner product. Using the norm defined by $\|f\| = \langle f, f\rangle^{1/2}$, we try to determine coefficients c_k, $k = 1, 2, \ldots, N$, such that

$$\left\| g - \sum_{k=1}^{N} c_k\varphi_k \right\|$$

is minimized for given g and N. The fact that the sequence φ_k is orthonormal makes the solution very simple—by adding and subtracting $\sum_{k=1}^{N} \langle g, \varphi_k\rangle^2$, we get

$$\begin{aligned} 0 \leq \left\| g - \sum_{k=1}^{N} c_k\varphi_k \right\|^2 &= \left\langle g - \sum_{k=1}^{N} c_k\varphi_k, g - \sum_{k=1}^{N} c_k\varphi_k \right\rangle \\ &= \|g\|^2 - 2\sum_{k=1}^{N} c_k\langle g, \varphi_k\rangle + \sum_{k=1}^{N} c_k^2 \\ &= \|g\| \ - \sum_{k=1}^{N} \langle g, \varphi_k\rangle^2 \\ &\quad + \sum_{k=1}^{N} [\langle g, \varphi_k\rangle^2 - 2c_k\langle g, \varphi_k\rangle + c_k^2] \\ &= \|g\|^2 - \sum_{k=1}^{N} \langle g, \varphi_k\rangle^2 + \sum_{k=1}^{N} [\langle g, \varphi_k\rangle - c_k]^2. \end{aligned} \tag{2}$$

But the first two terms in the last expression are independent of the choice of the c_k's, and the last sum is then minimized by taking $c_k = \langle g, \varphi_k\rangle$. The numbers $\langle g, \varphi_k\rangle$ are called the **Fourier coefficients** of g with respect to the orthonormal sequence $\{\varphi_k\}$. Notice that the simplicity of the answer to the problem depends very much on the orthogonality of the φ_k's.

We can summarize what has just been proved as follows.

9.1 Theorem

Let $\{\varphi_k\}_{k=1,2,\ldots}$ be an orthonormal sequence in a vector space with an inner product. Then, given an element g of the space, the distance

$$\left\| g - \sum_{k=1}^{N} c_k \varphi_k \right\|$$

is minimized for $N = 1, 2, \ldots$ by taking c_k to be the Fourier coefficient $\langle g, \varphi_k \rangle$.

The important thing about the conclusion of Theorem 9.1 is that the c_k's are uniquely determined, independently of N. In other words, if we wanted to improve the closeness of the approximation to g by increasing N, then Theorem 9.1 says that the c_k's already computed are to be left unchanged, and it is only necessary to compute additional coefficients $c_{N+1} = \langle g, \varphi_{N+1} \rangle$, etc.

As a by-product of the proof of Theorem 9.1, we have the following.

9.2 Bessel's Inequality

$$\|g\|^2 \geq \sum_{k=1}^{\infty} \langle g, \varphi_k \rangle^2.$$

Proof. The inequality

$$0 \leq \|g\|^2 - \sum_{k=1}^{N} \langle g, \varphi_k \rangle^2 + \sum_{k=1}^{N} [\langle g, \varphi_k \rangle - c_k]^2$$

was established in the last step of Equation (2). On taking $c_k = \langle g, \varphi_k \rangle$, the inequality becomes

$$0 \leq \|g\|^2 - \sum_{k=1}^{N} \langle g, \varphi_k \rangle^2$$

Bessel's inequality follows by letting N tend to infinity.

Example 1. The approximation of g by a sum $\sum_{k=1}^{N} c_k \varphi_k$ has been measured by a norm in Theorem 9.1. To see what this means for approximation by trigonometric polynomials, we use the inner product given by Equation

(1) on the space $C[-\pi, \pi]$ of continuous functions on $[-\pi, \pi]$. Given the orthonormal sequence $\varphi_1(x) = 1/\sqrt{2\pi}$, $\varphi_{2n}(x_n) = (\cos nx)/\sqrt{\pi}$, $\varphi_{2n+1}(x_n) = (\sin nx)/\sqrt{\pi}$, we try to minimize, for given g in $C[-\pi, \pi]$, the norm

$$\left\| g - \sum_{k=1}^{2N+1} c_k\varphi_k \right\|.$$

We have seen that this is done by taking $c_k = \langle g, \varphi_k\rangle$. But, by the definition of the inner product,

$$\langle g, \varphi_k\rangle = \begin{cases} \displaystyle\int_{-\pi}^{\pi} \frac{1}{\sqrt{2\pi}}\, g(t)\, dt, & k = 1 \\ \displaystyle\int_{-\pi}^{\pi} g(t)\frac{\cos nt}{\sqrt{\pi}}\, dt, & k = 2n \\ \displaystyle\int_{-\pi}^{\pi} g(t)\frac{\sin nt}{\sqrt{\pi}}\, dt, & k = 2n + 1. \end{cases}$$

Hence, the terms $c_k\varphi_k$ become

$$\langle g, \varphi_1\rangle\varphi_1(x) = \frac{1}{2\pi}\int_{-\pi}^{\pi} g(t)\, dt,$$

$$\langle g, \varphi_{2n}\rangle\varphi_{2n}(x) = \left[\frac{1}{\pi}\int_{-\pi}^{\pi} g(t)\cos nt\, dt\right]\cos nx,$$

$$\langle g, \varphi_{2n+1}\rangle\varphi_{2n+1}(x) = \left[\frac{1}{\pi}\int_{-\pi}^{\pi} g(t)\sin nt\, dt\right]\sin nx.$$

Then

$$\sum_{k=1}^{2N+1} c_k\varphi_k(x) = \frac{a_0}{2} + \sum_{k=1}^{N}(a_k\cos kx + b_k\sin kx),$$

where a_k and b_k are the trigonometric Fourier coefficients as defined in Section 5. The square of the norm to be minimized takes the form

$$\int_{-\pi}^{\pi}\left[g(x) - \frac{a_0}{2} - \sum_{k=1}^{N}(a_k\cos kx + b_k\sin kx)\right]^2 dx,$$

and Theorem 9.1 says that the minimum will be attained for any fixed N by taking a_k and b_k to be the Fourier coefficients of g.

The minimization of an integral of the form

$$\int_a^b\left[g(x) - \sum_{k=1}^{N} c_k\varphi_k(x)\right]^2 dx$$

is called a best **mean-square approximation** to g. In this sense we can say that the Fourier approximation provides the best mean-square approximation by a trigonometric polynomial.

In Section 5 we observed that the functions $\cos kx$ and $\sin kx$ are solutions of the differential equation $y'' = -k^2y$. The result can be stated by saying that $\cos kx$ and $\sin kx$ are **eigenfunctions** of the differential operator $d^2/(dx^2)$, corresponding to the **eigenvalue** $-k^2$. More generally, let L be a linear transformation defined on some vector space and having its range in the same space. We say that a *nonzero* vector f is an **eigenvector** of L, corresponding to the **eigenvalue** λ if $Lf = \lambda f$ for some number λ.

To see the connection between eigenvectors and orthogonal sets of functions, we need one more definition. Let $\mathcal{F}$ be a vector space with an inner product $\langle f, g\rangle$. Let L be a linear transformation from $\mathcal{F}$ into $\mathcal{F}$. Then L is **symmetric** with respect to the inner product if $\langle Lf, g\rangle = \langle f, Lg\rangle$ for all vectors f and g for which Lf and Lg are defined as elements of $\mathcal{F}$. A linear transformation with the same domain space and range space is sometimes called a **linear operator.**

Example 2. Let $\langle f, g\rangle$ be the inner product defined by Equation (1) on $C[-\pi, \pi]$. If we let $Lf = d^2f/dx^2$, then it is clear that Lf is in $C[-\pi, \pi]$ only for those f in $C[-\pi, \pi]$ that happen to have continuous second derivatives. Thus, for L to be symmetric, we must have $\langle Lf, g\rangle = \langle f, Lg\rangle$ for all twice continuously differentiable f and g on $[-\pi, \pi]$. Equivalently, we must have, because of the definition of the inner product,

$$\int_{-\pi}^{\pi} f''(x)g(x)\,dx = \int_{-\pi}^{\pi} f(x)g''(x)\,dx.$$

Integration by parts twice shows that

$$\int_{-\pi}^{\pi} f''(x)g(x)\,dx = f'(\pi)g(\pi) - f'(-\pi)g(-\pi) - f(\pi)g'(\pi) + f(-\pi)g'(-\pi) + \int_{-\pi}^{\pi} f(x)g''(x)\,dx.$$

Hence to make L symmetric we restrict its domain to some subspace of $C[-\pi, \pi]$ for which the nonintegrated terms will always add up to zero. This can be done in several ways. For example, we may restrict L to the subspace consisting of those functions h in $C[-\pi, \pi]$ for which $h(\pi) = h(-\pi) = 0$, or to the subspace for which $h'(\pi) = h'(-\pi) = 0$. With either of these restrictions L becomes symmetric. Notice that a restriction of the required type is a boundary condition, in that it specifies the values of f at the endpoints π and $-\pi$ of its domain of definition.

The connection between orthogonal functions and symmetric operators is as follows.

9.3 Theorem

Let L be a symmetric linear operator defined on a vector space $\mathcal{F}$ with an inner product. If f_1 and f_2 are eigenvectors of L corresponding to distinct eigenvalues λ_1 and λ_2, then f_1 and f_2 are orthogonal.

Proof. We assume that

$$Lf_1 = \lambda_1 f_1, \qquad Lf_2 = \lambda_2 f_2,$$

and prove that $\langle f_1, f_2 \rangle = 0$. We have

$$\langle Lf_1, f_2 \rangle = \langle \lambda_1 f_1, f_2 \rangle = \lambda_1 \langle f_1, f_2 \rangle$$

and

$$\langle f_1, Lf_2 \rangle = \langle f_1, \lambda_2 f_2 \rangle = \lambda_2 \langle f_1, f_2 \rangle.$$

Because L is symmetric, $\langle Lf_1, f_2 \rangle = \langle f_1, Lf_2 \rangle$, so $\lambda_1 \langle f_1, f_2 \rangle = \lambda_2 \langle f_1, f_2 \rangle$ or $(\lambda_1 - \lambda_2)\langle f_1, f_2 \rangle = 0$. Since λ_1 and λ_2 are not equal, we must have $\langle f_1, f_2 \rangle = 0$.

We consider the **Sturm-Liouville** differential operator of the form

$$Lf = (pf') + qf, \tag{3}$$

where p is a continuously differentiable function and q is assumed only continuous. (The operator d^2/dx^2 is a special case if we set $p(x) \equiv 1$, $q(x) \equiv 0$.) We want to see what boundary conditions should be imposed on the domain of L in order to make L symmetric with respect to the inner product

$$\langle f, g \rangle = \int_a^b f(x)g(x)\, dx.$$

The following formula simplifies the problem.

9.4 Lagrange Formula

If L is given by $Lf = (pf')' + qf$ on an interval $[a, b]$, then

$$\langle f_1, Lf_2 \rangle - \langle Lf_1, f_2 \rangle = [p(f_1 f_2' - f_1' f_2)]_a^b.$$

Proof. Starting with the definition of L, we rearrange $f_1(Lf_2) - (Lf_1)f_2$ as follows:

$$\begin{aligned} f_1(Lf_2) - (Lf_1)f_2 &= f_1[(pf_2')' + qf_2] - f_2[(pf_1')' + qf_1] \\ &= f_1(pf_2')' - f_2(pf_1')' \\ &= f_1[pf_2'' + p'f_2'] - f_2[pf_1'' + p'f_1'] \\ &= p'[f_1 f_2' - f_1' f_2] + p[f_1 f_2'' - f_1'' f_2] \\ &= [p(f_1 f_2' - f_1' f_2)]'. \end{aligned}$$

Integration of both sides from a to b gives the Lagrange formula.

Formula 9.4 shows that any condition on the coefficient p, or on the space containing f_1 and f_2, which makes $[p(f_1f_2' - f_1'f_2)]_a^b = 0$, will also make L symmetric.

Example 3. The operator L defined by $Lf(x) = (1 - x^2)f''(x) - 2xf'(x)$, has the form shown in Equation (3) if we set $p(x) = (1 - x^2)$ and $q(x) \equiv 0$. We shall consider L to be operating on twice continuously differentiable functions defined on $[-1, 1]$. To make L symmetric, we need to ensure that the right side of the Lagrange formula is always zero for $a = -1, b = 1$. But $p(x) = (1 - x^2)$, so $p(-1) = p(1) = 0$. Hence, L is symmetric on the space $C[-1, 1]$ without further restriction, and its domain consists of all twice continuously differentiable functions in $C[-1, 1]$.

The symmetric operator Lf defined in the previous example is usually associated with the differential equation

$$(1 - x^2)y'' - 2xy' + n(n + 1)y = 0. \tag{4}$$

This is called the **Legendre equation** of index n, and it is satisfied by the nth **Legendre polynomial** defined by

9.5 $$P_n(x) = \frac{1}{2^n n!}\frac{d^n}{dx^n}(x^2 - 1)^n, \qquad n = 0, 1, 2, \ldots.$$

That P_n satisfies Equation (4) can be verified by repeated differentiation. See Problem 8. The significance of the fact that P_n satisfies the Legendre equation comes from writing the equation in the form $Ly = -n(n + 1)y$, where L is the symmetric operator $Ly = (1 - x^2)y'' - 2xy'$ on $C[-1, 1]$. Then P_n can be looked at as an eigenfunction of L, corresponding to the eigenvalue $-n(n + 1)$. Hence, by Theorem 9.3, the Legendre polynomials are orthogonal, that is,

$$\int_{-1}^{1} P_n(x)P_m(x)\,dx = 0, \qquad n \neq m.$$

Furthermore, a fairly complicated calculation (see Problem 10) shows that

$$\int_{-1}^{1} P_n^2(x)\,dx = \frac{2}{2n + 1}, \qquad n = 0, 1, 2, \ldots.$$

Therefore, the normalized sequence $\{(\sqrt{2n + 1/2})P^n(x)\}$ $n = 0, 1, 2, \ldots$ is an orthonormal sequence in $C[-1, 1]$.

The Nth **Fourier-Legendre approximation** to a function g in $C[-1, 1]$

is the finite sum

$$\sum_{k=0}^{N} c_k P_k(x),$$

where

$$c_k = \frac{2k+1}{2}\int_{-1}^{1} g(x)P_k(x)\,dx. \tag{5}$$

Theorem 9.1 then implies that the best mean-square approximation to g by a linear combination of Legendre polynomials is given by the Fourier-Legendre approximation. In other words,

$$\int_{-1}^{1}\left[g(x) - \sum_{k=0}^{N} c_k P_k(x)\right]^2 dx$$

is minimized by computing c_k by Equation (5).

EXERCISES

1. (a) Verify that

$$\langle f, g\rangle = \int_a^b f(x)g(x)\,dx$$

defines an inner product on the space $C[a, b]$ of continuous functions defined on $[a, b]$.

(b) What condition is required of a continuous function w defined on $[a, b]$ in order that

$$\langle f, g\rangle = \int_a^b f(x)g(x)w(x)\,dx$$

define an inner product?

2. Let $\{\varphi_k\}$ be an orthonormal sequence of functions in $C[a, b]$, and let g be in $C[a, b]$. Show that if the real-valued function

$$\Delta(c_1, \ldots, c_N) = \int_a^b \left[g(x) - \sum_{k=1}^{N} c_k\varphi_k(x)\right] dx$$

has a local minimum as a function of $(c_1, \ldots, c_N)$, then

$$c_k = \int_a^b g(x)\varphi_k(x)\,dx.$$

[*Hint*. Differentiate the formula for Δ under the integral sign. This is justified under the assumptions made here. (See Problem 7 of Chapter 6, Section 2.)]

3. Let $\{\varphi_k\}$ be an orthonormal sequence of functions in $C[a, b]$ and let g be in $C[a, b]$. Use Bessel's inequality to show the following:

(a) $\displaystyle\sum_{k=1}^{\infty}\left[\int_a^b g(x)\varphi_k(x)\,dx\right]^2$ converges.

(b) If c_k is the kth Fourier coefficient of g with respect to $\{\varphi_k\}$, then $\lim_{k\to\infty} c_k = 0$.

(c) Find an example of a function whose trigonometric Fourier series has coefficients a_k and b_k such that $\sum_{k=1}^{\infty} (a_k^2 + b_k^2)$ converges, but such that $\sum_{k=1}^{\infty} a_k$ or $\sum_{k=2}^{\infty} b_k$ does not converge.

4. Let L be a linear transformation from a vector space $\mathcal{F}_1$ to an otherwise unrelated vector space $\mathcal{F}_2$. Can L have eigenvectors?

5. Find all eigenfunctions and corresponding eigenvalues of the differential operator d^2/dx^2 satisfying each of the following sets of boundary conditions. That is, solve $y'' = \lambda y$, subject to the boundary conditions

(a) $y(0) = y(\pi) = 0$.
(b) $y(0) = y(\pi) + y'(\pi) = 0$.
(c) $y(0) = y'(\pi) = 0$.

6. Let $C[a, b]$ be the continuous real-valued functions defined on $[a, b]$, $C'[a, b]$ the continuously differentiable functions on that interval, and $C''[a, b]$ the twice continuously differentiable functions there.

(a) Show that $C[a, b]$, $C'[a, b]$, and $C''[a, b]$ are vector spaces, each contained in the preceding one.

(b) Let $B_{x_0}[a, b]$ be the set of functions f contained in $C[a, b]$ and satisfying a condition of the form

$$cf(x_0) + df'(x_0) = 0,$$

where c and d are constants, not both zero. Show that $B_{x_0}[a, b]$ is a vector subspace of $C[a, b]$.

(c) Show that $C''[a, b] \cap B_a[a, b] \cap B_b[a, b]$ is a vector subspace of $C[a, b]$.

7. Show that the differential operator d^2/dx^2 is symmetric with respect to the inner product

$$\langle f, g\rangle = \int_{-1}^{1} f(x)g(x)\, dx,$$

and with each of the following sets of boundary conditions.

(a) $f(-1) = f(1) = 0$.
(b) $f'(1) = f'(-1) = 0$.
(c) $c_1 f(-1) + d_1 f'(-1) = c_2 f(1) + d_2 f'(1) = 0$, where $c_i^2 + d_i^2 > 0$.

8. Verify that the Legendre polynomial P_n defined by Formula 9.5 satisfies the Legendre equation: $(1 - x^2)y'' - 2xy' + n(n + 1)y = 0$. [*Hint*. Let $u = (x^2 - 1)^n$. Then $(x^2 - 1)u' = 2nxu$. Differentiate both sides $(n + 1)$ times with respect to x.]

9. Compute the Legendre polynomials P_0, P_1, and P_2.

10. (a) By using Formula 9.5 and repeated integration by parts, show that

$$\int_{-1}^{1} P_n^2(x)\,dx = \frac{(2n)!}{2^{2n}(n!)^2}\int_{-1}^{1}(1 - x^2)^n\,dx.$$

(b) Show that

$$\int_{-1}^{1}(1 - x^2)^n\,dx = 2\int_{0}^{\pi/2}\sin^{2n+1}\theta\,d\theta.$$

(c) Show that

$$\int_{0}^{\pi/2}\sin^{2n+1}\theta\,d\theta = \frac{2\cdot 4\cdot 6\cdot \ldots \cdot (2n)}{1\cdot 3\cdot 5\cdot \ldots \cdot (2n+1)}.$$

(d) Show that

$$\int_{-1}^{1} P_n^2(x)\,dx = \frac{2}{2n+1}.$$

11. Prove that P_n, the nth Legendre polynomial, has n distinct roots in the interval $[-1, 1]$. [*Hint.* Use Formula 9.5 and Rolle's theorem.]

12. The 3-dimensional Laplace equation in spherical coordinates (r, φ, θ) has the form

$$\frac{\partial}{\partial r}\left(r^2\frac{\partial u}{\partial r}\right) + \frac{1}{\sin\varphi}\frac{\partial}{d\varphi}\left(\sin\varphi\frac{\partial u}{\partial\varphi}\right) + \frac{1}{\sin^2\varphi}\frac{\partial^2 u}{\partial\theta^2} = 0.$$

(a) Show that, for solutions $u(r, \varphi, \theta) = v(r, \varphi)$ that are independent of θ, the equation has the form

$$r\frac{\partial^2}{\partial r^2}(rv) + \frac{1}{\sin\varphi}\frac{\partial}{\partial\varphi}\left(\sin\varphi\frac{\partial v}{\partial\varphi}\right) = 0.$$

(b) Show that the method of separation of variables applied to the equation of part (a) leads to the two ordinary differential equations

$$r^2G'' + 2rG' = \lambda G,$$

$$\frac{1}{\sin\varphi}\frac{d}{d\varphi}\left(\sin\varphi\frac{dH}{d\varphi}\right) = -\lambda H.$$

(c) Show that the equation for $G(r)$ has sotutions

$$r^{-(1/2)+\sqrt{\lambda+(1/4)}} \quad \text{and} \quad r^{-(1/2)-\sqrt{\lambda+(1/4)}}. \qquad [\textit{Hint.}\ \text{Let } r = e^t.]$$

(d) Show that the equation for H can be put in the form of the Legendre equation

$$(1 - x^2)H'' - 2xH' + \lambda H = 0.$$

[*Hint.* Let $x = \cos\varphi$.]

(e) By setting $\lambda = n(n + 1)$ find a sequence of solutions to the partial differential equation of part (a).

13. Show that in the case of the orthonormal sequence derived from the Legendre polynomials, the general expansion $\sum_{k=0}^{N}\langle g, \varphi_k\rangle\varphi_k$ reduces to $\sum_{k=0}^{N} c_k P_n(x)$, where c_k is given by Equation 5 of the text.

14. Show by using Theorem 8.3 of Chapter 2 that the normalized Legendre polynomials

$$\sqrt{\frac{2n+1}{2}}\,P_n(x)$$

are the same as the sequence of polynomials obtained by applying the Gram-Schmidt process to the functions $\{1, x, x^2, \ldots\}$. Use the inner product

$$\langle f, g\rangle = \int_{-1}^{1} f(x)g(x)\,dx$$

defined on $C[0, 1]$.

6

Multiple Integration

SECTION 1

ITERATED INTEGRALS

This chapter is devoted to the study of multiple integrals of functions with domains in $\mathcal{R}^n$. Such integrals occur in many branches of pure and applied mathematics, with interpretations such as volume, mass, probability, and flux. In this section we start with iterated integrals because they are computationally useful and because they provide a natural transition to the multiple integral from the ordinary definite integral,

$$\int_a^b f(x)\,dx,$$

of a real-valued function of one real variable. We begin with $n = 2$, that is, with the iterated integral of functions $\mathcal{R}^2 \xrightarrow{f} \mathcal{R}$.

Suppose $f(x, y)$ is a function defined on a rectangle $a \leq x \leq b$, $c \leq y \leq d$. By

$$\int_c^d f(x, y)\,dy$$

is meant simply the definite integral of the function of one variable obtained by holding x fixed; for example

$$\int_0^2 x^3y^2\,dy = \left[\frac{x^3y^3}{3}\right]_{y=0}^{y=2} = \frac{8}{3}x^3.$$

As this example shows, if the integral exists, it depends on x. Thus, we may set

$$F(x) = \int_c^d f(x, y)\,dy$$

and form the **iterated integral**

$$\int_a^b F(x)\,dx = \int_a^b \left[\int_c^d f(x, y)\,dy\right] dx.$$

A common notational convention, which we shall adopt, is to omit the brackets and write the iterated integral as

$$\int_a^b dx \int_c^d f(x, y)\,dy.$$

This notation has the advantage of emphasizing which variable goes with which integral sign, namely, x with $\int_a^b$ and y with $\int_c^d$.

Example 1. Consider $f(x, y) = x^2 + y$, defined on the rectangular region $0 \leq x \leq 1$, $1 \leq y \leq 2$.

$$\int_0^1 dx \int_1^2 (x^2 + y)\,dy = \int_0^1 \left[x^2 y + \frac{y^2}{2}\right]_{y=1}^{y=2} dx$$

$$= \int_0^1 [(2x^2 + 2) - (x^2 + \tfrac{1}{2})]\,dx$$

$$= \int_0^1 (x^2 + \tfrac{3}{2})\,dx = \tfrac{1}{3} + \tfrac{3}{2} = \tfrac{11}{6}.$$

To interpret this example geometrically, look at the surface defined by $z = x^2 + y$ shown in Fig. 1. For each x in the interval between 0 and 1,

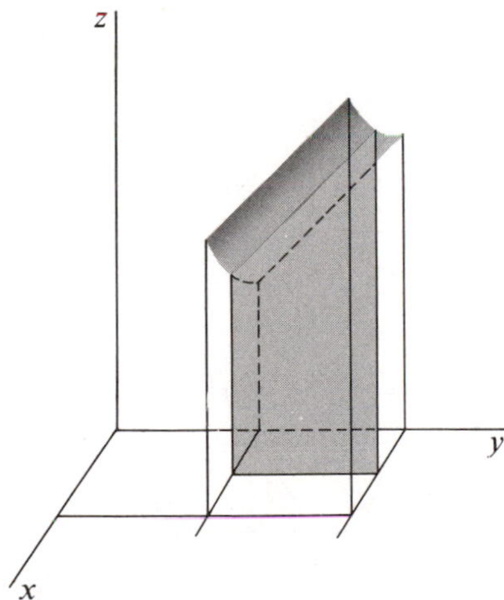

Figure 1

the integral

$$\int_1^2 (x^2 + y)\,dy = x^2 + \tfrac{3}{2}$$

is the area of the shaded cross section. It is customary to interpret the definite integral of an area-valued function as volume. Thus we can regard the iterated integral

$$\int_0^1 dx \int_1^2 (x^2 + y)\,dy = \tfrac{11}{6}$$

as the volume of the 3-dimensional region lying below the surface and above the rectangle $0 \le x \le 1$, $1 \le y \le 2$.

Example 2. We can perform the integration in Example 1 in the opposite order.

$$\begin{aligned}\int_1^2 dy \int_0^1 (x^2 + y)\,dx &= \int_1^2 \left[\frac{x^3}{3} + yx\right]_{x=0}^{x=1} dy \\ &= \int_1^2 (\tfrac{1}{3} + y)\,dy = \left[\frac{y}{3} + \frac{y^2}{2}\right]_1^2 \\ &= (\tfrac{2}{3} + 2) - (\tfrac{1}{3} + \tfrac{1}{2}) = \tfrac{11}{6}.\end{aligned}$$

This time

$$\int_0^1 (x^2 + y)\,dx = \tfrac{1}{3} + y$$

is the area of a cross section parallel to the xz-plane. See Fig. 2. The second integral again gives the volume of the 3-dimensional region lying below the surface $z = x^2 + y$ and above the rectangle $0 \le x \le 1$,

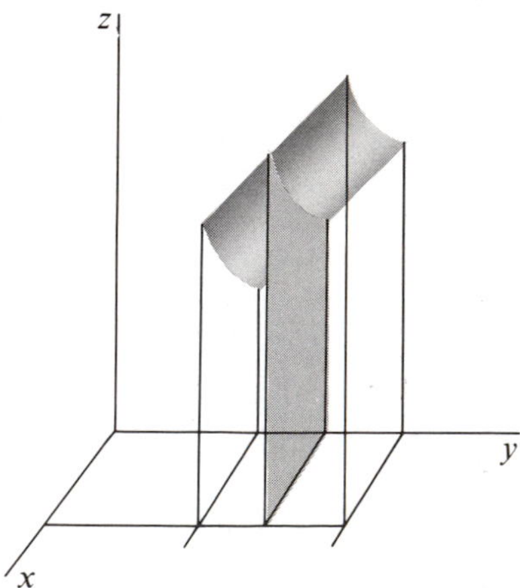

Figure 2

$1 \leq y \leq 2$. It is not surprising, therefore, that the two iterated integrals of Examples 1 and 2 are equal.

It is important to be able to integrate over subsets of the plane that are more general than rectangles. In such problems the limits in the first integration will depend on the remaining variable.

Example 3. Consider the iterated integral

$$\int_0^1 dx \int_0^{1-x^2} (x + y)\, dy = \int_0^1 \left[xy + \frac{y^2}{2}\right]_0^{1-x^2} dx$$
$$= \int_0^1 \left[x(1 - x^2) + \frac{(1 - x^2)^2}{2}\right] dx$$
$$= \int_0^1 \left[x - x^3 + \frac{1 - 2x^2 + x^4}{2}\right] dx = \tfrac{31}{60}.$$

For each x between 0 and 1, the number y lies between the values $y = 0$ and $y = 1 - x^2$. In other words, the point (x, y) runs along the line segment joining $(x, 0)$ and $(x, 1 - x^2)$. As x varies between 0 and 1, this line segment sweeps out the shaded region B as shown in Fig. 3.

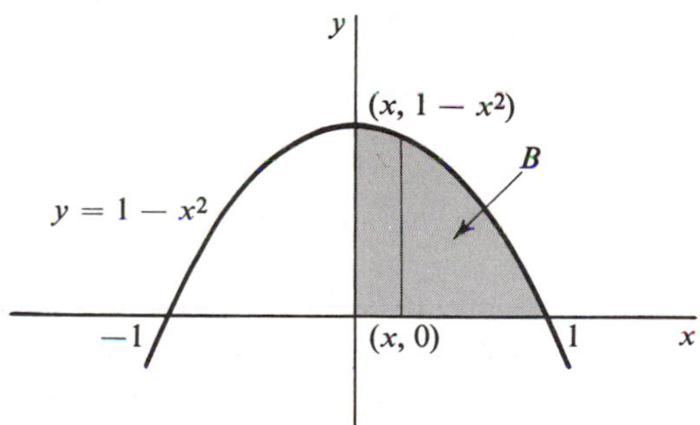

Figure 3

The integrand $f(x, y) = x + y$ has a graph (see Fig. 4), and the iterated integral is the volume under the graph and above the region B.

Suppose we are given an iterated integral over a plane region B in which the integrand is the constant function f defined by $f(x, y) = 1$, for all (x, y) in B. The integral may then be interpreted either as the volume of the slab of unit thickness and with base B or simply as the area of B. For example,

$$\int_0^1 dx \int_0^{1-x^2} dy = \tfrac{2}{3}$$

is the area of the region B shown in Fig. 3.

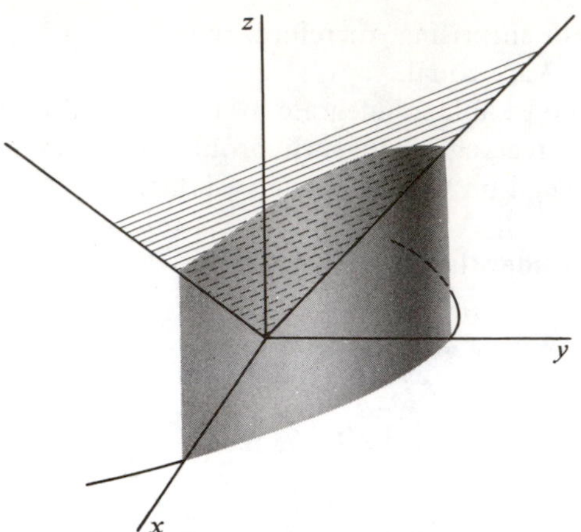

Figure 4

Example 4. Let f be defined by $f(x, y) = x^2y + xy^2$ over the region bounded by $y = |x|$, $y = 0$, $x = -1$, and $x = 1$. See Fig. 5. The two iterated integrals over the region are

$$\int_{-1}^{1} dx \int_{0}^{|x|} (x^2y + xy^2)\, dy$$

and

$$\int_{0}^{1} dy \left[\int_{-1}^{-y} (x^2y + xy^2)\, dx + \int_{y}^{1} (x^2y + xy^2)\, dx \right].$$

The second integral breaks into two pieces because, for fixed y between 0 and 1, the integration with respect to x is carried out over two separate intervals. Computation of the integral is straightforward. We get

$$\int_{0}^{1} \left[\frac{x^3y}{3} + \frac{x^2y^2}{2}\right]_{-1}^{-y} + \left[\frac{x^3y}{3} + \frac{x^2y^2}{2}\right]_{y}^{1} dy = \int_{0}^{1} \tfrac{2}{3}(y - y^4)\, dy = \tfrac{1}{3} - \tfrac{2}{15} = \tfrac{1}{5}.$$

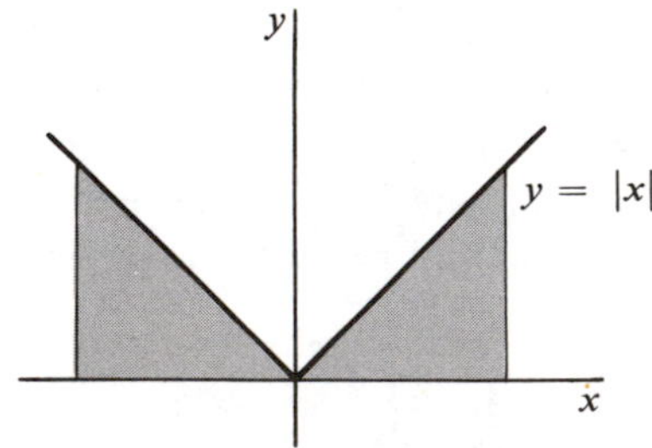

Figure 5

The iterated integral in the other order is

$$\int_{-1}^{1}\left[\frac{x^2y^2}{2}+\frac{xy^3}{3}\right]_0^{|x|}dx=\int_{-1}^{1}\left(\frac{x^4}{2}+\frac{x\,|x|^3}{3}\right)dx$$
$$=\int_{-1}^{1}\frac{x^4}{2}\,dx+\int_{-1}^{1}\frac{x\,|x|^3}{3}\,dx.$$

The functions $x^4/2$ and $x\,|x|^3/3$ are even and odd, respectively. It follows that

$$\int_{-1}^{1}\frac{x^4}{2}\,dx+\int_{-1}^{1}\frac{x\,|x|^3}{3}\,dx=\int_0^1 x^4\,dx=\tfrac{1}{5}.$$

The theorem which states that, under quite general hypotheses, the value of an iterated integral is independent of the order of integration will be proved in the next section. This will prove that different orders of integration in computing volume must lead to the same result.

Iterated integrals for functions defined on subsets of dimension greater than 2 can also be computed by repeated 1-dimensional integration.

Example 5.

$$\int_0^1 dx\int_{x^2}^{x} dy\int_x^{2x+y}(x+y+2z)\,dz=\int_0^1 dx\int_{x^2}^{x}(4x^2+6xy+2y^2)\,dy$$
$$=\int_0^1\left(\tfrac{23}{3}x^3-4x^4-3x^5-\frac{2x^6}{3}\right)dx$$
$$=\tfrac{23}{12}-\tfrac{4}{5}-\tfrac{1}{2}-\tfrac{2}{21}.$$

It is not possible to give a complete interpretation of this integral by drawing a picture. However, the region of integration B can be drawn and is shown in Fig. 6. It is bounded on the top by the surface $z = 2x + y$ and on the bottom by $z = x$. On the sides it is bounded by the surfaces obtained by projecting the curves $y = x^2$ and $y = x$ parallel to the z-axis. With the same limits of integration, the integral

$$\int_0^1 dx\int_{x^2}^{x} dy\int_x^{2x+y} dz$$

is the volume of B. For fixed x and y the first integral,

$$\int_x^{2x+y} dz,$$

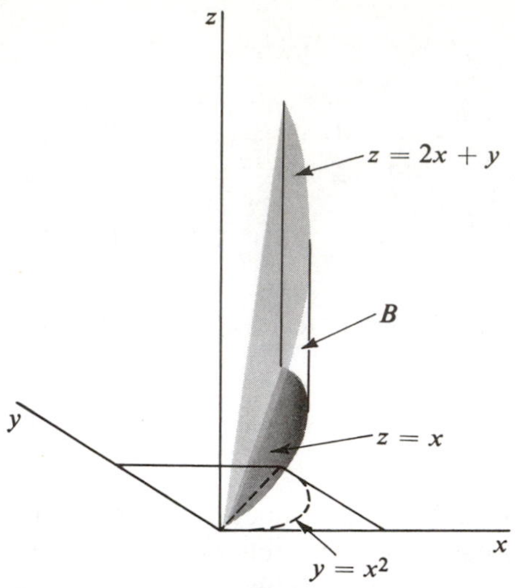

Figure 6

is the length of the vertical segment joining the point (x, y, x) to the point $(x, y, 2x + y)$. For fixed x, the integral

$$\int_{x^2}^{x} dy \int_{x}^{2x+y} dz$$

is the area of a cross section parallel to the yz-plane. Finally, the triply iterated integral is the volume.

Example 6. The n-fold iterated integral

$$\int_0^1 dx_1 \int_0^{x_1} dx_2 \ldots \int_0^{x_{n-1}} dx_n$$

can be thought of as the volume of the region in n-dimensional Euclidean space defined by the inequalities

$$0 \leq x_n \leq x_{n-1} \leq \ldots \leq x_2 \leq x_1 \leq 1.$$

To get some idea of what this region is like, consider the cases $n = 1$, $n = 2$, and $n = 3$. For $n = 1$, the integral

$$\int_0^1 dx_1 = 1$$

is simply the length of the unit interval $0 \leq x_1 \leq 1$. If $n = 2$, we have $0 \leq x_2 \leq x_1 \leq 1$. The region of integration is the intersection of the regions $0 \leq x_2$, $x_2 \leq x_1$, and $x_1 \leq 1$ shown in Fig. 7. For $n = 3$, we have simultaneously $0 \leq x_3$, $x_3 \leq x_2$, $x_2 \leq x_1$, and $x_1 \leq 1$. See Fig. 8. If we

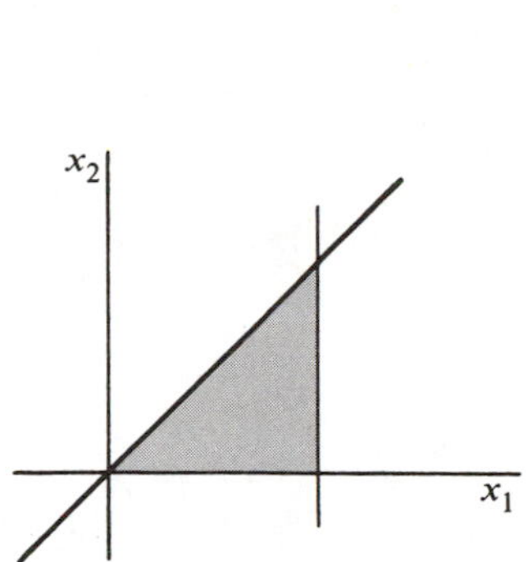

Figure 7

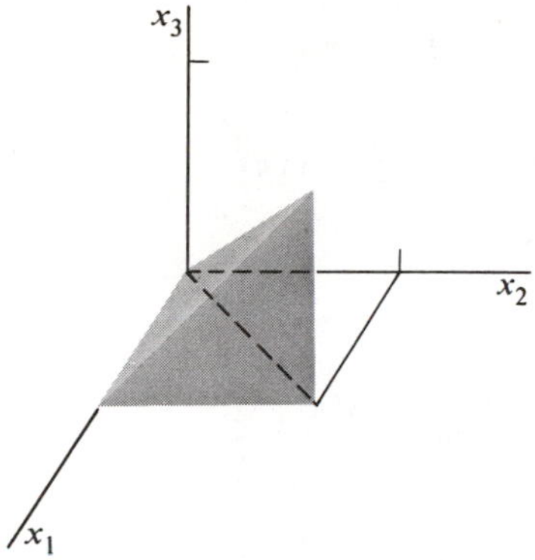

Figure 8

denote the n-fold integral by I_n, then $I_1 = 1$, $I_2 = \frac{1}{2}$, and $I_3 = \frac{1}{6}$. These numbers can be obtained either by direct computation or by observing that they are the length, area, and volume, respectively, of the regions of integration. Direct evaluation of I_n is straightforward:

$$\begin{aligned}
I_n &= \int_0^1 dx_1 \int_0^{x_1} dx_2 \ldots \int_0^{x_{n-1}} dx_n \\
&= \int_0^1 dx_1 \int_0^{x_1} dx_2 \ldots \int_0^{x_{n-2}} x_{n-1}\, dx_{n-1} \\
&= \int_0^1 dx_1 \int_0^{x_1} dx_2 \ldots \int_0^{x_{n-3}} \frac{x_{n-2}^2}{2}\, dx_{n-2} \\
&= \int_0^1 dx_1 \int_0^{x_1} dx_2 \ldots \int_0^{x_{n-4}} \frac{x_{n-3}^3}{3!}\, dx_{n-3} \\
&= \ldots = \int_0^1 \frac{x_1^{n-1}}{(n-1)!}\, dx_1 = \frac{1}{n!}
\end{aligned}$$

EXERCISES

Evaluate the following iterated integrals and sketch the region of integration for each.

1. $\displaystyle\int_{-1}^0 dx \int_1^2 (x^2y^2 + xy^3)\, dy.$

2. $\displaystyle\int_0^2 dy \int_1^3 |x - 2| \sin y\, dx.$ [*Ans.* 1 − cos 2.]

3. $\displaystyle\int_1^0 dx \int_2^0 (x + y^2)\, dy.$

4. $\int_0^{\pi/2} dy \int_{-y}^{y} \sin x\, dx.$ [*Ans.* 0.]

5. $\int_{-2}^{1} dy \int_0^{y^2} (x^2 + y)\, dx.$

6. $\int_{-1}^{1} dx \int_0^{|x|} dy.$ [*Ans.* 1.]

7. $\int_0^1 dx \int_0^{\sqrt{1-x^2}} dy.$

8. $\int_1^{-1} dx \int_x^{2x} e^{x+y} dy.$

9. $\int_0^{\pi/2} dy \int_0^{\cos y} x \sin y\, dx.$

10. $\int_1^2 dx \int_{x^2}^{x^3} x\, dy.$ [*Ans.* $\frac{49}{20}$.]

11. $\int_0^1 dz \int_0^z dy \int_0^y dx.$

12. $\int_0^2 dx \int_1^x dy \int_2^{x+y-1} y\, dz.$ [*Ans.* $\frac{2}{3}$.]

13. $\int_1^2 dy \int_0^1 dx \int_x^y dz.$

14. $\int_{-1}^{1} dx \int_0^{|x|} dy \int_0^1 (x + y + z)\, dz.$ [*Ans.* $\frac{5}{6}$.]

15. $\int_0^{\pi} \sin x\, dx \int_0^1 dy \int_0^2 (x + y + z)\, dz.$

16. Evaluate the integral $\int_0^1 dx \int_{-x}^{x} dy \int_{-x-y}^{x+y} dz \int_{-z}^{x} dw.$

17. Sketch the subset B defined by $0 \leq x \leq 1$, $0 \leq y \leq x$, and write down the integral over B in each of the two possible orders of $f(x, y) = x \sin y$. Evaluate both integrals.

18. Sketch the region defined by $x \geq 0$, $x^2 + y^2 \leq 2$, and $x^2 + y^2 \geq 1$. Write down the integral over the region in each of the two possible orders of $f(x, y) = x^2$. Evaluate both integrals. [*Ans.* $3\pi/8$.]

19. Consider two real-valued functions $c(x)$ and $d(x)$ of a real variable x. Suppose that, for all x in the interval $a \leq x \leq b$, we have $c(x) \leq d(x)$.

(a) Make a sketch of two such functions and of the subset B of the xy-plane consisting of all (x, y) such that $a \leq x \leq b$ and $c(x) \leq y \leq d(x)$.

(b) Express the area of B as an iterated integral.

(c) Set up the iterated integral of $f(x, y)$ over B.

20. Sketch the subset B of $\mathcal{R}^3$, defined by $0 \leq x \leq 1$, $0 \leq y \leq 1 + x$, and $0 \leq z \leq 2$. Write down the iterated integral with order of integration z, then y, and then x, of the function $f(x, y, z) = x^2 + z$ over the subset B. Compute the integral. [*Ans.* $\frac{25}{6}$.]

21. Sketch the region defined by $0 \leq x \leq 1$, $x^2 \leq y \leq \sqrt{x}$, and $0 \leq z \leq x + y$, and evaluate the iterated integral, in some order, of $f(x, y, z) = x + y + z$ over the region.

22. Let f be defined by $f(x, y, z) = 1$ on the hemisphere bounded by the plane $z = 0$ and the surface $z = \sqrt{1 - x^2 - y^2}$. Evaluate an iterated integral of f in some order over the region. [*Ans.* $2\pi/3$.]

23. Let f be defined by $f(x_1, \ldots, x_n) = x_1x_2 \ldots x_n$ on the cube $0 \leq x_1 \leq 1$, $0 \leq x_2 \leq 1, \ldots, 0 \leq x_n \leq 1$. Evaluate

$$\int_0^1 dx_1 \int_0^1 dx_2 \ldots \int_0^1 x_1x_2 \ldots x_n \, dx_n.$$

24. Show that if in the integral

$$\int_{a_1}^{b_1} dx_1 \int_{a_2}^{b_2} dx_2 \ldots \int_{a_n}^{b_n} f(x_1, \ldots, x_n) \, dx_n$$

the order of the limits of integration is interchanged on an even number of integral signs, then the value of the integral is unchanged. If the limits are interchanged on an odd number of integral signs, then the whole iterated integral changes sign.

25. Evaluate

$$\int_0^1 dx_1 \int_0^1 dx_2 \ldots \int_0^1 dx_{n-1} \int_0^{x_1} (x_1 + x_2) \, dx_n.$$

26. Prove that

$$\int_0^x dx_1 \int_0^{x_1} dx_2 \ldots \int_0^{x_{n-1}} f(x_n) \, dx_n = \frac{1}{(n-1)!} \int_0^x (x - t)^{n-1} f(t) \, dt.$$

SECTION 2

MULTIPLE INTEGRALS

Multiple integrals are closely related to the iterated integrals of the preceding section. Suppose we are given a real-valued function defined on a set B in $\mathcal{R}^n$. Our problem is to formulate a definition of the integral of f over B analogous to the definition of a one-variable integral, using sums rather than iterated integrals.

We first consider some simple sets in $\mathcal{R}^n$. **A closed coordinate rectangle** is a subset of $\mathcal{R}^n$ consisting of all points $\mathbf{x} = (x_1, \ldots, x_n)$ that satisfy a set of inequalities

$$a_i \leq x_i \leq b_i, \qquad i = 1, \ldots, n. \tag{1}$$

If in Formula (1) some of the symbols "$\leq$" are replaced by "$<$," the resulting set is still called a **coordinate rectangle.** In particular, if all the inequalities are of the form $a_i < x_i < b_i$, the set is open and is called an

open coordinate rectangle. A coordinate rectangle has its edges parallel to the coordinate axes. Throughout this section the word "rectangle" will be understood to mean "coordinate rectangle." Rectangles in $\mathcal{R}^2$ and $\mathcal{R}^3$ are illustrated in Fig. 9. A rectangle in $\mathcal{R}$ is just an interval.

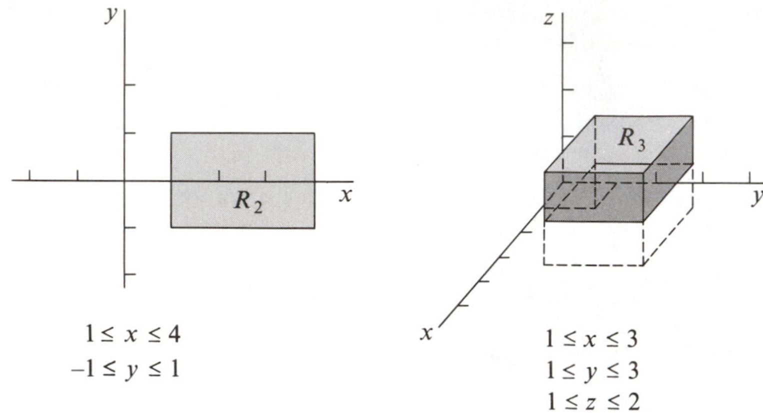

Figure 9

Let R be a rectangle (open, closed, or neither) defined by Formula (1) with replacement of any symbols "$\le$" by "$<$" permitted. The **volume** or **content** of R, written $V(R)$, is defined by

$$V(R) = (b_1 - a_1)(b_2 - a_2) \ldots (b_n - a_n). \tag{2}$$

In the examples shown in Fig. 9, $V(R_2) = (4 - 1)(1 - (-1)) = 6$ and $V(R_3) = (3 - 1)(3 - 1)(2 - 1) = 4$. If, for some i in Formula (1), $a_i = b_i$, then R is called **degenerate** and $V(R) = 0$. For rectangles in $\mathcal{R}^2$, content is the same thing as area, and we often write $A(R)$ instead of $V(R)$ to have the notation remind us of area rather than volume.

A subset B of $\mathcal{R}^n$ is called **bounded** if there is a real number k such that $|\mathbf{x}| < k$ for all $\mathbf{x}$ in B. A finite set of $(n - 1)$-dimensional planes in $\mathcal{R}^n$ (lines in $\mathcal{R}^2$) parallel to the coordinate planes will be called a **grid.** As illustrated in Fig. 10, a grid separates $\mathcal{R}^n$ into a finite number of closed, bounded rectangles $R_1, \ldots, R_r$ and a finite number of unbounded regions. A grid **covers** a subset B of $\mathcal{R}^n$ if B is contained in the union of the bounded rectangles $R_1, \ldots, R_r$. Obviously, a set can be covered by a grid if and only if the set is bounded. As a measure of the fineness of a grid, we take the maximum of the lengths of the edges of the rectangles $R_1, \ldots, R_r$. This number is called the **mesh** of the grid.

We now give the definition of multiple integral, also called the Riemann integral after Bernhard Riemann (1826–1866). Consider a function

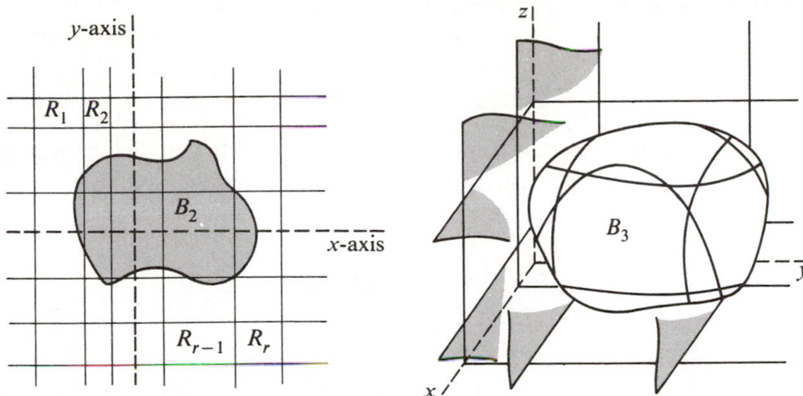

Figure 10

$\mathcal{R}^n \xrightarrow{f} \mathcal{R}$ and a set B such that

(a) B is a bounded subset of the domain of f.

(b) f is bounded on B.

Assertion (b) means that there exists a real number K such that $|f(\mathbf{x})| \leq K$, for all $\mathbf{x}$ in B. The multiple integral of f over B will be defined in terms of the function f_B, which is f altered to be zero outside B, that is

$$f_B(\mathbf{x}) = \begin{cases} f(\mathbf{x}), & \text{if } \mathbf{x} \text{ is in } B. \\ 0, & \text{if } \mathbf{x} \text{ is not in } B. \end{cases}$$

Let G be a grid that covers B and has mesh equal to $m(G)$. In each of the bounded rectangles R_i formed by G, $i = 1, \ldots, r$, choose an arbitrary point $\mathbf{x}_i$. The sum

$$\sum_{i=1}^{r} f_B(\mathbf{x}_i)V(R_i)$$

is called a **Riemann sum** for f over B. Its value, for given f and B, depends on G and $\mathbf{x}_1, \ldots, \mathbf{x}_r$. If, no matter how we choose grids G with mesh $m(G)$ tending to zero, it happens that

$$\lim_{m(G)\to 0} \sum_{i=1}^{r} f_B(\mathbf{x}_i)V(R_i)$$

exists and is always the same number, then this limit is the **integral of f over B** and is denoted by $\int_B f\,dV$. If the integral exists, f is said to be **integrable** over B.

The limit that defines the multiple integral is somewhat different from the limit of a vector function defined in Chapter 2, Section 2, although the idea behind it is similar. The defining equation

$$\lim_{m(G)\to 0} \sum_{i=1}^{r} f_B(\mathbf{x}_i)V(R_i) = \int_B f\,dV$$

means that, for any $\epsilon > 0$, there exists $\delta > 0$ such that if G is any grid that covers B and has mesh less than δ, and S is an arbitrary Riemann sum for f_B formed from G, then

$$\left| S - \int_B f\,dV \right| < \epsilon.$$

It should be emphasized that the integral is not defined for functions $\mathcal{R}^n \xrightarrow{f} \mathcal{R}$ and sets B unless the boundedness conditions on f and on B are satisfied. Without these conditions, even the Riemann sums may not be defined.

If f is a real-valued function of one real variable, that is, if $n = 1$, and if B is an interval $a \le x \le b$, the Riemann integral of f over B is the familiar definite integral

$$\int_a^b f(x)\,dx.$$

Other common notations for the integral of $\mathcal{R}^n \xrightarrow{f} \mathcal{R}$ over B are

$$\int_B f\,dA \quad \text{and} \quad \int_B f(x, y)\,dx\,dy, \qquad \text{if} \quad n = 2,$$

$$\int_B f(x, y, z)\,dx\,dy\,dz, \qquad \text{if} \quad n = 3,$$

$$\int_B f\,dx_1 \ldots dx_n, \qquad \text{for arbitrary } n.$$

In most applications the multiple integral can be replaced by iterated integrals as we shall see later. However, for theoretical purposes it is important to have some well-understood conditions under which multiple integrals exist. The following theorem, proved in the Appendix, gives such conditions.

2.1 Theorem

Let f be defined and bounded on a bounded set B in $\mathcal{R}^n$, and let the boundary of B be contained in finitely many smooth sets. If f is continuous on B except perhaps on finitely many smooth sets, then f is integrable over B. The value of $\int_B f\,dV$ is unchanged by changing values of f on any smooth set.

By a **smooth set** in $\mathcal{R}^n$ is meant the image of a closed bounded set under a continuously differentiable function $\mathcal{R}^m \xrightarrow{\phi} \mathcal{R}^n$, $m < n$. Thus, if $n = 2$ and $m = 1$, we may get a smooth curve. A smooth set in $\mathcal{R}^1$ will be understood to be just a point. To say that the value of $\int_B f\,dV$ is unchanged by changing the values of f on such a set means that f can be assigned arbitrary values on the set without affecting the existence or the value of the integral. For instance we can change the integrand on any finite set of points without changing the integral. This kind of modification is often convenient for removing discontinuities.

Example 1. Evaluate the multiple integral

$$\int_B (2x + y)\,dx\,dy,$$

where B is the rectangle $0 \leq x < 1$, $0 \leq y \leq 2$. The existence of the integral is ensured by Theorem 2.1. For this reason, any sequence of Riemann sums with mesh tending to zero may be used to evaluate it.

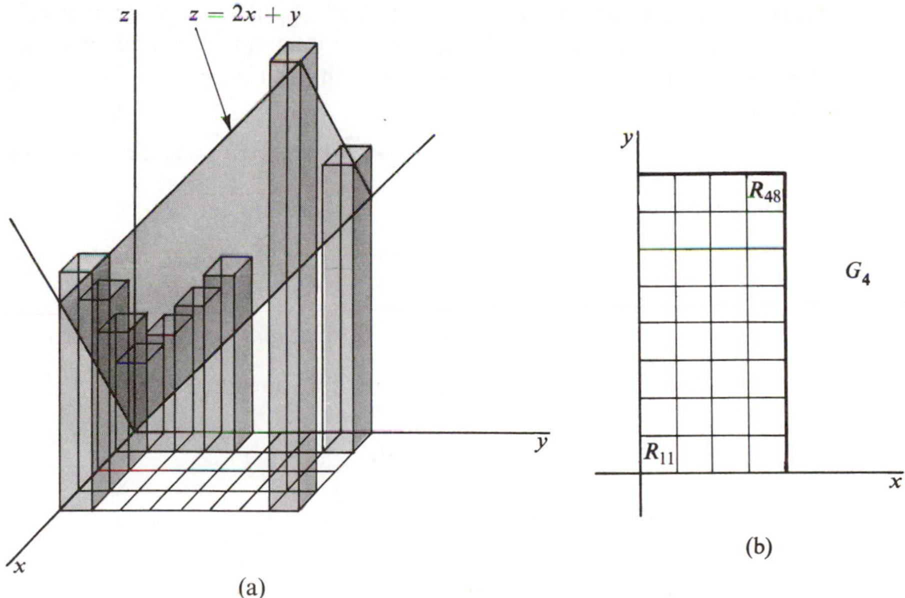

Figure 11

For each $n = 1, 2, \ldots,$ consider the grid G_n consisting of the lines $x = i/n$, $i = 0, \ldots, n$, and $y = j/n$, $j = 0, \ldots, 2n$. See Fig. 11(b). The mesh of G_n is $1/n$, and the area of each of the rectangles R_{ij} is $1/n^2$. Setting

$$\mathbf{x}_{ij} = (x_i, y_j) = \left(\frac{i}{n}, \frac{j}{n}\right),$$

we form the Riemann sum, illustrated in Figure 11(a),

$$\begin{aligned}\sum_{i=1}^{n}\sum_{j=1}^{2n}(2x_i + y_j)A(R_{ij}) &= \sum_{i=1}^{n}\sum_{j=1}^{2n}\left(\frac{2i}{n} + \frac{j}{n}\right)\frac{1}{n^2} \\ &= \frac{1}{n^3}\left(4n\sum_{i=1}^{n} i + n\sum_{j=1}^{2n} j\right) \\ &= \frac{1}{n^2}\left(\frac{4n^2 + n}{2} + \frac{4n^2 + 2n}{2}\right) \\ &= \frac{8n^2 + 3n}{2n^2} = 4 + \frac{3}{n}.\end{aligned}$$

Hence,

$$\int_B (2x + y)\,dx\,dy = \lim_{n\to\infty}\left(4 + \frac{3}{2n}\right) = 4.$$

A direct evaluation of a multiple integral will be very arduous for most functions we wish to integrate. Fortunately, in many instances the multiple integral can be easily evaluated by repeated application of ordinary 1-dimensional integration instead of by finding the limits of Riemann sums. The pertinent theorem, which we prove at the end of this section, is the following.

2.2 Theorem

Let B be a subset of $\mathcal{R}^n$ such that the iterated integral

$$\int dx_1 \int dx_2 \ldots \int f\,dx_n$$

exists over B. If, in addition, the multiple integral

$$\int_B f\,dV$$

exists, then the two integrals are equal.

Since the argument used to prove Theorem 2.2 applies equally well to any order of iterated integration, we have as an immediate corollary:

2.3 Theorem

If $\int_B f\,dV$ exists, and iterated integrals exist for some orders of integration, then all of these integrals are equal.

Example 2. Evaluate $\int_B (2x + y)\, dx\, dy$, where B is the rectangle $0 \leq x \leq 1, 0 \leq y \leq 2$. This is the same integral that occurs in Example 1. Theorem 2.2 is applicable, and we obtain

$$\begin{aligned}\int_B (2x + y)\, dx\ dy &= \int_0^1 dx \int_0^2 (2x + y)\, dy \\ &= \int_0^1 (4x + 2)\, dx \\ &= [2x^2 + 2x]_0^1 = 4.\end{aligned}$$

Example 3. Let R be the 3-dimensional rectangle defined by $-1 \leq x \leq 2, 0 \leq y \leq 1, 1 \leq z \leq 2$, and shown in Fig. 12. Consider $f(x, y, z) = xyz$. Then

$$\begin{aligned}\int_R f\, dV &= \int_R xyz\, dx\, dy\, dz \\ &= \int_{-1}^2 dx \int_0^1 dy \int_1^2 xyz\, dz = \int_{-1}^2 x\, dx \int_0^1 y\, dy \int_1^2 z\, dz \\ &= [(\tfrac{3}{2})(\tfrac{1}{2})(\tfrac{3}{2}) = \tfrac{9}{8}.\end{aligned}$$

Example 4. Let $f(x, y, z) = xyz$, and let the subset B of $\mathcal{R}^3$ be defined by $x^2 + y^2 + z^2 \leq 4, x \geq 0, y \geq 0, z \geq 0$. B is the interior and boundary of one-eighth of the spherical ball of radius 2 with center at the origin, shown in Fig. 13. The integral $\int_B f\, dV$ equals the triple iterated integral of the function $f(x, y, z) = xyz$ over B. For fixed x and y, the variable z

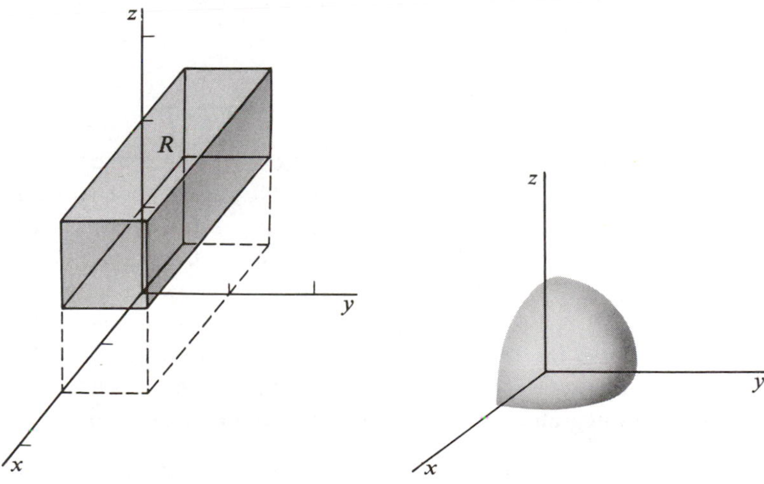

Figure 12 **Figure 13**

runs from 0 to $\sqrt{4 - x^2 - y^2}$, which are the limits of the first integration with respect to z. The result of this integration is a function of x and y that next must be integrated over the 2-dimensional subset obtained by projecting B on the xy-plane, that is, over the region $x^2 + y^2 \leq 4$, $x \geq 0$, $y \geq 0$. For fixed x, the variable y runs from 0 to $\sqrt{4 - x^2}$; hence, these are the limits on the integration with respect to y. Finally, x runs from 0 to 2, so we conclude that

$$\int_B f\,dV = \int_0^2 dx \int_0^{\sqrt{4-x^2}} dy \int_0^{\sqrt{4-x^2-y^2}} xyz\,dz.$$

Then

$$\int_B f\,dV = \frac{1}{2}\int_0^2 x\,dx \int_0^{\sqrt{4-x^2}} y(4 - x^2 - y^2)\,dy$$

$$= \frac{1}{2}\int_0^2 x\left(2(4 - x^2) - \frac{x^2}{2}(4 - x^2) - \frac{4 - x^2)^2}{4}\right)dx.$$

The last integral simplifies to

$$\int_0^2 (2x - x^3 + \tfrac{1}{8}x^5)\,dx = \tfrac{4}{3}.$$

If the constant function 1 is integrable over a subset B of $\mathcal{R}^n$, the **content** or **volume** of B is denoted by $V(B)$ and defined by

$$V(B) = \int_B 1\,dV = \int_B dV.$$

For sets B in $\mathcal{R}^2$, we write $A(B)$, for area, instead of $V(B)$. It follows from the last part of Theorem 2.1 that the content of a continuously differentiable k-dimensional ($k < n$) curve or surface S is zero, for

$$V(S) = \int_S dV = 0.$$

For some sets B, the integral $\int_B dV$ does not exist. If this happens, the content of B is not defined (see Exercise 21). Notice that for rectangles R, the content $V(R)$ has been defined twice: first as the product of the lengths of mutually perpendicular edges and second as an integral. That the two definitions agree follows immediately from Theorems 2.1 and 2.2.

Example 5. Let B be the region in $\mathcal{R}^2$ under the curve $y = f(x)$ from $x = a$ to $x = b$, where f is a nonnegative function. Assuming the existence of the following integrals, we obtain, using the iterated integral theorem (Theorem 2.2),

$$A(B) = \int_B dA = \int_a^b dx \int_0^{f(x)} dy = \int_a^b f(x)\,dx.$$

Hence, the above definition of content is consistent with the usual one for the area under the graph of a nonnegative integrable function of one variable. If f is integrable over B and also nonnegative on B, we could similarly show that the volume under the graph of f and above the set B is the double integral ${}_B\int f\,dA$.

Example 6. The volume above the disk D defined by $x^2 + y^2 \leq 1$ and under the graph of $f(x, y) = x^2 + y^2$ (see Fig. 14) is equal to

$$\int_D (x^2 + y^2)\,dx\,dy = \int_{-1}^{1} dx \int_{-\sqrt{1-x^2}}^{\sqrt{1-x^2}} (x^2 + y^2)\,dy$$

$$= 2\int_{-1}^{1} \left(x^2\sqrt{1 - x^2} + \tfrac{1}{3}(1 - x^2)\sqrt{1 - x^2}\right) dx$$

$$= \frac{4}{3}\int_{0}^{1} \left(\sqrt{1 - x^2} + 2x^2\sqrt{1 - x^2}\right) dx$$

$$= \frac{4}{3}\left(\frac{\pi}{4} + \frac{\pi}{8}\right) = \frac{\pi}{2}.$$

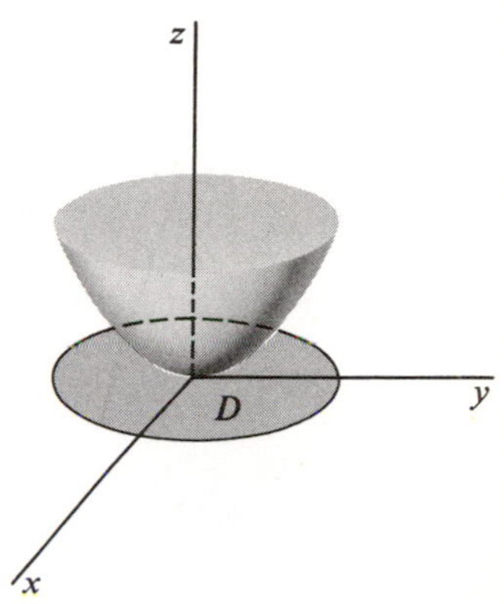

Figure 14

Example 7. Find the volume of the region B in $\mathcal{R}^3$ bounded by the four planes $x = 0$, $y = 0$, $z = 0$, and $x + y + z = 1$, shown in Fig. 15.

$$V(B) = \int_B dV = \int_0^1 dx \int_0^{1-x} dy \int_0^{1-x-y} dz = \tfrac{1}{6}.$$

The volume of the region B can be computed directly as a double integral. The projection of B on the xy-plane is the triangle D bounded by the lines $x = 0$, $y = 0$, $x + y = 1$. The set B itself can be described as the region under the graph of the function $f(x, y) = 1 - x - y$ and above D.

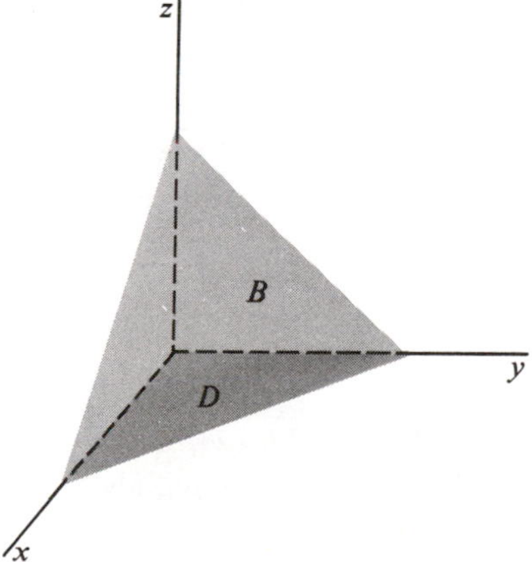

Figure 15

Hence, according to the remark at the end of Example 5,

$$V(B) = \int_D f\,dA = \int_0^1 dx \int_0^{1-x} (1 - x - y)\,dy = \tfrac{1}{6}.$$

Notice that what we have called the content of a subset of $\mathcal{R}^n$ is more properly called its **n-dimensional content.** For example, the square defined by the inequalities $0 \leq x \leq 2$, $0 \leq y \leq 2$ in $\mathcal{R}^2$ has 2-dimensional content 4, whereas the square defined in $\mathcal{R}^3$ by $0 \leq x \leq 2$, $0 \leq y \leq 2$, $z = 0$, and which looks the same, has 3-dimensional content 0. Thus the content of a set depends on the dimension of the containing Euclidean space with respect to which it is being measured, as well as on the shape of the set itself. Having already indicated that 2-dimensional content is called area, we remark that 1-dimensional content is length.

Some characteristic properties of the Riemann integral are summarized in the following four theorems.

2.4 Theorem. Linearity

If f and g are integrable over B and a and b are any two real numbers, then $af + bg$ is integrable over B and

$$\int_B (af + bg)\,dV = a\int_B f\,dV + b\int_B g\,dV.$$

2.5 Theorem. Positivity

If f is nonnegative and integrable over B, then

$$\int_B f\,dV \geq 0.$$

2.6 Theorem

If R is a rectangle, then $\int_R dV = V(R)$ (where the content $V(R)$ is defined by Equation (2)).

2.7 Theorem

If B is a subset of a bounded set C, then $\int_B f\,dV$ exists if and only if $\int_C f_B\,dV$ exists. Whenever both integrals exist, they are equal.

Proof of 2.4. Let $\epsilon > 0$ be given, and choose $\delta > 0$ so that if S_1 and S_2 are two Riemann sums for f_B and g_B, respectively, and whose grids have mesh less than δ, then

$$|a|\left|S_1 - \int_B f\,dV\right| < \frac{\epsilon}{2} \quad \text{and} \quad |b|\left|S_2 - \int_B g\,dV\right| < \frac{\epsilon}{2}.$$

Let S be any Riemann sum for $(af + bg)_B$ whose grid has mesh less than δ. Then

$$\begin{aligned} S &= \sum_i (af + bg)_B(\mathbf{x}_i)V(R_i) \\ &= a \sum_i f_B(\mathbf{x}_i)V(R_i) + b \sum_i g_B(\mathbf{x}_i)V(R_i) \\ &= aS_1 + bS_2. \end{aligned}$$

Hence,

$$\begin{aligned} &\left|S - a\int_B f\,dV - b\int_B g\,dV\right| \\ &\qquad = \left|aS_1 - a\int_B f\,dV + bS_2 - b\int_B g\,dV\right| \\ &\qquad \le |a|\left|S_1 - \int_B f\,dV\right| + |b|\left|S_2 - \int_B g\,dV\right| \\ &\qquad < \frac{\epsilon}{2} + \frac{\epsilon}{2} = \epsilon. \end{aligned}$$

Thus

$$\lim_{m(G)\to 0} \sum_i (af + bg)_B(\mathbf{x}_i)V(R_i) = a\int_B f\,dV + b\int_B g\,dV,$$

and the proof is complete.

Proof of 2.5. Since all the Riemann sums are nonnegative, the limit must also be nonnegative.

Proof of 2.6. This follows immediately from Theorems 2.1 and 2.2.

Proof of 2.7. The existence and the value of the integral $\int_B f\,dV$ depend only on the function f_B. Similarly, $\int_C f_B\,dV$ is defined by using $(f_B)_C$, which is equal to f_B.

Many of the important properties of the integral can be derived directly from the preceding four theorems, without reference to the original definition. The next two theorems are given as examples.

2.8 Theorem

If f and g are integrable over B and $f \leq g$ on B, then

$$\int_B f\,dV \leq \int_B g\,dV.$$

Proof. The function $g - f$ is nonnegative and, by Theorem 2.4, is integrable over B. Hence, by Theorems 2.4 and 2.5,

$$0 \leq \int_B (g - f)\,dV = \int_B g\,dV - \int_B f\,dV,$$

from which the conclusion follows.

The next theorem establishes an analog for the equation

$$\int_a^c f(x)\,dx = \int_a^b f(x)\,dx + \int_b^c f(x)\,dx$$

that holds for functions of one variable.

2.9 Theorem

If f is integrable over each of two disjoint sets B_1 and B_2, then f is integrable over their union and

$$\int_{B_1 \cup B_2} f\,dV = \int_{B_1} f\,dV + \int_{B_2} f\,dV.$$

Proof. By Theorem 2.7,

$$\int_{B_1} f\,dV + \int_{B_2} f\,dV = \int_{B_1 \cup B_2} f_{B_1}\,dV + \int_{B_1 \cup B_2} f_{B_2}\,dV.$$

Since B_1 and B_2 are disjoint, $f_{B_1 \cup B_2} = f_{B_1} + f_{B_2}$. Hence, by Theorem 2.4, $f_{B_1 \cup B_2}$ is integrable over $B_1 \cup B_2$, and

$$\int_{B_1 \cup B_2} f_{B_1}\,dV + \int_{B_1 \cup B_2} f_{B_2}\,dV = \int_{B_1 \cup B_2} f_{B_1 \cup B_2}\,dV.$$

Finally, by Theorem 2.7 again, f is integrable over $B_1 \cup B_2$ and

$$\int_{B_1 \cup B_2} f_{B_1 \cup B_2}\,dV = \int_{B_1 \cup B_2} f\,dV.$$

This completes the proof.

The next theorem will show that, for functions f and regions B for which $\int_B f\,dV$ exists, the value of the integral is completely determined by the properties stated in the four Theorems 2.4–2.7. The theorem is important because it enables us to identify the multiple integral with other integrals I, in particular, iterated integrals, which may be computed in any one of a number of orders. We shall use the symbol I to denote such an integral.

2.10 Theorem

Let I be an integral of bounded functions $\mathcal{R}^n \xrightarrow{f} \mathcal{R}$ over bounded sets B. Suppose that I has the following characteristic properties:

(a) If $I_B f$ and $I_B g$ are defined and a and b are real numbers, then $I_B(af + bg)$ is defined and

$$I_B(af + bg) = aI_B f + bI_B g.$$

(b) If f is nonnegative and $I_B f$ is defined, then $I_B f \geq 0$.

(c) If R is a rectangle, then $I_R 1 = V(R)$ (as defined by Equation (2)).

(d) If B is contained in a bounded set C, then $I_B f$ is defined if and only if $I_C f_B$ is defined. Whenever both exist, they are equal.

We can then conclude that, if $I_B f$ and $\int_B f\,dV$ both exist, they are equal. (No properties of the integral itself are assumed; we use only its definition.)

Proof. Suppose $\int_B f\,dV < I_B f$. Set

$$\epsilon = I_B f - \int_B f\,dV,$$

and choose $\delta > 0$ so that if S is any Riemann sum for f_B whose grid has mesh less than δ, then

$$\left| \int_B f\,dV - S \right| < \frac{\epsilon}{2}.$$

Let G be an arbitrary grid that covers B and has mesh less than δ, and denote the closed bounded rectangles formed by G by $R_1, \ldots, R_r$. Set

$$C = R_1 \cup \ldots \cup R_r,$$

$$\bar{f}_i = \text{least upper bound of } f_B \text{ in } R_i.$$

Consider the function g defined by

$$g = \sum_{i=1}^{r} \bar{f}_i \chi_{R_i}.$$

The function χ_{R_i} is the characteristic function of R_i. It is defined by

$$\chi_{R_i}(\mathbf{x}) = \begin{cases} 1, & \text{if } \mathbf{x} \text{ is in } R_i, \\ 0, & \text{otherwise.} \end{cases}$$

It follows immediately from properties (c), (d), and (a) that $I_C g$ is defined and that

$$I_C g = \sum_{i=1}^{r} \bar{f}_i V(R_i).$$

The definition of least upper bound implies that there exists a Riemann sum for f_B on the grid G that is arbitrarily close to $I_C g$. Hence,

$$\left| \int_B f\, dV - I_C g \right| \leq \frac{\epsilon}{2},$$

and so, by the definition of ϵ,

$$I_C g < I_B f. \tag{3}$$

By property (d), $I_B f = I_C f_B$. Moreover the function g has been constructed so that $f_B \leq g$. It follows from property (b) (as extended in Theorem 2.8) that

$$I_B f = I_C f_B \leq I_B g. \tag{4}$$

The inequalities (3) and (4) are contradictory; so we conclude

$$\int_B f\, dV \geq I_B f.$$

By an entirely analogous argument using the notion of greatest lower bound instead of least upper bound, we can obtain

$$\int_B f\, dV \leq I_B f,$$

and this completes the proof.

For an application of Theorem 2.10, take the functions $\mathcal{R}^2 \xrightarrow{f} \mathcal{R}$ and sets B for which the iterated integral $\int dx \int f\, dy$ over B is defined. Let

$$I_B f = \underset{(\text{over } B)}{\int dx \int f\, dy}.$$

Verification of the conditions of Theorem 2.10 is straightforward and reduces to a knowledge of the corresponding properties of the definite integral for functions of one variable. For example, for integration over intervals,

$$\int_\alpha^\beta (af + bg)\, dx = a\int_\alpha^\beta f\, dx + b\int_\alpha^\beta g\, dx.$$

$$\int_\alpha^\beta f\, dx \geq 0, \qquad \text{if } f \geq 0.$$

$$\int_\alpha^\beta dx = \beta - \alpha.$$

$$\int_\gamma^\delta f\, dx = \int_\alpha^\beta f_{[(\gamma,\delta)]}\, dx,$$

if $\alpha \leq \gamma \leq \delta \leq \beta$ and $[\gamma, \delta]$ is the interval $\gamma \leq x \leq \delta$.

It follows immediately from Theorem 2.10 that if both the iterated integral and the double integral of f exist over B, then they are equal. This proves Theorem 2.2 for two variables. The general case can be done by induction.

The possibility of changing order of integration has a number of consequences other than its obvious convenience for computing multiple integrals. One of these is the theorem for change of order in partial differentiation, proved in Section 5 of Chapter 3 by other means, and in a slightly stronger form in Exercise 8 of this section. Another consequence is the Leibnitz formula for interchanging differentiation and integration. The theorem is stated here and the proof is outlined in Exercise 7.

2.11 Leibnitz Rule

If $(\partial g/\partial y)(x, y)$ is continuous for $a \leq x \leq b$ and $c \leq y \leq d$, then

$$\frac{d}{dy}\int_a^b g(x, y)\, dx = \int_a^b \frac{\partial g}{\partial y}(x, y)\, dx.$$

EXERCISES

1. Make a drawing of the set B and compute $\int_B f\, dA$, where

   (a) $f(x, y) = x^2 + 3y^2$ and B is the disk $x^2 + y^2 \leq 1$. [*Ans.* π.]
   (b) $f(x, y) = 1/(x + y)$ and B is the region bounded by the lines $y = x$, $x = 1$, $x = 2$, $y = 0$. [*Ans.* log 2.]
   (c) $f(x, y) = x \sin xy$ and B is the rectangle $0 \leq x \leq \pi$, $0 \leq y \leq 1$. [*Ans.* π.]

(d) $f(x, y) = x^2 - y^2$ and B consists of all (x, y) such that $0 \le x \le 1$ and $x^2 - y^2 \ge 0$.

[*Ans.* $\frac{1}{3}$.]

2. Using the definition of the double integral as a limit of Riemann sums, compute $\int_B f(x, y)\, dx\, dy$, where

(a) $f(x, y) = x + 4y$ and B is the rectangle $0 \le x \le 2$, $0 \le y \le 1$.
[*Ans.* 6.]

(b) $f(x, y) = 3x^2 + 2y$ and B is the rectangle $0 \le x \le 2$, $0 \le y \le 1$.
[*Ans.* 10.]

The following formulas will be useful in doing this problem:

$$\sum_{i=1}^{n} i = \frac{n(n+1)}{2},$$

$$\sum_{i=1}^{n} i^2 = \frac{n(n+1)(2n+1)}{6},$$

$$\sum_{i=1}^{n} i^3 = \left(\sum_{i=1}^{n} i\right)^2.$$

3. Find the volume under the graph of f and above the set B, where

(a) $f(x, y) = x + y^2$ and B is the rectangle with corners (1, 1), (1, 3), (2, 3), and (2, 1). [*Ans.* $\frac{35}{3}$.]

(b) $f(x, y) = x + y + 2$ and B is the region bounded by the curves $y^2 = x$. and $x = 2$. [*Ans.* $\frac{128}{15}\sqrt{2}$.]

(c) $f(x, y) = |x + y|$ and B is the disk $x^2 + y^2 \le 1$. [*Ans.* $4\sqrt{2}/3$.]

4. Find by integration the area of the subset of $\mathcal{R}^2$ bounded by the curve $x^2 - 2x + 4y^2 - 8y + 1 = 0$. [*Ans.* 2π.]

5. Find an approximate value for each integral in Problem 1 by computing a Riemann sum with an appropriately fine grid.

6. Consider the rectangles

$$B_1 \text{ defined by } 0 < x \le 1, 0 \le y < 1$$
$$B_2 \text{ defined by } 1 \le x \le 2, -1 \le y \le 1$$

and the function

$$f(x, y) = \begin{cases} 2x - y, & \text{if } x < 1, \\ x^2 + y, & \text{if } x \ge 1. \end{cases}$$

Compute $\int_{B_1 \cup B_2} f(x, y)\, dx\, dy$. [*Ans.* $\frac{31}{6}$.]

7. (a) Prove the **Leibnitz rule** for differentiating an integral with respect to a parameter: If $g_y(x, y)$ is continuous on a rectangle $a \le x \le b$, $c \le y \le d$, then

$$\frac{d}{dy}\int_a^b g(t, y)\, dt = \int_a^b g_y(t, y)\, dt.$$

[*Hint.* Interchange the order of integration in $\int_c^y dy \int_a^b g_y(t, y)\, dt$, and then differentiate both sides with respect to y.]

(b) Use part (a) and the chain rule to show that if g_y is continuous, and h_1 and h_2 are differentiable, then

$$\frac{d}{dy}\int_{h_1(y)}^{h_2(y)} g(t, y)\, dt$$
$$= \int_{h_1(y)}^{h_2(y)} g_y(t, y)\, dt + h_2'(y)g(h_2(y), y) - h_1'(y)g(h_1(y), y).$$

8. Prove (compare 5.2, Chapter 3): If f_x, f_y, and f_{xy} are continuous on an open set, then $f_{xy} = f_{yx}$. [*Hint.* Apply the Leibnitz rule of Exercise 7(a) to the equation

$$f(x, y) - f(a, y) = \int_a^x f_x(t, y)\, dt,$$

and then differentiate both sides with respect to x.]

9. Given that $f(x, y, z) = xyz$ and that

$$\int_B f(x, y, z)\, dx\, dy\, dz = \int_0^2 dx \int_0^x dy \int_0^{x+y} xyz\, dz,$$

sketch the region B and evaluate the integral. [*Ans.* $\frac{68}{9}$.]

10. Compute the multiple integral of $f(x, y, z, w) = xyzw$ over the 4-dimensional rectangle

$0 \le x \le 1, \quad -1 \le y \le 2, \quad 1 \le z \le 2, \quad 2 \le w \le 3.$ [*Ans.* $\frac{45}{16}$]

11. Sketch the region B in $\mathcal{R}^3$ bounded by the surface $z = 4 - 4x^2 - y^2$ and the xy-plane. Set up the volume of B as a triple integral and also as a double integral. Compute the volume. [*Ans.* 4π.]

12. Write an expression for the volume of the ball $x^2 + y^2 + z^2 \le a^2$

(a) as a triple integral.
(b) as a double integral.

13. Sketch in $\mathcal{R}^3$ the two cylindrical solids defined by $x^2 + z^2 \le 1$ and $y^2 + z^2 \le 1$, respectively. Find the volume of their intersection. [*Ans.* $\frac{16}{3}$.]

14. The 4-dimensional ball B of radius 1 and with center at the origin is the subset of $\mathcal{R}^4$ defined by $x_1^2 + x_2^2 + x_3^2 + x_4^2 \le 1$. Set up an expression for the volume $V(B)$ as a fourfold iterated integral.

15. Use Theorem 2.8 to show that if f and $|f|$ are integrable over B, then $|\int_B f\, dV| \le \int_B |f|\, dV$.

16. Let $\mathcal{R}^n \xrightarrow{f} \mathcal{R}^m$ be defined on a set B in $\mathcal{R}^n$. We define

$$\int_B f\, dV = \left(\int_B f_1\, dV, \ldots, \int_B f_m\, dV\right),$$

provided that the integrals of the coordinate functions $f_1, \ldots, f_m$ of f all exist.

(a) Show that if $\mathcal{R}^n \xrightarrow{f} \mathcal{R}^m$ and $\mathcal{R}^n \xrightarrow{g} \mathcal{R}^m$ are both integrable over B, then

$$\int_B (af + bg)\, dV = a\int_B f\, dV + b\int_B g\, dV,$$

where a and b are constants.

(b) If $\mathbf{k}$ is a fixed vector in $\mathcal{R}^m$, and $\mathcal{R}^n \xrightarrow{f} \mathcal{R}^m$ is integrable over B, show that

$$\int_B \mathbf{k} \cdot f\, dV = \mathbf{k} \cdot \int_B f\, dV.$$

(c) Show that if $\mathcal{R}^n \xrightarrow{f} \mathcal{R}^m$ and $\mathcal{R}^n \xrightarrow{|f|} \mathcal{R}$ are integrable over B, then $|\int_B f\, dV| \le \int_B |f|\, dV$. [*Hint*: By the Cauchy-Schwarz inequality $f(\mathbf{x}) \cdot \int_B f\, dV \le |f(\mathbf{x})|\, |\int_B f\, dV|$, for all $\mathbf{x}$ in B. Integrate with respect to $\mathbf{x}$ and apply the result of part (b).]

17. Use the result of Exercise 16(c) to show that if $\mathcal{R}^n \xrightarrow{f} \mathcal{R}^m$ is continuous on a set B, and $\mathbf{x}_0$ is interior to B, then

$$\lim_{r \to 0} \frac{1}{V(B_r)} \int_{B_r} f\, dV = f(\mathbf{x}_0),$$

where B_r is a ball of radius r centered at $\mathbf{x}_0$.

18. Let R be a region in $\mathcal{R}^n$ having a finite volume. The vector

$$\mathbf{z}_0 = \frac{1}{V(R)} \int_R \mathbf{x}\, dV$$

is called the **centroid** of R, and the real number

$$I(\mathbf{z}) = \int_R |\mathbf{x} - \mathbf{z}|^2\, dV_{\mathbf{x}}$$

is called the **moment of inertia** of R about $\mathbf{z}$.

(a) Show that the centroid of a ball is its center.
(b) Show that $I(\mathbf{z})$ is minimized by taking $\mathbf{z}$ to be the centroid of R. [*Hint*. Show that

$$I(\mathbf{z}) = I(0) - V(R)\,|\mathbf{z}_0|^2 + V(R)\,|\mathbf{z} - \mathbf{z}_0|^2.]$$

19. Prove the analog for multiple integrals of Theorem 8.4 of Chapter 5: If R has finite volume, and the series $\sum_{k=1}^{\infty} f_k$ of continuous functions converges to f uniformly on R, then

$$\sum_{k=1}^{\infty} \int_R f_k\, dV = \int_R \left[\sum_{k=1}^{\infty} f_k\right] dV.$$

20. A function $\mathcal{R}^n \xrightarrow{p} \mathcal{R}$, integrable over a region R in $\mathcal{R}^n$ is called a **probability density** on R if

$$p(\mathbf{x}) \geq 0 \qquad \text{for all } \mathbf{x} \text{ in } R. \tag{1}$$

$$\int_R p\, dV = 1. \tag{2}$$

If E is an experiment with possible outcomes in $\mathcal{R}^n$ distributed according to the density p, then the **probability** that the outcome lies in a set B in $\mathcal{R}^n$ is defined by

$$Pr[E \text{ in } B] = \int_B p\, dV.$$

(a) For what constant k is the function $\mathcal{R}^2 \xrightarrow{p} \mathcal{R}$,

$$p(x, y) = \begin{cases} k(1 - x^2 - y^2), & x^2 + y^2 \leq 1 \\ 0, & x^2 + y^2 > 1, \end{cases}$$

a probability density? [*Ans.* $k = 2/\pi$.]

(b) If the outcomes of E are distributed according to the density of part (a), find the probability that E has an x-coordinate bigger than $\frac{1}{2}$.

21. Let B be the subset of $\mathcal{R}^2$ consisting of all points (x, y) such that $0 \leq y \leq 1$, and x is rational, $0 \leq x \leq 1$. What is the area of B?

22. On the rectangle $0 \leq x \leq 1$ and $0 \leq y \leq 1$, let $f(x, y) = 1$, if x is rational, and $f(x, y) = 2y$, if x is irrational. Show that

$$\int_0^1 dx \int_0^1 f(x, y)\, dy = 1,$$

but that f is not Riemann integrable over the rectangle.

23. Prove that

$$\int_0^1 dy \int_1^\infty (e^{-xy} - 2e^{-2xy})\, dx \neq \int_1^\infty dx \int_0^1 (e^{-xy} - 2e^{-2xy}) dy.$$

SECTION 3
CHANGE OF VARIABLE

The change-of-variable formula for 1-dimensional integrals is

$$\int_{\phi(a)}^{\phi(b)} f(x)\, dx = \int_a^b f(\phi(u))\phi'(u)\, du. \tag{1}$$

For example, taking $\phi(u) = \sin u$, we obtain

$$\int_0^1 \sqrt{1 - x^2\, dx} = \int_0^{\pi/2} \cos^2 u\, du = \frac{\pi}{4}.$$

In this section Equation (1) will be extended to dimensions higher than one. In n-dimensional space a change of variable is effected by a function $\mathcal{R}^n \xrightarrow{T} \mathcal{R}^n$. In what follows it will usually be more convenient to consider the domain space and range space of T as distinct. We therefore regard T as a transformation from one copy of $\mathcal{R}^n$, which we label $\mathcal{U}^n$, to another copy, which we continue to label $\mathcal{R}^n$, writing typically $T(\mathbf{u}) = \mathbf{x}$ where $\mathbf{u}$ is in $\mathcal{U}^n$ and $\mathbf{x}$ is in $\mathcal{R}^n$. The statement of the n-dimensional change-of-variable theorem follows.

3.1 Theorem

Let $\mathcal{U}^n \xrightarrow{T} \mathcal{R}^n$ be a continuously differentiable transformation. Let R be a set in $\mathcal{U}^n$ having a boundary consisting of finitely many smooth sets. Suppose that R and its boundary are contained in the interior of the domain of T and that

(a) T is one-to-one on R.

(b) $\det T'$, the Jacobian determinant of T, is different from zero on R.

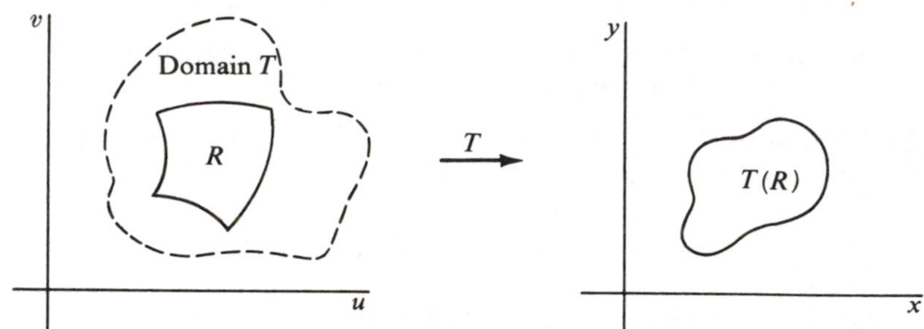

Figure 16

Then, if the function f is bounded and continuous on $T(R)$ (the image of R under T), we have

$$\int_{T(R)} f\,dV = \int_R (f \circ T)\,|\det T'|\,dV.$$

Either condition (a) or (b) is allowed to fail on a set of zero content.

The proof is in the Appendix. Before showing why the formula works, we give some examples of its application. Notice that the factor ϕ' that occurs in Equation (1) has been replaced in higher dimensions by the absolute value of the Jacobian determinant of T. Aside from the computation of $\det T'$, the application of the transformation formula is a matter of finding the geometric relationship between the subset R and its image $T(R)$ for various transformations T.

Example 1. The integral $\int_P (x + y)\,dx\,dy$, in which P is the parallelogram shown in Fig. 17, can be transformed into an integral over a rectangle.

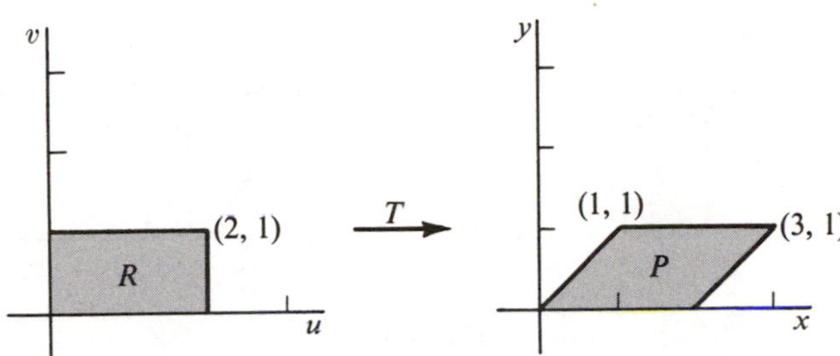

Figure 17

This is done by means of the transformation

$$\begin{pmatrix} x \\ y \end{pmatrix} = T\begin{pmatrix} u \\ v \end{pmatrix} = \begin{pmatrix} u + v \\ v \end{pmatrix}$$

The Jacobian determinant of T is

$$\det T' = \begin{vmatrix} 1 & 1 \\ 0 & 1 \end{vmatrix} = 1.$$

By the change-of-variable theorem,

$$\begin{aligned} \int_P (x + y)\,dx\,dy &= \int_R [(u + v) + v]1\,du\,dv \\ &= \int_0^2 du \int_0^1 (u + 2v)\,dv = 4. \end{aligned}$$

The transformation T is clearly one-to-one because it is a linear transformation with nonzero determinant. Notice that the region of integration in the given integral is in the range of the transformation rather than in its domain.

Example 2. The transformation

$$\begin{pmatrix} x \\ y \end{pmatrix} = \begin{pmatrix} u \cos v \\ u \sin v \end{pmatrix}$$

goes between the regions shown in Fig. 18. The Jacobian is

$$\det T' = \begin{vmatrix} \cos v & -u \sin v \\ \sin v & u \cos v \end{vmatrix} = u.$$

The transformation is one-to-one between R and $T(R)$. This can be seen geometrically because of the interpretation of v and u as angle and radius

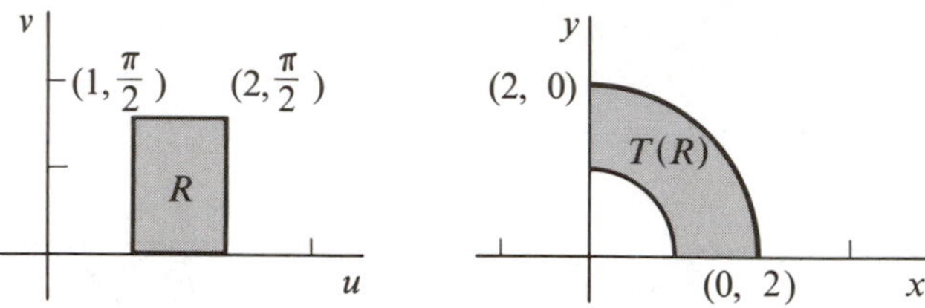

Figure 18

respectively, or directly from the relations

$$u = \sqrt{x^2 + y^2}, \qquad \cos v = \frac{x}{\sqrt{x^2 + y^2}},$$

together with the fact that $\cos v$ is one-to-one for $0 \leq v \leq \pi/2$. Given the integral of $x^2 + y^2$ over $T(R)$, we can transform as follows:

$$\int_{T(R)} (x^2 + y^2)\, dA = \int_R u^2 u\, dA = \int_1^2 u^3\, du \int_0^{\pi/2} dv = \frac{15\pi}{8}.$$

Example 3. Let B be the subset of 3-dimensional space $\mathcal{R}^3$ defined by the inequalities

$$x^2 + y^2 + z^2 \leq 1, \qquad \begin{cases} x \geq 0, \\ y \geq 0, \\ z \geq 0. \end{cases}$$

To transform the integral $\int_B (x^2 + y^2)\, dx\, dy\, dz$, we can define T by

$$\begin{pmatrix} x \\ y \\ z \end{pmatrix} = T\begin{pmatrix} u \\ v \\ w \end{pmatrix} = \begin{pmatrix} u \sin v \cos w \\ u \sin v \sin w \\ u \cos v \end{pmatrix}$$

Restricting (u, v, w) to the rectangle R defined by

$$\begin{cases} 0 \leq u \leq 1, \\ 0 \leq v \leq \pi/2, \\ 0 \leq w \leq \pi/2, \end{cases}$$

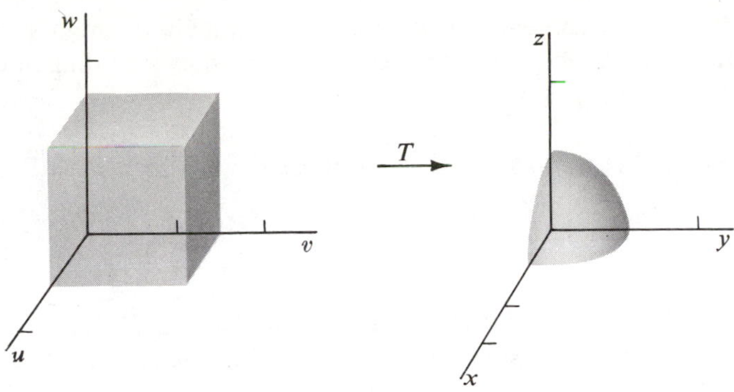

Figure 19

we get $T(R) = B$. The corresponding regions are shown in Fig. 19. Since

$$u = \sqrt{x^2 + y^2 + z^2}$$

$$\cos v = \frac{z}{\sqrt{x^2 + y^2 + z^2}},$$

$$\cos w = \frac{x}{\sqrt{x^2 + y^2}},$$

we conclude that the transformation T is one-to-one from R to B except on the boundary planes $u = 0$ and $v = 0$. The Jacobian determinant is

$$\det T' = \begin{vmatrix} \sin v \cos w & u \cos v \cos w & -u \sin v \sin w \\ \sin v \sin w & u \cos v \sin w & u \sin v \cos w \\ \cos v & -u \sin v & 0 \end{vmatrix}$$

$$= u^2 \sin v.$$

The transformed integral is

$$\int_B (x^2 + y^2)\, dx\, dy\, dz = \int_R (u^2 \sin^2 v \cos^2 w + u^2 \sin^2 v \sin^2 w) u^2 \sin v\, du\, dv\, dw$$

$$= \int_0^1 u^4\, du \int_0^{\pi/2} \sin^3 v\, dv \int_0^{\pi/2} dw$$

$$= \frac{1}{5} \cdot \frac{2}{3} \cdot \frac{\pi}{2} = \frac{\pi}{15}.$$

Notice that values of u and v appear in the transformed integral for which the Jacobian $u^2 \sin v$ is zero, that is, for $u = 0$ and $v = 0$. We have already remarked that the transformation fails to be one-to-one for points satisfying these conditions. However, the set of points on which this failure occurs has zero content, so the change-of-variable theorem still applies. Of course, the value of neither integral is affected by including or excluding these points.

Example 4. Let a function $\mathcal{U}^2 \xrightarrow{T} \mathcal{R}^2$ be defined by

$$\begin{pmatrix} x \\ y \end{pmatrix} = T\begin{pmatrix} u \\ v \end{pmatrix} = \begin{pmatrix} u^2 - v \\ u + v^2 \end{pmatrix}$$

The unit square R_{uv} defined by the inequalities $0 \leq u \leq 1$, $0 \leq v \leq 1$ is carried by T onto the subset R_{xy} shown in Fig. 20. Corresponding pieces

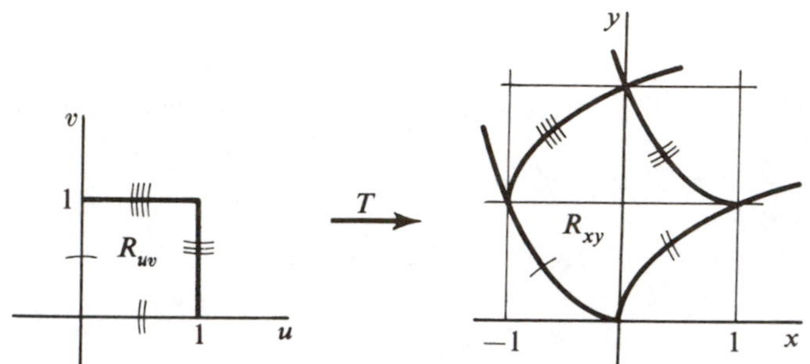

Figure 20

of the boundaries are indicated in the picture. The image of each of the four line segments that comprise the boundary of R_{uv} is computed as follows:

(a) If $u = 0$ and $0 \leq v \leq 1$, then $x = -v$ and $y = v^2$, that is, $y = x^2$ and $-1 \leq x \leq 0$.

(b) If $v = 0$ and $0 \leq u \leq 1$, then $x = u^2$ and $y = u$, that is, $x = y^2$ and $0 \leq y \leq 1$.

(c) If $u = 1$ and $0 \leq v \leq 1$, then $x = 1 - v$ and $y = 1 + v^2$, that is, $y - 1 = (x - 1)^2$ and $0 \leq x \leq 1$.

(d) If $v = 1$ and $0 \leq u \leq 1$, then $x = u^2 - 1$ and $y = u + 1$, that is, $(y - 1)^2 = x + 1$ and $1 \leq y \leq 2$.

It is not hard to verify that T is one-to-one on R_{uv}. Suppose

$$T\begin{pmatrix} u_1 \\ v_1 \end{pmatrix} = T\begin{pmatrix} u_2 \\ v_2 \end{pmatrix},$$

then

$$u_1^2 - v_1 = u_2^2 - v_2,$$
$$u_1 + v_1^2 = u_2 + v_2^2.$$

Obviously, if $u_1 = u_2$, then $v_1 = v_2$. Suppose $u_1 < u_2$. This implies

$$0 < u_2^2 - u_1^2 = v_2 - v_1,$$
$$0 < u_2 - u_1 = v_1^2 - v_2^2.$$

Hence, $v_1 < v_2$, whereas $v_2^2 < v_1^2$. This is impossible if both v_1 and v_2 are nonnegative; so the one-to-one-ness of T on R_{uv} is established. The Jacobian determinant of T is

$$\det T' = \begin{vmatrix} 2u & -1 \\ 1 & 2v \end{vmatrix} = 4uv + 1.$$

We therefore have as an application of the change-of-variable theorem

$$\begin{aligned}\int_{R_{xy}} x\,dx\,dy &= \int_{R_{uv}} (u^2 - v)(4uv + 1)\,du\,dv \\ &= \int_0^1 dv \int_0^1 (4u^3v - 4uv + u^2 - v)\,du \\ &= \int_0^1 (-2v^2 + \tfrac{1}{3})\,dv = -\tfrac{1}{3}.\end{aligned}$$

To understand why the change-of-variable formula works for a continuously differentiable vector function T, we need to know what effect T has on volume. We use the affine approximation to T that replaces $T(\mathbf{u})$ near $\mathbf{u}_0$ by $T(\mathbf{u}_0) + T'(\mathbf{u}_0)(\mathbf{u} - \mathbf{u}_0)$. The way in which T alters volume will be reflected in the way in which $d_{\mathbf{u}_0}T$ alters volume. In fact, translation of a subset by the vector $T(\mathbf{u}_0)$ leaves its volume unchanged, and the differential $d_{\mathbf{u}_0}T$, being a linear transformation, changes volume in a particularly simple way. Indeed, under a linear transformation volumes get multiplied by a constant factor, and the factor of proportionality is just the absolute value of the determinant of the transformation. For instance, suppose T is taken to be a linear transformation, and f is the constant function 1, that is, $f(\mathbf{u}) = 1$ for all $\mathbf{u}$ in $\mathcal{U}^n$. The change-of-variable theorem (Theorem 3.1) then implies the following.

3.2 Theorem

If T is a linear transformation from $\mathcal{R}^n$ to $\mathcal{R}^n$ having matrix A, then T multiplies volumes by the factor $|\det A|$.

Proof. By the change of variable theorem, setting $J = \det T'$, we get

$$V(T(R)) = \int_{T(R)} dV = \int_R |J|\,dV = |J|\,V(R.)$$

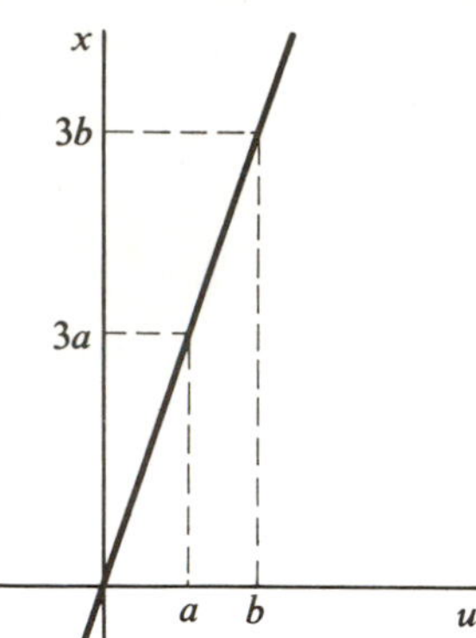

Figure 21

The last step is valid because for a linear transformation with matrix A, the Jacobian determinant is a constant det A. Then $|J| = |\det A|$, which can be taken outside the integral.

Example 5. The transformation from $\mathcal{U}$ to $\mathcal{R}$ given by $x = 3u$ is linear. It is therefore its own differential (see Chapter 3, Section 8, Exercise 9) and has Jacobian determinant $J = 3$. It is clear from Fig. 21 that lengths get multiplied by 3 under this transformation.

Example 6. The transformation T from $\mathcal{U}^2$ to $\mathcal{R}^2$ given by

$$T\begin{pmatrix} u \\ v \end{pmatrix} = \begin{pmatrix} u^2 \\ u + v \end{pmatrix}$$

has as its differential at $\mathbf{u}_0 = \begin{pmatrix} u_0 \\ v_0 \end{pmatrix} = \begin{pmatrix} 1 \\ 1 \end{pmatrix}$ the linear transformation

$$(d_{\mathbf{u}_0}T)\begin{pmatrix} u \\ v \end{pmatrix} = \begin{pmatrix} 2u_0 & 0 \\ 1 & 1 \end{pmatrix}\begin{pmatrix} u \\ v \end{pmatrix} = \begin{pmatrix} 2 & 0 \\ 1 & 1 \end{pmatrix}\begin{pmatrix} u \\ v \end{pmatrix}.$$

Near

$$\mathbf{u}_0 = \begin{pmatrix} 1 \\ 1 \end{pmatrix},$$

the function T is approximated by the affine transformation

$$A\begin{pmatrix} u \\ v \end{pmatrix} = T\begin{pmatrix} 1 \\ 1 \end{pmatrix} + T'(\mathbf{u}_0)\begin{pmatrix} u - 1 \\ v - 1 \end{pmatrix} = \begin{pmatrix} 1 \\ 2 \end{pmatrix} + \begin{pmatrix} 2 & 0 \\ 1 & 1 \end{pmatrix}\begin{pmatrix} u - 1 \\ v - 1 \end{pmatrix} = \begin{pmatrix} 2u - 1 \\ u + v \end{pmatrix}.$$

The square R in the uv-plane in Fig. 22 is carried by T onto the curved

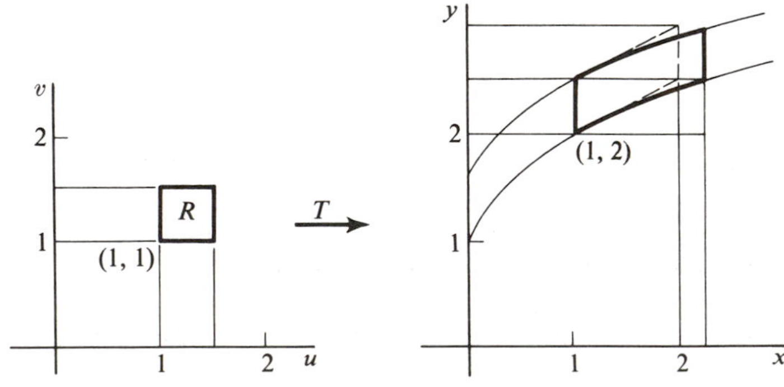

Figure 22

figure on the right. The affine approximation A carries R onto the parallelogram outlined with dashes. Notice that the area of the parallelogram is roughly equal to that of the curved figure. The exact area of the parallelogram is easily computed to be $\frac{1}{2}$, twice the area of R. The important point is that the affine approximation to T doubles the area of the square, while T itself approximately doubles that area. The magnification factor, 2, is given by the Jacobian determinant of T at $\mathbf{u}_0 = \begin{pmatrix} 1 \\ 1 \end{pmatrix}$. In fact

$$\det T' = \begin{pmatrix} 2 & 0 \\ 1 & 1 \end{pmatrix} = 2.$$

To find the exact area of the image $T(R)$, we use the change-of-variable theorem. Since u is positive on R, the transformation T is one-to-one there. The inverse function is given explicitly by

$$T^{-1}\begin{pmatrix} x \\ y \end{pmatrix} = \begin{pmatrix} \sqrt{x} \\ y - \sqrt{x} \end{pmatrix};$$

so T is one-to-one. Moreover, the Jacobian determinant $J = 2u$ is positive on R. Hence,

$$A(T(R)) = \int_{T(R)} dA = \int_R |J| \, dA$$

$$= \int_1^{3/2} dv \int_1^{3/2} 2u \, du = \tfrac{1}{2}u^2 \Big]_1^{3/2} = \frac{5}{8}.$$

To understand the change-of-variable theorem itself, let T be a continuously differentiable transformation and consider the corresponding regions R and $T(R)$. A 2-dimensional example is illustrated in Fig. 23.

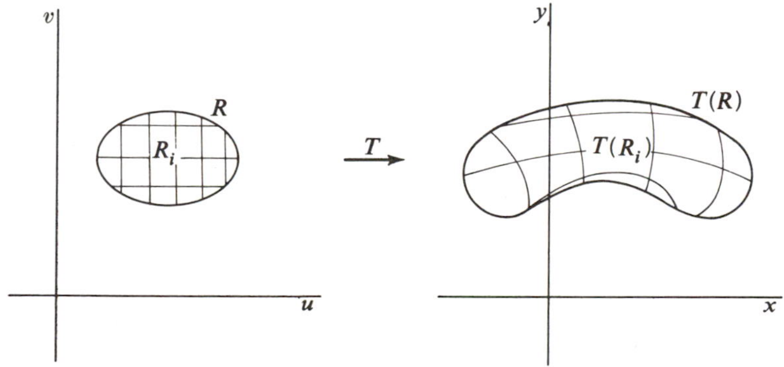

Figure 23

Decompose R into regions R_i by means of coordinate lines. Denoting approximate equality by the symbol $\approx$, we have

$$\int_{T(R)} f\,dV = \sum_i \int_{T(R_i)} f\,dV \approx \sum_i f_i V(T(R_i)), \tag{2}$$

where the number f_i is a value assumed by the function f in $T(R_i)$. We assume that $f_i V(T(R_i))$ is a reasonable approximation to $\int_{T(R_i)} f dV$. Next, approximate $V(T(R_i))$ by $|J_i|\ V(R_i)$, where J_i is a value assumed by the Jacobian determinant of T in R_i. Thus we are led to the approximation

$$\sum_i f_i V(T(R_i)) \approx \sum_i f_i\,|J_i|\ V(R_i). \tag{3}$$

But the number f_i is equally well a value of $f \circ T$ in R_i, which we can write $(f \circ T)_i$, getting from Formulas (2) and (3)

$$\int_{T(R)} f\,dV \approx \sum_i (f \circ T)_i\,|J_i|\ V(R_i).$$

Finally, the last sum can be used to approximate $\int_R (f \circ T)\,|J|\,dV$. To make this argument precise is difficult; so the proof given in the Appendix follows other lines.

The foregoing discussion shows that the Jacobian determinant can be interpreted as an approximate local magnification factor for volume. A one-to-one continuously differentiable transformation, looked at as a coordinate change, leads to another slightly different interpretation of J.

Example 7. Polar coordinate curves in $\mathcal{R}^2$ bound regions like S in Fig. 24. Since the Jacobian determinant of the polar coordinate transformation

$$\begin{pmatrix} x \\ y \end{pmatrix} = \begin{pmatrix} r\cos\theta \\ r\sin\theta \end{pmatrix} = T\begin{pmatrix} r \\ \theta \end{pmatrix}$$

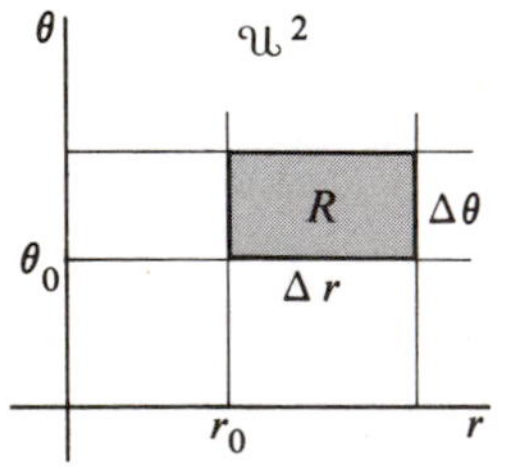

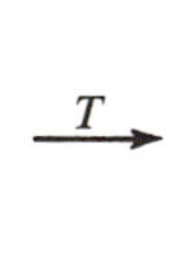

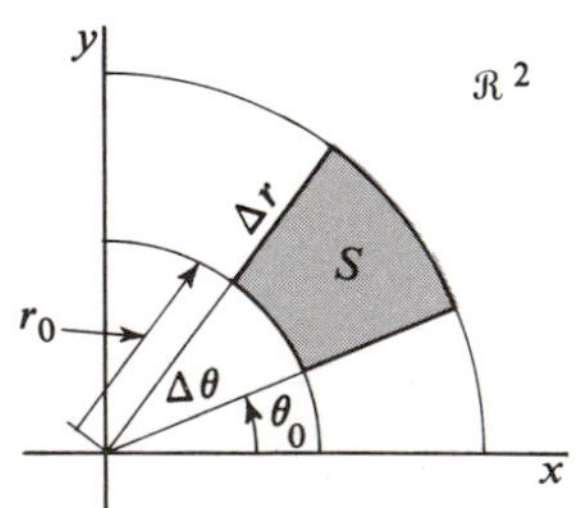

Figure 24

is

$$J = \begin{vmatrix} \cos\theta & -r\sin\theta \\ \sin\theta & r\cos\theta \end{vmatrix} = r,$$

we expect an approximation to the shaded area $S = T(R)$ in the xy-plane to be $r_0\,\Delta r\,\Delta\theta$. Computation of the exact area of S, using the change-of-variable theorem, gives

$$\begin{aligned} \int_S dA = \int_R r\,dA &= \int_{r_0}^{r_0+\Delta r} r\,dr \int_{\theta_0}^{\theta_0+\Delta\theta} d\theta \\ &= \left(\tfrac{1}{2}(r_0 + \Delta r)^2 - \tfrac{1}{2}r_0^2\right)\Delta\theta \\ &= r_0\,\Delta r\,\Delta\theta + \tfrac{1}{2}(\Delta r)^2\,\Delta\theta \\ &\approx r_0\,\Delta r\,\Delta\theta \qquad (\text{for small } \Delta r, \Delta\theta). \end{aligned}$$

Thus the significance of J in this case is that $J\,\Delta r\,\Delta\theta$ is an approximation to the area of a polar coordinate "rectangle," or region bounded by polar coordinate curves, with Δr and $\Delta\theta$ as the difference between pairs of values of r and θ.

The expression $r\,\Delta r\,\Delta\theta$ is called the **area element** in polar coordinates. More generally, if J is the Jacobian determinant of a coordinate change, then $|J|\,\Delta V$ is called its **volume element.**

Example 8. Spherical coordinates are introduced in $\mathcal{R}^3$ by means of the transformation

$$\begin{pmatrix} x \\ y \\ z \end{pmatrix} = \begin{pmatrix} r\sin\phi\cos\theta \\ r\sin\phi\sin\theta \\ r\cos\phi \end{pmatrix}.$$

Except for a notational change, the same transformation is considered in Example 3. The Jacobian determinant is $J = r^2\sin\phi$. This suggests the approximation $r^2\sin\phi\,\Delta r\,\Delta\phi\,\Delta\theta$ for the volume of the spherical coordinate "cube" C shown in Fig. 25. The spherical ball with center at the origin and radius 1 is defined in $\mathcal{R}^3$ by $x^2 + y^2 + z^2 \le 1$ and is denoted below by B_{xyz}. With respect to polar coordinates, the same ball is defined by the inequalities

$$0 \le r \le 1, \qquad 0 \le \phi \le \pi, \qquad 0 \le \theta \le 2\pi,$$

and is denoted below by $B_{r\phi\theta}$. Using the change-of-variable theorem, we

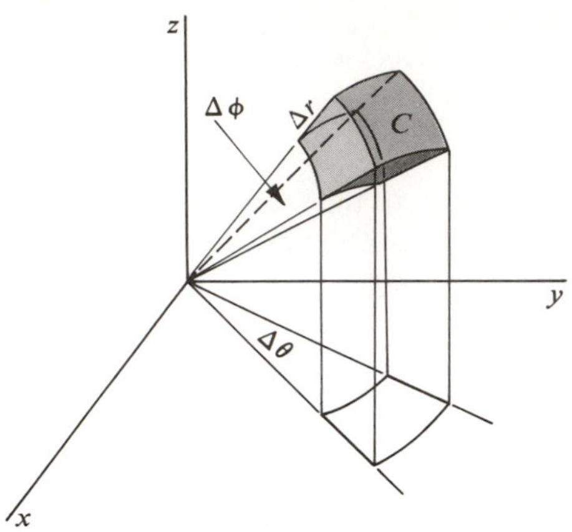

Figure 25

compute the volume of the ball to be

$$V(B_{xyz}) = \int_{B_{xyz}} dx\, dy\, dz = \int_{B_{r\phi\theta}} r^2 \sin\phi\, dr\, d\phi\, d\theta$$
$$= \int_0^1 r^2\, dr \int_0^{\pi} \sin\phi\, d\phi \int_0^{2\pi} d\theta$$
$$= \tfrac{1}{3}(1+1)2\pi = \frac{4\pi}{3}.$$

Notice (as in Example 3) that both conditions (a) and (b) of Theorem 3.1 fail to hold on $B_{r\phi\theta}$. However, except on a subset of $B_{r\phi\theta}$ having zero volume, the Jacobian $J = r^2 \sin\phi$ is positive, and the coordinate transformation is one-to-one. The change-of-variable formula is therefore applicable.

EXERCISES

1. Let

$$\begin{pmatrix} x \\ y \end{pmatrix} = T\begin{pmatrix} u \\ v \end{pmatrix} = \begin{pmatrix} u^2 - v^2 \\ 2uv \end{pmatrix}.$$

(a) Sketch the image under T of the square in $\mathcal{U}^2$ with vertices at (1, 1), $(1, \frac{3}{2})$, $(\frac{3}{2}, 1)$, $(\frac{3}{2}, \frac{3}{2})$.

(b) Sketch the image under $T'\begin{pmatrix} 1 \\ 1 \end{pmatrix}$ of the square in part (a).

(c) Sketch the translate of the image found in part (*b*) by the vector

$$T\begin{pmatrix} 1 \\ 1 \end{pmatrix} - T'\begin{pmatrix} 1 \\ 1 \end{pmatrix}\begin{pmatrix} 1 \\ 1 \end{pmatrix}.$$

Verify that this is the image of the square under the affine approximation to T at $\begin{pmatrix} 1 \\ 1 \end{pmatrix}$.

(d) Find the area of the region sketched in (c). [*Ans.* 2.]
(e) Find the area of the region sketched in (a). [*Ans.* $\frac{19}{6}$.]

2. Let

$$\begin{pmatrix} x \\ y \end{pmatrix} = T\begin{pmatrix} u \\ v \end{pmatrix} = \begin{pmatrix} u \cos v \\ u \sin v \end{pmatrix}.$$

(a) Sketch the image under T of the square S with vertices at $(0, 0)$, $(0, \pi/2)$, $(\pi/2, 0)$, and $(\pi/2, \pi/2)$.

(b) Sketch the image under $T'\begin{pmatrix} \pi/2 \\ 0 \end{pmatrix}$ of the square S. What is the area of the image?

(c) Sketch the image of S under the affine approximation to T at $(\pi/4, \pi/4)$. What is the area of the image?

(d) What is the area of the region sketched in (a)?

3. Let

$$T\begin{pmatrix} u \\ v \end{pmatrix} = \begin{pmatrix} u \cos v \\ u \sin v \end{pmatrix}.$$

Show that $T'\begin{pmatrix} u \\ v \end{pmatrix}$ transforms a rectangle of area A into a region having area uA.

4. Compute the area of the image of the rectangle in the uv-plane with vertices at $(0, 0)$, $(0, 1)$, $(2, 0)$, and $(2, 1)$ under the transformation

$$\begin{pmatrix} x \\ y \end{pmatrix} = \begin{pmatrix} 2 & 3 \\ 2 & 1 \end{pmatrix}\begin{pmatrix} u \\ v \end{pmatrix}.$$ [*Ans.* 8.]

5. Consider the transformation T defined by

$$\begin{pmatrix} x \\ y \end{pmatrix} = T\begin{pmatrix} u \\ v \end{pmatrix} = \begin{pmatrix} u^2 - v^2 \\ 2uv \end{pmatrix}.$$

Let R_{uv} be the quarter of the unit disk lying in the first quadrant, i.e., $u^2 + v^2 \leq 1$, $u \geq 0$, $v \geq 0$.

(a) Sketch the image region $R_{xy} = T(R_{uv})$.

(b) Compute $\displaystyle\int_{R_{xy}} \frac{dx\, dy}{\sqrt{x^2 + y^2}}$. [*Ans.* π.]

6. Let the transformation from the uv-plane to the xy-plane be defined by

$x = u + v$, $y = u^2 - v$. Let R_{uv} be the region bounded by (1) u-axis, (2) v-axis, and (3) the line $u + v = 2$.

(a) Find and sketch the image region R_{xy}.
(b) Compute the integral

$$\int_{R_{xy}} \frac{dx\,dy}{\sqrt{1 + 4x + 4y}}.$$ [*Ans.* 2.]

7. Let a transformation of the uv-plane to the xy-plane be given by

$$x = u, \qquad y = v(1 + u^2),$$

and let R_{uv} be the rectangular region given by $0 \le u \le 3$ and $0 \le v \le 2$.

(a) Find and sketch the image region R_{xy}.

(b) Find $\dfrac{\partial(x, y)}{\partial(u, v)}$.

(c) Transform $\int_{R_{xy}} x\,dx\,dy$ to an integral over R_{uv} and compute either one of them. [*Ans.* $\frac{99}{2}$.]

8. The transformation $u = x^2 - y^2$, $v = 2xy$ maps the region D (see sketch) onto a region R in the uv-plane, and it is one-to-one on D.

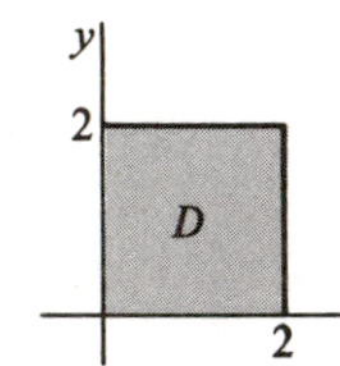

(a) Find R.
(b) Compute $\int_R 1\,du\,dv$ by integrating directly over R, and then by using the transformation formulas to integrate over D. [*Ans.* $\frac{128}{3}$.]
(c) Compute $\int_R v\,du\,dv$ both directly and by using the change-of-variable theorem. [*Ans.* 128.]

9. Let a transformation from the xy-plane to the uv-plane be given by

$$u = x$$
$$v = y(1 + 2x).$$

(a) What happens to horizontal lines in the xy-plane?
(b) If D is the rectangular region

$$0 \le x \le 3$$
$$1 \le y \le 3,$$

find the image region R of D.
(c) Find

$$\int_R du\,dv, \qquad \int_R v\,dv\,du, \qquad \text{and} \qquad \int_R u\,dv\,du$$

by direct integration, and then by reducing them to integrals over D. [*Ans.* 24, 228, 45.]

10. Compute the area bounded by the polar coordinate curves $\theta = 0$, $\theta = \pi/4$, and $r = \theta^2$. [*Ans.* $\pi^5/(2^{10} \cdot 10)$.]

11. Find the area bounded by the lemniscate $(x^2 + y^2)^2 = 2a^2(x^2 - y^2)$ by changing to polar coordinates. [*Ans.* $2a^2$.]

12. Compute the volume of the ellipsoid

$$\frac{x^2}{a^2} + \frac{y^2}{b^2} + \frac{z^2}{c^2} \leq 1.$$

[Use the transformation $(x, y, z) = (au, bv, cw)$ to transform the sphere $u^2 + v^2 + w^2 \leq 1$ onto the ellipsoid. Assume the volume of the sphere to be known.]

13. Evaluate the integral of $f(x, y, z) = a$ over the hemisphere $x^2 + y^2 + z^2 \leq 1$, $x \geq 0$ by changing to spherical coordinates. [*Ans.* $\frac{2}{3}\pi a$.]

14. (a) Compute the Jacobian of the cylindrical coordinate transformation

$$\begin{pmatrix} x \\ y \\ z \end{pmatrix} = \begin{pmatrix} r\cos\theta \\ r\sin\theta \\ z \end{pmatrix}.$$

(b) Use cylindrical coordinates to compute

$$\int_{\substack{0 \leq z \leq 1 \\ x^2+y^2 \leq 1,}} x^2\, dx\, dy\, dz.$$

15. Prove that the transformation

$$\begin{aligned} x_1 &= u_1 \\ x_2 &= u_1 + u_2 \\ x_3 &= u_1 + u_2 + u_3 \\ &\;\cdot \\ &\;\cdot \\ &\;\cdot \\ x_n &= u_1 + u_2 + \ldots + u_n \end{aligned}$$

leaves volumes of corresponding regions unchanged.

16. Cutting a solid of revolution R into thin cylindrical shells, with axis the same as the axis of revolution, leads intuitively to the following formula for the volume of the solid:

$$\int 2\pi r h(r)\, dr.$$

Here $h(r)$ is the thickness of the solid at a distance r from its axis, measured along a line parallel to the axis. Show that introducing cylindrical coordinates in the integral $\int_R dx\, dy\, dz$ leads to the same formula.

17. (a) Let a ball B of radius a have density ρ at each of its points equal to the distance of the point from a fixed diameter. Find the total mass of the ball.
[*Hint.* Compute the integral $\int_B \rho\, dV$ by using spherical coordinates.]

(b) Let a cylinder of height h and radius a have a density ρ equal at each point to the distance of the point from the axis of the cylinder. Find the total mass of the cylinder.

SECTION 4 IMPROPER INTEGRALS

The definition of the integral can be extended to functions that are unbounded and not necessarily zero outside some bounded set. We shall first consider some examples.

Example 1. The function $f(x, y) = 1/x^2y^2$, defined for $x \geq 1$ and $y \geq 1$, has the graph shown in Fig. 26. If B is the set of points (x, y) for

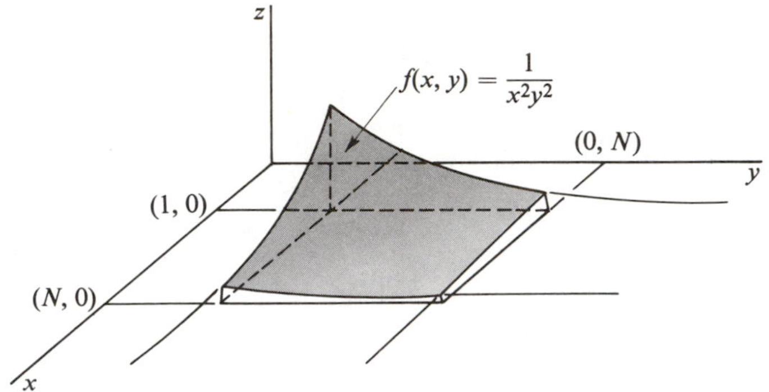

Figure 26

which $x \geq 1$ and $y \geq 1$, it is natural to define $\int_B f\, dA$ in such a way that it can be called the volume under the graph of f. We can approximate this volume by computing the volume lying above bounded subrectangles of B. To be specific, let B_N be the rectangle with corners at $(1, 1)$ and (N, N) and with edges parallel to the edges of B. For $N > 1$ we have

$$\int_{B_N} f\, dA = \int_1^N dx \int_1^N \frac{1}{x^2y^2}\, dy$$
$$= \left(\int_1^N \frac{dx}{x^2}\right)^2 = \left(1 - \frac{1}{N}\right)^2.$$

As N tends to infinity, the rectangles B_N eventually cover every point of B, and the regions above the B_N fill out the region under the graph of f. Then, by definition,

$$\int_B f\, dA = \lim_{N \to \infty} \int_{B_N} f\, dA = 1.$$

Example 2. Let B be the disk $x^2 + y^2 \leq 1$ in $\mathcal{R}^2$, and suppose

$$f(x, y) = -\log(x^2 + y^2), \qquad 0 < x^2 + y^2 \leq 1.$$

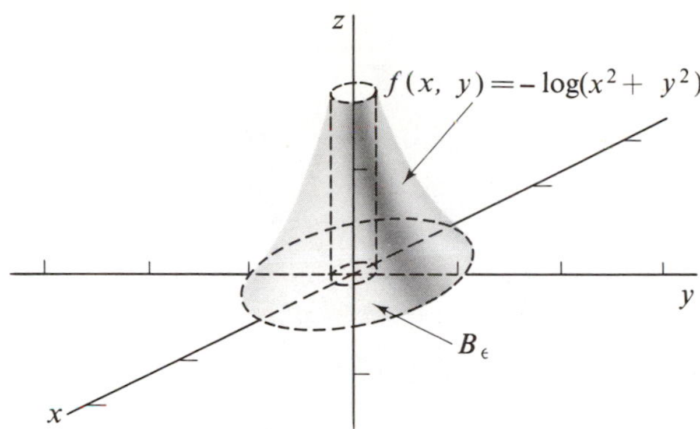

Figure 27

The graph of f is shown in Fig. 27. Since f is unbounded near (0, 0), we cut out from B a disk centered at (0, 0) and with radius ϵ. Call the part of B that is left B_ϵ. We have, using polar coordinates,

$$\int_{B_\epsilon} -\log(x^2+y^2)\,dx\,dy = \int_0^{2\pi} d\theta \int_\epsilon^1 -(\log r^2) r\,dr$$
$$= -2\pi[r^2 \log r - \tfrac{1}{2}r^2]_\epsilon^1$$
$$= \pi + 2\pi\epsilon^2 \log \epsilon - \pi\epsilon^2.$$

Since $\lim_{\epsilon \to 0} (2\pi\epsilon^2 \log \epsilon - \pi\epsilon^2) = 0$, we get, by definition,

$$\int_B -\log(x^2+y^2)\,dx\,dy = \lim_{\epsilon\to 0} \int_{B_\epsilon} -\log(x^2+y^2)\,dx\,dy = \pi.$$

In Example 2, the function $-\log(x^2 + y^2)$ becomes unbounded in the disk $x^2 + y^2 \leq 1$ only at the point (0, 0). It is of course important to find all such points in attempting to integrate an unbounded function. In general, we define an **infinite discontinuity point** for a function f to be a point $\mathbf{x}$ such that in any neighborhood of $\mathbf{x}$, $|f|$ assumes arbitrarily large values.

Example 3. Consider the function $g(x, y, z) = (x^2 + y^2 + z^2 - 1)^{-1/2}$ for (x, y, z) satisfying $1 < x^2 + y^2 + z^2 \leq 2$. The domain of g is the region between the concentric spheres shown in Fig. 28. Every point of the sphere $x^2 + y^2 + z^2 = 1$ is an infinite discontinuity point for g. To define the integral of g, we approximate its domain by shells B_ϵ determined by $1 + \epsilon \leq x^2 + y^2 + z^2 \leq 2$. These shells have the property of filling out the entire domain of g as ϵ tends to zero, although none of them

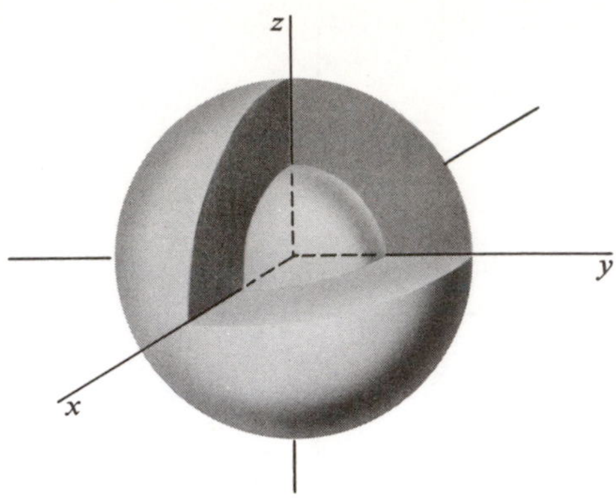

Figure 28

contains an infinite discontinuity point. Introducing spherical coordinates, we obtain

$$\int_{B_\epsilon} (x^2 + y^2 + z^2 - 1)^{-1/2}\, dx\, dy\, dz$$

$$= \int_{1+\epsilon}^{2} dr \int_0^{2\pi} d\theta \int_0^{\pi} (r^2 - 1)^{-1/2} r^2 \sin\phi\, d\phi$$

$$= 4\pi \int_{1+\epsilon}^{2} (r^2 - 1)^{-1/2} r^2\, dr$$

$$= 4\pi[\tfrac{1}{2} r\sqrt{r^2 - 1} + \tfrac{1}{2}\log(r + \sqrt{r^2 - 1})]_{1+\epsilon}^{2}.$$

It follows immediately that

$$\lim_{\epsilon \to 0} \int_{B_\epsilon} (x^2 + y^2 - z^2 + 1)^{1/2}\, dx\, dy\, dz = 4\pi\sqrt{3} + 2\pi \log(2 + \sqrt{3}).$$

Before collecting the ideas illustrated above into a general definition, we make two requirements about the integrand f and the set B over which it is to be integrated.

1. Let D be the set of points of B at which f is not continuous. The part of D lying in an arbitrary bounded rectangle is to be contained in finitely many smooth sets.

2. The part of B lying in an arbitrary bounded rectangle is to have a boundary consisting of finitely many smooth sets.

Both conditions are satisfied in the three examples considered so far, and we shall assume that they hold throughout the rest of the section.

In the examples, we have seen that the integral $\int_B f\, dV$ can sometimes be defined when either f or B is unbounded. The extended definition of the integral will be made in such a way that both phenomena can occur at once. We proceed as follows. An increasing family $\{B_N\}$ of subsets of B will be said to **converge to** B if every bounded subset of B on which f is bounded is contained in some one of the sets B_N. Notice that this notion of convergence depends not only on B but also on f. The index N can be chosen in any convenient way; it may, for example, tend to ∞ continuously or through integer values, or it may tend to some finite number. Throughout the rest of this section we shall assume that, in any increasing family $\{B_N\}$ converging to B, each of the sets B_N satisfies condition 2.

The **integral of f over B** is by definition

$$\lim_N \int_{B_N} f\,dV = \int_B f\,dV,$$

provided that the limit is finite and is the same for every increasing family of bounded sets B_N converging to B. It is assumed that the B_N are chosen so that the ordinary Riemann integrals $\int_{B_N} f\,dV$ (as defined in Section 2) exist. The integral thus obtained is called the **improper Riemann integral** when it is necessary to distinguish it from the Riemann integral of a bounded function over a bounded set.

Although the requirement that the value of the integral be independent of the converging family of sets used to define it is a natural one, we shall see later that it is sometimes interesting to disregard it. Nevertheless, the next theorem shows that for positive functions the limit of $\int_{B_N} f\,dV$ is always independent of the family of sets.

4.1 Theorem

Let f be nonnegative on B and suppose that

$$\lim_N \int_{B_N} f\,dV$$

is finite for some particular increasing family of sets B_N converging to B. Then $\int_B f\,dV$ is defined and has the same value,

$$\lim_N \int_{C_N} f\,dV,$$

for every other family $\{C_N\}$ converging to B.

Proof. Since f is bounded on each B_N, we have for each N an index K such that

$$B_N \subset C_K.$$

Similarly, there is an index M depending on K such that

$$C_K \subset B_M.$$

Then, because f is nonnegative,

$$\int_{B_N} f\,dV \leq \int_{C_K} f\,dV \leq \int_{B_M} f\,dV.$$

In addition,

$$\int_{C_N} f\,dV \leq \lim_N \int_{B_N} f\,dV$$

for all N. Because $\int_{C_N} f\,dV$ increases and is bounded above,

$$\lim_N \int_{C_N} f\,dV$$

exists. The double inequality shows that

$$\lim_N \int_{B_N} f\,dV = \lim_N \int_{C_N} f\,dV.$$

This completes the proof.

Example 4. Let f be defined on the infinite strip S in $\mathcal{R}^2$, shown in Fig. 29, by $f(x, y) = y^{-1/2}e^{-x}$. Clearly, f has an infinite discontinuity at

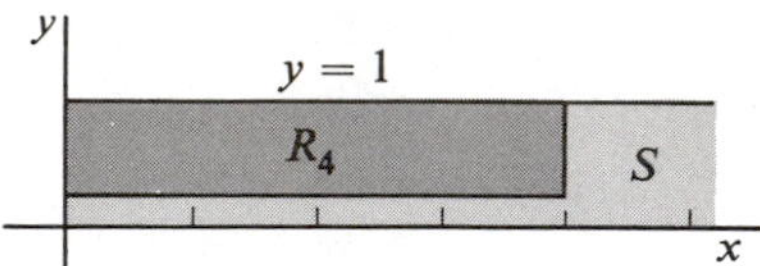

Figure 29

every point of the positive x-axis. We define R_N to be the rectangle in S bounded by the lines $x = N$ and $y = 1/N$, for $N > 1$. As N tends to infinity, R_N will converge to S. We have

$$\begin{aligned}\int_{R_N} f\,dA &= \int_0^N dx \int_{1/N}^1 y^{-1/2}e^{-x}\,dy \\ &= \left(\int_0^N e^{-x}\,dx\right)\left(\int_{1/N}^1 y^{-1/2}\,dy\right) \\ &= (1 - e^{-N})\left(2 - \frac{2}{\sqrt{N}}\right).\end{aligned}$$

Then

$$\int_S y^{-1/2}e^{-x}\,dx\,dy = \lim_{N\to\infty} (1 - e^{-N})\left(2 - \frac{2}{\sqrt{N}}\right) = 2.$$

Example 5. The integral of $1/x^a$ over the positive x-axis, denoted by $\int_0^\infty x^{-a}\,dx$, fails to exist for any a. Consider, for $N > 0$,

$$\int_{1/N}^{N} x^{-a}\,dx = \begin{cases} \dfrac{N^{1-a} - 1/N^{1-a}}{1-a}, & a \neq 1, \\ 2\log N, & a = 1. \end{cases}$$

As N tends to infinity, we get infinity for a limit in every case. However, it is easy to verify that

$$\int_1^\infty x^{-a}\,dx = \frac{1}{a-1}, \qquad \text{for } a > 1,$$

and

$$\int_0^1 x^{-a}\,dx = \frac{1}{1-a}, \qquad \text{for } a < 1.$$

The integral

$$\int_{-1}^{1} \frac{1}{x}\,dx$$

fails to exist if we require that its value be independent of the limit process by which it is computed. Indeed, if we integrate first over the intervals $[-1, -\delta]$ and $[\epsilon, 1]$ with $0 < \delta < 1$ and $0 < \epsilon < 1$, we get

$$\int_{-1}^{-\delta} \frac{1}{x}\,dx + \int_{\epsilon}^{1} \frac{1}{x}\,dx = \log|x|\Big]_{-1}^{-\delta} + \log x\Big]_{\epsilon}^{1}$$

$$= \log\frac{\delta}{\epsilon}.$$

As ϵ and δ tend to zero, $\log(\delta/\epsilon)$ can be made to tend to any number by controlling the limit of the ratio δ/ϵ. In particular, if we keep $\epsilon = \delta$, the limit is zero.

For a function f having a graph symmetric about some point $\mathbf{x}_0$, it is sometimes significant to define $\int_B f\,dV$ by a limit using sets B_N that are also symmetric about $\mathbf{x}_0$. This is what we have done in the previous example, and in general, we speak of computing a **principal value** (p.v.) of the integral. For the integral in the last part of Example 5 we would write

$$\text{p.v.}\int_{-1}^{1} \frac{dx}{x} = 0.$$

Example 6. Let (r, θ) be polar coordinates in $\mathcal{R}^2$, and set $f(r, \theta) = (\sin\theta)/r^2$ over the disk D of radius 1 centered at the origin. Clearly, f has an infinite discontinuity at the origin because, for instance, along the line $\theta = \pi/2$, f tends to ∞ and along the line $\theta = 3\pi/2$, f tends to $-\infty$. (See Fig. 30 for the graph of f.) However, $\int_D f\,dA$ fails to exist in the ordinary

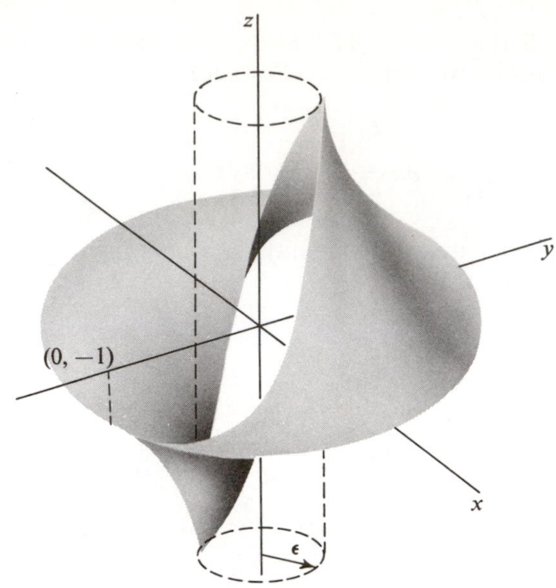

Figure 30

improper integral sense because the limit obtained from a sequence of regions in D will depend on the way in which the positive and negative values of f are balanced. A principal value of the integral can be determined by taking a limit over a family of annular regions. Let D_ϵ be the annulus $\epsilon \leq r \leq 1$. Then

$$\begin{aligned}\text{p.v.}\int_D f\,dV &= \lim_{\epsilon\to 0}\int_{D_\epsilon}\frac{\sin\theta}{r^2}\,r\,dr\,d\theta\\ &= \lim_{\epsilon\to 0}\int_0^{2\pi}\sin\theta\,d\theta\int_\epsilon^1\frac{dr}{r} = 0.\end{aligned}$$

The next theorem is a convenient test for the existence of an improper integral.

4.2 Theorem

Let f and g have the same infinite discontinuity points. If $|f| \leq g$ and $\int_B g\,dV$ exists, then so does $\int_B f\,dV$.

Proof. Let $\{B_N\}$ be an increasing family of sets converging to B. Since $f + |f| \leq 2\,|f| \leq 2g$, we have

$$\int_{B_N}(f + |f|)\,dV \leq 2\int_{B_N} g\,dV \leq 2\int_B g\,dV.$$

Then, because $f + |f| \geq 0$, the value of the integral $\int_{B_N} (f + |f|)\, dV$ increases as B_N increases, and we have

$$\lim_N \int_{B_N} (f + |f|)\, dV = l_1 \leq 2\int_B g\, dV.$$

Similarly,

$$\lim_N \int_{B_N} |f|\, dV = l_2 \leq \int_B g\, dV.$$

Finally,

$$\begin{aligned}\lim_N \int_{B_N} f\, dV &= \lim_N \left(\int_{B_N} (f + |f|)\, dV - \int_{B_N} |f|\, dV\right)\\ &= \lim_N \int_{B_N} (f + |f|)\, dV - \lim_N \int_{B_N} |f|\, dV = l_1 - l_2.\end{aligned}$$

Since the family B_N is arbitrary, $\int_B f\, dV$ is defined.

Example 7. Let B be the disk $x^2 + y^2 \leq 1$ in $\mathcal{R}^2$, and let f be defined by

$$f(x, y) = \begin{cases}(x^2 + y^2)^{-1/2}, & \text{for } x \geq 0 \text{ and } x^2 + y^2 > 0.\\ (x^2 + y^2)^{1/2}, & \text{for } x < 0.\end{cases}$$

Since $\int_B (x^2 + y^2)^{-1/2}\, dx\, dy$ exists, and $|f(x, y)| \leq (x^2 + y^2)^{-1/2}$ for $0 < x^2 + y^2 \leq 1$, it follows from Theorem 4.2 that $\int_B f\, dx\, dy$ exists. The computation of the integral of f is left as an exercise.

Example 8. Let $j(x) = (-1)^{n-1}/n$ for $n - 1 \leq x < n$ and $n = 1, 2, 3, \ldots$. The graph of j is shown in Fig. 31 as far out as $x = 4$. Then

$$\lim_{n\to\infty} \int_0^n j(x)\, dx = \sum_{n=1}^{\infty} \frac{(-1)^{n-1}}{n} = \log 2,$$

and we can write

$$\int_0^\infty j(x)\, dx = \log 2,$$

if it is understood that the passage to the limit has been carried out in this special way. This example shares with the principal-value examples the property that the value assigned to the integral depends on having taken

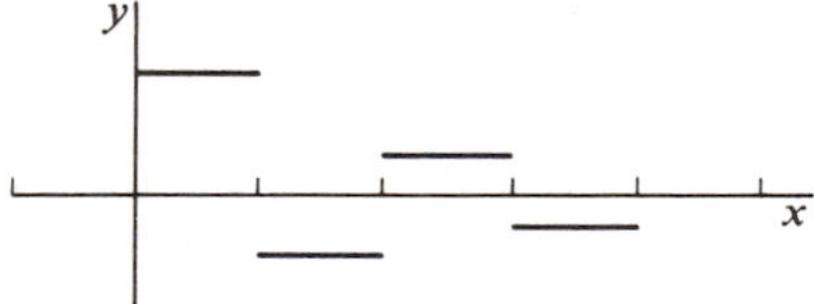

Figure 31

a limit over some particular sequence of regions. Such an integral is called **conditionally convergent.** For another example see Exercise 8.

EXERCISES

1. In each part determine whether the integral is defined or not. If it is defined, compute its value.

(a) $\displaystyle\int_0^\infty \frac{dx}{x^2+1}.$ [*Ans.* $\pi/2$.]

(b) $\displaystyle\int_{-\infty}^\infty \frac{dx}{x^2-1}.$

(c) $\displaystyle\int_0^1 \frac{dx}{\sqrt{1-x^2}}.$

(d) $\displaystyle\int_{x^2+y^2\le 1} \frac{dx\,dy}{\sqrt{x^2+y^2}}.$

(e) $\displaystyle\int_R \frac{(x-y)\,dx\,dy}{x^2+y^2}$, where R is the rectangle $\max(|x|, |y|) \le 1$.

(f) $\displaystyle\int_{x^2+y^2+z^2\ge 1} \frac{dx\,dy\,dz}{(x^2+y^2+z^2)^2}.$

(g) $\displaystyle\int_{x^2+y^2+z^2\ge 1} \frac{dx\,dy\,dz}{xyz}.$

(h) $\displaystyle\int_C e^{-x-y-z}\,dx\,dy\,dz$, where C is the infinite column

$$\max(|x|, |y|) \le 1, \qquad z \ge 0.$$

2. Prove that

(a)
$$\Gamma(n) = \int_0^\infty e^{-x}x^{n-1}\,dx = (n-1)!$$

for $n > 1$ an integer.

(b) Express $\int_T e^{-x}(x-y)^{-1/2}\,dx\,dy$ in terms of Γ, where T is the region $x \ge y \ge 0$. [*Ans.* $2\Gamma(\frac{3}{2})$.]

3. Let B be the ball $|\mathbf{x}| \le 1$ in $\mathcal{R}^n$. For what values of a does $\displaystyle\int_B \frac{dV}{|\mathbf{x}|^a}$ exist?

4. Compute: (a) p.v.$\displaystyle\int_{-\infty}^\infty \frac{x\,dx}{x^2+1}.$ (b) p.v.$\displaystyle\int_{-1}^1 \frac{x\,dx}{2x-1}.$

5. Compute the values of the function $g(y) = \displaystyle\int_{-1}^1 \frac{x\,dx}{x-y}$, taking a principal value of the integral when necessary.

6. Compute the integral of the function f in Example 7 in the text, and compute $\int_B (x^2+y^2)^{-1/2}\,dx\,dy$. [*Ans.* $\frac{4}{3}\pi$, 2π.]

7. Show that the integral of the function j in Example 8 of the text depends on the sequence of sets used to compute the limit. [*Suggestion.* Take each B_N to be a disconnected set of intervals.]

8. Let $f(x, y) = \sin(x^2 + y^2)$ over the quadrant Q defined by $x \geq 0$, $y \geq 0$. Show that $\int_Q f\,dA$ converges conditionally. (*Suggestion.* To get a limit, integrate over increasing squares. Then integrate over quarter disks.)

9. In what sense does each of the following integrals exist? The possibilities are ordinary Riemann integral, improper integral, conditionally convergent integral, or none of these.

(a) $\displaystyle\int_\pi^\infty \frac{\sin x}{x^2}\,dx.$ (c) $\displaystyle\int_0^\infty \sin x\,dx.$

(b) $\displaystyle\int_0^\infty \frac{\sin x}{x}\,dx.$ (d) $\displaystyle\int_0^1 \sin\frac{1}{x}\,dx.$

10. The integral $\displaystyle\int_0^\infty f(x)\,dx$ is said to be **Abel summable** to the value k if

$$\lim_{\epsilon\to 0+}\int_0^\infty e^{-\epsilon x} f(x)\,dx = k.$$

Find the Abel value of (a) $\displaystyle\int_0^\infty \sin x\,dx$, (b) $\displaystyle\int_0^\infty \cos x\,dx$.

11. (a) Compute $\displaystyle\int_{\mathscr{R}^2} e^{-x^2-y^2}\,dx\,dy$. (Use polar coordinates.)

(b) Use the result of part (a) to compute $\int_{-\infty}^{\infty} e^{-x^2}\,dx$.

(c) Compute $\int_{\mathscr{R}^n} \exp(-x_1^2 - \ldots - x_n^2)\,dV$.

12. (a) Show that the area bounded by the graph of $y = 1/x$, the x-axis, and the line $x = 1$ is infinite.

(b) Compute the volume swept out by rotating the region described in part (a) about the x-axis.

13. Let f be positive and unbounded on an unbounded set B in $\mathscr{R}^2$. Consider the region C between the graph of f and B, and show that if

$$\int_B f\,dA \quad \text{and} \quad \int_C dV$$

both exist, then they are equal.

14. Show that if $\int_B |f|\,\int dV$ exists, then so does $\int_B f\,dV$. Without conditions (1) and (2), this is false. For example, let B be the unit interval $0 \leq x \leq 1$ and f the function

$$f(x) = \begin{cases} 1, & \text{if } x \text{ is rational.} \\ -1, & \text{if } x \text{ is irrational.} \end{cases}$$

15. Show that if the ordinary Riemann integral $\int_B f\,dV$ exists, then it exists as an improper integral (given conditions (1) and (2)) and the two integrals are equal.

16. A nonnegative function $\mathcal{R}^n \xrightarrow{p} \mathcal{R}$, integrable over a region R in $\mathcal{R}^n$, is called a probability density if $\int_R p\,dV = 1$. The **mean** of p is defined to be the vector

$$M[p] = \int_R \mathbf{x}p(\mathbf{x})\,dV,$$

and the **variance** of p is the real number

$$\sigma^2[p] = \int_R |\mathbf{x} - M[p]|^2\,p(\mathbf{x})\,dV.$$

Show that each of the following functions is a probability density, and compute its mean and variance if they exist.

(a) $p(x, y) = (1/2\pi)e^{-(x^2+y^2)/2}$. [*Hint.* Use polar coordinates.]
(b) $p(x) = (\pi(1 + x^2))^{-1}$.

SECTION 5

ESTIMATES OF INTEGRALS

In many of the examples of this chapter, numerical evaluation of integrals has been made by using the fundamental theorem of calculus to arrive at a precise answer in terms of some elementary function. In practice, such a computation is very often not feasible, and then an estimate for the value of an integral may have to serve instead. The fundamental inequality used in making estimates is contained in Theorem 2.8. We repeat it here.

5.1 If f and g are integrable over B, and $f \leq g$ on B, then

$$\int_B f\,dV \leq \int_B g\,dV.$$

Example 1. The function of one variable defined by

$$f(x) = \frac{\sin x}{(1 + x^2)^2}, \qquad 0 \leq x \leq 1$$

has an integral which is difficult to compute. Comparing the graphs of x and $\sin x$ for $0 \leq x \leq 1$ shows that $0 \leq \sin x \leq x$ there, as we see from Fig. 32. Then, since $(1 + x^2)^2 > 0$,

$$0 \leq \frac{\sin x}{(1 + x^2)^2} \leq \frac{x}{(1 + x^2)}.$$

Hence, by 5.1,

$$0 \leq \int_0^1 \frac{\sin x}{(1 + x^2)^2}\,dx \leq \int_0^1 \frac{x\,dx}{(1 + x^2)^2}.$$

But the latter integral is easy to compute:

$$\int_0^1 \frac{x\,dx}{(1 + x^2)^2} = \left[\frac{-1}{2(1 + x^2)}\right]_0^1 = -\tfrac{1}{4} + \tfrac{1}{2} = 0.25;$$

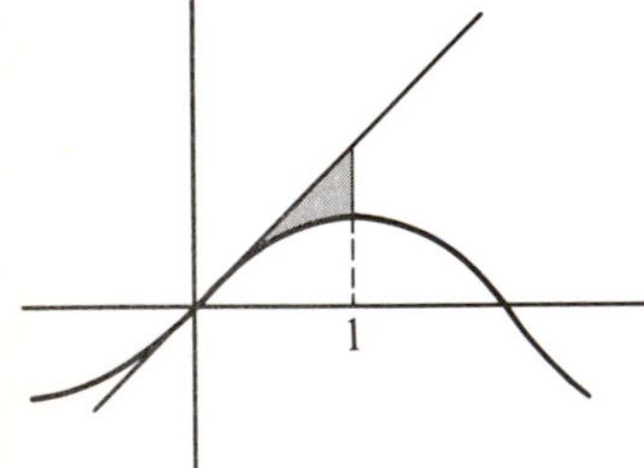

Figure 32

hence

$$0 \leq \int_0^1 \frac{\sin x}{(1 + x^2)^2}\, dx \leq 0.25.$$

Example 2. Suppose we want to estimate the integral of $f(x, y) = \cos(x + y/2)$ over the square $S: 0 \leq x \leq \frac{1}{2}, 0 \leq y \leq \frac{1}{2}$. The graph of f is

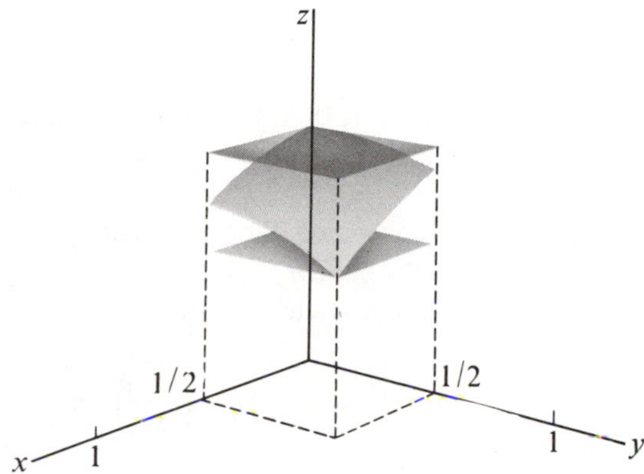

Figure 33

shown in Fig. 33, lying between the horizontal planes at $z = 1$ and $z = \cos \frac{3}{4} > 0.73$. Thus

$$0.73 \leq \cos\left(\frac{x + y}{2}\right) \leq 1.$$

By 5.1 we have

$$\int_S 0.73\, dx\, dy \leq \int_S \cos\left(\frac{x + y}{2}\right) dx\, dy \leq \int_S 1\, dx\, dy.$$

Evaluating the largest and smallest of these integrals is easy because they are the volumes of solid rectangles with base S and heights 0.73 and 1, respectively. Since the area of S is 0.25, we get

$$0.1825 \leq \int_S \cos(x + 2y)\, dx\, dy \leq 0.25.$$

The estimates in the two preceding examples are fairly rough because we have replaced the given integrand f by approximating functions that differ from f considerably over a relatively large part of the domain of integration. Of course, the estimates could be improved by choosing approximating functions that agree with f more closely. However, to be really useful, the approximating functions themselves should be easy to integrate. One way to achieve this is to choose for an approximating function a step function that is constant on each subrectangle of a rectangular grid in the domain of integration. The step function can be chosen so that its integral is a Riemann sum, and a sequence of these can be chosen so as to converge to the value of the integral.

Example 3. The integral

$$\int_R (1 + x + y)^3 \, dx \, dy,$$

where R is the rectangle $0 \le x \le 1$, $0 \le y \le \frac{1}{2}$, can be approximated by Riemann sums. We subdivide the rectangle by a grid with corners at points $(j/N, k/2M)$, where $j = 1, \ldots, N$, $k = 1, \ldots, M$. Evaluating the integrand at the upper right corner of each subrectangle gives function values of the form $(1 + j/N + k/2M)^3$. Each subrectangle has area $(1/N)\big(1/(2M)\big)$. The corresponding Riemann sum is

$$\sum_{k=1}^{N} \sum_{j=1}^{M} \left(1 + \frac{j}{N} + \frac{k}{2M}\right)^3 \left(\frac{1}{(2NM)}\right).$$

To simplify the expression, we can take $N = M$; hence the sum takes the form

$$S_N = \sum_{k=1}^{N} \sum_{j=1}^{N} \left(1 + \frac{j}{N} + \frac{k}{2N}\right)^3 \left(\frac{1}{2N^2}\right).$$

Evaluating sums like this for even moderately large values of N is best done on a computer. We get the following table:

N	S_N
1	7.8125
10	3.3215
20	3.1343
50	3.0248
100	2.9888
200	2.9707

Computation of the integral using iterated integrals and indefinite integration shows that, correct to two decimal places, the value is 2.95.

In Examples 1 and 2 estimates were made in such a way that the error involved in using the upper or lower estimates was limited by the difference of the two. In Example 3, however, we computed only a single Riemann sum corresponding to each subdivision of the domain of integration. To estimate the error in such a case we can sometimes use the following theorem.

5.2 Theorem

Let f be continuous on a closed bounded rectangle R in $\mathcal{R}^n$, defined by $a_i \leq x_i \leq b_i$, $i = 1, \ldots, n$, and let f have partial derivatives satisfying

$$\left| \frac{\partial f}{\partial x_i}(x) \right| \leq M_i, \qquad i = 1, \ldots, n.$$

Let R be subdivided by a grid G into rectangles of width $h_i = (b_i - a_i)/N_i$ in the x_i-coordinate, where N_i is an integer. Then the error in approximating $\int_R f\, dV$ by a Riemann sum based on G is at most

$$V(R) \sum_{i=1}^{n} h_i M_i,$$

where $V(R)$ is the volume of R.

Proof. Since $R = R_1 \cup \ldots \cup R_N$, and the R_k are disjoint, we can write

$$\int_R f\, dV = \sum_{k=1}^{N} \int_{R_k} f\, dV.$$

Then

$$\int_R f\, dV - S_N = \sum_{k=1}^{N} \int_{R_k} (f - f(x_k))\, dV,$$

because $f(x_k)$ is constant on R_k. Hence

$$\begin{aligned} \left| \int_R f\, dV - S_N \right| &\leq \sum_{k=1}^{N} \int_{R_k} |f - f(x_k)|\, dV \\ &\leq \sum_{k=1}^{N} V(R_k) \max_{x \text{ in } R_k} |f(x) - f(x_k)|. \end{aligned}$$

But by the mean-value theorem, for $\mathbf{x}$ and $\mathbf{y}$ in R_k,

$$\begin{aligned} |f(\mathbf{x}) - f(\mathbf{y})| &= |f'(\mathbf{z})(\mathbf{x} - \mathbf{y})| \\ &= \left| \sum_{i=1}^{n} \frac{\partial f}{\partial x_i}(\mathbf{z})(x_i - y_i) \right| \\ &\leq \sum_{i=1}^{n} h_i \left\{ \max_{x \text{ in } R} \left| \frac{\partial f}{\partial x_i}(x) \right| \right\}. \end{aligned}$$

By the previous inequality,

$$\begin{aligned} \left| \int_R f\, dV - S_N \right| &\leq \sum_{k=1}^{N} V(R_k) \sum_{i=1}^{n} h_i \left\{ \max_{x \text{ in } R} \left| \frac{\partial f}{\partial x_i}(x) \right| \right\} \\ &= V(R) \sum_{i=1}^{n} h_i M_i, \end{aligned}$$

as was to be shown.

Example 4. The integral

$$I = \int_R (1 + x + y)^3 \, dx \, dy$$

of Example 3 has integrand $f(x, y) = (1 + x + y)^3$. We have $f_x(x, y) = f_y(x, y) = 3(1 + x + y)^2$. The rectangle R determined by $0 \leq x \leq 1$ and $0 \leq y \leq \frac{1}{2}$ is such that

$$\max_{(x,y) \text{ in } R} |f_x(x, y)| = \max_{(x,y) \text{ in } R} |f_y(x, y)| = 3(\tfrac{5}{2})^2 = \tfrac{75}{4}.$$

Thus $M_1 = M_2 = \frac{75}{4}$. If we subdivide into K equal parts along the x-axis and L equal parts along the y-axis, we have $h_1 = 1/K$ and $h_2 = 1/2L$. Then a Riemann sum $S_{K,L}$ based on such a grid will satisfy

$$\begin{aligned} |I - S_{K,L}| &\leq (1)(\tfrac{1}{2})\left[\left(\frac{1}{K}\right)\left(\frac{75}{4}\right) + \left(\frac{1}{2L}\right)\left(\frac{75}{4}\right)\right] \\ &= \frac{75}{8}\left(\frac{1}{K} + \frac{1}{2L}\right). \end{aligned}$$

By choosing $L = 200$ and $K = 400$, we get the error bound

$$|I - S_{K,L}| < 0.05.$$

Integration over nonrectangular regions poses a problem in making estimates because there is likely to be error not only in approximating the function being integrated but also in fitting a grid to the region of integration. A bounded domain of integration B can always be extended to a rectangular one by the device of extending the integrand f to the function f_B, which is zero outside B. However, the error estimate in Theorem 5.2 will then usually no longer be applicable because f_B is, in general, discontinuous and so will fail to have the required partial derivatives. In some examples, the change-of-variable theorem can be used to transform the domain of integration into a rectangular one. Some care should then be used in the choice of transformation to ensure that it doesn't complicate the integrand unnecessarily.

Example 5. Let Q be the quarter disk in the first quadrant of $\mathcal{R}^2$ defined by $0 \leq x$, $0 \leq y$, $x^2 + y^2 \leq 1$. Using the polar coordinate transformation

$$x = u \cos v$$

$$y = u \sin v$$

for $0 \leq u \leq 1$ and $0 \leq v \leq \pi/2$, we can transform Q into the rectangle R shown in Fig. 34.

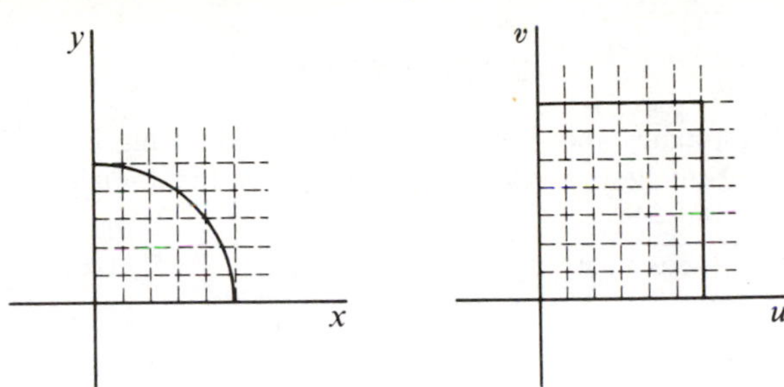

Figure 34

Given an integral of the form

$$I = \int_Q f(x, y)\, dx\, dy,$$

the change-of-variable theorem shows that

$$I = \int_R f(u \cos v, u \sin v) u\, du\, dv,$$

where u is the Jacobian determinant of the polar coordinate transformation. Since R is a rectangle, the estimates of this section apply readily. For example, if $f(x, y) = (x^2 + y^3)\sqrt{x^2 + y^2}$,

$$I = \int_R (u^2 \cos^2 v + u^3 \sin^3 v) u^2\, du\, dv.$$

Setting $g(u, v) = u^4 \cos^2 v + u^5 \sin^3 v$, we find

$$\left| \frac{\partial g}{\partial u}(u, v) \right| = |4u^3 \cos^2 v + 5u^4 \sin^3 v| \leq 9$$

and

$$\left| \frac{\partial g}{\partial v}(u, v) \right| = |-2u^4 \cos v \sin v + 3u^5 \sin^2 v \cos v| \leq 5$$

for $0 \leq u \leq 1$, $0 \leq v \leq \pi/2$. Now suppose $S_{N,M}$ is a Riemann sum based on a grid with N equal subdivisions along the u-axis and M equal subdivisions along the v-axis. Then Theorem 5.1 shows that

$$\begin{aligned} |I - S_{N,M}| &\leq (1)\left(\frac{\pi}{2}\right)\left(9\left(\frac{1}{N}\right) + 5\left(\frac{\pi/2}{M}\right)\right) \\ &\leq \frac{15}{N} + \frac{13}{M}. \end{aligned}$$

By taking $N = M = 60$, we get the error bound

$$|I - S_{60,60}| \leq \tfrac{15}{60} + \tfrac{13}{60} < 0.5.$$

EXERCISES

1. Find rough estimates above and below for the following integrals by using the largest and smallest values of the integrand on the domain of integration.

(a) $\int_0^{\pi/2} \sqrt{1 + \cos x}\, dx.$

(b) $\int_{-1}^{1} e^{x^2}\, dx.$

(c) $\int_0^1 dy \int_0^1 \cos xy\, dy.$

(d) $\int_{x^2+y^2 \le 1} e^{x+y}\, dx\, dy.$

(e) $\int_0^1 dx \int_0^1 dy \int_0^1 \sqrt{x + y + z}\, dz.$

(f) $\int_0^1 dx \int_1^2 dy \int_{-1}^1 (x^2 + y^2 + z^2)\, dz.$

2. Estimate each of the following integrals by computing a Riemann sum based on a subdivision of the domain of integration into four equal rectangles.

(a) $\int_0^1 (1 + x^2)\, dx.$ (b) $\int_0^1 dx \int_0^1 (x^2 + y^2)\, dy.$

3. For each integral in Problem 2, estimate the error in making the approximation. Use Theorem 5.1.

4. Find estimates for the error in approximating the integrals in Problem 2 if the number of equal subdivisions is 10 in part (a) and $10^2 = 100$ in part (b). Use Theorem 5.1.

5. Use a computer to make estimates by Riemann sums described in Problem 4.

6. If an integral is approximated by a Riemann sum with function values taken at the midpoint of each rectangle of the grid, then the integral is said to be approximated by the **midpoint rule.** It can be shown that, in this case, the error estimate of Theorem 5.1 can be changed to

$$\frac{V(R)}{24} \sum_{i=1}^{n} h_i^2 M_i^{(2)},$$

where $M_i^{(2)} \ge |(\partial^2 f/\partial x_i^2)(x)|$ for $i = 1, 2, \ldots, n$. If the integrals in Problem 2 are estimated using the midpoint rule for 10 equal subdivisions in part (a) and $10^2 = 100$ in part (b), find a bound for the error.

7. Transform the integral below by the change-of-variable theorem into one over a rectangle. Then use Theorem 5.1 to estimate the error in approximating by a Riemann sum with $10^2 = 100$ equal subdivisions.

$$\int_{x^2+y^2 \ge 1} (x + y^2)\, dx\, dy.$$

SECTION 6

NUMERICAL INTEGRATION

The estimates of the previous section are rather crude relative to the degree of accuracy often required in numerical work. The purpose of this section is to show how greater accuracy can be obtained. Theoretically, Theorem

5.2 may provide an arbitrarily high degree of accuracy. However, to achieve it, the amount of computing time required may be very great. Furthermore, so much arithmetic may have to be done that the accumulation of round-off error becomes unacceptably large. To get around these problems we need methods capable of giving good accuracy without so much arithmetic. One such method is Simpson's rule, which we first describe just for 1-dimensional integrals and then apply to multiple integrals.

Simpson's rule is motivated by the elementary observation that, for any quadratic polynomial $q(x) = Ax^2 + Bx + C$,

$$\int_a^b q(x)\,dx = \frac{b-a}{6}\left[q(a) + 4q\left(\frac{a+b}{2}\right) + q(b)\right]. \tag{1}$$

The proof of this formula is outlined in Example 5, Chapter 2, Section 2. To apply Equation (1) to approximate $\int_a^b f(x)\,dx$, we set

$$x_0 = a, \qquad x_1 = \frac{(a+b)}{2}, \qquad x_2 = b$$

and find a quadratic polynomial q such that

$$q(x_0) = f(x_0), \qquad q(x_1) = f(x_1), \qquad q(x_2) = f(x_2). \tag{2}$$

(Since $q(x)$ has the form $Ax^2 + Bx + C$ for some A, B, and C, to determine q we need only solve for A, B, and C in the equations $Ax_k^2 + Bx_k + C = f(x_2)$, $k = 1, 2, 3$. However this need not be done in practice.) The approximation we make is then

$$\int_a^b f(x)\,dx \approx \int_a^b q(x)\,dx.$$

Since by Equation (2), f and q are equal at x_0, x_1, and x_2, we can appeal to Equation (1) to get

6.1
$$\int_a^b f(x)\,dx \approx \frac{b-a}{6}\,[f(x_0) + 4f(x_1) + f(x_2)].$$

This formula is called **Simpson's three-point approximation.** Figure 35 shows that the approximation may be good or bad, depending on the shape of f. In Fig. 35(a), the graph of q lies close to that of f over $[a, b]$, and in addition there is cancellation between the shaded and unshaded areas because q lies alternately above and below f. In Fig. 35(b), the graphs are not particularly close, and there is no cancellation.

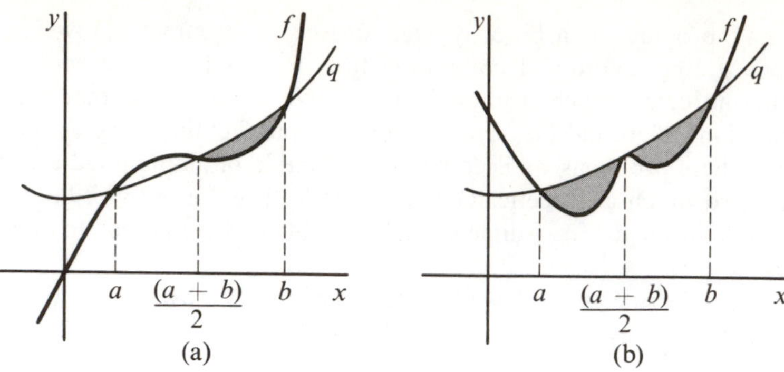

Figure 35

To improve the approximation, subdivide $[a, b]$ further into an *even* number of equal intervals by a new choice of points $x_0, x_1, \ldots, x_N$, N even:

$$a = x_0 \quad x_1 \quad x_2 \quad \cdots \quad x_{N-2} \quad x_{N-1} \quad x_N = b$$

Since

$$\int_a^b f(x)\,dx = \int_{x_0}^{x_2} f(x)\,dx + \int_{x_2}^{x_4} f(x)\,dx + \ldots + \int_{x_{N-2}}^{x_N} f(x)\,dx,$$

we can apply the three-point rule $N/2$ times to get

$$\begin{aligned}\int_a^b f(x)\,dx \approx\ & \frac{b-a}{6(N/2)}[f(x_0) + 4f(x_1) + f(x_2)] \\ & + \frac{b-a}{6(N/2)}[f(x_2) + 4f(x_3) + f(x_4)] \\ & \cdot \\ & \cdot \\ & \cdot \\ & + \frac{b-a}{6(N/2)}[f(x_{N-2}) + 4f(x_{N-1}) + f(x_N)].\end{aligned}$$

Combining terms gives

6.2 $$\int_a^b f(x)\,dx \approx \frac{b-a}{3N}[f(x_0) + 4f(x_1) + 2f(x_2) + \ldots + 4f(x_{N-1}) + f(x_N)].$$

Formula 6.2 is called **Simpson's rule.**

Example 1. Applying Simpson's rule to $I = \int_0^1 e^{x^2}\,dx$ for small values of N gives the approximation S_N:

$$\frac{1-0}{3N}[e^0 + 4e^{1/N} + 2e^{2/N} + \ldots + 4e^{(N-1)/N} + e^1].$$

N	S_N
2	1.4757
4	1.4637
10	1.4627
20	1.4626

Formula 6.2 says nothing about how accurate an approximation Simpson's rule gives. The information is contained in the following theorem.

6.3 Theorem

Let f be continuous on $[a, b]$ and be four times differentiable on (a, b). Let M be a number such that $|f^{(4)}(x)| \leq M$ for x in (a, b). Then the error in the Simpson approximation

$$\int_a^b f(x)\,dx$$

$$\approx \frac{b-a}{3N}[f(x_0) + 4f(x_1) + 2f(x_2) + \ldots + 4f(x_{N-1}) + f(x_N)],$$

using N intervals (N even), is at most

$$\frac{(b-a)^5 M}{180N^4}.$$

Proof. We first reduce the proof to the case $N = 2$, corresponding to the three-point rule. To do this we subdivide $[a, b]$ into $N/2$ *pairs* of intervals of length $(b-a)/(N/2)$. If we can show that the error over one pair is at most

$$\frac{((b-a)/N)^5 M}{90},$$

then the total error is at most $N/2$ times that or

$$\frac{(b-a)^5 M}{180N^4}.$$

Thus to prove the theorem we need only consider the three-point rule on $[a, b]$ and show that

$$\left| \int_a^b f(x)\,dx - \frac{h}{3}[f(a) + 4f(c) + f(b)] \right| \leq \frac{h^5 M}{90},$$

where $h = (b - a)/2$, and $c = (b + a)/2$. First let

$$E(t) = \int_{c-t}^{c+t} f(x)\,dx - \frac{t}{3}[f(c + t) + 4f(c) + f(c - t)]$$

and

$$F(t) = E(t) - \frac{t^5}{h^5} E(h).$$

Straightforward computation shows that

$$F'(t) = \tfrac{2}{3}[f(c + t) - 2f(c) + f(c - t)]$$
$$-\tfrac{1}{3}t[f'(c + t) - f'(c - t)] - \frac{5t^4}{h^5} E(h),$$

$$F''(t) = \tfrac{1}{3}[f'(c + t) - f'(c - t)]$$
$$- \tfrac{1}{3}t[f''(c + t) + f''(c - t)] - \frac{20t^3}{h^4} E(h),$$

$$F'''(t) = - \tfrac{1}{3}t[f'''(c + t) - f'''(c - t)] - \frac{60t^2}{h^5} E(h).$$

Now we apply the mean-value theorem to F at $t = 0$ and $t = h$. But since $F(0) = F(h) = 0$, we can conclude that $F'(t_1) = 0$ for some t_1 between 0 and h. Since $F'(0) = 0$, we can apply the mean-value theorem to F' at $t = 0$ and $t = t_1$ to conclude that $F''(t_2) = 0$ for some t_2 between 0 and t_1. But $F''(0) = 0$ also; so as before we conclude that $F'''(t_3) = 0$ for some t_3 between 0 and t_2. This implies that

$$E(h) = \frac{h^5}{180t_3}[f'''(c + t_3) - f'''(c - t_3)]. \tag{3}$$

Again using the mean-value theorem, we get

$$f'''(c + t_3) - f'''(c - t_3) = 2t_3 f^{(4)}(t_4) \tag{4}$$

for some t_4 between $c - t_3$ and $c + t_3$. Since $0 < t_3 < h$, the number t_4 lies in (a, b). We also have from (3) and (4)

$$E(h) = \frac{h^5}{90} f^{(4)}(t_4).$$

This completes the proof.

Example 2. We apply the previous theorem to

$$\int_0^1 e^{x^2}\,dx$$

We find $f^{(4)}(x) = 12e^{x^2} + 48x^2e^{x^2} + 16x^4e^{x^2}$; so on $[0, 1]$, we have $|f^{(4)}(x)| \leq 12e + 48e + 16e < 228$. Then the error in replacing the integral by the Simpson approximation using N points is at most

$$\frac{(1-0)^5 228}{180N^4} < \frac{1.3}{N^4}.$$

Thus for $N = 10$ the error is at most 0.00013, and for $N = 100$ it is at most 0.000000013.

To see how to apply Simpson's rule to a multiple integral over some rectangle, we apply Formula 6.2 repeatedly to an iterated integral in some order. Thus, if

$$\int_R f(x, y)\,dx\,dy = \int_c^d dy \int_a^b f(x, y)\,dx,$$

we consider for each y in $[c, d]$ the approximation

$$F(y) = \int_a^b f(x, y)\,dy \approx \frac{b-a}{3N}\sum_{j=0}^{J} A_j f(x_j, y), \tag{5}$$

where the numbers A_j follow the Simpson rule pattern $A_0 = 1$, $A_1 = 4$, $A_2 = 2, \ldots, A_{J-1} = 4$, $A_J = 1$, and J is even. The points x_j are evenly spaced at $a + j(b-a)/J$, for $j = 0, 1, \ldots, J$. We then apply Simpson's rule to the integral of F with K intervals in $[c, d]$, getting

$$\int_c^d F(y)\,dy \approx \frac{d-c}{3K}\sum_{k=0}^{K} B_k F(y_k), \tag{6}$$

where $B_0 = 1$, $B_1 = 4, \ldots, B_{K-1} = 4$, $B_K = 1$, as usual. At the risk of increasing the error, we substitute the approximation in Equation (5) into the one in (6) to get

$$\int_c^d dy \int_a^b f(x, y)\,dx \approx \frac{(b-a)}{3J}\frac{(d-c)}{3K}\sum_{k=0}^{K} B_k \sum_{j=0}^{J} A_j f(x_j, y_k)$$

or

6.4
$$\int_R f(x, y)\,dx\,dy \approx \frac{A(R)}{9JK}\sum_{k=0}^{K}\sum_{j=0}^{J} A_j B_k f(x_j, y_k),$$

where $x_j = a + j(b-a)/J$ and $y_k = c + k(d-c)/K$. The products A_jB_k are simply the products of the usual Simpson coefficients. For example, if $A_3 = 4$ and $B_2 = 2$, then $A_3B_2 = 8$.

Example 3. Consider

$$\int_R (x - y)^4 \, dx \, dy = \int_1^2 dy \int_1^3 (x - y)^4 \, dx.$$

Choosing $J = 4$ and $K = 2$ we find an approximation based on evaluation of the integrand at the points indicated in Fig. 36. The associated coeffi-

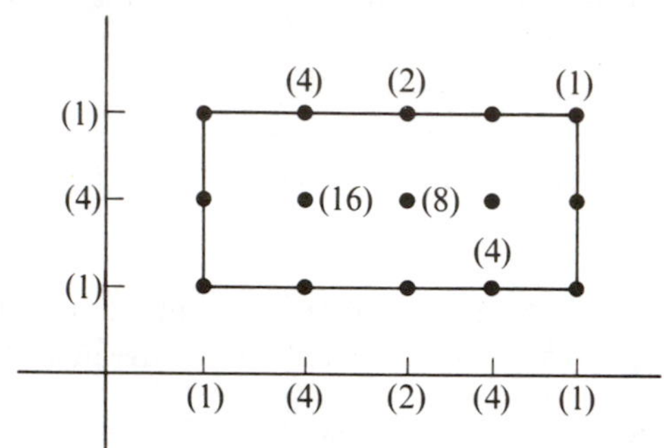

Figure 36

cient $A_j B_k$ is shown in a few cases.

Formula 6.4 gives

$$\int_R (x - y)^4 \, dx \, dy \approx \frac{2}{9(2)(4)} \sum_{k=0}^{2} \sum_{j=0}^{4} A_j B_k (x_j - y_k)^4,$$

where $x_j = 1 + j/2$ and $y_k = 1 + k/2$. The right side contains fifteen terms in the sum. Using a computer to evaluate it, we get $S_{8,4} = 2.13542$ for the Simpson approximation based on $J = 8$ and $K = 4$. Other values are listed below.

J, K	$S_{J,K}$
8, 4	2.13542
16, 8	2.13346
32, 16	2.13334
64, 32	2.13333

It so happens that the integral can be evaluated by iterated indefinite integration, and its value is $\frac{32}{15}$, which agrees with $S_{64,32}$ to as many decimal places as are given.

To approximate a triple integral of the form

$$\int_R f(x, y, z) \, dV = \int_{a_3}^{b_3} dz \int_{a_2}^{b_2} dy \int_{a_1}^{b_1} f(x, y, z) \, dx,$$

we apply the one-variable Simpson's rule three times to get the 3-dimensional Simpson rule:

6.5
$$\int_R f(x, y, z)\,dV \approx \frac{V(R)}{27JKL}\sum_{l=0}^{L}\sum_{k=0}^{K}\sum_{j=0}^{J} A_j B_k C_l f(x_j, y_k, z_l),$$

where $x_j = a_1 + j(b_1 - a_1)/J$, $y_k = a_2 + k(b_2 - a_2)/K$, and $z_l = a_3 + l(b_3 - a_3)/L$. Once again the A's, B's, and C's are the usual Simpson coefficients; hence, for example, with $A_2 = 2$, $B_3 = 4$, and $C_3 = 4$, we have $A_2B_3C_3 = 32$. Even in the case $J = K = L = 2$, the sum in 6.5 has twenty-seven terms, so it is desirable to have a computer to carry out the arithmetic.

Example 4. To approximate

$$\int_{1.1}^{2.3} dz \int_{1.5}^{1.8} dy \int_{1}^{3.6} \log(xyz)\,dx,$$

we apply Formula 6.4 to the solid rectangle $1 \le x \le 3.6$, $1.5 \le y \le 1.8$, $1.1 \le z \le 2.3$. The edges of the rectangle have lengths 2.6, 0.3, and 1.2, respectively; so we choose J, K, and L in proportion. For a first approximation we try $J = 16$, $K = 2$, and $L = 8$. The number of terms in the sum $S_{16,2,8}$ is then $(17)(3)(9) = 459$. We have

$$S_{16,2,8} = \frac{(3.6 - 1)(1.8 - 1.5)(2.3 - 1.1)}{27(16)(2)(8)}\sum_{l=0}^{8}\sum_{k=0}^{2}\sum_{j=0}^{16} A_j B_k C_l \log(x_j y_k z_l),$$

where $x_j = 1 + j(3.6 - 1)/8$, $y_k = 1.5 + k(1.8 - 1.5)/2$, and $z_l = 1.1 + l(2.3 - 1.1)/8$. We compute $S_{16,2,8} = 1.66797$, approximately.

To improve the accuracy in Example 4, we can refine the subdivision. However, since the arithmetic needed is fairly time-consuming, it is helpful to have an error estimate like 6.3 to use as a guide in choosing a grid.

6.6 Theorem

Let f be continuous on a closed, bounded rectangle R in $\mathcal{R}^n$, and have partial derivatives of order 4 satisfying

$$\left|\frac{\partial^4 f}{\partial x_i^4}(x)\right| \le M_i, \qquad i = 1, \ldots, n.$$

Let R be subdivided by a grid G into rectangles of width

$$h_i = \frac{b_i - a_i}{N_i}$$

in the x_i-coordinate, where each N_i is an even integer. Then the error in making a Simpson approximation to $\int_R f\,dV$ based on G is at most

$$\frac{V(R)}{180}\sum_{i=1}^{n} h_i^4 M_i.$$

Proof. The theorem contains Theorem 6.3 as the special case, $n = 1$, and is proved by reducing it to repeated application of 6.3. The case $n = 2$ presents the ideas adequately because it is the essential step in an inductive argument. Setting

$$E = \int_c^d dy \int_a^b f(x, z)\,dy - \frac{(b-a)(d-c)}{9JK}\sum_{j=0}^{J} A_j \sum_{k=0}^{K} B_k f(x_j, y_k),$$

we add and subtract an intermediate term to get

$$E = \int_c^d dy \int_a^b f(x, y)\,dy - \frac{b-a}{3J}\sum_{j=0}^{J} A_j \int_c^d f(x_j, y)\,dy$$

$$+ \frac{b-a}{3J}\sum_{j=0}^{J} A_j \int_c^d f(x_j, y)\,dy$$

$$- \frac{(b-a)(d-c)}{9JK}\sum_{j=0}^{J} A_j \sum_{k=0}^{K} B_k f(x_j, y_k).$$

Using the triangle inequality, and interchanging summation and integration, we have

$$|E| \leq \left| \int_c^d \left[\int_a^b f(x, y)\,dx - \frac{b-a}{3J}\sum_{j=0}^{J} A_j f(x_j, y) \right] dy \right|$$

$$+ \left| \frac{b-a}{3J}\sum_{j=0}^{J} A_j \left[\int_c^d f(x_j, y)\,dy - \frac{d-c}{3K}\sum_{k=0}^{K} B_k f(x_j, y_k) \right] \right|.$$

We now apply 6.3, the single-variable case, to the two expressions in square brackets. Thus

$$|E| \leq \int_c^d \left[\frac{(b-a)^5}{180J^4} M_1 \right] dy + \frac{b-a}{3J}\sum_{j=0}^{J} A_j \left[\frac{(d-c)^5}{180K^4} M_2 \right].$$

But now the expressions in square brackets are independent of y

and j, respectively. Furthermore, it is easy to check that $\sum_{j=0}^{J} A_j = 3J$. Hence

$$|E| \le (d-c)\frac{(b-a)^5}{180J^4}M_1 + (b-a)\frac{(d-c)^5}{180K^4}M_2$$
$$= \frac{(d-c)(b-a)}{180}\left[\frac{(b-a)^4}{J^4}M_1 + \frac{(d-c)^4}{K^4}M_2\right],$$

which is what we wanted to show.

Example 5. Returning to the integral

$$\int_{1.1}^{2.3} dz \int_{1.5}^{1.8} dy \int_{1}^{3.6} \log(xyz)\, dx$$

of Example 4, we find

$$V(R) = (2.3-1.1)(1.8-1.5)(3.6-1) < 1.$$

Also, from the integrand $f(x, y, z) = \log(xyz)$, we find

$$\frac{\partial^4 f}{\partial x^4}(x, y, z) = -6x^{-4}$$
$$\frac{\partial^4 f}{\partial y^4}(x, y, z) = -6y^{-4}$$
$$\frac{\partial^4 f}{\partial z^4}(x, y, z) = -6z^{-4}.$$

Since x, y, and z are all greater than or equal to 1 on R, the partial derivatives of order 4 are at most 6 in absolute value. Thus we can choose $M_1 = M_2 = M_3 = 6$. The numbers h_1, h_2, and h_3 we estimate more carefully, because they are to be raised to the fourth power. In fact, we take

$$h_1 = \frac{3.6-1}{J} = \frac{2.6}{J}$$
$$h_2 = \frac{1.8-1.5}{L} = \frac{0.3}{K}$$
$$h_3 = \frac{2.3-1.1}{L} = \frac{1.2}{L}.$$

Then Theorem 6.6 gives the error estimate

$$E < \frac{1}{180}\left(\left(\frac{2.6}{J}\right)^4(6) + \left(\frac{0.3}{K}\right)^4(6) + \left(\frac{1.2}{L}\right)^4(6)\right).$$

Computing this number for $J = 16$, $K = 2$, and $L = 8$ gives $E < 0.000029$. We recall that in Example 4 we computed the approximate value of the integral by Simpson's rule to be 1.66797. The error estimate now gives us a possible latitude of ± 0.000029, hence less than ± 0.00003. Thus Theorem 6.6 indicates that the true value of the integral lies between 1.66794 and 1.66800.

EXERCISES

1. In applying Simpson's rule to the approximation of $\int_0^6 x \sin x\, dx$, what is the smallest reasonable number of subintervals to use in order to take advantage of the shape of the graph of $x \sin x$?

2. Use Simpson's three-point rule to estimate each of the following integrals.

(a) $\displaystyle\int_1^3 x^4\, dx.$ (b) $\displaystyle\int_0^2 \sin x\, dx.$ (c) $\displaystyle\int_0^2 e^{-x^2}\, dx.$

3. Use the Simpson's rule with four intervals to estimate each of the following integrals.

(a) $\displaystyle\int_0^1 x \sin x\, dx,$ (b) $\displaystyle\int_1^{1.4} \log \sin x\, dx.$ (c) $\displaystyle\int_{-1}^0 e^x\, dx.$

4. Using a computer, apply Simpson's rule over 100 intervals to estimate each of the following integrals.

(a) $\displaystyle\int_0^2 e^{x^2}\, dx.$ (b) $\displaystyle\int_0^{2.1} \frac{dx}{x^3 + 1}.$ (c) $\displaystyle\int_{1.1}^{2.3} \frac{dx}{x}.$

5. Using Theorem 6.3, estimate the error in each approximation in Exercise 4.

6. Apply Formula 6.4 with $J = K = 2$ to estimate the following integrals.

(a) $\displaystyle\int_0^1 \int_1^2 (x + y)\, dx\, dy.$ (b) $\displaystyle\int_{-1}^1 dx \int_0^1 (x - y)\, dy.$

7. Using a computer and Formula 6.4 for $J = K = 8$, estimate the following integrals.

(a) $\displaystyle\int_0^1 \int_1^2 (x + e^y)^5\, dx\, dy.$ (b) $\displaystyle\int_{1.1}^{2.1} dx \int_0^{1.1} \log (x + y)\, dy.$

8. Use Theorem 6.6 to estimate the error in each approximation in Exercise 6.

9. Use Theorem 6.6 to estimate the error in each approximation in Exercise 7.

10. How large should J and K be chosen in Formula 6.4 to ensure that a Simpson approximation to

$$\int_0^1 dx \int_0^2 e^{xy}\, dy$$

gives an error no more than 0.0001? Remember that J and K must be even integers.

11. How large should J, K, and L be chosen in Formula 6.5 to ensure that a Simpson approximation to

$$\int_0^1 dx \int_0^{1.1} dy \int_0^{1.2} e^{xyz}\, dz$$

gives an error no more than 0.0001?

12. Let A be the annular region in $\mathcal{R}^2$ defined by $1 \le x^2 + y^2 \le 4$, that is, the region between concentric circles of radius 1 and 2 about $(0, 0)$. The integral

$$\int_A (x^2 + 2y^2)\, dx\, dy$$

cannot be well approximated directly by Simpson's rule because A is not rectangular.

(a) Transform the integral by the change-of-variable theorem into one over a rectangular region.
(b) Approximate the integral found in part (a), accurate to within 0.005.

7

Vector Field Theory

SECTION 1

GREEN'S THEOREM

The fundamental theorem of calculus says that if f' is integrable for $a \leq t \leq b$, then

$$\int_a^b f'(t)\,dt = f(b) - f(a). \tag{1}$$

In Section 2 of Chapter 4 the theorem has been extended to line integrals of a gradient ∇f by the equation

$$\int_a^b \nabla f(\mathbf{x}) \cdot d\mathbf{x} = f(\mathbf{b}) - f(\mathbf{a}). \tag{2}$$

The main theorems of the present chapter are also variations on the idea that an integral of some kind of derivative of a function can be evaluated by using the values of the function itself on a set of lower dimension. We begin with the version known as Green's theorem.

Let D be a plane region whose boundary is a single curve γ, parametrized by a function g in such a way that, as t increases from a to b, $g(t)$ traces γ once in the counterclockwise direction. An example is shown in Fig. 1. If F_1 and F_2 are real-valued functions defined on D, including its boundary, then Green's theorem says that

$$\int_D \left(\frac{\partial F_2}{\partial x_1} - \frac{\partial F_1}{\partial x_2}\right) dx_1\,dx_2 = \int_\gamma F_1\,dx_1 + F_2\,dx_2, \tag{3}$$

under appropriate smoothness conditions. The requirement that γ be traced counterclockwise is the analog of the fact that, in Equations (1) and (2), the differences on the right have to be taken in the proper order. The analogy of Equation (3) with Equations (1) and (2) can be further strengthened if we think of the integrand $(\partial F_2/\partial x_1) - (\partial F_1/\partial x_2)$ as a kind

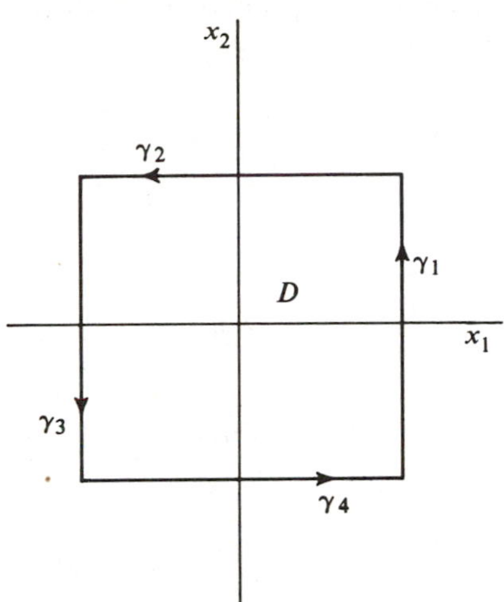

Figure 1

of derivative of the vector field $F = (F_1, F_2)$. Section 8 contains a justification of this viewpoint.

Example 1. Suppose D is the square defined by $-1 \leq x_1 \leq 1$, $-1 \leq x_2 \leq 1$, and let F_1 and F_2 be defined on D by $F_1(x_1, x_2) = -x_2 e^{x_1}$, and $F_2(x_1, x_2) = x_1 e^{x_2}$. Then

$$\frac{\partial F_2}{\partial x_1}(x_1, x_2) - \frac{\partial F_1}{\partial x_2}(x_1, x_2) = e^{x_2} + e^{x_1};$$

so

$$\int_D \left(\frac{\partial F_2}{\partial x_1} - \frac{\partial F_1}{\partial x_2}\right) dx_1\, dx_2 = \int_{-1}^{1} dx_1 \int_{-1}^{1} (e^{x_2} + e^{x_1})\, dx_2$$
$$= 4\left(e - \frac{1}{e}\right).$$

The boundary curve γ can be parametrized in four pieces γ_i, $i = 1, 2, 3, 4$, by

$$\begin{pmatrix} x_1 \\ x_2 \end{pmatrix} = \begin{cases} \begin{pmatrix} 1 \\ t \end{pmatrix} \\ \begin{pmatrix} -t \\ 1 \end{pmatrix} \\ \begin{pmatrix} -1 \\ -t \end{pmatrix} \\ \begin{pmatrix} t \\ -1 \end{pmatrix} \end{cases}, \qquad -1 \leq t \leq 1.$$

Notice that the traversal of γ is counterclockwise, as is shown in Fig. 1. On the first side of the square we have

$$\begin{aligned}\int_{\gamma_1} F_1\,dx_1 + F_2\,dx_2 &= \int_{\gamma} -x_2e^{x_1}\,dx_1 + x_1e^{x_2}\,dx_2 \\ &= \int_{-1}^{1}\left[(-te)\frac{dx_1}{dt} + e^t\frac{dx_2}{dt}\right]dt \\ &= \int_{-1}^{1} e^t\,dt = e - \frac{1}{e}.\end{aligned}$$

Similarly, the integrals over the other three sides are also equal to $(e - 1/e)$, so

$$\int_{\gamma} F_1\,dx_1 + F_2\,dx_2 = 4\left(e - \frac{1}{e}\right).$$

Equation (3) is thus verified for this particular example.

In the computation of a line integral, a given parametrization can always be replaced by an equivalent one for which the line integral will have the same value. In the previous example the boundary curve γ was given what appears to be the simplest parametrization, though any equivalent one would do. The question becomes more important if the boundary is presented without any parametrization, but merely as a set. It may be necessary to choose a parametrization, and if Green's theorem is to be applied, we shall see that this must be done so that the boundary is traced just once, and in the proper direction.

The importance of the clockwise versus counterclockwise traversal of the boundary becomes apparent when we observe that, for any line integral, a reversal of the direction of the path changes the sign of the integral. Thus, if γ is parametrized by $g(t)$ for $a \le t \le b$, we can denote by γ^- the curve parametrized by $g^-(t) = g(a + b - t)$ for $a \le t \le b$. It is clear that γ^- is the same set as γ, but is traced in the opposite direction, that is, from $g(b)$ to $g(a)$ instead of the other way around. Then, since $(g^-(t))' = -g'(a + b - t)$, we have

$$\int_{\gamma-} F \cdot d\mathbf{x} = -\int_a^b F(g(a + b - t)) \cdot g'(a + b - t)\,dt.$$

Changing variable by $t = a + b - u$ gives

$$\begin{aligned}\int_{\gamma-} F \cdot d\mathbf{x} &= -\int_a^b F(g(u)) \cdot g'(u)\,du \\ &= -\int_{\gamma} F \cdot d\mathbf{x}\end{aligned}$$

This proves the important formula

1.1
$$\int_{\gamma-} F \cdot d\mathbf{x} = -\int_{\gamma} F \cdot d\mathbf{x}.$$

Green's theorem can be proved most easily for regions D such that γ, the boundary of D, is crossed at most twice by a line parallel to a coordinate axis. Such a region is called **simple.** Thus a coordinate line intersects the boundary of a simple region either in a line segment or else in at most two points. In fact, using Equation 1.1, we can extend the theorem to finite unions of simple regions. A few such are shown in Fig. 2, where only D_1 is simple. In D_2 the boundary is shown traced not always counterclockwise, but rather with the region always to the left as a point traces the curve. For bounded regions with a single boundary curve, the two descriptions of the orientation of the boundary amount to the same thing.

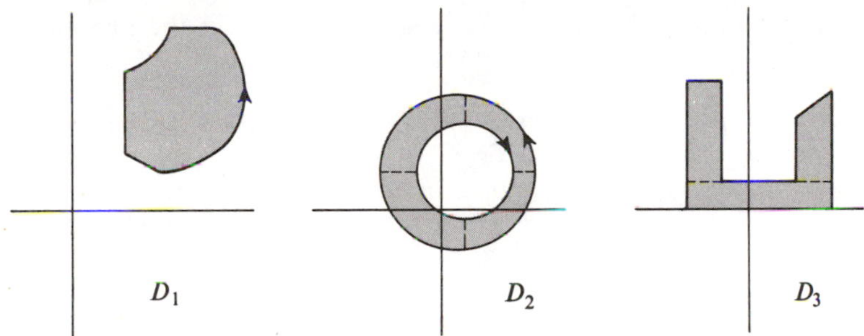

Figure 2

1.2 Green's Theorem

Let D be a bounded plane region which is a finite union of simple regions, each with a boundary consisting of a piecewise smooth curve. Let F_1 and F_2 be continuously differentiable real-valued functions defined on D together with γ, the boundary of D. Then

$$\int_D \left(\frac{\partial F_2}{\partial x_1} - \frac{\partial F_1}{\partial x_2}\right) dx_1\, dx_2 = \int_{\gamma} F_1\, dx_1 + F_2\, dx_2,$$

where γ is parametrized so that it is traced once, with D on the left.

Proof. Consider first the case in which D is a simple region, with boundary γ parametrized by

$$\begin{pmatrix} x_1 \\ x_2 \end{pmatrix} = \begin{pmatrix} g_1(t) \\ g_2(t) \end{pmatrix}, \qquad a \leq t \leq b.$$

Since

$$\int_\gamma F_1\,dx_1 + F_2\,dx_2 = \int_\gamma F_1\,dx_1 + \int_\gamma F_2\,dx_2,$$

we can work with each of the terms on the right separately. We have

$$\int_\gamma F_1(x_1, x_2)\,dx_1 = \int_a^b F_1(g_1(t), g_2(t))g_1'(t)\,dt.$$

The curve γ consists of the graphs of two functions $u(x_1)$ and $v(x_1)$, perhaps together with one or two vertical segments, as shown in Fig. 3. On a vertical segment, g_1 is constant, so $g_1' = 0$ there. On the remaining parts of γ we apply the change of variable $x_1 = g_1(t)$ so that, on the top curve, $g_2(t) = u(x_1)$, while on the bottom, $g_2(t) = v(x_1)$. It follows that

$$\int_\gamma F_1(x_1, x_2)\,dx_1 = \int_\beta^\alpha F_1(x_1, u(x_1))\,dx_1 + \int_\alpha^\beta F_1(x_1, v(x_1))\,dx_1,$$

where the integration from β to α occurs because the graph of u is traced from right to left. Reversing the limits in the first integral, we get

$$\begin{aligned}\int_\gamma F_1(x_1, x_2)\,dx_1 &= \int_\alpha^\beta [-F_1(x_1, u(x_1)) + F_1(x_1, v(x_1))]\,dx_1 \\ &= \int_\alpha^\beta \left[-\int_{v(x_1)}^{u(x_1)} \frac{\partial F_1}{\partial x_2}(x_1, x_2)\,dx_2\right] dx_1 \\ &= \int_D -\frac{\partial F_1}{\partial x_2}\,dx_1\,dx_2.\end{aligned}$$

A similar proof, referred to Fig. 4, shows that

$$\int_\gamma F_2(x_1, x_2)\,dx_2 = \int_D \frac{\partial F_2}{\partial x_1}\,dx_1\,dx_2.$$

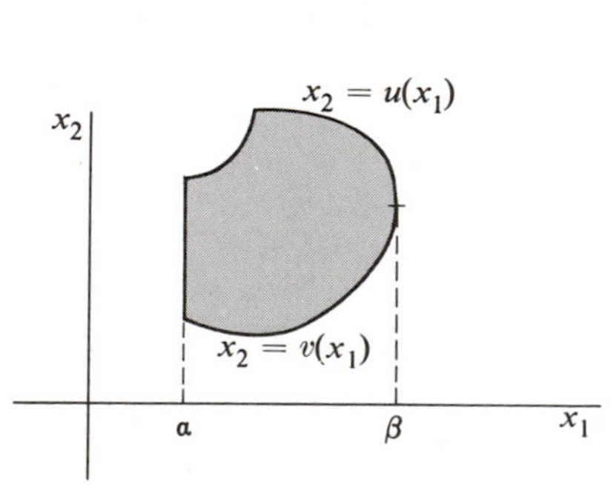

Figure 3

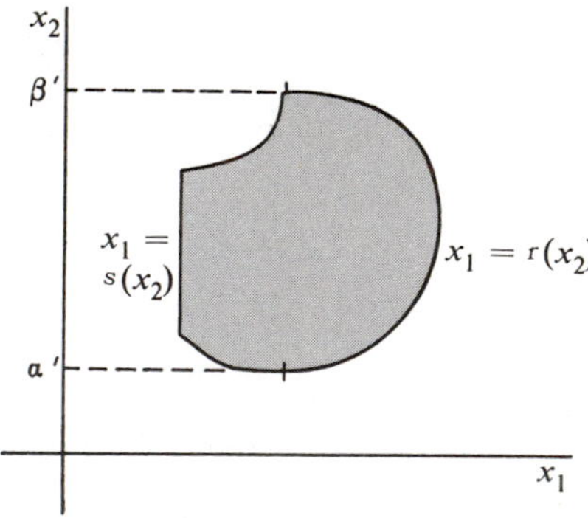

Figure 4

Combining this equation with the previous ones gives Green's theorem for the special class of simple regions.

We extend the theorem to a finite union, $D = D_1 \cup \ldots \cup D_K$, of simple regions each with a piecewise smooth boundary curve γ_k, $k = 1, \ldots, K$. Applying Green's theorem to each simple region D_k we get

$$\int_{D_k} \left(\frac{\partial F_2}{\partial x_1} - \frac{\partial F_1}{\partial x_2}\right) dx_1\, dx_2 = \int_{\gamma_k} F_1\, dx_1 + F_2\, dx_2$$

The sum of integrals over D_k is an integral over D; so

$$\int_{D} \left(\frac{\partial F_2}{\partial x_1} - \frac{\partial F_1}{\partial x_2}\right) dx_1\, dx_2$$

$$= \int_{\gamma_1} F_1\, dx_1 + F_2\, dx_2 + \ldots + \int_{\gamma_K} F_1\, dx_1 + F_2\, dx_2.$$

Now the boundary of D consists of pieces taken from several of the curves γ_k. In addition there may be parts of curves γ_k that are not a part of γ but which act as common boundary to two simple regions. The effect is illustrated in Fig. 5.

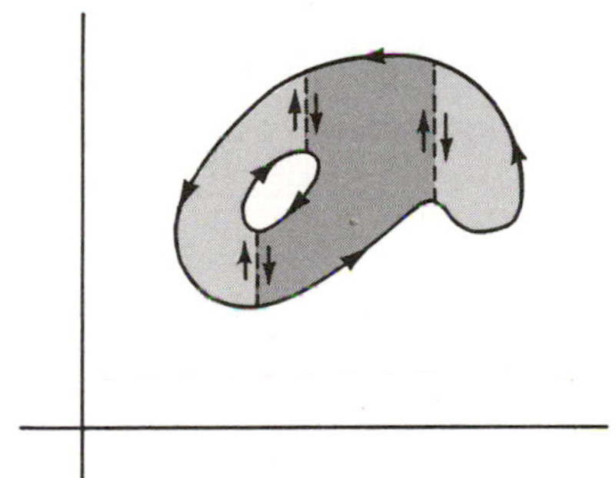

Figure 5

A piece δ of common boundary will be traced in one direction or the opposite, depending on which simple region it is associated with. But for a line integral we always have, by Equation 1.1,

$$\int_{\delta} F_1\, dx_1 + F_2\, dx_2 + \int_{\delta^-} F_1\, dx_1 + F_2\, dx_2 = 0.$$

Thus, while the parts of the curves γ_k that make up γ contribute to $\int_\gamma F_1\, dx_1 + F_2\, dx_2$, the other parts cancel, leaving

$$\int_{D} \left(\frac{\partial F_2}{\partial x_1} - \frac{\partial F_1}{\partial x_2}\right) dx_1\, dx_2 = \int_{\gamma} F_1\, dx_1 + F_2\, dx_2.$$

This completes the proof of Green's theorem.

The last part of the proof just given extends Green's theorem from simple regions to one like those shown in Fig. 6. The extension has an important consequence for line integrals $\int F_1\,dx_1 + F_2\,dx_2$ over two curves γ and δ, whenever the functions F_1 and F_2 are continuously differentiable in the region D between γ and δ as well as on the boundary curves. In Fig. 6(a) the curves are traced in the same direction (counterclockwise in the figure) and in Fig. 6(b) the curves go from one point to another in the

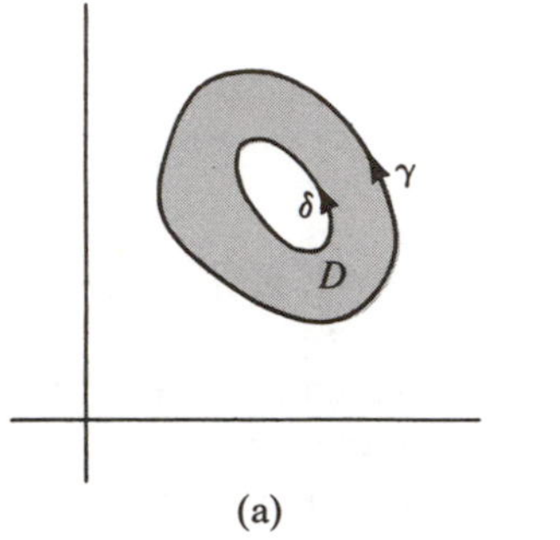

Figure 6

same direction. If it happens that the equation

$$\frac{\partial F_2}{\partial x_1} - \frac{\partial F_1}{\partial x_2} = 0 \tag{4}$$

holds throughout D, then we can conclude that

$$\int_\gamma F_1\,dx_1 + F_2\,dx_2 = \int_\delta F_1\,dx_1 + F_2\,dx_2.$$

The principle is illustrated in the next two examples.

Example 2. Let F_1 and F_2 be defined by

$$F_1(x_1, x_2) = \frac{-x_2}{x_1^2 + x_2^2}, \qquad F_2(x_1, x_2) = \frac{x_1}{x_1^2 + x_2^2},$$

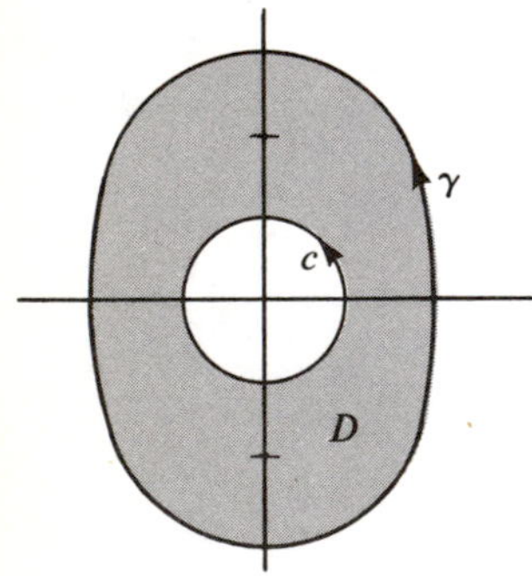

Figure 7

for $(x_1, x_2) \neq (0, 0)$. Direct computation shows that these functions satisfy Equation (4). If γ is the ellipse shown in Fig. 7 and defined by

$$\begin{pmatrix} x_1 \\ x_2 \end{pmatrix} = \begin{pmatrix} 2\cos t \\ 3\sin t \end{pmatrix}, \qquad 0 \leq t \leq 2\pi,$$

then the line integral $\int_\gamma F_1\,dx_1 + F_2\,dx_2$ would be troublesome to compute directly, even using tables. However, we can apply Green's theorem to the region D between γ and the circle c of radius 1 about the origin, parametrized by $(x_1, x_2) = (\cos t, \sin t)$, $0 \leq t \leq 2\pi$. Because Equation (4) is satisfied, Green's theorem yields

$$\int_{\gamma \cup c-} F_1\, dx_1 + F_2\, dx_2 = 0,$$

where c^- is c traced clockwise, so that D is on its left. The last equation can be written, by Equation 1.1,

$$\int_\gamma F_1\, dx_1 + F_2\, dx_2 = \int_c F_1\, dx_1 + F_2\, dx_2.$$

But on c we have $x_1^2 + x_2^2 = 1$, so

$$\int_\gamma F_1\, dx_1 + F_2\, dx_2 = \int_c - x_2\, dx_1 + x_1\, dx_2$$
$$= \int_0^{2\pi} (\sin^2 t + \cos^2 t)\, dt = 2\pi.$$

It is important to observe that Green's theorem could not have been applied directly to the entire interior of the ellipse because $(\partial F_2/\partial x_1)$ and $(\partial F_1/\partial x_2)$ fail to exist at the origin.

Example 3. The curve γ_1 given by $g(t) = (t, t^2)$, $0 \leq t \leq 1$, is shown in Fig. 8. Suppose that $F(x_1, x_2) = \big(F_1(x_1, x_2), F_2(x_1, x_2)\big)$ is a continuously differentiable vector field for $x_1^2 + x_2^2 < 4$ and satisfies Equation (4) there. The line integral of F over γ_1 could perhaps be computed directly in the form

$$\int_{\gamma_1} F \cdot d\mathbf{x} = \int_0^1 [F_1(t, t^2) + F_2(t, t^2)(2t)]\, dt.$$

However there are other possibilities. For example, the curve γ_2 can be parametrized by $g_2(t) = (t, t)$, $0 \leq t \leq 1$. Since we can apply Green's theorem to the region between γ_1 and γ_2, the fact that

$$\int_D \left(\frac{\partial F_2}{\partial x_1} - \frac{\partial F_1}{\partial x_2}\right) dx_1\, dx_2 = 0$$

would imply

$$\int_{\gamma_1} F \cdot d\mathbf{x} + \int_{\gamma_2^-} F \cdot d\mathbf{x} = 0,$$

where γ_2^- is given by $g_2^-(t) = (1 - t, 1 - t)$ for $0 \leq t \leq 1$. Then the line integrals over γ_1 and γ_2 are equal by Equation 1.1, and the latter integral can be written

$$\int_{\gamma_2} F \cdot d\mathbf{x} = \int_0^1 [F_1(t, t) + F_2(t, t)]\, dt.$$

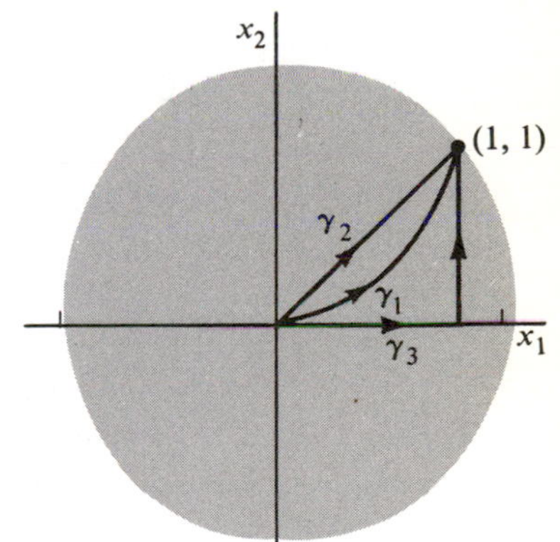

Figure 8

Another alternative would be to replace γ_1 by γ_3, where γ_3 is parametrized in two pieces by

$$g_3(t) = \begin{cases} \begin{pmatrix} t \\ 0 \end{pmatrix}, \\ \begin{pmatrix} 1 \\ t \end{pmatrix} \end{cases} \qquad 0 \le t \le 1.$$

Thus

$$\int_{\gamma_3} F \cdot d\mathbf{x} = \int_0^1 F_1(t, 0)\, dt + \int_0^1 F_2(1, t)\, dt.$$

This may be easier to compute than either of the integrals over γ_1 and γ_2, although all three are equal.

Line integrals around a *closed* path, sometimes called a **circuit,** are of sufficient importance that they are often distinguished from other integrals by means of an integral sign like $\oint$. In the plane this notation has the special advantage that $\circlearrowleft$ and $\circlearrowright$ can be used to indicate a counterclockwise or clockwise traversal of the path.

Example 4. Green's theorem has two distinct but closely related physical interpretations. We assume D to be a region in $\mathcal{R}^2$ whose boundary is a single counterclockwise-oriented curve γ. If γ has a smooth parametrization $g(t) = (g_1(t), g_2(t))$, $a \le t \le b$, and has a nonzero tangent at each point, we can form the unit tangent and normal vectors

$$\mathbf{t}(t) = \frac{g'(t)}{|g\,(t)|} = \left(\frac{g_1'(t)}{|g'(t)|}, \frac{g_2'(t)}{|g'(t)|}\right)$$

and

$$\mathbf{n}(t) = \left(\frac{g_2'(t)}{|g'(t)|}, \frac{-g_1'(t)}{|g'(t)|}\right).$$

An example is shown in Fig. 9. If $F = (F_1, F_2)$ is a continuously differentiable vector field defined on a region containing D and γ, then the line integral in Green's theorem can be written in the form

$$\oint_\gamma F_1\, dx_1 + F_2\, dx_2 = \int_a^b F(g(t)) \cdot \mathbf{t}(t)\, |g'(t)|\, dt$$
$$= \oint_\gamma F \cdot \mathbf{t}\, ds.$$

We define a real-valued function, curl F, called the **curl** of F by

$$\operatorname{curl} F(\mathbf{x}) = \frac{\partial F_2}{\partial x_1}(\mathbf{x}) - \frac{\partial F_1}{\partial x_2}(\mathbf{x}).$$

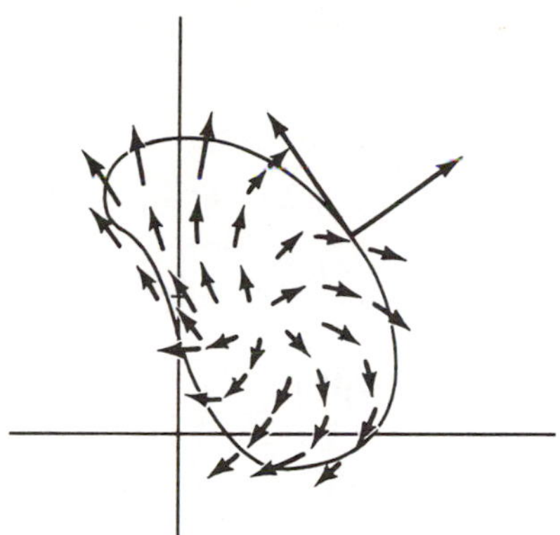

Figure 9

Green's theorem then becomes

$$\int_D \operatorname{curl} F \, dA = \oint_\gamma F \cdot \mathbf{t} \, ds,$$

sometimes called Stokes's theorem. Now interpret F as a force field in the plane. The line integral represents the work $W(\gamma)$ done in moving a particle around γ in the counterclockwise direction under the influence of F. Stokes's theorem says that $W(\gamma)$ is equal to the integral of the curl of F over D. In particular, if curl F is identically zero in D, then $W(\gamma) = 0$ for every smooth circuit γ contained in D, whether γ is oriented counterclockwise or not. For this conclusion to hold, it is of course necessary that curl F be defined throughout the inside of every circuit in D to which Stokes's theorem is applied. Conversely, it is possible to show that if $W(\gamma) = 0$ for every smooth circuit, then curl F is identically zero. See Section 2.

F can also be interpreted as the velocity field of a fluid flow in D. That is, the vector field F at each point of D represents the speed and direction of the flow at that point. In this case the line integral in Stokes's theorem is called the **circulation** of F around γ, and Stokes's theorem says that the circulation of F along γ is the integral of the curl of F over D. Thus to say that curl F is identically zero in D is to say that the circulation is zero around every smooth closed curve with its interior contained in D. A field F for which curl F is zero is called **irrotational** for this reason.

Now using the unit normal $\mathbf{n}(t)$, we can rewrite Green's theorem in another way. Instead of applying the fundamental Equation (3) to the field $F = (F_1, F_2)$, we apply it to the related pair of functions $(-F_2, F_1)$. Thus the line integral becomes

$$\begin{aligned}\oint_\gamma - F_2 \, dx_1 + F_1 \, dx_2 &= \int_a^b F\big(g(t)\big) \cdot \mathbf{n}(t) \, |g'(t)| \, dt \\ &= \oint_\gamma F \cdot \mathbf{n} \, ds.\end{aligned}$$

On the other hand, the area integral over D becomes

$$\int_D \left(\frac{\partial F_1}{\partial x_1} + \frac{\partial F_2}{\partial x_2}\right) dx_1\, dx_2.$$

We define a real-valued function div F, called the **divergence** of F, by

$$\text{div } F(\mathbf{x}) = \frac{\partial F_1}{\partial x_1}(\mathbf{x}) + \frac{\partial F_2}{\partial x_2}(\mathbf{x});$$

thus Green's theorem can be written

$$\int_D \text{div } F\, dA = \oint_\gamma F \cdot \mathbf{n}\, ds,$$

which, in this form, is called Gauss's theorem.

Using the fluid flow interpretation, in which F represents a velocity field, the line integral in Gauss's theorem is the integral of the outward normal coordinate of F over γ. Hence, this integral is called the total flow, or **flux,** of F across γ in the outward direction. Gauss's theorem shows that the flux, $\Phi(\gamma)$, across γ is equal to the integral of the divergence of F over the region bounded by γ. Thus div $F(\mathbf{x})$ measures the rate of change of the density of the fluid at the point $\mathbf{x}$. If div $F(\mathbf{x})$ is predominantly positive in D, then $\Phi(\gamma)$, the outward flow, will be positive, and a negative $\Phi(\gamma)$ indicates that more fluid is going into D than is going out. If div F is identically zero, then F is said to represent an **incompressible** flow.

EXERCISES

1. Use Green's theorem to compute the value of the line integral $\int_\gamma x_2\, dx_1 + x_1^2\, dx_2$, where γ is each of the following closed paths.

(a) The circle given by $g(t) = (\cos t, \sin t)$, $0 \le t \le 2\pi$.
(b) The square with corners at $(\pm 1, \pm 1)$, traced counterclockwise.
(c) The square with corners at $(0, 0)$, $(1, 0)$, $(1, 1)$, and $(0, 1)$, traced counterclockwise.

2. Using some one of the paths γ_1, γ_2, or γ_3 in Example 3 of the text, compute the line integral $\int_{\gamma_k} x_2\, dx_1 + x_1\, dx_2$.

3. Let γ be the curve parametrized by $g(t) = (2 \cos t, 3 \sin t)$, $0 \le t \le 2\pi$. Compute $\int_\gamma (2x_1 + x_2)\, dx_1 + (x_1 + 3x_2)\, dx_2$.

4. Evaluate the following line integrals by whatever method seems simplest

(a) $\int_\gamma e^x \cos y\, dx + e^x \sin y\, dy$, where γ is the triangle with vertices $(0, 0)$, $(1, 0)$, $(1, \pi/2)$, traced counterclockwise.
(b) Use the same integrand as in part (a), but change the path to the square with corners at $(0, 0)$, $(1, 0)$, $(1, 1)$, and $(0, 1)$, traced counterclockwise.
(c) $\int_c (x^2 - y^2)\, dx + (x^2 + y^2)\, dy$, where c is the circle of radius 1 centered at $(0, 0)$ and traced *clockwise*.

5. Show that if D is a simple region bounded by a piecewise smooth curve γ, then the area of D is given by

$$A(D) = \frac{1}{2}\oint (-y\,dx + x\,dy).$$

6. Let f be a real-valued function with continuous second-order derivatives in an open set D in $\mathcal{R}^2$. Let F be the vector field defined in D by $F(\mathbf{x}) = \nabla f(\mathbf{x})$, the gradient of f. Show that if $F(\mathbf{x}) = (F_1(\mathbf{x}), F_2(\mathbf{x}))$, then the equation $(\partial F_2/\partial x_1) - (\partial F_1/\partial x_2) = 0$ is satisfied in D.

7. Let $f(x_1, x_2) = \arctan(x_2/x_1)$, for $x_1 > 0$. Let the vector field $\nabla f(x_1, x_2)$ have coordinate functions $F_2(x_1, x_2)$ and $F_2(x_1, x_2)$. Show that F_1 and F_2 can be extended in a natural way to all $(x_1, x_2) \neq (0, 0)$. Let γ be a smooth curve parametrized to run from the point $(1, 0)$ to a fixed point (x_1, x_2), but on the way winding k times counterclockwise around the origin. Compute the line integral

$$\int_\gamma \nabla f \cdot d\mathbf{x},$$

(a) for $k = 0$; (b) for $k = 1$; (c) for k an arbitrary positive integer. (d) What interpretation can be given the line integral if k is a negative integer?

8. Let F be a continuously differentiable vector field defined everywhere but at two points $\mathbf{x}_1$ and $\mathbf{x}_2$ in $\mathcal{R}^2$, and satisfying $(\partial F_2/\partial x_1) - (\partial F_1/\partial x_2) = 0$. Let c_1 and c_2 be counterclockwise-oriented circles centered at $\mathbf{x}_1$ and $\mathbf{x}_2$, and with radii less than $|\mathbf{x}_1 - \mathbf{x}_2|$. Suppose

$$\int_{ck} F \cdot d\mathbf{x} = I_k, \qquad k = 1, 2.$$

Show that if γ is any closed smooth path that avoids $\mathbf{x}_1$ and $\mathbf{x}_2$, then

$$\int_\gamma F \cdot d\mathbf{x} = n_1 I_1 + n_2 I_2,$$

for some integers n_1 and n_2.

9. (a) Consider a particle moving in a plane vertical to the surface of the earth and subject to the gravitational field $G(x, y) = (0, mg)$, where m is the mass of the particle and g is the acceleration of gravity. Show that as the particle moves in the plane, the amount of work done is independent of the path between two points and depends only on the initial and final points. In particular, the work done in moving along a closed path is zero.
(b) Replace the field G by a field $F = (F_1, F_2)$ satisfying $(\partial F_2/\partial x_1) = (\partial F_1/\partial x_2)$ throughout the plane. Show that the same conclusions hold.

10. (a) Let $\mathcal{R} \xrightarrow{g} \mathcal{R}^2$ trace a simple closed curve γ in the counterclockwise direction. Show that, if a unit tangent vector to γ is given by $\mathbf{t}(t) = g'(t)/|g'(t)|$, then the *outward* pointing unit normal to γ at $g(t)$ is given by $\mathbf{n}(t) = (g_2'(t)/|g'(t)|, -g_1'(t)/|g'(t)|)$, where g_1 and g_2 are the coordinate functions of g.

(b) Show that Green's theorem can be written in the form

$$\int_D \left(\frac{\partial F_2}{\partial x_1} - \frac{\partial F_1}{\partial x_2}\right) dx_1\, dx_2 = \int_\gamma F \cdot \mathbf{t}\, ds,$$

where $F = (F_1, F_2)$, and ds denotes integration with respect to arc length.

(c) Show that Green's theorem can also be written in the form

$$\int_D \left(\frac{\partial F_1}{\partial x_1} + \frac{\partial F_2}{\partial x_2}\right) dx_1\, dx_2 = \int_\gamma F \cdot \mathbf{n}\, ds.$$

[*Hint*. In the previous formula, replace F_2 by F_1 and F_1 by $-F_2$.]

11. Assume that the vector field $F = (F_1, F_2)$ in Exercise 10(c) is a gradient field, that is, $F = \nabla f$ for some real-valued f. Show that Green's theorem can be written in the form

$$\int_D \Delta f\, dA = \int_\gamma \nabla f \cdot \mathbf{n}\, ds,$$

where $\Delta f = (\partial^2 f/\partial x_1^2) + (\partial^2 f/\partial x_2^2)$, the Laplacian of f.

12. (a) Show that if f is a continuous real-valued function defined in an open set D of $\mathcal{R}^2$, and $\int_M f\, dA = 0$ for every circular disk M in D, then f is identically zero in D. [*Hint*. Show that if $f(\mathbf{x}_0) \neq 0$ for some $\mathbf{x}_0$ in D, then there is a disk M centered at $\mathbf{x}_0$ such that $|f(\mathbf{x})| \geq \delta$ for some $\delta > 0$, and all $\mathbf{x}$ in M.]

(b) Use part (a) and Stokes's theorem to show that if curl F is continuous in an open set D, and the circulation of F is zero around every smooth circuit in D, then F is irrotational in D, that is, curl F is identically zero in D.

(c) Use part (a) and Gauss's theorem to show that if div F is continuous in D and $\Phi(\gamma) = 0$ for every smooth circuit γ in D, then F is incompressible.

13. The equations curl $F = 0$ and div $F = 0$ occur in complex variable theory in a slightly different form as the Cauchy-Riemann equations. Show that if $u(x, y)$ and $v(x, y)$ are the real and imaginary parts, respectively, of the following complex valued functions, then the vector field, given by $F(x, y) = (v(x, y), u(x, y))$, is irrotational and incompressible.

(a) $(x + iy)^2$.

(b) e^{x+iy}.

(c) $(\frac{1}{2}) \log (x^2 + y^2) + i \arctan y/x,\ x > 0$.

SECTION 2

CONSERVATIVE VECTOR FIELDS

The examples of the previous section show that, under certain conditions, it is possible to alter the path of integration in a line integral in the plane without affecting the value of the integral. Not all line integrals have this property, but those that do are particularly important, not only for the computational reasons already illustrated, but also because of their relation to the gradient. In fact, we have the following theorem, valid in $\mathcal{R}^n$, which is a converse to Theorem 2.5 of Chapter 4.

2.1 Theorem

Let F be a continuous vector field defined in a polygonally connected open subset D of $\mathcal{R}^n$. If the line integral

$$\int_\gamma F \cdot d\mathbf{x}$$

is independent of the piecewise smooth path γ from $\mathbf{x}_0$ to $\mathbf{x}$ in D, then the real-valued function defined by

$$f(\mathbf{x}) = \int_{\mathbf{x}_0}^{\mathbf{x}} F \cdot d\mathbf{x}$$

is continuously differentiable and satisfies the vector equation $\nabla f = F$ throughout D.

Proof. We have to show that, for each $\mathbf{x}$ in D, $\nabla f(\mathbf{x}) = F(\mathbf{x})$. Since $\mathbf{x}$ is an interior point of D, there is a ball of radius δ centered at $\mathbf{x}$ and contained in D. This implies that, for any unit vector $\mathbf{u}$ and for all real numbers t satisfying $|t| < \delta$, the vectors $\mathbf{x} + t\mathbf{u}$ are contained in D. Since the line integral is independent of the path, we choose an arbitrary piecewise smooth path from $\mathbf{x}_0$ to $\mathbf{x}$, lying in D, and extend it by a linear segment to the vector $\mathbf{x} + t\mathbf{u}$, $|t| < \delta$, as shown in Fig. 10. To show that f is continuously differentiable we observe that

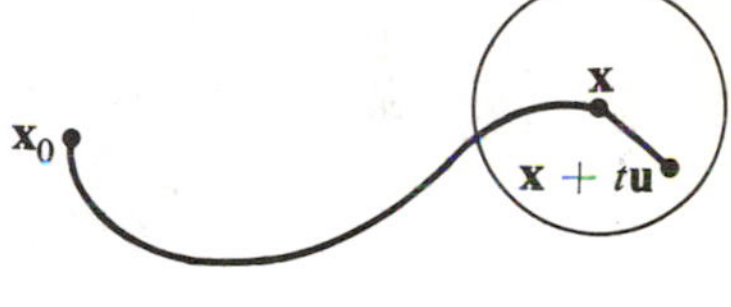

Figure 10

$$\begin{aligned} f(\mathbf{x} + t\mathbf{u}) - f(\mathbf{x}) &= \int_{\mathbf{x}_0}^{\mathbf{x}+t\mathbf{u}} F \cdot d\mathbf{x} - \int_{\mathbf{x}_0}^{\mathbf{x}} F \cdot d\mathbf{x} \\ &= \int_{\mathbf{x}}^{\mathbf{x}+t\mathbf{u}} F \cdot d\mathbf{x} \\ &= \int_0^t F(\mathbf{x} + v\mathbf{u}) \cdot \mathbf{u}\, dv. \end{aligned}$$

Then taking $\mathbf{u} = \mathbf{e}_j$, the jth natural basis vector in $\mathcal{R}^n$, we get

$$\begin{aligned} \frac{\partial f}{\partial x_j}(\mathbf{x}) &= \lim_{t \to 0} \frac{f(\mathbf{x} + t\mathbf{e}_j) - f(\mathbf{x})}{t} \\ &= \lim_{t \to 0} \frac{1}{t} \int_0^t F(\mathbf{x} + v\mathbf{e}_j) \cdot \mathbf{e}_j\, dv \\ &= \frac{d}{dt} \int_0^t F(\mathbf{x} + v\mathbf{e}_j) \cdot \mathbf{e}_j\, dv \Big|_{t=0} \\ &= F(\mathbf{x}) \cdot \mathbf{e}_j = F_j(\mathbf{x}), \end{aligned}$$

where F_j is the jth coordinate function of F. Since F was assumed continuous, so are the partial derivatives $\partial f / \partial x_j$; therefore f

is continuously differentiable on D. Finally, the equations $(\partial f/\partial x_j)(\mathbf{x}) = F_j(\mathbf{x})$, $j = 1, \ldots, n$, mean that $\nabla f = F$ in D.

A vector field F for which there is a real-valued function f such that $F = \nabla f$ is called a **conservative, or gradient, field.** In that case f is called the **potential** of F. The motivation for this terminology is discussed in the next example.

Example 1. Suppose that a continuous force field F, defined in a region D of $\mathcal{R}^3$, is such that the work done in moving a particle from one point to another under the influence of the field is independent of the path taken between the two points. Thus, if $\mathbf{x}_1$ and $\mathbf{x}_2$ are two points in the field and $W(\mathbf{x}_1, \mathbf{x}_2)$ represents the work done in going from $\mathbf{x}_1$ to $\mathbf{x}_2$, we can write

$$W(\mathbf{x}_1, \mathbf{x}_2) = \int_{\mathbf{x}_1}^{\mathbf{x}_2} F \cdot d\mathbf{x}.$$

If the particle follows a particular path given by $g(t)$, then the velocity and acceleration vectors are $\mathbf{v}(t) = g'(t)$, $\mathbf{a}(t) = g''(t)$, and we have $F(g(t)) = m\mathbf{a}(t)$, where m is the mass of the particle. Hence,

$$W(\mathbf{x}_1, \mathbf{x}_2) = \int_{t_1}^{t_2} m\mathbf{a}(t) \cdot \mathbf{v}(t)\, dt,$$

if $g(t_i) = \mathbf{x}_i$. But since $\mathbf{a}(t) = \mathbf{v}'(t)$, and $(d/dt)v^2(t) = 2\mathbf{v}(t) \cdot \mathbf{v}'(t)$, we have

$$\begin{aligned} W(\mathbf{x}_1, \mathbf{x}_2) &= \frac{m}{2} \int_{t_1}^{t_2} \frac{d}{dt} [v^2(t)]\, dt, \\ &= \frac{m}{2} \big(v^2(t_2) - v^2(t_1)\big). \qquad (1) \end{aligned}$$

The function $k(t) = (m/2)v^2(t)$ is called the **kinetic energy** of the particle at time t.

On the other hand, if we fix a point $\mathbf{x}_0$ in D, then by Theorem 2.1, the equation

$$u(\mathbf{x}) = -\int_{\mathbf{x}_0}^{\mathbf{x}} F \cdot d\mathbf{x}$$

defines a continuously differentiable function u in D. Using the independence of path to integrate from $\mathbf{x}_1$ to $\mathbf{x}_2$ via $\mathbf{x}_0$, we get

$$\begin{aligned} W(\mathbf{x}_1, \mathbf{x}_2) &= \int_{\mathbf{x}_1}^{\mathbf{x}_2} F \cdot d\mathbf{x} \\ &= \int_{\mathbf{x}_0}^{\mathbf{x}_2} F \cdot d\mathbf{x} - \int_{\mathbf{x}_0}^{\mathbf{x}_1} F \cdot d\mathbf{x} \\ &= -u(\mathbf{x}_2) + u(\mathbf{x}_1). \qquad (2) \end{aligned}$$

Comparison of Equations (1) and (2) shows that

$$u(\mathbf{x}_2) + \frac{m}{2} v^2(t_2) = u(\mathbf{x}_1) + \frac{m}{2} v^2(t_1).$$

In other words, along the path traced by $g(t)$, the sum $u(g(t)) + k(t)$ is a constant, independent of t, called the total energy of the path. For this reason, the function $u(\mathbf{x})$, which is a function of position in D, is called the **potential energy** of the field F. Notice that there is an arbitrary choice made in defining the potential in that the point $\mathbf{x}_0$ was picked to have zero potential. The choice of some other point $\mathbf{x}_0$ would change the function u at most by an additive constant equal to $W(\mathbf{x}_0, \mathbf{x}_1)$. It is the constant total energy which is "conserved" and which gives rise to the term "conservative field."

For a vector field F defined in a region D of $\mathcal{R}^n$, independence of path in the line integral $\int_\gamma F \cdot d\mathbf{x}$ has been defined to mean that

2.2
$$\int_{\gamma[x_1 x_2]} F \cdot d\mathbf{x} = \int_{\delta[\mathbf{x}_1 \mathbf{x}_2]} F \cdot d\mathbf{x}$$

where $\gamma[\mathbf{x}_1, \mathbf{x}_2]$ and $\delta[\mathbf{x}_1, \mathbf{x}_2]$ are any two piecewise smooth curves in D having initial point $\mathbf{x}_1$ and terminal point $\mathbf{x}_2$. An alternative formulation of the independence property is that

2.3
$$\int_\gamma F \cdot d\mathbf{x} = 0$$

for every piecewise smooth closed curve γ lying in D. The equivalence of the two properties follows from the observations that $\gamma[\mathbf{x}_1, \mathbf{x}_2]$ followed by $\delta[\mathbf{x}_1, \mathbf{x}_2]$ in reverse direction is a closed path, and that a closed path can be separated at points $\mathbf{x}_1$ and $\mathbf{x}_2$ into different paths joining $\mathbf{x}_1$ and $\mathbf{x}_2$. The details of the proof have already been illustrated in Section 1 and will be left as an exercise.

We can summarize what we have proved about gradient fields in Theorem 2.5 of Chapter 4, Theorem 2.1 of the present section, and in the previous remark, as follows.

2.4 Theorem

Let F be a continuous vector field defined in a polygonally connected open set D of $\mathcal{R}^n$. Then the following are equivalent:

1. F is the gradient of a function, f, continuously differentiable in D.

2. The line integral of F over a path from $\mathbf{x}_1$ to $\mathbf{x}_2$ is independent of the piecewise smooth curve γ from $\mathbf{x}_1$ to $\mathbf{x}_2$, and so can be written

$$\int_{\mathbf{x}_1}^{\mathbf{x}_2} F \cdot d\mathbf{x}.$$

3. For every piecewise smooth closed curve lying in D,

$$\oint F \cdot d\mathbf{x} = 0.$$

A more intrinsic criterion for deciding if a continuous vector field is a gradient field or not arises as follows. Suppose first that $\mathcal{R}^2 \xrightarrow{F} \mathcal{R}^2$ is continuous on an open set D, and that F is a gradient field, that is, there is a real-valued function f defined on D such that $\nabla f = F$. In terms of coordinate functions F_1 and F_2 of F, this means

$$\frac{\partial f}{\partial x_1} = F_1 \quad \text{and} \quad \frac{\partial f}{\partial x_2} = F_2.$$

If F itself is continuously differentiable, we can form the second partials,

$$\frac{\partial^2 f}{\partial x_2 \, \partial x_1} = \frac{\partial F_1}{\partial x_2} \quad \text{and} \quad \frac{\partial^2 f}{\partial x_1 \, \partial x_2} = \frac{\partial F_2}{\partial x_1},$$

and conclude from their equality that

$$\frac{\partial F_1}{\partial x_2} = \frac{\partial F_2}{\partial x_1} \tag{3}$$

throughout D. By the definition of curl F, Equation (3) can be written curl $F = 0$. The equation can also be expressed another way: We consider the more general field $\mathcal{R}^n \xrightarrow{F} \mathcal{R}^n$, which we assume continuously differentiable in an open subset D of $\mathcal{R}^n$. If F is a gradient field, there is an f such that $\nabla f = F$, or, in terms of coordinate functions

$$\frac{\partial f}{\partial x_j} = F_j, \qquad j = 1, \ldots, n.$$

Differentiating with respect to x_i, we get

$$\frac{\partial F_j}{\partial x_i} = \frac{\partial^2 f}{\partial x_i \, \partial x_j} = \frac{\partial^2 f}{\partial x_j \, \partial x} = \frac{\partial F_i}{\partial x_j}. \tag{4}$$

But the functions $\partial F_i / \partial x_j$ are the entries in the n-by-n Jacobian matrix of $\mathcal{R}^n \xrightarrow{F} \mathcal{R}^n$, and Equation (4) expresses the fact that this matrix is symmetric. Hence

2.5 Theorem

If $\mathcal{R}^n \xrightarrow{F} \mathcal{R}^n$ is a continuously differentiable gradient field, then F', the Jacobian matrix of F, is symmetric.

Example 2. The converse of Theorem 2.5 is false, as we see by looking at the example in $\mathcal{R}^2$,

$$F(x, y) = \left(\frac{-y}{x^2 + y^2}, \frac{x}{x^2 + y^2}\right),$$

defined for all $(x, y) \neq (0, 0)$. It is easy to check that $\partial F_1/\partial y = \partial F_2/\partial x$. But there is no continuously differentiable f such that $\nabla f = F$. The reason is that, for $x > 0$, the function $f(x, y) = \arctan(y/x)$ satisfies $\nabla f = F$, but this f cannot be extended to be a single-valued solution of the equation in the entire plane with the origin deleted.

Example 2 shows that the nature of the region D on which F is defined is significant in determining whether F is a gradient field. By making a special assumption about D we can obtain a partial converse to Theorem 2.5.

2.6 Theorem

Let R be an open coordinate rectangle in $\mathcal{R}^n$, and let F be a continuously differentiable vector field on R. If $F'(\mathbf{x})$, the Jacobian matrix of F, is symmetric on R, then F is a gradient field.

Proof. Pick a fixed point $\mathbf{x}_0$ in R and let $\mathbf{x}$ be any other point of R. We consider paths from $\mathbf{x}_0$ to $\mathbf{x}$, each consisting of a sequence of line segments parallel to the axes and such that each coordinate variable varies on at most one such segment. Three-dimensional examples are shown in Fig. 11. The reason for looking at such paths is to be able to approach $\mathbf{x}$ from any coordinate direction for the purpose of taking partial derivatives at $\mathbf{x}$. Choose one of these paths, call it $\gamma_{\mathbf{x}}$, and define a real-valued function f by

$$f(\mathbf{x}) = \int_{\gamma_{\mathbf{x}}} F \cdot d\mathbf{x}. \tag{5}$$

While the particular path $\gamma_{\mathbf{x}}$ is only one of several of the same type, we shall see that any of the other possible choices would lead to the same value for $f(\mathbf{x})$. The reason is that any one of these paths can be altered step by step into any one of the others by changes, each of

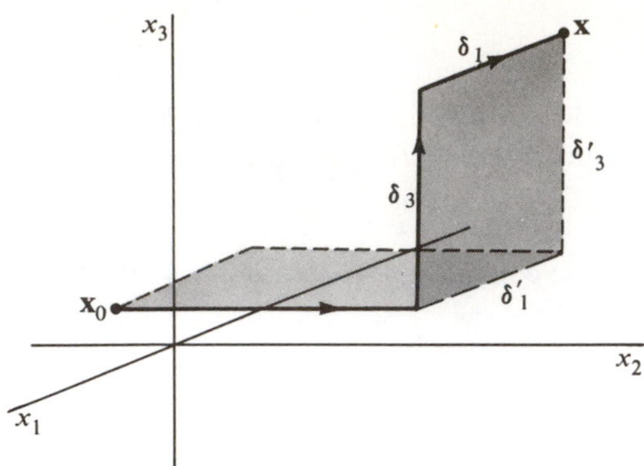

Figure 11

which leaves the value of the integral (5) unaltered. Each path can be described by a sequence of coordinate directions, only one of which is allowed to vary at a time. (For example, the dotted path in Fig. 11 corresponds to x_1, x_2, x_3, and the solid one to x_2, x_3, x_1.) Clearly, changing one such sequence into another can be accomplished by successively interchanging adjacent variables in pairs until the desired order is reached. But each interchange replaces a pair of segments (δ_i, δ_j) by another pair (δ'_i, δ'_j) lying in the same 2-dimensional plane. To see that the replacement leaves the value of the integral invariant, we form the circuit δ consisting of the segments δ_i and δ_j, followed by δ'_i and δ'_j in the reverse of their original directions. On these segments, x_i and x_j are the only variables that vary, so the circuit integral can be written

$$\oint_\delta F \cdot d\mathbf{x} = \oint_\delta F_i \, dx_j + F_j \, dx_j.$$

We apply Green's theorem to the 2-dimensional rectangle R_δ bounded by δ and get

$$\oint_\delta F \cdot d\mathbf{x} = \int_{R_\delta} \left(\frac{\partial F_i}{\partial x_i} - \frac{\partial F_i}{\partial x_j} \right) dx_i \, dx_j = 0,$$

since by the symmetry assumption, $\partial F_j / \partial x_i - \partial F_i / \partial x_j = 0$ in R. Thus

$$\oint_\delta F \cdot d\mathbf{x} = \int_{(\delta_i, \delta_j)} F \cdot d\mathbf{x} - \int_{(\delta'_j, \delta'_i)} F \cdot d\mathbf{x} = 0,$$

and so the change of path leaves the value of the integral invariant.

Once it has been established that $\mathbf{x}$ can be approached along a path of integration that varies only in an arbitrary coordinate, say the kth, we have, as in the proof of Theorem 2.1, the equation $\partial f / \partial x_k(\mathbf{x}) = F_k(\mathbf{x})$, for all k. Thus $\nabla f(\mathbf{x}) = F(\mathbf{x})$ for all $\mathbf{x}$ in R.

Example 3. Applying Theorem 2.4 to the field

$$F(x, y) = \left(\frac{-y}{x^2 + y^2}, \frac{x}{x^2 + y^2}\right), \qquad (x, y) \neq (0, 0),$$

of Example 2, we conclude that F, when restricted to any coordinate rectangle not containing the origin, is a gradient field. This is true, for example in any of the four half-planes bounded by a coordinate axis. A potential function f for the half-plane $x > 0$, can be computed by the line integral

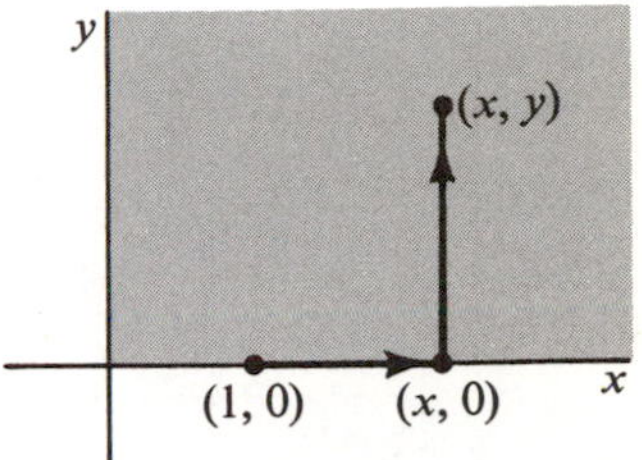

Figure 12

$$f(x, y) = \int_{(1,0)}^{(x,y)} \frac{-y\,dx}{x^2 + y^2} + \frac{x\,dy}{x^2 + y^2},$$

where the path of integration is any piecewise smooth curve from $(1, 0)$ to (x, y). A polygonal path from $(1, 0)$ to $(x, 0)$ and from $(x, 0)$ to (x, y) is particularly simple. On the first segment, the entire integral is zero because y is identically zero, and on the second segment, with x constant, the integral reduces to

$$\int_0^y \frac{x\,dy}{x^2 + y^2} = \arctan\left(\frac{y}{x}\right).$$

The most general potential of F in the right half-plane differs from this one by at most a constant. (Why?) The general solution there of $\nabla f = F$ is therefore

$$f(x, y) = \arctan \frac{y}{x} + c.$$

Compare the method of solution described in Problem 11 of Chapter 4, Section 2.

EXERCISES

1. Consider the approximation to the earth's gravitational field acting on a particle of mass 1 represented by the vector field $F(x, y, z) = (0, 0, -g)$.

(a) Find for F the potential function $u(x, y, z)$ that is zero when $(x, y, z) = (0, 0, 0)$.

(b) If a particle of mass 1 has at $(0, 0, 0)$ a velocity vector (v_1, v_2, v_3) with $v_3 > 0$, and no force but F acts on the particle, find the path of the particle.

(c) Verify that the sum of potential energy and kinetic energy remains constant for the path of part (b).

2. (a) Show that if F and G are gradient fields defined on the same domain D, then $F + G$ and cF are gradient fields, where c is a constant.

(b) Let $\mathcal{V}$ be the vector space of gradient fields defined on a domain D. Show that $\mathcal{V}$ has infinite dimension.

3. Use Theorem 2.4 or 2.6 to decide whether the following vector fields are gradient fields.

(a) $F(x, y) = (x - y, x + y)$, for (x, y) in $\mathcal{R}^2$.
(b) $G(x, y, z) = (y, z, x)$, for (x, y, z) in $\mathcal{R}^3$.
(c) $H(x, y) = \left(\frac{-y}{x^2 + y^2}, \frac{x}{x^2 + y^2}\right)$, $(x, y) \neq (0, 0)$.

4. Use Theorem 2.5 to show that the vector fields in Problems 3(a) and 3(b) are not gradient fields in any open subset of $\mathcal{R}^2$ or $\mathcal{R}^3$, respectively.

5. Show that the vector field of Problem 3(c) is a gradient field in the region $y > 0$ of $\mathcal{R}^2$ and find an explicit representation for its potential.

6. Consider the vector field defined in $\mathcal{R}^3$, with the z-axis deleted, by

$$F(x, y, z) = \left(\frac{-y}{x^2 + y^2}, \frac{x}{x^2 + y^2}, 0\right).$$

Is F a gradient field?

7. Find a potential for each of the following fields.

(a) $F(x, y, z) = (2xy, x^2 + z^2, 2yz)$.
(b) $G(x, y) = (y \cos xy, x \cos xy)$.
(c) $L(x_1, x_2) = \left(\frac{x_1}{x_1^2 + x_2^2}, \frac{x_2}{x_1^2 + x_2^2}\right)$, with $(x_1, x_2) \neq (0, 0)$.

8. Consider the vector field F which is the gradient of the Newtonian potential $f(\mathbf{x}) = -|\mathbf{x}|^{-1}$ for nonzero $\mathbf{x}$ in $\mathcal{R}^3$. Find the work done in moving a particle from $(1, 1, 1)$ to $(-2, -2, -2)$ along a smooth curve lying in the domain of F.

9. Give a detailed proof of the equivalence of Relations 2.2 and 2.3 of the text.

10. In $\mathcal{R}^n$, how many paths can there be from $\mathbf{x}_0$ to $\mathbf{x}$ of the special kind described in the proof of Theorem 2.6?

SECTION 3

SURFACE INTEGRALS

In Chapter 3, Section 4, we have defined integrals both of a real-valued function and of a vector over a smooth curve. Defining an integral over a surface S leads to a different geometric situation having, however, a close analogy with the line integral. To begin, we assume that S is parametrized by a continuously differentiable function $\mathcal{R}^2 \xrightarrow{g} \mathcal{R}^3$. We shall write g in the form

$$g(u, v) = \begin{pmatrix} g_1(u, v) \\ g_2(u, v) \\ g_3(u, v) \end{pmatrix}, \qquad (1)$$

with $\mathbf{u} = (u, v)$ in some set D in $\mathcal{R}^2$, which we assume bounded by a piecewise smooth curve. We further assume that, at each point $g(u, v)$ of S, the tangent vectors defined by the vector partial derivatives

$$\frac{\partial g}{\partial u}(u, v), \qquad \frac{\partial g}{\partial v}(u, v)$$

determine a 2-dimensional tangent plane to S; in other words, that the two tangents are linearly independent. If S satisfies all the above conditions, we shall refer to it as a **piece of smooth surface.**

On a smooth curve, the choice of a parametrization going from one endpoint to the other establishes an orientation for the curve. Analogously, on a piece of smooth surface, a one-to-one parametrization determines a particular normal vector

$$\frac{\partial g}{\partial u}(u, v) \times \frac{\partial g}{\partial v}(u, v) \tag{2}$$

at $g(u, v)$. See Fig. 13.

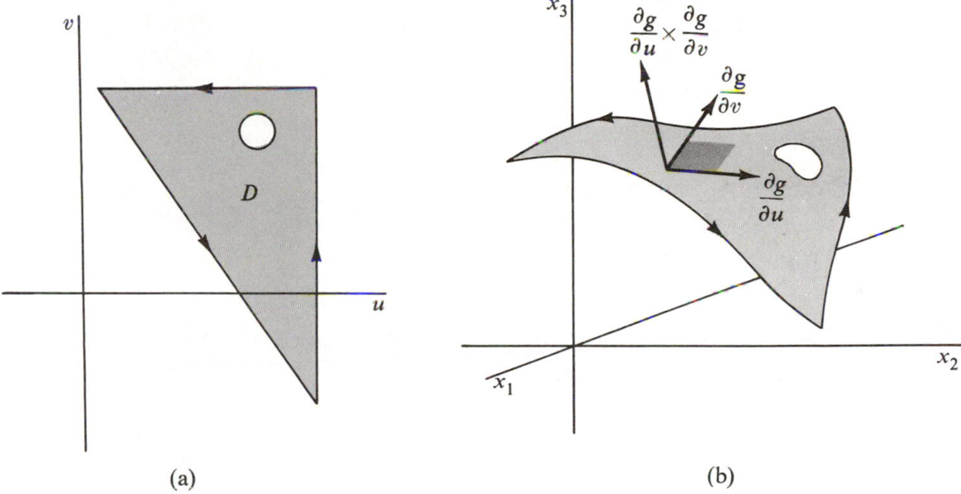

Figure 13

We recall that the length of the cross-product of two vectors **a** and **b** is the area of the parallelogram spanned by **a** and **b**. In particular,

$$\left| \frac{\partial g}{\partial u}(u, v) \times \frac{\partial g}{\partial v}(u, v) \right|$$

represents the area of the tangent parallelogram shown in Fig. 13(*b*). If we think of constructing such parallelograms at the points $g(u_k, v_k)$ corresponding to the corner points (u_k, v_k) of a grid over D, then it is natural to define the **area** of S by

3.1 $$\sigma(S) = \int_D \left| \frac{\partial g}{\partial u}(u, v) \times \frac{\partial g}{\partial v}(u, v) \right| du\, dv.$$

We assume that g is one-to-one. The integral over D exists because g was assumed continuously differentiable on D. Similarly, if p is a continuous real-valued function defined on S, we define the integral of p over S by

3.2 $$\int_S p\, d\sigma = \int_D p(g(u, v)) \left| \frac{\partial g}{\partial u}(u, v) \times \frac{\partial g}{\partial v}(u, v) \right| du\, dv.$$

This definition is the analog of that for a real-valued function over a smooth curve given by Equation (2) of Chapter 3, Section 4.

Example 1. Let S be parametrized by

$$g(u, v) = \begin{pmatrix} u \\ v \\ u^2 + v^2 \end{pmatrix},$$

for

$$1 \leq u^2 + v^2 \leq 4;$$

thus S is actually the graph of $x^2 + y^2$ for $1 \leq x^2 + y^2 \leq 4$. The surface is shown in Fig. 14. Then

$$\frac{\partial g}{\partial u}(u, v) = \begin{pmatrix} 1 \\ 0 \\ 2u \end{pmatrix}, \qquad \frac{\partial g}{\partial v}(u, v) = \begin{pmatrix} 0 \\ 1 \\ 2v \end{pmatrix}.$$

We have

$$\frac{\partial g}{\partial u}(u, v) \times \frac{\partial g}{\partial v}(u, v) = \left(\begin{vmatrix} 0 & 1 \\ 2u & 2v \end{vmatrix}, \begin{vmatrix} 2u & 2v \\ 1 & 0 \end{vmatrix}, \begin{vmatrix} 1 & 0 \\ 0 & 1 \end{vmatrix} \right).$$

Hence

$$\sigma(S) = \int_{1 \leq u^2 + v^2 \leq 4} \sqrt{4u^2 + 4v^2 + 1}\, du\, dv$$

$$= \int_0^{2\pi} d\theta \int_1^2 \sqrt{4r^2 + 1}\, r\, dr = \frac{\pi}{6}[17^{3/2} - 5^{3/2}].$$

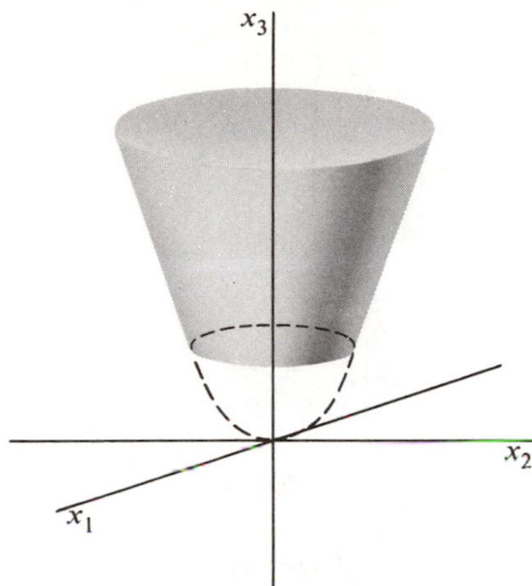

Figure 14

If S is given a nonuniform density equal at each point to the distance of that point from the axis of symmetry of S, then the density can be represented by the function $p(x, y, z) = \sqrt{x^2 + y^2}$. At a point $g(u, v)$ of S the density will be $p(g(u, v)) = \sqrt{u^2 + v^2}$. We can interpret the integral of p over S to be the total mass of the weighted surface, and we have

$$\int_S p\, d\sigma = \int_{1 \le u^2+v^2 \le 4} \sqrt{u^2 + v^2}\,\sqrt{4u^2 + 4v^2 + 1}\; du\, dv$$
$$= \int_0^{2\pi} d\theta \int_1^2 r^2\sqrt{4r^2 + 1}\; dr.$$

The main purpose of this section is the definition of the integral of a continuous vector field $\mathcal{R}^3 \xrightarrow{F} \mathcal{R}^3$ over a surface S. Continuing with the assumption that S is a piece of smooth surface represented by the function of Equation (1), we compare the normal vector $\partial g/\partial u \times \partial g/\partial v$ with the vector field F at a point $g(u, v)$ of S. These are shown in Fig. 15 at one point. If $\mathbf{n}$ is a *unit* normal to S at $g(u, v)$, then the dot-product $F \cdot \mathbf{n}$ at $g(u, v)$ is the coordinate of F in the direction of $\mathbf{n}$. But since

$$\mathbf{n} = \frac{\dfrac{\partial g}{\partial u} \times \dfrac{\partial g}{\partial v}}{\left|\dfrac{\partial g}{\partial u} \times \dfrac{\partial g}{\partial v}\right|},$$

it follows that

$$F(g(u, v)) \cdot \left(\frac{\partial g}{\partial u}(u, v) \times \frac{\partial g}{\partial v}(u, v)\right)$$

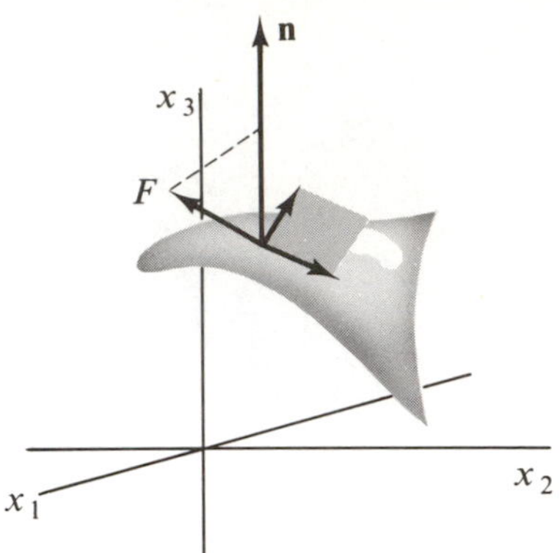

Figure 15

is equal to the coordinate of $F(g(u, v))$ in the direction of $\mathbf{n}$, multiplied by the area of the tangent parallelogram spanned by $\partial g/\partial u$ and $\partial g/\partial v$ at $g(u, v)$. We define the **surface integral** of F over S by

3.3 $$\int_D F(g(u, v)) \cdot \left(\frac{\partial g}{\partial u}(u, v) \times \frac{\partial g}{\partial v}(u, v)\right) du\, dv,$$

and denote it by $\int_S F \cdot dS$ or $\int_S F \cdot \mathbf{n}\, d\sigma$.

It is easy to check that the coordinates of the normal vector are given by Jacobian determinants as

$$\frac{\partial g}{\partial u} \times \frac{\partial g}{\partial v} = \left(\frac{\partial(x_2, x_3)}{\partial(u, v)}, \frac{\partial(x_3, x_1)}{\partial(u, v)}, \frac{\partial(x_1, x_2)}{\partial(u, v)}\right),$$

where x_1, x_2, and x_3 represent the coordinate functions of g. If the vector field F has coordinate functions F_1, F_2, F_3, then the surface integral is often written

$$\int_S F \cdot dS = \int_D F_1 \frac{\partial(x_2, x_3)}{\partial(u, v)} + F_2 \frac{\partial(x_3, x_1)}{\partial(u, v)} + F_3 \frac{\partial(x_1, x_2)}{\partial(u, v)}\, du\, dv$$

$$= \int_D F_1\, dx_2\, dx_3 + F_2\, dx_3\, dx_1 + F_3\, dx_1\, dx_2.$$

This last abbreviation is a particularly convenient one, and its significance is discussed in Section 7.

Example 2. Suppose that a continuous vector field $\mathcal{R}^3 \xrightarrow{F} \mathcal{R}^3$ describes the speed and direction of a fluid flow at each point of a region R in which it is defined. We shall define, using a surface integral, the flux, or rate of flow per unit of area and time across a piece of smooth surface S lying in R. If S is perfectly flat and F is a constant field, then the flux is equal to $F_{\mathbf{n}}\sigma(S)$, where $F_{\mathbf{n}}$ is the coordinate of F in the direction of a unit normal to S. Thus, in this case, the flux is equal to the volume of a tube of fluid illustrated in Fig. 16. Because $F_{\mathbf{n}} = F \cdot \mathbf{n}$, we get the formula

$$\Phi = F \cdot \mathbf{n}\sigma(S)$$

for the flux.

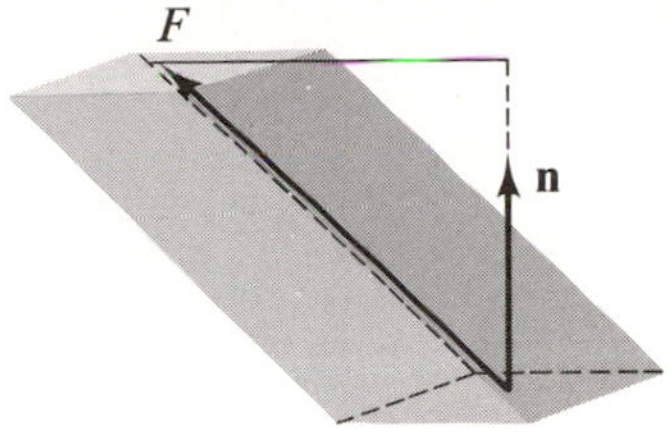

Figure 16

If S is a piece of smooth surface in R, we partition S along coordinate curves of the form $u = \text{const.}$ and $v = \text{const.}$ and assume that, within each part of S so formed, the field F is constant. Approximating S by tangent parallelograms spanned by vectors

$$\Delta u \frac{\partial g}{\partial u} \quad \text{and} \quad \Delta v \frac{\partial g}{\partial v}$$

leads to the picture shown in Fig. 17. The approximate flux across a typical subdivision S_k of S will have the form

$$\begin{aligned}\Phi_k &= F(g(\mathbf{u}_k)) \cdot \mathbf{n}_k\sigma(S_k)\\ &= F(g(\mathbf{u}_k)) \cdot \left(\frac{\partial g}{\partial u}(\mathbf{u}_k) \times \frac{\partial g}{\partial v}(\mathbf{u}_k)\right) \Delta u\, \Delta v.\end{aligned}$$

The sum

$$\sum_{k=1}^{N} \Phi_k = \sum_{k=1}^{N} F(g(\mathbf{u}_k)) \cdot \left(\frac{\partial g}{\partial u}(\mathbf{u}_k) \times \frac{\partial g}{\partial v}(\mathbf{u}_k)\right) \Delta u\, \Delta v$$

becomes a better approximation to what we would like to call the flux across S as the subdivision of S is refined by making finer the corresponding grid G in the parameter domain D. On the other hand, if F is

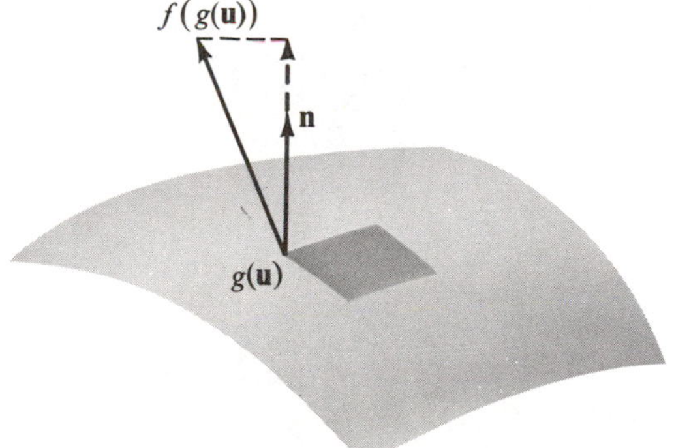

Figure 17

continuous on S and g is continuously differentiable on D, then

$$\lim_{m(G)\to 0} \sum_{k=1}^{N} \Phi_k = \int_D F(g(\mathbf{u}))\cdot \left(\frac{\partial g}{\partial u}(\mathbf{u}) \times \frac{\partial g}{\partial v}(\mathbf{u})\right) d\sigma$$

$$= \int_S F \cdot ds,$$

which is the previously defined integral of F over S. Consequently, we define the **flux** of F across S by

$$\Phi = \int_S F \cdot dS.$$

We remark that the sign of Φ would change if S were reparametrized so that the normal vector determined by the parametrization pointed in the opposite direction.

Example 3. Let a fluid flow outward from the origin in $\mathcal{R}^3$ be given by

$$F(x_1, x_2, x_3) = \left(\frac{x_1}{x_1^2 + x_2^2 + x_3^2}, \frac{x_2}{x_1^2 + x_2^2 + x_3^2}, \frac{x_3}{x_1^2 + x_2^2 + x_3^2}\right).$$

The flux of F across a sphere S_a centered at the origin and of radius a takes the form

$$\int_{S_a} F \cdot ds = \int_{S_a} \frac{x_1\, dx_2\, dx_3 + x_2\, dx_3\, dx_1 + x_3\, dx_1\, dx_2}{x_1^2 + x_2^2 + x_3^2}.$$

However, on S_a, the denominator is contantly equal to a^2, so the surface integral takes the simpler form

$$\frac{1}{a^2}\int_{S_a} x_1\, dx_2\, dx_3 + x_2\, dx_3\, dx_1 + x_3\, dx_1\, dx_2.$$

We can represent S_a parametrically by

$$\begin{pmatrix} x_1 \\ x_2 \\ x^3 \end{pmatrix} = \begin{pmatrix} a \sin\varphi \cos\theta \\ a \sin\varphi \sin\theta \\ a \cos\varphi \end{pmatrix}, \qquad \begin{matrix} 0 \le \varphi \le \pi \\ 0 \le \theta \le 2\pi, \end{matrix}$$

so that

$$\frac{\partial(x_2, x_3)}{\partial(\varphi, \theta)} = a^2 \sin^2\varphi \cos\theta$$

$$\frac{\partial(x_3, x_1)}{\partial(\varphi, \theta)} = a^2 \sin^2\varphi \sin\theta$$

$$\frac{\partial(x_1, x_2)}{\partial(\varphi, \theta)} = a^2 \sin\varphi \cos\varphi.$$

These are the coordinates of a normal vector pointing out from the sphere. Then

$$\int_{S_a} F \cdot dS = a\int_0^{2\pi} d\theta \int_0^{\pi} \sin^3 \varphi \cos^2 \theta + \sin^3 \varphi \sin^2 \theta + \sin \varphi \cos^2 \varphi \, d\varphi$$

$$= a\int_0^{2\pi} d\theta \int_0^{\pi} \sin \varphi \, d\varphi = 4\pi a.$$

For the purpose of computing a line integral over a piecewise smooth curve, it is natural to orient the smooth pieces of the curve so that the terminal point of one piece is the same as the initial point of the one that follows it. To integrate a vector field over a piecewise smooth surface, we need a notion of orientation for pieces of smooth surface S. If $\mathcal{R}^2 \xrightarrow{g} \mathcal{R}^3$ represents S parametricaly with g defined on D, then Fig. 13 shows how D and S may be related. The edge of S, corresponding under g to the boundary of D, we shall call the **border** of S. As a point $\mathbf{u}$ moves around the piecewise smooth boundary of D in the *counterclockwise* direction, its image $g(\mathbf{u})$ traces the border of S with what we shall call its **positive orientation.** A further geometric picture of the positive orientation can be had by observing that, as the normal vector $\partial g/\partial u \times \partial g/\partial v$ runs around the border in the positive direction, the surface S remains on its left.

A **piecewise smooth surface** is of course a finite union of pieces of smooth surface that are joined along common border curves. A piecewise smooth surface is **orientable** if, when the border curves of its pieces are positively oriented, borders common to two pieces are traced in opposite directions. If one of the two curves is reversed, their parametrizations must be equivalent. The surfaces pictured in Fig. 18 are assumed to be representable as piecewise smooth surfaces, and the first two are orientable. However, the joining together of two rectangular strips, one of them with a twist, gives a Möbius strip, which is the standard example of a non-orientable surface.

Figure 18

We define the integral of a continuous vector field over a piecewise smooth surface to be the sum of the integrals over each of its smooth pieces. Thus if $S = S_1 \cup S_2$,

$$\int_S F \cdot dS = \int_{S_1} F \cdot dS + \int_{S_2} F \cdot dS.$$

This definition holds even if S is not orientable, but in practice it is of little interest to integrate a vector field over a nonorientable surface. On the other hand,the integral of a real-valued function over a surface can be computed without regard to orientation. The reason is that in Formulas 3.1 and 3.2 the so-called element of surface area,

$$d\sigma = \left|\frac{\partial g}{\partial u} \times \frac{\partial g}{\partial v}\right| du\, dv,$$

does not change when the orientation is reversed by interchanging the roles of u and v. But in Formula 3.3, the vector surface element,

$$dS = \left(\frac{\partial g}{\partial u} \times \frac{\partial g}{dv}\right) du\, dv,$$

does change sign when u and v are interchanged. We observe that the surface element dS can also be written in the form

$$dS = \mathbf{n}\, d\sigma,$$

where $\mathbf{n}$ is the unit normal to the surface given by

$$\mathbf{n} = \frac{\dfrac{\partial g}{\partial u} \times \dfrac{\partial g}{\partial v}}{\left|\dfrac{\partial g}{\partial u} \times \dfrac{\partial g}{\partial v}\right|}.$$

EXERCISES

1. (a) Find the area of the spiral ramp represented by

$$g(u, v) = \begin{pmatrix} u \cos v \\ u \sin v \\ v \end{pmatrix}, \qquad 0 \le u \le 1, \qquad 0 \le v \le 3\pi.$$

$$\left[\textit{Ans.}\ \frac{3\pi}{2}(\sqrt{2} + \log(1 + \sqrt{2})).\right]$$

(b) Let the surface of part (a) have a density per unit of area at each point equal to the distance of that point from the central axis of the surface. Find the total mass of the weighted surface.

2. Compute $\int_S F \cdot dS$, where

(a) $F(x, y, z) = (x, y, z)$ and S is given by

$$g(u, v) = \begin{pmatrix} u - v \\ u + v \\ uv \end{pmatrix}, \qquad \begin{matrix} 0 \leq u \leq 1, \\ 0 \leq v \leq 2. \end{matrix}$$

(b) $F(x, y, z) = (x^2, 0, 0)$ and S is given by

$$g(u, v) = \begin{pmatrix} u \cos v \\ u \sin v \\ v \end{pmatrix}, \qquad \begin{matrix} 0 \leq u \leq 1, \\ 0 \leq v \leq 2\pi. \end{matrix}$$

3. Find the total mass of a spherical film having density at each point equal to the linear distance of the point from a single fixed point on the sphere.

4. Let $\mathbf{x} = g(u, v)$, for (u, v) in D, and $\mathbf{x} = h(s, t)$, for (s, t) in B, be parametrizations for the same piece of smooth surface S in $\mathcal{R}^3$. If there is a one-to-one transformation T, continuously differentiable both ways between D and B, such that the Jacobian determinant of T is positive, and such that $g(u, v) = h(T(u, v))$ for (u, v) in D, then g and h are called **equivalent** parametrizations of S.

(a) Show that equivalent parametrizations assign the same surface area to S. [*Hint.* Use the change-of-variable theorem.]
(b) Show that equivalent parametrizations assign the same value to the surface integral of a vector field over S.

5. Let the temperature at a point (x, y, z) of a region R be given by a continuously differentiable function $T(x, y, z)$. Then the vector field ∇T is called the **temperature gradient,** and under certain physical assumptions, $\nabla T(x, y, z)$ is proportional to the direction and rate of flow of heat per unit of area at (x, y, z).

(a) If $T(x, y, z) = x^2 + y^2$ for $x^2 + y^2 \leq 4$, find the total rate of flow of heat across the cylindrical surface $x^2 + y^2 = 1$, $0 \leq z \leq 1$.
(b) Give an example of a continuously differentiable vector field that cannot be a temperature gradient.

6. The Newtonian potential function $(x^2 + y^2 + z^2)^{-1/2}$ has as its gradient the attractive force field F of a charged particle at the origin acting on an oppositely charged particle at (x, y, z). The flux of the field across a piece of smooth surface is defined to be the surface integral of F over S. Show that the flux of F across a sphere of radius a centered at the origin is independent of a.

7. (a) If $\mathcal{R}^2 \xrightarrow{f} \mathcal{R}$ is continuously differentiable on a set D bounded by a piecewise smooth curve, show that the area of the graph of f is

$$\sigma(S) = \int_D \sqrt{1 + (f_x)^2 + (f_y)^2}\, dx\, dy.$$

(b) Find the area of the graph of $f(x, y) = x^2 + y$ for $0 \leq x \leq 1$, $0 \leq y \leq 1$. [*Ans.* $\sqrt{\frac{3}{2}} + \frac{1}{2}\log(\sqrt{2} + \sqrt{3})$.]

8. Show that if $\mathcal{R}^3 \xrightarrow{G} \mathcal{R}$ is continuously differentiable and implicitly determines a piece of smooth surface S on which $\partial G/\partial z \neq 0$, and which lies over a region D of the xy-plane, then

$$\sigma(S) = \int_D \sqrt{\left(\frac{\partial G}{\partial x}\right)^2 + \left(\frac{\partial G}{\partial y}\right)^2 + \left(\frac{\partial G}{\partial z}\right)^2} \left|\frac{\partial G}{\partial z}\right|^{-1} dx\, dy.$$

Assume that just one point of S lies over each point of D.

9. Prove that the border of a piece of smooth surface is a piecewise smooth curve.

10. For each of the following sets find a parametrization as a piecewise smooth orientable surface with outward pointing normal.

(a) The cylindrical can with bottom and no top given by $x^2 + y^2 = 1$, $0 \leq z \leq 1$ and $x^2 + y^2 \leq 1$, $z = 0$.
(b) The funnel given by $x^2 + y^2 - z^2 = 0$, $1 \leq z \leq 4$ and $x^2 + y^2 = 1$, $0 \leq z \leq 1$.
(c) The trough given by $y - z = 0$, $0 \leq x \leq 1$, $0 \leq z \leq 1$ and $y + z = 0$, $0 \leq x \leq 1$, $0 \leq z \leq 1$.

11. Let F be the vector field in $\mathcal{R}^3$ given by $F(x, y, z) = (x, y, 2z - x - y)$. Find the integral of F over the oriented surfaces of Problem 10.

12. Let F be a continuous fluid flow field and let M be a piecewise smooth Möbius strip lying in the domain of F. Is it possible to define the flux of F across M?

13. Parametrize the set of Problem 10(a) so that it is unoriented, with normals pointing out on the bottom and in on the sides. Compute the integral of $F(x, y, z) = (x, y, 2z - x - y)$ over the unoriented surface.

14. Prove that if F and G are continuous vector fields on a piece of smooth surface S, then

$$\int_S (aF + bG) \cdot dS = a\int_S F \cdot dS + b\int_S G \cdot dS,$$

where a and b are constants.

15. (a) Let F be a continuous vector field on a piece of smooth surface S. Show that

$$\left|\int_S F \cdot dS\right| \leq M\sigma(S),$$

where M is the maximum of $|F(\mathbf{x})|$ for $\mathbf{x}$ on S. [*Hint.* Write $\int F \cdot dS$ in the form $\int F \cdot \mathbf{n}\, d\sigma$.]

(b) Use part (a) to show that if S contracts to a point $\mathbf{x}_0$ in such a way that $\sigma(S)$ tends to zero, then $(1/\sigma(S))\int_S F \cdot dS$ tends to $F(\mathbf{x}_0) \cdot \mathbf{n}_0$ where $\mathbf{n}_0$ is a unit normal to S at $\mathbf{x}_0$.

16. Let f be a real-valued continuous differentiable function of one variable, nonnegative for $a \leq x \leq b$. The graph of f, rotated around the x-axis, generates a surface of revolution in $\mathcal{R}^3$.

    (a) Find a parametric representation for S in terms of f.

    (b) Prove that $\sigma(S) = 2\pi \int_a^b f(x)\sqrt{1 + (f'(x))^2}\, dx$.

17. The **solid angle** determined by one nappe of a solid cone $\mathcal{C}$ in $\mathcal{R}^3$, with vertex at the origin, is defined to be the area of the intersection of $\mathcal{C}$ with the unit sphere $|\mathbf{x}| = 1$. See Fig. 19.

    (a) Show that a suitable reduction of the above definition leads to the usual definition of the angle between two lines.

    (b) Compute the solid angle determined by the cone $x^2 + y^2 \leq 2z^2$, $0 \leq z$. [*Ans.* $2\pi(1 - 3^{-1/2})$.]

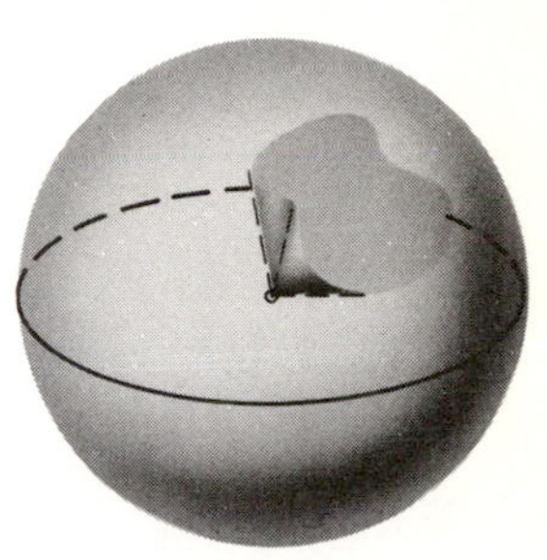

Figure 19

SECTION 4
STOKES'S THEOREM

An important extension of Green's theorem can be made as follows. Instead of considering a plane regon D bounded by a curve, we can think of lifting such a region, together with its boundary curve, into a 2-dimensional surface S in $\mathcal{R}^3$. Then S will have as its border a space curve γ corresponding to the boundary of D. The lifting is made precise by defining on D and its piecewise smooth boundary a function $\mathcal{R}^2 \xrightarrow{g} \mathcal{R}^3$ having S as the image of D. A typical picture is shown in Fig. 20. The

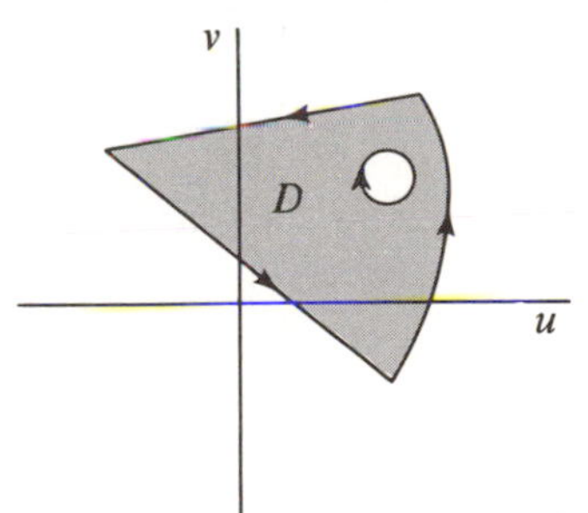

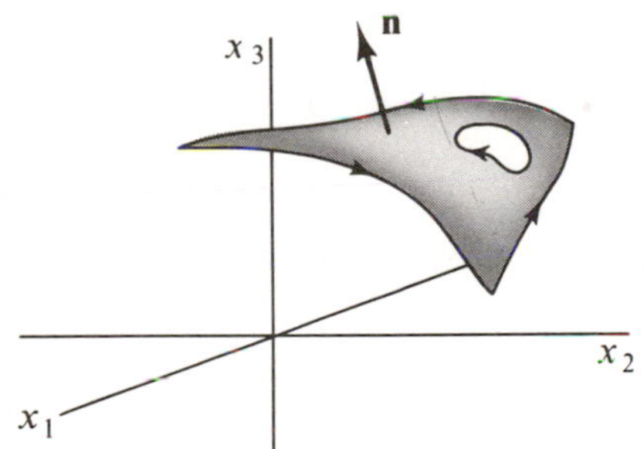

Figure 20

region D has its boundary oriented counterclockwise, and γ, the border curve of S, inherits what we have called the positive orientation with respect to S. If we parametrize the boundary of D by $\mathcal{R} \xrightarrow{h} \mathcal{R}^2$, for $a \leq t \leq b$, then the composition $g(h(t))$ will describe the border of S. We shall denote the positively oriented border of S by ∂S.

We can now relate the line integral of a vector field F around ∂S to the surface integral of an associated vector field over S. We assume that

$\mathcal{R}^3 \xrightarrow{F} \mathcal{R}^3$ is a continuously differentiable vector field whose domain contains S. Then the vector field **curl F** is defined by

$$\text{curl } F(\mathbf{x}) = \left(\frac{\partial F_3}{\partial x_2}(\mathbf{x}) - \frac{\partial F_2}{\partial x_3}(\mathbf{x}), \frac{\partial F_1}{\partial x_3}(\mathbf{x}) - \frac{\partial F_3}{\partial x_1}(\mathbf{x}), \frac{\partial F_2}{\partial x_1}(\mathbf{x}) - \frac{\partial F_1}{\partial x_2}(\mathbf{x})\right),$$

where F_1, F_2, and F_3 are the coordinate functions of F. If the domain of F is an open set, then the domain of curl F is the same set. The vector field curl F has been chosen so that if S is a piece of sufficiently smooth surface, then it will turn out that

$$\int_S \text{curl } F \cdot ds = \oint_{\partial S} F \cdot d\mathbf{x}. \tag{1}$$

Notice that if F were essentially a 2-dimensional vector field, with F_3 identically zero and F_1 and F_2 independent of x_3, then only the third coordinate function of curl F would be different from zero, and Equation (1) would reduce to Green's formula.

Example 1. Let S be the spiral surface parametrized by

$$\begin{pmatrix} x_1 \\ x_2 \\ x_3 \end{pmatrix} = \begin{pmatrix} u \cos v \\ u \sin v \\ v \end{pmatrix}, \qquad \text{for} \quad 0 \le u \le 1, \qquad 0 \le v \le \frac{\pi}{2}.$$

Then the border of S consists of three line segments and a spiral curve shown in Fig. 21 together with the domain D of the parametrization.

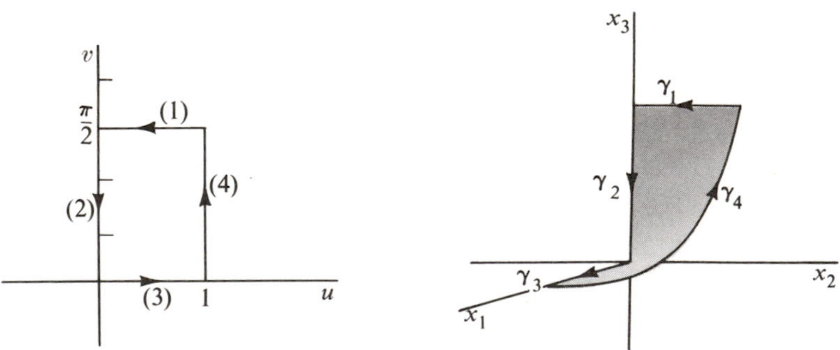

Figure 21

Restricting the parametrization of S to the boundary of D gives the following parametrizations of the smooth pieces of the border of S:

$$\gamma_1: \quad \begin{pmatrix} x_1 \\ x_2 \\ x_3 \end{pmatrix} = \begin{pmatrix} 0 \\ 1 - t \\ \pi/2 \end{pmatrix}, \quad 0 \le t \le 1$$

$$\gamma_2: \quad \begin{pmatrix} x_1 \\ x_2 \\ x_3 \end{pmatrix} = \begin{pmatrix} 0 \\ 0 \\ (\pi/2) - t \end{pmatrix}, \quad 0 \le t \le \frac{\pi}{2}$$

$$\gamma_3: \quad \begin{pmatrix} x_1 \\ x_2 \\ x_3 \end{pmatrix} = \begin{pmatrix} t \\ 0 \\ 0 \end{pmatrix}, \quad 0 \le t \le 1$$

$$\gamma_4: \quad \begin{pmatrix} x_1 \\ x_2 \\ x_3 \end{pmatrix} = \begin{pmatrix} \cos t \\ \sin t \\ t \end{pmatrix}, \quad 0 \le t \le \frac{\pi}{2}.$$

Now let F be the vector field $F(x_1, x_2, x_3) = (x_3, x_1, x_2)$. The line integrals of F over the γ' are all of the form

$$\int_{\gamma_i} x_3\, dx_1 + x_1\, dx_2 + x_2\, dx_3.$$

It is easy to see that the integrals over γ_1, γ_2, and γ_3 are all zero, while over γ_4 we get

$$\int_{\gamma_4} F \cdot d\mathbf{x} = \int_0^{\pi/2} (\cos^2 t + \sin t - t \sin t)\, dt$$

$$= \frac{\pi}{4}.$$

On the other hand, curl $F(x_1, x_2, x_3) = (1, 1, 1)$; so the integral of curl F over S is

$$\int_S \operatorname{curl} F \cdot dS = \int_D \left(\frac{\partial(x_2, x_3)}{\partial(u, v)} + \frac{\partial(x_3, x_1)}{(u, v)} + \frac{\partial(x_1, x_2)}{\partial(u, v)} \right) du\, dv$$

$$= \int_0^1 du \int_0^{\pi/2} (\sin v - \cos v + u)\, dv = \frac{\pi}{4}.$$

This verifies Equation (1) for our special example.

The proof that we give of Stokes's theorem depends on an application of Green's theorem to the region D on which the parametrization of S is defined. For this reason we need to assume enough about D to make Green's theorem hold on it. Also, if $\mathcal{R}^2 \xrightarrow{g} \mathcal{R}^3$ is the parametrization of S, we shall want the second-order partial derivatives of g to be continuous, that is, g should be twice continuously differentiable on D. These conditions can be relaxed, but to do so makes the proof much more difficult.

4.1 Stokes's Theorem

Let S be a piece of smooth surface in $\mathcal{R}^3$, parametrized by a twice continuously differentiable function g. Assume that D, the parameter domain of g, is a finite union of simple regions bounded by a piecewise smooth curve. If F is a continuously differentiable vector field defined on S, then

$$\int_S \operatorname{curl} F \cdot dS = \oint_{\partial S} F \cdot d\mathbf{x},$$

where ∂S is the positively oriented border of S.

Proof. Let F_1, F_2, F_3 be coordinate functions of F. We shall prove that

$$\oint_{\partial S} F_1\, dx_1 = \int_S -\frac{\partial F_1}{\partial x_2}\, dx_1\, dx_2 + \frac{\partial F_1}{\partial x_3}\, dx_3\, dx_1. \tag{2}$$

The proofs that

$$\oint_{\partial S} F_2\, dx_2 = \int_S -\frac{\partial F_2}{\partial x_3}\, dx_2\, dx_3 + \frac{\partial F_2}{dx_1}\, dx_1\, dx_2$$

and

$$\oint_{\partial S} F_3\, dx_3 = \int_S -\frac{\partial F_3}{\partial x_1}\, dx_3\, dx_1 + \frac{\partial F_3}{dx_2}\, dx_2\, dx_3$$

are similar, and addition of the three equations gives Stokes's formula. To prove the top equation, suppose that $h(t) = (u(t), v(t))$ is a counterclockwise-oriented parametrization of δ, the boundary of D. Then $g(h(t))$ is a piecewise smooth parametrization of the border of S, which by definition is then positively oriented. Writing g_1, g_2, g_3 for the coordinate functions of g, we have

$$\begin{aligned}
\oint_{\partial S} F_1\, dx_1 &= \int F_1(g(u, v)) \frac{d}{dt}\, g_1(u, v)\, dt \\
&= \int F_1(g(u, v)) \left[\frac{\partial g_1}{\partial u}(u, v)\frac{du}{dt} + \frac{\partial g_1}{\partial v}(u, v)\frac{dv}{dt}\right] dt \\
&= \oint_\delta F_1 \circ g \frac{\partial g_1}{\partial u}\, du + F_1 \circ g \frac{\partial g_1}{\partial v}\, dv.
\end{aligned}$$

This last integral is a line integral around the region D in $\mathcal{R}^2$, and we can apply Green's theorem to it, getting

$$\oint_{\partial S} F_1\, dx_1 = \int_D \left[\frac{\partial}{\partial u}\left(F_1 \circ g \frac{\partial g_1}{\partial v}\right) - \frac{\partial}{\partial v}\left(F_1 \circ g \frac{\partial g_1}{\partial u}\right)\right] du\, dv. \quad (3)$$

The fact that g is twice continuously differentiable ensures that the integral over D will exist. The same fact enables us to interchange the order of partial differentiation in a computation which shows that

$$\frac{\partial}{\partial u}\left(F_1 \circ g \frac{\partial g_1}{\partial v}\right) - \frac{\partial}{\partial v}\left(F_1 \circ g \frac{\partial g_1}{\partial u}\right) = -\frac{\partial F_1}{\partial x_2}\frac{\partial(g_1, g_2)}{\partial(u, v)} + \frac{\partial F_1}{\partial x_3}\frac{\partial(g_3, g_1)}{\partial(u, v)}. \quad (4)$$

Substitution of this identity into Equation (3) gives Equation (2), thus completing the proof.

Using Stokes's theorem we can derive an interpretation for the vector field curl F that gives some information about F itself. Let $\mathbf{x}_0$ be a point of an open set on which F is continuously differentiable. Let $\mathbf{n}_0$ be an arbitrary unit vector pointing away from $\mathbf{x}_0$, and construct a disk S_r of radius r centered at $\mathbf{x}_0$ and perpendicular to $\mathbf{n}_0$. This is shown in Fig. 22. Applying Stokes's theorem to F on the surface S_r and its border γ_r gives

$$\oint_{\gamma_r} F \cdot d\mathbf{x} = \int_S \operatorname{curl} F \cdot dS.$$

The value of the line integral is called the **circulation** of F around γ_r, and it measures the strength of the field tangential to γ_r. Thus, for small r, the circulation around γ_r is a measure of the tendency of the field near $\mathbf{x}_0$ to rotate around the axis determined by $\mathbf{n}_0$. On the other hand, the surface integral is, for small enough r, nearly equal to the dot product curl $F(\mathbf{x}_0) \cdot \mathbf{n}_0$, multiplied by the area of S_r. See Problem 15 of the previous section. It follows that the circulation around γ_r will tend to be larger if $\mathbf{n}_0$ points in the same direction as curl $F(\mathbf{x}_0)$. Thus we can think of curl $F(\mathbf{x}_0)$ as determining the axis about which the circulation of F is greatest near $\mathbf{x}_0$. Similarly, $|\operatorname{curl} F(\mathbf{x}_0)|$ measures the magnitude of the circulation around this axis near $\mathbf{x}_0$.

The extension of Stokes's theorem to piecewise smooth orientable surfaces is very simple, though a little care is needed in defining the border of such a surface. Figure 23 illustrates the method. The surfaces S_1 and S_2 have their borders joined so as to produce a piecewise smooth positively oriented surface which we denote by $S_1 \cup S_2$. Recall that the surface

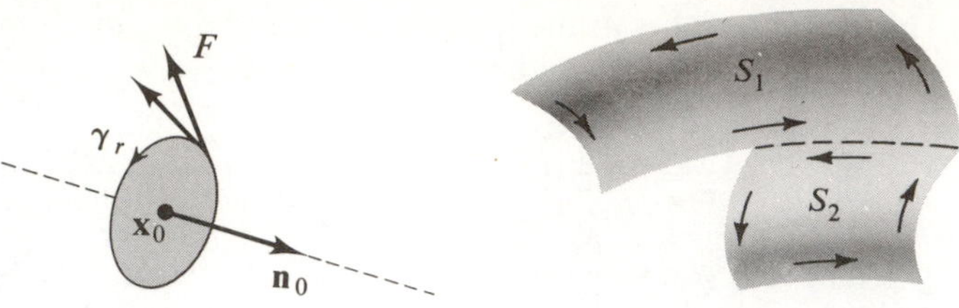

Figure 22

Figure 23

integral of a vector field F over $S_1 \cup S_2$ has already been defined by

$$\int_{S_1 \cup S_2} F \cdot dS = \int_{S_1} F \cdot dS + \int_{S_2} F \cdot dS.$$

The piece of common border curve, indicated by a broken line in Fig. 23, will be traced in opposite directions, depending on whether the parametrization induced by S_1 or by S_2 is used. Hence, the respective line integrals of F over the common border will have opposite sign, and when the line integrals over ∂S_1 and ∂S_2 are added, the integrals over the common part will cancel, leaving a line integral over the rest of the borders of S_1 and S_2. It is this remaining part that we call the positively oriented border of $S_1 \cup S_2$, and denote by $\partial(S_1 \cup S_2)$. (Thus in general $\partial(S_1 \cup S_2) \neq \partial S_1 \cup \partial S_2$.) With this understanding we write Stokes's theorem in the form

$$\int_S \text{curl } F \cdot dS = \oint_{\partial S} F \cdot d\mathbf{x},$$

for a piecewise smooth surface S.

Example 2. A sphere can be considered as a piecewise smooth surface on which all of the border curves cancel one another. In fact if we parametrize a sphere S_a in $\mathcal{R}^3$ by

$$g(u, v) = \begin{pmatrix} a \sin v \cos u \\ a \sin v \sin u \\ a \cos v \end{pmatrix}, \qquad \begin{matrix} 0 \leq u \leq 2\pi \\ 0 \leq v \leq \pi, \end{matrix}$$

then the positively oriented "border" of the sphere consists of the half-circle shown in Fig. 24 traced once in each direction. Thus the half-circle corresponds to the segments $u = 0$ and $u = 2\pi$ in the parameter domain. (What happens to the segments $v = 0$ and $v = \pi$?) The result is that a line integral over ∂S_a will be zero, and Stokes's theorem applied to a vector field F on S_a gives

$$\int_{S_a} \text{curl } F \cdot dS = 0.$$

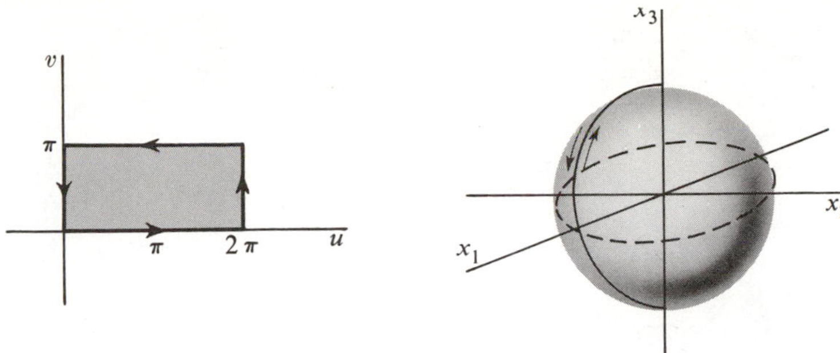

Figure 24

A surface like that in Example 2, in which the border is effectively nonexistent for the purpose of line integration over ∂S, is called a **closed surface.**

Stokes's theorem gives interesting information about gradient fields, that is, fields F such that $F = \nabla f$, or, using an alternative notation, $F = \text{grad}\, f$. If we assume that F is the continuously differentiable gradient field of f, we can form the vector field curl F. Since $F = (\partial f/\partial x_1, \partial f/\partial x_2, \partial f/\partial x_3)$, we get immediately from the definition of curl F and the equality of mixed partials that

4.2 $$\text{curl}\,(\text{grad}\, f)(\mathbf{x}) = 0,$$

for all $\mathbf{x}$ in the domain of f. We have already met the condition curl $F = 0$ in Theorems 2.5 and 2.6, where, for the 3-dimensional case that we consider here, it was stated in terms of the Jacobian matrix

$$F' = \begin{pmatrix} \dfrac{\partial F_1}{\partial x_1} & \dfrac{\partial F_2}{\partial x_1} & \dfrac{\partial F_3}{\partial x_1} \\[2ex] \dfrac{\partial F_1}{\partial x_2} & \dfrac{\partial F_2}{\partial x_2} & \dfrac{\partial F_3}{\partial x_2} \\[2ex] \dfrac{\partial F_1}{\partial x_3} & \dfrac{\partial F_2}{\partial x_3} & \dfrac{\partial F_3}{\partial x_3} \end{pmatrix}.$$

It is clear that the symmetry of F' about its main diagonal is equivalent to curl $F = 0$. Theorem 2.5 says in particular that if F is a gradient field, then curl F is identically zero. Theorem 2.6 gives only a partial converse, to the effect that if curl F is identically zero, then there is some rectangle in which F equals a gradient field. This is sometimes paraphrased by saying that F is *locally* a gradient field. And Example 2 of Section 2 shows that

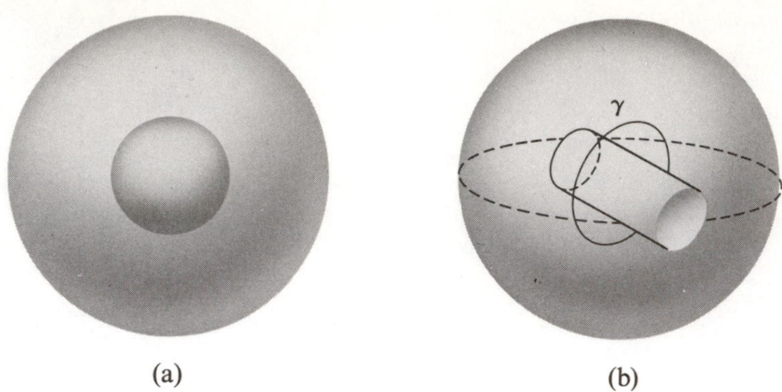

(a) (b)

Figure 25

the strict converse is false. Using Stokes's theorem, we can prove another partial converse, in which the local condition is replaced by a different kind of restriction on the domain of the given field.

For this purpose we shall define a **simply connected** open set B in $\mathcal{R}^n$. Roughly, a set B is simply connected if every closed curve γ in B can be contracted to a point in such a way as to stay within B during the contraction. As γ contracts to a point, it sweeps out a surface S lying in B, and γ is the border of S. The region between two spheres shown in Fig. 25(a) is simply connected. However, the open ball with a hole punched through it is not simply connected, because any surface whose border encircles the hole must lie at least partly outside B. In $\mathcal{R}^2$, the typical simply connected region is the inside of a closed curve, while the outside of such a curve is not simply connected. In Fig. 26(a), the curve γ is the border of the surface consisting of the part of the plane lying inside γ. However, the presence of the hole in Fig. 26(b) prevents a similar construction. More precisely, we shall say that an open set is simply connected if every piecewise smooth closed curve γ lying in B is the border of some piecewise smooth orientable surface S lying in B, and with parameter domain a disk in $\mathcal{R}^2$. We assume for applications that S is parametrized by

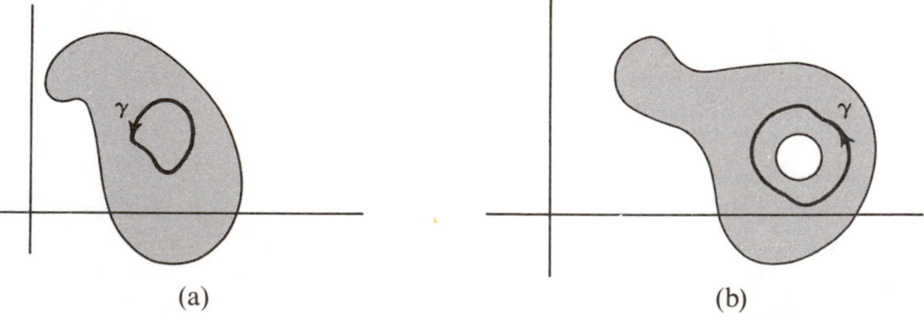

(a) (b)

Figure 26

twice continuously differentiable functions. All the differentiability conditions assumed above are usually relaxed to just continuity, but, since we shall have no use for the broader definition, we shall not introduce a special term for the one involving differentiability.

Now we can prove the following.

4.3 Theorem

Let F be a continuously differentiable vector field defined on an open set B in $\mathcal{R}^2$ or $\mathcal{R}^3$. If

1. B is simply connected, and
2. curl F is identically zero in B,

then F is a gradient field in B, that is, there is a real-valued function f such that $F = \nabla f$.

Proof. By Theorem 2.4 it is enough to show that $\oint_\gamma F \cdot d\mathbf{x} = 0$ for every piecewise smooth curve γ lying in B. Because B is simply connected, there is a piecewise smooth surface S of which γ is the border and to which we can apply Stokes's theorem in either 2 or 3 dimensions. Thus

$$\oint_{\partial S} F \cdot d\mathbf{x} = \int_S \operatorname{curl} F \cdot dS = 0.$$

EXERCISES

1. Compute curl F if

(a) $F(x_1, x_2, x_3) = (x_2 - x_3^2, x_3 - x_1^2, x_1 - x_2^2)$.
(b) $F(x, y, z) = (x, 2y, 3z)$.

2. Verify by computing both integrals that Stokes's theorem holds for the vector field $F(x_1, x_2, x_3) = (x_1, x_2, x_3)$ on a hemisphere centered at the origin in $\mathcal{R}^3$.

3. (a) Verify that if $F(x_1, x_2, x_3)$ is independent of x_3 and the third coordinate function of F is identically zero, then Stokes's formula, applied to a planar surface in the x_1x_2-plane, is the same as Green's formula.

(b) Consider the function $\mathcal{R}^2 \xrightarrow{g} \mathcal{R}^3$ defined by

$$g(u, v) = \begin{pmatrix} u \cos v \\ u \sin v \\ 0 \end{pmatrix}, \qquad \begin{matrix} 1 \le u \le 2, \\ 0 \le v \le 4\pi. \end{matrix}$$

If S is the image in $\mathcal{R}^3$ of g, give a precise description of the border of S.

(c) If F is a continuously differentiable vector field on S of the type described in part (a), use Stokes's theorem to compute the integral of F over the positively oriented border of S.

4. Show that the Stokes formula can be written in the form

$$\int_S \operatorname{curl} F \cdot \mathbf{n}\, d\sigma = \oint_{\partial S} F \cdot \mathbf{t}\, ds,$$

where $\mathbf{n}$ is a unit normal to S and $\mathbf{t}$ is a unit tangent to ∂S.

5. Use the result of Exercise 15 of the previous section and Stokes's theorem to prove that if F is a continuously differentiable vector field at $\mathbf{x}_0$, then

$$\lim_{r \to 0} \frac{1}{A(D_r)} \oint_c F \cdot \mathbf{t}\, ds = \operatorname{curl} F(\mathbf{x}_0) \cdot \mathbf{n}_0,$$

where D_r is a disk of radius r centered at $\mathbf{x}_0$, and $\mathbf{n}_0$ is a unit normal to the disk, and c is the boundary of D_r.

6. Prove that if F is a continuously differentiable vector field such that at each point $\mathbf{x}$ of a piece of smooth surface S, the vector curl $F(\mathbf{x})$ is tangent to S, then the integral of F around the border of S is zero.

7. Let F be a differentiable vector field defined in an open subset B of $\mathcal{R}^3$. Use the decomposition of a square matrix A into symmetric and skew-symmetric parts given by $A = \frac{1}{2}(A + A^t) + \frac{1}{2}(A - A^t)$ to show that for all $\mathbf{y}$ in $\mathcal{R}^3$

$$F'(\mathbf{x})\mathbf{y} = S(\mathbf{x})\mathbf{y} + \tfrac{1}{2} \operatorname{curl} F(\mathbf{x}) \times \mathbf{y},$$

where $S(\mathbf{x})$ is a symmetric matrix.

8. Let F be the gradient field of the Newtonian potential $f(x, y, z) = (x^2 + y^2 + z^2)^{-1/2}$. Show that near each point of the domain of F the circulation of F is zero.

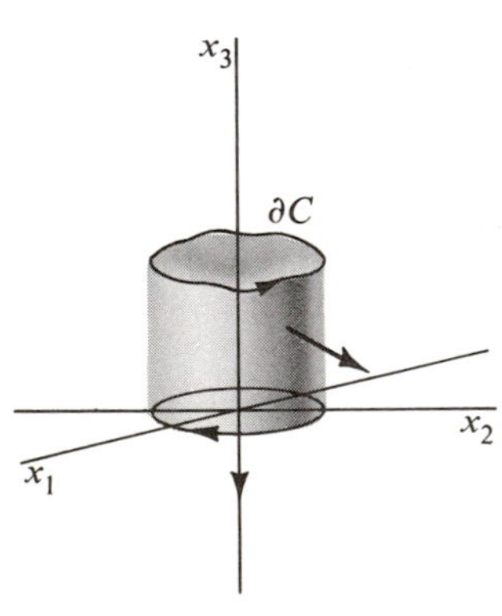

Figure 27

9. Carry out the computation of the identity in Equation (4) of the proof of Stokes's theorem.

10. Consider the cylindrical can C of radius 1 having an unspecified smooth border and an orientation as shown in Fig. 27. Let $F(x_1, x_2, x_3) = (2x_2^2, x_1^2, 3x_3^2)$. What is the value of the line integral of F over the border of C?

11. Compute the integral of curl F, where $F(x_1, x_2, x_3) = (x_2^3, -x_1^3, x_3^3)$, over the hemisphere $x_1^2 + x_2^2 + x_3^2 = 1$, $x_3 \geq 0$, by considering an integral over the disk that makes the hemisphere closed.

12. Show that the open subset of $\mathcal{R}^2$ consisting of $\mathcal{R}^2$ with the origin deleted is not simply connected by finding a vector field F for which curl F is identically zero, but such that F is not a gradient field. [*Hint.* See Problem 7 of Section 1.]

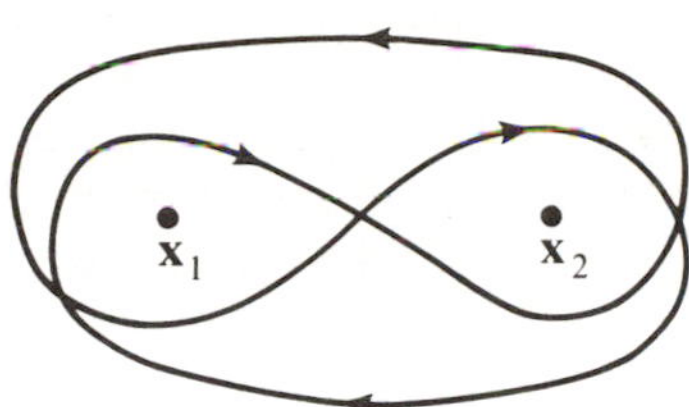

Figure 28

13. The open set in $\mathcal{R}^2$ consisting of $\mathcal{R}^2$ with two points $\mathbf{x}_1$ and $\mathbf{x}_2$ deleted is not simply connected. However, show that if F is any continuously differentiable vector field in such a region such that curl $F = 0$ there, then the integral of F over the smooth curve shown in Fig. 28 is equal to zero.

14. In the open subset B of $\mathcal{R}^3$ consisting of $\mathcal{R}^3$ with the origin deleted, show that a circle centered at the origin is the border of a piecewise smooth surface lying in B.

SECTION 5

GAUSS'S THEOREM

The Gauss theorem (described below) has many applications, some of which are discussed in Section 6. Like Stokes's theorem, it can be looked at as an extension of Green's theorem. We begin with a region R in $\mathcal{R}^3$ having as boundary a piecewise smooth surface S. Each piece of S will be parametrized by a continuously differentiable function $\mathcal{R}^2 \xrightarrow{g} \mathcal{R}^3$ such that the normal vector $\partial g/\partial u \times \partial g/\partial v$ points away from R at each point of S. The boundary surface S is then said to have **positive orientation**, and we denote the positively oriented boundary of R by ∂R. To state the theorem, we consider a vector field F, continuously differentiable on R and its boundary. We define the **divergence** of F to be the real-valued function div F defined on R by

$$\text{div } F(\mathbf{x}) = \frac{\partial F_1}{\partial x_1}(\mathbf{x}) + \frac{\partial F_2}{\partial x_2}(\mathbf{x}) + \frac{\partial F_3}{\partial x_3}(\mathbf{x}),$$

where F_1, F_2, F_3 are the coordinate functions of F. Then the Gauss (or divergence) formula is

$$\int_R \text{div } F\, dV = \int_{\partial R} F \cdot dS.$$

The Gauss formula is like Stokes's formula, Green's formula, and the formula

$$\int_a^b \text{grad } f \cdot d\mathbf{x} = f(\mathbf{b}) - f(\mathbf{a}),$$

in that it relates an integral of some kind of derivative of a function to the behavior of that function on a boundary. In each case the orientation of the boundary is important. For example, if we apply Gauss's theorem to the region R in $\mathcal{R}^3$ given by $1 \leq |\mathbf{x}| \leq 2$, then the oriented boundary, ∂R, must be such that its normal vectors on the outer sphere point away from the origin, and on the inner sphere point toward it, as shown in Fig. 29. We shall say that ∂R is **positively oriented** with respect to R if the normal vectors given by the parametrization of ∂R point away from R.

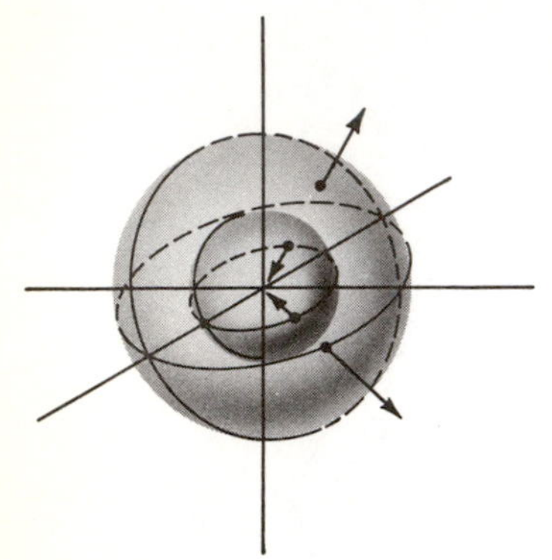

Figure 29

We shall prove Gauss's theorem for the case in which R is a finite union of simple regions, where a **simple region** in $\mathcal{R}^3$ is one whose boundary is crossed by a line parallel to a coordinate axis at most twice. The region between two spheres, shown in Fig. 29, is a union of eight simple regions, one in each octant.

5.1 Gauss's Theorem

Let R be a finite union of simple regions in $\mathcal{R}^3$, having a positively oriented piecewise smooth boundary ∂R. If F is a continuously differentiable vector field on R and ∂R, then

$$\int_R \operatorname{div} F \, dV = \int_{\partial R} F \cdot dS.$$

Proof. In terms of coordinate functions of F, Gauss's formula reads

$$\int_R \left(\frac{\partial F_1}{\partial x_1} + \frac{\partial F_2}{\partial x_2} + \frac{\partial F_3}{\partial x_3}\right) dx_1\, dx_2\, dx_3$$

$$= \int_{\partial R} F_1\, dx_2\, dx_3 + F_2\, dx_3\, dx_1 + F_3\, dx_3\, dx_2.$$

We assume first that R is a simple region and prove only the equation

$$\int_R \frac{\partial F_2}{\partial x_2}\, dx_1\, dx_2\, dx_3 = \int_{\partial R} F_2\, dx_3\, dx_1,$$

the proofs for the terms containing F_1 and F_3 being similar. Addition of the resulting equations will then prove the theorem for simple regions. Because R is simple, ∂R consists of the graphs of two functions, $s(x_1, x_3)$ and $r(x_1, x_3)$, perhaps together with pieces parallel to the x_2-axis as shown in Fig. 30. Let

$$g(u, v) = \begin{pmatrix} g_1(u, v) \\ g_2(u, v) \\ g_3(u, v) \end{pmatrix}, \qquad (u, v) \text{ in } D,$$

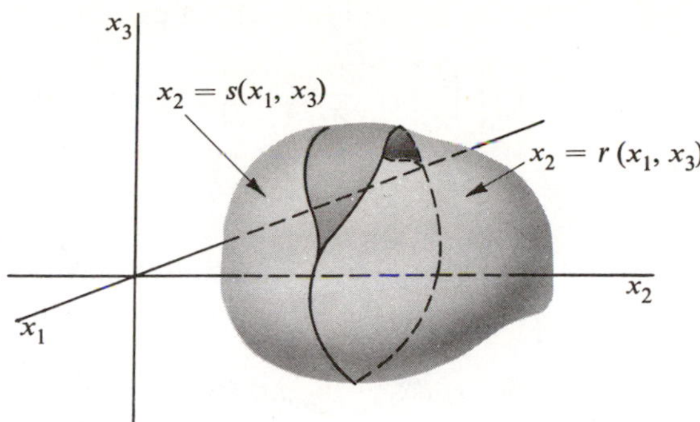

Figure 30

be a parametrization for ∂R that orients it positively. Then, by the definition of the surface integral,

$$\int_{\partial R} F_2\, dx_3\, dx_1 = \int_D F_2(g_1, g_2, g_3) \frac{\partial(g_3, g_1)}{\partial(u, v)}\, du\, dv, \tag{1}$$

and, on the sections of ∂R that are parallel to the x_2-axis, the normal vector to ∂R is perpendicular to the x_2-axis. Hence $\partial(g_3, g_1)/\partial(u, v)$, the second coordinate of the normal, is equal to zero, thus eliminating the part of the integral that is not on the graph of r or s. We now apply the change-of-variable theorem to the two remaining parts of the integral in Equation (1). The appropriate transformations are

$$\begin{pmatrix} x_3 \\ x_1 \end{pmatrix} = \begin{pmatrix} g_3(u, v) \\ g_1(u, v) \end{pmatrix},$$

with (u, v) in either D_r or D_s, where D_r and D_s are the parts of D corresponding to the graphs of r and s. The Jacobian determinant $\partial(g_3, g_1)/\partial(u, v)$ is positive on the graph of r and negative on the graph of s, because it represents the x_2-coordinate of the outward normal. On D_r we have $g_2(u, v) = r(x_1, x_3)$, while on D_s, $g_2(u, v) = s(x_1, x_3)$. Using these facts, we get from the change of variable theorem and Equation (1),

$$\int_{\partial R} F_2\, dx_3\, dx_1 = \int_{R_2} F_2(x_1, s(x_1, x_3), x_3)(-1)\, dx_1\, dx_3 + \int_{R_2} F_2(x_1, r(x_1, x_3), x_3)\, dx_1\, dx_3,$$

where R_2 is the plane region got by projecting R on the x_1x_3-plane. These last integrals are not surface integrals, but rather integrals over a set. Then, by the fundamental theorem of calculus,

$$\int_{\partial R} F_2\, dx_3\, dx_1 = \int_{R_2}\left[\int_{s(x_1,x_3)}^{r(x_1,x_3)} \frac{\partial F_2}{\partial x_2}(x_1, x_2, x_3)\, dx_2\right] dx_1\, dx_3$$
$$= \int_R \frac{\partial F_2}{\partial x_2}\, dx_1\, dx_2\, dx_3.$$

Similar arguments involving F_1 and F_3 complete the proof for simple regions, since the addition of the three resulting equations gives

$$\int_{\partial R} F_1\, dx_2\, dx_3 + F_2\, dx_3\, dx_1 + F_3\, dx_1\, dx_2$$
$$= \int_R \left(\frac{\partial F_1}{\partial x_1} + \frac{\partial F_2}{\partial x_2} + \frac{\partial F_3}{\partial x_3}\right) dx_1\, dx_2\, dx_3.$$

This is the Gauss formula in coordinate form.

The extension of Gauss's theorem to a finite union R of simple regions is essentially the same as the analogous extension of Green's theorem. In the present case, when two simple regions have a common boundary surface, the respective outward normals will be negatives of one another. The corresponding surface integrals are then negatives of one another, and so cancel out. The remaining surface integrals extend over the surface ∂R.

Example 1. Problem 6 of Section 3 consists of showing that the flux of the gradient field F of the potential function

$$f(x, y, z) = (x^2 + y^2 + z^2)^{-1/2}$$

across a sphere of radius a, centered at the origin, is independent of a. Using Gauss's theorem we can prove something more general, and with a minimum of calculation. Thus, let S_1 and S_2 be any two piecewise smooth closed surfaces, one contained in the other, both containing the origin, and bounding a region R between them; for example, R might be the region between two spheres. A routine calculation of the gradient shows that $F(x, y, z) = (x^2 + y^2 + z^2)^{-3/2}(-x, -y, -z)$, and then that the divergence of this field is zero, i.e., $\operatorname{div} F = 0$ everywhere except at the origin. In particular, $\operatorname{div} F = 0$ throughout R. Applying Gauss's theorem to R gives

$$\int_{\partial R} F \cdot dS = \int_R \operatorname{div} F\, dV = 0.$$

But ∂R consists of S_1 with inward pointing normal and S_2 with outward pointing normal; so, with the understanding that S_1^- stands for the inner surface with reversed normal, we get

$$\int_{\partial R} F \cdot dS = -\int_{S_1^-} F \cdot dS + \int_{S_2} F \cdot dS = 0.$$

Thus the integrals over the outward-oriented surfaces are equal. To find the actual value, it is enough to compute it for one surface, say a sphere. The result is -4π.

The divergence of a vector field F at a point $\mathbf{x}$ can be interpreted as a measure of the tendency of the field to radiate outward from $\mathbf{x}$, hence the term "divergence." To see this, consider a ball B_r of radius r, centered at an interior point $\mathbf{x}_0$ of the set on which F is continuously differentiable. Dividing both sides of Gauss's formula by $V(B_r)$ gives

$$\frac{1}{V(B_r)} \int_{B_r} \operatorname{div} F \, dV = \frac{1}{V(B_r)} \int_{\partial B} F \cdot \mathbf{n} \, d\sigma,$$

where $\mathbf{n}$ is the outward unit normal. As r tends to zero, the left side tends to div $F(\mathbf{x}_0)$. See Problem 17 of Chapter 6, Section 2. On the right side, the integral is the average rate of flow per unit of volume across the sphere of radius r centered at $\mathbf{x}_0$. Hence the limit as r tends to zero is, by definition, the rate per unit of volume of flow outward from $\mathbf{x}_0$. Because of its connection with the interpretation of divergence, Gauss's theorem is often called the divergence theorem.

EXERCISES

1. Compute the divergence of the following vector fields:

(a) $F(x_1, x_2, x_3) = (x_1^2, x_2^2, x_3^2)$.
(b) $F(x, y, z) = (\sin xy, 0, 0)$.
(c) $F(x_1, x_2, x_3) = (x_2, x_3, x_1)$.

2. Prove the following identities for any twice continuously differentiable vector field F or real-valued function f.

(a) div (curl F)($\mathbf{x}$) $= 0$.
(b) curl (grad f)($\mathbf{x}$) $= 0$.

3. (a) Show that for $f(x, y, z) = (x^2 + y^2 + z^2)^{-1/2}$ the equation

div (grad f)($\mathbf{x}$) $= 0$ holds for all $\mathbf{x} \neq 0$.

(b) Show by example that

div (grad f)($\mathbf{x}$) $\neq 0$

may hold for some twice continuously differentiable function f.

(c) If the operator Δ is defined by

$$\Delta f = \operatorname{div} (\operatorname{grad} f),$$

find a formula for Δf in terms of partial derivatives of f. (A function such that $\Delta f(\mathbf{x}) = 0$ for all $\mathbf{x}$ in the domain of f is called **harmonic**, and Δ is called the Laplace operator.)

4. The trace of a square matrix is defined as the sum of the elements on its main diagonal. If $\mathcal{R}^n \xrightarrow{F} \mathcal{R}^n$ is a differentiable vector field, we define div F to be the real-valued function given by

$$\text{div } F(\mathbf{x}) = \text{tr } F'(\mathbf{x}),$$

where tr A stands for the trace of A. Show that in the 2- and 3-dimensional cases this definition agrees with those previously given.

5. Use Gauss's theorem to compute

$$\int_S F \cdot dS$$

over the sphere of radius 1 centered at the origin in $\mathcal{R}^3$, and with outward pointing normal, where F is

(a) $F(x_1, x_2, x_3) = (x_1^2, x_2^2, x_3^2)$.
(b) $F(x, y, z) = (xz^2, 0, z^3)$.

6. Show that for a region R to which Gauss's theorem applies, the volume of R is given by

$$V(R) = \frac{1}{3} \int_{\partial R} x_1\, dx_2\, dx_3 + x_2\, dx_3\, dx_1 + x_3\, dx_1\, dx_2.$$

7. (a) Use Gauss's theorem to prove that if F is a continuously differentiable vector field with zero divergence in a region R, then the integral of F over ∂R is zero.
(b) Write an intuitive argument, based on the interpretation of the divergence, for the assertion in part (a).

8. Let S be the ellipsoid

$$\frac{x^2}{a^2} + \frac{y^2}{b^2} + \frac{z^2}{c^2} = 1,$$

and let $D(x, y, z)$ be the distance from the origin to the tangent plane to S at (x, y, z).

(a) Show that if $F(x, y, z) = (x/a^2, y/b^2, z/c^2)$, then $F \cdot \mathbf{n} = D^{-1}$, where $\mathbf{n}$ is the outward unit normal to S at (x, y, z).

(b) Show that $\displaystyle\int_S D^{-1}\, d\sigma = \frac{4\pi}{3}\left(\frac{bc}{a} + \frac{ca}{b} + \frac{ab}{c}\right)$.

9. A vector field $\mathcal{R}^3 \xrightarrow{F} \mathcal{R}^3$ defined in a region R is called **irrotational** if curl $F(\mathbf{x}) = 0$ for all $\mathbf{x}$ in R, and **incompressible** if div $F(\mathbf{x}) = 0$ for all $\mathbf{x}$ in R. Assume F continuously differentiable.

(a) Show that if F is irrotational, then the circulation of F is zero around every sufficiently small circular path in R.
(b) Show that if F is incompressible, then the flux of F is zero across every sufficiently small sphere with its interior in R.

SECTION 6

THE OPERATORS ∇, $\nabla\times$, AND $\nabla\cdot$

To facilitate the application of the Gauss and Stokes theorems, it is helpful to extend the use of the symbol ∇, called "del," that is used in denoting the gradient field of a real-valued function. In terms of the natural basis $\mathbf{e}_1$, $\mathbf{e}_2$, $\mathbf{e}_3$ for $\mathcal{R}^3$, we recall that

$$\nabla f = \frac{\partial f}{\partial x_1}\mathbf{e}_1 + \frac{\partial f}{\partial x_2}\mathbf{e}_2 + \frac{\partial f}{\partial x_3}\mathbf{e}_3. \tag{1}$$

This equation defines ∇ as an operator from real-valued differentiable functions $\mathcal{R}^3 \xrightarrow{f} \mathcal{R}$, to vector fields $\mathcal{R}^3 \xrightarrow{F} \mathcal{R}^3$. If we write

$$\nabla = \frac{\partial}{\partial x_1}\mathbf{e}_1 + \frac{\partial}{\partial x_2}\mathbf{e}_2 + \frac{\partial}{\partial x_3}\mathbf{e}_3, \tag{2}$$

then Equation (1) follows by application of both sides of Equation (2) to f.

The formalism described above makes the following definitions natural. If F is a differentiable vector field given by

$$F(\mathbf{x}) = F_1(x)\mathbf{e}_1 + F_2(\mathbf{x})\mathbf{e}_2 + F_3(\mathbf{x})\mathbf{e}_3,$$

then the operator $\nabla\times$ is defined by taking the formal cross-product of ∇ and F to get

$$\nabla \times F = \begin{vmatrix} \frac{\partial}{\partial x_2} & \frac{\partial}{\partial x_3} \\ F_2 & F_3 \end{vmatrix}\mathbf{e}_1 + \begin{vmatrix} \frac{\partial}{\partial x_3} & \frac{\partial}{\partial x_1} \\ F_3 & F_1 \end{vmatrix}\mathbf{e}_2 + \begin{vmatrix} \frac{\partial}{\partial x_1} & \frac{\partial}{\partial x_2} \\ F_1 & F_2 \end{vmatrix}\mathbf{e}_3$$

$$= \left(\frac{\partial F_3}{\partial x_2} - \frac{\partial F_2}{\partial x_3}\right)\mathbf{e}_1 + \left(\frac{\partial F_1}{\partial x_3} - \frac{\partial F_3}{\partial x_1}\right)\mathbf{e}_2 + \left(\frac{\partial F_2}{\partial x_1} - \frac{\partial F_1}{\partial x_2}\right)\mathbf{e}_3.$$

Thus $\nabla \times F$ is the vector field that we have called the curl of F and written curl F. Similarly, for a differentiable vector field F, we define the operator $\nabla\cdot$ by taking the formal dot-product of ∇ and F to get

$$\nabla \cdot F = \frac{\partial F_1}{\partial x_1} + \frac{\partial F_2}{\partial x_2} + \frac{\partial F_3}{\partial x_3}. \tag{3}$$

This real-valued function we have called the divergence of F and have written div F. The meaning of the notation just introduced is easy to remember if Equation (2) is kept in mind.

Using the ∇-notation, Stokes's formula becomes

6.1 $$\int_S (\nabla \times F)\cdot \mathbf{n}\, d\sigma = \int_{\partial S} F \cdot \mathbf{t}\, ds,$$

and Gauss's formula becomes

6.2
$$\int_R \nabla \cdot F \, dV = \int_{\partial R} F \cdot \mathbf{n} \, d\sigma.$$

To exploit these formulas fully we need some identities involving ∇. In the formulas below, f and g are real-valued differentiable functions, F and G are differentiable vector fields, and a and b are constants.

$$\nabla (af + bg) = a \nabla f + b \nabla g \tag{4}$$

$$\nabla (fg) = f \nabla g + g \nabla f \tag{5}$$

$$\nabla \times (aF + bG) = a \nabla \times F + b \nabla \times G \tag{6}$$

$$\nabla \times (fF) = f \nabla \times F + \nabla f \times F \tag{7}$$

$$\nabla \cdot (aF + bG) = a \nabla \cdot F + b \nabla \cdot G \tag{8}$$

$$\nabla \cdot (fF) = f \nabla \cdot F + \nabla f \cdot F \tag{9}$$

$$\nabla \cdot (F \times G) = (\nabla \times F) \cdot G - F \cdot (\nabla \times G). \tag{10}$$

Each of these formulas is an immediate consequence of the coordinate definitions of the operators. Using the same kind of proof establishes that if f and F are twice differentiable, then

$$\nabla \cdot (\nabla \times F) = 0, \tag{11}$$

$$\nabla \times (\nabla f) = 0, \tag{12}$$

$$\nabla \cdot \nabla f = \nabla^2 f, \tag{13}$$

where $\nabla^2 f$ is the Laplacian of f, defined by

$$\nabla^2 f = \frac{\partial^2 f}{\partial x_1^2} + \frac{\partial^2 f}{\partial x_2^2} + \frac{\partial x^2 f}{\partial x_3^2}.$$

(Equations (11), (12), and (13) are the same as those in Problems 2 and 3 of the previous section, where the alternative symbol Δ was used for the Laplace operator.)

The formulas given above can be used to derive many special cases of the Gauss and Stokes theorems. A particularly important kind arises if the vector field F is assumed to be a gradient ∇f, or a multiple $f \nabla g$. If we set $F = \nabla f$ in Formula 6.2, the result is

$$\int_R \nabla \cdot \nabla f \, dV = \int_{\partial R} \nabla f \cdot \mathbf{n} \, d\sigma. \tag{14}$$

But by Equation (13), $\nabla \cdot \nabla f = \nabla^2 f$, and by Equation 2.1 of Chapter 4,

$\nabla f \cdot \mathbf{n} = (\partial/\partial \mathbf{n})f$. Thus we have

$$\int_R \nabla^2 f \, dV = \int_S \frac{\partial f}{\partial \mathbf{n}} \, d\sigma. \tag{15}$$

If we replace F in Formula 6.2 by $f \nabla g$, instead of by ∇f, we have from Equation (9)

$$\nabla \cdot (f \nabla g) = f \nabla \cdot \nabla g + \nabla f \cdot \nabla g;$$

and so Gauss's formula yields

6.3

$$\int_R f \nabla^2 g \, dV + \int_R \nabla f \cdot \nabla g \, dV = \int_S f \frac{\partial g}{\partial \mathbf{n}} \, d\sigma.$$

This is called **Green's first identity.** Because of the symmetry in the middle term, interchange of f and g and subtraction of the corresponding terms gives **Green's second identity.**

6.4

$$\int_R (f \nabla^2 g - g \nabla^2 f) \, dV = \int_S \left(f \frac{\partial g}{\partial \mathbf{n}} - g \frac{\partial f}{\partial \mathbf{n}} \right) d\sigma.$$

Example 1. Let R be a polygonally connected region in $\mathcal{R}^3$ with a piecewise smooth boundary surface S. If h is a real-valued function defined in R, we consider a **Poisson equation**

$$\nabla^2 u = h,$$

subject to a preassigned boundary condition, $u(\mathbf{x}) = \varphi(\mathbf{x})$ for $\mathbf{x}$ on S. We suppose that there is at least one solution $u(\mathbf{x})$ defined in R and satisfying the boundary condition. We can show, using Green's first identity, that *such a solution must be unique*. Let us suppose that there were two solutions u_1 and u_2; then the function u defined by $u(\mathbf{x}) = u_1(\mathbf{x}) - u_2(\mathbf{x})$ would satisfy the **Laplace equation** $\nabla^2 u = 0$ in R, together with the boundary condition $u(\mathbf{x}) = 0$ on S. Setting $f = g = u$ in Formula 6.3 gives

$$\int_R u \nabla^2 u \, dV + \int_R |\nabla u|^2 \, dV = \int_S u \frac{\partial u}{\partial \mathbf{n}} \, d\sigma.$$

But the first and last terms are zero because $\nabla^2 u = 0$ in R and $u = 0$ on S. It follows from $\int_R |\nabla u|^2 \, dV = 0$ that $\nabla u = 0$ identically on R and S. Hence u must be a constant in the polygonally connected region R. Finally, u must in fact be identically zero because $u(\mathbf{x}) = 0$ for $\mathbf{x}$ on S. We remark that the Laplace equation is the special case of the Poisson equation obtained by taking h identically zero; thus we have proved a uniqueness theorem for the Laplace equation also.

Example 2. As an application of Green's second identity, we consider the following problem. Let D be a differential operator acting on functions f defined in a region R in $\mathcal{R}^3$ having a piecewise smooth boundary S. If we let $\mathscr{F}$ be the vector space of continuous functions defined on R together with S, we can define an inner product on $\mathscr{F}$ by

$$\langle f, g \rangle = \int_R fg\,dV.$$

It is easy to check that the integral is indeed an inner product. With respect to the inner product, it is often important to know when the operator D is **symmetric** in the sense that

$$\langle Df, g \rangle = \langle f, Dg \rangle,$$

whenever Df and Dg are defined and continuous. See Exercise 17 of Chapter 5, Section 2 for the analog for finite dimensional spaces. In general, to solve such a problem it is necessary to impose appropriate boundary conditions on the functions f in $\mathscr{F}$. Thus we can show that the Laplace operator ∇^2 is a symmetric operator acting on the subspace $\mathscr{F}_0$ of $\mathscr{F}$ consisting of continuous functions f that satisfy a fixed boundary condition of the form

$$a_1 f(\mathbf{x}) + a_2 \frac{\partial f}{\partial \mathbf{n}}(\mathbf{x}) = 0. \tag{16}$$

In Equation (16), a_1 and a_2 are constants, not both zero, and $\partial/\partial\mathbf{n}$ denotes differentiation with respect to the outward unit normal on S.

We suppose that f and g are two functions satisfying Equation (16) such that $\nabla^2 f$ and $\nabla^2 g$ are continuous. Then Formula 6.4, Green's second identity, shows that $\langle \nabla^2 f, g \rangle = \langle f, \nabla^2 g \rangle$, if and only if

$$\int_S \left(f \frac{\partial f}{\partial \mathbf{n}} - g \frac{\partial f}{\partial \mathbf{n}} \right) d\sigma = 0. \tag{17}$$

Since f and g both satisfy Equation (16), we have

$$g a_1 f + g a_2 \frac{\partial f}{\partial \mathbf{n}} = 0,$$

$$f a_1 g + f a_2 \frac{\partial g}{\partial \mathbf{n}} = 0,$$

and subtraction of one from the other gives

$$a_2 \left[g(\mathbf{x}) \frac{\partial f}{\partial \mathbf{n}}(\mathbf{x}) - f(\mathbf{x}) \frac{\partial g}{\partial \mathbf{n}}(\mathbf{x}) \right] = 0, \qquad \text{for all } \mathbf{x} \text{ on } S.$$

If $a_2 \neq 0$, this implies $g(\partial f/\partial\mathbf{n}) - f(\partial g/\partial\mathbf{n}) = 0$ identically on S; so Equation (17) is satisfied. If, on the other hand, $a_2 = 0$, then a_1 is not zero, and the boundary condition implies that both f and g are identically zero on S. But then Equation (17) is still satisfied. Thus we have shown that ∇^2 is symmetric on $\mathcal{F}_0$. Notice that $\nabla^2 f$ is not defined for all f in $\mathcal{F}_0$, but only for sufficiently differentiable f.

From Formula 6.2 we can derive some equations for vector-valued integrals. For the definition and properties of the integral, see Problem 16 of Chapter 6, Section 2. Let $\mathbf{k}$ be an arbitrary constant vector and let $F(\mathbf{x}) = f(\mathbf{x})\mathbf{k}$, where f is real-valued and continuously differentiable on a region R. Then, because $\nabla \cdot f\mathbf{k} = \nabla f \cdot \mathbf{k}$ (Verify!), Formula 6.2 becomes

$$\int_R \nabla f \cdot \mathbf{k}\, dV = \int_{\partial R} f\mathbf{k} \cdot \mathbf{n}\, d\sigma.$$

Since $\mathbf{k}$ is constant,

$$\mathbf{k} \cdot \int_R \nabla f\, dV = \mathbf{k} \cdot \int_{\partial R} f\mathbf{n}\, d\sigma,$$

and, from the fact that $\mathbf{k}$ is arbitrary, we conclude that

$$\int_R \nabla f\, dV = \int_{\partial R} f\mathbf{n}\, d\sigma. \tag{18}$$

Similarly, replacing F in Formula 6.2 by $\mathbf{k} \times F$, where $\mathbf{k}$ is a constant vector, we can show that

$$\int_R \nabla \times F\, dV = \int_{\partial R} \mathbf{n} \times F\, d\sigma. \tag{19}$$

The proof is left as an exercise.

We conclude the section with a description of the expressions for gradient, divergence, and curl with respect to an *orthogonal* curvilinear coordinate system. Thus we assume that all coordinate curves intersect at right angles, which means that the natural tangent vectors $\mathbf{c}_1$, $\mathbf{c}_2$, $\mathbf{c}_3$ are perpendicular at each point. If we are given a real-valued function f or a vector field F defined in a region D of $\mathcal{R}^3$, then for practical convenience we may want to substitute the curvilinear coordinate variables u_1, u_2, u_3 using a transformation

$$(x_1, x_2, x_3) = T(u_1, u_2, u_3).$$

This leads us to consider a new real-valued function $\bar{f}$, defined in the region of the u-space corresponding to D by

$$f \circ T = \bar{f}.$$

For the vector field, we make an additional change and express it in terms of the unit vectors

$$\frac{1}{h_1}\mathbf{c}_1, \frac{1}{h_2}\mathbf{c}_2, \frac{1}{h_3}\mathbf{c}_3,$$

where $h_i = |\mathbf{c}_i|$. We get coordinate functions $\bar{F}_1, \bar{F}_2, \bar{F}_3$ satisfying

$$F \circ T = \frac{1}{h_1} \bar{F}_1 \mathbf{c}_1 + \frac{1}{h_2} \bar{F}_2 \mathbf{c}_2 + \frac{1}{h_3} \bar{F}_3 \mathbf{c}_3.$$

Rather lengthy computation using the chain rule gives the formulas:

6.5
$$(\nabla f) \circ T = \frac{1}{h_1^2} \frac{\partial \bar{f}}{\partial u_1} \mathbf{c}_1 + \frac{1}{h_2^2} \frac{\partial \bar{f}}{\partial u_2} \mathbf{c}_2 + \frac{1}{h_3^2} \frac{\partial \bar{f}}{\partial u_3} \mathbf{c}_3.$$

6.6
$$(\nabla \cdot F) \circ T = \left[\frac{\partial h_2 h_3 \bar{F}_1}{\partial u_1} + \frac{\partial h_1 h_3 \bar{F}_2}{\partial u_2} + \frac{\partial h_1 h_2 \bar{F}_3}{\partial u_3} \right]$$

6.7
$$(\nabla \times F) \circ T = \frac{\pm 1}{h_1 h_2 h_3} \left[\left(\frac{\partial h_3 \bar{F}_3}{\partial u_2} - \frac{\partial h_2 \bar{F}_2}{\partial u_3} \right) \mathbf{c}_1 + \left(\frac{\partial h_1 \bar{F}_1}{\partial u_3} - \frac{\partial h_3 \bar{F}_3}{\partial u_1} \right) \mathbf{c}_2 + \left(\frac{\partial h_2 \bar{F}_2}{\partial u_1} - \frac{\partial h_1 \bar{F}_1}{\partial u_2} \right) \mathbf{c}_3 \right]$$

In the last formula, the plus sign is chosen if the **c**'s form a right-handed system, and the minus sign is chosen otherwise. Of course the **c**'s and the h's may vary from point to point. The formulas are no *more* complicated than they are because we have assumed the **c**'s perpendicular. As a result, the product $h_1 h_2 h_3$ can be interpreted as the volume of the rectangular box spanned by $\mathbf{c}_1$, $\mathbf{c}_2$, and $\mathbf{c}_3$. In the case of rectangular coordinates in $\mathcal{R}^3$, the transformation T becomes the identity, and the resulting formulas reduce to the corresponding definitions in terms of rectangular coordinates.

Example 3. Introducing spherical coordinates in $\mathcal{R}^3$, we have from Chapter 4, Section 5, Example 4, that

$$h_1 = 1, \qquad h_2 = r, \qquad h_3 = r \sin \varphi.$$

Then at a point (x, y, z) in $\mathcal{R}^3$ with spherical coordinates (r, φ, θ) we have

$$\nabla f(x, y, z) = \frac{\partial \bar{f}}{\partial r}(r, \varphi, \theta)\mathbf{c}_1 + \frac{1}{r^2} \frac{\partial \bar{f}}{\partial \varphi}(r, \varphi, \theta)\mathbf{c}_2 + \frac{1}{r^2 \sin^2 \varphi} \frac{\partial \bar{f}}{\partial \theta}(r, \varphi, \theta)\mathbf{c}_3.$$

and

$$\nabla \cdot F(x, y, z) = \frac{1}{r^2 \sin \varphi}\left[\sin \varphi \frac{\partial r^2 \bar{F}_1(r, \varphi, \theta)}{\partial r} + r \frac{\partial \sin \varphi \bar{F}_2(r, \varphi, \theta)}{\partial \varphi} + r \frac{\partial \bar{F}_3(r, \varphi, \theta)}{\partial \theta}\right]$$

Using the fact that the Laplacian is defined by $\nabla^2 f = \nabla \cdot \nabla f$, we find

$$\nabla^2 f(x, y, z) = \frac{1}{r^2 \sin^2 \varphi}\left[\sin^2 \varphi \frac{\partial}{\partial r}\left(r^2 \frac{\partial \bar{f}(r, \varphi, \theta)}{\partial r}\right) + \sin \varphi \frac{\partial}{\partial \varphi}\left(\sin \varphi \frac{\partial \bar{f}(r, \varphi, \theta)}{\partial \varphi}\right) + \frac{\partial^2 \bar{f}(r, \varphi, \theta)}{\partial \theta^2}\right]$$

In checking this derivation, it is important to remember that Equation 6.6 deals with coordinate functions relative to *unit* vectors in the direction of the **c**'s and not relative to the **c**'s themselves.

EXERCISES

1. Verify the identities (4) through (10) of the text.

2. If $F = (F_1, F_2, F_3)$ is a vector field, define the operator $F \cdot \nabla$ to be

$$F_1 \frac{\partial}{\partial x_1} + F_2 \frac{\partial}{\partial x_2} + F_3 \frac{\partial}{\partial x_3}.$$

(a) If $F(\mathbf{x}) = \mathbf{x}$, compute $(\mathbf{x} \cdot \nabla)G(\mathbf{x})$, where $G(x_1, x_2, x_3) = (x_1^2, x_1x_2, x_3)$.
(b) Show that in general $(\nabla \cdot F)G \neq (F \cdot \nabla)G$.
(c) If $\mathbf{k}$ is a constant vector, show that

$$\nabla \times (\mathbf{k} \times F) = \mathbf{k}(\nabla \cdot F) - (\mathbf{k} \cdot \nabla)F,$$

where F is a differentiable vector field.
(d) Use part (c) to show that if F and G are differentiable, then

$$\nabla \times (F \times G) = F(\nabla \cdot G) - (F \cdot \nabla)G - G(\nabla \cdot F) + (G \cdot \nabla)F.$$

3. Prove that if $\mathbf{k}$ is a constant vector, then

$$\nabla \times \frac{\mathbf{k} \times \mathbf{x}}{|\mathbf{x}|} = \frac{\mathbf{k}}{|\mathbf{x}|} + \frac{\mathbf{k} \cdot \mathbf{x}}{|\mathbf{x}|^3}\mathbf{x}.$$

4. Prove that

(a) $\nabla\left(\frac{1}{|\mathbf{x}|}\right) = \frac{-\mathbf{x}}{|\mathbf{x}|^3}, \quad \mathbf{x} \neq 0.$

(b) $\nabla^2\left(\frac{1}{|\mathbf{x}|}\right) = 0, \quad \mathbf{x} \neq 0.$

5. If $T(\mathbf{x})$ is the steady-state temperature at a point $\mathbf{x}$ of an open set R in $\mathcal{R}^3$, then the flux of the temperature gradient across any smooth closed surface in R is zero. Use this fact and Equation (15) to show that a steady-state temperature function that is twice continuously differentiable is harmonic, i.e., $\nabla^2 T \equiv 0$. [*Hint.* Suppose $\nabla^2 T(\mathbf{x}_0) > 0$. Show that $\nabla^2 T(\mathbf{x}) > 0$ in some ball centered at $\mathbf{x}_0$.]

6. The boundary condition (16) of Example 2 may be generalized to

$$\phi_1(\mathbf{x}) f(\mathbf{x}) + \phi_2(\mathbf{x}) \frac{\partial f}{\partial \mathbf{n}}(\mathbf{x}) = 0,$$

where ϕ_1 and ϕ_2 are continuous functions satisfying $\phi_1^2(\mathbf{x}) + \phi_2^2(\mathbf{x}) > 0$. Show that ∇^2, the Laplace operator, is still symmetric with this more general condition.

7. Show that if $\nabla \cdot F$ is identically zero in a ball B in $\mathcal{R}^3$, then the vector field G defined by

$$G(\mathbf{x}) = \int_0^1 [F(t\mathbf{x}) \times (t\mathbf{x})]\, dt$$

satisfies $\nabla \times G(\mathbf{x}) = F(\mathbf{x})$ in B. The proof consists of the following steps.

(a) If $G(\mathbf{x}) = \int_0^1 [F(t\mathbf{x}) \times (t\mathbf{x})]\, dt$, then show that

$$\nabla \times G(\mathbf{x}) = \int_0^1 \nabla \times [F(t\mathbf{x}) \times (t\mathbf{x})]\, dt.$$

[*Hint.* Apply Leibnitz rule of Problem 7(a), Chapter 6, Section 2.]

(b) Show that $\nabla \times [F(t\mathbf{x}) \times (t\mathbf{x})] = 2tF(t\mathbf{x}) + t^2(d/dt)F(t\mathbf{x})$ by using the identity of Exercise 2(d).

(c) Show that $\nabla \times G(\mathbf{x}) = F(\mathbf{x})$.

8. A vector field $\mathcal{R}^3 \xrightarrow{F} \mathcal{R}^3$ defined in a region R is called **solenoidal** if, in some neighborhood B of every point of R, the field F can be represented as the curl of another field G_B. The field F is called **incompressible** if the divergence of F is zero at every point of R. Show that a continuously differentiable field is solenoidal if and only if it is incompressible. See Exercise 9 of the previous section for the interpretation of incompressibility.

9. Consider the Newtonian potential function $N(\mathbf{x}) = |\mathbf{x}|^{-1}$ and its associated gradient field $\nabla N(\mathbf{x})$. (See Exercise 4.) Show that $N(\mathbf{x})$ can be interpreted as the work done in moving a particle from ∞ to $\mathbf{x}$ along some smooth path through the field ∇N.

10. Let R be a bounded region in $\mathcal{R}^3$ and let $p(\mathbf{x})$ be the density of material at the point $\mathbf{x}$ in R. Then the integral

$$N_p(\mathbf{x}) = \int_R \frac{p(\mathbf{y})}{|\mathbf{x} - \mathbf{y}|}\, dV_{\mathbf{y}}$$

is called the **Newtonian potential** of the material distributed with density p throughout R.

(a) Show that if p is continuous on an open set R with a smooth boundary, then, for $\mathbf{x}$ *not* in R,

$$\nabla^2 N_p = 0.$$

That is, show that for $\mathbf{x}$ not in R, $N_p(\mathbf{x})$ is harmonic.

(b) Show that under the above assumptions on p and R, the integral $N_p(\mathbf{x})$ exists as an improper integral when $\mathbf{x}$ is in R.

11. (a) Use Equations 6.5 and 6.6 to show that the Laplacian in cylindrical coordinates has the form

$$\nabla^2 f(x, y, z) = \frac{1}{r^2}\left[r\frac{\partial}{\partial r}\left(r\frac{\partial \bar{f}(r, \theta, z)}{\partial r}\right) + \frac{\partial^2 \bar{f}(r, \theta, z)}{\partial \theta^2} + r^2\frac{\partial^2 \bar{f}(r, \theta, z)}{\partial z^2}\right].$$

(b) Show that if $f(x, y, z)$ is independent of z, then in cylindrical coordinates

$$\nabla^2 f(x, y, z) = \frac{\partial^2 \bar{f}(r,\theta)}{\partial r^2} + \frac{1}{r^2}\frac{\partial^2 \bar{f}(r, \theta)}{\partial \theta^2} + \frac{1}{r}\frac{\partial \bar{f}(r, \theta)}{\partial r}.$$

Compare with Chapter 4, Section 5, Exercise 13.

(c) Show that if $f(x, y, z)$ can be written as a function $\bar{f}(\sqrt{x^2 + y^2})$, then

$$\nabla^2 f(x, y, z) = \frac{\partial^2 \bar{f}(r)}{\partial r^2} + \frac{1}{r}\frac{\partial \bar{f}(r)}{\partial r}.$$

12. Show that if $f(x, y, z)$ can be written as a function $\bar{f}(\sqrt{x^2 + y^2 + z^2})$, then

$$\nabla^2 f(x, y, z) = \frac{\partial^2 \bar{f}(r)}{\partial r^2} + \frac{2}{r}\frac{\partial \bar{f}(r)}{\partial r}.$$

SECTION 7 DIFFERENTIAL FORMS

Having defined the line integral in Chapter 3, we observed that it could be abbreviated

$$\int_\gamma F_1\,dx_1 + F_2\,dx_2 + F_3\,dx_3$$

in the 3-dimensional case. Our purpose here is to show that the integrand $F_1\,dx_1 + F_2\,dx_2 + F_3\,dx_3$ has an interpretation which leads to another way of looking at the line integral. From there we can go naturally to a definition of surface integral.

We shall denote by dx_k the function that assigns to a vector $\mathbf{a}$ in $\mathcal{R}_n$ its kth coordinate. Thus if $\mathbf{a} = (a_1, \ldots, a_k, \ldots, a_n)$ is in $\mathcal{R}^n$, then $dx_k(\mathbf{a}) = a_k$. In particular, if $\mathbf{a} = (-1, 1, 3)$, then $dx_1(\mathbf{a}) = -1$, $dx_2(\mathbf{a}) = 1$, and $dx_3(\mathbf{a}) = 3$. Geometrically, $dx_k(\mathbf{a})$ is the length, with appropriate sign, of the projection of $\mathbf{a}$ on the kth coordinate axis, as shown in Fig. 31. Linear combinations of the functions dx_k with constant coefficients produce new functions

$$c_1\,dx_1 + \ldots + c_n\,dx_n.$$

Going one step further, if $F_1, \ldots, F_n$ are real-valued functions defined in a

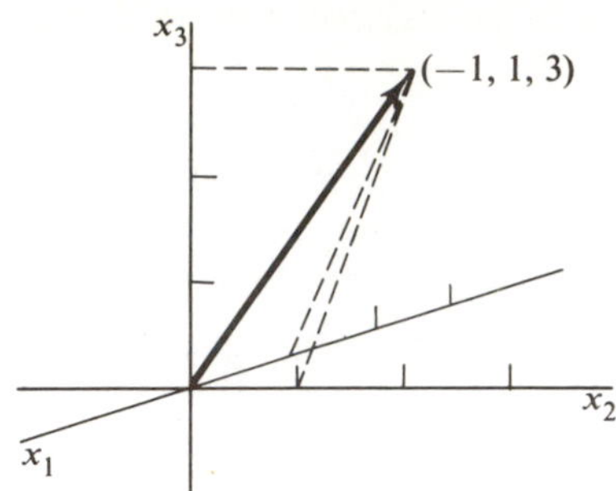

Figure 31

region D of $\mathcal{R}^n$, we can form for each $\mathbf{x}$ in D the linear combination

$$\omega_{\mathbf{x}} = F_1(\mathbf{x})\, dx_1 + \ldots + F_n(\mathbf{x})\, dx_n, \tag{1}$$

where $\omega_{\mathbf{x}}$ acts on vectors $\mathbf{a}$ in $\mathcal{R}^n$ by

$$\omega_{\mathbf{x}}(\mathbf{a}) = F_1(\mathbf{x})\, dx_1(\mathbf{a}) + \ldots + F_n(\mathbf{x})\, dx_n(\mathbf{a}). \tag{2}$$

For example, if in $\mathcal{R}^2$ we have $\omega_{(x,y)} = x^2\, dx + y^2\, dy$, then $\omega_{(x,y)}(a, b) = ax^2 + by^2$, and $\omega_{(-1,3)}(a, b) = a + 9b$. A function ω_x as defined by Formula (2) defines a **differential 1-form** or, briefly, a **1-form.**

Example 1. Let $\mathcal{R}^3 \xrightarrow{f} \mathcal{R}$ be a differentiable function in a region D of $\mathcal{R}^3$. Then $d_{\mathbf{x}}f$, the differential of f at $\mathbf{x}$, is a 1-form in D, for $d_{\mathbf{x}}f$ acting on $\mathbf{a}$ can be written, if $\mathbf{a} = (a_1, a_2, a_3)$,

$$\begin{aligned} d_{\mathbf{x}}f(\mathbf{a}) &= \frac{\partial f}{\partial x_1}(\mathbf{x})a_1 + \frac{\partial f}{\partial x_2}(\mathbf{x})a_2 + \frac{\partial f}{\partial x_3}(\mathbf{x})a_3 \\ &= \frac{\partial f}{\partial x_1}(\mathbf{x})\, dx_1(\mathbf{a}) + \frac{\partial f}{\partial x_2}(\mathbf{x})\, dx_2(\mathbf{a}) + \frac{\partial f}{\partial x_3}(\mathbf{x})\, dx_3(\mathbf{a}). \end{aligned}$$

Thus the coefficient functions $F_k(\mathbf{x})$ in Formula (1) have the form $F_k(\mathbf{x}) = (\partial f/\partial x_k)(\mathbf{x})$. However, not every 1-form is the differential of a function. Recall that, for sufficiently differentiable f, we have $\partial^2 f/(\partial x_2\, \partial x_1) = \partial^2 f/(\partial x_1\, \partial x_2)$, but that $\partial F_1/\partial x_2 \neq \partial F_2/\partial x_1$, unless F_1 and F_2 are specially related, as they are by the requirement that $\nabla f = F$ for some F.

Example 2. If $\omega_{\mathbf{x}}$ is a 1-form defined in a region D of $\mathcal{R}^3$, and γ is a differentiable curve lying in D and given by $\mathcal{R} \xrightarrow{g} \mathcal{R}^3$ for $a \leq t \leq b$, we can at each point $\mathbf{x} = g(t)$ of γ apply the linear function $\omega_{g(t)}$ to the tangent vector $g'(t)$ at $\mathbf{x}$. The result is a real number which we can express as

$$\begin{aligned}\omega_{g(t)}(g'(t)) &= F_1(g(t))\,dx_1(g'(t)) + F_2(g(t))\,dx_2(g'(t)) + F_3(g(t))\,dx_3(g'(t))\\ &= F_1(g(t))g_1'(t) + F_2(g(t))g_2'(t) + F_3(g(t))g_3'(t)\\ &= F(g(t)) \cdot g'(t).\end{aligned}$$

If we write

$$\int_a^b \omega_{g(t)}(g'(t))\,dt = \int_a^b F(g(t)) \cdot g'(t)\,dt,$$

we see that the right-hand integral is the line integral of the field F over γ.

The previous example leads us to the natural definition of the integral of a 1-form over a smooth curve γ. If $\omega_{\mathbf{x}}$ is defined by Formula (1) in a region D of $\mathcal{R}^n$, and γ, lying in D, is parametrized by $\mathcal{R} \xrightarrow{g} \mathcal{R}^n$ for $a \leq t \leq b$, we can do either of two things. We can use the coefficient functions $F_1, \ldots, F_n$ of $\omega_{\mathbf{x}}$ to form a vector field F in D and define

$$\int_\gamma \omega_{\mathbf{x}} = \int_a^b F(g(t)) \cdot g'(t)\,dt,$$

or we can form a partition P of the interval $a \leq t \leq b$ at points $a = t_0 < t_1 < \ldots < t_K = b$ and define

$$\int_\gamma \omega_{\mathbf{x}} = \lim_{m(P) \to 0} \sum_{k=1}^{K} \omega_{g(t_k)}(g'(t_k))(t_k - t_{k-1}).$$

It is clear that the two formulas give the same definition of $\int_\gamma \omega_{\mathbf{x}}$, the *integral of the* 1*-form* $\omega_{\mathbf{x}}$ *over* γ.

Next we define a product of 1-forms which is different from ordinary pointwise multiplication of functions. We first define the product of the basic 1-forms dx_1, dx_2, dx_3 in $\mathcal{R}^3$. The product $dx_1 \wedge dx_2$ is defined so that it is a function on ordered pairs of vectors in $\mathcal{R}^3$. Geometrically, $dx_1 \wedge dx_2(\mathbf{a}, \mathbf{b})$ will be the area of the parallelogram spanned by the projections of $\mathbf{a}$ and $\mathbf{b}$ into the x_1x_2-plane. The sign of the area is determined so that if the projections of $\mathbf{a}$ and $\mathbf{b}$ have the same orientation as the positive x_1 and x_2 axes, then the area is positive; it is negative when these orientations are opposite. Such a projection is shown in Fig. 32. Thus, if $\mathbf{a} = (a_1, a_2, a_3)$ and $\mathbf{b} = (b_1, b_2, b_3)$, then

$$dx_1 \wedge dx_2\,(\mathbf{a}, \mathbf{b}) = \det \begin{pmatrix} a_1 & b_1 \\ a_2 & b_2 \end{pmatrix} = a_1b_2 - a_2b_1,$$

and the determinant automatically gives the area the correct sign. We can use the basic 1-forms dx_1 and dx_2 to write the last equation as

$$dx_1 \wedge dx_2\,(\mathbf{a}, \mathbf{b}) = \det \begin{pmatrix} dx_1\,(\mathbf{a}) & dx_1\,(\mathbf{b}) \\ dx_2\,(\mathbf{a}) & dx_2\,(\mathbf{b}) \end{pmatrix}.$$

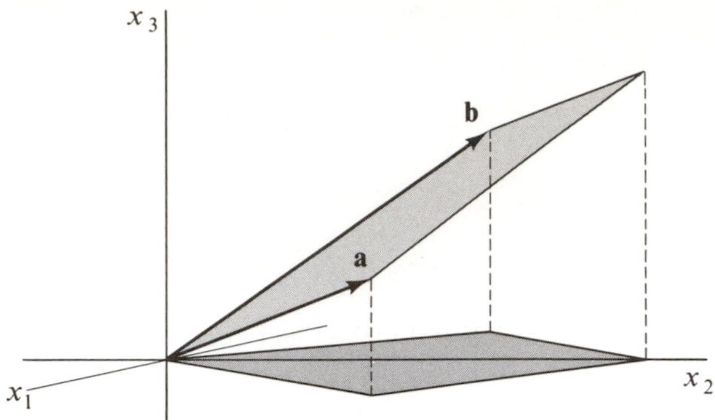

Figure 32

The definitions of all possible products of dx_1, dx_2, and dx_3 in either order can thus all be written in one formula:

$$dx_i \wedge dx_j\,(\mathbf{a}, \mathbf{b}) = \det \begin{pmatrix} dx_i\,(\mathbf{a}) & dx_i\,(\mathbf{b}) \\ dx_j\,(\mathbf{a}) & dx_j\,(\mathbf{b}) \end{pmatrix}, \tag{3}$$

with a similar geometric interpretation for each one. For example, we have

$$dx_2 \wedge dx_3\,(\mathbf{a}, \mathbf{b}) = \det \begin{pmatrix} a_2 & b_2 \\ a_3 & b_3 \end{pmatrix},$$

which is the signed area of the projection of the **ab**-parallelogram on the x_2x_3-plane.

As a consequence of Equation (3) and the properties of determinants, the following relations hold:

7.1 $$dx_i \wedge dx_j = -dx_j \wedge dx_i.$$

7.2 $$dx_i \wedge dx_i = 0.$$

The first equation holds because the interchange of adjacent rows of a determinant changes its sign, the second, because a determinant with two equal rows is zero. Similarly, we have, on interchanging columns,

7.3 $$dx_i \wedge dx_j\,(\mathbf{b}, \mathbf{a}) = -dx_i \wedge dx_j\,(\mathbf{a}, \mathbf{b}).$$

If we now ask for the most general linear combination of the functions $dx_i \wedge dx_j$, it is clear from 7.1 and 7.2 that it can be written in the form

$$c_1\, dx_2 \wedge dx_3 + c_2\, dx_3 \wedge dx_1 + c_3\, dx_1 \wedge dx_2.$$

Furthermore, if $F = (F_1, F_2, F_3)$ is a vector field in a region D of $\mathcal{R}^3$, we can define for each $\mathbf{x}$ in D the function

$$\tau_{\mathbf{x}} = F_1(\mathbf{x})\, dx_2 \wedge dx_3 + F_2(\mathbf{x})\, dx_3 \wedge dx_1 + F_3(\mathbf{x})\, dx_1 \wedge dx_2$$

of ordered pairs $(\mathbf{a}, \mathbf{b})$ of vectors in $\mathcal{R}^3$. The function $\tau_{\mathbf{x}}$ is called a **differential 2-form** or **2-form.**

Example 3. The 2-form

$$\tau = 2\, dx_2 \wedge dx_3 + dx_3 \wedge dx_1 + 5\, dx_1 \wedge dx_2$$

is the same function at every point of $\mathcal{R}^3$ because its coefficients are constant. (We have written τ instead of $\tau_{\mathbf{x}}$, and we shall do this whenever explicit mention of the variable $\mathbf{x}$ is not needed.) Letting $\mathbf{a} = (1, 2, 3)$ and $\mathbf{b} = (0, 1, 1)$, we have

$$\tau(\mathbf{a}, \mathbf{b}) = 2 \det \begin{pmatrix} 2 & 1 \\ 3 & 1 \end{pmatrix} + \det \begin{pmatrix} 3 & 1 \\ 1 & 0 \end{pmatrix} + 5 \det \begin{pmatrix} 1 & 0 \\ 2 & 1 \end{pmatrix} = 2.$$

We recall, and it is easy to verify directly, that the vector $\mathbf{a} \times \mathbf{b}$ with coordinates

$$\det \begin{pmatrix} 2 & 1 \\ 3 & 1 \end{pmatrix}, \quad \det \begin{pmatrix} 3 & 1 \\ 1 & 0 \end{pmatrix}, \quad \det \begin{pmatrix} 1 & 0 \\ 2 & 1 \end{pmatrix}$$

is perpendicular to $(1, 2, 3)$ and to $(0, 1, 1)$, and that its length is equal to the area of the parallelogram P spanned by $\mathbf{a}$ and $\mathbf{b}$. Thus $\tau(\mathbf{a}, \mathbf{b}) = (2, 1, 5) \cdot (\mathbf{a} \times \mathbf{b})$ and is thus the coordinate of $(2, 1, 5)$ in the direction perpendicular to P, multiplied by the area of P. If we interpret $(2, 1, 5)$ as the constant velocity vector of a fluid flow in space, $\tau(\mathbf{a}, \mathbf{b})$ will be the total flow across P in one unit of time. Such a flow is shown in Fig. 33. For a flow F of constant speed and direction across a flat surface S, the **flux** is defined to be the normal coordinate $F_{\mathbf{n}}$ of F, times the area of S, and so is equal to $(F \cdot \mathbf{n})\sigma(S)$.

We can summarize what has just been done by pointing out that we have defined a multiplication, called the **exterior product** of basic 1-forms dx_1, dx_2, etc. The resulting products written $dx_i \wedge dx_j$ are *basic* 2-forms in that every 2-form is a linear combination of them. In $\mathcal{R}^2$ there is only one basic 2-form, $dx_1 \wedge dx_2$, while in $\mathcal{R}^3$ there are three of them. (How many in $\mathcal{R}^1$?) From here we can proceed to define the exterior product of any two 1-forms to be the 2-form got by multiplying the 1-forms as if they

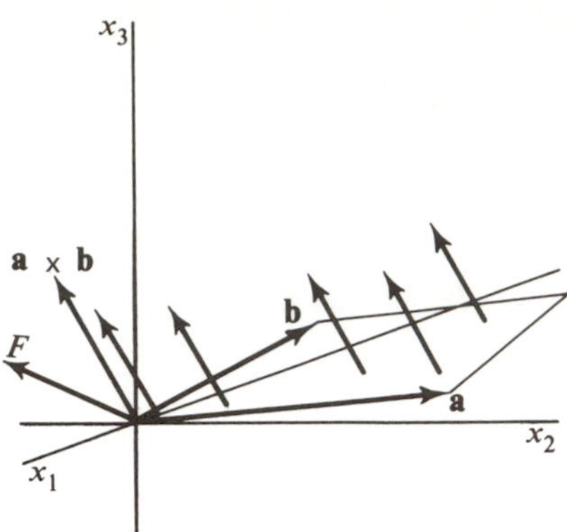

Figure 33

were polynomials in the variables dx_1, dx_2, etc., and then using the rules 7.1 and 7.2 to simplify. In practice, the wedges are often omitted from the notation.

Example 4. To save writing subscripts, we shall denote basic 1-forms by dx, dy, dz, even though the latter notation is less convenient for formulating the general definition. We have

$$\begin{aligned}(x\,dx + y^2\,dy) \wedge (dx + x\,dy) \\ &= x\,dx \wedge dx + y^2\,dy \wedge dx + x^2\,dx \wedge dy + xy^2\,dy \wedge dy \\ &= 0 - y^2\,dx \wedge dy + x^2\,dx \wedge dy + 0 \\ &= (x^2 - y^2)\,dx \wedge dy.\end{aligned}$$

In three variables we have the example

$$\begin{aligned}(dx + dy + dz) \wedge (x\,dx + z\,dy) \\ &= x\,dx \wedge dx + x\,dy \wedge dx + x\,dz \wedge dx \\ &\quad + z\,dx \wedge dy + z\,dy \wedge dy + z\,dz \wedge dy \\ &= -z\,dy \wedge dz + x\,dz \wedge dx + (z - x)\,dx \wedge dy.\end{aligned}$$

Three-forms arise in attempting to define the product of a 2-form and a 1-form. The meaning of the basic 3-form $dx_1 \wedge dx_2 \wedge dx_3$ is that of a signed volume function. Thus if $\mathbf{a} = (a_1, a_2, a_3)$, $\mathbf{b} = (b_1, b_2, b_3)$, and $\mathbf{c} = (c_1, c_2, c_3)$, we define

$$dx_1 \wedge dx_2 \wedge dx_3(\mathbf{a}, \mathbf{b}, \mathbf{c}) = \det \begin{pmatrix} a_1 & b_1 & c_1 \\ a_2 & b_2 & c_2 \\ a_3 & b_3 & c_3 \end{pmatrix},$$

which is the 3-dimensional-oriented volume of the parallelepiped spanned by the vectors **a**, **b**, and **c**. Higher-dimensional forms are defined in Exercise 11.

A differential form of unspecified dimension will be called a p-form and denoted ω^p, σ^p, etc. As we did with the 1-form, we shall define the integral of a p-form in $\mathcal{R}^n$ over the image S of a set in $\mathcal{R}^p$ under a continuously differentiable function g. To simplify the discussion, we suppose that $\mathcal{R}^p \xrightarrow{g} \mathcal{R}^n$ is differentiable on a closed bounded rectangle R in $\mathcal{R}^p$. If ω^p is a p-form with coefficient functions defined on the image $S = \partial R$ in $\mathcal{R}^n$, we can apply ω^p at a point $g(\mathbf{u})$ to the derived vectors $(\partial g/\partial u_k)(\mathbf{u})$ for $k = 1, \ldots, p$. The result looks like

$$\omega^p_{g(\mathbf{u})}\left(\frac{\partial g}{\partial u_1}(\mathbf{u}), \ldots, \frac{\partial g}{\partial u_p}(\mathbf{u})\right). \tag{4}$$

Furthermore, if G is a grid over R with corner points $\mathbf{u}_1, \ldots, \mathbf{u}_N$, and rectangles $R_1, \ldots, R_N$, we can form the sum

$$\sum_{k=1}^{N} \omega^p_{g(\mathbf{u}_k)}\left(\frac{\partial g}{\partial u_1}(\mathbf{u}_k), \ldots, \frac{\partial g}{\partial u_p}(\mathbf{u}_k)\right) V(R_k),$$

where $V(R_k)$ is the p-dimensional volume of R_k. We define the integral of ω^p over S by

$$\int_S \omega^p = \lim_{m(G)\to 0} \sum_{k=1}^{N} \omega^p_{g(\mathbf{u}_k)}\left(\frac{\partial g}{\partial u_1}(\mathbf{u}_k), \ldots, \frac{\partial g}{\partial u_p}(\mathbf{u}_k)\right) V(R_k),$$

provided the limit exists in the Riemann sum sense. If g is continuously differentiable, and ω^p has continuous coefficient functions, then the function of $\mathbf{u}$ given by Formula (4) will be continuous on R, and $\int_S \omega^p$ will exist and be equal to the Riemann integral of that function. Thus

$$\int_S \omega^p = \int_R \omega^p_g\left(\frac{\partial g}{\partial u_1}, \ldots, \frac{\partial g}{\partial u_p}\right) dV.$$

Example 5. If

$$\omega^2_{\mathbf{x}} = F_1(\mathbf{x})\, dx_2 \wedge dx_3 + F_2(\mathbf{x})\, dx_3 \wedge dx_1 + F_3(\mathbf{x})\, dx_1 \wedge dx_2$$

has continuous coefficients in $\mathcal{R}^3$, and $\mathcal{R}^2 \xrightarrow{g} \mathcal{R}^3$ is continuously differentiable on R, with coefficient functions g_1, g_2, g_3, then

$$\omega_g\left(\frac{\partial g}{\partial u_1}, \frac{\partial g}{\partial u_2}\right) = F_1 \circ g\, \frac{\partial(g_2, g_3)}{\partial(u_1, u_2)} + F_2 \circ g\, \frac{\partial(g_3, g_1)}{\partial(u_1, u_2)} + F_3 \circ g\, \frac{\partial(g_1, g_2)}{\partial(u_1, u_2)}.$$

Hence, in this case,

$$\int_S \omega^p = \int_S F_1\, dx_2 \wedge dx_3 + F_2\, dx_3 \wedge dx_1 + F_3\, dx_1 \wedge dx_2$$

$$= \int_R \left[F_1 \circ g \frac{\partial(g_2, g_3)}{\partial(u_1, u_2)} + F_2 \circ g \frac{\partial(g_3, g_1)}{\partial(u_1, u_2)} + F_3 \circ g \frac{\partial(g_1, g_2)}{\partial(u_1, u_2)} \right] du_1\, du_2.$$

Except for smoothness conditions on S, assumed here for convenience, this is the same as the definition of the surface integral given in Section 3.

EXERCISES

1. Find the value of each of the following differential forms acting on the indicated vector or ordered set of vectors.

(a) $dx_1 + 2dx_2$; $(1, 1)$.
(b) $3dx - dy + dz$; $(1, -1, 0)$.
(c) $dx_2 + dx_3$; $(1, 3, -5)$.
(d) $dx_1 + 2\, dx_2 + \ldots + n\, dx_n$; $(1, -1, 1, -1, \ldots)$.
(e) $2dx_1 \wedge dx_2$; $((1, 1), (1, -1))$.
(f) $dy \wedge dy + 2dx \wedge dz$; $((1, 2, 1), (-1, 2, 3))$.
(g) $dx \wedge dy \wedge dz$; $((1, 1, 1), (1, 2, 1), (2, 1, 2))$.
(h) $dx_1 \wedge dx_2$; $((-2, -3, 0), (2, 0, 2))$.

2. Multiply out and simplify the following products.

(a) $(dx_1 + dx_2) \wedge (dx_1 - dx_2)$.
(b) $(2dx + 3dy - 2dz) \wedge dx$.
(c) $(dx + dy) \wedge (dx + dy)$.
(d) $(x^2\, dx + z^2\, dz) \wedge (dx - 2dy)$.
(e) $(\sin z\, dx + \cos x\, dy) \wedge (dx + dz)$.

3. Compute $\int_\gamma y\, dx + x\, dy$, where γ is given by $g(t) = (\cos t, \sin t)$, $0 \leq t \leq \pi/2$.

4. Compute $\int_\gamma x_1\, dx_1 + x_2\, dx_2 + x_3\, dx_3$, where γ is given by

(a) $g(t) = (-t, t^2, t)$ for $-1 \leq t \leq 1$.
(b) $h(t) = (t, t, t)$ for $0 \leq t \leq 1$.
(c) $k(t) = (t^2, t^2, t^2)$ for $-1 \leq t \leq 1$.

5. Compute $\int_\gamma \omega$ where $\omega = x_1\, dx_1 + x_n^2\, dx_2 + \ldots + x_n^2\, dx_n$ and γ is given by $g(t) = (t, t, t, \ldots)$ for $0 \leq t \leq 1$.

6. Let ω and $\bar{\omega}$ be 1-forms defined in a region D of $\mathcal{R}^n$, and let f and g be real-valued functions defined in D.

(a) Show that $f\omega + g\bar{\omega}$ defines a 1-form in D.
(b) Show that if ω, ν, and μ are 1-forms in D, then

$$(f\omega + g\nu) \wedge \mu = f\omega \wedge \mu + g\nu \wedge \mu.$$

7. Prove that if ω and $\bar{\omega}$ are 1-forms and if γ_1 and γ_2 are curves over which ω and $\bar{\omega}$ are integrable, then

$$\int_{\gamma_1} (a\omega + b\bar{\omega}) = a\int_{\gamma_1} \omega + b\int_{\gamma_1} \bar{\omega},$$

where a and b are constant and

$$\int_{\gamma_1 \cup \gamma_2} \omega = \int_{\gamma_1} \omega + \int_{\gamma_2} \omega.$$

8. Let f be a continuously differentiable real-valued function defined on a region D of $\mathcal{R}^n$, and let γ be a continuously differentiable curve lying in D with parametrization $\mathcal{R} \xrightarrow{g} \mathcal{R}^n$, for $a \leq t \leq b$. Show that if $g(a) = \mathbf{a}$ and $g(b) = \mathbf{b}$, then

$$\int_\gamma d_{\mathbf{x}} f = f(\mathbf{b}) - f(\mathbf{a}).$$

9. (a) Prove that if $\omega_{\mathbf{x}}$ is a 1-form in a region D of $\mathcal{R}^n$, then for each fixed $\mathbf{x}_0$ in D, $\omega_{\mathbf{x}_0}$ is a real-valued linear function on all of R^n.

(b) Prove the converse to part (a), namely, that if $\omega_{\mathbf{x}_0}$ is a real-valued linear function defined on all of $\mathcal{R}^n$, then $\omega_{\mathbf{x}_0}(\mathbf{a}) = \sum_{k=1}^{n} c_k(\mathbf{x}_0)\, dx_k\, (\mathbf{a})$, for all $\mathbf{a}$ in $\mathcal{R}^n$, where the $c_k(\mathbf{x}_0)$ are real numbers.

10. (a) Let τ be a real-valued function of pairs $(\mathbf{a}, \mathbf{b})$ of vectors in $\mathcal{R}^3$ such that $\tau(\mathbf{a}, \mathbf{b}) = -\tau(\mathbf{b}, \mathbf{a})$ and such that τ is linear in $\mathbf{a}$ and in $\mathbf{b}$. Show that there is a vector $\mathbf{c}_\tau$ in $\mathcal{R}^3$ such that $\tau(\mathbf{a}, \mathbf{b}) = \det(\mathbf{a}, \mathbf{b}, \mathbf{c}_\tau)$ for all $\mathbf{a}, \mathbf{b}$ in $\mathcal{R}^3$. [*Hint.* Show first that the result holds if $\mathbf{a}$ and $\mathbf{b}$ are in a basis for $\mathcal{R}^3$.]

(b) Use part (a) to show that if $\tau(\mathbf{a}, \mathbf{b}) = -\tau(\mathbf{b}, \mathbf{a})$, and τ is bilinear, then τ is a 2-form in $\mathcal{R}^3$.

11. For an ordered p-tuple $(\mathbf{a}_1, \mathbf{a}_2, \ldots, \mathbf{a}_p)$ of vectors in $\mathcal{R}^n$ where $p \geq 1$, define

$$dx_{k_1} \wedge dx_{k_2} \wedge \ldots \wedge dx_{k_p}(\mathbf{a}_1, \ldots, \mathbf{a}_p) = \det\left(dx_{k_i}(\mathbf{a}_j)\right)_{\substack{i=1,\ldots,p \\ j=1,\ldots,p}}.$$

This equation defines the basic p-forms in $\mathcal{R}^n$, of which the general p-forms are linear combinations.

(a) Compute $dx_2 \wedge dx_3 \wedge dx_4 + 2\, dx_1 \wedge dx_2 \wedge dx_4(\mathbf{a}, \mathbf{b}, \mathbf{c})$, where $\mathbf{a} = (1, -1, 0, 2)$, $\mathbf{b} = (-1, 1, 1, 1)$, and $\mathbf{c} = (0, 1, 2, 0)$.

(b) Prove that the interchange of adjacent factors in a basic p-form changes the sign of the form.

(c) Prove that a basic p-form with a repeated factor is zero.

(d) Prove that the general p-form can be written

$$\omega^p = \sum_{i_1 < \ldots < i_p} f_{i_1,\ldots,i_p}\, dx_{i_1} \wedge \ldots \wedge dx_{i_p},$$

where $1 \leq i_k \leq n$ for $k = 1, \ldots, p$.

(e) Prove that if $p > n$, ω^p is identically zero.

(f) Prove that there are $\binom{n}{p}$ terms in the p-form of part (d).

12. If ω^p and ω^q are p- and q-forms in $\mathcal{R}^n$ with

$$\omega^p = \sum_{i_1<\ldots<i_p} f_{i_1,\ldots,i_p}\, dx_{i_1} \wedge \ldots \wedge dx_{i_p}$$

and

$$\omega^p = \sum_{j_1<\ldots<j_q} g_{j_1,\ldots,j_q}\, dx_{j_1} \wedge \ldots \wedge dx_{j_q},$$

define their exterior product $\omega^p \wedge \omega^q$ by

$$\omega^p \wedge \omega^q = \sum f_{i_1\ldots i_p} g_{j_1\ldots j_q}\, dx_{i_1} \wedge \ldots \wedge dx_{i_p} \wedge dx_{j_1} \wedge \ldots \wedge dx_{j_q}.$$

(a) Prove that if $\omega^p \wedge \omega^q$ is reduced to standard form by using Equations 7.1 and 7.2, then it has $\binom{n}{p+q}$ terms.

(b) Prove that $\omega^p \wedge \omega^q = (-1)^{pq}\omega^q \wedge \omega^p$.

13. Show that the definition of the integral of a p-form agrees with that given for a 1-form when $p = 1$.

14. Compute $\int_S \omega^2$ if $\omega^2 = dx \wedge dy + dx \wedge dz$, and S is the image of $0 \le u \le 1$, $0 \le v \le \pi/2$ under $g(u, v) = (u \cos v, u \sin v, v)$.

15. (a) If ω^3 is the 3-form $f\, dx_1 \wedge dx_2 \wedge dx_3$ in $\mathcal{R}^3$, and $\mathcal{R}^3 \xrightarrow{g} \mathcal{R}^3$ is differentiable, show that

$$\omega^3_g\left(\frac{\partial g}{\partial u_1}, \frac{\partial g}{\partial u_2}, \frac{\partial g}{\partial u_3}\right) = f \circ g\, \frac{\partial(g_1, g_2, g_3)}{\partial(u_1, u_2, u_3)}.$$

(b) Compute the integral $\int_S dx \wedge dy \wedge dz$, where S is the image of $0 \le u \le 1$, $0 \le v \le 1$, $0 \le w \le 1$ under $g(u, v, w) = (u^2, v^2, w^2)$.

SECTION 8

THE EXTERIOR DERIVATIVE

The fundamental theorem of calculus states that if $(d/dx)f$ is integrable on $[a, b]$, then

$$\int_a^b \frac{df}{dx}\, dx = f(b) - f(a). \tag{1}$$

The Stokes and Gauss formulas,

$$\int_S \operatorname{curl} F \cdot dS = \int_{\partial S} F \cdot \mathbf{t}\, ds \tag{2}$$

$$\int_R \operatorname{div} F\, dV = \int_{\partial R} F \cdot \mathbf{n}\, d\sigma, \tag{3}$$

are similar in that they express the integral of a kind of derivative of a function in terms of the function itself on a set of lower dimension. Using differential forms, we shall define exterior differentiation, which unifies the above formulas.

The operation of **exterior differentiation** is defined inductively on differential forms as follows. Let f be a real-valued differentiable function on $\mathcal{R}^n$. Then, by definition,

$$df = \frac{\partial f}{\partial x_1} dx_1 + \ldots + \frac{\partial f}{\partial x_n} dx_n.$$

Thus the exterior derivative of f is the particular 1-form that at each point of the domain of f is equal to what we have earlier called the differential of f. To continue, if

$$\omega^1 = f_1\, dx_1 + \ldots + f_n\, dx_n$$

is a 1-form with differentiable coefficients, then in terms of the 1-forms $df_1, \ldots, df_n$, we define

$$d\omega^1 = (df_1) \wedge dx_1 + \ldots + (df_n) \wedge dx_n.$$

Thus $d\omega^1$ is a 2-form. In general, if ω^p is a p-form, then $d\omega^p$ is the $(p + 1)$-form got by replacing each coefficient function of ω^p by the 1-form that is its exterior derivative. To keep the terminology consistent, we may refer to a real-valued function as a 0-form.

Example 1. If $f(x_1, x_2) = x_1^2 + x_2^3$, then df is given by

$$d(x_1^2 + x_2^3) = 2x_1\, dx_1 + 3x_2^2\, dx_2.$$

If $\omega^1_{(x_1,x_2)} = x_1x_2\, dx_1 + (x_1^2 + x_2^2)\, dx_2$, then $d\omega^1$ is given by

$$\begin{aligned} d\big(x_1x_2\, dx_1 &+ (x_1^2 + x_2^2)\, dx_2\big) \\ &= (x_2\, dx_1 + x_1\, dx_2) \wedge dx_1 + (2x_1\, dx_1 + 2x_2\, dx_2) \wedge dx_2 \\ &= x_1\, dx_1 \wedge dx_2. \end{aligned}$$

If $\omega^2_{(x,y,z)} = xz\, dx \wedge dy + y^2z\, dx \wedge dz$, then $d\omega^2$ is given by

$$\begin{aligned} d(xz\, dx \wedge dy &+ y^2z\, dx \wedge dz) \\ &= (z\, dx + x\, dz) \wedge dx \wedge dy + (2yz\, dy + y^2\, dz) \wedge dx \wedge dz \\ &= (x - 2yz)\, dx \wedge dy \wedge dz. \end{aligned}$$

Using the exterior derivative we can state the general Stokes formula in the form

8.1
$$\int_B d\omega^p = \int_{\partial B} \omega^p,$$

where B is $(p + 1)$-dimensional and ∂B is its p-dimensional boundary. We shall interpret the formula in several specific cases.

If ω^1 is a 1-form in $\mathcal{R}^2$, then we can write ω in the form

$$\omega^1 = F_1\,dx_1 + F_2\,dx_2,$$

and then

$$\begin{aligned} d\omega^1 &= \left(\frac{\partial F_1}{\partial x_1}\,dx_1 + \frac{\partial F_1}{\partial x_2}\,dx_2\right) \wedge dx_1 + \left(\frac{\partial F_2}{\partial x_1}\,dx_1 + \frac{\partial F_2}{\partial x_2}\,dx_2\right) \wedge dx_2 \\ &= \left(\frac{\partial F_2}{\partial x_1} - \frac{\partial F_1}{\partial x_2}\right) dx_1 \wedge dx_2. \end{aligned} \tag{4}$$

Substitution of $d\omega^1$ and ω^1 into Equation 8.1 gives

$$\int_B \left(\frac{\partial F_2}{\partial x_1} - \frac{\partial F_1}{\partial x_2}\right) dx_1 \wedge dx_1 = \int_{\partial B} F_1\,dx_1 + F_2\,dx_2. \tag{5}$$

This is almost Green's formula of Section 1, if B is a suitable set in $\mathcal{R}^2$, and ∂B stands for its counterclockwise-oriented boundary curve. However, the left-hand integral in Equation (5) has been defined as an integral over a parametrized set in the previous section, whereas the corresponding integral of Green's formula,

$$\int_D \left(\frac{\partial F_2}{\partial x_1} - \frac{\partial F_2}{\partial x_2}\right) dx_1\,dx_2 = \oint_\gamma F_1\,dx_1 + F_2\,dx_2,$$

is an integral over a set without a parametrization. The difference between the two is such that the parametrized set B may be covered more than once by its parametrization, while the integral over the set D covers each part of the set only once. See Exercise 9.

If ω^1 is a 1-form in $\mathcal{R}^3$ given by

$$\omega^1 = F_1\,dx_1 + F_2\,dx_2 + F_3\,dx_3,$$

then a straightforward calculation like that in Equation 4 yields

$$\begin{aligned} d\omega^1 = &\left(\frac{\partial F_3}{\partial x_2} - \frac{\partial F_2}{\partial x_3}\right) dx_2 \wedge dx_3 \\ &+ \left(\frac{\partial F_1}{\partial x_3} - \frac{\partial F_3}{\partial x_1}\right) dx_3 \wedge dx_1 + \left(\frac{\partial F_2}{\partial x_1} - \frac{\partial F_1}{\partial x_2}\right) dx_1 \wedge dx_2. \end{aligned} \tag{6}$$

Thus the 2-form $d\omega^1$ has as coefficient functions the coordinates of the vector field curl F where $F = (F_1, F_2, F_3)$. It is immediate that the general Stokes formula becomes precisely the Stokes formula of Section 4 if we make B and ∂B stand for a piece of smooth surface S and its positively oriented border ∂S.

The Gauss formula of Section 5 comes from considering a 2-form

$$\omega^2 = F_1\,dx_2 \wedge dx_3 + F_2\,dx_3 \wedge dx_1 + F_3\,dx_1 \wedge dx_2.$$

A short computation shows that

$$d\omega^2 = \left(\frac{\partial F_1}{\partial x_1} + \frac{\partial F_2}{\partial x_2} + \frac{\partial F_3}{\partial x_3}\right) dx_1 \wedge dx_2 \wedge dx_3. \tag{7}$$

This 3-form has as coefficient the divergence of the field $F = (F_1, F_2, F_3)$, and substitution into the general Stokes formula gives the Gauss, or divergence, formula of Section 5, except that, as with Green's formula, the volume integral of Gauss's formula is not identical with the integral of a 3-form. See Exercise 10.

The correspondence between a vector field $F = (F_1, F_2, F_3)$ in $\mathcal{R}^3$, and a differential form with coefficient functions F_1, F_2, F_3, has been described in Equations (6) and (7) and can be summarized as follows:

$$\text{if} \quad \omega^2 \leftrightarrow F, \qquad \text{then} \quad d\omega^2 \leftrightarrow \text{div}\, F,$$

$$\text{if} \quad \omega^1 \leftrightarrow F, \qquad \text{then} \quad d\omega^1 \leftrightarrow \text{curl}\, F.$$

Finally, if $\omega^0 \leftrightarrow f$, then $d\omega^0 \leftrightarrow \text{grad}\, f$, where f is real-valued.

EXERCISES

1. Compute $d\omega$, where ω is

(a) $(x_1^2 + x_2^2)\, dx_1 - x_2\, dx_2$.
(b) $\sin x_1\, dx_3$.
(c) $x_1\, dx_1 \wedge dx_2 + x_2\, dx_2 \wedge dx_3$.
(d) $yz\, dx + zx\, dy + xy\, dz$.
(e) $x_1\, dx_1 \wedge dx_2 \wedge dx_3$.

2. (a) If $\mathcal{R}^2 \xrightarrow{f_1} \mathcal{R}$ and $\mathcal{R}^2 \xrightarrow{f_2} \mathcal{R}$ are differentiable functions with the same domain, show that

$$df_1 \wedge df_2 = \frac{\partial(f_1, f_2)}{\partial(x_1, x_2)}\, dx_1 \wedge dx_2.$$

(b) Generalize part (a) to the case of three functions from $\mathcal{R}^3$ to $\mathcal{R}$.

3. Show that if ω^p and ω^q are p- and q-forms, respectively, with differentiable coefficients, then

$$d(\omega^p \wedge \omega^q) = d\omega^p \wedge \omega^q + (-1)^p \omega^p \wedge (d\omega^q),$$

(a) for ω^p and ω^q defined in $\mathcal{R}^2$.
(b) for ω^p and ω^q defined in $\mathcal{R}^n$.

4. If ω^0 is a 0-form in $\mathcal{R}^n$ with coefficients that are twice continuously differentiable, show that $d(d\omega^0) = 0$.

5. (a) Let ω^1 be a 1-form in $\mathcal{R}^3$ with twice continuously differentiable coefficients. Show that $d(d\omega^1) = 0$.

(b) Use the correspondence between a vector field $F = (F_1, F_2, F_3)$ and the 1-form with coefficients F_1, F_2, F_3, to show that the result of part (a) is equivalent to the relation div (curl F) = 0.

(c) Show that $d(d\omega^0) = 0$ is equivalent to curl (grad f) = 0.

6. (a) Find an example of a 1-form ω^1 such that there is no 0-form ω^0 for which $d\omega^0 = \omega^1$.

(b) Show similarly that if ω^2 is a 2-form, there may not exist a 1-form ω^1 such that $d\omega^1 = \omega^2$.

(c) Interpret parts (a) and (b) in terms of gradient and curl.

7. Prove that if ω^p is a p-form with twice continuously differentiable coefficients, then $d(d\omega^p) = 0$.

8. (a) Let $F = (F_1, \ldots, F_n)$ be a vector field in $\mathcal{R}^n$, and consider the 1-form $\omega^1 = F_1\,dx_1 + \ldots + F_n\,dx_n$. Show that the condition $d\omega^1 = 0$ is equivalent to the requirement that the Jacobian matrix F' be symmetric.

(b) Show that if ω^1 is a 1-form with continuously differentiable coefficients in a rectangle R in $\mathcal{R}^n$, then $d\omega^1 = 0$ in R implies that there is a function $\mathcal{R}^n \xrightarrow{f} \mathcal{R}$ such that $df = \omega^1$. [*Hint.* Use Theorem 2.6.]

9. In the general Stokes formula, Equation 8.1, let B be a disc in $\mathcal{R}^2$ parametrized by $(x_1, x_2) = (u \cos v,\ u \sin v)$, $0 \le u \le 1$, $0 \le v \le 4\pi$. Notice that the disc is covered twice by this parametrization. What is the correct parametrization of ∂B to make 8.1 hold for this example? Take ω^p to be a 1-form with continuously differentiable coefficients.

10. In the general Stokes formula, Equation 8.1, let ω^p be a 2-form with continuously differentiable coefficients in the ball $|\mathbf{x}| \le 1$ of $\mathcal{R}^3$. Show that it is possible to parametrize the ball B in such a way that the corresponding parametrization of ∂B has an inward pointing normal. Explain the apparent contradiction to the requirement in the Gauss theorem of Section 5 that ∂B have an outward-pointing normal.

Appendix

SECTION 1

INTRODUCTION

The theorems and techniques of calculus depend on both algebraic and topological properties of $\mathcal{R}^n$ and of functions from $\mathcal{R}^n$ to $\mathcal{R}^m$. The algebraic properties are taken up in Chapter 1, while topological matters are treated in the section of Chapter 3 on limits and continuity. However, at certain points of the later development of the subject, notably in existence theorems for integrals, we need more facts about continuity than we have treated at the beginning of Chapter 3. These topics are discussed briefly in the present section, but without proofs. For example, consider the following theorem.

1.1 Theorem

Let $\mathcal{R}^n \xrightarrow{f} \mathcal{R}$ be continuous on a closed, bounded subset K of $\mathcal{R}^n$. Then f attains both its maximum and minimum on K.

This theorem is used to guarantee the existence of a maximum or a minimum value for a real-valued function. The usual techniques of calculus are useful for finding maxima and minima, but they fail to solve the problem of existence. The following theorem is basic to this entire discussion.

1.2 Theorem

Let K be a closed, bounded subset of $\mathcal{R}$. Then K contains a largest element, called the ***supremum*** of K, and a smallest element, called the ***infimum*** of K.

Theorem 1.2 can be used to prove Theorem 1.1 as well as the following result, used to establish the existence of the integral of a continuous function.

1.3 Theorem

Let $\mathcal{R}^n \xrightarrow{f} \mathcal{R}$ be continuous on a closed, bounded subset K of $\mathcal{R}^n$. Then f is uniformly continuous on K, that is, given $\epsilon > 0$, there is a $\delta > 0$ such that $|f(x) - f(y)| < \epsilon$ whenever x and y are in K and $|x - y| < \delta$.

Proofs of these three theorems, and also of Theorem 1.4, can be found in any textbook on real-variable theory. One of the best is *Principles of Mathematical Analysis*, Second Ed., Walter Rudin, McGraw-Hill, 1964.

Next we describe an intrinsic criterion for the convergence of a sequence $\mathbf{x}_1, \mathbf{x}_2, \mathbf{x}_3, \ldots$ of vectors in $\mathcal{R}^n$. We say that a sequence is a **Cauchy sequence,** if, given $\epsilon > 0$ there is a number N such that

$$|\mathbf{x}_k - \mathbf{x}_l| < \epsilon$$

whenever k, $l > N$. While it is quite easy to show directly from the definitions that every convergent sequence is a Cauchy sequence, the converse given below depends on Theorem 1.2.

1.4 Theorem

If $\mathbf{x}_1, \mathbf{x}_2, \mathbf{x}_3, \ldots$ is a Cauchy sequence of vectors in $\mathcal{R}^n$, then the sequence has a limit $\mathbf{x}$ in $\mathcal{R}^n$, so that

$$\lim_{k \to \infty} \mathbf{x}_k = \mathbf{x}.$$

Theorem 1.4 is needed to prove Theorem 5.1 which justifies the modified Newton method for solving vector equations.

The definitions of limit point, limit, continuity, interior point, open set, boundary, closed set, and differentiability have been given directly or indirectly in terms of the Euclidean norm on $\mathcal{R}^n$. Relative to the norm, the distance between two points $\mathbf{x} = (x_1, \ldots, x_n)$ and $\mathbf{y} = (y_1, \ldots, y_n)$ is the number

$$|\mathbf{x} - \mathbf{y}| = [(x_1 - y_1)^2 + \ldots + (x_n - y_n)^2]^{1/2}.$$

The same definitions can be made provided any norm is given. A norm on $\mathcal{V}$ is a real-valued function $\| \quad \|$ with domain equal to $\mathcal{V}$ and with the three properties of Euclidean length:

1.5 Positivity $\|\mathbf{x}\| > 0$, except that $\|0\| = 0$.

1.6 Homogeneity $\|a\mathbf{x}\| = |a| \, \|\mathbf{x}\|$.

1.7 Triangle inequality $\|\mathbf{x} + \mathbf{y}\| \leq \|\mathbf{x}\| + \|\mathbf{y}\|$.

One purpose of this section is to prove that, in a finite-dimensional vector space, the limit definitions are independent of the choice of norm. That is, if $\mathbf{x}_0$ is a limit point of a set S with respect to one norm, then it is a limit point of S with respect to every norm, and the same goes for the other definitions referred to above. It follows, in particular, that these basic limit concepts are not dependent upon a Euclidean inner product. An example of a norm on $\mathcal{R}^n$ different from the Euclidean norm is the so-called **box norm**, or **maximum norm,** defined, for any $\mathbf{x} = (x_1, \ldots, x_n)$, by

$$\|\mathbf{x}\| = \max \{|x_1|, \ldots, |x_n|\}.$$

If $\mathbf{x}_0 = (2, 1)$, the set of all $\mathbf{x}$ in $\mathcal{R}^2$ such that $\|\mathbf{x} - \mathbf{x}_0\| < 0.5$ is the parallelogram shown in Fig. 1. We have purposely drawn nonperpendicular coordinate axes with different scale to emphasize the fact that this norm is not Euclidean.

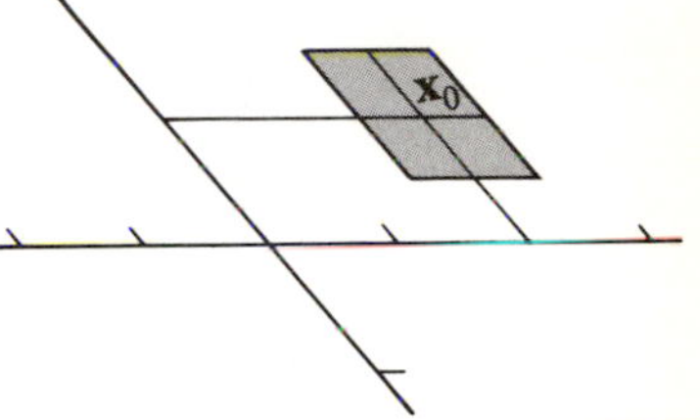

Figure 1

For any two norms $\| \quad \|_1$ and $\| \quad \|_2$ on a vector space $\mathcal{V}$, we define $\| \quad \|_1$ to be **equivalent** to $\| \quad \|_2$ if there exist positive real numbers k and K such that, for any $\mathbf{x}$ in $\mathcal{V}$,

$$k \, \|\mathbf{x}\|_1 \leq \|\mathbf{x}\|_2 \leq K \, \|\mathbf{x}_1\|. \tag{1}$$

It is easy to check that this is a true equivalence relation, that is, it satisfies the three requirements:

Reflexivity Every norm is equivalent to itself.

Symmetry $\| \quad \|_1$ is equivalent to $\| \quad \|_2$ if and only if $\| \quad \|_2$ is equivalent to $\| \quad \|_1$.

Transitivity If $\| \quad \|_1$ is equivalent to $\| \quad \|_2$ and if $\| \quad \|_2$ is equivalent to $\| \quad \|_3$, then $\| \quad \|_1$ is equivalent to $\| \quad \|_3$.

More important is the fact that equivalent norms result in the same definitions of limit point, limit, interior point, and differentiability. Continuity, open set, boundary, and closed set, which are defined in terms of the preceding concepts, are therefore also independent of a choice between equivalent norms.

To verify the above contention, let $\| \quad \|_1$, and $\| \quad \|_2$ be equivalent norms, and suppose that $\mathbf{x}_0$ is a limit point of S with respect to $\| \quad \|_1$.

Then, for any $\epsilon_1 > 0$, there exists a point $\mathbf{x}$ in S such that $0 < \|\mathbf{x} - \mathbf{x}_0\|_1 < \epsilon_1$. Thus, if $\epsilon_2 > 0$ is given arbitrarily, we may set $\epsilon_1 = \epsilon_2/K$ and obtain, by Inequality (1),

$$0 < \|\mathbf{x} - \mathbf{x}_0\|_2 \leq K \|\mathbf{x} - \mathbf{x}_0\|_1 < K\epsilon_1 = \epsilon_2.$$

Hence, $\mathbf{x}_0$ is a limit point of S with respect to $\| \quad \|_2$. Suppose, next, that with respect to $\| \quad \|_1$ we have

$$\lim_{\mathbf{x} \to \mathbf{x}_0} f(\mathbf{x}) = \mathbf{y}_0.$$

Then, as we have just proved, $\mathbf{x}_0$ is a limit point of the domain of f with respect to both norms. For any $\epsilon_1 > 0$, there exists $\delta_1 > 0$ such that if $\mathbf{x}$ is in the domain of f and $0 < \|\mathbf{x} - \mathbf{x}_0\|_1 < \delta_1$, then $\|f(\mathbf{x}) - \mathbf{y}_0\|_1 < \epsilon_1$. Let $\epsilon_2 > 0$ be given arbitrarily, set $\epsilon_1 = \epsilon_2/K$, and then choose $\delta_2 = \delta_1 k$. If $0 < \|\mathbf{x} - \mathbf{x}_0\|_2 < \delta_2$, it follows by Inequality (1) that

$$0 < \|\mathbf{x} - \mathbf{x}_0\|_1 \leq \frac{1}{k} \|\mathbf{x} - \mathbf{x}_0\|_2 < \frac{\delta_2}{k} = \delta_1.$$

Hence,

$$\|f(\mathbf{x}) - \mathbf{y}_0\|_2 \leq K \|f(\mathbf{x}) - \mathbf{y}_0\|_1 < K\epsilon_1 = \epsilon_2,$$

and we conclude that $\lim f(\mathbf{x}) = \mathbf{y}_0$ with respect to $\| \quad \|_2$. The arguments for the definitions of interior point and differentiability are similar, and we omit the details.

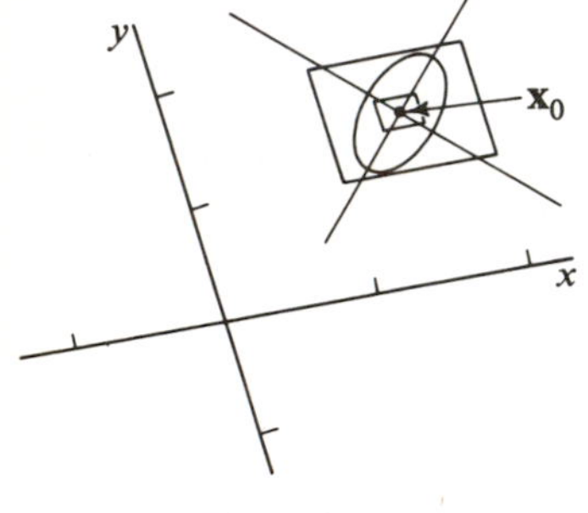

Figure 2

With respect to a given norm on a vector space $\mathcal{V}$, the **ϵ-ball with center $\mathbf{x}_0$** is the set of all $\mathbf{x}$ in $\mathcal{V}$ such that $\|\mathbf{x} - \mathbf{x}_0\| < \epsilon$. In general, an ϵ-ball doesn't look very round. For example, with respect to the box norm on $\mathcal{R}^2$, every ϵ-ball is a parallelogram (see Fig. 1). It follows directly from the Inequalities (1) that two norms are equivalent if and only if any ϵ-ball about $\mathbf{x}_0$ with respect to one norm is contained in some δ-ball about $\mathbf{x}_0$ with respect to the other norm, and vice versa (see Fig. 2). We now turn to the principal theorem.

1.8 Theorem

Any two norms on a finite-dimensional vector space $\mathcal{V}$ are equivalent.

Proof. Let $\| \quad \|$ be an arbitrary norm on $\mathcal{V}$. Choose a basis $\{\mathbf{x}_1, \ldots, \mathbf{x}_n\}$ for $\mathcal{V}$, and define a Euclidean norm on $\mathcal{V}$ by setting

$$|\mathbf{x}| = \sqrt{x_1^2 + \ldots + x_n^2},$$

for any $\mathbf{x} = x_1\mathbf{x}_1 + \ldots + x_n\mathbf{x}_n$. We shall show that $\| \quad \|$ is equivalent to $| \quad |$, that is, there exist positive real numbers k and K such that $k\,|\mathbf{x}| \leq \|\mathbf{x}\| \leq K\,|\mathbf{x}|$, for all $\mathbf{x}$ in $\mathcal{V}$. By the transitivity property of the equivalence relation between norms, it then follows that any two norms on $\mathcal{V}$ are equivalent.

For any $\mathbf{x} = x_1\mathbf{x}_1 + \ldots + x_n\mathbf{x}_n$, we have

$$\|\mathbf{x}\| \leq \sum_{i=1}^{n} |x_i|\,\|\mathbf{x}_i\| \leq \left(\sum_{i=1}^{n}\|\mathbf{x}_i\|\right)\max_i\,\{|x_i|\}$$
$$\leq \left(\sum_{i=1}^{n}\|\mathbf{x}_i\|\right)\sqrt{\sum_{i=1}^{n} x_i^2} = K\,|\mathbf{x}|,$$

where $K = \sum_{i=1}^{n} \|\mathbf{x}_i\| > 0$. We now prove that k exists. We contend that, as a function of $\mathbf{x}$, the real-valued function $\|\ \ \|$ is continuous with respect to the Euclidean norm $|\ \ |$. For instance, if $\epsilon > 0$ is given, we pick $\delta = \epsilon/K$. Then, if $|\mathbf{x} - \mathbf{x}_0| < \delta$,

$$|\,\|\mathbf{x}\| - \|\mathbf{x}_0\|\,| \leq \|\mathbf{x} - \mathbf{x}_0\| \leq K\,|\mathbf{x} - \mathbf{x}_0| < \epsilon.$$

Let k be the minimum value of the function $\|\ \ \|$ restricted to the Euclidean unit sphere $|\mathbf{x}| = 1$. Then, for any $\mathbf{x} \neq 0$, it follows that $\|\mathbf{x}/|\mathbf{x}|\,\| \geq k$ and, hence, that

$$\|\mathbf{x}\| \geq k\,|\mathbf{x}|, \qquad \text{for any } \mathbf{x} \text{ in } \mathcal{V}.$$

This completes the proof of the equivalence of norms.

SECTION 2

THE CHAIN RULE

In Chapter 4 we have presented two different versions of the chain rule, one in the section on the gradient and another in Section 3 devoted entirely to the chain rule. Both of these theorems contain assumptions that certain functions are continuously differentiable. Furthermore, in Theorem 3.1 we assumed that the composition $g \circ f$ was defined on an open set. Neither of these assumptions need to be made in order for the chain rule formula to hold. Of course, to conclude that $g \circ f$ is continuously differentiable, we need to assume that both f and g have continuous derivative matrices. The virtue of the next theorem is that it contains a minimum of assumptions, and the proof is, in style, very much what would be used to prove the theorem for real functions of a real variable.

2.1 Theorem. The Chain Rule

If $\mathcal{R}^n \xrightarrow{f} \mathcal{R}^m$ is differentiable at $\mathbf{x}_0$ and $\mathcal{R}^m \xrightarrow{g} \mathcal{R}^p$ is differentiable at $f(\mathbf{x}_0)$, then $g \circ f$ is differentiable at $\mathbf{x}_0$ and

$$(g \circ f)'(\mathbf{x}_0) = g'(f(\mathbf{x}_0))f'(\mathbf{x}_0).$$

Proof. The first thing to prove is that $\mathbf{x}_0$ is an interior point of the domain of $g \circ f$. Since g is differentiable at $f(\mathbf{x}_0)$, the point $\mathbf{y}_0 = f(\mathbf{x}_0)$ is by definition an interior point of the domain of g. Hence, there exists a positive real number δ' such that a point $\mathbf{y}$ is in the domain of g whenever $|\mathbf{y} - \mathbf{y}_0| < \delta'$. The function f, being differentiable at $\mathbf{x}_0$, is also continuous there. Furthermore, $\mathbf{x}_0$ is by definition an interior point of the domain of f. It follows that there exists a positive number δ such that if $|\mathbf{x} - \mathbf{x}_0| < \delta$, then $\mathbf{x}$ is in the domain of f and

$$|f(\mathbf{x}) - \mathbf{y}_0| = |f(\mathbf{x}) - f(\mathbf{x}_0)| < \delta'.$$

But δ' has been chosen just so that, if the last inequality holds, then $f(\mathbf{x})$ is in the domain of g. Thus any point $\mathbf{x}$ in $\mathcal{R}^n$ that satisfies $|\mathbf{x} - \mathbf{x}_0| < \delta$ lies in the domain of the composite function $g \circ f$, and the vector $\mathbf{x}_0$ is therefore an interior point of that domain.

It remains to prove that the composite linear function with matrix $g'(\mathbf{y}_0)f'(\mathbf{x}_0)$ satisfies the criterion for being the differential of $g \circ f$ at $\mathbf{x}_0$. That is, we must prove that if

$$g \circ f(\mathbf{x}) - g \circ f(\mathbf{x}_0) - g'(\mathbf{y}_0)f'(\mathbf{x}_0)(\mathbf{x} - \mathbf{x}_0) = |\mathbf{x} - \mathbf{x}_0|\, Z(\mathbf{x} - \mathbf{x}_0), \tag{1}$$

then

$$\lim_{\mathbf{x}\to\mathbf{x}_0} Z(\mathbf{x} - \mathbf{x}_0) = 0.$$

Since f and g are differentiable at $\mathbf{x}_0$ and $\mathbf{y}_0$, respectively, there are functions Z_1 and Z_2 such that

$$f(\mathbf{x}) - f(\mathbf{x}_0) = f'(\mathbf{x}_0)(\mathbf{x} - \mathbf{x}_0) + |\mathbf{x} - \mathbf{x}_0|\, Z_1(\mathbf{x} - \mathbf{x}_0),$$
$$g(\mathbf{y}) - g(\mathbf{y}_0) = g'(\mathbf{y}_0)(\mathbf{y} - \mathbf{y}_0) + |\mathbf{y} - \mathbf{y}_0|\, Z_2(\mathbf{y} - \mathbf{y}_0),$$

and

$$\lim_{\mathbf{x}\to\mathbf{x}_0} Z_1(\mathbf{x} - \mathbf{x}_0) = \lim_{\mathbf{y}\to\mathbf{y}_0} Z_2(\mathbf{y} - \mathbf{y}_0) = 0.$$

Using the f-equation to substitute into the g-equation, we get

$$\begin{aligned} g \circ f(\mathbf{x}) - g \circ f(\mathbf{x}_0) = {} & g'(\mathbf{y}_0)\big(f'(\mathbf{x}_0)(\mathbf{x} - \mathbf{x}_0) + |\mathbf{x} - \mathbf{x}_0|\, Z_1(\mathbf{x} - \mathbf{x}_0)\big) \\ & + |f(\mathbf{x}) - f(\mathbf{x}_0)|\, Z_2\big(f(\mathbf{x}) - f(\mathbf{x}_0)\big). \end{aligned}$$

From this it follows, by the linearity of the matrix multiplier $g'(\mathbf{y}_0)$, that

$$\begin{aligned} & g \circ f(\mathbf{x}) - g \circ f(\mathbf{x}_0) - g'(\mathbf{y}_0)f'(\mathbf{x}_0)(\mathbf{x} - \mathbf{x}_0) \\ & \quad = |\mathbf{x} - \mathbf{x}_0|\, g'(\mathbf{y}_0)[Z_1(\mathbf{x} - \mathbf{x}_0)] + |f(\mathbf{x}) - f(\mathbf{x}_0)|\, Z_2\big(f(\mathbf{x}) - f(\mathbf{x}_0)\big). \end{aligned} \tag{2}$$

We note that, since f is differentiable at $\mathbf{x}_0$,

$$\begin{aligned}|f(\mathbf{x}) - f(\mathbf{x}_0)| &= |f'(\mathbf{x}_0)(\mathbf{x} - \mathbf{x}_0) + |\mathbf{x} - \mathbf{x}_0|\, Z_1(\mathbf{x} - \mathbf{x}_0)| \\ &\leq k\,|\mathbf{x} - \mathbf{x}_0| + |Z_1(\mathbf{x} - \mathbf{x}_0)|\,|\mathbf{x} - \mathbf{x}_0|,\end{aligned}$$

where in the last step we have used the triangle inequality, and Theorem 7.3 of Chapter 3 provides the constant k. This inequality enables us to estimate the right side of Equation (2), showing that its norm is less than or equal to $|\mathbf{x} - \mathbf{x}_0|$ multiplied by a function that tends to zero as $\mathbf{x}$ tends to $\mathbf{x}_0$. In fact, we have the upper estimate

$$|\mathbf{x} - \mathbf{x}_0|\,\{|g'(\mathbf{y}_0)[Z_1(\mathbf{x} - \mathbf{x}_0)]| + [k + |Z_1(\mathbf{x} - \mathbf{x}_0)|]\,|Z_2(f(\mathbf{x}) - f(\mathbf{x}_0))|\}.$$

But as $\mathbf{x}$ tends to $\mathbf{x}_0$, $f(\mathbf{x})$ tends to $f(\mathbf{x}_0)$; so both $Z(\mathbf{x}_1 - \mathbf{x}_0)$ and $Z_2(f(\mathbf{x}) - f(\mathbf{x}_0))$ tend to zero. This shows that $Z(\mathbf{x} - \mathbf{x}_0)$ itself, defined by Equation (1), tends to zero, thus completing the proof.

SECTION 3

ARC LENGTH FORMULA

To appreciate fully the significance of the connection between arc length and the formula $\int_a^b |f'(t)|\, dt$ that is used to compute it, it is necessary to know that the formula doesn't always work. The reason is that there is a continuous curve γ in $\mathcal{R}^2$ which has length 2, but such that if the integral formula is applied to it the result is 1. The construction of such an example is fairly complicated, and showing that the relevant integral has value 1 is itself nontrivial. For the curve γ we can take the graph of the so-called Cantor function. (See R. P. Boas, *A Primer of Real Functions*, John Wiley & Sons, 1960, p. 131, or B. R. Gelbaum and J. M. H. Olmstead, *Counter-examples in Analysis*, Holden-Day, Inc., 1964, p. 97.) Once the Cantor function is understood, it is fairly easy to show that its graph has length 2, as defined by the least upper bound of the lengths of inscribed polygons.

For a piecewise smooth curve γ, given by a piecewise continuously differentiable function $\mathcal{R} \xrightarrow{g} \mathcal{R}^n$ for $a \leq t \leq b$, we have the following theorem, stated in Chapter 3. (The graph of the Cantor function is not, of course, a piecewise smooth curve.)

Theorem

Let γ be a piecewise smooth curve as described above. Then $l(\gamma)$, the length of γ, is finite, and

$$l(\gamma) = \int_a^b |g'(t)|\, dt.$$

Proof. We show first that $l(\gamma) \leq \int_a^b |g'(t)|\, dt$, noting that since $|g'|$ is continuous on each of finitely many closed intervals, $|g'|$ is bounded, and the integral will be finite. The inequality will be proved if we can show that

$$\sum_{k=1}^{K} |g(t_k) - g(t_{k-1})| \leq \int_a^b |g'(t)|\, dt, \tag{1}$$

when $a = t_0 < t_1 < \ldots < t_K = b$ is an *arbitrary* partition P of $[a, b]$. But, by the triangle inequality for the norm, we only increase the sum on the left if we add to the partition all end points of the finitely many intervals on which g' is continuous. So we assume this has been done. Then for each interval $[t_{k-1}, t_k]$ we have

$$\begin{aligned} |g(t_k) - g(t_{k-1})| &= \left| \int_{t_{k-1}}^{t_k} g'(t)\, dt \right| \\ &\leq \int_{t_{k-1}}^{t_k} |g'(t)|\, dt. \end{aligned}$$

The equality holds by Exercise 16 of Section 2, Chapter 3, and the inequality holds by Exercise 17 of the same section. Summing over $k = 1, \ldots, K$, gives (1).

To prove the reverse inequality, $\int_a^b |g'(t)|\, dt \leq l(\gamma)$, we shall show that for any $\eta > 0$, we can *find* a partition P of $[a, b]$ such that

$$\int_a^b |g'(t)|\, dt - \eta \leq \sum_{k=1}^{K} |g(t_k) - g(t_{k-1})|. \tag{2}$$

This will show that no number smaller than the integral is an upper bound for the sums on the right. We take as an initial partition P_0 all the finitely many endpoints of closed intervals on which g' is continuous. On each such interval, g' is uniformly continuous by Theorem 1.3 of this Appendix. This means that given $\epsilon > 0$, there is a $\delta > 0$ such that $|g'(t) - g'(u)| < \epsilon$, if $|t - u| < \delta$ and $t_{k-1} \leq t$, $u \leq t_k$. Since there are only finitely many intervals $[t_{k-1}, t_k]$, we can choose a single positive δ that will work for all of them. Now make a partition P fine enough that $\max\,(t_k - t_{k-1}) < \delta$, still including in P all the points of P_0. On each interval of the new partition we have $|g'(t)| \leq |g'(t_k)| + \epsilon$, by the uniform continuity of g'. Thus,

$$\begin{aligned} \int_{t_{k-1}}^{t} |g'(t)|\, dt &\leq [|g'(t_k)| + \epsilon](t_k - t_{k-1}) \\ &= |g'(t_k)|\,(t_k - t_{k-1}) + \epsilon(t_k - t_{k-1}). \end{aligned} \tag{3}$$

But we also have the identity

$$|g'(t_k)|(t_k - t_{k-1}) = \left| \int_{t_{k-1}}^{t_k} (g'(t) + g'(t_k) - g'(t))\, dt \right|$$
$$\leq \left| \int_{t_{k-1}}^{t_k} g'(t)\, dt \right| + \left| \int_{t_{k-1}}^{t_k} (g'(t_k) - g'(t))\, dt \right|$$
$$\leq |g(t_k) - g(t_{k-1})| + \int_{t_{k-1}}^{t_k} |g'(t_k) - g'(t)|\, dt,$$

where in the last step we have again used the results of Exercises 16 and 17 of Section 2, Chapter 3. Again using the uniform continuity of g', together with the previous inequality, we get

$$|g'(t_k)|(t_k - t_{k-1}) \leq |g(t_k) - g(t_{k-1})| + \epsilon(t_k - t_{k-1}).$$

Applying this inequality to Equation (3) gives

$$\int_{t_{k-1}}^{t_k} |g'(t)|\, dt \leq |g(t_k) - g(t_{k-1})| + 2\epsilon(t_k - t_{k-1}).$$

summing over k gives

$$\int_a^b |g'(t)|\, dt \leq \sum_{k=1}^{K} |g(t_k) - g(t_{k-1})| + 2\epsilon(b - a).$$

If ϵ is chosen so that $2\epsilon(b - a) < \eta$, then the desired Inequality (2) will be satisfied for the partition P constructed above.

SECTION 4 CONVERGENCE OF FOURIER SERIES

Theorem 5.2 of Chapter 5 asserts that the Fourier series of a piecewise smooth function f converges pointwise to the average of the right and left limits of f at each point. The assumption that f is piecewise smooth means that the interval $[-\pi, \pi]$ can be broken into finitely many subintervals, on each of which f and f' can be extended to be continuous. At the endpoints of an interval $[x_k, x_{k+1}]$ we require f to be continuous if it is given the respective values $f(x_k+) = \lim_{u\to 0+} f(x_k + u)$ and $f(x_{k+1}-) = \lim_{u\to 0-} f(x_k + u)$. Similarly, we require f' to be continuous on the closed interval if it is given the values of the right and left derivatives, respectively:

$$f^+(x_k) = \lim_{u\to 0+} \frac{f(x_k + u) - f(x_k+)}{u},$$

$$f^-(x_{k+1}) = \lim_{u\to 0+} \frac{f(x_{k+1} + u) - f(x_{k+1}-)}{u}.$$

We restate the convergence theorem as follows.

4.1 Theorem

Let f be piecewise smooth on $[-\pi, \pi]$. Then the Fourier series of f converges at each point x of the interval to $(\frac{1}{2})[f(x-)+f(x+)]$. In particular, if f is continuous at x, then the series converges to $f(x)$.

Proof. We have to show that if a_k and b_k are the Fourier coefficients of f, then

$$\lim_{N\to\infty} s_N(x) = \lim_{N\to\infty}\left[\frac{a_0}{2}+\sum_{k=1}^{N} a_k \cos kx + b_k \sin kx\right]$$

$$= \tfrac{1}{2}[f(x-)+f(x+)], \quad \text{for all } x \text{ in } [-\pi, \pi].$$

Replacing a_k and b_k by their definitions, we get

$$s_N(x) = \frac{1}{2\pi}\int_{-\pi}^{\pi} f(t)\,dt$$
$$+\frac{1}{\pi}\sum_{k=1}^{N}\left[\cos kx\int_{-\pi}^{\pi} f(t)\cos kt\,dt + \sin kx\int_{-\pi}^{\pi} f(t)\sin kt\,dt\right]$$
$$= \frac{1}{\pi}\int_{-\pi}^{\pi} f(t)\left[\tfrac{1}{2}+\sum_{k=1}^{N}\cos k(t-x)\right]dt.$$

But trigonometric identities (See Exercise 5 of Chapter 5, Section 5) show that

$$\tfrac{1}{2}+\sum_{k=1}^{N}\cos ku = \frac{\sin (N+\frac{1}{2})u}{2\sin(\frac{1}{2})u}. \tag{1}$$

Hence,

$$s_N(x) = \frac{1}{\pi}\int_{-\pi}^{\pi} f(t)\frac{\sin (N+\frac{1}{2})(t-x)}{2\sin(\frac{1}{2})(t-x)}\,dt.$$

We now extend f outside the interval $[-\pi, \pi]$ so that it has period 2π, and make the change of variable $t = x + u$. Then the new interval of integration is $[-\pi - x, \pi - x]$. But since the integrand has period 2π, the value of the integral remains unchanged if we shift back to the interval $[-\pi, \pi]$. Thus we have

$$s_N(x) = \frac{1}{\pi}\int_{-\pi}^{\pi} f(x+u)\frac{\sin (N+\frac{1}{2})u}{2\sin(\frac{1}{2})u}\,du.$$

We shall show that

$$\lim_{N\to\infty} s_N^{+}(x) = (\tfrac{1}{2})f(x+), \tag{2}$$

where

$$s_N^+(x) = \frac{1}{\pi}\int_0^\pi f(x+u)\frac{\sin(N+\frac{1}{2})u}{2\sin(\frac{1}{2})u}\,du. \tag{3}$$

A similar proof would show that $\lim_{N\to\infty} s_N^-(x) = (\frac{1}{2})f(x-)$ where $s_N^-(x) = (1/\pi)\int_{-\pi}^0 f(x+u)\left(\sin(N+\frac{1}{2})u/2\sin(\frac{1}{2})u\right)du$, and addition of the two equations will finish the proof.

To prove Equation (2), we observe from Equation (1) that

$$\frac{1}{\pi}\int_0^\pi \frac{\sin(N+\frac{1}{2})u}{2\sin(\frac{1}{2})u}\,du = \frac{1}{2}.$$

Multiplying both sides by $f(x+)$ and subtracting from Equation (3) gives

$$s_N^+(x) - \tfrac{1}{2}f(x+) = \frac{1}{\pi}\int_0^\pi \frac{\sin(N+\frac{1}{2})u}{2\sin(\frac{1}{2})u}[f(x+u)-f(x+)]\,du. \tag{4}$$

The proof will be complete if we show that this last integral tends to zero as N tends to infinity. But $g(u) = [f(x+u) - f(x+)]/\sin(\frac{1}{2})u$ is piecewise continuous, so the result is a consequence of

4.2 Riemann's Lemma

If g is continuous on the closed interval $[a, b]$, then

$$\lim_{k\to\infty}\int_a^b g(u)\sin ku\,du = \lim_{k\to\infty}\int_a^b g(u)\cos ku\,du = 0.$$

Proof. By Bessel's inequality, proved in Chapter 4, Section 7, we have

$$\int_a^b g^2(u)\,du \geq \sum_{k=1}^{\infty}(a_k^2 + b_k^2).$$

Hence, the series on the right converges. We conclude that a_k and b_k tend to zero as k tends to infinity.

SECTION 5 PROOF OF THE INVERSE AND IMPLICIT FUNCTION THEOREMS

We start with a theorem that not only guarantees the existence of an inverse function, but also proves the convergence of the modified Newton iteration given in Equation 9.2 of Chapter 3.

5.1 Theorem

Let $\mathcal{R}^n \xrightarrow{f} \mathcal{R}^n$ be continuously differentiable in some closed ball of radius r centered at $\mathbf{x}_0$. Suppose $[f'(\mathbf{x}_0)]^{-1}$ exists and that for some number K, with $0 < K < 1$, the maximum absolute value of the entries in the matrix

$$I - [f'(\mathbf{x}_0)]^{-1}f'(\mathbf{x})$$

is less than K/n whenever $|\mathbf{x}_0 - \mathbf{x}| < r$. If

$$|[f'(\mathbf{x}_0)]^{-1}f(\mathbf{x}_0)| < (1 - K)r, \tag{1}$$

then the equation

$$\mathbf{x}_{k+1} = \mathbf{x}_k - [f'(\mathbf{x}_0)]^{-1}f(\mathbf{x}_k)$$

defines a sequence $\mathbf{x}_0, \mathbf{x}_1, \mathbf{x}_2, \ldots$ converging to a vector $\bar{\mathbf{x}}$ such that $f(\bar{\mathbf{x}}) = 0$. Furthermore, $\bar{\mathbf{x}}$ is the only solution of $f(\mathbf{x}) = 0$ contained in $B_r(\mathbf{x}_0)$.

Proof. Let r be chosen as in the hypotheses. Let $g(\mathbf{x}) = \mathbf{x} - [f'(\mathbf{x}_0)]^{-1}f(\mathbf{x})$. Then for $\mathbf{x}$ and $\mathbf{x}'$ in $B_r(\mathbf{x}_0)$, we have $g(\mathbf{x}') - g(\mathbf{x}) = (\mathbf{x}' - \mathbf{x}) - [f'(\mathbf{x}_0)]^{-1}\big(f(\mathbf{x}') - f(\mathbf{x})\big)$. Applying the mean-value theorem to each coordinate function f_j of f, we get

$$f_j(\mathbf{x}') - f_j(\mathbf{x}) = \sum_{i=1}^{n} \frac{\partial f_j}{\partial x_i}(\mathbf{y}_j)(x_i' - x_i),$$

where $\mathbf{x}' = (x_1', \ldots, x_m')$, $\mathbf{x} = (x_1, \ldots, x_m)$, and $\mathbf{y}_j$ is on the segment joining $\mathbf{x}$ and $\mathbf{x}'$. Writing $F(\mathbf{x}, \mathbf{x}')$ for the matrix

$$\left(\frac{\partial f_j}{\partial x_i}\right)_{i,\,j=1,\ldots,n},$$

we get

$$f(\mathbf{x}') - f(\mathbf{x}) = F(\mathbf{x}, \mathbf{x}')(\mathbf{x}' - \mathbf{x});$$

hence

$$\begin{aligned} g(\mathbf{x}') - g(\mathbf{x}) &= (\mathbf{x}' - \mathbf{x}) - [f'(\mathbf{x}_0)]^{-1}F(\mathbf{x}, \mathbf{x}')(\mathbf{x}' - \mathbf{x}) \\ &= \big(I - [f'(\mathbf{x}_0)]^{-1}F(\mathbf{x}, \mathbf{x}')\big)(\mathbf{x}' - \mathbf{x}). \end{aligned} \tag{2}$$

By assumption, the maximum absolute entry in the matrix $A = I - [f'(\mathbf{x}_0)]^{-1}F(\mathbf{x}, \mathbf{x}')$ is less than or equal to K/n for some positive $K < 1$. But for any matrix $A = (a_{ij})_{i,j=1,\ldots,n}$ and any vector $\mathbf{y} = (y_1, \ldots, y_n)$, we have

$$\begin{aligned} |A\mathbf{y}| &= \sqrt{\sum_{i=1}^{n}\left(\sum_{j=1}^{n} a_{ij}y_j\right)^2} \\ &\le \sqrt{\sum_{i=1}^{n}\left(\sum_{j=1}^{n} a_{ij}^2 \sum_{j=1}^{n} y_j^2\right)}, \end{aligned}$$

by the Cauchy-Schwarz inequality. Hence

$$|A\mathbf{y}| \le |\mathbf{y}|\sqrt{\sum_{i=1}^{n}\sum_{j=1}^{n} a_{ij}^2} \le |\mathbf{y}|\,\sqrt{n^2 \max_{i,j} a_{ij}^2} = n\,|\mathbf{y}|\,\max_{i,j}|a_{ij}|.$$

Applying this result to the matrix $I - [f'(\mathbf{x}_0)]^{-1}F(\mathbf{x}, \mathbf{x}')$, we get from (2):

$$|g(\mathbf{x}') - g(\mathbf{x})| \leq \frac{K}{n} n\,|\mathbf{x}' - \mathbf{x}| = K\,|\mathbf{x}' - \mathbf{x}|. \tag{3}$$

Now define for $k \geq 1$, and for $\mathbf{x}_k$ in the domain of g,

$$\mathbf{x}_{k+1} = g(\mathbf{x}_k).$$

Then

$$\begin{aligned} |\mathbf{x}_{k+1} - \mathbf{x}_k| &= |g(\mathbf{x}_k) - g(\mathbf{x}_{k-1})| \\ &\leq K\,|\mathbf{x}_k - \mathbf{x}_{k-1}| \leq K^k\,|\mathbf{x}_1 - \mathbf{x}_0|. \end{aligned} \tag{4}$$

It follows by repeated application of (4) that for $k \geq m \geq 0$ we have

$$\begin{aligned} |\mathbf{x}_{k+1} - \mathbf{x}_m| &\leq |\mathbf{x}_{k+1} - \mathbf{x}_k| + |\mathbf{x}_k - \mathbf{x}_{k-1}| + \ldots + |\mathbf{x}_{m+1} - \mathbf{x}_m| \\ &\leq (K^k + K^{k-1} + \ldots + K^m)\,|\mathbf{x}_1 - \mathbf{x}_0| \\ &\leq K^m\left(\frac{1 - K^{k-m+1}}{1 - K}\right)|\mathbf{x}_1 - \mathbf{x}_0| \\ &\leq \frac{K^m}{1 - K}|\mathbf{x}_1 - \mathbf{x}_0| = \frac{K^m}{1 - K}|[f'(\mathbf{x}_0)]^{-1}f(\mathbf{x}_0)| \\ &\leq K^m r \end{aligned}$$

For $m = 0$ we get

$$|\mathbf{x}_{k+1} - \mathbf{x}_0| \leq r;$$

so, for $k = 1, 2, 3, \ldots$, $\mathbf{x}_k$ lies in $N_r(\mathbf{x}_0)$ and, hence, in the subset of the domain of f on which the hypotheses of the theorem are satisfied. Thus each x_k is defined and satisfies the Inequality (4).

Returning to (4) we see that

$$|\bar{\mathbf{x}}_{k+1} - \mathbf{x}_m| \leq K^m r$$

implies that $\mathbf{x}_0, \mathbf{x}_1, \mathbf{x}_2, \ldots$ is a Cauchy sequence which necessarily converges to some vector $\bar{\mathbf{x}}$ satisfying $|\bar{\mathbf{x}} - \mathbf{x}_0| \leq r$. Since

$$\mathbf{x}_{k+1} = g(\mathbf{x}_k)$$

and g is continuous, we have

$$\bar{\mathbf{x}} = \lim_{k\to\infty} \mathbf{x}_{k+1} = \lim_{k\to\infty} g(\mathbf{x}_k) = g(\bar{\mathbf{x}}),$$

that is,

$$\bar{\mathbf{x}} = \bar{\mathbf{x}} - [f'(\mathbf{x}_0)]^{-1}f(\bar{\mathbf{x}}).$$

Thus

$$f(\bar{\mathbf{x}}) = 0.$$

To show that the solution $\bar{\mathbf{x}}$ is unique in $B_r(\mathbf{x}_0)$, observe that the inequality (3), rewritten in terms of f, is

$$|(\mathbf{x}' - \mathbf{x}) - [f'(\mathbf{x}_0)]^{-1}(f(\mathbf{x}') - f(\mathbf{x}))| \leq K|\mathbf{x}' - \mathbf{x}|.$$

From this it follows by the triangle inequality that

$$|\mathbf{x}' - \mathbf{x}| - |[f'(\mathbf{x}_0)]^{-1}(f(\mathbf{x}') - f(\mathbf{x}))| \leq K|\mathbf{x}' - \mathbf{x}|.$$

Hence

$$(1 - K)|\mathbf{x}' - \mathbf{x}| \leq |[f'(\mathbf{x}_0)]^{-1}(f(\mathbf{x}') - f(\mathbf{x}))|. \tag{5}$$

This shows that, for $\mathbf{x}$ and $\mathbf{x}'$ in $B_r(\mathbf{x}_0)$, $f(\mathbf{x}') - f(\mathbf{x}) = 0$ implies $\mathbf{x}' = \mathbf{x} = 0$, that is, that f is one-to-one on $B_r(\mathbf{x}_0)$.

5.2 Inverse Function Theorem

Let $\mathcal{R}^n \xrightarrow{f} \mathcal{R}^n$ be a continuously differentiable function such that $f'(\mathbf{x}_0)$ has an inverse. Then there is an open set N containing $\mathbf{x}_0$ such that $f(N)$ is open and such that f, when restricted to N, has a continuously differentiable inverse f^{-1}. In addition,

$$[f^{-1}]'(\mathbf{y}_0) = [f'(\mathbf{x}_0)]^{-1},$$

where $\mathbf{y}_0 = f(\mathbf{x}_0)$.

Proof, Part 1. We apply the previous theorem to the function $f(\mathbf{x}) - \mathbf{y}$, where $\mathbf{y}$ is any vector such that the Inequality (1) is satisfied. That is, we assume that $\mathbf{y}$ is in the set S of vectors $\mathbf{z}$ satisfying

$$|[f'(\mathbf{x}_0)]^{-1}(f(\mathbf{x}_0) - \mathbf{z})| < (1 - K)r.$$

By the approximation theorem, there is a vector $\bar{\mathbf{x}}$, unique in $B'(\mathbf{x}_0)$, such that $f(\bar{\mathbf{x}}) - \mathbf{y} = 0$.

Furthermore, Inequality (1) shows that the function f^{-1} defined by $f^{-1}(\mathbf{y}) = \bar{\mathbf{x}}$ is continuous because

$$(1 - K)|f^{-1}(\mathbf{y}') - f^{-1}(\mathbf{y})| \leq |[f'(\mathbf{x}_0)]^{-1}(\mathbf{y}' - \mathbf{y})|.$$

We define the set $N = f^{-1}(S)$. Clearly, $N \subset B_r(\mathbf{x}_0)$, and it is easily seen that N and $f(N)$ are open sets because f and f^{-1} are continuous.

Proof, Part 2. For any $\mathbf{x}$ *in* N, *the inverse of* $f'(\mathbf{x})$ *satisfies the condition for being the derivative of* f^{-1} at $\mathbf{y} = f(\mathbf{x})$.

For $\mathbf{x}'$ and $\mathbf{x}$ in N, let $\mathbf{y}' = f(\mathbf{x}')$ and $\mathbf{y} = f(\mathbf{x})$. Since f is differentiable at $\mathbf{x}$,

$$f(\mathbf{x}') - f(\mathbf{x}) = f'(\mathbf{x})(\mathbf{x}' - \mathbf{x}) + |\mathbf{x}' - \mathbf{x}|\, Z(\mathbf{x}' - \mathbf{x}),$$

where $\lim_{\mathbf{x}' \to \mathbf{x}} Z(\mathbf{x}' - \mathbf{x}) = 0$. Alternatively, we can write

$$\mathbf{y}' - \mathbf{y} = f'(\mathbf{x})(f^{-1}(\mathbf{y}') - f^{-1}(\mathbf{y})) + |\mathbf{x}' - \mathbf{x}|\, Z(\mathbf{x}' - \mathbf{x}).$$

Applying $[f'(\mathbf{x})]^{-1}$ to both sides gives

$$f^{-1}(\mathbf{y}') - f^{-1}(\mathbf{y}) - [f'(\mathbf{x})]^{-1}(\mathbf{y}' - \mathbf{y}) = |\mathbf{x}' - \mathbf{x}|\, [f'(\mathbf{x})]^{-1}Z(\mathbf{x}' - \mathbf{x}).$$

Lemma 5.3 shows that $|\mathbf{x}' - \mathbf{x}|/|\mathbf{y}' - \mathbf{y}|$ remains bounded as $\mathbf{y}'$ tends to $\mathbf{y}$. Since $[f'(\mathbf{x})]^{-1}Z(\mathbf{x}' - \mathbf{x})$ tends to zero as $\mathbf{x}'$ tends to $\mathbf{x}$, the inequality $|\mathbf{y}' - \mathbf{y}| \geq M\, |\mathbf{x}' - \mathbf{x}|$ resulting from Lemma 5.3 shows that the right side of the above equation tends to zero when divided by $|\mathbf{y}' - \mathbf{y}|$ and as $\mathbf{y}'$ tends to $\mathbf{y}$. This completes the proof of part 2.

Proof, Part 3. f^{-1} is continuously differentiable. Part 2 shows that f^{-1} is differentiable on $f(N)$ and that

$$[f^{-1}]'(\mathbf{y}) = [f'(\mathbf{x})]^{-1}.$$

Theorem 8.3 of Chapter 1 shows that the inverse of a matrix A has as entries continuous functions of the entries in A. Since $f'(f^{-1}(\mathbf{y}))$ has continuous entries, so does its inverse. This completes the proof of part 3 and so of the inverse function theorem.

5.3 Lemma

There is a neighborhood N of x_0 and a number $M > 0$ such that

$$M\, |x' - x| \leq |f(x') - f(x)|$$

for all x and x' in N.

Proof. By Theorem 7.3 of Chapter 3 there is a constant k such that

$$|[f'(x_0)]^{-1}(f(x') - f(x))| \leq k\, |f(x') - f(x)|.$$

Combining this inequality with inequality (5) above gives the desired result with $M = (1 - K)/k$.

5.4 Implicit-Function Theorem

Let $\mathcal{R}^{n+m} \xrightarrow{F} \mathcal{R}^m$ be a continuously differentiable function. Suppose that for some $\mathbf{x}_0$ in $\mathcal{R}^n$ and $\mathbf{y}_0$ in $\mathcal{R}^m$

1. $F(\mathbf{x}_0, \mathbf{y}_0) = 0$.
2. $F_{\mathbf{y}}(\mathbf{x}_0, \mathbf{y}_0)$ has an inverse.

Then there exists a continuously differentiable function $\mathcal{R}^n \xrightarrow{f} \mathcal{R}^m$ defined on some neighborhood N of $\mathbf{x}_0$ such that $f(\mathbf{x}_0) = \mathbf{y}_0$ and $F(\mathbf{x}, f(\mathbf{x})) = 0$, for all $\mathbf{x}$ in N.

Proof. The proof consists in reducing the theorem to an application of the inverse function theorem. For that purpose we extend F to a function $\mathcal{R}^{n+m} \xrightarrow{H} \mathcal{R}^{n+m}$ by setting $H(\mathbf{x}, \mathbf{y}) = (\mathbf{x}, F(\mathbf{x}, \mathbf{y}))$. In terms of the coordinate functions $F_1, \ldots, F_m$ of F, the coordinate functions of H are given by

$$
\begin{aligned}
H_1(\mathbf{x}, \mathbf{y}) &= x_1 \\
H_2(\mathbf{x}, \mathbf{y}) &= x_2 \\
&\vdots \\
H_n(\mathbf{x}, \mathbf{y}) &= x_n \\
H_{n+1}(\mathbf{x}, \mathbf{y}) &= F_1(x_1, \ldots, x_n, y_1, \ldots, y_m) \\
&\vdots \\
H_{n+m}(\mathbf{x}, \mathbf{y}) &= F_m(x_1, \ldots, x_n, y_1, \ldots, y_m).
\end{aligned}
$$

The Jacobian matrix of H at $(\mathbf{x}_0, \mathbf{y}_0)$ is

$$
\begin{bmatrix}
1 & 0 & \ldots & 0 & 0 & \ldots & 0 \\
0 & 1 & \ldots & 0 & 0 & \ldots & 0 \\
\vdots & & & & & & \vdots \\
0 & 0 & & 1 & 0 & & 0 \\
\frac{\partial F_1}{\partial x_1} & \frac{\partial F_1}{\partial x_2} & \ldots & \frac{\partial F_1}{\partial x_n} & \frac{\partial F_1}{\partial y_1} & \ldots & \frac{\partial F_1}{\partial y_m} \\
\vdots & & & & & & \\
\frac{\partial F_m}{\partial x_1} & \frac{\partial F_m}{\partial x_2} & \ldots & \frac{\partial F_m}{\partial x_n} & \frac{\partial F_m}{\partial y_1} & \ldots & \frac{\partial F_m}{\partial y_m}
\end{bmatrix},
$$

where all the partial derivatives are evaluated at $(\mathbf{x}_0, \mathbf{y}_0)$. By assumption 2, the m columns on the right of the matrix are independent. Since they are also independent of the n independent columns on the left, all columns of the matrix are independent, and therefore the differential of H at $(\mathbf{x}_0, \mathbf{y}_0)$ has an inverse.

The function H is certainly continuously differentiable; so we can apply the inverse function theorem at the point $(\mathbf{x}_0, \mathbf{y}_0)$ to get a function H^{-1} that is inverse to H from some open set N' in $\mathcal{R}^{n+m}$ containing $H(\mathbf{x}_0, \mathbf{y}_0)$ to an open set about $(\mathbf{x}_0, \mathbf{y}_0)$. Since $H(\mathbf{x}_0, \mathbf{y}_0) = (\mathbf{x}_0, F(\mathbf{x}_0, \mathbf{y}_0)) = (\mathbf{x}_0, 0)$, the set N of all points $\mathbf{x}$ in $\mathcal{R}^n$ such that $(\mathbf{x}, 0)$ is in N' is an open set and contains $\mathbf{x}_0$.

Let G_1 be the function that selects the first n variables of a point in $\mathcal{R}^{n+m}$, and G_2 the function that selects the last m variables. Thus, $G_1(\mathbf{x}, \mathbf{y}) = \mathbf{x}$ and $G_2(\mathbf{x}, \mathbf{y}) = \mathbf{y}$. Since $H(\mathbf{x}, \mathbf{y}) = (\mathbf{x}, F(\mathbf{x}, \mathbf{y}))$, the function H is the identity on $\mathbf{x}$. The same must therefore be true of H^{-1}. Hence,

$$G_1 = G_1 \circ H^{-1}.$$

We define f by

$$f(\mathbf{x}) = G_2 H^{-1}(\mathbf{x}, 0), \qquad \text{for every } \mathbf{x} \text{ in } N.$$

Then

$$H^{-1}(\mathbf{x}, 0) = (G_1 H^{-1}(\mathbf{x}, 0), G_2 H^{-1}(\mathbf{x}, 0)) = (\mathbf{x}, f(\mathbf{x})).$$

Applying H to both sides, we get

$$(\mathbf{x}, 0) = H(\mathbf{x}, f(\mathbf{x})) = (\mathbf{x}, f(\mathbf{x})),$$

for every $\mathbf{x}$ in N. The two parts of the first and last pairs must be equal. Hence,

$$0 = F(\mathbf{x}, f(\mathbf{x})), \qquad \text{for every } \mathbf{x} \text{ in } N.$$

Finally, f, being the composition of two continuously differentiable functions, is itself continuously differentiable by the chain rule.

SECTION 6 — PROOF OF LAGRANGE'S THEOREM

The proof makes use of the implicit function theorem, and of the fact that, for a linear function L defined in $\mathcal{R}^n$, the dimension of the domain is equal to the dimension of the range plus the dimension of the null space. We begin by restating the theorem.

Lagrange's Theorem

Let the function $\mathcal{R}^n \xrightarrow{G} \mathcal{R}^m$, $n > m$, be continuously differentiable and have coordinate functions $G_1, G_2, \ldots, G_m$. Suppose the equations

$$\begin{aligned} G_1(x_1, \ldots, x_n) &= 0 \\ G_2(x_1, \ldots, x_n) &= 0 \\ &\vdots \\ G_m(x_1, \ldots, x_n) &= 0 \end{aligned}$$

implicitly define a surface S in $\mathcal{R}^n$, and that at a point $\mathbf{x}_0$ of S the matrix $G'(\mathbf{x}_0)$ has some m columns linearly independent.

If $\mathbf{x}_0$ is an extreme point of a differentiable function $\mathcal{R}^n \xrightarrow{f} \mathcal{R}$, when restricted to S, then $\mathbf{x}_0$ is a critical point of the function

$$f + \lambda_1 G_1 + \ldots + \lambda_m G_m$$

for some constants $\lambda_1 \ldots, \lambda_m$.

Proof. The implicit function theorem ensures that there is a parametric representation for S in a neighborhood of $\mathbf{x}_0$. For instance, suppose that for some choice of m variables, say $x_1, \ldots, x_m$, the columns of the matrix

$$\begin{bmatrix} \frac{\partial G_1}{\partial x_1} & \frac{\partial G_1}{\partial x_2} & \cdots & \frac{\partial G_1}{\partial x_m} \\ \cdot & & & \cdot \\ \cdot & & & \cdot \\ \cdot & & & \cdot \\ \frac{\partial G_m}{\partial x_1} & \frac{\partial G_m}{\partial x_2} & \cdots & \frac{\partial G_m}{\partial x_m} \end{bmatrix}_{\mathbf{x}_0} \tag{1}$$

are independent. Then the matrix has an inverse. Write $\mathbf{x}_0 = (a_1, \ldots, a_n)$, and set $\mathbf{u}_0 = (a_1, \ldots, a_m)$ and $\mathbf{v}_0 = (a_{m+1}, \ldots, a_n)$ By the implicit function theorem, there is a differentiable function $\mathcal{R}^{n-m} \xrightarrow{h} \mathcal{R}^m$ defined on a neighborhood N of $\mathbf{v}_0$ such that $h(\mathbf{v}_0) = \mathbf{u}_0$ and $G(h(\mathbf{v}), \mathbf{v}) = 0$ for all $\mathbf{v}$ in N. The function $\mathcal{R}^{n-m} \xrightarrow{H} \mathcal{R}^n$ defined by

$$H(\mathbf{v}) = (h(\mathbf{v}), \mathbf{v}) \qquad \text{for all } \mathbf{v} \text{ in } N,$$

is a parametric representation of a part of S containing $\mathbf{x}_0 = H(\mathbf{v}_0)$. The surface S has a tangent $\mathcal{T}$ of dimension $n - m$ at $\mathbf{x}_0$. The reason is that, first of all, the derivative of H at $\mathbf{v}_0$ is the $(n - m)$-by-$(n - m)$ matrix

$$\begin{bmatrix} \frac{\partial h_1}{\partial x_{m+1}} & \frac{\partial h_1}{\partial x_{m+2}} & \cdots & \frac{\partial h_1}{\partial x_n} \\ \cdot & & & \\ \cdot & & & \\ \cdot & & & \\ \frac{\partial h_m}{\partial x_{m+1}} & & \cdots & \frac{\partial h_m}{\partial x_n} \\ 1 & 0 & \cdots & 0 \\ 0 & 1 & \cdots & 0 \\ \cdots & & & \\ 0 & 0 & \cdots & 1 \end{bmatrix}_{\mathbf{v}_0},$$

where $h_1, \ldots, h_m$ are the coordinate functions of H. In addition, the columns of this matrix are independent because the columns of 0's and 1's in it are independent.

Now compose H with f. Since $\mathbf{x}_0$ is an extreme point of f in S, the point $\mathbf{v}$ is an extreme point of $f \circ H$. Hence,

$$(f \circ H)'(\mathbf{v}_0) = f'(\mathbf{x}_0)H'(\mathbf{v}_0) = 0. \tag{2}$$

Because G is constantly zero on S,

$$(G \circ H)'(\mathbf{v}_0) = G'(\mathbf{x}_0)H'(\mathbf{v}_0) = 0. \tag{3}$$

Looking at (2) and (3) together, we see that $d_{\mathbf{x}_0} f$ and $d_{\mathbf{x}_0} G$ are both zero on the range of $d_{\mathbf{v}_0} H$, which set is the tangent $\mathcal{T}$. Thus the matrix

$$\begin{bmatrix} \dfrac{\partial f}{\partial x_1} & \cdots & \dfrac{\partial f}{\partial x_n} \\ \dfrac{\partial G_1}{\partial x_1} & \cdots & \dfrac{\partial G_1}{\partial x_n} \\ \vdots & & \vdots \\ \dfrac{\partial G_m}{\partial x_1} & \cdots & \dfrac{\partial G_m}{\partial x_n} \end{bmatrix}_{\mathbf{x}_0}$$

defines a linear function $\mathcal{R}^n \xrightarrow{L} \mathcal{R}^{m+1}$ that is identically zero on $\mathcal{T}$. Since the dimension of $\mathcal{T}$ is $n - m$, we have

$$n - m \leq \text{dimension of null space of } L.$$

It is always true for a linear function L that

$$n = \text{dimension of null-space of } L + \text{dimension of range of } L;$$

so

$$n \geq n - m + \text{dimension of range of } L,$$

that is,

$$m \geq \text{dimension of range of } L.$$

Recall that the range of L is a subspace of $\mathcal{R}^{m+1}$. Then we can define a linear function $\mathcal{R}^{m+1} \xrightarrow{\Lambda} \mathcal{R}$ such that Λ is zero on the range of L, but not identically zero on $\mathcal{R}^{m+1}$. In other words, there is a nonzero Λ such that $\Lambda \circ L = 0$. In matrix form $\Lambda = (\lambda_0, \lambda_1, \ldots, \lambda_m)$, and so

$$(\lambda_0, \lambda_1, \ldots, \lambda_m)\begin{bmatrix} \dfrac{\partial f}{\partial x_1} & \cdots & \dfrac{\partial f}{\partial x_n} \\ \dfrac{\partial G_1}{\partial x_1} & & \dfrac{\partial G_1}{\partial x_n} \\ \vdots & & \vdots \\ \dfrac{\partial G_m}{\partial x_1} & \cdots & \dfrac{\partial G_m}{\partial x_n} \end{bmatrix}_{\mathbf{x}_0} = 0. \tag{4}$$

It cannot happen that $\lambda_0 = 0$, for then the rows of (1) are dependent, contradicting the fact that (1) has an inverse. Taking $\lambda_0 = 1$ (if $\lambda_0 \neq 1$, divide through by λ_0), the condition (4) becomes

$$\frac{\partial f}{\partial x_j}(\mathbf{x}_0) + \lambda_1 \frac{\partial G_1}{\partial x_j}(\mathbf{x}_0) + \ldots + \lambda_m \frac{\partial G_m}{\partial x_j}(\mathbf{x}_0) = 0,$$

for $j = 1, \ldots, m$. In other words, $(f + \lambda_1 G_1 + \ldots + \lambda_m G_m)'(\mathbf{x}_0) = 0$. This completes the proof.

SECTION 7

PROOF OF TAYLOR'S THEOREM

The method of proof consists of reducing the problem to the one-variable case and then making an estimate of the size of the integral formula for the remainder.

7.1 Taylor's Theorem

Let $\mathcal{R}^n \xrightarrow{f} \mathcal{R}$ have all derivatives of order N continuous in a neighborhood of $\mathbf{x}_0$. Let $T_N(\mathbf{x} - \mathbf{x}_0)$ be the Nth degree Taylor expansion of f about $\mathbf{x}_0$. That is,

$$T_N(\mathbf{x} - \mathbf{x}_0) = f(\mathbf{x}_0) + d_{\mathbf{x}_0} f(\mathbf{x} - \mathbf{x}_0) + \ldots + \frac{1}{N!} d^N_{\mathbf{x}_0} f(\mathbf{x} - \mathbf{x}_0).$$

Then

$$\lim_{\mathbf{x} \to \mathbf{x}_0} \frac{(f(\mathbf{x}) - T_N(\mathbf{x} - \mathbf{x}_0))}{|\mathbf{x} - \mathbf{x}_0|^N} = 0, \tag{1}$$

and T^N is the only Nth-degree polynomial having this property.

Proof. Let $\mathbf{y} = \mathbf{x} - \mathbf{x}_0$ and define

$$F(t) = f\big(\mathbf{x}_0 + t(\mathbf{x} - \mathbf{x}_0)\big) = f(\mathbf{x}_0 + t\mathbf{y}).$$

Then for $k = 0, 1, \ldots, N$, we can apply the chain rule to get

$$F^{(k)}(t) = d^k_{\mathbf{x}_0+t\mathbf{y}} f(\mathbf{y}). \tag{2}$$

To see this, notice that for $k = 0$ the formula is true by definition. Assuming it to hold for some $k < N$, we have

$$\begin{aligned}
F^{(k+1)}(t) &= \frac{d}{dt}\, d^k_{\mathbf{x}_0+t\mathbf{y}} f(\mathbf{y}) \\
&= \frac{d}{dt}\left(y_1 \frac{\partial}{\partial x_1} + \ldots + y_n \frac{\partial}{\partial x_n}\right)^k_{\mathbf{x}_0+t\mathbf{y}} f \\
&= d_{\mathbf{x}_0+t\mathbf{y}}\left[\left(y_1 \frac{\partial}{\partial x_1} + \ldots + y_n \frac{\partial}{\partial x_n}\right)^k f\right](\mathbf{y}) \\
&= \left(y_1 \frac{\partial}{\partial x_1} + \ldots + y_n \frac{\partial}{\partial x_n}\right)^{k+1}_{\mathbf{x}_0+t\mathbf{y}} f \\
&= d^{k+1}_{\mathbf{x}_0+t\mathbf{y}} f(\mathbf{y}).
\end{aligned}$$

This completes the proof of Equation 2 by induction. In particular,

$$F^{(k)}(0) = d^k_{\mathbf{x}_0} f(\mathbf{y}).$$

From Theorem 3.1 of Chapter 5 obtain

$$\begin{aligned}
F(1) - F(0) - \frac{1}{1!} F'(0) - \ldots - \frac{1}{N!} F^{(N)}(0) \\
= \frac{1}{(N-1)!} \int_0^1 (1-t)^{N-1} [F^{(N)}(t) - F^{(N)}(0)]\, dt.
\end{aligned}$$

In terms of f, this is

$$\begin{aligned}
f(\mathbf{x}) - T_N(\mathbf{x} - \mathbf{x}_0) &= f(\mathbf{x}) - f(\mathbf{x}_0) - \frac{1}{1!} d_{\mathbf{x}_0} f(\mathbf{y}) - \ldots - \frac{1}{N!} d^N_{\mathbf{x}_0} f(\mathbf{y}) \\
&= \frac{1}{(N-1)!} \int_0^1 (1-t)^{N-1} [d^N_{\mathbf{x}_0+t\mathbf{y}} f(\mathbf{y}) - d^N_{\mathbf{x}_0} f(\mathbf{y})]\, dt.
\end{aligned}$$

We now estimate this difference.

$$\begin{aligned}
|f(\mathbf{x}) - T_N(\mathbf{y})| &\le \frac{1}{(N-1)!} \max_{0 \le t \le 1} |d^N_{\mathbf{x}_0+t\mathbf{y}} f(\mathbf{y}) - d^N_{\mathbf{x}_0} f(\mathbf{y})| \\
&\le \max_{0 \le t \le 1} \left| \left(y_1 \frac{\partial}{\partial x_1} + \ldots + y_n \frac{\partial}{\partial x_n}\right)^N_{\mathbf{x}_0+t\mathbf{y}} f \right. \\
&\qquad \left. - \left(y_1 \frac{\partial}{\partial x_1} + \ldots + y_n \frac{\partial}{\partial x_n}\right)^N_{\mathbf{x}_0} f \right| \\
&= \max_{0 \le t \le 1} \left| \sum_{k_1+\ldots+k_n=N} \binom{N}{k_1 \ldots k_n} y_1^{k_1} \ldots y_n^{k_n} \right. \\
&\qquad \left. \cdot \left(\frac{\partial^N f}{\partial x_1^{k_1} \ldots \partial x_n^{k_n}}(\mathbf{x}_0 + t\mathbf{y}) - \frac{\partial_N f}{\partial x_1^{k1} \ldots \partial x_n^{k_n}}(\mathbf{x}_0) \right) \right|.
\end{aligned}$$

Then since $|y_i| \leq |\mathbf{y}|$, we have

$$\frac{|y_1^{k_1} \ldots y_n^{k_n}|}{|\mathbf{y}|^N} \leq 1,$$

and so

$$\frac{|f(\mathbf{x}) - T_N(\mathbf{y})|}{|\mathbf{y}|^N} \leq \tag{3}$$

$$\sum_{k_1+\ldots+k_n=N} \binom{N}{k_1 \ldots k_n} \cdot \max_{0 \leq t \leq 1} \left| \frac{\partial^N f}{\partial x_1^{k_1} \ldots \partial x_n^{k_n}} (\mathbf{x}_0 + t\mathbf{y}) - \frac{\partial^N f}{\partial x_1^{k_1} \ldots \partial x_n^{k_n}} (\mathbf{x}_0) \right| .$$

By assumption, the derivatives of f through order N are continuous functions at $\mathbf{x}_0$. Then as $\mathbf{y}$ tends to zero, each term in the last sum tends to zero, which proves Equation (1). The inequality (3) shows that if f is a polynomial of degree N, then it equals its Nth-degree Taylor expansion. For then all terms on the right are zero.

The proof that T_N is the only Nth-degree polynomial satisfying Equation (1) goes as follows. Let T_N and T'_N be two such polynomials. By (1),

$$\lim_{\mathbf{y} \to 0} \frac{T_N(\mathbf{y}) - T'_N(\mathbf{y})}{|\mathbf{y}|^N} = 0.$$

Suppose that $T_N - T'_N$ were not identically zero, and let

$$P_k(\mathbf{y}) + R(\mathbf{y}) = T_N(\mathbf{y}) - T'_N(\mathbf{y}),$$

where P_k is the polynomial consisting of the terms of lowest degree (say k) that actually occur in $T_N - T'_N$. Then, there is a vector $\mathbf{y}_0$ such that $P_k(\mathbf{y}_0) \neq 0$. On the other hand, since $k \leq N$,

$$\begin{aligned} 0 &= \lim_{t \to 0} \frac{T_N(t\mathbf{y}_0) - T'_N(t\mathbf{y}_0)}{|t\mathbf{y}_0|^k} \\ &= \lim_{t \to 0} \frac{P_k(t\mathbf{y}_0) + R(t\mathbf{y}_0)}{|t\mathbf{y}_0|^k} \\ &= \frac{P_k(\mathbf{y}_0)}{|\mathbf{y}_0|^k} + \lim_{t \to 0} \frac{R(t\mathbf{y}_0)}{|t|^k \, |\mathbf{y}_0|^k} . \end{aligned}$$

However, because all the terms of R have degree greater than k, the last limit is zero. But then $P_k(\mathbf{y}_0) = 0$, which is a contradiction.

SECTION 8

EXISTENCE OF THE RIEMANN INTEGRAL

Theorem 2.1 of Chapter 6 is as follows.

Theorem

> Let f be defined and bounded on a bounded set B in $\mathcal{R}^n$, and let the boundary of B be contained in finitely many smooth sets. If f is continuous on B, except perhaps on finitely many smooth sets, then f is integrable over B. The value of $\int_B f\,dV$ is unchanged by changing the values of f on any smooth set.

We recall that a **smooth set** in $\mathcal{R}^n$ is the image of a closed bounded set under a continuously differentiable function $\mathcal{R}^m \xrightarrow{g} \mathcal{R}^n$, with $m < n$. We first show why smooth sets are negligible in the domain of integration of f.

8.1 Theorem

> Let $\mathcal{R}^m \xrightarrow{g} \mathcal{R}^n$ be continuously differentiable. Then for every closed, bounded subset K in the domain of g, there is a constant M such that
>
> $$|g(\mathbf{y}) - g(\mathbf{x})| \leq M\,|\mathbf{x} - \mathbf{y}|$$
>
> for all $\mathbf{x}$ and $\mathbf{y}$ in K.

Proof. Denote by $K \times K$ the subset of $\mathcal{R}^{2m}$ consisting of all $2m$-tuples $(\mathbf{x}, \mathbf{y})$ such that $\mathbf{x}$ and $\mathbf{y}$ are each in K. It is easy to see that $K \times K$ is closed and bounded in $\mathcal{R}^{2m}$. Now consider the equation

$$g(\mathbf{y}) - g(\mathbf{x}) - g'(\mathbf{x})(\mathbf{x} - \mathbf{y}) = |\mathbf{x} - \mathbf{y}|\, Z(\mathbf{y} - \mathbf{x}).$$

This equation defines Z except at 0; so we define $Z(0) = 0$. We first show that $Z(\mathbf{x} - \mathbf{y})$ thus defined is a continuous function on $K \times K$.

To solve for $Z(\mathbf{y} - \mathbf{x})$, we divide both sides of the defining equation by $|\mathbf{x} - \mathbf{y}|$. It is then apparent, because both g and g' are continuous, that $Z(\mathbf{x} - \mathbf{y})$ is continuous except perhaps when $\mathbf{x} = \mathbf{y}$. If both $\mathbf{x}$ and $\mathbf{y}$ tend to some point $\mathbf{z}$ in the domain of g, then $(\mathbf{x}, \mathbf{y})$ tends to $(\mathbf{z}, \mathbf{z})$, and we want to show that $Z(\mathbf{x} - \mathbf{y})$ tends to zero. We apply the mean-value theorem, Theorem 8.2 of Chapter 2, to the coordinate functions g_k of g, for $k = 1, \ldots, n$. We have for each k,

$$g_k(\mathbf{y}) - g_k(\mathbf{x}) = g_k'(\mathbf{x}_k)(\mathbf{x} - \mathbf{y}),$$

for $\mathbf{x}$ and $\mathbf{y}$ is some sufficiently small neighborhood of $\mathbf{z}$ and for some $\mathbf{x}_k$ on the segment joining $\mathbf{x}$ and $\mathbf{y}$. Then

$$\begin{aligned}
||\mathbf{x} - \mathbf{y}|\, Z(\mathbf{x} - \mathbf{y})| &= |g(\mathbf{y}) - g(\mathbf{x}) - g'(\mathbf{x})(\mathbf{y} - \mathbf{x})| \\
&\leq \max_{1 \leq k \leq n} |g_k(\mathbf{y}) - g_k(\mathbf{x}) - g_k'(\mathbf{x})(\mathbf{y} - \mathbf{x})| \\
&\leq \max_{1 \leq k \leq n} |g_k'(\mathbf{x}_k)(\mathbf{y} - \mathbf{x}) - g_k'(\mathbf{x})(\mathbf{y} - \mathbf{x})| \\
&= \max_{1 \leq k \leq n} \left| \sum_{j=1}^{m} \left(\frac{\partial g_k}{\partial x_j}(\mathbf{x}_j) - \frac{\partial g_k}{\partial x_j}(\mathbf{x}) \right)(y_j - x_j) \right| \\
&\leq \max_{1 \leq k \leq n} \left| \sum_{j=1}^{m} \frac{\partial g_k}{\partial x_j}(\mathbf{x}_k) - \frac{\partial g_k}{\partial x_j}(\mathbf{x}) \right| |\mathbf{y} - \mathbf{x}|.
\end{aligned}$$

Hence

$$|Z(\mathbf{y} - \mathbf{x})| \leq \max_{1 \leq k \leq n} \left| \sum_{j=1}^{m} \frac{\partial g_k}{\partial x_j}(\mathbf{x}_k) - \frac{\partial g_k}{\partial x_j}(\mathbf{x}) \right|.$$

Since the partial derivatives are assumed continuous, and since each $\mathbf{x}_k$ tends to $\mathbf{z}$ as $\mathbf{x}$ and $\mathbf{y}$ do, it follows that $Z(\mathbf{y} - \mathbf{x})$ tends to zero as $\mathbf{x}$ and $\mathbf{y}$ tend to $\mathbf{z}$. This shows that $Z(\mathbf{x} - \mathbf{y})$ is continuous on $K \times K$.

Since $|Z(\mathbf{x} - \mathbf{y})|$ is continuous, it attains its maximum value M'; so $|Z(\mathbf{y} - \mathbf{x})| \leq M'$ for all $\mathbf{x}$ and $\mathbf{y}$ in K. Hence

$$|g(\mathbf{y}) - g(\mathbf{x}) - g'(\mathbf{x})(\mathbf{y} - \mathbf{x})| \leq M' |\mathbf{y} - \mathbf{x}|.$$

The inequality $|A| - |B| \leq |A - B|$ shows that

$$|g(\mathbf{y}) - g(\mathbf{x})| \leq M' |\mathbf{y} - \mathbf{x}| + |g'(\mathbf{x})(\mathbf{y} - \mathbf{x})|. \tag{1}$$

But

$$|g'(\mathbf{x})(\mathbf{y} - \mathbf{x})| \leq \max_{1 \leq k \leq n} \sum_{j=1}^{m} \frac{\partial g_k}{\partial x_j}(\mathbf{x})\, |\mathbf{y} - \mathbf{x}|,$$

and the continuity of the partial derivatives on K implies the existence of a constant M'' such that

$$|g'(\mathbf{x})(\mathbf{y} - \mathbf{x})| \leq M'' |\mathbf{y} - \mathbf{x}|.$$

This inequality, together with (1), implies that

$$|g(\mathbf{y}) - g(\mathbf{x})| \leq (M' + M'')\, |\mathbf{y} - \mathbf{x}|,$$

for all $\mathbf{x}$ and $\mathbf{y}$ in K, as was to be shown.

8.2 Theorem

If S is a smooth set in $\mathcal{R}^n$, then S can be covered by finitely many coordinate rectangles of arbitrarily small total content. The covering can be done in such a way that no point of S lies on the boundary of the union of the set of covering rectangles.

Proof. The case in which S is just a point is trivially true, so we assume $m \geq 1$. The smooth subset S is the image under a continuously differentiable function $\mathcal{R}^m \xrightarrow{g} \mathcal{R}^n$, $m < n$, of a closed bounded set K in $\mathcal{R}^m$. We enclose K in a cube of side length s, and subdivide the cube into smaller cubes of side length s/N, where N is an integer bigger than 1. There are N^m of these little cubes. On each of the little cubes that contain any points of K we have by Theorem 8.1

$$\|g(\mathbf{x}) - g(\mathbf{y})\| \leq M\,\|\mathbf{x} - \mathbf{y}\| \leq M\frac{s}{N},$$

where M is a constant depending only on K and g. This means that the image under g of the part of K in each little cube is contained in a cube of side length $M(s/N)$. Then the surface S is contained in N^m cubes each of volume $(Ms/N)^n$. The total volume of the cubes containing S is at most $N^m(Ms/N)^n = (Ms)^n/N^{n-m}$. Since $n > m$, the total volume can be made arbitrarily small by making N large. By enlarging the side length of each covering rectangle to $(Ms + 1)/N$, the last condition of the theorem can be met.

As a corollary, we get the fact that *a smooth set has zero content.*

Now we can prove the existence theorem for integrals stated at the beginning of the section. Suppose that f and B are as described in the hypotheses. We must produce a number which we shall prove is the Riemann integral of f over B. Let f_B be the function f extended to be zero outside B. For an arbitrary grid G covering B, let R_k be the kth bounded rectangle of G, and let $\underline{f}_k$ be the *infimum* of f_B on R_k. Define

$$\underline{S}(G) = \sum_{k=1}^{N} \underline{f}_k V(R_k).$$

Similarly, define

$$\bar{S}(G) = \sum_{k=1}^{N} \bar{f}_k V(R_k),$$

where $\bar{f}_k$ is the *supremum* of f_B on R_k. Then clearly

$$\underline{S}(G) \leq \sum_{k=1}^{N}(f_B(\mathbf{x}_k))V(R_k) \leq \bar{S}(G), \tag{2}$$

if the Riemann sum is an arbitrary one formed from the grid G. Furthermore, if G' is a grid consisting of a subdivision of the rectangles of a grid G, we have

$$\underline{S}(G) \le \underline{S}(G') \le \bar{S}(G') \le \bar{S}(G).$$

In particular, if G and G'' are two grids, and G' contains all the rectangles of both of them, then

$$\underline{S}(G) \le \underline{S}(G') \le \bar{S}(G') \le \bar{S}(G''). \tag{3}$$

We define

$$I_B f = \textit{supremum} \text{ of } \underline{S}(G),$$

where the *supremum* is taken over all grids G covering B. We have from relation (3)

$$\underline{S}(G) \le I_B f \le \bar{S}(G)$$

or

$$-\bar{S}(G) \le -I_B f \le -\underline{S}(G).$$

This inequality added to (2) gives

$$\left| \sum_{k=1}^{N} (f_B(\mathbf{x}_k)) V(R_k) - \bar{S}_B f \right| \le \bar{S}(G) - \underline{S}(G),$$

in which the Riemann sum has been formed from the grid G.

Now all we have to do is show that $\bar{S}(G) - \underline{S}(G)$ can be made arbitrarily small if the mesh of G is made small enough. Then according to the definition of the integral, we will have shown that the integral of f over B exists and is $I_B f$. Let ϵ be a positive number. By Theorem 6.2, we can cover the boundary of B, the smooth surfaces containing the discontinuity points of f, and any other smooth surface on which we would like to disregard the values of f, with finitely many open rectangles $R'_1, \ldots, R'_l$, of total content less than ϵ. On the part of B not covered by these rectangles, f is continuous; so by Theorem 1.3 there is a $\delta > 0$ such that $\bar{f}_k - \underline{f}_k < \epsilon$ over any rectangle R_k belonging to a grid with mesh less than δ. By making the mesh still smaller, say, less than δ', we can arrive at a mesh size such that the rectangles $R'_1, \ldots, R'_l$, are always contained in finitely many rectangles $R''_1, \ldots, R''_m$ of any grid with mesh less than δ' and such that the total content of the latter rectangles is less than 2ϵ. Suppose that the remaining rectangles of such a grid G are $R_1, \ldots, R_n$, that $|f| < M$ on B, and that B is contained in a rectangle of volume C. Then

$$\begin{aligned} \bar{S}(G) - \underline{S}(G) &= \sum_{k=1}^{m} (\bar{f}''_k - \underline{f}''_k) V(R''_k) + \sum_{k=1}^{n} (\bar{f}_k - \underline{f}_k) V(R_k) \\ &< (2M)(2\epsilon) + \epsilon C = \epsilon(4M + C). \end{aligned}$$

Thus we have made

$$\left| \sum_{k=1}^{N} f_B(\mathbf{x}_k)V(R_k) - I_B f \right| < \epsilon(4M + C)$$

for any grid of small enough mesh. Since ϵ can be made arbitrarily small, the proof is complete.

SECTION 9

THE CHANGE-OF-VARIABLE FORMULA FOR INTEGRALS

This section contains a proof of the change-of-variable theorem (Theorem 3.1) of Chapter 6.

Theorem

Let $\mathcal{R}^n \xrightarrow{T} \mathcal{R}^n$ be a continuously differentiable transformation. Let R be a set in $\mathcal{R}^n$ having a boundary consisting of finitely many smooth sets. Suppose that R and its boundary are contained in the interior of the domain of T and that

1. T is one-to-one on R.

2. J, the Jacobian determinant of T, is different from zero on R, except perhaps on finitely many smooth sets.

Then, if the function f is bounded and continuous on $T(R)$ (the image of R under T),

$$\int_{T(R)} f\, dV = \int_R (f \circ T)\, |J|\, dV.$$

If f should be discontinuous on a smooth set S contained in R, then the theorem can be applied to R with S deleted. The subsequent inclusion of S and $T(S)$ in the domains of integration will affect neither integral since these sets have zero content.

Proof.† We first consider the special case in which f is the constant function 1, and T is linear, then, by Theorem 3.9 of Chapter 2, T can, except for trivial interchanges of variables, be written as the product of elementary linear transformations of two types: numerical multiplication of a coordinate,

$$M(x_1, \ldots, x_k, \ldots, x_n) = (x_1, \ldots, ax_k, \ldots, x_n); \qquad (1)$$

† The proof we give is contained in one by J. Schwartz, "The formula for change of variable in a multiple integral," *American Math. Monthly*, vol. 61, no. 2 (February, 1954). See also D. E. Varberg, "Change of variables in multiple integrals," *American Math. Monthly*, vol. 78, no. 1 (January, 1971).

addition of a multiple of one coordinate to another,

$$A(x_1, \ldots, x_k, \ldots, x_n) = (x_1, \ldots, x_k + rx_j, \ldots, x_n). \qquad (2)$$

By looking at the matrices of these transformations, it is easy to see that det $M = a$ and det $A = 1$. Once the special case of the theorem has been verified for each of these two types, it follows for arbitrary nonsingular linear transformations by successive application of the product rule for determinants. Let R_k be the projection of R on the subspace perpendicular to the kth coordinate axis. For each point $(x_1, \ldots, x_{k-1}, x_{k+1}, \ldots, x_n)$ in R_k, let I_k be the set of all x_k such that $(x_1, \ldots, x_n)$ is in R. For the linear transformation (1), we have by iterated integration

$$\int_R |J|\, dV = \int_{R_k} dV_{n-1} \int_{I_k} |a|\, dx_k.$$

If we denote by $|a|\, I_k$ the set of all numbers of the form $|a|\, x_k$, where x_k is in I_k, we obtain by 1-dimensional change of variable

$$\int_{R_k} dV_{n-1} \int_{I_k} |a|\, dx_k = \int_{R_k} dV_{n-1} \int_{|a|I_k} du_k$$

$$= \int_{M(R)} dV.$$

For the linear transformation (2), we denote by $I_k + rx_j$ the set of all numbers $x_k + rx_j$, where x_k is in I_k. Then iterated integration and 1-dimensional change of variable yield

$$\int_R |J|\, dV = \int_R dV = \int_{R_k} dV_{n-1} \int_{I_k} dx_k$$

$$= \int_{R_k} dV_{n-1} \int_{I_j + rx} dx_k = \int_{A(R)} dV.$$

This completes the proof of the theorem for linear transformations T and constant functions f.

In proving the general theorem we shall use the following norm for the matrix $A = (a_{ij})$ of a linear function:

$$\|A\| = \max_{1 \leq i \leq n} \sum_{j=1}^{n} |a_{ij}| .$$

If we similarly define a norm for vectors $\mathbf{x} = (x_1, \ldots, x_n)$ by

$$\|\mathbf{x}\| = \max_{1 \leq j \leq n} |\mathbf{x}_j| ,$$

then it follows immediately that

$$\|A\mathbf{x}\| \leq \|A\| \|\mathbf{x}\| .$$

The reason for using this norm is that, if $\mathbf{x}$ and $\mathbf{y}$ are vectors in a coordinate cube of side length $2s$, then $\|\mathbf{x} - \mathbf{y}\| \leq 2s$. Similarly, if $\mathbf{p}$ is the center of such a cube, then the closed cube consists of all vectors $\mathbf{x}$ such that $\|\mathbf{x} - \mathbf{p}\| \leq s$.

Suppose now that C is a cube of side length $2s$ contained in R and with center $\mathbf{p}$. We have by the mean-value theorem, for $\mathbf{x}$ in C,

$$T_k(\mathbf{x}) - T_k(\mathbf{p}) = T_k'(\mathbf{y}_k)(\mathbf{x} - \mathbf{p}),$$

where the T_k are the coordinate functions of T, and $\mathbf{y}_k$ is some point on the segment joining $\mathbf{x}$ to $\mathbf{p}$. Then

$$\|T(\mathbf{x}) - T(\mathbf{p})\| \leq \max_{\mathbf{y} \text{ in } C} \|T'(\mathbf{y})\| \|\mathbf{x} - \mathbf{p}\| .$$

This implies that $T(C)$ is contained in the cube of vectors $\mathbf{z}$ defined by

$$\|\mathbf{z} - T(\mathbf{p})\| \leq s \max_{\mathbf{y} \text{ in } C} \|T'(\mathbf{y})\| .$$

Raising both sides to the nth power, we conclude that

$$V(T(C)) \leq \left\{ \max_{\mathbf{y} \text{ in } C} \|T'(\mathbf{y})\| \right\}^n V(C).$$

Notice that if L is an arbitrary one-to-one linear transformation with matrix A, and S is a set bounded by finitely many smooth sets, then $V(L(S)) = |\det A|\, V(S)$. This follows from the special case of the change-of-variable theorem that we have just proved for linear transformations.

Now we take $S = T(C)$ and $A = [T'(\mathbf{x})]^{-1}$. Then, applying (3) with T replaced by $(d_{\mathbf{x}}T)^{-1} \circ T$, we get

$$\begin{aligned} |\det [T'(\mathbf{x})]^{-1}|\, V(T(C)) &= V((d_{\mathbf{x}}T)^{-1} \circ T(C)) \\ &\leq \left\{ \max_{\mathbf{y} \text{ in } C} \|[T'(\mathbf{x})]^{-1} T'(\mathbf{y})\| \right\}^n V(C) \end{aligned}$$

or

$$V(T(C)) \leq |\det T\,(\mathbf{x})| \left\{ \max_{\mathbf{y} \text{ in } C} \|[T'(\mathbf{x})]^{-1} T'(\mathbf{y})\| \right\}^n V(C). \tag{4}$$

Let the cube C be divided into a finite set $C_1, \ldots, C_N$ of nonoverlapping cubes with centers $\mathbf{x}_1, \ldots, \mathbf{x}_N$, and suppose that δ is the maximum side-length of all of them. Apply (4) to each C_k, taking $\mathbf{x} = \mathbf{x}_k$ in each case. Addition gives

$$V(T(C)) \leq \sum_{n=1}^{n} |\det T'(\mathbf{x}_k)| \left\{ \max_{\mathbf{y} \text{ in } C_k} \|[T'(\mathbf{x}_k)]^{-1} T'(\mathbf{y})\| \right\}^n V(C_k).$$

Since T is continuously differentiable, $T'(\mathbf{y})$ is a continuous function of $\mathbf{y}$, and $[T'(\mathbf{x}_k)]^{-1}T'(\mathbf{y})$ approaches the identity matrix as y tends to $\mathbf{x}_k$. Then there is a function $h(\delta)$, tending to zero as δ tends to zero, such that

$$\left\{\max_{\mathbf{y} \text{ in } C_k} \|[T'(\mathbf{x}_k)]^{-1}T'(\mathbf{y})\|\right\}^n \leq 1 + h(\delta).$$

This gives

$$V(T(C)) \leq [1 + h(\delta)]\sum_{k=1}^{n} |\det T'(\mathbf{x}_k)|\, V(C_k).$$

As δ approaches zero, the sum on the right approaches $\int_C |\det T'|\, dV$ Then the last inequality becomes

$$V(T(C)) \leq \int_C |\det T'|\, dV. \tag{5}$$

Having proved this last inequality, we use it to prove the formula for more general sets than cubes. We shall assume $f \geq 0$. The general case follows by considering the positive and negative parts of f separately and adding the resulting formula for each part. Let G be a cubical grid covering R and having mesh δ. Let $C_1, \ldots, C_N$ be the cubes of G that are contained in R. If we let R_N be the part of R that is not contained in any of the cubes C_k, then $R = C_1 \cup \ldots \cup C_N \cup R_N$. Whenever $\mathbf{y}_k$ is a point of C_k and $\mathbf{x}_k = T(\mathbf{y}_k)$, we shall write f_k for $(f \circ T)\mathbf{y}_k$ and $f(\mathbf{x}_k)$. Then, because of (5), we have

$$\sum_{k=1}^{N} f_k \int_{T(C_k)} dV \leq \sum_{k=1} f_k \int_{C_k} |J|\, dV.$$

From this it follows that

$$\begin{aligned} D &= \int_{T(R)} f\, dV - \int_R (f \circ T)\,|J|\, dV \\ &\leq \int_{T(R)} f\, dV - \sum_{k=1}^{N} f_k \int_{T(C_k)} dV + \sum_{k=1}^{N} f_k \int_{C_k} |J|\, dV - \int_R (f \circ T)\,|J|\, dV. \end{aligned}$$

Since $T(R) = T(C_1) \cup \ldots \cup T(C_N) \cup T(R_N)$, we have

$$\begin{aligned} D \leq &\int_{T(R_N)} f\, dV + \sum_{k=1}^{N} \int_{T(C_k)} (f - f_k)\, dV \\ &+ \sum_{k=1}^{N} \int_{C_k} (f_k - f \circ T)\,|J|\, dV - \int_{R_N} (f \circ T)\,|J|\, dV. \end{aligned}$$

Because T is continuously differentiable, it follows from Theorem 8.1 of this Appendix that there is a constant B for which

$$\|T(\mathbf{x}) - T(\mathbf{y})\| \leq B\|\mathbf{x} - \mathbf{y}\|, \qquad \text{for all } \mathbf{x} \text{ and } \mathbf{y} \text{ in } R. \tag{6}$$

Now let ϵ be a positive number. Since f is uniformly continuous on $T(R)$ (apply Theorem 1.3 of this Appendix to $T(R)$ together with its boundary), we can choose δ, the mesh of G, small enough so that

$$|(f \circ T)\mathbf{y} - f_k| \leq \epsilon, \qquad \text{for } \mathbf{y} \text{ in } C_k, \qquad k = 1, \ldots, N.$$

By using (6) and, if necessary, taking δ still smaller, we can get

$$|f(\mathbf{x}) - f_k| \leq \epsilon, \qquad \text{for } \mathbf{x} \text{ in } T(C_k), \qquad k = 1, \ldots, N.$$

Then

$$D \leq \int_{T(R_N)} f\,dV - \int_{R_N} (f \circ T)\,|J|\,dV + \epsilon\{V(T(R)) + V(R)\}.$$

Since R is assumed to have a volume, there is a mesh such that $V(R_N) \leq \epsilon$. Again using (6) and, if necessary, decreasing the mesh again, we can get $V(T(R_N)) \leq B^n\epsilon$. Then

$$D \leq \epsilon\{MB^n + M + V(T(R)) + V(R)\},$$

where M is a number such that $f \leq M$ on $T(R)$ and $(f \circ T)\,|J| \leq M$ on R. Since ϵ is arbitrary, we must have $D \leq 0$, that is,

$$\int_{T(R)} f\,dV \leq \int_R (f \circ T)\,|J|\,dV.$$

If we apply this last inequality to the situation in which T is replaced by T^{-1}, we get

$$\int_R (f \circ T)\,|J|\,dV \leq \int_{T(R)} (f \circ T \circ T^{-1})\,|J \circ T^{-1}|\,|J^{-1}|\,dV.$$

where J^{-1} is the Jacobian determinant of the transformation T^{-1}. But $(J \circ T^{-1})(J^{-1}) = 1$; so

$$\int_{T(R)} f\,dV \leq \int_R (f \circ T)\,|J|\,dV \leq \int_{TR} f\,dV,$$

and the desired equality has been proved.

If J is zero on some piece of smooth surface S in R, then the above proof breaks down because T^{-1} may fail to be continuously differentiable. However, by Theorem 8.2 of this Appendix, S can be enclosed in the interior of a union U of finitely many rectangles of arbitrarily small content v, and the image surface $T(S)$ will be contained in an image region $T(U)$ having content at most $B^n v$, where B is the constant of relation (6). Then, applying the change-of-variable formula to the region R with U deleted, we get

$$\left| \int_{T(R)} f\,dV - \int_R (f \circ T)\,|J|\,dV \right| \leq \left| \int_{T(U)} f\,dV \right| + \left| \int_U (f \circ T)\,|J|\,dV \right|$$
$$\leq MB^n v + Mv,$$

where $|f|$ and $|f \circ T|\,|J|$ are both less than M. Letting v tend to zero, we get the final equality.

Index

D

E

N

O

P

T

U

V

W

Z